Lecture Notes in Mathematics

Edited by A. Dold and B. Eckmann

597

Geometry and Topology

III Latin American School of Mathematics

Proceedings of the School held at the
Instituto de Matemática Pura e Aplicada
CNPq, Rio de Janeiro, July 1976

Edited by
Jacob Palis and Manfredo do Carmo

Springer-Verlag
Berlin · Heidelberg · New York 1977

Editors
Jacob Palis
Manfredo do Carmo
Instituto de Matemática
Pura e Aplicada
Rua Luiz de Camoes, 68
2000 Rio de Janeiro – RJ, Brasil

AMS Subject Classifications (1970): 49F XX, 53 XX, 55 XX, 57 XX, 58 XX, 70 XX

ISBN 3-540-08345-6 Springer-Verlag Berlin · Heidelberg · New York
ISBN 0-387-08345-6 Springer-Verlag New York · Heidelberg · Berlin

Printed in Germany

Printing and binding: Beltz Offsetdruck, Hemsbach/Bergstr.
2141/3140-543210

Preface

The III Latin American School of Mathematics (III ELAM) aimed to stimulate the development of the areas of geometry and topology in Latin America and to expand the interchange of ideas and contacts among mathematicians of this region initiated with the support of the OAS in the I and II Schools, held respectively in Brazil and Mexico. The III ELAM congregated more than 250 mathematicians (researches and students) most of them from Latin America. The members of the Organizing Committee were: C. Camacho, P. Schweitzer, M. do Carmo and J. Palis (Coordinator).

These Proceedings reflect one of the two main activities of the meeting, namely a series of research talks covering topics in Dynamical Systems, Differential Geometry, Foliations, Singularities of Mappings and Algebraic Topology. The other activity was a series of introductory courses in these subjects given by M. Peixoto, M. do Carmo, P.Schweitzer, J. Sotomayor and J. Adem, whose lecture notes are being or have been informally published by the Organizing Committee.

We hope that these Proceedings can contribute to the further progress of these areas of Mathematics within and outside Latin America.

To many of our colleagues, to several universities and research institute of Latin America and to the director of the host institution L. Dias we express our gratitude for their most valuable help.

We acknowledge the financial support of the Brazilian agencies CNPq, FINEP, FAPESP and CAPES, as well as the OAS.

Jacob Palis and Manfredo do Carmo

Rio de Janeiro, April, 1977

CONTENTS

José Adem and Kee Yuen Lam	Evaluation of some Maunder cohomology operations	1
Jorge Soto Andrade	Constructions géometrique de certaines series discretes	32
J.L. Arraut and N.Moreira dos Santos	The point spectrum of the adjoint of an automorphism of a vector bundle	56
Thomas Banchoff and Clint McCrory	Whitney duality and singularities of projections	68
César Camacho and A. Lins Neto	Orbit preserving diffeomorphisms and the stability of Lie group actions and singular foliations	82
A.G. Colares and Manfredo P. do Carmo	On minimal immersions with parallel normal curvature tensor	104
Shiing-shen Chern	Circle bundles	114
D.B.A. Epstein	A topology for the space of foliations	132
D.B.A. Epstein and H. Rosenberg	Stability of compact foliations	151
S. Feder and W. Iberkleid	Secondary operations in K- theory and the generalized vector field problem	161
R. Gérard	On holomorphic solutions of certain kind of Pfaffian systems with singularities	176
Herman Gluck and David Singer	Scattering problems in differential geometry	191
C. Gutierrez and W. de Melo	The connected components of Morse-Smale vector fields on two manifolds	230
Gilbert Hector	Feuilletages en cylindres	252
Michael R. Herman	1) Mesure de Lebesgue et nombre de rotation	271
	2) The Godbillon-Vey invariant of foliations by planes of T^3	294
M. Herrera	On the construction of the trace in Serre duality	308
Peter Hilton	Localization theories for groups and homotopy types	319
Henrich Kollmer	On hyperblic attractors of codimension one	330
A. Lins, W. de Melo and C.C. Pugh	On Liénard's equation	335
S. Lojasiewicz	On the Weierstrass preparation theorem	358

Ricardo Mañé	1) Reduction of semilinear parabolic equations to finite dimensional C^1 flows	361
	2) Axiom A for endomorphisms	379
	3) Characterizations of AS diffeomorphisms	389
Airton S. de Medeiros	Structural stability of integrable differential forms	395
W. de Melo	Accessibility of an optimum	429
J. Moser	1) The scattering problem for some particle systems on the line	441
	2) Proof of a generalized form of a fixed point theorem	464
Jacob Palis Jr.	Some developments on stability and bifurcations of dynamical systems	495
Lucio L. Rodríguez	Convexity and tightness of manifolds with boundary	510
Otto Raul Ruiz M.	Existence of brake orbits in Finsler mechanical systems	542
José Carlos de Souza Kiihl	On the complex projective bundle construction	568
Floris Takens	Symmetries, conservation laws and variational principles	581
Geovan Tavares dos Santos	Classification of generic quadratic vector fields with no limit cycles	605
Juan Tirao	The classifying ring of SL(2,C)	641
Friedrich Tomi	On the finite solvability of Plateau's problem	679
A.J. Tromba	The set of curves of uniqueness for Plateau's problem has a dense interior	696
C.T.C. Wall	Geometric properties of generic differentiable manifolds	707
E.C. Zeeman	A catastrophe model for the stability of ships	775
Eduard Zehnder	1) Note on smoothing symplectic and volume preserving diffeomorphism	828
	2) A simple proof of a generalization of a theorem by C.L. Siegel	855

EVALUATION OF SOME MAUNDER COHOMOLOGY OPERATIONS

By

José Adem and Kee Yuen Lam

Introduction

In this paper we present an explicit evaluation of primary, secondary and tertiary Maunder cohomology operations on complex projective spaces. These operations are related to divisibility properties of the Chern character, as was first discovered by Adams for primary operations ([2]) and later extended by Maunder to higher order operations ([14]).

If η is the canonical complex line-bundle over the N-dimensional complex projective space CP^N, we have

$$ch(\eta-1)^n = (e^{\omega}-1)^n = \omega^n + \Sigma_{t=1}^{\infty}\{n+t,n\}\omega^{n+t},$$

where the rational numbers $\{n+t,n\}$ are the Taylor coefficients and $\omega \in H^2(CP^N;Z)$ is the integral cohomology generator. According with Adams (loc. cit.) the number

$$(*) \qquad m(t)\{n+t,n\}$$

is an integer, where the numerical function $m(t)$ is defined by

$$m(t) = \prod_p p^{[t/(p-1)]}$$

where p runs over the primer numbers.

Roughly speaking, to compute the higher order Maunder operations (mod 2) on the class ω^n is equivalent to determine the highest power of two contained in the integers (*). In section 3 we give explicit expressions that allow us to compute the 2-integers $2^t\{n+t,n\}$ mod 8 and this is enough for our purpose.

We obtain simple formulas in terms of binomial coefficients to evaluate the Maunder secondary and tertiary operations on the class ω^n (see (4.6), (4.17)). Their duals, denoted by $\Phi_{2r}^{(2)}$ for secondary and by $\Phi_{2r}^{(3)}$ for tertiary operations, are also evaluated (see (6.5), (7.12)). The secondary operations $\Phi_{2r}^{(2)}$ have been used to solve several problems. Through "accompany relations" they can be combined to yield non-trivial operations on real projective spaces, that are applied to solve non-immersion problems ([5],[6]). We do not know how to apply this procedure to the $\Phi_{2r}^{(3)}$. In general, examples of non-trivial tertiary operations on real projective spaces promise to be interesting.

We do not give any application of our results, with exception of this one, presented right here. Let $q = 2^{n+2}$ and $r = 2^n$, with $n \geq 2$. Our evaluation of tertiary operations in (7.12), gives

$$\Phi_{2r}^{(3)}\omega^q = \omega^{q+r} \qquad \text{(mod 2)},$$

in $H^*(CP^{q+r}/CP^{q-1};Z_2)$. Now, a result of Gitler and Milgram ([11]), shows that the operation $\Phi_{2r}^{(3)}$ on an integral class of dimension $\ell \leq 2r-7$ is a primary operation. This operation in our case becomes zero and we have proved that the stunted projective space CP^{q+r}/CP^{q-1} cannot be desuspended $2q-2r+7$ times.

Finally, we want to indicate that Dennis Phee Hurley has obtained in [16], evaluation results for a full family of <u>modified Maunder dual operations</u> of all orders. His work agrees with our theorem (7.12) and can be regarded as a generalization of the results given in [10], for secondary operations. However, it seems that he cannot obtain our theorem (4.17).

1. The Maunder operations

As defined in [14;p.753], let $C(3,r)$ be the chain complex $(3 \le r)$

$$C_3 \xrightarrow{d_3} C_2 \xrightarrow{d_2} C_1 \xrightarrow{d_1} C_0, \tag{1.1}$$

where each C_i is a free-graded left module over the mod 2 Steenrod algebra A, on generators described as follows:

C_0 with A-basis $\{c_0\}$,

C_1 with A-basis $\{c_1, c_{10}, c_{11}\}$,

C_2 with A-basis $\{c_2, c_{20}, c_{21}, c_{22}\}$,

C_3 with A-basis $\{c_3\}$,

where $\dim c_0 = 0$, $\dim c_i = 2r+i-1$ for $i = 1,2,3$, and $\dim c_{ni} = 3n-2i$, for $n = 1,2$. On these generators, the maps d_i are given by

$$(1.2)\quad \begin{cases} d_1c_1 = (\chi Sq^{2r})c_0, \\ d_1c_{10} = Sq^{01}c_0, \\ d_1c_{11} = Sq^1c_0. \end{cases}$$

$$(1.3)\quad \begin{cases} d_2c_2 = Sq^1c_1 + (\chi Sq^{2r})c_{11} + (\chi Sq^{2r-2})c_{10}, \\ d_2c_{20} = Sq^{01}c_{10}, \\ d_2c_{21} = Sq^1c_{10} + Sq^{01}c_{11} \\ d_2c_{22} = Sq^1c_{11}, \end{cases}$$

$$(1.4)\quad d_3c_3 = Sq^1c_2 + (\chi Sq^{2r})c_{22} + (\chi Sq^{2r-2})c_{21} + (\chi Sq^{2r-4})c_{20}$$

where $Sq^{01} = Sq^2Sq^1 + Sq^1Sq^2$ and $\chi : A \longrightarrow A$ is the canonical antiautomorphism. Each map d_i is of degree zero and $d_2d_3 = d_1d_2 = 0$ follows from the relations

(1.5) $$Sq^1Sq^1 = 0,$$

(1.6) $$Sq^{01}Sq^{01} = 0,$$

(1.7) $$Sq^1Sq^{01} + Sq^{01}Sq^1 = 0,$$

(1.8) $$Sq^1(\chi Sq^{2r}) + (\chi Sq^{2r-2})Sq^{01} + (\chi Sq^{2r})Sq^1 = 0.$$

Associated with $C(3,r)$ we have a pyramid of operations $\Phi^{s,t}$, for $3 \geq s > t \geq 0$, where each $\Phi^{s,t}$ is of $(s-t)$th order. We recall that this set of operations verify the Maunder's axioms and that they satisfy the relations with the Chern character as stated in [14;Th.2].

Let $w \in H^2(CP^N;Z_2)$ be the mod 2 cohomology generator of the N-dimensional complex projective space, where N is a large enough number, and let w^n be the n-fold product of w. Sometimes it is necessary to work in a $(2n-1)$-connected complex (see [14;Th.1]). In this case, through the collapsing map, w^n is regarded as an element of $H^*(CP^N/CP^{n-1};Z_2)$. Set

(1.9) $$\varepsilon : C_0 \longrightarrow H^*(CP^N/CP^{n-1};Z_2),$$

as the A-map defined by $\varepsilon(c_0) = w^n$. For ε (or equivalently for w^n), we shall make explicit the primary, secondary and tertiary operations associated with the chain complex (1.1).

In order to motivate the computation developed in section 3, we advance the following. Let η be the canonical complex line bundle over CP^N, and $\mu = \eta - 1$. The n-fold product μ^n can be regarded as an element of the Grothendieck ring $\tilde{K}_c(CP^N/CP^{n-1})$ (see [3;p.622]). If $\omega \in H^2(CP^N;Z)$ represents the integral cohomology generator then, for the Chern character, we have $ch\mu = e^{\omega}-1$ and

$$(1.10) \qquad ch\mu^n = (e^{\omega}-1)^n.$$

2. A binomial identity

In this section we state the Jensen identity concerning binomial coefficients $\binom{p}{q}$. Let α,β,γ and n be positive integers. Then we have

$$(2.1) \qquad \sum_{k=0}^{n} \binom{\alpha+\beta k}{k}\binom{\gamma+\beta(n-k)}{n-k} = \sum_{i=0}^{n} \beta^i \binom{\alpha+\gamma+n\beta-i}{n-i}.$$

For a proof see [9].

3. Computation of $2^t\{n+t,n\}$ mod 8

Write $(e^{\omega}-1)^n = \sum_{t=0}^{\infty} \{n+t,n\}\omega^{n+t}$ where $\{n+t,n\}$ are the Taylor coefficients. The value of $2^t\{n+t,n\}$ mod 2^r must be computed in order to evaluate the Maunder operations on CP^{∞}. In this section we compute them mod 8. Computation mod 16 is also available. In general the task becomes more complicated modulo higher powers of 2.

All computations and divisibility arguments take place in $Z_{(2)}$, the ring of rational numbers with odd denominators. If $a,b \in Z_{(2)}$, $a \equiv b \pmod{2^r}$ means that $a-b = 2^r c$ for some $c \in Z_{(2)}$. This is the only kind of congruence considered in our paper.

First we assert that $2^t\{n+t,n\}$ is in $Z_{(2)}$ for all n and t. When $n = 1$, $2^t\{1+t,1\} = 2^t/(t+1)!$ is indeed in $Z_{(2)}$, because of the well known fact:

(3.1) <u>The highest power of</u> 2 <u>dividing</u> m! <u>is</u> $2^{m-\alpha(m)}$, <u>where</u> $\alpha(m)$ <u>is the number of</u> 1's <u>in the dyadic expansion of</u> m.

The general assertion that $2^t\{n+t,n\} \in Z_{(2)}$ now follows inductively from

<u>Proposition</u> (3.2). <u>For all</u> $n,t > 0$ <u>we have</u>

$$2^t\{n+1+t,\ n+1\} = \sum_{i=0}^{t} 2^i\{n+i,n\}2^{t-i}\{1+t-i,1\}. \tag{3.3}$$

The proof is by equating coefficients of ω^{n+1+t} in $(e^{\omega}-1)^{n+1} = (e^{\omega}-1)^n(e^{\omega}-1)$.

To compute $2^t\{n+t,n\}$ mod 8, we first do so for $n = 1$ in theorem (3.4) below. We then attack the general case inductively on n, using arguments based on (3.2) and the binomial identity (2.1).

<u>Theorem</u> (3.4). <u>For every</u> $t > 0$ <u>the value of</u> $2^t/(t+1)!$ mod 8 <u>is given by</u>

$$\frac{2^t}{(t+1)!} = \binom{1+2t}{t} - 2\binom{2t-2}{t-1} + 4\binom{2t-6}{t-3} \pmod 8. \tag{3.5}$$

<u>Proof</u>. The formula is directly verified for $t = 0,1,2$ and 3, so we take $t \geq 4$, in which case the last term on the right hand side is always divisible by 8 (see(3.1)) and may as well be dropped.

Introduce the function $F(t) = \prod_{i=0}^{t} (2i+1)$ and note the identity

$$(3.6) \qquad \binom{2t}{t} = \frac{2^t}{t!} F(t-1).$$

Writing $\binom{1+2t}{t} = \frac{2t+1}{t+1}\binom{2t}{t}$ and using (3.6) to substitute for the binomial coefficients, we reduce (3.5) to the equivalent form

$$(3.7) \qquad \frac{2^t}{(t+1)!}(F(t)-1) - \frac{2^t}{(t-1)!}F(t-2) \equiv 0 \qquad (\text{mod } 8).$$

Since $F(t+2) \equiv -F(t)$ (mod 4), we have

$$F(t) \equiv \begin{cases} 1 & \\ & (\text{mod } 4) \\ -1 & \end{cases} \qquad \begin{array}{l} t = 4h \text{ or } 4h+3 \\ \\ t = 4h+1 \text{ or } 4h+2 \end{array}$$

From this and (3.1), we easily check that (3.7) is true when $t \equiv 0,3$ (mod 4) or when $\alpha(t+1) \geq 3$; for in these cases each individual term of (3.7) is a multiple of 8. The only remaining possibility is when $t = 2^r+1$, in which case $2^t/(t+1)!$ and $2^t/(t-1)!$ are both even, so we may replace $F(t)$, $F(t-2)$ by their mod 4 values in (3.7), reducing it to

$$(3.8) \qquad 2\frac{2^t}{(t+1)!} - \frac{2^t}{(t-1)!} \equiv 0 \quad (\text{mod } 8) \qquad t = 2^r+1.$$

Now (3.8) is true because its left hand side equals $2^t/(t+1)!$ times a multiple of 4. This ends the proof of our theorem.

<u>Theorem</u> (3.9). <u>The value of</u> $2^t\{n+t,n\}$ mod 8 <u>is given by</u>

$$(3.10) \qquad 2^t\{n+t,n\} \equiv \binom{n+2t}{t} - 2\binom{n+2t-\varepsilon}{t-1} + 4A \qquad (\text{mod } 8),$$

where $\varepsilon = 1$ or 3 according as n is even or odd, and A depends on the mod 4 value of n as follows:

n(4)	1	2	3	4
A	$\binom{n+2t-7}{t-3}$	$\binom{n+2t-3}{t-2} + \binom{n+2t-6}{t-2}$	$\binom{n+2t-3}{t-2} + \binom{n+2t-7}{t-2}$	0

Proof. By induction on n. When n = 1 this is theorem (3.4). Supposing inductively that the theorem holds for n = 4m+1, we shall derive in succession that it also holds for 4m+2, 4m+3, 4m+4 and 4m+5. The argument is given only for passing from 4m+1 to 4m+2, since the other steps are entirely similar, and equally tedious.

Take, then, n = 4m+1, and assume it has been proved that

$$2^t\{n+t,n\} \equiv \binom{n+2t}{t} - 2\binom{n+2t-3}{t-1} + 4\binom{n+2t-7}{t-3} \pmod 8.$$

By formula (3.3) we have

$$(3.11)\qquad 2^t\{n+1+t,\ n+1\} = \sum_{i=0}^{t} 2^i\{n+i,n\}2^{t-i}\{1+t-i,1\} \equiv$$

$$\sum_{i=0}^{t} \left[\binom{n+2i}{i} - 2\binom{n+2i-3}{i-1} + 4\binom{n+2i-7}{i-3}\right]\left[\binom{1+2t-2i}{t-i} - 2\binom{2t-2i-2}{t-i-1} + 4\binom{2t-2i-6}{t-i-3}\right].$$

We can expand the product in the above expression, drop all multiples of 8, and end up with 6 summation terms. Each term can be written in the form $\Sigma 2^k \rho_k$ according to (2.1). For example,

$$\sum_{i=0}^{t}\binom{n+2i}{i}\binom{1+2t-2i}{t-1} = \sum_{k=0}^{t} 2^k\binom{n+1+2t-k}{t-k} = \binom{n+1+2t}{t} + 2\binom{n+2t}{t-1} + 4\binom{n+2t-1}{t-2} + \cdots$$

If we do this for all the six terms, and ignore multiples of 8 along the way, we end up with the following huge sum

$$\{\binom{n+2t+1}{t} + 2\binom{n+2t}{t-1} + 4\binom{n+2t-1}{t-2}\} - 2\{\binom{n+2t-2}{t-1} + 2\binom{n+2t-3}{t-2}\} \tag{3.12}$$

$$+ 4\binom{n+2t-6}{t-3} - 2\{\binom{n+2t-2}{t-1} + 2\binom{n+2t-3}{t-2}\} + 4\binom{n+2t-5}{t-2} + 4\binom{n+2t-6}{t-3}.$$

In this sum, the second and the third { } can be combined and simplified mod 8, the last term cancels a previous term mod 8. Using the identity $\binom{a}{b} = \binom{a-1}{b} + \binom{a-1}{b-1}$, we get the result

$$\binom{n+1+2t}{t} - 2\binom{n+1+2t-1}{t-1} + 4[\binom{n+1+2t-3}{t-2} + \binom{n+1+2t-6}{t-2}],$$

which is the mod 8 value of $2^t\{n+1+t,n+1\}$, as is to be proved. This establishes the theorem for $n+1 = 4m+2$, and we can go on to $4m+j$, $3 \leq j \leq 5$ by the same steps of argument, completing the induction.

Of course, if we are only interested in the mod 4 or mod 2 values of $2^t\{n+t,n\}$, things will be greatly simplified. We shall just record

<u>Corollary</u> (3.13). $$2^t\{n+k,n\} \equiv \binom{n+2t}{t} \pmod 2.$$

<u>Corollary</u> (3.14). $$2^t\{n+t,n\} \equiv \binom{n+2t}{t} - 2\binom{n+2t-1}{t-1} \pmod 4.$$

Now, as a check to the correctness of theorem (3.9), we propose to show, in a somewhat different way.

<u>Proposition</u> (3.15). *If* $n \equiv 0 \pmod{2^{\ell-1}}$, *then*

$$2^t\{n+t,n\} \equiv \binom{n+2t}{t} - 2\binom{n+2t-1}{t-1} \pmod{2^{\ell}}.$$

Proof. By induction on ℓ. If $\ell = 1$, see corollary (3.13). Assume the proposition true for n, we will show its truth for 2n. For this purpose write the inductive hypothesis as

$$\left(\frac{e^{2\omega}-1}{2\omega}\right)^{n} \equiv \sum_{t=0}^{\infty} \left[\binom{n+2t}{t} - 2\binom{n+2t-1}{t-1}\right]\omega^{t} \qquad (\text{mod } 2^{\ell}).$$

Now, if $a \equiv b$, mod 2^{ℓ}, $\ell > 0$, then $a^2 \equiv b^2$ mod $2^{\ell+1}$ because $a^2 - b^2 = (a-b)^2 + 2b(a-b)$. This is true even if a,b are power series. Hence we get

$$\left(\frac{e^{2\omega}-1}{2\omega}\right)^{2n} \equiv \left(\sum_{t=0}^{\infty} \left[\binom{n+2t}{t} - 2\binom{n+2t-1}{t-1}\right]\omega^{t}\right)^{2} \qquad (\text{mod } 2^{\ell+1}).$$

The coefficient of ω^t in the right hand side is

$$\sum_{i=0}^{t} \left[\binom{n+2i}{i} - 2\binom{n+2i-1}{i-1}\right]\left[\binom{n+2t-2i}{t-i} - 2\binom{n+2t-2i-1}{t-i-1}\right].$$

If we treat this summation in the same way as we treated (3.11), by applying (2.1), and dropping multiples of $2^{\ell+1}$ along the way, we obtain the simple result

$$\binom{2n+2t}{t} - 2\binom{2n+2t-1}{t-1}$$

which is congruent mod $2^{\ell+1}$ to $2^t\{2n+t,2n\}$, the coefficient of ω^t in the left hand side. Thus our induction is completed.

4. Evaluation of the Maunder operations on CP^N

Now we go back to the evaluation on ε of the primary, secondary and tertiary operations, associated with the chain complex $C(3,r)$. In Maunder's notations these are: $\Phi^{1,0}(\varepsilon)$, $\Phi^{2,0}(\varepsilon)$ and $\Phi^{3,0}(\varepsilon)$, where ε is the A-map (1.9) of degree 2n.

If λ is of degree q, we recall that $\Phi^{K,0}(\lambda)$ is an equivalence class of maps $\xi : C_s \longrightarrow H^*(CP^N;Z_2)$ of degree $q-K+1$, where two maps are considered equivalent if they differ by a map in the indeterminacy of $\Phi^{K,0}$. Also, $\Phi^{K,0}(\lambda)$ is defined only for those A-maps λ such that each $\Phi^{t,0}(\lambda)$, for $K > t > 0$, is the equivalence class containing the zero map (see [13],[14]). For our particular case of the first three operations on CP^N, we will try to present these facts in a simpler equivalent form.

The primary operation is $\Phi^{1,0}(\varepsilon) = \varepsilon d_1$ and from (1.2), it follows that this operation is given by

$$(4.1)\qquad \begin{cases} (\chi Sq^{2r})w^n, \\ Sq^{01}w^n = 0, \\ Sq^1 w^n = 0. \end{cases}$$

Consequently, we can regard $\Phi^{1,0}(\varepsilon)$ as represented only by $(\chi Sq^{2r})w^n$.

We make the first use of the computation developed in section 3, as follows. From Adams's theorem [2;Th.2] together with (1.10) and the congruence (3.13), we get

$$(4.2)\qquad (\chi Sq^{2r})w^n = \binom{n+2r}{r} w^{n+r},$$

and this completes the evaluation of the primary operation $\Phi^{1,0}(\varepsilon)$.

Before continuing with the next case we will make some remarks and establish some notation. In general, a secondary cohomology operation is constructed from a given set of relations on primary operations ([1]). Let $\Psi^{(2)}_{2r}$ be a secondary operation constructed with the relation (1.8). We can write

(4.3) $$\Psi^{(2)}_{2r} = \chi\Phi^{(2)}_{2r},$$

where $\Phi^{(2)}_{2r}$ represents the <u>dual</u> operation ([13],[15]) associated with the relation

(4.4) $$Sq^{2r}Sq^{1} + Sq^{01}Sq^{2r-2} + Sq^{1}Sq^{2r} = 0.$$

Clearly, (4.4) is the image of the relation (1.8) under the antiautomorphism χ.

The secondary operation $\Phi^{2,0}(\varepsilon)$ is defined if $\Phi^{1,0}(\varepsilon) = 0$, and this condition is equivalent with $(\chi Sq^{2r})w^{n} = 0$ (see (4.1)). Hence, from (4.2), it follows that $\Phi^{2,0}(\varepsilon)$ is defined if $\binom{n+2r}{r} \equiv 0 \pmod 2$.

The operation $\Phi^{2,0}(\varepsilon)$ is an equivalence class of maps $\xi : C_2 \longrightarrow H^*(CP^N;Z_2)$ where each ξ is of degree $2n-1$. For the A-basis of C_2 we have $\xi(c_{20}) = \xi(c_{21}) = \xi(c_{22}) = 0$, since these elements are of odd dimension, and $\xi(c_2) \in H^{2n+2r}(CP^N;Z_2)$.

The indeterminacy of $\Phi^{2,0}(\varepsilon)$ is the set of all maps

$$C_2 \xrightarrow{d_2} C_1 \xrightarrow{\varepsilon_1} H^*(CP^N;Z_2),$$

where ε_1 is any A-map of degree $2n-1$. Here, for dimensional reasons: $\varepsilon_1(c_1) = 0$ and $\varepsilon_1 d_2(c_2)$ is the only nontrivial possible value of $\varepsilon_1 d_2$ in the A-basis of C_2 (see(1.3)). If we set $\varepsilon_1(c_{10}) = w^{n+1}$ and

$\varepsilon_1(c_{11}) = w^n$, then $\varepsilon_1 d_2(c_2) = (\chi Sq^{2r})w^n + (\chi Sq^{2r-2})w^{n+1}$. But we assume that $(\chi Sq^{2r})w^n = 0$, therefore, the only relevant term in the indeterminacy is

$$(4.5) \qquad (\chi Sq^{2r-2})w^{n+1} = \binom{n+2r-1}{r-1} w^{n+r},$$

where the term to the right follows from (4.2).

The operation $\Phi^{2,0}(\varepsilon)$ is equivalent to the secondary operation $(\chi\Phi^{(2)}_{2r})w^n$ of (4.3), constructed with the relation (1.8). This is not hard to see and we refer the reader to [1], [13] and [14] for the construction of a proof.

Therefore, the evaluation of $\Phi^{2,0}(\varepsilon)$ has been reduced to the calculation of $(\chi\Phi^{(2)}_{2r})w^n$ and this is given in the following

Theorem (4.6). *Suppose that*

$$(4.7) \qquad \binom{n+2r}{r} \equiv 0 \quad \text{and} \quad \binom{n+2r-1}{r-1} \equiv 0 \qquad (\text{mod } 2).$$

Then, $(\chi\Phi^{(2)}_{2r})w^n$ *is defined and with zero indeterminacy, we have*

$$(4.8) \qquad (\chi\Phi^{(2)}_{2r})w^n = \begin{cases} \frac{1}{2}\binom{n+2r}{r} w^{n+r}, & \text{if } n \text{ even}, \\ \left[\frac{1}{2}\binom{n+2r}{r} + \binom{n+2r-3}{r-1}\right] w^{n+r}, & \text{if } n \text{ odd}. \end{cases}$$

Proof. From (4.2) and (4.5) it follows that the conditions (4.7) are enough in order to have the operation defined and with zero indeterminacy. In the same form used to prove (4.2), we now apply the Maunder's result [14;Th.2] together with the mod 4 congruence (3.14). This allows us to compute $\xi(c_2)$ (see above) and, consequently, a representative of the operation. The stated expressions follow easily from (3.14) and this ends the proof.

The tertiary operation $\Phi^{3,0}(\varepsilon)$ is defined if each of the operations $\Phi^{1,0}(\varepsilon)$ and $\Phi^{2,0}(\varepsilon)$ is the equivalence class containing the zero map. To simplify matters we assume that the operations, in the equivalent form (4.1) and (4.7), are zero with zero indeterminacy.

The indeterminacy of $\Phi^{3,0}(\varepsilon)$ is the set of equivalence classes of maps $\Phi^{3,1}(\lambda)$, as λ runs over all the maps $\lambda : C_1 \longrightarrow H^*(CP^N;Z_2)$ of degree 2n–1, such that $\Phi^{3,1}(\lambda)$ is defined. We recall that the chain complex

$$C_3 \xrightarrow{d_3} C_2 \xrightarrow{d_2} C_1 \xrightarrow{\lambda} H^*(CP^N;Z_2),$$

together with the cycle $z = d_3c_3$, are used to construct this operation.

It is not difficult to see that $\Phi^{3,1}(\lambda)$, for all possible λ's, can be identified with the pair of operations $(\chi\Phi^{(2)}_{2r})w^n$ and $(\chi\Phi^{(2)}_{2r-2})w^{n+1}$.

Extending the notation used in (4.3), set

$$(\chi\Phi^{(3)}_{2r})w^n = \Phi^{3,0}(\varepsilon). \tag{4.9}$$

Theorem (4.10). *Suppose that the following arithmetic conditions are fulfilled.*

If n *is even*:

$$\left\{\begin{array}{ll} \binom{n+2r}{r} \equiv 0 & \text{mod } 4, \\ \binom{n+2r-1}{r-1} + 2\binom{n+2r-4}{r-2} \equiv 0 & \text{mod } 4, \\ \binom{n+2r-2}{r-2} \equiv 0 & \text{mod } 2. \end{array}\right. \tag{4.11}$$

If n *is odd*:

$$(4.12)\quad \begin{cases} \binom{n+2r}{r} + 2\binom{n+2r-3}{r-1} \equiv 0 & \text{mod } 4, \\ \binom{n+2r-1}{r-1} \equiv 0 & \text{mod } 4, \\ \binom{n+2r-2}{r-2} \equiv 0 & \text{mod } 2. \end{cases}$$

Then the tertiary operation $(\chi\Phi^{(3)}_{2r})w^n$ *is defined and with zero indeterminacy, we have*

$$(\chi\Phi^{(3)}_{2r})w^n = \left\{ {n \atop r} \right\} w^{n+r},$$

where the coefficient $\left\{ {n \atop r} \right\}$ *depends of the* mod 4 *value of* n *as follows*:

n	$\left\{ {n \atop r} \right\}$
4k	$\frac{1}{4}\binom{n+2r}{r}$
4k+2	$\frac{1}{4}\binom{n+2r}{r} + \binom{n+2r-4}{r-3} + \binom{n+2r-6}{r-2}$
4k+1	$\frac{1}{4}\left[\binom{n+2r}{r} - 2\binom{n+2r-3}{r-1}\right] + \binom{n+2r-7}{r-3}$
4k+3	$\frac{1}{4}\left[\binom{n+2r}{r} - 2\binom{n+2r-3}{r-1}\right] + \binom{n+2r-3}{r-2} + \binom{n+2r-7}{r-2}$

Proof. From the previous considerations it follows that all we need in order to have the tertiary operation defined and with zero indeterminacy, is to have the operations $(\chi\Phi^{(2)}_{2r})w^n$ and $(\chi\Phi^{(2)}_{2r-2})w^{n+1}$ defined, with zero value and zero indeterminacy. With the expressions (4.2) and (4.8) this follows immediately from the conditions (4.11) and (4.12) stated.

As before, if we regard our operation as an equivalence class of maps $\zeta : C_3 \longrightarrow H^*(CP^N;Z_2)$, again we use Maunder's result [14;Th.2] together with the mod 8 congruence (3.9), to compute $\zeta(c_3)$ and this gives us the values of $\{{n \atop r}\}$.

If n is odd, the values of $\{{n \atop r}\}$ in the table follow straightforward from (3.9).

If $n = 4k$, we will first show that $T = \binom{n+2r-4}{r-2} \equiv 0 \mod 2$ $(r \neq 2)$. For r odd this is trivial. Let $r = 2\ell$ even. From (3.1) it follows that the highest power of two in a binomial coefficient $C = \binom{a}{b}$ is $\nu(C) = \alpha(b) + \alpha(a-b) - \alpha(a)$. Then, we have
$\nu(T) = \alpha(2\ell-2) + \alpha(4k-2\ell-2) - \alpha(4k+4\ell-4) = \alpha(\ell-1) + \alpha(2k-\ell-1) - \alpha(k+\ell-1)$.
Now, from (4.11) we have the condition $S = \binom{n+2r-2}{r-2} \equiv 0 \mod 2$, and $\nu(S) = \alpha(\ell-1) + \alpha(2k+\ell) - \alpha(k+\ell-1) - 1 \geq 1$, where we are using the identity $\alpha(4k+4\ell-2) = \alpha(k+\ell-1) + 1$. In general, $\alpha(t) + 1 \geq \alpha(t+1)$, then $\alpha(2k+\ell-1) \geq \alpha(2k+\ell) - 1$, therefore $\nu(T) \geq \nu(S) \geq 1$, and this proves that $T \equiv 0 \mod 2$.

Hence, the second row of condition (4.11) reduces to

$$(4.13) \qquad \binom{n+2r-1}{r-1} \equiv 0 \qquad \mod 4.$$

Consequently,

$$\{{n \atop r}\} = \frac{1}{4}[\binom{n+2r}{r} - 2\binom{n+2r-1}{r-1}] = \frac{1}{4}\binom{n+2r}{r} \qquad \text{mod } 2.$$

If $n = 4k+2$, directly from (3.9) we get

$$\{{n \atop r}\} = \frac{1}{4}[\binom{n+2r}{r} - 2\binom{n+2r-1}{r-1}] + \binom{n+2r-3}{r-2} + \binom{n+2r-6}{r-2}.$$

From (4.11), we have $-\binom{n+2r-1}{r-2} \equiv 2\binom{n+2r-4}{r-2}$ mod 4. Then, $-\frac{1}{2}\binom{n+2r-1}{r-2} \equiv \binom{n+2r-4}{r-2}$ mod 2. Replacing this in $\{{n \atop r}\}$ and using the identity $\binom{n+2r-4}{r-2} + \binom{n+2r-3}{r-2} \equiv \binom{n+2r-4}{r-3}$ mod 2, the value given in the table follows. This ends the proof of (4.10).

As we will see, by simultaneously considering the mod 4 value of n and the parity of r, the expressions of (4.10) can be made simpler and more handy. For that purpose let us write the following.

(4.14) $$\binom{n+2r}{r} \equiv \binom{n+2r-1}{r-1} \equiv 0 \qquad \text{mod } 4,$$

(4.15) $$\binom{n+2r-2}{r-2} \equiv 0 \quad \text{mod } 2, \qquad (r \neq 2)$$

(4.16) $$\binom{n+2r-3}{r-1} \equiv 0 \quad \text{mod } 2.$$

<u>Theorem</u> (4.17). <u>Given two positive integers</u> n <u>and</u> r, <u>assume they fulfill the hypotheses indicated below</u>. <u>Then</u>, <u>the tertiary operation</u> $(\chi\Phi_{2r}^{(3)})w^n$ <u>is defined and with zero indeterminacy</u>, <u>we have</u>

$$(\chi\Phi_{2r}^{(3)})w^n = \{{n \atop r}\}w^{n+r},$$

<u>where the coefficient</u> $\{{n \atop r}\}$ <u>is given in the following table</u>, <u>according with the</u> mod 4 <u>value of</u> n <u>and the parity of</u> r.

Case	Hypotheses	$\left\{ {n \atop r} \right\}$
$n = 4k$ $r = 2\ell$	(4.14) and (4.15)	$\frac{1}{4}\binom{n+2r}{r}$
$n = 4k$ $r = 2\ell+1$	(4.14)	0
$n = 4k+2$ $r = 2\ell$	(4.11)	$\frac{1}{4}\binom{n+2r}{r} + \binom{n+2r-6}{r-2}$
$n = 4k+2$ $r = 2\ell+1$	(4.14)	$\binom{n+2r-4}{r-3}$
$n = 4k+1$ $r = 2\ell$	(4.14) and (4.15)	$\frac{1}{4}\binom{n+2r}{r} + \frac{1}{2}\binom{n+2r-3}{r-1}$
$n = 4k+1$ $r = 2\ell+1$	(4.14) and (4.15)	$\frac{1}{4}\binom{n+2r}{r} + \binom{n+2r-7}{r-3}$
$n = 4k+3$ $r = 2\ell$	(4.14) and (4.15)	$\frac{1}{4}\binom{n+2r}{r} + \binom{n+2r-7}{r-2}$
$n = 4k+3$ $r = 2\ell+1$	(4.14), (4.15) and (4.16)	$\frac{1}{4}\binom{n+2r}{r} + \frac{1}{2}\binom{n+2r-3}{r-1}$

Proof. For each case we will indicate the main facts from which the proof is directly obtained. To simplify statements, we name our binomial coefficients as follows:

$$P = \binom{n+2r}{r}, \quad Q = \binom{n+2r-1}{r-1}, \quad R = \binom{n+2r-3}{r-1}, \quad S = \binom{n+2r-2}{r-2}, \quad T = \binom{n+2r-4}{r-2}.$$

Without explicit mention, we use the following: if $C = \binom{a}{b}$ with a even and b odd, then $C \equiv 0 \pmod 2$.

<u>Case</u> $4 = 4k$ <u>and</u> $r = 2\ell$. The argument used to set (4.13) proves that $\nu(T) \geq \nu(S)$, but, by (4.16), $S \equiv 0\ (2)$, then $T \equiv 0\ (2)$, and in (4.11) this implies that $Q \equiv 0\ (4)$. Now, the proof of this case follows easily.

<u>Case</u> $n = 4k$ and $r = 2\ell+1$. If we only suppose n even and r odd, we get

$$\nu(P) = \nu(Q) + 1. \tag{4.18}$$

From r odd, we have $T \equiv 0\ (2)$ and from (4.11), it follows that $Q \equiv 0\ (4)$. Therefore, $P \equiv 0\ (8)$, and that is enough to establish this case.

<u>Case</u> $n = 4k+2$ <u>and</u> $r = 2\ell$. Here, we only use $\binom{n+2r-4}{r-3} \equiv 0\ (2)$, that follows from the fact that n and r are even.

<u>Case</u> $n = 2k+2$ <u>and</u> $r = 2\ell+1$. From n even and r odd, we have $T \equiv S \equiv 0\ (2)$ and $\binom{n+2r-6}{r-2} \equiv 0\ (2)$. From (4.11), we have $Q \equiv 0\ (4)$ and since (4.18) holds, this implies that $P \equiv 0\ (8)$.

<u>Case</u> $n = 4k+1$ <u>and</u> $r = 2\ell$. We have $R \equiv 0\ (2)$ and $\binom{n+2r-7}{r-3} \equiv 0\ (2)$. Then, from (4.12), $P \equiv 0\ (4)$, and this is sufficient for this case.

<u>Case</u> $n = 4k+1$ <u>and</u> $r = 2\ell+1$. We prove that $\nu(R) \geq \nu(Q)$. Then, from $Q \equiv 0\ (4)$, we get $R \equiv 0\ (4)$.

Case $n = 4k+3$ and $r = 2\ell$. As before, we can prove that $\nu(R) \geq \nu(Q)$. From (4.12), we have $Q \equiv 0$ (4), therefore $R \equiv 0$ (4). Now, $S \equiv 0$ (2) implies that $\binom{n+2r-3}{r-2} \equiv 0$ (2), and that is enough for this case.

Case $n = 4k+3$ and $r = 2\ell+1$. We have $\nu(P) = \nu(Q)$ and $Q \equiv 0$ (4). Then, from (4.12), we get that $R \equiv 0$ (2). Also $\binom{n+2r-3}{r-2} \equiv \binom{n+2r-7}{r-2} \equiv 0$ (2). The result for this last case follows, and this ends the proof of the theorem.

5. S-duals of stunted projective spaces

Two spaces K and L are S-homotopy-equivalent or of the same S-type if some suspension of K is homotopy-equivalent to some suspension of L. Let K be a finite CW-complex embedded in an m-sphere S^m. An m-dual of K is defined to be any subcomplex of $S^m - K$, which is an S-deformation retract of $S^m - K$. Two complexes K and L are S-duals if K and some m-dual of L are of the same S-type (cf.[17]).

For the fields of real, complex and quaternion numbers, Atiyah proves in [8;Th.(6.1)] a result which asserts that the spaces of certain pairs, formed with a stunted projective space and a stunted quasi-projective space (cf.[12;p.117]), are S-duals. Atiyah's theorem, translated to our notation and for the case of complex projective spaces, becomes

Theorem (5.1) (Atiyah) The stunted projective spaces

$$CP^{q+r}/CP^{q-1} \quad \text{and} \quad CP^{M-q-1}/CP^{M-q-r-2}$$

are S-*duals*, *where* M *is an integer such that* $M \cdot J(\eta) = 0$ *in* $J(CP^{q+r})$.

In this theorem, η is the canonical complex line bundle over CP^{q+r} and $J(CP^{q+r})$ is the group introduced by Atiyah in [8;p.292].

Later, it was proved by Adams and Walker in [4], that the integer M was equal to the so-called Atiyah-Todd number M_{q+r+1}, previously defined, for another purpose, in [7;Th.(1.7)]. If $\nu(M_k)$ is the exponent of the highest power of 2 dividing M_k, we have $\nu(M_k) \geq k-1$ (loc. cit.). Therefore,

(5.2)
$$M = M_{q+r+1} = 2^{q+r}N,$$

where the integer $N > 0$.

6. *The dual secondary operations*

Let us consider the stunted projective spaces of (5.1), that are S-duals and denote, respectively, by w^q and $w^{M-q-r-1}$ their first non-trivial cohomology class with Z_2 coefficients. Through the Alexander and Pontrjagin duality isomorphism, a cohomology operation acting on w^q is equivalent with a *dual* cohomology operation acting on $w^{M-q-r-1}$ (see [15]).

For primary operations this means that χSq^{2r} and Sq^{2r} are duals. Actually, we have

$$(\chi Sq^{2r})w^{M-q-r-1} \neq 0 \text{ if and only if } Sq^{2r}w^q \neq 0.$$

This follows directly from the well known formulas (cf.(4.2))

$$(\chi Sq^{2r})w^{M-q-r-1} = \binom{M-q+r-1}{r} w^{M-q-1} \text{ and } Sq^{2r}w^q = \binom{q}{r} w^{q+r},$$

where *both binomial coefficients have the same power of two*. In symbols

$$\nu \binom{M-q+r-1}{r} = \nu \binom{q}{r}. \tag{6.1}$$

In fact, we need to prove that

$$\alpha(r) + \alpha(M-q-1) - \alpha(M-q+r-1) = \alpha(r) + \alpha(q-r) - \alpha(q).$$

From (5.2), we have

$$\alpha(M-q-1) = \alpha(2^{q+r}(N-1)) + 2^{q+r} - 1 - q) = \alpha(2^{q+r}(N-1)) + q + r - \alpha(q),$$

and also,

$$\alpha(M-q+r-1) = \alpha(2^{q+r}(N-1) + q + r - \alpha(q-r).$$

Therefore, (6.1) follows.

We regard now the secondary operations $\chi\Phi_{2r}^{(2)}$ and $\Phi_{2r}^{(2)}$ associated, respectively, with the relations (1.8) and (4.4). Let $n = M - q - r - 1$. Suppose that $(\chi\Phi_{2r}^{(2)})w^n$ is defined and has zero indeterminacy, so that (4.7) holds. As before, we get

$$\binom{n+2r}{r} \equiv \binom{q}{r} \quad \text{and} \quad \binom{n+2r-1}{r-1} \equiv \binom{q}{r-1} \quad (\text{mod } 2). \tag{6.2}$$

Consequently, $Sq^{2r}w^q = Sq^{2r-2}w^q = 0$, and this implies that $\Phi_{2r}^{(2)}w^q$ is defined and has zero indeterminacy. Its value can be obtained from (4.6), as stated in the following

Theorem (6.3). Suppose that

$$(6.4)\qquad \binom{q}{r} \equiv 0 \quad \text{and} \quad \binom{q}{r-1} \equiv 0 \qquad (\text{mod } 2).$$

Then, $\Phi_{2r}^{(2)}w^q$ is defined and with zero indeterminacy, we have

$$(6.5)\qquad \Phi_{2r}^{(2)}w^q = \begin{cases} \frac{1}{2}\binom{q}{r}w^{q+r}, & \text{if } q+r \text{ odd}, \\ \left[\frac{1}{2}\binom{q}{r} + \binom{q+2}{r-1}\right]w^{q+r}, & \text{if } q+r \text{ even}. \end{cases}$$

Proof. Assume the results of [15;Th.(4.1)]. Clearly, (6.2) and (6.4) imply that, both $(\chi\Phi_{2r}^{(2)})w^n$ and $\Phi_{2r}^{(2)}w^q$, are defined and have zero indeterminacy. We have

$$(\chi\Phi_{2r}^{(2)})w^n \neq 0 \text{ if and only if } \Phi_{2r}^{(2)}w^q \neq 0.$$

Then, (6.5) follows from (4.8) together with (6.1) and $\nu\binom{n+2r-3}{r-1} = \nu\binom{q+2}{r-1}$. This last equality, like (6.1), is verified, and the proof of the theorem is finished.

7. The dual Maunder operations

The results of the preceding section have been generalized by Maunder, for higher order cohomology operations ([13]). Here, we will use his formulation to study the dual tertiary operations that correspond to the operations $(\chi\Phi_{2r}^{(3)})w^n$ that we evaluated in (4.17).

Before describing the chain complex $C^*(3,r)$ dual to $C(3,r)$, we consider the following construction. Let E and F be free left A-modules with A-basis, respectively, $\{e_1,\ldots,e_s\}$ and $\{f_1,\ldots,f_t\}$ and let $d : E \longrightarrow F$ be an A-map such that $d(e_p) = \Sigma\alpha_{pq}f_q$, where $\alpha_{pq} \in A$. Then, construct $\delta : F^* \longrightarrow E^*$ as follows: F^* and E^* are free left A-modules with A-basis, respectively, $\{f_1^*,\ldots,f_t^*\}$ and $\{e_1^*,\ldots,e_s^*\}$, and the A-map δ is given by $\delta(f_q^*) = \Sigma\chi(\alpha_{qp})e_q^*$.

Now, apply this construction to $C(3,r)$ to get the chain complex $C^*(3,r)$, formed by

$$C_0^* \xrightarrow{\delta_1} C_1^* \xrightarrow{\delta_2} C_2^* \xrightarrow{\delta_3} C_3^*.$$

According to Maunder, we grade the new elements c_i^* so that $\dim c_i^* = K - \dim c_i$, where K is a convenient integer. Taking $K = 2r + 2$, we get $\dim c_0^* = 2r + 2$, $\dim c_i^* = 3 - i$ for $i = 1,2,3$, and $\dim c_{ni}^* = 2r + 2 - 3n + 2i$, for $n = 1,2$. On these generators, the maps δ_i are given by

$$(7.1)\qquad \begin{cases} \delta_3 c_2^* = Sq^1 c_3^* \\ \delta_3 c_{20}^* = Sq^{2r-4} c_3^* \\ \delta_3 c_{21}^* = Sq^{2r-2} c_3^* \\ \delta_3 c_{22}^* = Sq^{2r} c_3^* \end{cases}$$

$$(7.2)\qquad \begin{cases} \delta_2 c_1^* = Sq^1 c_2^* \\ \delta_2 c_{10}^* = Sq^{2r-2} c_2^* + Sq^{01} c_{20}^* + Sq^1 c_{21}^* \\ \delta_2 c_{11}^* = Sq^{2r} c_2^* + Sq^{01} c_{21}^* + Sq^1 c_{22}^* \end{cases}$$

$$(7.3)\qquad \delta_1 c_0^* = Sq^{2r} c_1^* + Sq^{01} c_{10}^* + Sq^1 c_{11}^*$$

Let $\lambda : C_3^* \longrightarrow H^*(CP^{q+r}/CP^{q-1};Z_2)$ be the A-map defined by $\lambda(c_3^*) = w^q$. As in section 4, we consider here the cohomology operations $\chi\Phi^{1,0}(\lambda)$, $\chi\Phi^{2,0}(\lambda)$ and $\chi\Phi^{3,0}(\lambda)$, associated with the chain complex $C^*(3,r)$. From (7.1) it follows that the primary operation $\chi\Phi^{1,0}(\lambda) = \lambda\delta_3$, is given by

(7.4)
$$\begin{cases} Sq^1 w^q = 0 \\ Sq^{2r-4} w^q = \binom{q}{r-2} w^{q+r-2} \\ Sq^{2r-2} w^q = \binom{q}{r-1} w^{q+r-1} \\ Sq^{2r} w^q = \binom{q}{r} w^{q+r} \end{cases}$$

Suppose that $\chi\Phi^{1,0}(\lambda) = 0$. From (7.4), this is equivalent to the conditions

(7.5)
$$\binom{q}{r} \equiv \binom{q}{r-1} \equiv \binom{q}{r-2} \equiv 0 \quad (\mathrm{mod}\ 2).$$

Then, the secondary operation $\chi\Phi^{2,0}(\lambda)$ is defined and its indeterminacy is the set of all maps

$$C_1^* \xrightarrow{\delta_2} C_2^* \xrightarrow{\zeta} H^*(CP^{q+r}/CP^{q-1};Z_2)$$

where ζ is any A-map of degree $2q-1$. The only non-trivial map is $\zeta(c_2^*) = w^q$ with $\zeta(c_{2i}^*) = 0$, for $i = 0,1,2$, The operations given by $\zeta\delta_2$ are $Sq^{2r}w^q$ and $Sq^{2r-2}w^q$ that already have been considered in (7.5).

The operation $\chi\Phi^{2,0}(\lambda)$ is equivalent to the three secondary operations constructed with the cycles $z_1 = \delta_2 c_1^*$, $z_2 = \delta_2 c_{10}^*$ and $z_3 = \delta_3 c_{11}^*$. These are the secondary Bockstein and the operations $\Phi_{2r}^{(2)} w^q$ and $\Phi_{2r-2}^{(2)} w^q$ already considered in (6.3). Then, assuming that $\chi\Phi^{2,0}(\lambda) = 0$, we have

(7.6)
$$\Phi_{2r-2}^{(2)} w^q = 0 \text{ and } \Phi_{2r}^{(2)} w^q = 0.$$

Hence, if $\chi\Phi^{1,0}(\lambda) = \chi\Phi^{2,0}(\lambda) = 0$, the operation $\chi\Phi^{3,0}(\lambda)$ is defined and its indeterminacy is $\chi\Phi^{3,1}(\xi)$, where $\xi : C_2^* \longrightarrow H^*(CP^{q+r}/CP^{q-1};Z_2)$ is of degree $2q-1$. The only non-trivial map ξ is given by $\xi(c_2^*) = w^q$ and $\xi(c_{2i}^*) = 0$, for $i = 0,1,2$, and it can be easily shown that the indeterminacy becomes the operation $\Phi_{2r}^{(2)}w^q$, already considered zero in (7.6).

The results of Maunder (see [13;Th.4.3.1]) assert that the tertiary operation

$$\chi\Phi^{3,0}(\lambda) = \Phi_{2r}^{(3)}w^q,$$

associated with the chain complex $C^*(3,r)$, can be computed by means of $\chi\Phi_{2r}^{(3)}w^n$, where $n = M - q - r - 1$. As in (4.17), we will distinguish several cases. To state our hypotheses let us list the following congruences.

(7.7) $\binom{q}{r} \equiv \binom{q}{r-1} \equiv 0 \mod 4$, (7.8) $\binom{q}{r} \equiv 0 \mod 4$,

(7.9) $\binom{q}{r-1} + 2\binom{q+2}{r-2} \equiv 0 \mod 4$,

(7.10) $\binom{q}{r-2} \equiv 0 \mod 2$, (7.11) $\equiv \binom{q+2}{r-1} \equiv 0 \mod 2$.

Theorem (7.12). *Let* q *and* r *be two integers fulfilling the hypotheses indicated below. Then the tertiary operation* $\Phi_{2r}^{(3)}w^q$ *is defined and with zero indeterminacy, we have*

$$\Phi_{2r}^{(3)}w^q = \left[{n \atop r}\right]w^{q+r},$$

where the coefficients $\left[{n \atop r}\right]$ *are given in the following table.*

Case	Hypotheses	$\left[{q \atop r} \right]$
$q+r = 4k$, $r = 2\ell$	(7.7) and (7.10)	$\frac{1}{4}\binom{q}{r} + \binom{q+5}{r-2}$
$q+r = 4k$, $r = 2\ell+1$	(7.7), (7.10) and (7.11)	$\frac{1}{4}\binom{q}{r} + \frac{1}{2}\binom{q+2}{r-1}$
$q+r = 4k+2$, $r = 2\ell$	(7.7) and (7.10)	$\frac{1}{4}\binom{q}{r} + \frac{1}{2}\binom{q+2}{r-1}$
$q+r = 4k+2$, $r = 2\ell+1$	(7.7) and (7.10)	$\frac{1}{4}\binom{q}{r} + \binom{q+4}{r-3}$
$q+r = 4k+1$, $r = 2\ell$	(7.8), (7.9) and (7.10)	$\frac{1}{4}\binom{q}{r} + \binom{q+4}{r-2}$
$q+r = 4k+1$, $r = 2\ell+1$	(7.7)	$\binom{q+1}{r-3}$
$q+r = 4k+3$, $r = 2\ell$	(7.7) and (7.10)	$\frac{1}{4}\binom{q}{r}$
$q+r = 4k+3$, $r = 2\ell+1$	(7.7)	0

Proof. We begin with the following auxiliary result. As before, let $n = M - q - r - 1$, where $M = M_{q+r+1}$ is the Atiyah-Todd number, considered in (5.2). Then, if θ and ε are two non-negative integers such that $2^{q+r} - 1 \geq q + \theta - r$, $q + \theta - \varepsilon \geq 0$, then

$$\text{(7.13)} \qquad \nu\binom{n+2r-\theta}{r-\varepsilon} = \nu\binom{q+\theta-\varepsilon}{r-\varepsilon}.$$

This is a generalization of (6.1) and the proof is the same.

Clearly, from the above considerations ((7.5), (7.6), etc.), the operation $\Phi^{(3)}_{2r}w^q$ is defined and has zero indeterminacy if and only if the conditions (7.6) are satisfied with zero indeterminacy. If we take these conditions from (6.3) and simplify them, with the same type of argument used in the proof of (4.17), we get the hypotheses indicated in the table.

Another form is to dualize the table of (4.17) directly from the relation

$$(7.14) \qquad (\chi\Phi^{(3)}_{2r})w^n = \Phi^{(3)}_{2r}w^q,$$

where $n = M - q - r - 1$. In this way we will obtain the hypotheses and the values of the operation.

Since the expressions depend on the mod 4 value of n and the parity of r, write n in the form $n = 4A + 3 - (q+r)$. Then, n and q+r (mod 4) are related as follows:

n	4K	4K+1	4K+2	4K+3
q+r	4k+3	4k+2	4k+1	4k

These four cases, combined with the parity of r, add to eight cases and each of them can be translated directly with the use of (7.13).

For example, if $q+r = 4k$ and $r = 2\ell$ then, n as in (7.14) is of the form $n = 4K+3$ and $r = 2\ell$. Using (7.13) for the expressions in the penult row of the table in (4.17), we get $\binom{n+2r}{r} \equiv \binom{q}{r} \equiv 0 \mod 4$, $\binom{n+2r-1}{r-1} \equiv \binom{q}{r-1} \equiv 0 \mod 4$ and $\binom{n+2r-2}{r-2} \equiv 0 \mod 2$. Also, $\frac{1}{4}\binom{n+2r}{r} \equiv \frac{1}{4}\binom{q}{r} \mod 2$ and $\binom{n+2r-7}{r-2} \equiv \binom{q+5}{r-2} \mod 2$. Therefore, the first row of our table follows.

All the other cases are alike, excepting $q+r = 4k+1$ and $r = 2\ell$. Here, the congruence (7.9) is of another type and we have to prove its equivalence with the corresponding congruence of (4.11). If any of the binomial coefficients contains a power of two higher than one, the result is trivial in this case. Otherwise, we can easily argue directly and the equivalence is established.

Centro de Investigación del IPN, México, D. F.
University of British Columbia, Vancouver, B. C.

References

[1] J. F. Adams, On the non-existence of elements of Hopf invariant one, Ann. of Math. 72(1960), 20-104.

[2] J. F. Adams, On Chern characters and the structure of the unitary group, Proc. Camb. Phil. Soc. 57(1961), 189-208.

[3] J. F. Adams, Vector fields on spheres, Ann. of Math. 75(1962), 603-632.

[4] J. F. Adams and G. Walker, On complex Stiefel manifolds, Proc. Camb. Phil. Soc. 61(1965), 81-103.

[5] J. Adem and S. Gitler, Secondary characteristic classes and the immersion problem, Bol. Soc. Mat. Mexicana 8(1963), 53-78.

[6] J. Adem and S. Gitler, Non-immersion theorems for real projective spaces, Bol. Soc. Mat. Mexicana 9(1964), 37-49.

[7] M. F. Atiyah and J. A. Todd, On complex Stiefel manifolds, Proc. Camb. Phil. Soc. 56(1960), 342-353.

[8] M. F. Atiyah, Thom complexes, Proc. London Math. Soc. (3) 11 (1961), 291-310.

[9] A. García-Máynez, Sobre algunas identidades de Jensen y Gould, Bol. Soc. Mat. Mexicana 14(1969), 70-74.

[10] S. Gitler, M. Mahowald and R. J. Milgram, Secondary cohomology operations and complex vector bundles, Proc. Amer. Math. Soc. 22(1969), 223-229.

[11] S. Gitler and R. J. Milgram, Unstable divisibility of the Chern character, Bol. Soc. Mat. Mexicana 16(1971), 1-14.

[12] I. M. James, Spaces associated with Stiefel manifolds, Proc. London Math. Soc. (3) 9(1959), 115-140.

[13] C. R. F. Maunder, Cohomology operations of the Nth kind, Proc. London Math. Soc. (3) 13(1963), 125-154.

[14] C. R. F. Maunder, Chern characters and higher-order cohomology operations, Proc. Camb. Phil. Soc. 60(1964), 751-764.

[15] F. P. Peterson and N. Stein, The dual of a secondary cohomology operation, Illinois J. Math., 4(1960), 397-404.

[16] D. Phee Hurley, Una familia de operaciones cohomológicas de orden superior, Doctoral dissertation, Centro de Investigación del IPN, 1973.

[17] E. H. Spanier and J. H. C. Whitehead, Duality in homotopy theory, Mathematika, 2(1955), 56-80.

CONSTRUCTION GEOMETRIQUE DE CERTAINES SERIES DISCRETES

par Jorge SOTO ANDRADE

La construction de toutes les représentations irréductibles (complexes) des groupes classiques est un problème difficile qui reste ouvert depuis longtemps. L'aspect le plus délicat en est la construction des représentations des séries discrètes (dites aussi représentations paraboliques (cf. [5])). Tout particulièrement, on manque de méthodes géométriques générales de construction de ces séries.

A défaut de méthodes générales, le but de ce travail est de montrer comment dans un cas spécial intéressant, celui du groupe $G = GSp(4,k)$ des similitudes symplectiques en 4 variables sur un corps fini k (de caractéristique quelconque), on arrive à obtenir une réalisation géométrique d'une série discrète, à savoir celle associée à l'unique extension quadratique K de k. Rappelons que le groupe G possède en outre une série discrète associée à l'unique extension quartique de k et une série discrète exceptionnelle paramétrée par $Car(k^{\times})$. Toutes les représentations irréductibles de ce groupe n'ont d'ailleurs été construites qu'en 1974 (cf. [1], [2]), par une décomposition de sa "représentation de Weil", bien que les caractères de groupe symplectique associé $Sp(4,k)$ étaient connus depuis 1968 [6]. Cette méthode ne fournit cependant pas directement une construction géométrique des séries discrètes. Signalons enfin que l'on a pas encore une construction générale explicite de toutes les représentations des groupes symplectiques de rang supérieur.

Le plan de l'exposé est le suivant.

Dans le paragraphe 1 , nous rappelons la construction de la représentation de Weil de $H = GL(2,k)$, dont nous nous servons dans le paragraphe 2 pour construire les représentations $(Bin(\pi_1, \pi_2), \tau)$ (où π_1 et π_2 sont des représentations de la série discrète de H . Ces représentations s'obtiennent à partir de la représentation évidente de G (que nous appelons bi-naturelle et notons (Bin, τ) ci-dessous) dans l'espace des fonctions complexes de paires de vecteurs dans $E = k^4$. Elles ne sont autres que les composantes isotypiques de Bin suivant une double action de H dans Bin , donnée par l'action naturelle évidente et par une représentation de Weil. Nous donnons ensuite une réalisations plus maniable et géométrique des représentations $Bin(\pi_1, \pi_2)$, comme de sous-représentations de la représentation naturelle $(Hyp(\pi_1), \tau)$ de G (avec $\pi_2 = \pi_\Lambda$ $(\Lambda \in Car(K^\times)$, $\Lambda \neq \Lambda^q))$ dans l'espace $Hyp(\pi_1)$ des fonctions complexes de bases des plans hyperboliques dans E (pour la forme bilinéaire alternée J qui définit G) qui sont π_1 - homogènes par rapport à l'action naturelle de H sur ces bases. Les représentations en question s'obtiennent en imposant une condition supplémentaire d'harmonicité où intervient le caractère Λ . Dans le paragraphe 3 , nous établissons l'irréductibilité des représentations $Hyp(\pi_1, \Lambda)$, en étudiant l'algèbre commutante $A(\pi_1)$ de $(Hyp(\pi_1), \tau)$. Le point essentiel y est la commutativité de cette algèbre. Enfin , nous décrivons l'entrelacement entre ces représentations, qui fournissent (pour $\pi_1 \not\approx \pi_\Lambda$) la série discrète de G associée à l'unique extension quadratique K de k .

Le passage de G au groupe symplectique $Sp(4,k)$, par restriction, ne présente pas de difficultés. Nous travaillons plutôt avec G car, de manière générale, la structure des représentations des groupes des similitudes associés est plus simple que celle des groupes classiques eux-mêmes (cf. [2] , [3]) . D'autre part, on remarquera que l'essentiel de nos méthodes est transposable au cas continu (notemment au cas réel ...), au prix de quelques difficultés techniques. Nous espérons y revenir ailleurs.

§ 1. La représentation de Weil de $H = GL(2,k)$

Pour plus de détails sur les résultats que nos rappelons ici, nous renvoyons à [3] .

1.1 Définitions.

Notons k le corps fini à q éléments (de caractéristique quelconque) . Posons $H = GL(2,k)$. Définissons les générateurs $\underline{h}(t)$ $(t \in k^\times)$, $\underline{h}'(r)$ $(r \in k^\times)$ $\underline{u}(s)$ $(s \in k^+)$ et w de H par

(1) $$\underline{h}(t) = \begin{pmatrix} t & 0 \\ 0 & t^{-1} \end{pmatrix} \qquad (t \in k^\times) ,$$

(2) $$\underline{h}'(r) = \begin{pmatrix} 1 & 0 \\ 0 & r \end{pmatrix} \qquad (r \in k^\times) ,$$

(3) $$\underline{u}(s) = \begin{pmatrix} 1 & s \\ 0 & 1 \end{pmatrix} \qquad (s \in k^+) ,$$

(4) $$w = \begin{pmatrix} 0 & 1 \\ 1 & 0 \end{pmatrix} .$$

Soit (E,Q) un espace quadratique (non-dégénéré) sur k . Notons B la forme bilinéaire associée, définie par

(5) $$B(x,y) = Q(x+y) - Q(x) - Q(y) \qquad (x,y \in E) .$$

Notons X l'ensemble des caractères non-triviaux de k^+ . Le groupe multiplicatif $k^\times$ agit transitivement sur X par

(6) $$\psi^t(r) = \psi(tr) \qquad (t \in k^\times,\ r \in k^+, \psi \in X) .$$

THEOREME 1.-

Supposons que la dimension de l'espace quadratique (E,Q) soit paire, égale à $2m$. Soit U un espace vectoriel complexe quelconque. On peut définir une représentation (V_Q, ρ_Q) de H, appelée représentation de Weil de H associée à (E, Q) (et à U), en posant

$$(7) \qquad V_Q = V^{E \times X}$$

et en se donnant $\rho_Q = \rho$ sur les générateurs ci-dessus de H par les formules suivantes

$$(8) \qquad [\rho(\underline{h}(t))\, f](x,\psi) = f(tx,\psi) \qquad (t \in k^{\times}),$$

$$(9) \qquad [\rho(\underline{h}'(r^{-1}))\, f]\,(x,\psi) = f(x,\psi^{r}) \qquad (r \in k^{\times})$$

$$(10) \qquad [\rho(\underline{u}(s))\, f](x,\psi) = \psi(sQ(x))\, f(x,\psi) \qquad (s \in k^{+})$$

$$(11) \qquad [\rho(w)\, f\,](x,\psi) = \varepsilon(Q)\, q^{-m} \sum_{y \in E} \psi(B(x,y))\, f(y,\psi),$$

(ou $\varepsilon(Q)$ dénote le signe de la forme quadratique Q, égal à 1 si Q est déployée et à -1 sinon), pour $f \in V_Q$, $x \in E$, $\psi \in X$.

D'autre part, si γ est une similitude de (E,Q), de multiplicateur $m(\gamma)$, alors γ induit un automorphisme $\sigma(\gamma)$ de (V_Q,ρ_Q) défini par

$$(12) \qquad [\sigma(\gamma)\, f](x,\psi) = f(\gamma\, x,\, \psi^{m(\gamma)^{-1}}),$$

pour $f \in V_Q$, $x \in E$, $\psi \in X$.

1.2 Décomposition dans le cas du plan non-déployé.

La dernière assertion du théorème 1 montre que la représentation de Weil (V_Q,ρ_Q) se décompose suivant les représentations irréductibles du groupe de similitudes $GO(Q)$ de (E,Q). En décomposant les deux représentations de Weil associées aux deux plans quadratiques sur k, on obtient toutes les représentations irréductibles de H. Le plan quadratique non-déployé fournit la série discrète de représentations de H (formée des représentations irréductibles qui n'admettent pas de vecteur fixe par le sous-groupe U des $\underline{u}(s)$ $(s \in k^{+})$).

Nous réalisons le plan quadratique non-déployé comme (K,N) , où K désigne l'unique extension quadratique de K et N la norme de K sur k .

Le groupe $GO(N)$ est alors le produit semi-direct du groupe des γ_a (où $\gamma_a(b) = ab$, pour $b \in K$, $a \in K^\times$) et du groupe $\{1,F\}$ où F désigne la conjugation ($F(a) = a^q$ ($a \in K$)) , avec les relations

$$F^2 = 1 \qquad \text{et} \qquad F\gamma_a = \gamma_{a^q} F \qquad (a \in K^\times) .$$

DEFINITION 1.-

Pour tout caractère $\Lambda \in Car(K^\times)$, *notons* V_Λ *le sous-espace de la représentation de Weil* (V , ρ_N) *associée à* (K,N) (et à $U = \mathbb{C}$) *formé des* $f \in V_N$ *telles que*

$$(13) \qquad f(ab, \psi^{N(a)^{-1}}) = \Lambda(a) f(b,\psi) \qquad (a \in K^\times,\ b \in K, \psi \in X) .$$

Nous notons de plus π_Λ la restriction de ρ_N au sous-espace stable V_Λ de V_N

THEOREME 2.-

Les représentations (V_Λ, π_Λ) *sont irréductibles pour tout* $\Lambda \in Car(K^\times)$. *Le seul cas d'isomorphie entre elles est*

$$(V_\Lambda, \pi_\Lambda) \simeq (V_{\Lambda^q}, \pi_{\Lambda^q}) \qquad (\Lambda \in Car(K^\times)) ;$$

l'isomorphisme est donné par

$$[F(f)](a,\psi) = f(a^q,\psi) \qquad (f \in V_N,\ a \in K,\ \psi \in X)$$

on a

$$\dim V_\Lambda = q - 1 \qquad (\Lambda \neq \Lambda^q) ,$$

$$\dim V_\Lambda = q \qquad (\Lambda = \Lambda^q)$$

La série de types d'isomorphie des (V_Λ, π_Λ) *pour* $\Lambda \in Car(K^\times) - Car(k^\times)$ *(c'est-à-dire* $\Lambda \neq \Lambda^q$) *est la série discrète de* H . *Elle est formée de* $(1/2)q(q - 1)$ *types d'isomorphie.*

REMARQUES.- 1) On obtient une base de V_Λ $(\Lambda = \Lambda^q)$ en associant à chaque $\phi \in X$ le vecteur v_ϕ de V de support minimal tel que $v_\phi(1,\phi) = 1$.

2) La restriction de π_Λ au sous-groupe B de H formé des matrices triangulaires supérieures est encore irréductible et son type d'isomorphie ne dépend que de la restriction de Λ à $k^\times$.

§ 2. La représentation binaturelle de $G = GSp(4,k)$.

2.1 Définitions.-

Nous notons G le groupe des similitudes symplectiques $GSp(4,k)$ en quatre variables sur le corps fini $k = \mathbf{F}_q$ à q élements. De maniere plus précise, nous réalisons G comme le groupe de similitudes de la forme bilinéaire alternée non-dégénérée J dans $E = k^4$, telle que

(1) $$J(u,v) = u_1v_3 - u_3v_1 + u_2v_4 - u_4v_2$$

pour $u = (u_1,u_2,u_3,u_4)$, $v = (v_1,v_2,v_3,v_4)$ dans E . Pour qu'un k-automorphisme g de E appartienne alors à G , il faut et il suffit que

(2) $$J(ug,vg) = m(g)J(u,v) \qquad (u,v \in E)$$

pour un scalaire convenable $m(g)$, appelé le multiplicateur de g . Ci-dessus, nous faisons agir g à droite dans E , simplement par multiplication matricielle (ce choix s'avère plus commode dans la suite) .

DEFINITION 1.-

Notons X l'ensemble des caracteres non-triviaux du groupe additif k^+ . Posons $M = E \times E$ et $\tilde{M} = M \times X$.

On appelle représentation binaturelle de G la représentation (Bin,τ) , d'espace $\mathrm{Bin} = \mathbf{C}^{\tilde{M}}$ et d'action τ donnée par

$$[\tau(g)f](\xi) = f(\xi g) \qquad (g \in G, f \in \mathrm{Bin}, \xi \in \tilde{M}) ,$$

ou $$\xi g = (x_1 g, x_2 g \,;\, \psi^{m(g)^{-1}})$$

pour $$\xi = (x,\psi) = (x_1,x_2;\psi) \in \tilde{M} .$$

2.2 Décomposition de (Bin, τ).-

La représentation bi-naturelle de G peut se décomposer à l'aide d'une double action du groupe $H = GL(2,k)$ dans l'espace Bin. La première action est évidente, de nature géométrique, mais la seconde l'est un peu moins etant donnée par une représentation de Weil (*cf.* 1.1)

DEFINITION 2.-

Pour tout $h = \begin{pmatrix} a & b \\ c & d \end{pmatrix} \in H$, $\xi = (x_1, x_2 ; \psi) \in \widetilde{M}$, $f \in V^{\widetilde{M}}$ (où V est un espace vectoriel complexe quelconque), on pose

$$h\xi = (hx, \psi^{\det(h)^{-1}}) = (ax_1 + bx_2, cx_1 + dx_2 ; \psi^{\det(h)^{-1}})$$

et

$$(\sigma_1(h)f)(\xi) = f(h\xi)$$

Nous avons ainsi une action à droite (ou anti-représentation) σ_1 de H dans $V^{\widetilde{M}}$.

Pour définir la seconde action, nous remarquons que la forme bilinéaire alternée J peut etre regardée comme une forme quadratique, de rang 8, sur l'espace $M = E \times E = k^8$. Il lui est associée la forme bilinéaire symétrique B définie

$$B(x,y) = J(x_1, y_2) + J(y_1, x_2) \tag{3}$$

pour $x = (x_1, x_2) \in M$ et $y = (y_1, y_2) \in M$. D'après le paragraphe 1, il lui est aussi associée une représentation de Weil ρ_J de H dans l'espace $V^{\widetilde{M}}$

DEFINITION 3.-

Pour tout $h \in H$ et $f \in V^M$, on pose,

$$\sigma_2(h)f = \rho_J(h)^{-1} f .$$

Il est clair que les anti-répresentations σ_1 et σ_2 commutent entre elles et aussi au prolongement évident de τ à l'espace $V^{\widetilde{M}}$, noté encore τ.

En effet, tout $h \in H$ et tout $g \in G$ définissent des similitudes orthogonales de l'espace quadratique (M,J) , donc, d'après 1.1 , les opérateurs $\sigma_1(h)$ et $\tau(g)$ commutent aux opérateurs $\rho_J(h')$. $(h' \in H)$. Il est enfin trivial que $\sigma_1(h)$ et $\tau(g)$ commutent, pour tout $h \in H$ et $g \in G$.

Il s'ensuit que (Bin,τ) se décompose suivant les composantes isotypiques de l'anti-représentation $\sigma_1 \otimes \sigma_2$ de $H \times H$ dans Bin . Il sera commode d'écrire les types de ces composantes sous la forme $\pi_1 \otimes \check{\pi}_2$ ou mieux $[\pi_2, \pi_1]$, où π_1 et π_2 désignent des représentations irréductibles de H ($\check{\pi}_2$ dénote la contragrédiente de π_2) et où l'on a posé la

DEFINITION 4.-

Si (V_1, π_1) et (V_2, π_2) sont deux représentations de H , on note $([V_2,V_1] , [\pi_2, \pi_1])$ la représentation de H donnée par

$$[V_2,V_1] = \mathrm{Hom}_{\mathbb{C}}(V_2,V_1) ,$$

$$[\pi_2,\pi_1](h,h') \psi = \pi_1(h') \circ \psi \circ \pi_2(h)^{-1}$$

pour $\psi \in [V_2,V_1]$, $h, h' \in H$.

Nous réalisons ces composantes isotypiques de la manière suivante.

DEFINITION 5.-

Soient (V_1, π_1) et (V_2, π_2) deux représentations irréductibles de H . On note $(\mathrm{Bin}(\pi_1,\pi_2) ,\tau)$ la représentation de G dont l'espace $\mathrm{Bin}(\pi_1,\pi_2)$ est formé des fonctions F de $\tilde{M} = M \times X$ dans $[V_2,V_1]$ telles que

$$\sigma_2(h)\, \sigma_1(h')F = [\pi_2, \pi_1] (h,h')F \qquad (h,h' \in H) \tag{4}$$

et dont l'action τ est donnée par

$$(\tau(g)F)(\xi) = F(\xi g) \tag{5}$$

pour $g \in G$, $F \in \mathrm{Bin}(\pi_1, \pi_2)$, $\xi \in \tilde{M}$.

La condition (4) équivaut aux conditions

(6) $$F(h'\xi) = \pi_1(h') \circ F(\xi) \qquad (h' \in H \;,\; \xi \in \widetilde{M}) \;,$$

et

(7) $$[\rho_J(h)F](\xi) = F(\xi) \circ \pi_2(h) \qquad (h \in H \;,\; \xi \in \widetilde{M}) \;.$$

Bien entendu, come σ_1 et ρ_J coïncident sur le centre de H, pour que $\mathrm{Bin}(\pi_1, \pi_2)$ soit non-nul, il faut que π_1 et π_2 coïncident aussi sur le centre de H.

Nous verrons qu'en prenant π_1 et π_2 dans la série discrète de H (et coïncidant sur le centre de H), on obtient une famille de représentations irréductibles de G qui constitue la série discrète de G associée au tore T_1 de G, formé des $g \in G$ tels que $P_{13}g = P_{13}$ et $P_{24}g = P_{24}$ (où l'on note P_{ij} le plan de E engendré par les vecteurs e_i et e_j de la base canonique de E, $(1 \leqslant i, j \leqslant 4)$). Nous suivons ici la terminologie de [4].

2.3 Une autre réalisation de $\mathrm{Bin}(\pi_1,\pi_2)$

Nous supposons dorénavant que π_1 et π_2 sont des représentations irréductibles de la série discrète de H qui coïncident sur le centre de H. Nous réalisons en autre ces représentations par leurs modèles de Weil. (cf. 1.2.)

Prenons $\pi_2 = \pi_\Lambda$ $(\Lambda \in \mathrm{Car}(K^\times) - \mathrm{Car}(k^\times))$. Pour tout $\psi \in X$, notons v_ψ l'element de l'espace $V_2 = V_\Lambda$ de π_2 de support minimal tel que $v_\psi(1,\psi) = 1$.

Notons en outre Hyp l'ensemble des $x = (x_1,x_2) \in M$ tels que $J(x) \neq 0$. L'ensemble Hyp est alors l'ensemble des bases des plans hyperboliques pour J dans E et l'on a

(8) $$|\mathrm{Hyp}| = |H|^{-1}(q^4-1)(q^4-q^3) = q^2\cdot(q^2+1)$$

(avec $q = |k|$).

PROPOSITION 1.-

En associant à chaque $F \in \mathrm{Bin}(\pi_1, \pi_2)$ la fonction f de Hyp dans V_1 définie par

(9) $$f(x) = [F(x,\psi)](v_{\psi^{J(x)}}) \qquad (x \in \mathrm{Hyp})$$

pour un choix arbitraire de $\psi \in X$, on établit un isomorphisme de $(\mathrm{Bin}(\pi_1, \pi_2), \tau)$ sur la représentation $(\mathrm{Hyp}(\pi_1,\Lambda), \tau)$ d'espace $\mathrm{Hyp}(\pi_1,\Lambda)$ formé de toutes les fonctions f de Hyp dans V_1 telles que

(10) $$f(hx) = \pi_1(h)(f(x)) \qquad (h \in H,\ x \in \mathrm{Hyp})$$

et

(11) $$H_{r,s} f = -q^3 \Lambda(r,s) f \qquad (r \in k^\times,\ s \in k^+),$$

où

$$(H_{r,s} f)(x) = \sum_{\substack{y \in \mathrm{Hyp} \\ J(y) = rJ(x) \\ B(x) = sJ(x)}} f(y) \qquad (x \in \mathrm{Hyp})$$

et

$$\Lambda(r,s) = \sum_{\substack{z \in K \\ N(z) = r \\ Tr(z) = s}} \Lambda(z) \qquad (r \in k^\times,\ s \in k^+)\ ;$$

l'action τ dans $\mathrm{Hyp}(\pi_1,\Lambda)$ est donnée par

$$[\tau(g) f](x) = f(xg) \qquad (x \in \mathrm{Hyp},\ g \in G).$$

Démonstration :

La condition (7) pour F, apliquée a $h = \underline{u}(t)$ $(t \in k^+)$ entraîne aussitôt que l'aplication linéaire $F(\xi) \in [V_2, V_1]$ doit s'annuler sur les composantes U-isotypiques de V_2 de type autre que $\psi^{J(x)}$. C'est-à-dire, $F(\xi)$ s'annule sur tous les v_ϕ tels que $\phi \neq \psi^{J(x)}$. En particulier, comme π_2 est dans la série discrète, on en tire que $F(x,\psi) = 0$ dès que $J(x) = 0$

D'autre part, la condition (7) pour $h = \underline{h}'(t^{-1})$ $(t \in k^\times)$ donne, avec $\theta = \psi^{J(x)}$,

$$F(x,\psi^t)(v_{\theta^t}) = F(x,\psi)(\pi_1(\underline{h}'(t^{-1}))\, v_{\theta^t}) = F(x,\psi)(v_\theta)$$

pour tout $x \in \mathrm{Hyp}$, $\psi \in X$, $t \in k^\times$ ce qui justifie la définition (9) . Pour achever la démonstration, nous n'avons qu'à montrer que la condition (7) pour F et pour $h = w$ équivaut à la condition (11) pour la restriction f de F donnée par (9) . Or on a, pour $F \in \mathrm{Bin}(\pi_1,\pi_2)$, $x \in \mathrm{Hyp}$, $\psi \in X$ et $\theta = \psi^{r'}$ $(r' \in k^\times)$,

$$(12) \qquad [(\rho_J(w)F)(x,\psi)]\,(v_\theta) = q^{-4} \sum_{\substack{y \in \mathrm{Hyp} \\ J(y) = r'}} \psi(B(x,y))\, f(y) \quad ;$$

d'autre part

$$\pi_2(w)v_\theta = -q^{-1} \sum_{t \in k^\times} \Lambda^{-1}(t) \Big[\sum_{\substack{b \in K \\ N(b) = t}} \theta(\mathrm{Tr}(b))\, \Lambda(b)\Big]\, v_{\theta^t} \quad ,$$

d'où

$$F(x,\psi)(\pi_2(w)v_\theta) = -q^{-1}\, \Lambda(r') \Big[\sum_{\substack{a \in K^\times \\ N(a) = r'J(x)}} \psi(\mathrm{Tr}(a))\, \Lambda^{-1}(a)\Big] f(x) \quad ,$$

Par conséquent, (12) équivaut à

$$(13) \qquad \sum_{\substack{y \in \mathrm{Hyp} \\ J(y) = r'}} \psi(B(x,y)) f(y) = -q^3\, \Lambda(r') \sum_{\substack{a \in K^\times \\ N(a) = r'J(x)}} \psi(r(a))\Lambda^{-1}(a)\; f(x)$$

pour tout $\psi \in X$, $x \in \mathrm{Hyp}$ et $r' \in k^\times$. Or, dans le cas $\psi = 1$, l'égalité (13) est encore vérifiée: le membre de gauche est nul car f est π_1 - homogène et la restriction de π_1 à $SL(2,k)$ ne contient pas de sous-représentation unité, et le membre de droite est aussi nul car $\Lambda \neq \Lambda^q$. Comme les fonctions $\psi \in \mathrm{Car}(k^+)$ engendrent linéairement toute fonction sur $k^\times$, il s'ensuit que (13) équivaut à

$$(14) \quad \sum_{\substack{y \in Hyp \\ J(y) = r' \\ B(x,y) = s'}} f(y) = -q^3 \Lambda(r') \Big[\sum_{\substack{a \in K^\times \\ N(a) = r'J(x) \\ Tr(a) = s'}} \Lambda^{-1}(a) \Big] f(x)$$

pour tout $r' \in k^\times$, $s' \in k^+$. En posant $r' = rJ(x)$ et $s' = sJ(x)$, la relation (11) résulte aussitot.

C. Q. F. D.

§ 3. L'entrelacement des $Bin(\pi_1, \pi_2)$.

3.1 L'algebre commutante $A(\pi)$.

Nous posons dorénavant $\pi_1 = \pi$ et $\pi_2 = \pi_\Lambda$. Dans le paragraphe précédent nous avons alors réalisé $Bin(\pi, \pi_\Lambda)$ comme une sous-représentation $Hyp(\pi,\Lambda)$ de la représentation $Hyp(\pi)$ de G donnée par la

DEFINITION 1.-

Soit (V,π) une représentation irréductible de H . On note $(Hyp(\pi), \tau)$ la représentation de G dont l'espace $Hyp(\pi)$ est formé des fonctions f de Hyp dans V telles que

$$f(hx) = \pi(h)(f(x)) \qquad (h \in H, \quad x \in Hyp)$$

et dont l'action τ est donnée par

$$[\tau(g) f](x) = f(xg) \qquad (g \in G, \quad f \in Hyp(\pi), x \in Hyp) .$$

Dans la suite, nous suppons donc que π appartient à la série discrète de H . L'irréductibilité et le non-entrelacement des représentations $Hyp(\pi,\Lambda)$ (quand π_Λ parcourt la série discrète de H et π est fixée) résultera de l'étude de l'algèbre commutante $A(\pi)$ de $(Hyp(\pi),\tau)$.

Le résultat suivant, de vérification immédiate, décrit de manière générale l'espace $E(\pi,\pi')$ des opérateurs d'entrelacement de $Hyp(\pi)$ dans $Hyp(\pi')$.

PROPOSITION 1.-

<u>L'espace d'opérateurs d'entrelacement</u> $E(\pi,\pi')$ <u>est isomorphe a l'espace</u> $N(\pi,\pi')$ <u>formé de tous les noyaux</u> $K : Hyp \times Hyp \to Hom_{\mathbb{C}}(V,V')$ <u>invariants par</u> G et (π,π') - <u>homogènes</u>, <u>c'est-à-dire vérifiant</u>

(1) $$K(xg,yg) = K(x,y) \qquad (x,y \in Hyp\ ,\ g \in G)\ ,$$

et

(2) $$K(hx,h'y) = \pi'(h)K(x,y)\pi(h')^{-1} \qquad (x,y \in Hyp,\ h,h' \in H)\ .$$

<u>A un noyau</u> $K \in N(\pi,\pi')$ <u>est associé, par cet isomorphisme, l'opérateur d'entrelacement</u> $\phi \in E(\pi,\pi')$ <u>donné par</u>

$$(\phi f)(x) = \sum_{y \in Hyp} K(x,y)\, f(y) \qquad (f \in Hyp(\pi),\ x \in Hyp)\ .$$

Dans la suite de ce numéro, nous posons $N(\pi) = N(\pi,\pi)$ et nous écrirons toujours $A(\pi)$ a la place de $E(\pi,\pi)$.

DEFINITION 2.-

<u>Pour</u> $x \in Hyp$, <u>notons</u> $P(x)$ <u>le plan hyperbolique engendré par</u> x . <u>Pour</u> $x,y \in Hyp$, <u>définissons la matrice</u> $\mathbf{B}(x,y) \in M_2(k)$ <u>par</u>

$$\mathbf{B}(x,y)_{ij} = J(x_i,y_j) \qquad (1 \leqslant i,\ j \leqslant 2)\ .$$

On a alors

(3) $$\mathbf{B}(h'x,hy) = h'B(x,y)h^* \qquad (h,h' \in H,\ x,y \in Hyp)\ ,$$

(4) $$\mathbf{B}(y,x) = -\mathbf{B}(x,y)^* \qquad (x,y \in Hyp)$$

(où l'on note a^* la matrice transposée de $a \in M_2(k)$) .

PROPOSITION 2.

La liste des $H \times H \times G$ - orbites dans $Hyp^2 = Hyp \times Hyp$ (suivant l'action $(x,y) \to (hxg, h'xg)$ $(x,y \in Hyp \; ; \; h,h' \in H, g \in G)$) est la suivante:

i) l'orbite dégénérée 0_o formée de tous les $(x,y) \in Hyp^2$ tels que $P(x) = P(y)$;

ii) l'orbite semi-dégénérée de tous les $(x,y) \in Hyp^2$ tels que $P(x) \cap P(y)$ est une droite, notée 0_1 ;

iii) les orbites régulières 0_s $(s \in k^\times, s \neq 1)$ formées, chacune, de tous les $(x,y) \in Hyp^2$ tels que $P(x) \cap P(y) = \underline{0}$ et $I(x,y) = s$, où

$$I(x,y) = J(x)^{-1} J(y)^{-1} \det(\mathbf{B}(x,y)) \;;$$

iv) l'orbite régulière 0^1 formée de tous les $(x,y) \in Hyp^2$ tels que $P(x) \cap P(y) = \underline{0}$ et $\mathrm{rang}(\mathbf{B}(x,y)) = 1$;

v) l'orbite régulière 0° formé de tous les $(x,y) \in Hyp^2$ tels que $P(x)$ soit orthogonal à $P(y)$.

La vérification, facile, est laissée au lecteur. Notons néanmois qu'il n'existe pas de $(x,y) \in Hyp^2$ tel que $P(x) \cap P(y) = \underline{0}$ et que

$$\det(\mathbf{B}(x,y) = J(x)J(y) \quad !$$

REMARQUE.- L'invariant I vaut 1 sur les orbites 0_o et 0_1 et il s'annule sur les orbites 0^1 et 0°.

PROPOSITION 3.

Soit $\phi \in A(\pi)$ et $K \in N(\pi)$ son noyau. Posons $R(x, y) = B(x, y)w^{-1}J(y)^{-1}$ pour $x,y \in Hyp$. Alors on a

i) K est nul sur les orbites 0° et 0^1 ;

ii) $K(x,y) = \lambda_o \pi(x|y)$ pour $(x,y) \in 0_o$, où l'on note $(x|y)$ l'unique $h \in H$ tel que $x = hy$; et où $\lambda_o \in \mathbb{C}$,

iii) $K(x,y) = \lambda_t \pi(R(x,y))$ pour $(x,y) \in 0_t$ $(t \in k^{\times})$, avec $\lambda_t \in \mathbb{C}$.

iv) $$(\phi f)(x) = \mu_o f(x) + \sum_{\substack{P \in Hyp \\ I(P(x),P) \neq 0}} \mu_{I(P(x),P)} f(pr_P(x))$$

pour tout $f \in Hyp(\pi)$, $x \in Hyp$, où l'on note HYP l'ensemble des tous les plans hyperboliques dans E et où l'on a posé

$$pr_P(x) = R(x,y)y \qquad (x \in Hyp)$$

pour une base arbitraire $y \in P$.

Démonstration :

Pour établir i) à iii), on considère les stabilisateurs dans $H \times H \times G$ des $(x,y) \in Hyp^2$. Notons e_1,e_2,e_3,e_4 la base canonique de E. Pour le représentant $(x^o,y^o) = ((e_1,e_3), (e_2,e_4))$ de 0° on trouve aussitôt, d'après les conditions (1) et (2)

$$\pi(h)K(x^o,y^o) = K(x^o,y^o)\pi(h')$$

pour tous les $h,h' \in H$ de même determinant, d'où $K(x^o,y^o) = \underline{0}$.

De manière analogue, en considérant le représentant $(x^1,y^1) = ((e_1,e_3), (e_2+e_3,e_4))$ de 0^1, on obtient, en particulier,

$$\pi(u)K(x^1,y^1) = K(x^1,y^1)\pi(u') \tag{5}$$

pour tous les u, u' dans le sous-groupe unipotent supérieur U de H. Comme π est dans la série discrète, sa restriction à U ne contient pas la sous-représentation unité, et il s'ensuit donc de (5) que $K(x^1,y^1) = \underline{0}$. Cela établit i).

Pour montrer ii), on remarque, de même, que

$$\pi(h)K(x,x) = K(x,x)\pi(h) \qquad (x \in Hyp, \ h \in H) .$$

Par le lemme de Schur, $K(x,x)$ est donc une homothétie, d'où ii).

Pour établir iii), considérons tout d'abord le cas de 0^1. Posons $(x^1, y^1) = ((e_1,e_3), (e_1 + e_2,e_3))$. L'endomorphisme $K(x^1,y^1)$ doit satisfaire alors à la (seule) condition

$$\pi(b)K(x^1,y^1) = K(x^1,y^1)\pi(b)$$

pour tout $b \in B$ (où l'on note B le sous-groupe triangulaire supérieur de H). Comme la restriction de π à B est encore irréductible, $K(x^1,y^1)$ doit être une homothétie. Pour $(x,y) \in 0_1$, il existe $h,h' \in H$, $g \in G$ tels que $(x, y) = (h'x^1g, hy^1g)$ et donc

$$K(x,y) = \lambda_1 \pi(h'h^{-1}) = \lambda_1 \pi(h'\mathbf{B}(x^1,y^1)h^{*}w^{-1} \det(h)^{-1})$$

$$= \lambda_1 \pi(R(x,y))$$

(où λ_1 désigne le rapport de l'homothétie $K(x^1,y^1)$).

Soit enfin $(x,y) \in 0_s$ $(s \in k^{\times}, s \neq 1)$. Pour $h,h' \in H$, la condition $(h'x,hy) = (xg,yg)$ pour $g \in G$ convenable, équivaut aussitôt (à l'aide du théorème de Witt) à

$$h' = \det(h)\mathbf{B}(x,y)h^{*-1}\mathbf{B}(x,y)^{-1} ,$$

c'est-à-dire à

$$h' = (\mathbf{B}(x,y)w^{-1})h(\mathbf{B}(x,y)w^{-1})^{-1}$$

et par suite, puisque

$$\pi(h')K(x,y) = K(x,y)\pi(h)$$

et que π et la representation conjuguée π' : $h \mapsto$ (h') sont irréductibles, on en déduit que $K(x,y)$ doit être proportionnel à l'opérateur d'entrelacement évident $(\mathbb{B}(x,y)w^{-1})$ de π dans π'. On a donc

$$K(x,y) = \lambda(x,y)\pi(\mathbb{B}(x,y)w^{-1})$$

pour $\lambda(x,y) \in \mathbb{C}$ convenable. Or comme K satisfait à (1) et (2), on a

$$\lambda(h'xg, hyg) = \lambda(x,y)\pi(m(g)\det(h))^{-1} \qquad (h', h \in H),$$

d'où

$$\lambda(x,y) = \lambda_s \pi(J(y))^{-1} \qquad ((x,y) \in 0_s),$$

ce qui achève d'établir iii)

La formule iv) est une conséquence immédiate de i), ii) et iii).

C.Q.F.D.

REMARQUES.-

(1) Aussi pour $(x,y) \in 0_o$, on a $K(x,y) = \lambda_o \pi(R(x,y))$.

(2) On a

$$R(h'xg, hyg) = \pi(h)R(x,y)\pi^{-1}(y)$$

pour $x, y \in Hyp$, $h,h' \in H$ et $g \in G$.

TEOREME 1.-

<u>L'algebre commutante</u> $A(\pi)$ <u>est commutative et de dimension</u> q <u>sur</u> $\mathbb{C}$.

<u>Démonstration</u> :

Le fait que $\dim A(\pi) = q$ résulte aussitôt de la proposition ci-dessus. Montrons qu'elle est commutative.

Notons K_s le noyau dans $N(\pi)$ de support 0_s avec $\lambda_s = 1$ $(s \in k^\times)$. Il suffit de prouver que $K_r K_s = K_s K_r (r,s \in k^\times)$, et donc que

$$(6) \qquad (K_r K_s)(x,z) = (K_s k_r)(x,z)$$

pour tout $(x,z) \in Hyp^2$. Mais comme, à priori, $K_s K_r$ et $K_r K_s$ appartiennent à $N(\pi)$, il suffit de montrer (6) pour un (x,z) dans chaque 0_t $(t \in k^+)$, et encore, comme $(K_s K_r)(x,z)$ et $(K_r K_s)(x,z)$ sont des multiples scalaires de $\pi(R(x,z))$, il suffit de prouver que les traces des deux membres de (6) coïncident.

Or on a

$$(7) \qquad \begin{aligned} Tr(\pi(R(z,x)) &= Tr(\pi(-\mathbf{B}(x,z)^* w^{-1} J(x)^{-1})) \\ &= Tr(\pi(w^{-1}\mathbf{B}(x,z)J(z)^{-1})) \\ &= Tr(\pi(R(x,z))) \end{aligned}$$

pour tout $(x,z) \in 0_t$ tel que $J(x) = J(z)$ (en se servant du fait que h^* est conjugué à h, por tout $h \in H$). On en tire que, por tout $(x,z) \in 0_t$, avec $J(x) = J(z)$, on a

$$\begin{aligned} Tr((K_s K_r(x,z)) &= Tr((K_s K_r)(z,x)) \\ &= \sum_{\substack{(z,y)\in 0_s \\ (y,x)\in 0_r}} Tr(\pi(\mathbf{B}(z,y)w^{-1}J(y)^{-1}\mathbf{B}(y,x)w^{-1}J(x)^{-1})) \\ &= \sum_{\substack{(x,y)\in 0_r \\ (y,z)\in 0_s}} Tr(\pi(w\mathbf{B}(x,y)J(y)^{-1}w\mathbf{B}(y,z)J(z)^{-1})) \\ &= Tr((K_r K_s)(x,z)) , \end{aligned}$$

comme voulu.

COROLLAIRE.-

La représentation $(\mathrm{Hyp}(\pi), \tau)$ est somme directe de q représentations irréductibles non-isomorphes deux à deux.

3.2 Irreductibilité et dimensions des $\mathrm{Hyp}(\pi,\Lambda)$.

Posons $\pi = \pi_\Psi$ ($\Psi \in \mathrm{Car}(K^\times)-\mathrm{Car}(k^\times)$). On vérifie sans difficulté que les représentations $(\mathrm{Bin}(\pi,\pi_2), \tau)$ non-nulles, pour π_2 dans la série principale de H (cf. [3]), s'obtiennent avec $\pi_2 = \pi_{\alpha,\beta}$, où $\alpha,\beta \in \mathrm{Car}(k^\times)$, $\alpha\beta = \Psi$ sur $k^\times$, et avec π_2 égal à la représentation de Steinberg associée à $\alpha \in \mathrm{Car}(k^\times)$ tel que $\alpha^2 = \Psi$ sur $k^\times$. En faisant le décompte dans les différents cas (car k = 2 ou non, Ψ carré dans $\mathrm{Car}(K^\times)$ ou non), on trouve que $(\mathrm{Bin}(\pi),\tau)$ a, en tout, au moins q composantes non-nulles. Comme il en est de même pour $(\mathrm{Hyp}(\pi),\tau)$, le corollaire ci-dessus montre alors que les sous-représentations $\mathrm{Hyp}(\pi,\tau)$ sont toutes irréductibles.

PROPOSITION 4.- L'unique projecteur équivariant P_Λ de $\mathrm{Hyp}(\pi)$ sur sa composante isotypique irréductible $\mathrm{Hyp}(\pi,\Lambda)$ est donné par

$$P_\Lambda = (q+1)^{-1}\mathrm{Id} - q^{-3}(q^2-1)^{-1} \sum_{a \in K^\times} \Lambda^{-1}(a) H_{N(a),\, Tr(a)}$$

De plus, on a

$$\dim \mathrm{Hyp}(\pi,\Lambda) = (q-1)^2(q^2+1) \qquad (\pi_\Lambda \not\simeq \pi),$$
$$\dim \mathrm{Hyp}(\pi,\Lambda) = q(q-1)(q^2+1) \qquad (\pi_\Lambda \simeq \pi) .$$

Démonstration

On vérifie sans difficulté que P_Λ est égal à l'identité sur $\mathrm{Hyp}(\pi,\Lambda)$ et nul sur les autres composantes isotypiques irréductibles de $\mathrm{Hyp}(\pi)$. D'autre part, on trouve aussitôt

$$\mathrm{Tr}(H_{r,s}) = \sum_{P \in HYP} \sum_{h \in L(r,s;\, x_P)} \mathrm{Tr}(\pi(h)) ,$$

où l'on a choisi une base quelconque x_P de P , pour tout $P \in HYP$ et où l'on a posé.

$$L(r,s\, ;\, x_P) = \{h \in H | J(hx_P) = rJ(x_P) \text{ et } B(x_P, hx_P) = sJ(x_P)\} .$$

Par conséquent ,

$$\mathrm{Tr}(H_{r,s}) = |HYP| \sum_{\substack{h \in H \\ \det(h) = r \\ \mathrm{Tr}(h) = s}} \mathrm{Tr}(\pi(h)) ,$$

d'où, avec $\pi = \pi_\Psi$ ($\Psi \in \mathrm{Car}(K^\times) - \mathrm{Car}(k^\times)$, $\Psi = \Lambda$ sur $k^\times$)

(8) $$\mathrm{Tr}(H_{t^2,2t}) = -q^3(q^2+1)(q-1)\Psi(t) \qquad (t \in k^\times) ,$$

(9) $$\mathrm{Tr}(H_{r,s}) = 0$$

si (r,s) n'est pas de la forme $(N(a)\, , \mathrm{Tr}(a))$ pour $a \in K^\times$, et

(10) $$\mathrm{Tr}(H_{N(a),\mathrm{Tr}(a)}) = -q^3(q^2+1)(\Psi(a) + \Psi^q(a))$$

pour $a \in K^\times - k^\times$. Il en résulte aussitôt que

$$\mathrm{Tr}\Big(\sum_{a \in K^\times} \Lambda^{-1}(a) H_{N(a),\mathrm{Tr}(a)}\Big) = q^3(q-1)^2(q^2+1) \quad (\Lambda \neq \Psi, \Psi^q)$$
$$= -q^4(q-1)^2(q^2+1) \quad (\Lambda = \Psi) ,$$

d'où les valeurs annoncées de $\dim \mathrm{Hyp}(\pi, \Lambda) = \mathrm{Tr}(P_\Lambda)$.

C. Q.F.D.

THEOREME 2.-

Les représentations $(\mathrm{Hyp}(\pi,\Lambda),\tau)$ son irréductibles por tout π dans la série discrète de H et tout $\Lambda \in \mathrm{Car}(K^\times) - \mathrm{Car}(k^\times)$ tel que π et Λ coïncident sur le centre de H. Elles appartiennent à la serie discrète de G, sauf quand $\pi_\Lambda \simeq \pi$ et leur nombre, pour π fixe, est de

i) $(1/2)(q-1)$ si car $k \neq 2$ et $\pi = \pi_\Psi$ avec Ψ non-carré dans $\mathrm{Car}(K^\times)$;

ii) $(1/2)(q-3)$ si car $k = 2$ et $\pi = \pi_\Psi$ avec Ψ carré dans $\mathrm{Car}(K^\times)$,

iii) $(1/2)(q-2)$ si car $k = 2$.

Cela est clair d'après ce qui précède. Le fait que $\mathrm{Hyp}(\pi,\Lambda)$ est dans la série discrète de G, si $\pi_\Lambda \not\simeq \pi$, résulte de sa dimension (si l'on compare avec la classification de [4]) ou encore du fait que dans ce cas $\mathrm{Hyp}(\pi,\Lambda)$ n'admet pas de vecteur invariant par les radicaux unipotents des sous-groupes paraboliques maximaux de G (cf. [4]).

REMARQUE.-

La représentation $\mathrm{Hyp}(\pi_\Lambda,\Lambda)$ appartient à la série induite à partir du sous-groupe parabolique P_1 de G (cf. [1], [2] ou [4]).

3.3 Entrelacement des $\mathrm{Hyp}(\pi,\Lambda)$

Pour terminer, nous décrivons l'entrelacement entre $\mathrm{Hyp}(\pi,\Lambda)$ et $\mathrm{Hyp}(\pi',\Lambda)$ où (V,π) et (V,π') sont dans la série discrète de H et $\Lambda, \Lambda' \in \mathrm{Car}(K^\times) - \mathrm{Car}(k^\times)$ sont tels que π_Λ et π (resp. $\pi_{\Lambda'}$ et π') coïncident sur le centre de H mais $\pi \not\simeq \pi_\Lambda$ (resp. $\pi' \not\simeq \pi_{\Lambda'}$) .

A l'aide des propositions 1 et 2 du numéro 1 , on obtient aussitôt la

PROPOSITION 5.-

Si π et π' ne son pas isomorphes, alors

$$\dim E(\pi,\pi') = 1$$

Démonstration:

En effet, dans ce cas, on trouve, en procédant comme dans la démostration de la proposition 3 du numéro 1 , que tout noyau $K \in N(\pi,\pi')$ est porté par la seule orbite 0_1 . Compte tenu de la remarque 2 au théorème 2 de 1.2 , l'assertion est alors claire.

C. Q. F. D.

Pour achever la description de l'entrelacement, il suffit de remarquer (en posant $\pi = \pi_\Lambda$ et $\pi' = \pi_{\Lambda'}$) que l'on a bien un isomorphisme de $Hyp(\pi, \Lambda')$ sur $Hyp(\pi', \Lambda)$. Cet isomorphisme est plus évident si l'on réalise ces représentations sous la forme $Bin(\pi,\pi')$ et $Bin(\pi',\pi)$. Il est alors induit par l'isomorphisme bien connu R de la representation de Weil ρ_J de H. avec sa representation naturelle $h \mapsto \sigma_1(h^*)$ (cf. 2.2., déf. 2) donné par

$$(R(F))(x_1,x_2;\psi) = q^{-2} \sum_{u \in E} F(x_1,u;\psi)\psi(x_2,u)$$

pour $F \in Bin(\pi,\pi')$, $x_1,x_2 \in E$, $\psi \in X$. De manière précise l'isomorphisme S de $Bin(\pi,\pi')$ sur $Bin(\pi',\pi)$ est le composé $F \mapsto T\circ[R(F)]$ $(F \in Bin(\pi,\pi'))$

avec $\qquad T(\phi) = \beta\circ\phi^*\circ\alpha$

où α (res β) est un isomorphisme involutif de (V,π) sur $(V^*,\hat{\pi})$ (res. de $(V'^*,\hat{\pi}')$ sur (V',π') et ϕ^* est l'application transposée de $\phi \in \mathrm{Hom}_{\mathbb{C}}(V', V)$ et où l'on a posé $\hat{\pi}(h) = \pi(h^*)^*$ $(h \in H)$ et de meme pour $\hat{\pi}'$.

Nous résumons nos résultats dans le

THEOREME 3.-

Pour chaque paire de représentations non-isomorphes $(\pi,\pi') = (\pi_\Lambda, \pi_{\Lambda'})$ dans la série discrète de H, qui coïncident sur le centre de H, on a

$$(\mathrm{Hyp}(\pi,\Lambda'),\tau) \simeq (\mathrm{Bin}(\pi,\pi'),\tau) .$$

Ces représentations sont toutes irréductibles et non-isomorphes deux-à-deux (pour des paires non-isomorphes) sauf dans le cas

$$(\mathrm{Hyp}(\pi,\Lambda'),\tau) \simeq (\mathrm{Hyp}(\pi',\Lambda),\tau) .$$

Elles forment une série de $(1/8)q(q-1)(q-3)$ (resp. $(1/8)q(q-1)(q-2)$) types d'isomorphie si car $k \neq 2$ (resp. si car $k = 2$) de dimension commune $(q-1)^2(q^2+1)$; c'est la série discrète de G associée au tore T_1 (isomorphe au groupe des $(a,b) \in H \times H$ tels que $\det(a) = \det(b)$).

B I B L I O G R A P H I E

1 J. SOTO ANDRADE, C. R. Acad. Sc. Paris, 278, 1974., p. 321-324.

2 J. SOTO ANDRADE, Sem. Th. Groupes (1972/73) P. CARTIER, 2e. partie, Ch. VII, Les représentations de $Sp(4, \mathbb{F}_q)$ et $GSp(4, \mathbb{F}_q)$, IHES, Bures-sur-Yvette, 1975.

3 J. SOTO ANDRADE, Sem. Th. Groupes (1972/73) P. CARTIER 1e. partie, Ch. III, Les représentations de $GL(2, \mathbb{F}_q)$ et $SL(2, \mathbb{F}_q)$, IHES, Bures-sur-Yvette, à paraître.

4 T. A. SPRINGER, Seminar on Algebraic Groups and related Finite Groups, Characters of special groups, Lecture Notes in Maths., 131, Springer Verlag, 1970 .

5 T. A. SPRINGER, Caractères de groupes de Chevalley finis, Sem. BOURBAKI, 25eme. année, 1972/73, exposé 429 .

6 Bh. SRINIVASAN, The characters of the finite symplectic group $Sp(4,q)$, Trans. A. M. S., 131, 1968, p. 488-525 .

DEPARTAMENTO DE MATEMATICAS
FACULTAD DE CIENCIAS
UNIVERSIDAD DE CHILE
SANTIAGO.

The Point Spectrum of the Adjoint of an Automorphism of a Vector Bundle

J.L. Arraut and N. Moreira dos Santos

1. Introduction

We shall be studying the eigenvalues (point spectrum) of the adjoint operator to an automorphism of a vector bundle, on the space of bounded continuous sections over a recurrent orbit, or over its closure. We prove a theorem that combined with Mather's theorem [1] gives some information on the spectrum of the adjoint operator on the closure of a recurrent orbit.

Let M be a complete metric space, $\xi = (E,\pi,M)$ a q-dimensional complex vector bundle with a hermitian structure on it, and $F = (\tilde{f},f)$ an automorphism of ξ, that is $\tilde{f}\colon E \to E$ and $f\colon M \to M$ are homeomorphisms such that $\pi\tilde{f} = f\pi$ and the restriction $\tilde{f}\colon E_x \to E_{f(x)}$ is a linear isomorphism for each $x \in M$ (here, as usual, $E_x = \pi^{-1}(x)$ is the fiber over $x \in M$). Let Λ be a subset of M invariant under f, that is, $f(\Lambda) = \Lambda$. The set $\Gamma_b^o(\Lambda,\xi)$ of all bounded continuous sections s of ξ over Λ is a Banach space under the sup norm $| \ |_o$. The automorphism F induces its adjoint operator

$$F_*\colon \Gamma_b^o(\Lambda,\xi) \to \Gamma_b^o(\Lambda,\xi)$$

given by $F_*s = \tilde{f}\circ s\circ f^{-1}$, which is a linear homeomorphism.

A point $x \in M$ is recurrent if there exists a sequence (n_j) of non zero integers such that $f^{n_j}(x) \to x$ as $j \to \infty$.
The orbit $\mathcal{O}(x)$ of a point $x \in M$ relative to $f\colon M \to M$ is the set

$$\mathcal{O}(x) = [f^n(x);\ n \in \mathbb{Z}].$$

We denote by $\overline{\Theta(x)}$ the closure of $\Theta(x)$.

Here we study the point spectrum $\sigma_p(F_*)$ of F_* when Λ is the orbit of a recurrent point $x \in M$ of $f: M \to M$. If $\lambda \in \sigma_p(F_*)$, let us consider the eigenspaces

$$P^k(\lambda;F_*) = \ker(F_*-\lambda I)^k \quad \text{where} \quad k \geq 1 \quad \text{is an integer}$$

and

$$P(\lambda;F_*) = [s \in \Gamma_b^o(\Lambda,\xi);\ (F_*-\lambda I)^k\ s = 0 \quad \text{for some integer,} \quad k \geq 1].$$

Given $\lambda_i \in \sigma_p(F_*)$, $1 \leq i \leq \ell$ we also consider the eigenspaces

$$P^k(\lambda_1,\ldots,\lambda_\ell;\ F_*) = P^k(\lambda_1;F_*) \oplus\ldots\oplus P^k(\lambda_\ell;F_*)$$

and

$$P(\lambda_1,\ldots,\lambda_\ell;\ F_*) = P(\lambda_1;F_*) \oplus\ldots\oplus P(\lambda_\ell;F_*).$$

The automorphism $f = (f \times id,\ f)$ of the product bundle $\xi_o = (M \times C,\ \pi,\ M)$ plays a special role due to the fact that the point spectrum $\sigma_p(f_*)$ is a subgroup of the unit circle $S^1 = [\alpha \in C;\ |\alpha| = 1]$ and we have a natural action

$$\sigma_p(f_*) \times \sigma_p(F_*) \to \sigma_p(F_*) \tag{1}$$

given by $(\alpha,\lambda) \mapsto \alpha\lambda$. Let us denote by

$$[\lambda] = [\alpha\lambda;\ \alpha \in \sigma_p(f_*)]$$

the orbit of λ under this action. We remark that the sections of ξ_o are naturally identified with the bounded continuous functions $\varphi: \Lambda \to C$ and $\Gamma_b^o(\Lambda,\xi)$ is a module over the ring $\Gamma_b^o(\Lambda,\xi_o)$. Under this identification we have $f_*(\varphi) = \varphi\circ f^{-1}$. For each $y \in \Lambda$ we have the <u>evaluation mapping</u> at y

$$ev_y: \Gamma_b^o(\Lambda,\xi) \to E_y$$

given by $ev_y(s) = s(y)$.

We prove

Theorem 1 - Let Λ be the orbit of a recurrent point x of $f:M \to M$ and $\lambda_1,\ldots,\lambda_\ell \in \sigma_p(F_*)$. If $[\lambda_i] \neq [\lambda_j]$ for all $i \neq j$, $1 \leq i,j \leq \ell$, then the restriction

$$ev_y: P(\lambda_1,\ldots,\lambda_\ell;\ F_*) \to E_y$$

of the evaluation mapping is injective for every $y \in \Lambda$.

The following are immediate consequences of the above theorem:

i) the eigenspaces $P(\lambda_1,\ldots,\lambda_\ell;\ F_*)$ are finite dimensional and their dimension are bounded above by the dimension q of the fiber of $\xi = (E,\pi,M)$;

ii) the action (1) has at most q distinct orbits.

We have the following corollary

Corollary 1 - Let Λ be the closure of the orbit of a recurrent point x of $f: M \to M$ and λ be any complex number. The following statements are equivalent,

(i) $F_*-\lambda I: \Gamma_b^0(\Lambda,\xi) \to \Gamma_b^0(\Lambda,\xi)$ is surjective

(ii) $F_*-\lambda I: \Gamma_b^0(\Lambda,\xi) \to \Gamma_b^0(\Lambda,\xi)$ is a linear homeomorphism.

J. Mather's theorem in [1] can be stated as

Theorem - Let $\Lambda = \overline{\mathcal{O}(x)}$ be the closure of the orbit of a non periodic point $x \in M$ of $f: M \to M$ and $\sigma(F_*)$ be the spectrum of $F_*: \Gamma_b^0(\Lambda,\xi) \to \Gamma_b^0(\Lambda,\xi)$. If $\lambda \in \sigma(F_*)$ and $\alpha \in \mathbb{C}$, $|\alpha| = 1$ then $\alpha k \in \sigma(F_*)$.

As a consequence of this theorem and Corollary 1, we have,

Corollary 2 - Let $\Lambda = \overline{\mathcal{O}(x)}$ be the closure of the orbit of a recurrent point $x \in M$ of $f: M \to M$ and λ be any complex number. The following statements are equivalent,

(i) $F_*-\lambda I: \Gamma_b^o(\Lambda,\xi) \to \Gamma_b^o(\Lambda,\xi)$ is surjective

(ii) the spectrum of $F_*: \Gamma_b^o(\Lambda,\xi) \to \Gamma_b^o(\Lambda,\xi)$ does not intercept the circle of radius $r = |\lambda|$.

The results in this paper were presented at the Symposium held at the University of Warwick in 1974 and reported in [2].

2. The Point Spectrum of the Adjoint Operator

In this section we prove Theorem 1 and Corollaries 1 and 2. To simplify notation we denote by s^k the sections in $P^k(\lambda;F_*)$ and if i is a non negative integer we put $s^{k-i} = (F_*-\lambda I)^i\, s^k$. Thus $s^{k-i} \in P^{k-i}(\lambda;F_*)$. The binominal formula gives

$$F_*^n = (F_* - \lambda I + \lambda I)^n = \sum_{i=0}^{n} \binom{n}{i} \lambda^{n-i}(F_*-\lambda I)^i$$

and noticing that $s^{k-i} = 0$ for $i \geq k$, we get

$$F_*^n(s^k) = \sum_{i=0}^{k-1} \binom{n}{i} \lambda^{n-i}\, s^{k-i} \tag{2}$$

for all integer $n \geq k$.

Lemma 2.1 - Let Λ be the orbit $\mathcal{O}(x)$ of a recurrent point x of $f: M \to M$ and $\lambda \in \sigma_p(F_*)$. If y is any point in $\mathcal{O}(x)$, then the evaluation mapping

$$ev_y: P^k(\lambda;F_*) \to E_y$$

is injective for each positive integer k. The same is true if Λ is the closure of $\mathcal{O}(x)$.

Proof: If $s^1 \in P^1(\lambda;F_*)$ and $s^1(y) = 0$ then $F_*^n\, s^1 = \lambda^n\, s^1$ and

$$F_*^n\, s^1(f^n(y)) = \tilde{f}^n\, s^1(y) = \lambda^n\, s^1(f^n(y)) = 0 \tag{3}$$

for all integer n. By continuity of s^1 we have then $s^1 = 0$.

Suppose now, by induction, that the lemma is true for $k-1$. If $s^k \in P^k(\lambda;F_*)$ and $s^k(y) = 0$ then $F_*^n\, s^k(f^n(y)) = \tilde{f}^n s^k(y) = 0$. From this and (2) we get

$$s^1(f^n(y)) = -\sum_{i=0}^{k-2} \frac{\binom{n}{i}}{\binom{n}{k-1}} \lambda^{k-i-1}\, s^{k-i}\,(f^n(y)) \tag{4}$$

As x is a recurrent point and $y \in \mathcal{O}(x)$ there exists a sequence $f^{n_j}(y) \to y$ as $n_j \to \infty$. We remark that

$$P^k(\lambda;F_*) = P^k(\lambda^{-1};F_*^{-1}) \tag{5}$$

for all positive integer k.

From (5), substituting f by f^{-1} and λ by λ^{-1} in (4) if necessary, we may assume that $n_j \geq k$. Taking limits in (4) as $j \to \infty$ we get $s^1(y) = 0$ and by the above $s^1 = 0$. Thus $(F_*-\lambda I)^{k-1}\, s^k = s^1 = 0$. The induction assumption gives then $s^k = 0$, proving the lemma.

Lemma 2.2 - Let Λ be the orbit of a point $x \in M$ relative to $f\colon M \to M$. Then the point spectrum $\sigma_p(f_*)$ is a subgroup of the unit circle. The same is true if $\Lambda = \overline{\mathcal{O}(x)}$.

Proof: We first notice that since $|f_*(\varphi)|_o = |\varphi\circ f^{-1}|_o = |\varphi|_o$ for all bounded continuous functions $\varphi\colon \Lambda \to C$ thus $\sigma_p(f_*)$ is contained in the unit circle S^1. Clearly $1 \in \sigma_p(f_*)$ and if $\lambda, \mu \in \sigma_p(f_*)$ then there exists bounded continuous functions $\varphi\colon\Lambda \to C$, $\psi\colon \Lambda \to C$ such that $\varphi, \psi \neq 0$ and $\varphi\circ f^{-1} = \lambda\varphi$, $\bar{\psi}\circ f^{-1} = \bar{\mu}\bar{\psi}$. By Lemma 2.1, $(\varphi\bar{\psi})(x) = \varphi(x)\bar{\psi}(x) \neq 0$. Since $(\varphi\bar{\psi})\circ f^{-1} = \lambda\bar{\mu}(\varphi\bar{\psi})$ then $\lambda\bar{\mu} = \lambda\mu^{-1} \in \sigma_p(f_*)$, proving the lemma.

Lemma 2.3 - Let Λ be the orbit of a point $x \in M$ relative to $f\colon M \to M$. If $\alpha \in \sigma_p(f_*)$ and $\lambda \in \sigma_p(F_*)$, then $\alpha\lambda \in \sigma_p(F_*)$.

Proof: By assumption there exists $s \in \Gamma_b^o(\Lambda,\xi)$ and a bounded continuous function $\varphi\colon \Lambda \to C$ such that $F_* s = \lambda s$ and $\varphi\circ f^{-1} = \alpha\varphi$. Since $F_*(\varphi s) = (\varphi\circ f^{-1})F_* s = \alpha\lambda(\varphi s)$ and by Lemma 2.1 $(\varphi s)(x) = \varphi(x)s(x) \neq 0$ then $\alpha\lambda \in \sigma_p(F_*)$, proving the lemma.

Lemma 2.4 - Let $\Lambda = \mathcal{O}(x)$ be the orbit of a recurrent point $x \in M$ of $f\colon M \to M$ and $\alpha \in C$, $|\alpha| = 1$. A necessary and sufficient condition for $\alpha \in \sigma_p(f_*)$ is that $\alpha^{n_j} \to 1$ whenever $f^{n_j}(x) \to x$.

Proof: If $\alpha \in \sigma_p(f_*)$ then there exists a bounded continuous function $\varphi\colon \Lambda \to C$ such that $\varphi\circ f^{-1} = \alpha\varphi$ and $\varphi(x) = 1$. From (5) we get $\varphi\circ f^n = \alpha^{-n}\varphi$ and $\varphi(f^n(x)) = \bar{\alpha}^n$ for all integer n. Thus, by continuity of φ, if $f^{n_j}(x) \to x$ then $\alpha^{n_j} \to 1 = \varphi(x)$. Conversely, suppose that $\alpha^{n_j} \to 1$ whenever $f^{n_j}(x) \to x$. It is easy to see that the function $\varphi\colon \Lambda \to C$, $\varphi(f^n(x)) = \bar{\alpha}^n$ is well defined, bounded and $\varphi\circ f^{-1} = \alpha\varphi$. To show continuity of φ, we notice that if $f^{n_j}(x) \to y = f^{\ell}(x)$ then $f^{n_j-\ell}(x) \to x$ and by assumption $\bar{\alpha}^{n_j-\ell} \to 1$, that is $\bar{\alpha}^{n_j} \to \bar{\alpha}^{\ell}$ showing $\varphi(f^{n_j}(x)) \to \varphi(y)$, proving the lemma.

An interesting problem is to find conditions in which $\varphi\colon \Lambda \to C$ is uniformly continuous. In particular, is it true when the closure $\overline{\mathcal{O}(x)}$ is a minimal set?

Lemma 2.5 - Let Λ be the orbit $\mathcal{O}(x)$ of a recurrent point x of $f\colon M \to M$ and $\lambda_1,\ldots,\lambda_\ell \in \sigma_p(F_*)$ be eigenvalues such that $[\lambda_i] \neq [\lambda_j]$ whenever $i \neq j$, $1 \le i,j \le \ell$. Then the evaluation mapping

$$ev_y\colon P^1(\lambda_1;F_*) \oplus\ldots\oplus P^1(\lambda_\ell;F_*) \to E_y$$

is injective for every $y \in \Lambda$.

Proof: The lemma is true for $\ell = 1$ by Lemma 2.1. Suppose, by induction, that the lemma is true if we have less than ℓ eigenvalues. We show that it is true for ℓ eigenvalues. As in the proof of Lemma 2.1, we may assume the existence of a sequence of positive integers (n_j) such that $f^{n_j}(y) \to y$ as $n_j \to \infty$.

Suppose $\sum_{i=1}^{\ell} s_i(y) = 0$ where $s_i \in P^1(\lambda_i;F_*)$. (6)

We consider two cases,

1) $|\lambda_1| = \dots = |\lambda_\ell|$.

Since, by assumption $\alpha = \frac{\lambda_{\ell-1}}{\lambda_\ell}$ is not an eigenvalue of f_* thus by Lemma 2.4 there exists an infinite sequence of integers (n_j) such that $f^{n_j}(y) \to y$ as $j \to \infty$ and $(\frac{\lambda_i}{\lambda_\ell})^{n_j} \to c_i$ where $c_{\ell-1} \neq 1$. From (3) and (6) we have

$$s_\ell(f^n(y)) = - \sum_{i=1}^{\ell-1} (\frac{\lambda_i}{\lambda_\ell})^n s_i(f^n(y)). \tag{7}$$

Taking limit in (7) as $n_j \to \infty$ and using (6) we get

$$\sum_{i=1}^{\ell-1} (c_i-1)\, s_i(y) = 0.$$

As $c_{\ell-1} \neq 1$, the induction assumption gives $s_i = 0$ for $1 \leq i \leq \ell$.

2) $|\lambda_1| \leq \dots \leq |\lambda_{p-1}| < |\lambda_p| \leq \dots \leq |\lambda_\ell|$.

Since $|\lambda_{p-1}| < |\lambda_p|$ we may choose an infinite sequence of positive integers (n_j) such that $f^{n_j}(y) \to y$ as $j \to \infty$ and $(\frac{\lambda_i}{\lambda_\ell})^{n_j} \to c_i$ where $c_i = 0$ for $1 \leq i \leq p-1$. Taking limit in (7) as $j \to \infty$ we get

$$s_\ell(y) + \sum_{i=p}^{\ell-1} c_i s_i(y) = 0$$

and by induction assumption $s_i = 0$ for $1 \leq i \leq \ell$, proving the lemma.

Lemma 2.6 - Let Λ be the orbit of a recurrent point $x \in M$ of $f\colon M \to M$ and $\lambda_1,\dots,\lambda_\ell \in \sigma_p(F_*)$ such that

(i) $|\lambda_p| = |\lambda_\ell|$ for all $1 \leq p \leq \ell$.

(ii) $[\lambda_i] \neq [\lambda_j]$ for all $i \neq j$.

Then

$$ev_y: P^k(\lambda_1,\ldots,\lambda_\ell;F_*) \to E_y$$

is injective for all positive integer k and all $y \in \Lambda$.

Proof: It is true for $k = 1$ by Lemma 2.5. By induction assumption, the lemma is true for $k-1$. Suppose $\sum_{p=1}^{\ell} s_p^k(y) = 0$ where $s_p^k \in P^k(\lambda_p;F_*)$ for $1 \le p \le \ell$. Using (2) we get

$$\sum_{p=1}^{\ell} \left[\sum_{i=0}^{k-1} \binom{n}{i} \lambda_p^{n-i} s_p^{k-i} (f^n(y))\right] = 0 \tag{8}$$

for all positive integer $n \ge k$. From this we have

$$s_\ell^1(f^n(y)) + \sum_{p=1}^{\ell-1} \left(\frac{\lambda_p}{\lambda_\ell}\right)^{n-k+1} s_p^1(f^n(y)) =$$

$$= - \sum_{p=1}^{\ell} \left[\sum_{i=0}^{k-2} \frac{\binom{n}{i}}{\binom{n}{k-1}} \left(\frac{\lambda_p}{\lambda_\ell}\right)^n \frac{\lambda_\ell^{k-1}}{\lambda_p^i} s_p^{k-i}(f^n(y))\right]. \tag{9}$$

As in the proof of Lemma 2.1, we may assume the existence of a sequence of positive integers (n_j) such that $f^{n_j}(y) \to y$ as $n_j \to \infty$ and $\left(\frac{\lambda_p}{\lambda_\ell}\right)^{n_j-k+1} \to c_p$, $|c_p| = 1$ for $1 \le p \le \ell$.

Taking limit in (9) as $n_j \to \infty$ we get

$$s_\ell^1(y) + \sum_{p=1}^{\ell-1} c_p s_p^1(y) = 0.$$

From this we see that $s_p^1 = 0$ for $1 \le p \le \ell$. Thus $s_p^k \subset P^{k-1}(\lambda_p;F_*)$ for $1 \le p \le \ell$. By induction assumption we have $s_p^k = 0$ for $1 \le p \le \ell$, proving the lemma.

We recall from Calculus that if $p(x)$ is any polynomial and α is a complex number such that $|\alpha| < 1$, then

$$\lim_{n\to\infty} p(n) \alpha^n = 0. \tag{10}$$

Proof of Theorem 1

We have to show, under the assumption of the theorem, that,

$$ev_y: P^k(\lambda_1,\ldots,\lambda_\ell;F_*) \to E_y \tag{11}$$

is injective for all positive integers ℓ and k.

Lemma 2.1 syas (11) is true for all k if $\ell = 1$. Assume, by induction, that (11) is true if the number of eigenvalues is less than ℓ. By Lemma 2.6 we may assume the existence of an integer m such that $|\lambda_1| \leq \ldots \leq |\lambda_{m-1}| < |\lambda_m| = \ldots = |\lambda_\ell|$. Choose an infinite sequence of positive integers (n_j) such that $f^{n_j}(y) \to y$ and $(\frac{\lambda_p}{\lambda_\ell})^{n_j} \to c_p$, $|c_p| = 1$ for $m \leq p \leq \ell$. This can be done as in the proof of Lemma 2.1. Suppose $\sum_{p=1}^{\ell} s_p^k(y) = 0$ where $s_p^k \in P^k(\lambda_p;F_*)$. Taking limit in (9) as $j \to \infty$ we get

$$s_\ell^1(y) + \sum_{p=m}^{\ell} c_p^{-k+1} s_p^1(y) = 0$$

and by induction assumption, $s_p^1 = 0$ for $m \leq p \leq \ell$. Thus, from (8) we get

$$\sum_{p=1}^{m-1} \lambda_p^{n-k+1} s_p^1(f^n(y)) = - \sum_{p=1}^{\ell} [\sum_{i=0}^{k-2} \frac{\binom{n}{i}}{\binom{n}{k-1}} \lambda_p^{n-i} s_p^{k-i}(f^n(y))]. \tag{12}$$

If we multiply (12) by λ_{m-1}^{-n+k-1} and take limit as $j \to \infty$, we show the existence of the limit

$$\lim_{j\to\infty} \sum_{p=m}^{k-2} \frac{\binom{n_j}{i}}{\binom{n_j}{k-1}} (\frac{\lambda_p}{\lambda_{m-1}})^{n_j} \frac{\lambda_{m-1}^{k-1}}{\lambda_p^i} s_p^{k-i} (f^{n_j}(y))] \tag{13}$$

for some subsequence (n_j) of the above sequence because $|\lambda_p| \leq |\lambda_{m-1}|$ for $1 \leq p \leq m-1$ and $\lim_{n\to\infty} \frac{\binom{n}{i}}{\binom{n}{k-1}} = 0$ for $0 \leq i \leq k-2$.

We show that $s_p^k = 0$ for $m \leq p \leq \ell$. Assume we have shown that $s_p^t = 0$ for $1 \leq t < r$ and $m \leq p \leq \ell$. By (10) we have

$$\lim_{j\to\infty} \frac{\binom{n_j}{k-1}}{\binom{n_j}{k-r}} (\frac{\lambda_{m-1}}{\lambda_\ell})^{n_j} = 0 \tag{14}$$

because $|\lambda_{m-1}| < |\lambda_\ell|$. If we multiply (13) and (14) we get

$$\sum_{p=m}^{\ell} \lambda_p^{-k+r} s_p^r(y) = 0$$

and by induction assumption, $s_p^r = 0$ for $m \leq p \leq \ell$. Now $s_p^k = 0$

for $m \le p \le \ell$ and $\sum_{p=1}^{\ell} s_p^k(y) = 0$ gives

$$\sum_{p=1}^{m-1} s_p^k(y) = 0 \quad \text{where} \quad m-1 < \ell$$

and again, by induction assumption, $s_p^k = 0$, $1 \le p \le \ell$, proving the theorem.

Remarks:

1. Let $\Lambda = \overline{\mathcal{O}(x)}$ where $x \in M$ is a recurrent point of $f\colon M \to M$ and λ be an eigenvalue of $F_*\colon \Gamma_b^o(\mathcal{O}(x),\xi) \to \Gamma_b^o(\mathcal{O}(x),\xi)$.

Then the set $P_\Lambda(\lambda,F_*)$ of all uniformly continuous sections $s \in P(\lambda;F_*)$ is a linear subspace of $P(\lambda;F_*)$. Notice that $P_\Lambda(\lambda,F_*)$ is naturally identified with the set of continuous sections over Λ. Thus Theorem 1 is true for

$$P_\Lambda(\lambda_1,\dots,\lambda_\ell;F_*) = P_\Lambda(\lambda_1;F_*) \oplus\dots\oplus P_\Lambda(\lambda_\ell;F_*)$$

and all $y \in \mathcal{O}(x)$, provided that $\frac{\lambda_i}{\lambda_j}$ is not an eigenvalue of $f_*\colon \Gamma_b^o(\mathcal{O}(x),\xi_o) \to \Gamma_b^o(\mathcal{O}(x),\xi_o)$ for $i \ne j$, $1 \le i,j \le \ell$.

2. Notice that if $|\lambda_i| \ne |\lambda_j|$ then $[\lambda_i] \ne [\lambda_j]$. Thus Theorem 1 is true if $|\lambda_i| \ne |\lambda_j|$, $i \ne j$, $1 \le i,j \le \ell$.

Proof of Corollary 1

Clearly (ii) implies (i). Assume now that $F_*-\lambda I\colon \Gamma_b^o(\Lambda,\xi) \to \Gamma_b^o(\Lambda,\xi)$ is surjective. We show that it is also injective. Suppose that $\ker(F_*-\lambda I) \ne \{0\}$. By Theorem 1, $\ker(F_*-\lambda I)$ is finite dimensional. Thus $F_*-\lambda I$ has a bounded right inverse $\Omega\colon \Gamma_b^o(\Lambda,\xi) \to \Gamma_b^o(\Lambda,\xi)$. Denote by $L(\Gamma_b^o(\Lambda,\xi))$ the Banach space of all bounded linear operators $T\colon \Gamma_b^o(\Lambda,\xi) \to \Gamma_b^o(\Lambda,\xi)$. Because the set of all linear homeomorphisms is open in $L(\Gamma_b^o(\Lambda,\xi))$, the continuity of the mapping

$$f\colon \mathbb{C} \to L(\Gamma_b^o(\Lambda,\xi)), \quad f(z) = (F_*-zI)\Omega$$

and the fact that $f(\lambda) = I$ gives a neighborhood V of λ in C such that $f(z): \Gamma_b^o(\Lambda,\xi) \to \Gamma_b^o(\Lambda,\xi)$ is a linear homeomorphism for all $z \in V$. Thus $F_*-zI: \Gamma_b^o(\Lambda,\xi) \to \Gamma_b^o(\Lambda,\xi)$ is surjective and

$$\dim \ker(F_*-zI) = \dim \ker(F_*-\lambda I) \geq 1$$

for all $z \in V$, contradicting Theorem 1.

3. Some Problems

We think that the following are interesting problems

1) Under which conditions is Lemma 2.4 true if Λ is the closure $\overline{\mathfrak{G}(x)}$ of a recurrent orbit? Is it true if $\Lambda = \overline{\mathfrak{G}(x)}$ is a minimal set?

2) Is Theorem 1 true if $\Lambda = \overline{\mathfrak{G}(x)}$ and $\sigma_p(f_*)$ is the point spectrum of $f_*: \Gamma_b^o(\Lambda,\xi_o) \to \Gamma_b^o(\Lambda,\xi_o)$? Is it true under the above conditions if Λ is a minimal set?

3) Let M be a compact manifold, $f: M \to M$ a diffeomorphism, $\xi = (TM,\pi,M)$ the tangent bundle of M, $F = (Tf,f)$ and $\Lambda = M$. Suppose there exists a positive integer k such that

$$\dim P^{\ell}(1;F_*) = \dim M \quad \text{for all} \quad \ell \geq k.$$

Is $f: M \to M$ minimal?

4) If Λ is the orbit of a recurrent point $x \in M$ of a homeomorphism $f: M \to M$, under what conditions on $F = (\tilde{f},f)$ do there exist $\lambda_1,\dots,\lambda_\ell \in \sigma_p(F_*)$ such that

$$ev_x: P^k(\lambda_1,\dots,\lambda_\ell;F_*) \to E_x \quad \text{is surjective?}$$

Bibliography

[1] Mather, J. (1968), Indag. Math., 30 pp. 479-483.

[2] Arraut, J.L., Moreira dos Santos, N. (1974), Lecture Notes in Math., Springer-Verlag, (468) pp. 20-21.

Departamento de Matemática
Pontifícia Universidade Católica
Rio de Janeiro - Brasil

WHITNEY DUALITY AND SINGULARITIES OF PROJECTIONS*

Thomas Banchoff and Clint McCrory

When we look at a transparent smooth surface, the most apparent features are the fold curves; that is, the points of the surface where the tangent plane contains the horizontal line of sight. If we look with both eyes, then we see two different fold curves, and they intersect at precisely those points where the tangent plane is horizontal; i.e., at the critical points of the height function in the vertical direction. This relationship between intersections of fold curves and critical points of a function is the prototype of a cycle level version of the Whitney duality theorem relating the tangential and normal characteristic classes of a manifold. We will establish this cycle level duality theorem in this paper.

The idea of using singularities of projections to Euclidean spaces in order to define characteristic homology classes is contained in the work of Pontrjagin [3, p. 265] and Thom [5]. In his proof of the Whitney duality theorem, Chern in [1] proves formulas in the intersection ring of the Grassmann manifold. We show in the final section of this paper that some of his formulas have a precise cycle level interpretation.

§1. Surfaces immersed in 3-space.

A characteristic cycle of a smooth surface M is the singularity set of a stable mapping of M to a Euclidean space. (See [4] for the definition of a stable mapping.) We will use the geometry of projections to obtain intersection formulas for characteristic cycles.

Let $f\colon M^2 \to E^3$ be an immersion of a smooth surface (compact, without boundary) into Euclidean 3-space, with basis $\{e_1, e_2, e_3\}$.

*Research was supported by the National Science Foundation under MCS 76-06324 and MCS 76-09817.

Let $\pi_{12}: E^3 \to E^2_{12}$ be the orthogonal projection to the plane spanned by $\{e_1, e_2\}$. The <u>singularity set</u> $\Sigma(\pi_{12} \circ f) = \{p \in M^2 \mid \text{rank } d(\pi_{12} \circ f)_p < 2\}$ is precisely the set of points where the tangent plane $T_p f(M)$ to $f(M)$ at $f(p)$ contains the line spanned by the vector e_3; i.e. $T_p f(M)$ is orthogonal to the plane E^2_{12}. [2] For almost all choices of basis, the map $\pi_{12} \circ f$ is stable, and so the singularity set is a finite collection of disjoint closed curves on M^2, the <u>fold curves</u> of $\pi_{12} \circ f$. If $\pi_{13}: E^3 \to E^2_{13}$ is another such projection, then the two cycles $\Sigma(\pi_{12} \circ f)$ and $\Sigma(\pi_{13} \circ f)$ are homologous. Moreover, the homology class is independent of the immersion f. The first <u>Stiefel-Whitney class</u> w_1 of M^2 is defined to be the cohomology class Poincaré dual to the common homology class of these characteristic cycles. Thus, $w_1 = \mathcal{D}[\Sigma(\pi_{12} \circ f)] = \mathcal{D}[\Sigma(\pi_{13} \circ f)]$.

We now consider orthogonal projection $\pi_1: E^3 \to E^1_1$ onto the line spanned by the vector e_1. The singularity set $\Sigma(\pi_1 \circ f) = \{p \in M^2 \mid \text{rank } d(\pi_1 \circ f) < 1\}$ is precisely the set of points where the tangent plane $T_p f(M)$ is orthogonal to the line E^1_1. For almost all choices of the basis $\{e_1, e_2, e_3\}$, $\pi_1 \circ f$ is stable, and so $\Sigma(\pi_1 \circ f)$ is a finite collection of <u>critical points</u> of $\pi_1 \circ f$. If $\tilde{\pi}_1$ is another such projection, then the two characteristic cycles $\Sigma(\pi_1 \circ f)$ and $\Sigma(\tilde{\pi}_1 \circ f)$ are homologous, and the second Stiefel-Whitney class w_2 of M^2 is defined to be the Poincaré dual of their common homology class. Thus, $w_2 = \mathcal{D}[\Sigma(\pi_1 \circ f)]$.

A stable map $\pi_1 \circ f: M^2 \to E^1_1$ is just a nondegenerate Morse function. If $\pi_1 \circ f$ has m_0 minima, m_2 maxima, and m_1 nondegenerate saddle points as critical points, then $\Sigma(\pi_1 \circ f)$

represents the trivial homology class if and only if there are an even number of critical points; i.e., if $m_0 + m_1 + m_2 \equiv 0 \pmod 2$. This happens precisely when the Euler characteristic $\chi(M^2) = m_0 - m_1 + m_2$ is even, so $w_2 = 0$ if and only if $\chi(M^2)$ is even.

The observation which relates these characteristic cycles of M^2 is that for an immersion f,

$$\Sigma(\pi_{12}\circ f) \cap \Sigma(\pi_{13}\circ f) = \Sigma(\pi_1\circ f). \tag{1}$$

This is just a statement of the fact that the plane $T_p f(M)$ is orthogonal to both E^2_{12} and E^2_{13} if and only if it is orthogonal to their common line E^1_1. Moreover, for almost all choices of basis $\{e_1,e_2,e_3\}$, the intersection of the two cycles $\Sigma(\pi_{12}\circ f)$ and $\Sigma(\pi_{13}\circ f)$ on M is transverse, so their set-theoretic intersection $\Sigma(\pi_1\circ f)$ represents the intersection product of their homology classes. Thus, we may interpret (1) in terms of cohomology using the fact that the intersection product of homology classes is Poincaré dual to the cup product of the corresponding cohomology classes. Thus

$$\begin{aligned} w_1 \smile w_1 &= \mathscr{D}[\Sigma(\pi_{12}\circ f)] \smile \mathscr{D}[\Sigma(\pi_{13}\circ f)] \\ &= \mathscr{D}[\Sigma(\pi_{12}\circ f) \cap \Sigma(\pi_{13}\circ f)] \\ &= \mathscr{D}[\Sigma(\pi_1\circ f)] \\ &= w_2. \end{aligned}$$

§2. Surfaces immersed in 4-space.

We now consider an immersion $f\colon M^2 \to E^4$ of a surface into Euclidean 4-space with basis $\{e_1,e_2,e_3,e_4\}$. As before we may consider compositions $\pi_{12}\circ f$, $\pi_{13}\circ f$, $\pi_1\circ f$ from the surface into lower dimensional subspaces, but now we must also treat the composition $\pi_{123}\circ f$, where $\pi_{123}: E^4 \to E^3_{123}$ is orthogonal projection to the 3-space spanned by $\{e_1, e_2, e_3\}$. The singularity set $\Sigma(\pi_{123}\circ f) = \{p \in M^2 \mid \text{rank } d(\pi_{123}\circ f)_p < 2\}$ is precisely the set of points where the tangent plane $T_p f(M)$ is orthogonal to the 3-space E^3_{123}, so the projection $\pi_{123}\circ f$ fails to be an immersion at p. For almost all choices of basis, $\pi_{123}\circ f$ is stable, and its singularity set is a finite collection of pinch points, (Whitney umbrella points).[2] Again the homology class of $\Sigma(\pi_{123}\circ f)$ is independent of the projection, and this class is Poincaré dual to the second normal Stiefel-Whitney class $\bar{w}_2$ of M. Thus, we define $\bar{w}_2 = \mathcal{D}[\Sigma(\pi_{123}\circ f)]$.

Let $D(\pi_{123}\circ f) = \{x \in E^3 \mid (\pi_{123}\circ f)^{-1}(x)$ has more than one element$\}$, the set of double points of the image of $\pi_{123}\circ f$. If $\pi_{123}\circ f$ is stable, $D(\pi_{123}\circ f)$ is a finite collection of disjoint curves in E^3 whose set of boundary points is precisely the image $f(\Sigma(\pi_{123}\circ f))$. Furthermore, the restriction $f|\Sigma(\pi_{123}\circ f)$ is injective. Thus, $\Sigma(\pi_{123}\circ f)$ has an even number of points; i.e. $\bar{w}_2 = 0$. But we will show that the characteristic cycle $\Sigma(\pi_{123}\circ f)$ still enters into the relation between the characteristic cycles $\Sigma(\pi_{12}\circ f)$, $\Sigma(\pi_{13}\circ f)$, and $\Sigma(\pi_1\circ f)$.

If rank $d(\pi_{123}\circ f)_p = 2$, then p is in $\Sigma(\pi_{12}\circ f) \cap \Sigma(\pi_{13}\circ f)$ if and only if p is in $\Sigma(\pi_1\circ f)$. If rank $d(\pi_{123}\circ f)_p < 2$, then p is in $\Sigma(\pi_{12}\circ f) \cap \Sigma(\pi_{13}\circ f)$. Thus,

$$\Sigma(\pi_{12}\circ f) \cap \Sigma(\pi_{13}\circ f) = \Sigma(\pi_1\circ f) \cup \Sigma(\pi_{123}\circ f). \qquad (2)$$

Moreover, for almost all choices of basis $\{e_1, e_2, e_3, e_4\}$, the

0-cycles $\Sigma(\pi_1 \circ f)$ and $\Sigma(\pi_{123} \circ f)$ are disjoint, the intersection of the 1-cycles $\Sigma(\pi_{12} \circ f)$ and $\Sigma(\pi_{13} \circ f)$ on M is transverse, and we may translate (2) into a statement about cohomology classes:

$$\begin{aligned} w_1 \smile w_1 &= \mathscr{D}[\Sigma(\pi_{12} \circ f)] \smile \mathscr{D}[\Sigma(\pi_{13} \circ f)] \\ &= \mathscr{D}[\Sigma(\pi_{12} \circ f) \cap \Sigma(\pi_{13} \circ f)] \\ &= \mathscr{D}[\Sigma(\pi_1 \circ f) \cup \Sigma(\pi_{123} \circ f)] \\ &= \mathscr{D}[\Sigma(\pi_1 \circ f)] + \mathscr{D}[\Sigma(\pi_{123} \circ f)] \\ &= w_2 + \overline{w}_2 \end{aligned}$$

This last statement is the degree 2 homogeneous part of the Whitney duality theorem for surfaces.

§3. Schubert cycles of planes in 4-space.

The cycle-level interpretation of Whitney duality (2) is a reflection of the geometry of the classical Schubert cycles in a Grassmann manifold.

Let $G = G_2(E^4)$ be the Grassmann manifold of 2-dimensional linear subspaces of E^4. G is a compact smooth 4-manifold without boundary. Let $\{e_1, e_2, e_3, e_4\}$ be a basis for E^4. Let $\pi_1 \colon E^4 \to E^1_1$, $\pi_{12} \colon E^4 \to E^2_{12}$, $\pi_{13} \colon E^4 \to E^2_{13}$, and $\pi_{123} \colon E^4 \to E^3_{123}$ be as above. Let

$$\begin{aligned} S_1 &= \{X \in G \mid \dim(\pi_1 X) < 1\}, \\ S_{12} &= \{X \in G \mid \dim(\pi_{12} X) < 2\}, \\ S_{13} &= \{X \in G \mid \dim(\pi_{13} X) < 2\}, \\ S_{123} &= \{X \in G \mid \dim(\pi_{123} X) < 2\}. \end{aligned}$$

S_1, S_{12}, S_{13} and S_{123} are <u>Schubert cycles</u> of G. In fact, S_1 and S_{123} are homeomorphic with the real projective plane, and S_{12} [or S_{13}] is a 3-circuit (pseudomanifold) with an isolated singular point, corresponding to $X = E^2_{34}$ [or $X = E^2_{24}$]. Furthermore, S_{12} and S_{13} are homologous. Let W_1, W_2, $\overline{W}_2$ be the cohomology classes of G Poincaré dual to the homology classes of the circuits S_{12}, S_1, S_{123}, respectively.

Now let $f\colon M^2 \to E^4$ be an immersion of a surface in 4-space. For each $p \in M$, let $g(p) \in G$ be the translation to the origin of the tangent plane $T_p f(M)$. (The map $g\colon M \to G$ is a generalization of the classical <u>Gauss map</u> for a surface in E^3.) By the definition of the singularity set Σ,

$$\begin{aligned}
\Sigma(\pi_1 \circ f) &= g^{-1}(S_1),\\
\Sigma(\pi_{12} \circ f) &= g^{-1}(S_{12}),\\
\Sigma(\pi_{13} \circ f) &= g^{-1}(S_{13}),\\
\Sigma(\pi_{123} \circ f) &= g^{-1}(S_{123}).
\end{aligned}$$

If $\pi_1 \circ f$ is stable, g is <u>transverse</u> to S_1, so the homology class of $\Sigma(\pi_1 \circ f)$ is Poincaré dual in M^2 to the pull-back by g of the cohomology class in G Poincaré dual to the homology class of S_1:

$$\mathscr{D}[\Sigma(\pi_1 \circ f)] = g^* \mathscr{D}[S_1].$$

Similar statements are true for S_{12} and S_{123}. Therefore,

$$w_1 = g^* W_1 \ , \quad w_2 = g^* W_2 \ , \quad \overline{w}_2 = g^* \overline{W}_2.$$

It follows that the relation $w_1 \smile w_1 = w_2 + \overline{w}_2$ for <u>every</u> surface

M^2 is a consequence of the single statement $W_1 \smile W_1 = W_2 + \bar{W}_2$.

We have the intersection formula

$$S_{12} \cap S_{13} = S_1 \cup S_{123}. \qquad (3)$$

The 3-circuits S_{12} and S_{13} cross transversely in G, except on a set of dimension 1 (containing their respective singular points), so (3) translates, by Poincaré duality, into $W_1 \smile W_1 = W_2 + \bar{W}_2$. The cycle-level Whitney duality theorem (2) is the pull-back, by the Gauss map g, of the intersection formula (3) for Schubert cycles.

This formula can be visualized by identifying G with the space of projective lines in projective 3-space $P^3 = G_1(E^4)$. P^3 can be thought of as E^3 with a (projective) plane added at infinity. The lines spanned by the basis vectors e_1, e_2, e_3, e_4 correspond to four independent points p_1, p_2, p_3, p_4 in P^3. Thus, the planes E_{12} and E_{13} correspond to <u>lines</u> P_{12} and P_{13}, with intersection $P_{12} \wedge P_{13}$ a <u>point</u> P_1 and span $P_{12} \vee P_{13}$ a <u>plane</u> P_{123}.

Now S_{12} is the set of lines in P^3 which meet the line P_{12}, and S_{13} is the set of lines meeting P_{13}. S_1 comprises the lines passing through the point P_1, and S_{123} is the set of lines lying in the plane P_{123}. But a line meets P_{12} and P_{13} if and only if it either passes through $P_{12} \wedge P_{13} = P_1$ or it lies in $P_{12} \vee P_{13} = P_{123}$. This is a restatement of (3). Using local coordinates, it is easily checked that S_{12} and S_{13} cross transversely, except on the set of lines which simultaneously pass through the point P_1 and lie in the plane P_{123}; i.e. except on the set $S_1 \cap S_{123}$.

§4. Higher dimensions.

For an n-dimensional manifold M^n, the tangential and normal Stiefel-Whitney classes can also be described in terms of singularities of projections. We now explain how this description is related to the obstruction theory definition in terms of singularities of systems of vector fields on M^n. Consider an immersion $f: M^n \to E^{2n}$ into Euclidean 2n-space with orthonormal basis $\{e_1, e_2, \ldots, e_{2n}\}$. For any fixed vector x in E^{2n} and any point $p \in M$, we may write $x = \tau_p f(x) + \nu_p f(x)$ where $\tau_p f(x)$ is a tangent vector in $T_p f(x)$ and $\nu_p f(x)$ is in the normal space $T_p^{\perp} f(M)$. We may then define $2n$ tangent vector fields $v_1 f, \ldots, v_{2n} f$ on M by $v_k f(p) = \tau_p f(e_k)$. The set of zeroes $Z(v_1 f) = \{p \in M^2 |\ v_1 f(p) = 0\}$ is precisely the singularity set $\Sigma(\pi_1 \circ f)$. If $\pi_1 \circ f$ is stable, $Z(v_1 f)$ is a finite set, and each zero has multiplicity $(-1)^r$, where r is the Morse index of the corresponding nondegenerate critical point of $\pi_1 \circ f$. More generally, for $1 \leq i \leq n$, $Z(v_1 f, \ldots, v_{n-i+1} f) = \{p |\ v_1 f(p), \ldots, v_{n-i+1} f(p)$ are linearly dependent$\} = \{p |\ \text{rank}\ d(\pi^{n-i+1} \circ f)_p < n - i + 1\} = \Sigma(\pi^{n-i+1} \circ f)$, where $\pi^{n-i+1}: E^{2n} \to E^{n-i+1}_{12\ldots n-i+1}$ denotes the orthogonal projection to the plane spanned by $\{e_1, \ldots, e_{n-i+1}\}$. For almost all choices of basis, $\Sigma(\pi^{n-i+1} \circ f)$ is a cycle (mod 2) in a single homology class. By definition, the ith tangential Stiefel-Whitney class w_i of M is the Poincaré dual of this homology class $w_i = \mathscr{D}[\Sigma(\pi^{n-i+1} \circ f)]$.

Similarly, we may define 2n normal vector fields $\bar{v}_1 f, \ldots, \bar{v}_{2n} f$

on M by $\bar{v}_k f(p) = \nu_p f(e_k)$, and $Z(\bar{v}_{2n}f) = \{p \mid \bar{v}_{2n}f(p) = 0\} = \{p \mid e_{2n} \in T_p f(M)\} = \Sigma(\pi^{2n-1} \circ f)$. More generally, for $1 \leq j \leq n$, $Z(\bar{v}_{n+j}f, \ldots, \bar{v}_{2n}f) = \{p \mid \bar{v}_{n+j}f(p), \ldots, \bar{v}_{2n}f(p)$ are linearly dependent$\}$ $=$ $\{p \mid T_p f(M)$ contains a vector in the span of $\{e_{n+j}, \ldots, e_{2n}\}\} = \{p \mid \text{rank } d(\pi^{n+j-1} \circ f)_p < n\} = \Sigma(\pi^{n+j-1} \circ f)$. For almost all choices of basis of E^{2n}, $\Sigma(\pi^{n+j-1} \circ f)$ is a cycle in a single homology class. The jth normal Stiefel-Whitney cohomology class is defined to be the Poincaré dual of $\bar{w}_j = \mathscr{D}[\Sigma(\pi^{n+j-1} \circ f)]$

The generalization of the intersection formula (2) to n-manifolds is a cycle-level description of $w_i \smile \bar{w}_j$. Let $A = A^{n-i+1} = \text{span}\{e_1, \ldots, e_{n-i+1}\}$ and $B = B^{n+j-1} = \text{span}$ $\{e_1, \ldots, e_{n-i}, e_{n+1}, \ldots, e_{n+i+j-1}\}$. Let $A' = (A')^{n-i}$ be the intersection of A and B, $A' = \text{span}\{e_1, \ldots, e_{n-i}\}$, and let $B' = (B')^{n+j}$ be the span of A and B, $B' = \text{span}\{e_1, \ldots, e_{n-i+1}, e_{n+1}, \ldots, e_{n+i+j-1}\}$:

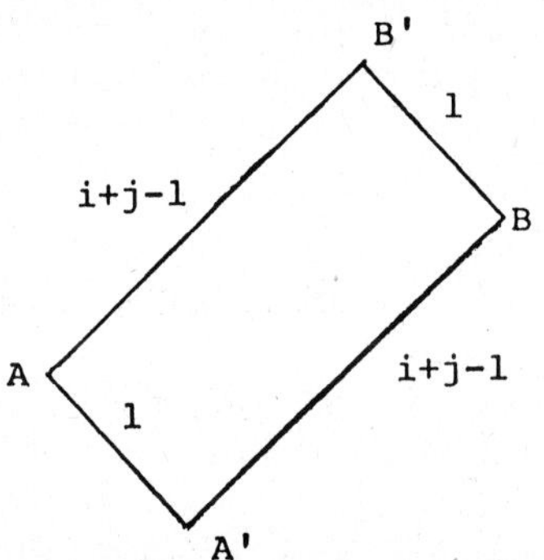

Let π_A π_B $\pi_{A'}$ $\pi_{B'}$ denote the orthogonal projections of E^{2n} to A, B, A', B', respectively.

Now let $f\colon M^n \to E^{2n}$ be an immersion. If $\text{rank } d(\pi_{B'} \circ f)_p = n$, then p is in $\Sigma(\pi_A \circ f) \cap \Sigma(\pi_B \circ f)$ if and only if p is in $\Sigma(\pi_{A'} \circ f) \cap \Sigma(\pi_B \circ f)$. If $\text{rank } d(\pi_{B'} \circ f)_p < n$, then p is in

$\Sigma(\pi_A \circ f) \cap \Sigma(\pi_{B'} \circ f)$. Therefore,

$$\Sigma(\pi_A \circ f) \cap \Sigma(\pi_B \circ f) = (\Sigma(\pi_{A'} \circ f) \cap \Sigma(\pi_B \circ f)) \cup (\Sigma(\pi_A \circ f) \cap \Sigma(\pi_{B'} \circ f)). \tag{4}$$

For almost all choices of basis $\{e_1,\ldots,e_{2n}\}$, $\pi_A \circ f$, $\Sigma(\pi_{A'} \circ f) \cap \Sigma(\pi_B \circ f)$ and $\Sigma(\pi_A \circ f) \cap \Sigma(\pi_{B'} \circ f)$ are $n - (i+j)$ cycles, whose respective homology classes do not depend on the choice of basis. (Notice that $[\Sigma(\pi_{A'} \circ f) \cap \Sigma(\pi_B \circ f)]$ is <u>not</u> the intersection product of $[\Sigma(\pi_{A'} \circ f)]$ and $[\Sigma(\pi_B \circ f)]$, since this product would have dimension $n - (i+j) - 1$. $\Sigma(\pi_{A'} \circ f)$ and $\Sigma(\pi_B \circ f)$ are not in general position because $A' \subset B$. A similar comment holds for $\Sigma(\pi_A \circ f) \cap \Sigma(\pi_{B'} \circ f)$.) We define $y_{i-1,j+1} = \mathscr{D}[\Sigma(\pi_{A'} \circ f) \cap \Sigma(\pi_B \circ f)]$ and $y_{i,j} = \mathscr{D}[\Sigma(\pi_A \circ f) \cap \Sigma(\pi_{B'} \circ f)]$. For almost all choices of basis, $\Sigma(\pi_A \circ f)$ and $\Sigma(\pi_B \circ f)$ intersect transversely, except possibly on a set of dimension $n - (i+j) - 1$, and so (4) translates to an equation in cohomology:

$$w_i \smile \bar{w}_j = y_{i-1,j+1} + y_{i,j}. \tag{5}$$

Summing this equation over all i and j with $i + j = k$, we obtain $\sum_{i=1}^{k-1} w_i \smile \bar{w}_{k-i} = y_{0,k} + y_{k,0} = w_k + \bar{w}_k = w_k \smile \bar{w}_0 + w_0 \smile \bar{w}_k$. ($\bar{w}_0 = w_0 = 1$.) Thus

$$\sum_{i+j=k} w_i \smile \bar{w}_j = 0, \tag{6}$$

the degree k homogeneous part of the <u>Whitney duality theorem</u> for n-manifolds:

$$w \smile \bar{w} = 1.$$

In fact, (6) is also true on the cycle-level. For $1 \leq i \leq k$, let $A^{n-i+1} = \text{span}\{e_1,\dots,e_{n-i+1}\}$, and for $1 \leq j \leq k$, let $B^{n+j-1} = \text{span}\{e_1,\dots,e_{n-k+j},e_{n+1},\dots,e_{n+k-1}\}$. Let π^{n-i+1}: $E^{2n} \to A^{n-i+1}$ and $\overline{\pi}^{n+j-1}$: $E^{2n} \to B^{n+j-1}$ be orthogonal projections. For almost all choices of basis, $\Sigma(\pi^{n-i+1}\circ f)$ and $\Sigma(\overline{\pi}^{n+j-1}\circ f)$ are cycles in fixed homology classes which intersect transversely, except possibly on a set of dimension $n - (i+j) - 1$. Let $|\Sigma(\pi^{n-i+1}\circ f)|\cdot|\Sigma(\overline{\pi}^{n+j-1}\circ f)|$ denote their intersection cycle. Then

$$\sum_{i+j=k} |\Sigma(\pi^{n-i+1}\circ f)|\cdot|\Sigma(\overline{\pi}^{n+j-1}\circ f)| = 0 \qquad (k > 0), \tag{7}$$

summed as cycles mod 2 (with cancellation mod 2).

This intersection formula is proved by applying (4) to the individual steps in the ladder:

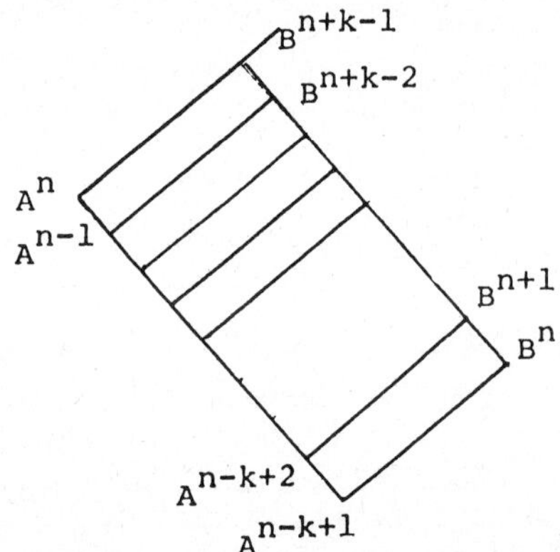

(4) and (7) can be interpreted as intersection formulas for Schubert cycles, generalizing (3). The formula (5) is a consequence of the relation

$$W_i \smile \overline{W}_j = Y_{i-1,j+1} + Y_{i,j} \tag{8}$$

between Schubert cohomology classes. This relation, due to Chern [1, p. 370], is a special case of an intersection identity modeled after a classical formula of Pieri. Formula (4) corresponds to a precise cycle-level version of (8).

REFERENCES

[1] S. Chern, On the multiplication in the characteristic ring of a sphere bundle, Annals of Math. 49(1948), 362-372.

[2] J. Mather, Generic projections, Annals of Math. 98(1973), 226-245.

[3] L.S. Pontrjagin, Vector fields on manifolds, Mat. Sbornik (N.S.) 24(66)(1949), 129-162. Amer. Math. Soc. Translations, series one, volume 7 (1962).

[4] J. Sotomayor, Singularidades de Aplicações Diferenciáveis, III ELAM, (1976) IMPA.

[5] R. Thom, Les singularités des applications différentiables, Ann. Inst. Fourier, Grenoble 6 (1955-56), 43-87.

Department of Mathematics
Brown University
Providence, Rhode Island 02912

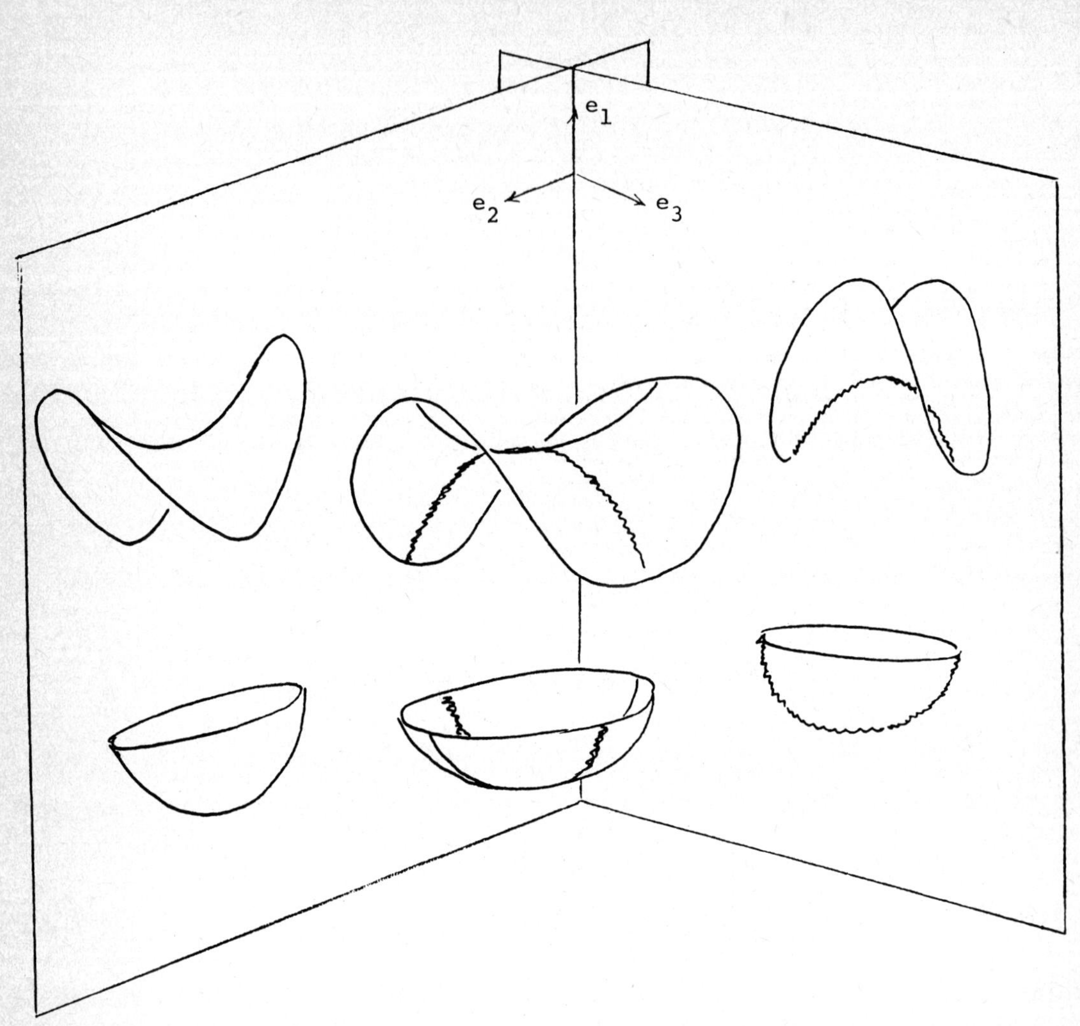
e1
e2
e3

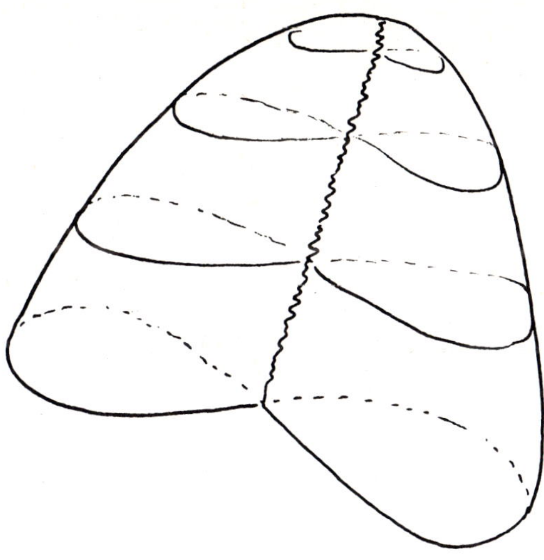

Whitney umbrella, or pinch point

Orbit Preserving Diffeomorphisms and the Stability of Lie Group Actions and Singular Foliations

César Camacho   Alcides Lins Neto

In certain situations concerning differentiable Lie group actions and foliations with singularities one is led to consider pairs (X,f) where X is a vector field on a manifold M and $f: M \to M$ is a diffeomorphism sending orbits of X to orbits of X. We call such an f an orbit preserving diffeomorphism.

Severe obstructions occur for f even when X is a linear vector field in R^n. More precisely, let $(\mu_i)_{i=1}^{n} = (\alpha_j \pm \beta_j)_{j=1}^{k}$ be the set of eigenvalues of X and $E = \bigoplus_{j=1}^{k} E_j$ the eigenspace decomposition. Suppose that for every $j = 1,\dots,k$ $\alpha_j > 0$ (or $\alpha_j < 0$) and let $|\alpha_1| < |\alpha_j|$ for all $j = 2,\dots,k$.

Define $\Sigma = \{x = (x_1,\dots,x_k) \in E_1 \times\dots\times E_k \mid |x_1| = 1\}$. Putting $E = E_2 \times\dots\times E_k$ one obtains that $\Sigma = \{-1,1\} \times E$ or $\Sigma = S^1 \times E$ according to whether $\beta_1 = 0$ or not.

Let $\pi: R^n - E \to \Sigma$ be the projection along the orbits of X. We shall see in (1.2) that E is f-invariant. Consequently π is well defined and onto. Therefore the map $\bar{f} = \pi \circ f: \Sigma \to \Sigma$ is a diffeomorphism.

<u>Theorem 1</u> - Let (X,f), X linear, $f \in C^r$, $r \geq 1$, be as above. Suppose that $|\alpha_1| < |\alpha_2| < \dots < |\alpha_k|$. Then

A. If $\beta_1 = 0$, $\bar{f} = (\bar{f}_2,\dots,\bar{f}_k)$ is written

$$\bar{f}_j(x_2,\dots,x_k) = A_j x_j + \varphi_j(x_2,\dots,x_{j-1}) \qquad 2 \leq j \leq k$$

B. If $\beta_1 \neq 0$, $\bar{f} = (\bar{f}_\theta,\bar{f}_2,\dots,\bar{f}_k)$ is written

$$\bar{f}_\theta(\theta, x_2, \ldots, x_k) = \theta + \theta_o$$

$$\bar{f}_j(\theta, x_2, \ldots, x_k) = A_j x_j + \varphi_j(\theta, x_2, \ldots, x_{j-1}) \qquad 2 \leq j \leq k$$

where $\theta_o \in \mathbb{R}$, $A_j \in \mathbb{R}$ if $\beta_j = 0$ and $A_j = \begin{pmatrix} a_j & b_j \\ -b_j & a_j \end{pmatrix}$ if $\beta_j \neq 0$.

We say that the eigenvalues of X satisfy nonresonant conditions if for any sequence $(m_1, m_2, \ldots, m_n)$ of non negative integers with $\sum_{j=1}^{n} m_j \geq 2$ one has $\mu_i \neq \sum_{j=1}^{n} m_j \mu_j$ $i=1,\ldots,n$. The following generalizes N. Kopell theorem ([3]).

Theorem 2 - Let X be linear and (X,f) as above. Suppose the eigenvalues of X, $\mu_1, \ldots, \mu_n$ satisfy nonresonant conditions and $|Re(\mu_1)| < |Re(\mu_j)| \leq |Re(\mu_n)|$ for $1 < j < n$. Assume furthermore that $f \in C^K$, $K > |Re(\mu_n)|/|Re(\mu_1)|$. Then

A. If $\mu_1 \in \mathbb{R}$ then $\bar{f}$ is linear

B. If $\mu_1 \notin \mathbb{R}$ and $\bar{f} = (\bar{f}_\theta, \bar{f}_2, \ldots, \bar{f}_k)$ then $\bar{f}_\theta(\theta, x_2, \ldots, x_k) = \theta + \theta_o$ and $\bar{f}_j$ is linear, $2 \leq j \leq k$.

In particular it follows that f cannot leave fixed X-orbits outside the invariant subspaces of X unless it induces the identity map in the orbit space of X.

At the end of §1. we study orbit preserving diffeomorphisms of vector fields on 2-manifolds. There we prove the following corollary of Theorem 1.

Corollary I - Let M be a compact 2-manifold and $\mathfrak{X}^r(M)$ the space of vector fields with the uniform C^r topology. There is an open and dense subset $\mathcal{G} \subset \mathfrak{X}^r(M)$, $r \geq 2$, such that if $Y \in \mathcal{G}$ and $f\colon M \to M$ is a C^r diffeomorphism preserving the orbits of Y, then the induced map $\tilde{f}\colon M/Y \to M/Y$ in the orbit space of Y satisfies $\tilde{f}^2 = $ identity.

This theorem was generalized to n-manifolds by Paulo Sad in his thesis concerning centralizers of vector fields [10].

In §2. we apply Theorem 2 to the stability of actions

$\varphi: \mathbb{R}^2 \times S^n \to S^n$ of the group $\mathbb{R}^2$ on the n-sphere S^n. Such an action is said to be non trivial if there is no open subset of S^n where the φ-orbits have dimension less than two.

The following theorem solves Problem 4 of [4].

Theorem 3 - There are C^2-structurally stable non trivial $\mathbb{R}^2$-actions on the n-sphere.

Section 3 is devoted to the stability of foliations with singularities defined by 1-forms ω with the integrability condition $\omega \wedge d\omega = 0$. A singularity of ω is a point at which ω vanishes. Let $\mathcal{J}^r(M)$ be the space of integrable 1-forms on a manifold M endowed with the C^r-uniform topology. Then ω is structurally stable if there is a neighborhood $N(\omega)$ in $\mathcal{J}^r(M)$ and for any $\eta \in N(\omega)$ a homeomorphism $h: M \to M$ sending leaves of ω to leaves of η and singularities of ω to singularities of η.

Theorem 4 - On any compact orientable 3-manifold there exists a C^2-structurally stable foliation whose singular subset is a union of finitely many closed curves.

Theorem 5 - Let M^2, N^n be compact manifolds and $\pi_1: M^2 \times N^n \to M^2$, $\pi_1(x,y) = x$. There is an open and dense subset $\mathcal{F} \subset \mathcal{J}^r(M^2)$ of the space of 1-forms on M^2 such that for any $\omega \in \mathcal{F}$, $\pi_1^*\omega$ is C^2-structurally stable.

We wish to thank conversations on the subject with A. de Medeiros and P. Sad.

§1. Diffeomorphisms preserving the orbits of linear vector fields.

Let X and Y be vector fields in $\mathbb{R}^n$. Denote the orbit of X through x by $\mathcal{O}_X(x)$ and the parametrized orbit by $X_t(x)$. We say that the germ $f: (\mathbb{R}^n,p) \to (\mathbb{R}^n,q)$ sends orbits of X to orbits of Y if for any representative $\tilde{f}: U \to \mathbb{R}^n$ of f, $\tilde{f}(\mathcal{O}_X(x) \cap U) \subset \subset \mathcal{O}_Y(\tilde{f}(x))$ for $x \in U$. To simplify notation we shall denote any representative of f by f again.

(1.1) Proposition - Let X and Y be C^1 vector fields in $\mathbb{R}^n$ and $X(p) = Y(q) = 0$. Suppose that $f: (\mathbb{R}^n,p) \to (\mathbb{R}^n,q)$ is a germ of a C^1 diffeomorphism sending orbits of X to orbits of Y. Let $A = DX(p)$, $B = DY(q)$ and $J = Df(p)$. Suppose further that A and B are nonsingular. Then $A = \mu J^{-1}BJ$ where $\mu \in \mathbb{R}$. If $X = Y$ and $p = q$, then $\mu = \pm 1$. If $X = Y$ and there is an eigenvalue λ of A such that $-\lambda$ is not eigenvalue of A, then $\mu = 1$.

Proof: We can suppose $p = q = 0$. By hypothesis we have

$$Df_x(X(x)) = \mu(x)\, Y(f(x)) \qquad (*)$$

where $\mu: (\mathbb{R}^n,0) \to (\mathbb{R},0)$ is continuous outside $0 \in \mathbb{R}^n$. Putting $J(x) = Df(x)$ obtain from $(*)$

$$J(tx)\left(\frac{1}{t} X(tx)\right) = \mu(tx)\left(\frac{1}{t} Y(f(tx))\right).$$

Since $\lim_{t\to 0} J(tx) = J$, $\lim_{t\to 0} \left(\frac{1}{t} X(tx)\right) = A\cdot x$, $\lim_{t\to 0} \frac{1}{t} Y(f(tx)) = B\cdot J\cdot x$ and A, B, J are non singular, there exists $\lim_{t\to 0} \mu(tx) = \mu(x)$ and $J\cdot A(x) = \mu(x) B\cdot J(x)$.

Let us show that μ is constant. Set $R = J.A$, $S = B\cdot J$. If $x,y \in \mathbb{R}^n$ are linearly independent, so are Sx and Sy, and $\mu(x)\, S\cdot x + \mu(y)\, S\cdot y = R(x+y) = \mu(x+y)\,(S\cdot x+S\cdot y)$, therefore $\mu(x) = \mu(x+y) = \mu(y)$ which implies that μ is constant.

Suppose that $X = Y$ and assume by contradiction that $|\mu| > 1$. Let $\tilde{A}$, $\tilde{J}$ be the complexifications of A, J. Let ν be an eigenvalue of $\tilde{A}$ with minimum norm and $v \in \mathbb{C}^n$ an eigenvector of $\tilde{A}$ with $\tilde{A}(v) = \nu v$. We have $\nu\tilde{J}(v) = \tilde{J}\tilde{A}(v) = \mu\tilde{A}(\tilde{J}v)$ which implies $\tilde{A}(\tilde{J}(v)) = \mu^{-1}\nu\tilde{J}(v)$. Therefore $\mu^{-1}\nu$ is an eigenvalue of $\tilde{A}$ with $|\mu^{-1}\nu| < |\nu|$ which is absurd. The case $|\mu| < 1$ is analogous, hence $\mu = \pm 1$. If there is an eigenvalue λ of A such that $-\lambda$ is not eigenvalue of A then by an analogous argument, one shows that $\mu = 1$.

Assume now that X is a linear vector field in $\mathbb{R}^n$ with eigenvalues $(\mu_i)_{i=1}^n = (\alpha_j \pm i\beta_j)_{j=1}^k$. Suppose that $\alpha_j < 0$ (or $\alpha_j > 0$) for $j = 1,\ldots,k$ $\alpha_i \neq \alpha_i'$ if $i \neq i'$. Let E_j, $j = 1,\ldots,k$ be the invariant subspaces of X and <u>assume that</u> λ_1 <u>is the weakest eigenvalue of</u> X, i.e.

$$|\alpha_1| < |\alpha_j| \qquad j = 2,\ldots,k.$$

Call $E = \bigoplus_{j \neq 1} E_j$ the complementary subspace of E_1.

(1.2) <u>Lemma</u> - If $f\colon (\mathbb{R}^n,0) \to (\mathbb{R}^n,0)$ is a germ of a C^r, $r \geq 1$, diffeomorphism sending orbits of X to orbits of X, then E is invariant by f. That is, if $f\colon U \to \mathbb{R}^n$ is a representative of f, then $f(U \cap E) \subset E$.

<u>Proof</u>: Let $J = Df(0)$, $A = DX(0)$. By (1.1) $AJ = JA$. Since λ_1 is the weakest eigenvalue of A it follows that $J(E) = E$. Therefore $f(U \cap E)$ is tangent to E at $0 \in \mathbb{R}^n$. Suppose that $0 \in \mathbb{R}^n$ is a sink and let $\rho(t)$ be the angle between $X(X_t(x))$ and E_1. One has $\lim_{t\to\infty} \rho(t) = 0$ if and only if $X_t(x) \cap E = \emptyset$. Hence $f(U \cap E) \subset E$.

(1.3) <u>Proof of Theorem 1</u>: Let $\mathbb{R}^n = E_1 \oplus\ldots\oplus E_k$, where $\dim E_i = 1$ or 2 and E_i is invariant by X. Write $\lambda_j = \alpha_j + i\beta_j$ for $j = 1,\ldots,k$. Since $Df(0)$ commutes with X (see (1.1)), we can write $f = (f_1,\dots,f_k)$, where $f_j\colon \mathbb{R}^n \to E_j$ is of the form

$$f_j(x_1,\dots,x_k) = A_j x_j + R_j(x_1,\dots,x_k),$$

and

$$A_j = a_j \in \mathbb{R} \text{ if } \lambda_j \in \mathbb{R},\quad A_j = \begin{bmatrix} a_j & b_j \\ -b_j & a_j \end{bmatrix} \text{ if } \lambda_j \notin \mathbb{R},\quad \lim_{x\to 0} \frac{R_j(x)}{\|x\|} = 0.$$

If $\lambda_1 \in \mathbb{R}$, then $\Sigma = \{(\pm 1, x_2,\dots,x_k) \mid x_j \in E_j\}$ and in this case we denote $p \in \Sigma$ by $(\tau_2,\dots,\tau_k)$ and $\bar{f} = (\bar{f}_2,\dots,\bar{f}_k)$. If $\lambda_1 \notin \mathbb{R}$, then $\Sigma = \{(x_1,\dots,x_k) \mid \|x_1\|^2 = 1$ and $x_j \in E_j$ for $j = 2,\dots,k\}$ and we denote $p \in \Sigma$ by $(\theta, \tau_2,\dots,\tau_k)$ and $\bar{f} = (\bar{f}_\theta, \bar{f}_2,\dots,\bar{f}_k)$, where $e^{i\theta} \in S^1$.

Suppose first that $\lambda_1 \in \mathbb{R}$. Then the trajectory of X through the point $\tau = (1,\tau_2,\dots,\tau_k)$ has the following expression: $x_1(t,\tau) = e^{\lambda_1 t}$ and if $2 \le j \le k$, $x_j(t,\tau) = e^{\alpha_j t}\tau_j$ if $\lambda_j \in \mathbb{R}$ and

$x_j(t,\tau) = e^{\alpha_j t} B(\beta_j t)\cdot\tau_j$, where $B(t) = \begin{bmatrix} \cos(t) & \sin(t) \\ -\sin(t) & \cos(t) \end{bmatrix}$ if $\lambda \notin \mathbb{R}$.

This curve in terms of x_1 is $t = \ell n(x_1)^{1/\lambda_1}$, $x_j(x_1,\tau) = (x_1)^{\bar{\alpha}_j}\tau_j$ if $\lambda_j \in R$ and $x_j(x_1,\tau) = (x_1)^{\bar{\alpha}_j} B(\bar{\beta}_j\, \ell n(x_1))\cdot\tau_j$ if $\lambda_j \notin \mathbb{R}$, where $\bar{\alpha}_j = \alpha_j/\alpha_1$ and $\bar{\beta}_j = \beta_j/\alpha_1$. Since $|\mathrm{Re}(\lambda_1)| < \dots < |\mathrm{Re}(\lambda_k)|$ by hypothesis, we have $1 < \bar{\alpha}_2 < \dots < \bar{\alpha}_k$. Now suppose that $a_1 = \frac{\partial f_1}{\partial x_1}(0) > 0$ and let $E_+ = \{(x_1,\dots,x_k) \mid x_1 > 0\}$, $\Sigma_1 = \{(1,\tau_2,\dots,\tau_k) \mid \tau_j \in E_j\}$. In this case $f(E_+) \subset E_+$, therefore $\bar{f}(\Sigma_1) = \Sigma_1$ and $\bar{f}(\Sigma_{-1}) = \Sigma_{-1}$. Since f sends the curve $x(x_1,\tau) = (x_1, x_2(x_1,\tau),\dots,x_k(x_1,\tau))$ to the curve $x(x_1, \bar{f}(\tau))$, we can write

a) $f_j(x(x_1,\tau)) = [f_1(x(x_1,\tau))]^{\bar{\alpha}_j}\cdot\bar{f}_j(\tau)$, if $\lambda_j \in \mathbb{R}$, or

b) $f_j(x(x_1,\tau)) = [f_1(x(x_1,\tau))]^{\bar{\alpha}_j}\cdot B(\bar{\beta}_j\, \ell n(f_1(x(x_1,\tau))))\cdot\bar{f}_j(\tau)$,
if $\lambda_j \notin \mathbb{R}$.

Let us consider case b), since the other is similar. We have:

$$A_j \cdot (x_1^{\bar{\alpha}_j} B(\bar{\beta}_j \ell n(x_1))) \cdot \tau_j + R_j(x(x_1,\tau)) =$$

$$= [a_1 x_1 + R_1(x(x_1,\tau))]^{\bar{\alpha}_j} \cdot B(\bar{\beta}_j \ell n(a_1 x_1 + R_1)) \cdot \bar{f}_j(\tau).$$

Multiplying both members by $x_1^{-\bar{\alpha}_j} B(-\bar{\beta}_j \ell n(x_1))$, we get

(1) $$A_j \cdot \tau_j + x_1^{-\bar{\alpha}_j} B(-\bar{\beta}_j \ell n(x_1)) R_j =$$

$$= [a_1 + \frac{R_1}{x_1}]^{\bar{\alpha}_j} \cdot B(\bar{\beta}_j \ell n(a_1 + \frac{R_1}{x_1})) \cdot \bar{f}_j(\tau).$$

Differentiating both members with respect to τ_j, we get:

(2) $$A_j + B(-\bar{\beta}_j \ell n(x_1)) \cdot D_j R_j(x(x_1,\tau)) \cdot B(\bar{\beta}_j \ell n(x_1)) =$$

$$= [a_1 + \frac{R_1}{x_1}]^{\bar{\alpha}_j} \cdot B(\bar{\beta}_j \ell n(a_1 + \frac{R_1}{x_1})) \cdot \frac{\partial \bar{f}_j}{\partial \tau_j} + (x_1)^{\bar{\alpha}_j - 1} Y_j,$$

where

$$Y_j(v) = [a_1 + \frac{R_1}{x_1}]^{\bar{\alpha}_j - 1} (D_j R_1 \circ B(\bar{\beta}_j \ell n(x_1))(v)) \cdot \bar{X}_j \cdot B(\bar{\beta}_j \ell n(a_1 + \frac{R_1}{x_1})) \cdot \bar{f}_j,$$

where $\bar{X}_j = \begin{bmatrix} \bar{\alpha}_j & \bar{\beta}_j \\ -\bar{\beta}_j & \bar{\alpha}_j \end{bmatrix}$. Now we have $\lim_{x_1 \to 0} x(x_1,\tau) = 0$,

$\lim_{x_1 \to 0} D_j R_j(x(x_1,\tau)) = \lim_{x_1 \to 0} D_j R_1(x(x_1,\tau)) = 0$ and $\lim_{x_1 \to 0} \frac{R_1(x(x_1,\tau))}{x_1} = 0$,

therefore $\lim_{x_1 \to 0} [a_1 + \frac{R_1}{x_1}]^{\bar{\alpha}_j} = (a_1)^{\bar{\alpha}_j}$, $\lim_{x_1 \to 0} B(\bar{\beta}_j \ell n(a_1 + \frac{R_1}{x_1})) =$

$= B(\bar{\beta}_j \ell n(a_1))$. Since $B(\bar{\beta}_j \ell n(x_1))$ is bounded, Y_j is bounded.

Hence, taking the limit as $x_1 \to 0$ of both members of (2), we get,

$$\frac{\partial \bar{f}_j}{\partial \tau_j} = (a_1)^{-\bar{\alpha}_j} B(-\bar{\beta}_j \ell n(a_1)) \cdot A_j = \bar{A}_j = \begin{bmatrix} \bar{a}_j & \bar{b}_j \\ -\bar{b}_j & \bar{a}_j \end{bmatrix}.$$

Differentiating both members of (1) with respect to τ_ℓ, $\ell \neq j$, we get:

(3) $$x_1^{\bar{\alpha}_\ell - \bar{\alpha}_j} B(-\bar{\beta}_j \ell n(x_1)) D_\ell R_j B(\bar{\beta}_\ell \ell n(x_1)) =$$

$$= [a_1 + \frac{R_1}{x_1}]^{\bar{\alpha}_j} \cdot B(\bar{\beta}_j \ell n(a_1 + \frac{R_1}{x_1})) \frac{\partial \bar{f}_j}{\partial \tau_\ell} + x_1^{\bar{\alpha}_\ell - 1} Y_\ell,$$

where Y_ℓ is bounded. If $\ell > j$, then $\bar{\alpha}_\ell - \bar{\alpha}_j > 0$, hence taking the limit as $x_1 \to 0$ of both members of (3), we get $\frac{\partial \bar{f}_j}{\partial \tau_\ell} = 0$. This proves A of Theorem 1.

Suppose now that $\lambda_1 \notin \mathbb{R}$. The orbit of X through the point $\tau = (\tau_1, \tau_2, \ldots, \tau_k)$, where $\tau_1 = (\cos\theta_o, \sin\theta_o)$, is given by $x_j(t) = e^{\alpha_j t}\tau_j$ if $\lambda_j \in \mathbb{R}$ and $x_j(t) = e^{\alpha_j t} B(\beta_j t)\tau_j$ if $\lambda_j \notin \mathbb{R}$. If $r = \|x_1\|$, then $r = e^{\alpha_1 t}$, therefore $t = \frac{1}{\alpha_1} \ln(r)$ and we can parametrize the orbit in terms of r as: $x_1(r,\tau) = r\, B(\bar{\beta}_1 \ln(r)) \cdot \tau_1$, $x_j(r,\tau) = (r)^{\bar{\alpha}_j} \tau_j$ if $\lambda_j \in \mathbb{R}$ or $x_j(r,\tau) = (r)^{\bar{\alpha}_j} B(\bar{\beta}_j \ln(r)) \cdot \tau_j$ if $\lambda_j \notin \mathbb{R}$.

Let $f = (f_1, f_2, \ldots, f_k)$, where $f_1 = (f_{11}, f_{12})$. Let $\rho = \sqrt{f_{11}^2 + f_{12}^2}$. Then $\rho(x) = \mu r + R(x)$, where $\mu = \sqrt{a_1^2 + b_1^2}$, and $\lim_{x\to 0} \frac{R(x)}{\|x\|} = 0$. Since $\bar{f}$ sends orbits of X into orbits of X, we have:

$$\rho(x(r,\tau))\, B(\bar{\beta}_1 \ln(\rho)) \cdot \begin{pmatrix} \cos(\bar{f}_\theta(\theta_o,\tau)) \\ \sin(\bar{f}_\theta(\theta_o,\tau)) \end{pmatrix} = f_1(x(r,\tau)) \quad \text{or}$$

$$(\mu r + R)\, B(\bar{\beta}_1 \ln(\mu r + R)) \cdot \begin{pmatrix} \cos(\bar{f}_\theta) \\ \sin(\bar{f}_\theta) \end{pmatrix} =$$

$$= A_1 \left(r B(\bar{\beta}_1 \ln(r)) \cdot \begin{pmatrix} \cos\theta_o \\ \sin\theta_o \end{pmatrix} \right) + R_1(x(r,\tau)).$$

Multiplying both members by $r^{-1} B(-\bar{\beta}_1 \ln(r))$, we get

$$\left(\mu + \frac{R}{r}\right) B\left(\bar{\beta}_1 \ln\left(\mu + \frac{R}{r}\right)\right) \cdot \begin{pmatrix} \cos(\bar{f}_\theta) \\ \sin(\bar{f}_\theta) \end{pmatrix} =$$

$$= A_1 \cdot \begin{pmatrix} \cos\theta_o \\ \sin\theta_o \end{pmatrix} + B(-\bar{\beta}_1 \ln(r)) \frac{R_1}{r}.$$

Now $\lim_{r\to 0} \frac{R(x(r,\tau))}{r} = \lim_{r\to 0} \frac{R(x(r,\tau))}{\|x(r,\tau)\|} \cdot \frac{\|x(r,\tau)\|}{r} = 0$ (because $\frac{\|x(r,\tau)\|}{r}$ is bounded) and similarly $\lim_{r\to 0} \frac{R_1(x(r,\tau))}{r} = 0$. Hence $\mu\, B(\bar{\beta}_1 \ln(\mu)) \begin{pmatrix} \cos(\bar{f}_\theta) \\ \sin(\bar{f}_\theta) \end{pmatrix} = A_1 \cdot \begin{pmatrix} \cos\theta_o \\ \sin\theta_o \end{pmatrix}$. This implies that $\bar{f}_\theta(\theta_o, \tau_2, \ldots, \tau_k) = \theta_o + \varphi_o$.

The proof that $\frac{\partial \bar{f}_j}{\partial \tau_\ell} = 0$ if $\ell > j$ and $\frac{\partial \bar{f}_j}{\partial \tau_j} = \bar{A}_j$ (constant) is analogous to the proof of A. QED

(1.4) Corollary - Let $n = 2$, X, Σ, $f \in C^1$ and $\bar{f}: \Sigma \to \Sigma$ be as above and let λ_1, λ_2 be the eigenvalues of X.

A - Suppose that $\lambda_1, \lambda_2 \in \mathbb{R}$, $0 < |\lambda_1| < |\lambda_2|$. Then $\bar{f}$ is affine, that is, $\bar{f}(x_2) = ax_2 + b$ where $a, b \in \mathbb{R}$.

B - Suppose that $\lambda_1 = \bar{\lambda}_2 \notin \mathbb{R}$. Then $\bar{f}: S^1 \to S^1$ is a rotation.

Remark: Corollary (1.4) can be proved even if $\lambda_1 = \lambda_2 \neq 0$. If

$$X = \begin{bmatrix} \lambda_1 & 0 \\ 0 & \lambda_1 \end{bmatrix}$$

and f is C^1 we have two cases:

C - There exists $x \neq 0$, such that $f(\mathcal{O}(x)) \subset \mathcal{O}(x)$. In this case we take $E = \{\lambda x \mid \lambda \in R\}$, $v \notin E$, $\Sigma_v = E+v$ and $\Sigma = \Sigma_v \cup \Sigma_{-v}$. Then $\bar{f}: \Sigma \to \Sigma$ is affine.

D - For any $x \neq 0$, $f(\mathcal{O}(x)) \not\subset \mathcal{O}(x)$. In this case we take $\Sigma = S^1$. Then $\bar{f}: S^1 \to S^1$ is a rotation.

Similarly if $X = \begin{bmatrix} \lambda & 1 \\ 0 & \lambda \end{bmatrix}$ and $f \in C^2$.

Now consider the following situation: Let X be a C^2 vector field in $\mathbb{R}^2$ such that $X(p_o) = 0$ and let λ_1, λ_2 be the eigenvalues of $DX(p_o)$. Assume that $\mathrm{Re}(\lambda_i) > 0$ (or $\mathrm{Re}(\lambda_i) < 0$) $i = 1,2$. Let $S^1 \subset \mathbb{R}^2$ be a circle transverse to X, such that for any $x \in S^1$ the α-limit (or ω-limit) set of x is p_o. Let $f: (\mathbb{R}^2, p_o) \to (\mathbb{R}^2, p_o)$ be a germ of C^1 diffeomorphism which sends orbits of X into orbits of X. Then the pair (X,f) induces a C^1 diffeomorphism $\bar{f}: S^1 \to S^1$. From Hartman's theorem of C^1 linearization of C^2 hyperbolic sources (or sinks) [2] and Theorem 1, one obtains the following corollary.

(1.5) Corollary - A - Let $0 < |\lambda_1| < |\lambda_2|$. In this case there exist $p_1, p_2 \in S^1$, such that if $S^1 - \{p_1, p_2\} = S_+ \cup S_-$ (where S_+ and S_- are connected), then $\bar{f}(S_+) = S_+$ or S_-. Suppose that $\bar{f}(S_+) = S_+$ and $\bar{f}$ has two fixed points in S_+, then $\bar{f}$ is the identity in S_+.

B - Let $\lambda_1 = \bar{\lambda}_2 \notin \mathbb{R}$. If $\bar{f}$ has one fixed point, then $\bar{f}$ is the identity in S^1.

(1.6) Proof of Theorem 2: Let us suppose $\lambda_1 \in \mathbb{R}$. In this case by (1) of the proof of Theorem 1 we get:

$$a_1^{\bar{\alpha}_j} B(\bar{\beta}_j \ln(a_1))\, \bar{f}_j(\tau) = A_j \tau_j + \lim_{x_1 \to 0} [x_1^{-\bar{\alpha}_j} B(-\bar{\beta}_j \ln(x_1)) R_j(x(x_1,\tau))]$$

if $\lambda_j \notin R$, or $a_1^{\bar{\alpha}_j} \bar{f}_j(\tau) = a_j \tau_j + \lim_{x_1 \to 0} [x_1^{-\bar{\alpha}_j} R_j(x(x_1,\tau))]$, if $\lambda_j \in R$. Since f_j is C^K we have $R_j(x) = P_j(x) + Q_j(x)$, where P_j is a polynomial of degree K, $\lim_{x \to 0} \frac{P_j(x)}{\|x\|} = 0$ and $\lim_{x \to 0} \frac{Q_j(x)}{\|x\|^K} = 0$. Since $K > |\mathrm{Re}(\lambda_n)|/|\mathrm{Re}(\lambda_1)| = \bar{\alpha}_n > \bar{\alpha}_{n-1} > \ldots > \bar{\alpha}_2 > 1$, it is easy to see that $\lim_{x_1 \to 0} x_1^{-\bar{\alpha}_j} Q_j(x(x_1,\tau)) = 0$. Let us show that

$\lim_{x_1 \to 0} [x_1^{-\bar{\alpha}_j} P_j(x_1,\tau))] = 0$. In order to simplify the notation we suppose that $\lambda_i \in \mathbb{R}$, $i=1,\ldots,n$. In this case $x = (x_1,\ldots,x_n)$ and $P(x) = $
$= \sum_{2 \le |\sigma| \le K} a_\sigma x^\sigma$, where $\sigma = (\sigma_1,\ldots,\sigma_n)$, $|\sigma| = \sigma_1 + \ldots + \sigma_n$ and $x^\sigma = x_1^{\sigma_1} \ldots x_n^{\sigma_n}$. We have

$$x_1^{-\alpha_j} P_j(x(x_1,\tau)) = x_1^{-\bar{\alpha}_j} P_j(x_1, x_1^{\bar{\alpha}_2}\tau_2, \ldots, x_1^{\bar{\alpha}_n}\tau_n) = \sum_{2 \le |\sigma| \le K} a_\sigma x_1^{\bar{\alpha}\cdot\sigma - \bar{\alpha}_j},$$

where $\bar{\alpha}\cdot\sigma = \sigma_1 + \bar{\alpha}_2\sigma_2 + \ldots + \bar{\alpha}_n\sigma_n$. Since $\alpha_1,\ldots,\alpha_n$ do not satisfy resonant relations, $\bar{\alpha}\cdot\sigma - \bar{\alpha}_j \neq 0$, therefore $\lim_{x_1 \to 0} (x_1^{\bar{\alpha}\cdot\sigma - \bar{\alpha}_j}) = 0$ or $+\infty$. Since $\lim_{x_1 \to 0} [x_1^{-\bar{\alpha}_j} P_j(x(x_1,\tau))] = a_1^{\bar{\alpha}_j} \bar{f}_j(\tau) - a_j \tau_j$ exists, it is clear that $a_\sigma = 0$ if $\bar{\alpha}\cdot\sigma - \bar{\alpha}_j < 0$, which implies that $a_1^{\bar{\alpha}_j} \bar{f}_j(\tau) = a_j \tau_j$. QED. The proof in case B is analogous.

§1.a - Diffeomorphisms preserving the orbits of a vector field on a 2-manifold.

Let X be a C^r-vector field $r \geq 2$ on a compact 2-manifold M without boundary and $M \xrightarrow{\pi} M/X$ the orbit space of M. In what follows the set of nonwandering points $\Omega(X)$ will be a finite union of orbits.

Definition - Let $h: M/X \to M/X$ be a homeomorphism. We say that h is C^r compatible with X if for any $\gamma_1, \gamma_2 \subset \Omega(X)$ with $h(\pi(\gamma_1)) = \pi(\gamma_2)$ there are points $x_i \in \gamma_i$, neighborhoods $V_i \ni x_i$ and a C^r diffeomorphism $g: V_1 \to V_2$ such that the following diagram commutes:

$$\begin{array}{ccc} V_1 & \xrightarrow{g} & V_2 \\ \downarrow \pi & & \downarrow \pi \\ \pi(V_1) & \xrightarrow{h} & \pi(V_2) \end{array}$$

Examples of this are C^r diffeomorphisms preserving the orbits of X.

Let $M\text{-}S^r(M)$ be the set of Morse-Smale C^r vector fields on M. Here we show the following

Corollary I - There is an open and dense subset $\mathcal{G} \subset M\text{-}S^r(M)$, $r \geq 2$, such that if $X \in \mathcal{G}$ and $h: M/X \to M/X$ is a homeomorphism C^1-compatible with X, then h^2 is the identity.

We remark that Peixoto's Theorem [5] implies that $\mathcal{G}$ is open and dense in $\mathfrak{X}^r(M)$ when M is orientable.

Proof of Corollary A: To fix notation, if p is a singularity of X denote with $W^s(p)$, $W^u(p)$ the stable and unstable manifolds of p. If p is a sink (or a source) with eigenvalues $\lambda_2 < \lambda_1 < 0$ (or $0 < \lambda_1 < \lambda_2$) and eigenvectors v_1, v_2 there is a unique invariant submanifold of dimension 1 tangent to v_2. We call it "strong stable" (or unstable) manifold of p and denote it by $W^{ss}(p)$ (or $W^{uu}(p)$).

Define the subset $\mathcal{G}$ by the following five conditions:

(C.1) For any singularity p of X, the eigenvalues of $DX(p)$ are different.

(C.2) If p, q are singularities of X then $\operatorname{Spec}(DX(p)) \neq \lambda \operatorname{Spec}(DX(q))$ for $\lambda \in \mathbb{R}$. Given any two closed orbits γ, γ' of X, the associated eigenvalues of their Poincaré maps are different.

(C.3) If p is a sink (resp. source) with real eigenvalues and q is a saddle, then $W^{ss}(p) \cap W^{u}(q) = \emptyset$ (resp. $W^{uu}(p) \cap W^{s}(q) = \emptyset$).

(C.4) If p is a sink and q is a source with real eigenvalues, then $W^{ss}(p) \cap W^{uu}(q) = \emptyset$.

The next condition will be imposed to the following situation. Let $V \subset M$ be an open subset such that $\partial V = \gamma_1 \cup \gamma_2$ where $\gamma_1, \gamma_2 \subset \Omega(X)$ and for any $x \in V$, $\omega(x) = \gamma_1$, $\alpha(x) = \gamma_2$. Suppose further that if γ_i is a singularity then the eigenvalues of γ_i are not real. Let $\Sigma_i \subset V$ be a transverse circle contained in a small neighborhood of γ_i $i=1,2$. (See picture below).

A

Let $h: M/X \to M/X$ be C^r compatible with X inducing C^r diffeomorphisms $\bar{f}_i: \Sigma_i \to \Sigma_i$ and $P: \Sigma_2 \to \Sigma_1$ the map defined by $P(x) =$

$= \mathcal{O}(x) \cap \Sigma_1$. It is easy to see using Hartman's Theorem on C^1-linearizations of C^2-sinks or sources plus minor arguments that there are C^1-diffeomorphisms $g_i : \Sigma_i \to S^1$ such that $g_i \circ \bar{f}_i \circ g_i^{-1}$ is a rotation $i = 1,2$. Then $g_1 \circ P \circ g_2^{-1}$ is a C^1-diffeomorphism of S^1.

On the other hand, let $\mathcal{B}^r \subset \mathrm{Diff}^r(S^1)$ be the set of C^r diffeomorphisms $f: S^1 \to S^1$ such that the only linear rotation commuting with f is the identity. It is easy to see that $\mathrm{int}(\mathcal{B}^r)$ is dense in $\mathrm{Diff}^r(S^1)$.

(C.5) $g_1 \circ P \circ g_2^{-1} \in \mathrm{int}(\mathcal{B}^1)$.

It can be checked, although is not immediate, that $\mathcal{G}^r$ is open and dense in $\mathfrak{X}^r(M)$.

Suppose $X \in \mathcal{G}^r$ and $h: M/X \to M/X$ is C^1-compatible with X. Since X is Morse-Smale the union of all stable manifolds of sinks is open and dense in M. So, to show that h^2 = identity it is enough to show that h^2 is the identity in a neighborhood of any sink γ in M/X. This is obtained with elementary arguments making repeated use of Corollary 1.5 of Theorem 1. We leave this to the reader.

<u>Remark:</u> In many situations one obtains indeed that h = identity. For instance when one of the following conditions hold: (a) There is a sink or source γ and a neighborhood $U \ni \gamma$ such that h is the identity in U/X. (b) X has a singularity with non real eigenvalues. (c) X has more than one sink or more than one source. (d) The local diffeomorphisms defined by $h: M/X \to M/X$ are C^0 near the identity.

§2. Stability of $\mathbb{R}^2$-actions.

In this section we prove the existence of structurally stable $\mathbb{R}^2$-actions on the n-sphere S^n.

Throughout this section we make repeated use of the following hypothesis,

(*): Let X be a C^r-vector field on an n-manifold and p a singularity of X. Let $(\mu_i)_{i=1}^{n}$ be the set of eigenvalues of $DX(p)$. Suppose:

(1) $\mathrm{Re}(\mu_i) < 0$ (or $\mathrm{Re}(\mu_i) > 0$) for all $i = 1,\ldots,n$.

(2) $\mu_i \neq \sum_{j=1}^{n} m_j\mu_j$ for all $i = 1,\ldots,n$ and integers $m_j \geq 0$, $\sum_{j=1}^{n} m_j \geq 2$.

(3) $|\mathrm{Re}(\mu_1)| \leq |\mathrm{Re}(\mu_j)| \leq |\mathrm{Re}(\mu_n)|$ for $j = 1,\ldots,n$ and $r > \dfrac{|\mathrm{Re}(\mu_n)|}{|\mathrm{Re}(\mu_1)|}$.

It is clear that conditions (1), (2), (3) hold for any vector field sufficiently close to X. Moreover by Sternberg linearization theorem [7] these conditions are sufficient to obtain a C^r-linearization of X in a neighborhood of p.

(2.1) Lemma - There exists a vector field Z on the m-sphere S^m and a neighborhood $N^r(Z)$ in the C^r-topology, $r \geq 2$, such that (i) the set of nonwandering points of Z consists of two points p_1, p_2 which are hyperbolic. (ii) For any $W \in N^r(Z)$ and $V \in \mathfrak{X}^r(S^m)$ with $[V,W] = 0$ one has $V = cW$, $c \in \mathbb{R}$.

Proof: Define Z on S^m as the gradient of a function $f\colon S^m \to \mathbb{R}$ with two singularities, a maximum p_1 and a minimum p_2 such that

(P_1) There is a system of coordinates in a neighborhood $U_i \ni p_i$ in which Z is expressed as a linear vector field satisfying

(1)-(3) of (*).

(P_2) For any trajectory τ of Z such that $\tau \cap U_1$ is contained in an eigenspace of Z/U_1, the smallest invariant linear subspace of Z/U_2 containing $\tau \cap U_2$ is U_2.

By Theorem 2 a vector field V on S^m with $[V,Z] = 0$ is necessarily linear in U_1. Given an eigenspace E of Z/U_1 there is by commutativity a linear combination $Y = \alpha V - \alpha c Z$, $\alpha \neq 0$, such that $Y(x) = 0$ for $x \in E$. Since Z satisfies (P_2) one obtains $Y \equiv 0$ in U_2 which implies $Y \equiv 0$ in S^m. So $V = cZ$. Since any W sufficiently near to Z is C^r-linearizable near p_i, the same arguments above show $V = cW$.

(2.2) Proof of Theorem 3 - Consider S^n as the union of $T_1 = D^{n-1} \times S^1$ and $T_2 = S^{n-2} \times D^2$ joined along the boundary by a diffeomorphism $f: \partial T_1 \to \partial T_2$ of the form $f(x,\theta) = (f_1(x), f_2(\theta))$ where $(x,\theta) \in \partial T_1 = \partial T_2 = S^{n-2} \times S^1$. We define an action $\varphi: \mathbb{R}^2 \times S^n \to S^n$ by two commutative vector fields X and Y on S^n as follows.

Let $(x,\theta) = (x_1,\ldots,x_{n-1},\theta) \in D^{n-1} \times S^1 = T_1$. Then define $Y(x,\theta) = \frac{\partial}{\partial\theta}$ and $X(x,\theta) = X(x) = \sum_{i=1}^{n-1} X_i(x) \frac{\partial}{\partial x_i}$ on T_1. Take $X(0) = 0$ and assume that X at $0 \in D^{n-1}$ satisfies (*) with the further hypothesis that for any $x \in D^{n-1}$ the α-limit set of x is $0 \in D^{n-1}$ and in a neighborhood of ∂D^{n-1}, $X_i(x) = x_i$ for all $i = 1,\ldots,n-1$.

Let $(z,y) = (z,y_1,y_2) \in S^{n-2} \times D^2 = T_2$ and extend Y to T_2 by putting $Y(z,y) = -y_2 \frac{\partial}{\partial y_1} + y_1 \frac{\partial}{\partial y_2}$.

Let Z be the vector field on S^{n-2} given by Lemma (2.1). Then extend X to T_2 as

$$X(z,y) = \lambda(y_1^2+y_2^2)\left(-y_1 \frac{\partial}{\partial y_1} - y_2 \frac{\partial}{\partial y_2}\right) + \alpha(y_1^2+y_2^2)\, Z(z)$$

where $\alpha(t)$, $\lambda(t)$ are positive C^∞ functions; $\alpha(t) = 1$ for $0 \le t \le 1/4$ $\alpha(t) = 0$ for $3/4 \le t \le 1$ and $\lambda(t) = 1$ for $3/4 \le t \le$ ≤ 1; $\lambda(t) = \lambda_o$ for $0 \le t \le 1/4$. Where λ_o is chosen so that p_1 is an attractor of X and X at p_i satisfies (*).

Clearly $[X,Y] = 0$. We proceed to show that with a convenient choice of $p_2 \in S^{n-2} \times \{0\}$ φ is structurally stable.

The set $\mathrm{Sing}(\varphi)$ of singular points of the action φ is the union of $\gamma(\varphi) = 0 \times S^1 \subset T_1$ and $\Gamma(\varphi) = S^{n-2} \times \{0\} \subset T_2$.

By definition X is normally hyperbolic at $\gamma(\varphi)$. So given $\epsilon > 0$ there is a neighborhood $N^r(X)$ and for $\tilde{X} \in N^r(X)$ a curve $\tilde{\gamma}$, ϵ-close to $\gamma(\varphi)$ and points $\tilde{p}_i$ ϵ-close to p_i $i=1,2$, such that $\tilde{X}$ is normally hyperbolic at $\tilde{\gamma}$ and hyperbolic at $\tilde{p}_i$. Define $\tilde{\Gamma} = \overline{W^u(\tilde{p}_2)}$. Then $\tilde{\Gamma}$ is homeomorphic to S^{n-2}.

Let $\tilde{\varphi}$ be an action near φ defined by $\tilde{X}$, $\tilde{Y}$. Then $\tilde{\Gamma}$ is invariant by $\tilde{\varphi}$. Moreover for $N^r(X)$ small enough, $\tilde{X}$ is transverse to ∂T_2 and $\tilde{X}$, $\tilde{Y}$ are linearly independent on ∂T_2. So we obtain that $\mathrm{Sing}(\tilde{\varphi}) \subset \tilde{\gamma} \cup \tilde{\Gamma}$. We prove now that $\tilde{\Gamma}$ is a C^r-submanifold and $\mathrm{Sing}(\tilde{\varphi}) = \tilde{\gamma} \cup \tilde{\Gamma}$.

In fact, we know by hypothesis that there are neighborhoods $p_i \in U_i \subset$ $\subset \Gamma(\varphi)$, $i = 1,2$ where X is linear and for any trajectory τ of X such that $\tau \cap U_1$ is contained in an eigenspace of X/U_1 then the smallest invariant linear subspace of X/U_2 containing $\tau \cap U_2$ is U_2. Let $\tilde{p}_i = p_i$ and take $\tilde{X}$ such that $\tilde{X}$ at p_1 and $\tilde{X}/W^u(p_2)$ at p_2 satisfy (*). Then by Theorem 2 there are neighborhoods $V_i \ni p_i$ such that $\tilde{X}$, $\tilde{Y}$ are simultaneously C^r-linearizable in V_1 and $V_2' = V_2 \cap$ $\cap W^u(p_2)$. Assume for simplicity that $\tilde{X}$, $\tilde{Y}$ are linear in V_1 and V_2'. Then, given any 1-dimensional real eigenspace E for $\tilde{X}/V_1$, $\tilde{Y}/V_1$ one has $E \subset \mathrm{Sing}(\tilde{\varphi})$. So $E \subset \tilde{\Gamma}$ and for some $\alpha,\beta \in R$, $(\alpha\tilde{X}+\beta\tilde{Y})(z) = 0$

for any $z \in E$. Let $\tilde{\tau}$ be the $\tilde{X}$-orbit of z. Then the smallest $\tilde{X}$-invariant linear subspace of $\tilde{X}/V_2'$ containing $\tilde{\tau}$ has dimension $n-2$. So $(\alpha\tilde{X}+\beta\tilde{Y})/W^u_{\tilde{X}}(p_2) \equiv 0$ which shows that $\tilde{\Gamma} \subset \mathrm{Sing}(\tilde{\varphi})$ and $\overline{W^u_{\tilde{X}}(p_2)} \cap V_1$ is an invariant subspace of X, showing that $\tilde{\Gamma}$ is a C^r-submanifold.

On the other hand some linear combination $W = \tilde{X} + \delta\tilde{Y}$ has $\tilde{\gamma}$ as a hyperbolic closed expansor. Commutativity of the flows W_t and $\tilde{X}_s$ imply that $\tilde{\gamma}$ is invariant by $\tilde{\varphi}$. This proves the assertion.

The disc $\Sigma_\theta = D^{n-1} \times \theta \subset D^{n-1} \times S^1$ is transverse to $\tilde{Y}$ for $\tilde{Y}$ close enough to Y. Therefore there is an open subset $V \subset \Sigma_\theta$ containing $0 \times \theta \in D^{n-1} \times \theta$ where the first return map $f\colon V \to \Sigma_\theta$ along the positive trajectories of $\tilde{Y}$ is well defined. On the other hand there are functions a, b such that $\zeta = a\tilde{X} + b\tilde{Y}$ is tangent to Σ_θ. The map f sends orbits of ζ to orbits of ζ. The subset $\mathcal{O} = W^s_{\tilde{X}}(p_2) - \{p_2\}$ is a cylindrical orbit of $\tilde{\varphi}$ intersecting ∂T_2 along a closed curve near $W^s_X(p_2) \cap \partial T_2$. Then $f(\mathcal{O} \cap V) \subset \mathcal{O} \cap \Sigma_\theta$. Taking p_2 such that $\mathcal{O} \cap \Sigma_\theta$ is far away from the eigenspaces of $D\zeta(0)$ one obtains that f induces the identity map in the orbit space of ζ. So $\tilde{\varphi}/T_1$ is a product foliation.

Since $\tilde{\varphi}$ is linearizable at V_1, there is an invariant fibration $\tilde{\pi}\colon V_1 \to V_1 \cap \tilde{\Gamma}$. This fibration extends to a neighborhood of $\tilde{\Gamma}$ by putting $\tilde{\pi}^{-1}(z) = \tilde{X}_{-t}(\pi^{-1}(\tilde{X}_t(z))) \cap T_2$ where $X_t(z) \in V_1$ and $\tilde{\pi}^{-1}(\tilde{p}_2) = W^s_{\tilde{X}}(p_2) \cap T_2$. Assume for simplicity that this neighborhood is T_2 and $\tilde{\Gamma} = \Gamma$.

Now, given a conjugacy $h\colon \Gamma \to \Gamma$ between X and $\tilde{X}$ h induces a homeomorphism $h\colon \partial T_2 \to \partial T_2$, $\partial T_2 \approx S^{n-2} \times S^1$, by $h(x,\theta) = \pi^{-1}(h(x)) \cap S^{n-2} \times \{\theta\}$. Extend h to T_2 by $h(X_t(x,\theta)) = \tilde{X}_t(h(x,\theta))$ for $(x,\theta) \in \partial T_2$ and $t > 0$. Clearly h is a homeomorphism and on ∂T_1 it preserves the fibers $\partial(D^{n-1} \times \theta)$. So we extend h to T_1 writting $h(X_{-t}(x,\theta)) = \zeta_{-t}(h(x,\theta))$. This finish the proof of the theorem.

The following is an interesting question.

Problem - Are there structurally stable R^2-actions on any manifold?

§3. Stability of Foliations.

In this section we show stability theorems for integrable differential 1-forms ω in some compact manifolds.

Theorem 4 - On any compact orientable 3-manifold, there exists a C^2 structurally stable integrable 1-form whose singular subset is a union of a finite number of closed curves.

In order to construct such stable forms we prove the following lemma.

(3.1) Lemma - Let ω_o be a linear 1-form on $S^1 \times S^1$ with rational rotation number. There is a C^2-structurally stable integrable 1-form ω on $D^2 \times S^1$ such that $\omega/\partial(D^2 \times S^1) = \omega_o$.

Proof: The foliation of $D^2 \times S^1$ with leaves $D^2 \times \{\theta\}$ is clearly stable, so we will assume from now on that the leaves of ω_o are transverse to $\partial D^2 \times \{\theta\}$, $\theta \in S^1$.
Let $\mathfrak{F}_o$ be the foliation on $\partial D^2 \times [0,1]$ induced by the projection $p: \partial D^2 \times [0,1] \to \partial D^2$, $p(x,t) = x$. The foliation ω_o is obtained from $\mathfrak{F}_o$ after identification of $\partial D^2 \times \{0\}$ with $\partial D^2 \times \{1\}$ by a linear rotation $R(e^{i\theta}) = e^{i(\theta + 2\pi\, m/n)}$, $m \in \mathbb{Z}$, $n \in \mathbb{Z}_+$ relatively prime.
Define the 1-form ω_1 on D^2 in polar coordinates as

$$\omega_1 = -\beta(r)\,\delta(\theta)\,r^2\,dr + r\,\alpha(r)\,d\theta$$

where α, β, δ are C^∞ functions defined as follows. $\alpha(r) = 1/2 - r$ if $r > 1/4$ and $\alpha(r) = 1$ if $r < 1/8$; $\beta(r) = 1$ if $|r-1/2| < 1/8$

and $\beta(r) = 0$ if $|r-1/2| > 1/4$; δ is periodic of period $2\pi\, m/n$ and $\delta(\theta) = 0$, $|\delta'(\theta)| < 1/2$ at points $\theta = k\pi\, m/n$, $k \in \mathbb{Z}$.

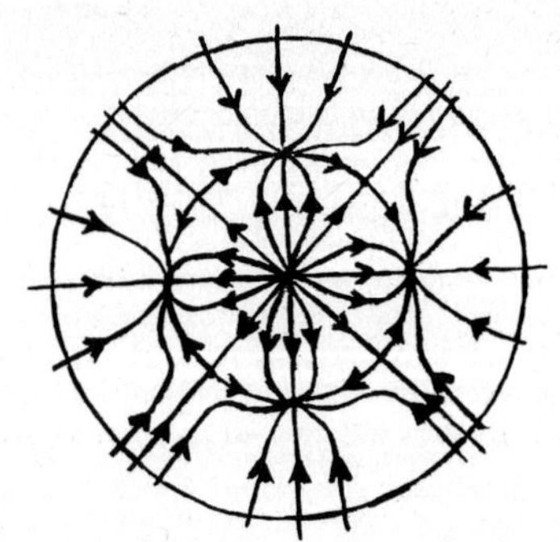

Let $p: D^2 \times [0,1] \to D^2$, $p(x,t) = x$. Since ω_1 is invariant by the rotation $R(re^{i\theta}) = r\, e^{i(\theta+2\pi\, m/n)}$, $p^*\omega_1$ will induce a 1-form ω on $D^2 \times S^1 = D^2 \times [0,1]/\sim$ where $(x,0) \sim (R(x),1)$.

We proceed to show that ω is structurally stable. Let $\mathrm{Sing}(\omega)$ be the set of zeros of ω and $\mathrm{attr}(\omega)$ (resp. $\mathrm{sadd}(\omega)$) the union of singular curves which are normally attractors (resp. saddle) (see [1]). Let $\mathrm{Sing}(\omega) = \bigcup_{i=1}^{3} \gamma_\omega^i$, $U_i \supset \gamma_\omega^i$, disjoint neighborhoods and Ω a volume form. Define the vector field X_ω in U_i by $d\omega = i_{X_\omega}(\Omega)$. Then the γ_ω^i are periodic orbits of X_ω and one can consider small cross sections to X_ω, Σ_i, passing through points $x_i \in \gamma_\omega^i$ where the first return map $\Sigma_i \supset V_i \xrightarrow{f_\omega^i} \Sigma_i$, $V_i \ni x_i$, along the positive trajectories of X_ω is well defined.

Let $N(\omega)$ be a neighborhood of ω such that for any $\eta \in N(\omega)$, $\mathrm{Sing}(\eta)$ is homeomorphic to $\mathrm{Sing}(\omega)$, and for all $1 \le i \le 3$, $\gamma_\eta^i = \mathrm{Sing}(\eta) \cap U_i$ is normally hyperbolic and $\Sigma_i \supset V_i \xrightarrow{f_\eta^i} \Sigma_i$ is well defined.

Since the eigenvalues of ω_1 at any singular point in the circle $r = 1/2$ are real and different, one obtains from (1.4) that for $N(\omega)$ small enough any f_η^i induces the identity map in the leaf space of

η/V_i. Now, the union of leaves L of η such that $\lim(L) \cap$ $\cap \operatorname{attr}(\eta) \neq \phi$ is dense in $D^2 \times S^1$. Therefore $\eta/\partial D^2 \times S^1$ is equivalent to ω_o and there is a homeomorphism $h: \mathcal{L}(\omega) \to \mathcal{L}(\eta)$ between the leaf spaces of ω and η.

Let $S(\eta)$ be the union of leaves L such that $\lim(L) \cap$ $\cap \operatorname{sadd}(\eta) \neq \phi$. Assume $S(\eta) = S(\omega)$ and $\operatorname{Sing}(\eta) = \operatorname{Sing}(\omega)$. Let $\mathcal{N}$ be a vector field transverse to the leaves of ω and tangent to the fibers $D^2 \times \{\theta\}$, $\theta \in S^1$. In each connected component R of $M - \overline{S(\eta)}$ an integral of $\mathcal{N}/R$ intersects precisely once any leaf of η/R. Define $H: D^2 \times S^1 \to D^2 \times S^1$ putting $H(x) = x$ for $x \in S(\omega) \cup$ $\cup \operatorname{Sing}(\omega)$. Let $\pi_\eta: D^2 \times S^1 \to \mathcal{L}(\eta)$ be the quotient map. For $x \in R$ let $H(x)$ be the intersection of the integral $\mathcal{N}_x$ of $\mathcal{N}/R$ with the leaf $\pi_\eta^{-1} \circ h \circ \pi_\omega(x)$ of η/R. A simple verification shows that H is the required equivalence.

(3.2) Proof of Theorem 4: Let M be a compact, orientable 3-manifold.

By Wallace Theorem ([8]) there is a collection of embedded tori $T_i, T'_i \approx D^2 \times S^1$, $i = 1,\ldots,n$ in M and $S^1 \times S^2$ respectively and a diffeomorphism $g: M - \bigcup_{i=1}^{n} T_i \to S^1 \times S^2 - \bigcup_{i=1}^{n} T'_i$. It was proved in [9] (see also [6]) that the T'_i can be taken transversal to the fibers $\{\theta\} \times S^2$.

Define on M an integrable 1-form ω putting first $\omega = g^*(d\theta)$ on $M - \bigcup_{i=1}^{n} T_i$ and then extending each $\omega_o = \omega/\partial T_i$ to T_i by Lemma (3.1). We assert that ω is structurally stable.

By the stability of $\omega | T_i$ there is a neighborhood $N(\omega)$ such that for any $\eta \in N(\omega)$ the form $\delta = (g^{-1})^*(\eta | M - \bigcup_{i=1}^{n} T_i)$ on $N = S^2 \times S^1 - \bigcup_{i=1}^{n} T'_i$ is close to $d\theta$ and any leaf of δ is a 2-sphere with a finite number of holes with boundary on $\bigcup_{i=1}^{n} \partial T'_i$. Let L_o be a leaf of δ and assume for simplicity that L_o is also a leaf of $d\theta | N$. Then N can be considered as the quotient of $L_o \times [0,1]$ by an equivalence

relation identifying points of $L_o \times 0$ with points of $L_o \times 1$. Suppose $\mathrm{Sing}(\eta) = \mathrm{Sing}(\omega)$ and $S(\eta) = S(\omega)$ as in (3.1) and let $\mathcal{N}$ be a vector field on M transverse to the leaves of ω, tangent to ∂T_i for all i and such that if $e_i\colon D^2 \times S^1 \to M$ is the embedding $T_i = e_i(D^2 \times S^1)$ then $\mathcal{N}$ leaves invariant the fibers $e_i(D^2 \times \{\theta\})$. Call $\mathcal{N}^o$ the vector field on $L_o \times [0,1]$ inducing $dg \circ \mathcal{N} \circ g^{-1}$ on N. Then any leaf of δ intersects any orbit of $\mathcal{N}^o$ in precisely one point. This provides us with a homeomorphism $h\colon \mathcal{L}(\delta) \to \mathcal{L}(d\theta|N)$ between the leaf spaces of δ and $d\theta|N$. Define a topological equivalence $H\colon N \to N$ between δ and $d\theta|N$ by putting $H(x) = x$ for $x \in L_o$ and $H(x) = \pi_{d\theta}^{-1} \circ h \circ \pi_\delta(L_x(\delta)) \cap \mathcal{N}_x^o$ for $x \in N - L_o \approx$ $\approx L_o \times (0,1)$. Here $\pi_\delta\colon N \to \mathcal{L}(\delta)$, $\pi_{d\theta}\colon N \to \mathcal{L}(d\theta|N)$ denote quotient maps and $\mathcal{N}_x^o$ the orbit of x by $\mathcal{N}^o$.

The homeomorphism $g^{-1} \circ H \circ g$ of $M - \bigcup_{i=1}^{n} T_i$ in itself is an equivalence between η and ω and can be extended as in (3.1) to an equivalence in M.

(3.3) <u>Proof of Theorem 5</u>: Let Ω be a volume form on M^2. Define a continuous isomorphism $d\colon \mathfrak{X}^r(M^2) \to \mathcal{J}^r(M^2)$ by putting $d(X) = i_X(\Omega)$. Then the subset $\mathfrak{F}$ is defined by $\mathfrak{F} = d(\mathfrak{G})$. Any singularity $y_o \in M^2$ of ω gives rise to a singular submanifold $Y_o = \pi_1^{-1}(y_o)$ of $\pi_1^*\omega$ of codimension two. In a neighborhood of Y_o the foliation $\mathfrak{S}$ defined by $\{Y \in \mathfrak{X}^r(M);\ i_Y(d\omega) = 0\}$ has codimension two is tangent to the leaves of $\pi_1^*\omega$ and Y_o is a leaf of $\mathfrak{S}$. Let M_x be the fiber $M \times \{x\}$, $x \in N$. Then $\mathfrak{S}$ induces a diffeomorphism $\bar{f}$ of a neighborhood of $(x,y_o) \in M_x$ sending leaves of $\pi_1^*\omega/M_x$ to leaves of $\pi_1^*\omega/M_x$. Similarly, given a closed integral c of ω one obtains a local diffeomorphism of a neighborhood of $x \times c \subset$ $\subset M_x$ preserving the leaves of $\pi_1^*\omega/M_x$.

Since the same construction holds for η near $\pi_1^*\omega$ we conclude that there is a neighborhood $N(\pi_1^*\omega)$ such that for any $\eta \in N(\pi_1^*\omega)$ there is a homeomorphism $h: M_x/\eta/M_x \to M_x/\eta/M_x$ C^1-compatible with η/M_x. By Corollary I obtain that h is the identity. So any leaf of η intersects with M_x in a connected subset. The actual construction of a homeomorphism $H: M \to M$ sending leaves of $\pi_1^*\omega$ to leaves of η follows easily as in the case of vector fields on 2-manifolds.

References

[1] C. Camacho, Structural stability of foliations with singularities, School of Topology, PUC, 1976, Springer Lect. Notes.

[2] P. Hartman, On local homeomorphisms of euclidean spaces, Bol. Soc. Mat. Mexicana 5 (1960).

[3] N. Kopell, Commuting diffeomorphisms, Global Anal. Proc. Symp. vol. 14 (1970). Am. Math. Soc.

[4] J. Palis and C. Pugh, Fifty problems in dynamical systems, Dynamical Systems - Warwick 1974, Springer Lect. Notes nº 468.

[5] M. Peixoto, Structural stability on two-dimensional manifolds, Topology 1, (1962).

[6] H. Rosenberg and R. Roussarie, Les feuilles exceptionnelles ne sont pas exceptionnelles, Conm. Math. Helv. (1971).

[7] S. Sternberg, Local contractions and a theorem of Poincaré, Am. Jr. of Math. vol. 79, (1957).

[8] A.H. Wallace, Modifications and cobounding manifolds, Canad. J. Math. 12, (1960).

[9] J. Wood, Foliations on 3-manifolds, Ann. of Math. 89, (1969).

[10] P. Sad, Centralizers of vector fields, To be published.

Instituto de Matemática Pura e Aplicada
Rio de Janeiro - Brazil

ON MINIMAL IMMERSIONS WITH PARALLEL NORMAL CURVATURE TENSOR

by

A.G. Colares and M.P. do Carmo

In this paper we first prove the following theorem on reduction of codimension of minimal immersions:

Theorem 1 - *Let* $x: M^n \to X$ *be a minimal immersion of an* n-*dimensional connected manifold* M^n *into an* $(n+\ell)$-*dimensional space* X *of constant curvature. Assume that the curvature tensor of the normal connexion is parallel in the normal bundle and the first normal space of the immersion has constant dimension* k. *Then there exists an* $(n+k)$-*dimensional totally geodesic submanifold* X' *of* X *such that* x *is a minimal immersion of* M^n *in* X'.

By using Theorem 1 and a result of I. Itoh [2] we then obtain the following

Theorem 2 - *Let* $x: M^2 \to S^n$ *be a minimal immersion of a compact 2-dimensional manifold* M^2 *into the* n-*sphere* $S^n \subset R^{n+1}$. *Assume that the curvature tensor of the normal connexion is parallel in the normal bundle. Then* $x(M^2)$ *is a Veronese surface in* S^4 *or* x *is a minimal immersion of* M^2 *in* S^3.

Our interest in Theorem 1 comes from the fact that it gives an indication that the normal curvature tensor of an immersion is intimately related with the problem of reducing the codimension of the immersion. Actually, it is likely that if $(\nabla^{\perp})^m R^{\perp} = 0$ and the mth-normal space has constant dimension k, then the codimension of a minimal immersion reduces to k.

Besides the conditions on the normal curvature tensor, some extra conditions on the immersion are necessary to reduce the codimension. In §2 we construct an example of a 2-dimensional manifold which can be immersed with flat normal bundle $(R^{\perp} \equiv 0)$ and arbitrarily high codimension. It would be interesting to find geometrical conditions which can replace minimality in Theorem 1.

Theorem 1 should be compared with Theorem 10 of Yau [4].

§1. Preliminaries - Let $x: M^n \to X$ be an isometric immersion of an n-dimensional connected Riemannian manifold into an $(n+\ell)$-dimensional space X of constant curvature. Given $p \in M$, we identify the tangent space M^n_p to M^n at p with $dx(M^n_p)$. We write X_p for the tangent space to X at $x(p)$. The normal space $(M^n_p)^{\perp}$ is the subspace of X_p consisting of all vectors $\xi(p) \in X_p$ which are normal to M^n_p, with respect to the metric g of X. Let $\bar{\nabla}$ (resp. ∇) be the covariant differentiation of the Riemannian connexion in X (resp. M^n) and $\nabla^{\perp}$ the covariant differentiation in the normal bundle of x. Given $\xi(p) \in (M^n_p)^{\perp}$ we define the second fundamental form of x relative to $\xi(p)$

$$A_{\xi(p)}: M^n_p \to M^n_p,$$

by the Weingarten equation:

$$\bar{\nabla}_X \xi = -A_{\xi} X + \nabla^{\perp}_X \xi, \tag{1.1}$$

where $X \in M^n_p$ and ξ is any extension of $\xi(p)$. We will denote the curvature tensor of $\bar{\nabla}$ by $\bar{R}$ and that of $\nabla^{\perp}$ by $R^{\perp}$, that is,

$$\bar{R}_{XY} = \bar{\nabla}_X \bar{\nabla}_Y - \bar{\nabla}_Y \bar{\nabla}_X - \bar{\nabla}_{[X,Y]} \tag{1.2}$$

and

$$R^{\perp}_{XY} = \nabla^{\perp}_X \nabla^{\perp}_Y - \nabla^{\perp}_Y \nabla^{\perp}_X - \nabla^{\perp}_{[X,Y]}, \tag{1.3}$$

for tangent vectors X, Y to M^n. If $R^{\perp} \equiv 0$ we will say that the normal bundle is <u>flat</u>. It is a result of Cartan that the normal bundle is flat if and only if at each point all the second fundamental forms are simultaneously diagonalizable. This comes from the Ricci equation

$$g(R^{\perp}_{XY}\xi,\eta) = g([A_{\xi},A_{\eta}]X,Y), \tag{1.4}$$

which holds for ambient spaces of constant curvature, where ξ and η are normal vector fields. When $n = 2$ and $n+\ell = 4$, the function $K_N = g(R^{\perp}_{XY}\xi,\eta)$ does not depend on the choice of the orthonormal set X, Y, ξ, η and is called the <u>normal curvature</u> of the immersion. We say that a normal vector field ξ is <u>parallel</u> in the normal bundle if $\nabla^{\perp}_{X}\xi = 0$ for any tangent vector X. A subbundle $\tilde{N}$ of the normal bundle is said to be <u>parallel</u> in the normal bundle if it is invariant by parallel translation with respect to $\nabla^{\perp}$; that is, if $\xi \in \tilde{N}$ then $\nabla^{\perp}_{Z}\xi \in \tilde{N}$, for any tangent vector Z. We say that the curvature tensor of the normal connexion is <u>parallel</u> in the normal bundle if

$$\nabla^{\perp}_{Z}R^{\perp}_{XY} = 0, \quad \text{for every } X,Y,Z.$$

This means that for any normal vector field ξ,

$$(\nabla^{\perp}_{Z}R^{\perp}_{XY})\xi = \nabla^{\perp}_{Z}(R^{\perp}_{RY}\xi) - R^{\perp}_{XY}\nabla^{\perp}_{Z}\xi = 0 \tag{1.5}$$

If the ambient space X has constant curvature c, then, for any vector fields $\bar{X}, \bar{Y}, \bar{Z}$, we have

$$\bar{R}_{\bar{X}\bar{Y}}\bar{Z} = c(g(\bar{Z},\bar{Y})\bar{X} - g(\bar{Z},\bar{X})\bar{Y}).$$

Hence, for X, Y tangent and ξ normal to M^n,

$$\bar{R}_{XY}\xi = 0. \tag{1.6}$$

We say that an immersion $x: M^n \to X$ is <u>minimal</u> if

$$\frac{1}{n}\sum_{\alpha=1}^{p} (\text{trace } A_{\xi_\alpha})\xi_\alpha = 0, \tag{1.7}$$

for any orthonormal vector fields ξ_α. When $A_{\xi_\alpha} = 0$ for every ξ_α

we say that x is <u>totally geodesic</u>.

§2. <u>Minimal immersions with parallel normal curvature tensor</u>

Here $x: M^n \to X$ is an isometric immersion (not necessarily minimal), X, Y and Z are tangent vectors to M^n, and ξ and η are normal vector fields to M^n.

The first normal space $N_1(p)$ at $p \in M^n$ is defined to be the orthogonal complement in $(M_p^n)^\perp$ of

$$\{\xi(p) \in (M_p^n)^\perp ;\ A_{\xi(p)} = 0\}.$$

<u>Lemma 1</u> - <u>Let</u> $x: M^n \to X$ <u>be an isometric immersion of an</u> n-<u>dimensional Riemannian manifold</u> M^n <u>in an</u> $(n+\ell)$-<u>dimensional space</u> X <u>of constant curvature. Assume that the curvature tensor of the normal connexion is parallel in the normal bundle. For each</u> $q \in M^n$ <u>let</u>

$$\tilde{N}(q) = \{\xi(q) \in (M_q^n)^\perp ;\ R^\perp_{XY}\xi = 0,\ \ \forall\ X,Y\}.$$

<u>Then,</u> $\tilde{N}$ <u>is parallel in the normal bundle</u>.

<u>Proof</u>: Given $p \in M^n$, let $m = \dim \tilde{N}(p)$. Choose an orthonormal basis $\xi_1(p),\ldots,\xi_m(p)$ for $\tilde{N}(p)$ and extend these to an orthonormal basis $\xi_1(p),\ldots,\xi_m(p),\ \xi_{m+1}(p),\ldots,\xi_\ell(p)$ of $(M_p^n)^\perp$. Now, extend these to vector fields ξ_α, $1 \le \alpha \le \ell$ in a normal neighborhood U of p in M^n by parallel translation with respect to $\nabla^\perp$ along geodesics of M^n starting at p.

We first show that the dimension of $\tilde{N}$ is locally constant, that is,

$$\xi_\alpha(q) \in \tilde{N}(q),\ \ \forall\ q \in U \Leftrightarrow \alpha \le m. \tag{2.1}$$

Given $q \in U$, let $\beta(t)$ be the geodesic from p to q in U with $\beta(0) = p$ and $\beta(1) = q$. For each pair X Y of tangent vector fields set

$$f_\alpha(t) = g(R^\perp_{XY}\,\xi_\alpha(\beta(t)),\ R^\perp_{XY}\,\xi_\alpha(\beta(t))).$$

By the hypothesis of the lemma and (1.5),

$$R^\perp_{XY}(\nabla^\perp_{\beta'}\,\xi_\alpha(\beta)) = \nabla^\perp_{\beta'}(R^\perp_{XY}\,\xi_\alpha(\beta)).$$

Hence, since $\xi_\alpha(\beta)$ is parallel,

$$f'_\alpha = 2g(\bar{\nabla}_{\beta'}\,R^\perp_{XY}\,\xi_\alpha(\beta),\ R^\perp_{XY}\,\xi_\alpha(\beta)) = 0.$$

Thus, f_α is constant for every α. Therefore, $f_\alpha(1) = 0$ if and only if $f_\alpha(0) = 0$. That is,

$$R^\perp_{XY}\,\xi_\alpha(q) = 0 \Leftrightarrow R^\perp_{XY}\,\xi_\alpha(p) = 0 \Leftrightarrow \xi_\alpha(p) \in N(p),$$

proving (2.1).

If follows that the dimension of $\tilde{N}$ is everywhere constant, hence $\tilde{N}$ is a subbundle of the normal bundle.

To prove that $\tilde{N}$ is parallel in the normal bundle, let $\xi \in \tilde{N}$, i.e., $R^\perp_{XY}\xi = 0$. By (1.5), this implies that $R^\perp_{XY}\,\nabla^\perp_Z\xi = 0$, i.e., $\nabla^\perp_Z\xi \in \tilde{N}$, for every tangent vector Z. This completes the proof of the lemma.

Proof of Theorem 1: Case I: The normal bundle $N(M^n)$ is flat - Let N_1 be the first normal space of x. Since $\dim N_1$ in constant, $N' = (N_1)^\perp$, where $\perp$ is the orthogonal complement in the normal space, is a subbundle of the normal bundle. We want to show that N' is parallel in the normal bundle.

Given $p \in M^n$, choose orthonormal vector fields $\xi_1,\ldots,\xi_k$ spanning N_1 at each point in a neighborhood V of p in M^n. Extend these to $\xi_1,\ldots,\xi_\ell$ so that the latter span the normal space at each point of V. Consider the vector fields $\xi_{k+1},\ldots,\xi_\ell$ which generate the subbundle N' over V.

It suffices to show that N' over V is parallel in the normal bundle. Given $q \in V$, let $X_1,\ldots,X_n$ be coordinate vector fields in a neighborhood $V' \subset V$ of q which diagonalize, at q, all the second fundamental forms of x. This is possible because the normal bundle is flat. On the other hand, because the ambient space has constant curvature, $\bar{R}_{X_j X_i} \xi_\alpha = 0$, by (1.6), $j,i = 1,\ldots,n$, $\alpha = 1,\ldots,\ell$. But

$$\bar{R}_{X_j X_i} \xi_\alpha = \bar{\nabla}_{X_j} \bar{\nabla}_{X_i} \xi_\alpha - \bar{\nabla}_{X_i} \bar{\nabla}_{X_j} \xi_\alpha =$$

$$= \bar{\nabla}_{X_j} (\nabla^\perp_{X_i} \xi_\alpha - A_{\xi_\alpha} X_i) - \bar{\nabla}_{X_i} (\nabla^\perp_{X_j} \xi_\alpha - A_{\xi_\alpha} X_j).$$

When $\xi_\alpha \in N'$, i.e., $\alpha = k+1,\ldots,\ell$, this gives

$$0 = \bar{R}_{X_j X_i} \xi_\alpha = \nabla^\perp_{X_j} \nabla^\perp_{X_i} \xi_\alpha - A_{\nabla^\perp_{X_i} \xi_\alpha} X_j - \nabla^\perp_{X_i} \nabla^\perp_{X_j} \xi_\alpha + A_{\nabla^\perp_{X_j} \xi_\alpha} X_i =$$

$$= R^\perp_{X_j X_i} \xi_\alpha - A_{\nabla^\perp_{X_i} \xi_\alpha} X_j + A_{\nabla^\perp_{X_j} \xi_\alpha} X_i.$$

Because the normal bundle is flat, $R^\perp_{X_j X_i} \xi_\alpha = 0$. Hence we have

$$A_{\nabla^\perp_{X_j} \xi_\alpha} X_i - A_{\nabla^\perp_{X_i} \xi_\alpha} X_j = 0. \tag{2.2}$$

Since $X_1,\ldots,X_n$ diagonalize all the second fundamental forms at q, we have, at q,

$$A_{\nabla^\perp_{X_j} \xi_\alpha} X_i = a^i_j X_i \quad \text{and} \quad A_{\nabla^\perp_{X_i} \xi_\alpha} X_j = b^j_i X_j,$$

for some numbers a^i_j, b^j_i. Hence, (2.2) implies that

$$A_{\nabla^\perp_{X_j}(q) \xi_\alpha} X_i(q) = 0, \quad j \neq i.$$

Since the immersion is minimal,

$$A_{\nabla^\perp_{X_j}(q) \xi_\alpha} X_i(q) = 0, \quad \text{for every } j,i,$$

and so

$$A_{\nabla^\perp_{X_j}(q) \xi_\alpha} = 0, \quad \text{for every } j.$$

This implies that $\nabla^\perp_{X_j}(q) \xi_\alpha \in N'(q)$, by the definition of N'.

Since $q \in V$ is arbitrary, we have proved that

$$\nabla^{\perp}\xi_{\alpha} \in N' \quad \text{on} \quad V, \quad \alpha = k+1,\ldots,\ell,$$

i.e., that N' over V is parallel in the normal bundle. It follows that N' is parallel in the normal bundle. Therefore, N_1 is also parallel in the normal bundle. By a theorem of Erbacher ([1], pg.333), there exists a totally geodesic submanifold X' of X such that

$$\dim X' = n+k \quad \text{and} \quad x(M^n) \subset X',$$

completing the proof of Case I.

Case II: the normal bundle is not flat. Set

$$\tilde{N} = \{\xi \in N(M^n);\ R^{\perp}\xi = 0\}.$$

By Lemma 1, $\tilde{N}$ is parallel in the normal bundle. Let N' be the orthogonal complement of N_1 in the normal bundle $N(M^n)$. Clearly $N' \subset \tilde{N}$. Observe that, by Lemma 1, $\tilde{N}$ is parallel and, by the Ricci equation (1.4), all second fundamental forms A_{ξ}, $\xi \in \tilde{N}$, can be simultaneously diagonalized. Thus we can apply the argument of Case I, with $\tilde{N}$ is place of $N(M^n)$ to conclude that N', hence N_1, is parallel in the normal bundle. Again, as in Case I, by Erbacher's theorem, $x(M^n) \subset X'$, where X' is a totally geodesic submanifold of X with dimension $n+k$. This completes Case II and so Theorem 1 is proved.

Example: We now describe the example mentioned in the introduction. Let $e_1,\ldots,e_n$ be the canonical basis in R^n and let $P \subset R^n$ be the plane generated by e_1,e_2. Choose n-2 points p_i, $i = 3,\ldots,n$, in P isolated by neighborhoods $U_i \subset P$. At each p_i, erect, in the space generated by P and e_i, a 2-dimensional graph M_i of a real function over U_i. By joining $\bigcup_i M_i$ with $P - \bigcup_i U_i$ and smoothing corners we obtain a surface $M^2 \subset R^n$. Such a surface has flat normal bundle and the codimension cannot be reduced.

§3. An application of Theorem 1 - Here, $x: M^2 \to S^n$, is a minimal immersion of a 2-dimensional connected manifold M^2 in the n-sphere $S^n \subset \mathbb{R}^{n+1}$.

For a 2-dimensional Riemannian manifold M^2 there exists isothermal parameters defined locally on M^2; that is, at each point of M^2 there exists a coordinate system $z = x+iy$ with range in an open set of the complex plane such that $\frac{\partial}{\partial x}$ and $\frac{\partial}{\partial y}$ are orthogonal and have the same length λ. Let $x: M^2 \to X$ be a minimal immersion and (h^α_{ij}) be the matrix of the second fundamental form A_{ξ_α} with respect to the orthonormal vector fields $\frac{\partial}{\partial x}/\lambda$ and $\frac{\partial}{\partial y}/\lambda$. It is known [3] that the form of degree four ψ given by

$$\psi = \sum_\alpha (h^\alpha_{11} - ih^\alpha_{12})^2 \lambda^4 dz^4$$

is holomorphic and globally defined when M^2 is orientable. If $\psi \not\equiv 0$ the zeros of ψ are isolated points in M^2. This implies that if x is not totally geodesic the common zeros of A_{ξ_α} are isolated.

Proof of Theorem 2: Case I: Suppose that the normal bundle is flat. Then, for the first normal space N_1 we have, since x is minimal, $\dim N_1(p) \leq 2$. On the other hand, since the normal bundle is flat, all the second fundamental forms of x are simultaneously diagonalizable at p and so $\dim N_1(p) \leq 1$. Actually, if x is not totally geodesic, then $\dim N_1(p) = 1$, except at isolated points. In fact, if $\dim N_1(p) = 0$ for some $p \in M^2$, then all the second fundamental forms vanish at p; by the last observation in §1, p is an isolated zero.

Now let $\{p_1,\ldots,p_n\}$ be the subset of M^2 constituted by the common zeros of the second fundamental forms of x. Then,

$$\dim N_1 = 1 \quad \text{on} \quad M^2 - \{p_1,\ldots,p_n\},$$

and so we can apply Theorem 1 to conclude that

$$x(M^2 - \{p_1,\ldots,p_n\}) \subset S^3.$$

Hence, also $x(M^2) \subset S^3$. Clearly, x is minimal as an immersion of M^2 in S^3, proving the theorem in this case.

Case II: Suppose the normal bundle is not flat. We first prove that $\dim N_1 = 2$ at every point of M^2.

In fact, if $\dim N_1(p) \leq 1$ for some $p \in M^2$, the second fundamental forms of x would be linearly dependent at p and so, by the Ricci equation (1.4),

$$g(R^{\perp}_{XY}\xi,\eta)(p) = 0,$$

for all tangent vectors X, Y and normal vectors ξ, η at p. By an argument similar to that of Lemma 1 and the fact that M^2 is connected, we prove that

$$(R^{\perp}_{XY}\xi,\eta) \equiv 0,$$

for all tangent fields X, Y and all normal fields ξ, η. This contradicts the fact that the normal bundle is not flat and proves that $\dim N_1 = 2$ everywhere. Thus we can apply Theorem 1 above to conclude that x is in fact a minimal immersion of M^2 in $S^4 \subset R^5$. Moreover, by the above argument, we see that the normal curvature of such an immersion is a non-zero constant. In this situation there exists a theorem of T. Itoh [2] which states that $x(M^2)$ is a Veronese surface in S^4. This completes the proof of Theorem 2.

REFERENCES

[1] Erbacher, J. - Reduction of codimension of isometric immersions, J. Dif. Geom. V (1971).

[2] Itoh, T. - Minimal surfaces in 4-dimensional Riemannian manifold of constant curvature, Kodai Math. Sem. Rep., 23 (1971).

[3] Yau, S.T. - Submanifolds with constant mean curvature I, Am. J. Math. v. 96 (1974).

[4] Yau, S.T. - Submanifolds with constant mean curvature II, Am. J. Math., v. 97 (1975), 76-100.

Departamento de Matemática
Universidade Federal do Ceará
Fortaleza, Brasil

Instituto de Matemática Pura e Aplicada
Rua Luiz de Camões, 68
Rio de Janeiro, Brasil

Circle Bundles

Shiing-shen Chern[1)]

It has been 41 years since Hassler Whitney coined the name "sphere bundle" in 1935. Since then the subject has been studied as a branch of algebraic topology, and justifiably so. It seems therefore opportune to see what it has done to mathematics in general. I will take the simplest case, that of circle bundles. For various reasons (among which is the fact that the circle bundles over a given space form an abelian group) they occupy a special position among sphere bundles.

1. Generalities.

By a circle is meant the unit circle in the complex plane, i.e.,

$$S^1 = \{z \in \mathbb{C} \mid |z| = 1\} . \tag{1}$$

A circle bundle is a family of circles parametrized by a space such that it is locally a product. The situation is best illustrated by the diagram

$$\begin{array}{ccc} S^1 & \xrightarrow{\text{inc}} & B \\ \pi\downarrow & & \downarrow\pi \\ x & \longrightarrow & M \end{array} , \qquad \text{inc} = \text{inclusion} \tag{2}$$

where B is the union of all the circles, M is the parameter space or base space, and π, the projection, assigns to a point of a circle, the point of the base space. The local product structure of B is expressed as follows: M has an open covering $\mathfrak{U} = \{U,V,W,\ldots\}$ such that $\pi^{-1}(U)$ is a product $U \times S^1$ and has therefore the local coordinates (x,z_U), $x \in U$, $z_U \in S^1$. If $U \cap V \neq \emptyset$, the points of $\pi^{-1}(U \cap V)$ have two sets of local coordinates (x,z_U), (x,z_V), $x \in U \cap V$, $z_U, z_V \in S^1$, relative to U and V respectively, and z_U, z_V are

---

1) Work done under partial support by NSF grant MCS-74-23180.

related by

$$z_U = g_{UV}(x) z_V, \qquad x \in U \cap V , \tag{3}$$

where $g_{UV}(x)$ is a complex number of absolute value 1. In order that this definition is consistent, the functions $g_{UV}(x)$, $x \in U \cap V$, must satisfy the conditions

$$(4) \qquad \begin{cases} g_{UU}(x) = 1 & x \in U \\ g_{UV}(x) g_{VU}(x) = 1 , & x \in U \cap V, \\ g_{UV}(x) g_{VW}(x) g_{WU}(x) = 1 , & x \in U \cap V \cap W \end{cases}$$

In this paper we suppose that our spaces are C^∞-manifolds and all functions (such as g_{UV}), real or complex-valued, are C^∞.

The functions g_{UV}, defined for any two members U,V of the covering, with $U \cap V \neq \emptyset$, and satisfying the relations (4), are called the *transition functions*. Equation (3) was called by H. Weyl the *gauge transformation*.

Two circle bundles B and B', with the transition functions $\{g_{UV}\}$ and $\{g'_{UV}\}$, can be added, their sum $\tilde{B} = B + B'$ having the transition functions $\{\tilde{g}_{UV}\}$, given by

$$\tilde{g}_{UV} = g_{UV} g'_{UV} . \tag{5}$$

In this way all the circle bundles over M, relative to the same covering $\mathfrak{U}$, form an abelian group.

The notion of a circle bundle is closely related to one-dimensional cohomology. In fact, consider the *nerve* $N(\mathfrak{U})$ of the covering $\mathfrak{U}$. $N(\mathfrak{U})$ is a complex whose vertices are $U,V,W,\ldots$ and which contains a simplex $UV\ldots W$, whenever $U \cap V \cap \ldots \cap W \neq \emptyset$. The transition function g_{UV} is an assignment to every oriented 1-simplex UV and is hence a one-dimensional cochain. The third equation of (4) implies that it is a cocycle.

The mathematical concept which expresses this situation precisely is *sheaf cohomology*. The above description of a circle bundle depends on the covering $\mathfrak{U}$ and on the choice of the local coordinate z_U with respect to U; it will be called a *coordinate bundle*. Given M, let $\mathfrak{C}^*$ be the sheaf of germs of C^∞ complex-valued functions of absolute value 1. Then the equivalence classes of circle bundles over M can be identified with the one-dimensional cohomology group $H^1(M; \mathfrak{C}^*)$, whose group addition in terms of coordinate bundles is the one described above.

Using elementary notions of sheaf cohomology, the group $H^1(M; \mathfrak{C}^*)$ can be identified with the cohomology group $H^2(M;\mathbb{Z})$ with integer coefficients $\mathbb{Z}$. In fact, let $\mathfrak{A}$ be the sheaf of germs of C^∞ real-valued functions. Then the following sequence of sheaves

$$(6) \qquad 0 \to \mathbb{Z} \to \mathfrak{A} \xrightarrow{\ell} \mathfrak{C}^* \to 0$$

is exact, where ℓ maps a germ of functions f to the germ $\ell(f) = \exp(2\pi i f)$. This follows, because the logarithm of a complex number of absolute 1 is purely imaginary and is locally well-defined. From the sheaf exact sequence follows the exact sequence of cohomology groups:

$$(7) \qquad H^1(M;\mathfrak{A}) \to H^1(M; \mathfrak{C}^*) \xrightarrow{c_1} H^2(M;\mathbb{Z}) \to H^2(M;\mathfrak{A}).$$

Since $\mathfrak{A}$ is a fine sheaf, the groups at the two ends of this sequence are zero and we have the isomorphism

$$(8) \qquad H^1(M; \mathfrak{C}^*) \cong H^2(M;\mathbb{Z}) .$$

The left-hand side is the group of circle bundles of M. In view of this isomorphism it is a discrete group. If $B \in H^1(M; \mathfrak{C}^*)$, the image $c_1(B) \in H^2(M;\mathbb{Z})$ is called the *characteristic class* of B.

In applications it is important to relate the local and global properties and it is advantageous to make use of the coordinate bundle. From (3) we get

$$d \log z_U = d \log g_{UV} + d \log z_V . \tag{9}$$

Notice that log is a multiple-valued function, but the operator d log is well-defined. Since z_U, z_V, g_{UV} all have absolute value 1, every term in (9) is purely imaginary. A *connection* or *gauge potential* is a real-valued 1-form θ_U, defined in every U of the covering $\mathfrak{U}$, such that

$$\theta_U = \theta_V - i d \log g_{UV}, \quad \text{in} \quad U \cap V \neq \emptyset . \tag{10}$$

By an extension argument it is easily proved that a connection always exists.

From (9) and (10) we derive

$$\varphi \underset{\text{def}}{=} i d \log z_U + \theta_U = i d \log z_V + \theta_V . \tag{11}$$

The common expression is therefore a real-valued form in B, which is independent of the neighborhood U and is hence globally defined in B. Writing $z_U = e^{ix_U}$, where x_U is real and is defined up to an integral multiple of 2π, we have

$$\varphi = i d \log z_U + \theta_U = -dx_U + \theta_U . \tag{12}$$

The connection is equally well defined by the form φ, in B having the local expression (12). In applications this gives a useful criterion to verify whether a one-form in B defines a connection.

Let γ be a curve in M, with parameter t. To each point $t \in \gamma$ we associate a point of the circle $\pi^{-1}(t)$. Such an association is called a *section*. This will be given by $z_U(t)$ in U; if $t \in U \cap V$, $z_U(t)$ and $z_V(t)$ are related by (3). The section is called *parallel*, if

(13) $$d \log z_U - i\theta_U(t) = 0,$$

where $\theta_U(t)$ is the restriction of θ_U to γ. Integrating the differential equation (13), we get the following expression of a parallel section:

(14) $$z_U = \exp(i \int_\gamma \theta_U(t)).$$

Up to a constant factor of θ_U, $\int \theta_U$ is called the *phase* in physics and the right-hand side of (14) the *phase factor*. In gauge theory in physics it is the phase factor, not the phase, that is physically meaningful. It is of interest to note that also in circle bundles it is the phase factor that is geometrically meaningful.

From (10) we get by exterior differentiation

(15) $$d\theta_U = d\theta_V \ .$$

The real two-form

(16) $$\Theta \underset{\text{def}}{=} \Theta_U = d\theta_U,$$

which is independent of U, is called the *curvature* of the connection. The following theorem, which is easy to prove, is fundamental ([3], [5]):

Theorem. *The closed 2-form* $\frac{1}{2\pi}\Theta$ *determines an element of* $H^2(M;\mathbb{Z})$ *in the sense of DeRham's theory. This cohomology class is the characteristic class of the circle bundle.*

In spite of its simplicity this is a profound theorem. While curvature is a local property, this relates it to the globally defined characteristic class. It shows that the deRham class of the form $\frac{1}{2\pi}\Theta$ is independent of the choice of the connection. As the differential form is a continuous object and the characteristic class has integer coefficients, the theorem expresses a relation between the continuous and the discrete. For example, let Y be a 2-dimensional cycle of M with integer coefficients. Then the integral

$$\frac{1}{2\pi}\int_Y \Theta$$

is an integer.

2. Electricity and magnetism.

Hermann Weyl's gauge theory of electricity and magnetism is a forerunner of circle bundles. Gauge theory plays an important role also in other field theories in physics.

In the field theory of electricity and magnetism M is a 4-dimensional Lorentzian manifold. In a circle bundle B over M there is given the gauge potential $\{\theta_U\}$. The curvature form

$$\Theta = \Theta_U = d\theta_U \tag{16}$$

is also called the *magnetic strenth* or *Faraday*. Maxwell's equations are

$$\begin{aligned} d\Theta &= 0, \\ \delta\Theta &= -4\pi J \ , \end{aligned} \tag{17}$$

where $\delta = *d*$ is the codifferential and J is called the *source vector*.

The fundamental problem in electromagnetic field theory is whether Maxwell's equations describe all the electrical and magnetic phenomena. The answer turns out to be negative. An experiment proposed by Y. Aharanov and D. Bohm in 1959 and carried out by R. G. Chambers in 1960 gives an electric-magnetic field in a non-simply connected domain whose Faraday is zero, but the phase factor can be observed. This question has recently been critically analyzed by T. T. Wu and C. N. Yang [10]. Their conclusion is: Electricity and magnetism is gauge-invariant manifestation of a non-integrable phase factor. The Faraday underdescribes e and m; phase overdescribes e and m. Complete description of e and m is phase factor. In mathematical terms *e and m field theory is a connection in a circle bundle*.

Actually in most physical applications the bundle is a trivial one. In a famous paper in 1931 [4] Dirac associated the non-trivial circle bundles with possible *magnetic monopoles*. If M is the 4-dimensional number space R^4 and Y is the trajectory of a magnetic monopole, then the second homology group $H_2(R^4 - Y;\mathbb{Z})$ is free cyclic. Let Z be a cycle which defines a generator. Then $\frac{1}{2\pi}\int_Z \Theta$ is an integer, called *Dirac's quantization*. On the possible existence of a magnetic monopole, Dirac said, "No change whatever in the formalism (from trivial circle bundle to any circle bundle)... Under these circumstances one would be surprised if Nature had made no use of it." (Phrase between parentheses added by the author.)

It might be observed that mathematically the existence of non-trivial circle bundles is also a sophisticated fact.

It will be incomplete in leaving the subject of fiber bundles and physics without mentioning the celebrated Yang-Mills theory (1954). The latter is a connection in an SU(2)-bundle and was originally introduced in the study of the isotopic spin.

3. Differential geometry.

Although never very clearly formulated, circle bundles present themselves in classical surface theory in differential geometry. Let M be an oriented surface in the ordinary euclidean space E^3. All the unit tangent vectors of M form a circle bundle B over M, the fiber consisting of all unit tangent vectors with the same origin. Let $x \in M$, ν the unit normal vector to M at x, and $\underline{\xi}$ a unit tangent vector at x. Let $\eta = \nu \times \xi$. (Vector product in the classical sense!) Then

$$\varphi = -d\xi \cdot \eta = +(d\xi, \xi, \nu) \tag{18}$$

is a one-form in B. It is easy to verify that φ has locally the expression (12). Hence it defines a connection in B in the sense of §1. By a theorem in surface theory, this connection depends only on the induced riemannian metric on M.

Using standard formulas in surface theory, the curvature form of this connection is

$$\Theta = +KdA, \tag{19}$$

where dA is the element of area and K is the guassian curvature of M. From our general theorem $+\frac{1}{2\pi}\Theta = \frac{1}{2\pi}KdA$ represents the characteristic class of the unit tangent bundle B of M. Our theorem in §1 contains as a special case the Gauss-Bonnet formula

$$\frac{1}{2\pi}\iint_M KdA = \chi , \tag{20}$$

where M is a closed orientable surface with the Euler characteristic χ.

Actually Gauss proved the formula for a geodesic triangle, thus involving the angles of the triangle, and Bonnet proved it for a simply-connected domain with an arbitrary smooth boundary curve. The formula (20) relating the integral of the gaussian curvature with the topological invariant χ of a closed surface was first given by von Dyck.

Although simpler, the circle bundle of unit normal vectors of a closed curve M in E^3 has been considered only recently. Since the curve M can be oriented, the normal bundle is oriented accordingly. A unit normal vector ν determines uniquely the unit normal vector $\nu^{\perp}$ perpendicular to ν such that $\nu, \nu^{\perp}$ define the orientation on the fiber. The form

$$(21) \qquad \varphi = -d\nu \cdot \nu^{\perp}$$

defines a connection in the normal bundle. Here the curvature form vanishes identically, because M is one-dimensional.

To define a global invariant we consider a smooth unit normal vector field $\nu(x)$, $x \in M$, i.e., assign to each x a unit normal vector $\nu(x)$ at x. Then we have the integral

$$(22) \qquad \frac{1}{2\pi}\int_M \varphi(x) = -\frac{1}{2\pi}\int_M d\nu(x)\cdot\nu^{\perp}(x) .$$

It is easily seen that change of the normal vector field modifies the integral by an additive integer. The integral

$$(23) \qquad \tau = \frac{1}{2\pi}\int_M \varphi(x) \quad \mathrm{mod}\ 1$$

is called the *total twist* of M. If the curvature of the curve M is never zero, say positive, we can choose $\nu(x)$ to be the corresponding unit principal normal vector at x. Then, by Frenet's formula, $\int_M \varphi(x)$ is the integral of the torsion of M. But the total twist is defined for any immersed curve. This is the simplest instance of global invariants defined mod 1.

The total twist of a closed curve has some interesting properties:

1) (T. Banchoff and J. White [2]) τ is invariant under conformal transformations in E^3.

2) (W. Scherrer [7]) A surface S is a sphere, if and only if the total twist of every closed curve is zero.

Generalizations of the total torsion have been considered by B. Segre. Scherrer's Theorem was generalized by D. Ferus to codimension two submanifolds of any dimension. The total twist of a compact orientable n-dimensional manifold in an euclidean space of dimension 2n+1 has been studied by J. White [8].

4. Value distributions [9].

Perhaps the best-known circle bundle is the *Hopf bundle*

$$S^3 = \{(z_0,z_1) \mid |z_0|^2 + |z_1|^2 = 1\} \tag{24}$$
$$\pi \downarrow$$
$$P_1(C) \quad ,$$

where $P_1(\mathbb{C})$ is the complex projective line and π maps the point $(z_0,z_1) \in S^3$ into the point of $P_1(\mathbb{C})$ having z_0,z_1 as homogeneous coordinates. This mapping is essential, i.e., not homotopic to a constant. Since $P_1(\mathbb{C})$ is homeomorphic to the two-sphere S^2, this was the first example of an essential mapping of a sphere into a sphere of lower dimension.

The Hopf bundle plays a fundamental role in complex function theory. The main reason lies in the identification of the element of area of $P_1(\mathbb{C})$ with the curvature form of a connection in the bundle (24). To explain it, let C_2 be the two-dimensional complex space, with the points $Z = (z_0,z_1)$. In C_2 we introduce the hermitian scalar product

$$(Z,W) = \overline{(W,Z)} = z_0\bar{w}_0 + z_1\bar{w}_1, \qquad W = (w_0,w_1). \tag{25}$$

Then S^3 consists of all Z_0 such that

$$(Z_0,Z_0) = 1 . \tag{26}$$

A unitary frame consists of points Z_α , $\alpha = 0,1$, satisfying

$$(Z_\alpha, Z_\beta) = \delta_{\alpha\beta}\ , \qquad \alpha,\beta,\ \gamma = 0,1. \tag{27}$$

The space of all unitary frames can be identified with the unitary group $U(2)$ in two variables. We can write

$$dZ_\alpha = \sum_\beta \omega_{\alpha\beta}\, Z_\beta\ , \tag{28}$$

where as a consequence of (27), $\omega_{\alpha\beta}$ are skew-hermitian:

$$\omega_{\alpha\beta} + \bar{\omega}_{\beta\alpha} = 0\ . \tag{29}$$

Exterior differentiation of (28) gives the Maurer-Cartan equations for $U(2)$:

$$d\omega_{\alpha\beta} = \sum_\gamma \omega_{\alpha\gamma} \wedge \omega_{\gamma\beta}\ . \tag{30}$$

The form

$$-i\ \omega_{00} = -i(dZ_0, Z_0) \tag{31}$$

in S^3 is real-valued and defines a connection in the Hopf bundle (because it is locally of the form (12)). By the Maurer-Cartan equations (30) its curvature form is

$$\Theta = -id\omega_{00} = -i\ \omega_{01} \wedge \omega_{10} = +\ i\omega_{01} \wedge \bar{\omega}_{01}\ . \tag{32}$$

Notice that

$$dA \underset{\text{def}}{=} \frac{i}{2\pi}\, \omega_{01} \wedge \bar{\omega}_{01} \tag{33}$$

is also the element of area of $P_1(\mathbb{C})$, normalized so that the total area is 1.

Hence curvature and area are related by

$$\frac{1}{2\pi}\Theta = dA . \tag{34}$$

Complex function theory studies holomorphic or covering maps

$$f\colon V \to P_1(\mathbb{C}), \tag{35}$$

where V is a Riemann surface. If V is compact, the two fundamental theorems are:

1) (First main theorem). Counted with multiplicities, every point of $P_1(\mathbb{C})$ is covered the same number n (= an integer) of times, and n is equal to the area of the image $f(V)$.

2) (Second main theorem or Riemann-Hurwitz formula). The Euler characteristic of V is given by

$$\chi(V) = 2n - w, \tag{36}$$

where w is the number of branch points of the covering. It follows that $n \geq \frac{1}{2}\chi(V)$, which gives a lower bound of n when V is given.

The famous value distribution theory of Picard-Borel-Nevanlinna studies holomorphic maps f when V is non-compact, and in particular, the question of the size of $f(V)$ and the uniformity that a point of $P_1(\mathbb{C})$ is covered. A natural step is to suppose V to be exhausted by a family of domains D_t with boundary, and to consider the geometric invariants of the mapping as functions of t. The historical example concerns with the case $V = \mathbb{C}$, in which D_t is the family of concentric disks with radius $r = e^t \geq r_0$, a positive constant. Then $t = \log r$ is a harmonic function on $\mathbb{C}$. In general, it is important to have this property preserved: V is said to have a *harmonic exhaustion* if the parameter t defining the domains D_t is a harmonic function. Let $a \in P_1(\mathbb{C})$ and let $n(a,t)$ be the number of times that a is covered by $f(D_t)$. Let

(37) $$N(a,t) = \int_{t_0}^{t} n(a,t)dt, \qquad T(t) = \int_{t_0}^{t} A(f(D_t))dt,$$

where $A(f(D_t))$ is the area of $f(D_t)$. $T(t)$ is Nevanlinna's *order function.*

To study how well $T(t)$ approximates $N(a,t)$ we consider the Hopf bundle over $f(D_t)$. The fact that the total area is the same as the total curvature makes it possible to apply Stokes' Theorem on the basis of the formula (34). An important consequence of the first main theorem so obtained is the remarkable inequality

(38) $$N(a,t) < T(t) + \text{const.}$$

Also, the derivation of the second main theorem involves a circle bundle, the canonical circle bundle.

5. Harmonic analysis on nilmanifolds.

The group of all matrices of the form

(39) $$\begin{pmatrix} 1 & y & z \\ 0 & 1 & x \\ 0 & 0 & 1 \end{pmatrix}$$

is called the Heisenberg group, to be denoted by N. If we denote the above element by (x,y,z), then the group multiplication is given by

(4) $$(x,y,z)(a,b,c) = (x + a,\ y + b,\ z + c + ya).$$

Let Γ be the discrete subgroup of N, consisting of the matrices in which x,y,z are integers. Let

$$\Gamma\backslash N = \{\Gamma n,\ n \in N\}$$

be the homogeneous space of right cosets.

The center of N is $C = \{(0,0,c)\}$. The orbits of $\Gamma\backslash N$ under the action of C are circles. They give the fibering

$$\begin{array}{c} \Gamma\backslash N \\ \pi \downarrow \\ T^2 \end{array} , \tag{41}$$

whose base space is a two-dimensional torus. This circle bundle is clearly non-trivial, for $\Gamma\backslash N$ has Γ as its fundamental group, which can not be the fundamental group of a three-dimensional torus.

$\Gamma\backslash N$ is called a nilmanifold, in the sense that it is the homogenous space of the nilpotent Lie group N. Its harmonic analysis is of great importance and interest.

Let $L^2(\Gamma\backslash N)$ be the Hilbert space of all square-integrable functions on $\Gamma\backslash N$. Then N acts on $L^2(\Gamma\backslash N)$ as a unitary operator and we have the direct sum decomposition

$$L^2(\Gamma\backslash N) = \bigoplus_{n\in\mathbb{Z}} H_n , \tag{42}$$

where a function $f \in H_n$, if and only if it satisfies the functional equation

$$f(x,y,z + t) = e^{2\pi int} f(x,y,z). \tag{43}$$

Let C_n be the subspace of continuous functions in H_n. Then C_n is dense in H_n, and the functions of C_n are closely related to the Jacobi theta functions.

Let $\zeta = x + iy$, and $n \in \mathbb{Z}^+$. A continuous function $G(\zeta)$ is called a

Jacobi theta function or order n, or simply $G \in \Theta_n$, if it satisfies the functional equations

$$(44A) \qquad G(\zeta + i) = G(\zeta)$$

$$(44B) \qquad G(\zeta + i) = \{\exp \pi i n(-2\zeta - i)\}G(\zeta).$$

To $G_n(\zeta) \in \Theta_n$ we associate the function $M_n(G_n) \in C_n$ by

$$(45) \qquad M_n(G_n)(x,y,z) = e^{2\pi i n z} e^{-\pi n y^2} G_n(\zeta) .$$

This mapping has the following remarkable property:

The mapping

$$(46) \qquad M_n: \Theta_n \to C_n$$

is a linear isomorphism.

The possible fruitfulness of harmonic analysis on nilmanifolds was first indicated by P. Cartier (1965) and W. Weil (1964). Since then much work has been done by L. Auslander and his coworkers [1]. Frequently classical theta identities follow from group-theoretic arguments.

6. Complex line bundles [3], [6].

An account of circle bundles will not be complete without a mention of the complex line bundles, which are closely related. These are bundles whose fibers are $\mathbb{C}$ and whose group is the miltiplicative group $\mathbb{C}^*$ of non-zero complex numbers. The local fiber coordinate is $z_U \in \mathbb{C}$ and the gauge transformation is

$$(47) \qquad z_U = g_{UV} z_V ,$$

where $g_{UV}: U \cap V \to \mathbb{C}^*$. Complex line bundles are particularly significant on complex manifolds, when we suppose in addition that the transition functions are holomorphic. The bundle is then called *holomorphic*. All the holomorphic line bundles over a complex manifold M form a group which is isomorphic to $H^1(M,\mathfrak{O}^*)$, where $\mathfrak{O}^*$ is the sheaf of germs of non-zero holomorphic functions.

To describe the group $H^1(M,\mathfrak{O}^*)$ consider on M the exact sequence of sheaves

(48) $$0 \longrightarrow \mathbb{Z} \xrightarrow{j} \mathfrak{O} \xrightarrow{e} \mathfrak{O}^* \longrightarrow 0,$$

where $\mathfrak{O}$ is the sheaf of germs of holomorphic functions and e is defined on a germ of functions by

$$e(f(z)) = \exp(2\pi i f(z)).$$

From (48) follows the exact cohomology sequence:

(49) $$H^1(M;\mathbb{Z}) \xrightarrow{j} H^1(M,\mathfrak{O}) \xrightarrow{e} H^1(M,\mathfrak{O}^*) \xrightarrow{\delta} H^2(M;\mathbb{Z}) \to \dots$$

The image of $H^1(M;\mathfrak{O}^*)$ under δ is best described by differential geometry. We introduce in the coordinate bundle L an hermitian structure, i.e., C^∞ functions $a_U > 0$ in each U, such that

(50) $$a_U|z_U|^2 = a_V|z_V|^2, \quad \text{whenever} \quad U \cap V \neq \emptyset .$$

It follows that

$$d \log a_U + d \log g_{UV} + d \log \bar{z}_U = d \log a_V + d \log \bar{z}_V,$$

and hence that

(51) $$\partial \log a_U + d \log g_{UV} = \partial \log a_V.$$

Thus the form $\frac{i}{2\pi}\partial\bar{\partial}\log a_U$, of bidegree (1,1), in independent of U. It can be proved that it represents the characteristic class $c_1(L)$ in the sense of deRham's theory. In fact, we have

(52) $$\delta H^1(M,\mathfrak{D}^*) \cong H^2_{(1,1)}(M;\mathbb{Z}) ,$$

where the right-hand side is the subgroup of $H^2(M;\mathbb{Z})$, whose classes can be represented by a form of bidegree (1,1).

From the exactness of (49) it also follows that the subgroup of all holomorphic line bundles with characteristic class zero is isomorphic to $H^1(M,\mathfrak{D})/jH^1(M;\mathbb{Z})$. If M is a compact Kähler manifold, this is a complex torus; the proof of this uses the theory of elliptic operators.

The consideration of the group $H^1(M;\mathfrak{D}^*)$ provides the ideal background for some of the fundamental results in algebraic geometry. Let M be a non-singular projective variety. Let

$\mathfrak{G}$ = group of divisors,
$\mathfrak{G}_a$ = " " " which are homologous to zero,
$\mathfrak{G}_\ell$ = " " " which are linearly equivalent to zero.

Then

(53) $$\begin{cases} \mathfrak{G}/\mathfrak{G}_\ell \cong H^1(M;\mathfrak{D}^*), \\ \mathfrak{G}_a/\mathfrak{G}_\ell \cong H^1(M,\mathfrak{D})/jH^1(M;\mathbb{Z}) = \text{Picard torus} \\ \mathfrak{G}/\mathfrak{G}_a \cong H^2_{(1,1)}(M;\mathbb{Z}) \text{ (Lefschetz-Hodge Theorem).} \end{cases}$$

Bibliography

1. L. Auslander and R. Tolimieri, Abelian harmonic analysis, theta functions, and function algebras on a nilmanifold, Springer Lecture Notes, No. 436, (1975).

2. T. Banchoff and J. White, The behavior of the total twist and the self-linking number of a closed space curve under inversions, Math. Scandinavica, 36(1975), 254-262.

3. S. S. Chern, Complex manifolds without potential theory, van Nostrand, Princeton, 1967.

4. P. A. M. Dirac, Quantised singularities in the electromagnetic field, Proc. R. Soc. London A133 (1931), 60-72.

5. J. W. Milnor and J. D. Stasheff, Characteristic classes, Annals of Math. Studies 76, Princeton Univ. Press, 1974.

6. J. Morrow and K. Kodaira, Complex manifolds, Holt, Rinehart, and Winston, 1971.

7. W. Scherrer, Eine Kennzeichnung der Kugel, Vierteljahrschift der Naturforsch. Ges. Zurich 85(1940), Beiblatt 32, 40-46.

8. J. H. White, Twist invariants and the Pontryagin numbers of immersed manifolds, Proc. of Symp. in Pure Math., Vol. 27, Differential Geometry, Part 1, 429-437.

9. H. Wu, The equidistribution theory of holomorphic curves, Annals of Math. Studies, No. 64, Princeton Univ. Press, 1970.

10. T. T. Wu and C. N. Yang, Concept of nonintegrable phase factors and global formulation of gauge fields, Physical Review D, 12 (1975), 3845-3857.

University of California
Berkeley

A TOPOLOGY FOR THE SPACE OF FOLIATIONS

D. B. A. Epstein

§1. Introduction.

In foliation theory, just as in the theory of differential equations, stability properties under a small perturbation are of interest. Such properties were considered by Reeb in his thesis [3]. To provide a proper framework for the consideration of such questions we propose to topologize the set of C^r-foliations of codimension k on a smooth manifold M, where $1 \leq r \leq \infty$.

One standard way to define a topology is as follows. Given a foliation $\mathcal{F}$, we associate to each point x of M the tangent space to $\mathcal{F}$ at x. This gives a section to the bundle over M whose fibre over x is the Grassmannian of all (n-k)-planes of the tangent space to M at x. Any topology which one puts on the space of all sections of this bundle will topologize the space of foliations. For example we can topologize the set of C^r-foliations by taking the C^{r-1}-topology on the space of foliations. This works very nicely when $r = \infty$. Unfortunately, when r is finite a degree of differentiability is lost.

To see more clearly the problem when using a topology on the space of sections, we formulate the axioms which should be satisfied by a topology on the set of C^r-foliations of M of codimension k.

Axiom 1. Let $\mathcal{F}$ be a C^r-foliation of M. Let $\phi: M \to M$ be a C^r-diffeomorphism which is C^r-close to the identity in the fine C^r-topology. Then $\phi\mathcal{F}$ is close to $\mathcal{F}$.

Axiom 2. Let $D(t) = \{x | x \in R^k \text{ and } |x| < t\}$. Let $\mathcal{F}$ be a foliation of M. Let $h: I \times D(1) \to M$ be a C^r-map such that $h|\{t\} \times D(1)$ is an embedding transverse to $\mathcal{F}$ for each $t \in I$ and such that for each $x \in D(1)$, $h(I \times \{x\})$ lies on a single leaf of $\mathcal{F}$. We insist that our topology should have the following property. If $\mathcal{G}$ is sufficiently close to $\mathcal{F}$, then there is a C^r-map

$$k : I \times D(\tfrac{1}{2}) \to M$$

such that:

i) $k|0 \times D(\frac{1}{2}) = h|0 \times D(\frac{1}{2})$;

ii) for each $x \in D(\frac{1}{2})$, $k(I \times \{x\})$ lies on a single leaf of $\mathcal{G}$;

iii) $k(\{t\} \times D(\frac{1}{2})) \subset h(\{t\} \times D(\frac{3}{4}))$ and $k|\{t\} \times D(\frac{1}{2})$ is an embedding for each t;

iv) for each $t \in I$, $\mathcal{G}$ is transverse to $h(\{t\} \times D(\frac{3}{4}))$;

v) k is C^r-near to $h|I \times D(\frac{1}{2})$.

Note that the first four properties determine k uniquely. The condition on the topology of the space of foliations is that k should exist and be C^r-close to h.

Axiom 2 says that if $\mathcal{G}$ is near $\mathcal{F}$, then holonomy maps arising from $\mathcal{G}$ are near holonomy maps arising from $\mathcal{F}$.

Axiom 2 can be strengthened to Axiom 2', which is sometimes needed. In this version, we have a family of maps $h_i: I \times D(1) \to M$ $(i \in I)$ as in Axiom 2, such that the family of subsets $\{\text{Im } h_i\}$ is locally finite. Axiom 2' asserts the simultaneous existence of maps $k_i: I \times D(\frac{1}{2}) \to M$ $(i \in I)$, with the same properties i)-v) with respect to h_i as k has with respect to h, if $\mathcal{G}$ is sufficiently near to $\mathcal{F}$.

The topology on the space of sections of the Grassmannian bundle does not satisfy even a weakened form of Axiom 2. To see this, we construct a C^∞ function $\phi : R^2 \to R$ with compact support in a disk of radius $\frac{1}{2}$. and such that $|\partial\phi/\partial y\ (0)| \neq 0$. Let $\psi_\varepsilon : R^2 \to R^2$ be the map $\psi_\varepsilon(x,y) = (x, y + \varepsilon\phi(x,y/\varepsilon))$ for $\varepsilon > 0$. Then ψ_ε has the Jacobian matrix

$$\begin{bmatrix} 1 & 0 \\ \varepsilon\dfrac{\partial\phi}{\partial x} & 1 + \dfrac{\partial\phi}{\partial y} \end{bmatrix}$$

Therefore ψ_ε is a diffeomorphism. Let $\mathcal{F}_\varepsilon$ be the image under ψ_ε of the foliation $\mathcal{F}_o$ of R^2 by horizontal lines (y = constant).

If we are topologizing C^1-foliations, there is really no choice but the C^o-topology on the space of sections of the Grassmannian. Explicitly, this means that we measure the distance of $\mathcal{F}_\varepsilon$ from $\mathcal{F}_o$ by $\varepsilon \sup |\partial\phi/\partial x|$, which tends to zero as ε tends to zero. However, the holonomy map from $x = 1$ to $x = 0$ is given by

$$y \to y + \varepsilon\phi(0, y/\varepsilon)$$

whose derivative at $y = 0$ is $1 + \partial\phi/\partial y(0)$. Therefore the holonomy map associated to $\mathcal{F}_\varepsilon$ does not tend to the identity in the C^1-topology.

The topology we are going to introduce was first considered by M.W. Hirsch [1]. However, Hirsch did not prove that neighbourhoods of a foliation $\mathcal{F}$ can be computed by using <u>any</u> family of admissible coordinate charts for $\mathcal{F}$. It is this result which is our main objective. (Also the reader should be cautioned that Hirsch asserts without proof that his topology is the same as the C^{r-1}-topology on the space of sections of the Grassmannian.

The above example shows that this assertion is false.)

§2. C^r-P-structures.

Let P be a pseudogroup of diffeomorphisms between open subsets of R^n. (For the definition of a pseudogroup see Kobayashi and Nomizu [2]). Let M be a C^∞-manifold of dimension n. Suppose that every element of P is $C^r (r \geq 1)$. A C^r-P-structure on M is a maximal atlas $\{\phi_i : U_i \to R^n\}_{i \in I}$ where ϕ_i is a C^r-diffeomorphism onto an open cube of R^n and $\phi_i \phi_j^{-1}$ belongs to P.

If S is a C^r-P-structure, we define a neighbourhood scheme for S as a collection $(I, \{\phi_i\}, \{K_i\})$, where I is an indexing set, $\phi_i : U_i \to R^n$ is in S and $K_i \subset U_i$ for each $i \in I$, where:

2.1) $\phi_i K_i$ is a closed cube in R^n with sides parallel to the axes (all cubes in this paper have sides parallel to the axes);

2.2) the family $\{K_i\}$ is locally finite;

2.3) the family $\{\text{int } K_i\}$ covers M

If $\alpha = (\alpha_1, \ldots, \alpha_n)$ is a multi-index, with each $\alpha_i \geq 0$ and an integer, we define $|\alpha| = \Sigma \alpha_i$ and put $D^\alpha = \partial_1^{\alpha_1} \ldots \partial_n^{\alpha_n}$, the usual composition of partial derivatives in R^n. Let $\varepsilon_i > 0$ for each $i \in I$. We write $N(I, \{\phi_i\}, \{K_i\}, \{\varepsilon_i\})$ for the set of C^r-P-structures T for which there exist charts $\phi_i' : U_i' \to R^n$ $(i \in I)$ such that:

2.4) $K_i \subset U_i'$;

2.5) there exists subsets K_i' of U_i' such that $\phi_i' K_i'$ is a cube in R^n and $K_i \subset K_i'$; and

2.6) $|D^\alpha(\phi_i' \phi_i^{-1} - \text{id})| < \varepsilon_i$ on $\phi_i K_i$ for $|\alpha| \leq r$.

Then $(I, \{\phi_i'\}, \{K_i'\}$ is again a neighbourhood scheme. By abuse of language we will say that a neighbourhood scheme $(I, \{\phi'_i\}, \{K_i'\})$ as above is in $N(I, \{\phi_i\}, \{K_i\}, \{\varepsilon_i\})$, although this last set really consists of C^r-P-structures and not neighbourhood schemes.

Lemma 2.7. Let $(I, \{\phi_i'\}, \{K_i'\})$ be a neighbourhood scheme in $N(I, \{\phi_i\}, \{K_i\}, \{\varepsilon_i\})$. Then there exists positive numbers δ_i $(i \in I)$ such that

$$N(I, \{\phi_i'\}, \{K_i'\}, \{\delta_i\}) \subset N(I, \{\phi_i\}, \{K_i\}, \{\varepsilon_i\}).$$

Proof. Let $(I, \{\phi_i''\}, \{K_i''\})$ be a neighbourhood scheme in $N(I, \{\phi_i'\}, \{K_i'\}, \{\delta_i\})$, where $\phi_i'' : U_i'' \to R^n$ and $K_i \subset K_i' \subset K_i'' \subset U_i''$. Then on $\phi_i K_i$,

$$\phi_i'' \phi_i^{-1} - id = (\phi_i'' (\phi_i')^{-1} - id) \circ (\phi_i' \phi_i^{-1}) + (\phi_i' \phi_i^{-1} - id).$$

As δ_i tends to zero, the first term on the right tends to zero in the C^r-topology. Therefore for δ_i sufficiently small, and $|\alpha| \leq r$, we have $|D^\alpha (\phi_i'' \phi_i^{-1} - id)| < \varepsilon_i$ on $\phi_i K_i$.

Proposition 2.8. As $(I, \{\phi_i\}, \{K_i\})$ vary over all neighbourhood schemes for all C^r-P-structures on M, and $\{\varepsilon_i\}$ varies over all families of positive numbers, the sets $N(I, \{\phi_i\}, \{K_i\}, \{\varepsilon_i\})$ define open neighbourhoods for a topology on the set of C^r-P-structures on M. (We call this the fine C^r-topology.)

Proof. Let $(I, \{\phi_i\}, \{K_i\})$ and $(J, \{\phi_j\}, \{K_j\})$ be neighbourhood schemes. Let T be a C^r-P structure which is in both $N_1 = N(I, \{\phi_i\}, \{K_i\}, \{\varepsilon_i\})$ and in $N_2 = N(J, \{\phi_j\}, \{K_j\}, \{\varepsilon_j\})$. Then there are a neighbourhood schemes for T, $(I, \{\phi_i'\}, \{K_i'\})$ in N_1 and $\{J, \{\phi_j'\}, \{K_j'\})$ in N_2.

There is no loss of generality in assuming I and J disjoint. By Lemma 2.7 we may choose $\delta_i (i \in I)$ and $\delta_j (j \in J)$ small enough so that

$$N(I \cup J, \{\phi_i'\} \cup \{\phi_j'\}, \{K_i'\} \cup \{K_j'\}, \{\delta_i\} \cup \{\delta_j\})$$

is contained in both N_1 and in N_2. This proves the proposition.

Definition 2.9. We may also define the compact C^r-topology by an analogous definition, lemma and proposition. In this definition one obtains a neighbourhood by fixing also a finite subset of I for each neighbourhood, and then requiring 2.4, 2.5 and 2.6 only for i in this finite subset.

§3. Foliations

We now specialize to the case which really interests us, namely C^r-foliations of codimension k with $r \geq 1$. Here P is the pseudogroup of C^r-diffeomorphisms between open subsets of R^n, which are locally of the form $(x,y) \to (h(x,y), g(y))$ where $x, h(x,y) \in R^{n-k}$ and $y, g(y) \in R^k$.

Let $\mathcal{F}$ be the foliation of R^n of codimension k given by $R^n = R^{n-k} \times R^k$. For $i = 1,2,\ldots, N$ let $\phi_i : U_i \to \mathrm{Im}\,\phi_i \subset R^n$ be in P. Let K_i be a compact subset of U_i $(1 \leq i \leq N)$ and suppose that $0 < b < 1/8$ and that $[-1-2b, 1+2b]^n \subset \bigcup_{i=1}^N \mathrm{int}\, K_i$.

Lemma 3.1 Suppose we are in the situation described above. Then given $\varepsilon > 0$ there exists $\delta > 0$ with the following property. Suppose $\mathcal{G}$ is a codimension k C^r-foliation of a neighbourhood of $\bigcup_{i=1} K_i$ which has admissible coordinate charts $\psi_i : U_i' \to R^n$ where $K_i \subset U_i'$, such that for any multi-index α with $|\alpha| \leq r$, and for any i with $1 \leq i \leq N$, we have

$$|D^\alpha(\psi_i \phi_i^{-1} - \mathrm{id})| < \delta \text{ on } \phi_i K_i.$$

Then there exists a C^r-chart $\psi : (-1-b,1+b)^n \to R^n$ which is admissible for $\mathcal{G}$ and such that $|D^\alpha(\psi - id)| < \varepsilon$ if $|\alpha| \leq r$. Moreover we may assume that $\psi(x,y) = (x,g(x,y))$ for some C^r-function

$$g: (-1-b,1+b)^{n-k} \times (-1-b,1+b)^k \to R^k.$$

and $g(0,y) = y$. A chart ψ with these properties is uniquely determined on $(-1-\frac{b}{2},\ 1+\frac{b}{2})^n$ for δ sufficiently small.

Proof.

Since the derivatives of ϕ_i are bounded on K_i and the derivatives of ϕ_i^{-1} are bounded on $\phi_i K_i$, there is no loss of generality in changing the condition on ψ_i to $|D^\alpha(\psi_i - \phi_i)| < \delta$ on K_i. For each point (x_o,y_o) of $[-1-2b,\ 1+2b]^n$ we choose a cubical neighbourhood $N(x_o,y_o)$ such that $N(x_o,y_o) \subset \text{int } K_i$ for some i. We choose $N(x_o,y_o)$ small enough so that on $N(x_o,y_o)$

$$\phi_i(x,y) = (h(x,y),g(y))$$

where g is a diffeomorphism of a cubical neighbourhood of y_o in R^k onto an open subset of R^k. Let $\phi_i(x_o,y_o) = (u_o,v_o)$. Let $\phi_i^{-1}(u,v) = (\alpha(u,v),\beta(v))$ on $\phi_i N(x_o,y_o)$. Let $\rho(u,v) = (\alpha(u,v),v)$. If $N(x_o,y_o)$ is small enough, then ρ is a diffeomorphism whose domain is $\phi_i N(x_o,y_o)$.

Let $M > 0$ be large enough so that any closed cube of sidelength $1/2^{M-4}$ which meets $[-1-2b,1+2b]^n$ is contained in some neighbourhood of the form $N(x_o,y_o)$. We also assume that $1/2^{M-5} < b$. We now cut R^n up into cubes of sidelength $1/2^M$ with sides parallel to the axes, so that each coordinate of each vertex has the form $m/2^M$ for some integer m. All the cubes referred to in the rest of the proof will be formed from unions of these cubes.

Let $v_1,\ldots,v_p$ be the vertices of cubes of sidelength $1/2^M$ which meet $[-1-2b,1+2b]^n$. We arrange this list so that as i increases $(|\Pi_1 v_i|,|\Pi_2 v_i|)$ increases. Here $\Pi_1 : R^{n-k} \times R^k \to R^{n-k}$ and $\Pi_2 : R^{n-k} \times R^k \to R^k$ are the standard projections; if $x = (x_1,\ldots,x_q) \in R^q$, we write $|x| = \sup\{|x_i| : 1 \leq i \leq q\}$; and pairs of real numbers are ordered lexicographically. Let C_j^r be the cube with centre v_j and sidelength $2r/2^M$.

Each C_j^8 is contained in some $N(x_o,y_o)$ which is contained in int K_i for some i $(1 \leq i \leq N)$. Here $N(x_o,y_o)$ and the index i have the same relationship as in the discussion above. We write C^r for C_j^r, ϕ for ϕ_i and ψ for ψ_i; fixing i and j for the moment.

3.2. We choose δ sufficiently small so that $\psi C^r \subset \text{int}\, \phi C^{r+1}$ and $\phi C^r \subset \text{int}\, \psi C^{r+1}$ for $1 \leq r < 8$. Let g', h', α' and β' be defined by

$$\psi(x,y) = (h'(x,y),\ g'(x,y)) \text{ on } C^8$$

$$\text{and} \qquad \psi^{-1}(u,v) = (\alpha'(u,v),\ \beta'(u,v)) \text{ on } \phi C^7 \text{ (by 3.2)}$$

Let $\rho'(u,v) = (\alpha'(u,v),v)$ on ϕC^7. We choose δ sufficiently small so that ρ' is a diffeomorphism on ϕC^7. Moreover $\rho'|\text{int}\, \phi C^7$ is in our pseudogroup.

As δ tends to zero, $\rho'\psi|C^6$ tends to $\rho\phi|C^6$. The first coordinate of $\rho'\psi(x,y)$ is $\alpha'(h'(x,y),g'(x,y)) = x$ for $(x,y) \in C^6$. The same calculation applies to $\rho\phi(x,y)$ for $(x,y) \in C^8$.

3.3. We are now in the following situation. $\mathcal{G}$ is a C^r-foliation of a neighbourhood of $[-1-2b,1+2b]^n$. After changing the indexing on the ϕ's and ψ's and altering them as above, we have $\phi_i : C_i^8 \to R^n$ $(1 \leq i \leq P)$ each ϕ_i being a chart for $\mathcal{F}$, the foliation by horizontal planes. We also have $\psi_i : C_i^6 \to R^n$ $(1 \leq i \leq P)$, a family of charts for $\mathcal{G}$. We have $\phi_i(x,y) = (x,g_i(y))$ and $\psi_i(x,y) = (x,g_i'(x,y))$. For $1 \leq j < 6$ we have $\psi_i C_i^j \subset \text{int } \phi_i C_i^{j+1}$ and $\phi_i C_i^j \subset \text{int } \psi_i C_i^{j+1}$.

Given $\varepsilon > 0$ we are looking for a $\delta > 0$ (δ a universal constant, depending on ε and on the $\{\phi_i\}$, but not on any of the data about $\mathcal{G}$) such that if $|D^\alpha(\phi_i - \psi_i)| < \delta$ on C_i^6 for $1 \leq i \leq P$ and $|\alpha| \leq r$, then there is a chart $\psi(-1-b,1+b)^n \to R^n$ for $\mathcal{G}$ such that $\psi(x,y) = (x,g(x,y))$, $\psi(0,y) = (0,y)$ and $|D^\alpha(\psi - id) < \varepsilon$ for $|\alpha| \leq r$.

The next step is to change ϕ_i and ψ_i by composing on the left with $1 \times g_i^{-1} = \phi_i^{-1}$. This is possible since $\psi_i C_i^6 \subset \phi_i C_i^7$. We may therefore assume without loss of generality that ϕ_i is the identity map for each i. We than have $C_i^j \subset \text{int } \psi_i C_i^{j+1}$ and $\text{int } \psi_i C_i^j \subset C_i^{j+1}$ for $1 \leq j < 6$. Therefore we can define $\beta_i' \colon C_i^5 \to R^k$ by $\psi_i^{-1}(u,v) = (u,\beta_i'(u,v))$. As δ tends to zero, β_i' tends to Π_2 in the C^r-topology.

Let $c > 0$ be choosen with $c < 1/(64.P.n)$. We choose $\delta > 0$ sufficiently small so that $|\partial\beta_i'/\partial x_j| < c$ on C_i^5 for $1 \leq j \leq n-k$. Locally, a leaf is obtained as follows: we fix v and take the graph of the function $x \to y = \beta_i'(x,v)$. This serves to determine the leaves locally on C_i^4, since $C_i^4 \subset \text{int } \psi_i^{-1} C_i^5$.

Let L be a leaf of $\mathcal{G}$. Let L'be a component of $L \cap \bigcup_{i=1}^P C_i^4$ which meets $\bigcup_{i=1}^P C_i^1$. To see what L' looks like, we choose a point $x \in L' \cap \bigcup_{i=1}^P C_i^1$. Let v_i be a vertex such that $|x - v_i| \leq 1/2^M$. Let σ be a permutation of $\{1,\ldots,P\}$ such that if $\sigma r < \sigma s$ then $(|\Pi_1(v_r - v_i)|, |\Pi_2(v_r - v_i)|) \leq (|\Pi_1(v_s - v_i)|, |\Pi_2(v_s - v_i)|)$.

In particular we must have $\sigma i = 1$. Let $C^j = \bigcup\{C_r^4 : \sigma r \leq j\}$. Let L^j be the component of $C^j \cap L'$ containing x.

Using the remark above about the graph of β_i' being equal to part of a leaf, and the inequality on the derivatives of β_i', we will prove the following two statements by induction on j.

3.4. $\Pi_2 L^j$ has diameter less than $cjn/2^{M-5} < 1/2^{M+1}$ so that $|\Pi_2 L^j - \Pi_2 v_i| < 1/2^{M-1}$.

3.5. L^j is the graph of a C^r-function from $\Pi_1 C^j$ to $\Pi_2 C_i^2$.

Let $\sigma r = j+1$ and suppose the induction statements are true for j. We have two cases: either $\Pi_1 C_r^4 \subset \Pi_1 C^j$ or $\Pi_1 C_r^4 \not\subset \Pi_1 C^j$. If $\Pi_1 C_r^4 \subset \Pi_1 C^j$, then the boundary of L^j is contained in the boundary of C^{j+1} by 3.5. Therefore $L^j = L^{j+1}$. If $\Pi_1 C_r^4 \not\subset \Pi_1 C^j$ then our ordering ensures that $\Pi_2 v_r = \Pi_2 v_i$. Let $\Pi_1 C_r^4 \cap \Pi_1 C^j = K$. Then K is connected because of the order in which we are adding on these cubes. By 3.4 and 3.5, $L^j \cap C_r^4$ is connected. Therefore $g_r'(L^j \cap C_r^4)$ is a single point $\{v\}$ say. Then L^{j+1} is equal to the union of L^j and the graph of $x \to \beta_r'(x,v)$ for $x \in \Pi_1 C_r^4$. This completes the induction. We have therefore proved the following statement.

3.6. If L is a leaf of $\mathcal{G}$ and L' is a component of $L \cap \bigcup_{j=1}^P C_j^4$ which meets $\bigcup_{j=1}^P C_j^1$, then L' is the graph of a C^r-function from $\Pi_1(\bigcup_{j=1}^P C_j^4)$ to $\Pi_2 C_i^2$ for any i such that $C_i^1 \cap L' \neq \emptyset$.

Hence we have a unique function $\psi: \bigcup_{j=1}^P C_j^1 \to R^n$ such that $\psi(x,y) = (x, \theta(x,y))$, where θ is constant on any component L' as above, and $\theta(0,y) = y$.

Let $\gamma_i(y) = g_i'(\Pi_1 v_i, y)$ for $y \in \operatorname{int} \Pi_2 C_i^6$ $(1 \leq i \leq P)$.

We may assume that γ_i is a diffeomorphism and that ψ_i and γ_i^{-1} move points by less than $1/P.2^{M+2}$. As δ tends to zero, $\gamma_i | \Pi_2 C_i^5$ and $\gamma_i^{-1} | \Pi_2 C_i^5$ tend to the identity in the C^r-topology.

By our ordering of the v_i's, there is an integer Q such that $\Pi_1 v_i = 0$ for $1 \leq i \leq Q$ and $\Pi_1 v_i \neq 0$ for $Q < i \leq P$. If $j \leq Q$, then $\psi | C_j^1 = (1 \times \gamma_j^{-1})\, \psi_j | C_j^1$. This is because the two sides are equal on the y-axis, ψ_j is locally constant on a leaf, and we can apply 3.6. If $j > Q$ so that $\Pi_1 v_j \neq 0$, let i be an integer less than j such that $v_i \in C_j^1$, $\Pi_2 v_i = \Pi_2 v_j$ and $|\Pi_1 v_i| < |\Pi_1 v_j|$. The integers i and j continue to have this relationship for the remainder of the proof.
Let $\theta_i(y) = \theta(\Pi_1 v_i, y)$ for $y \in \Pi_2(\bigcup_{i=1}^{P} C_i^1)$. Then $(1 \times \theta_i)(1 \times \gamma_i^{-1})\psi_i$ is defined on C_j^3 if C_j^3 meets $[-1-\frac{3h}{2}, 1+\frac{3h}{2}]^n$. To see this, note that $C_j^3 \subset C_i^4$ which is sent by ψ_i into C_i^5 and then by $1 \times \gamma_i^{-1}$ into C_i^6. Finally $h/2 > 16/2^M$ so that $C_i^6 \subset [-1-2h, 1+2h]^n \subset \bigcup_{r=1}^{P} C_r^1$.

If $C_j^3 \cap [-1-\frac{3h}{2}, 1+\frac{3h}{2}]^n \neq \emptyset$, then $\psi | C_j^1 = (1 \times \theta_i)(1 \times \gamma_i^{-1})\psi_i | C_j^1$.

This is because both sides are equal on (x,y) when $x = \Pi_1 v_i$ and $(x,y) \in C_j^2$, both sides are locally constant on a leaf, both sides are defined on C_j^2, and we can apply 3.6.

$$\text{Let } D_j = C_j^1 \cap \left[-1 - \frac{3h}{2} + \frac{j}{P.2^M},\ 1 + \frac{3h}{2} - \frac{j}{P.2^M}\right]^n .$$

We prove by induction that $\psi | D_j$ is C^r and tends to the identity in the C^r-topology as δ tends to zero. When $j \leq Q$, this follows from $\psi|C_j^1 = (1 \times \gamma_j^{-1})\psi_j|C_j^1$. If $j > Q$, let i be as above. Then, since γ_i^{-1} and ψ_i move points by less than $1/P.2^{M+2}$, $\Pi_2(1 \times \gamma_i^{-1})\psi_i D_j \subset \Pi_2 D_i$ (recall that $\Pi_2 C_i^1 = \Pi_2 C_j^1$). By induction $\theta_i|\Pi_2 D_i$ tends to the identity in the C^r-topology as δ tends to zero. It follows that $\psi|D_j = (1 \times \theta_i)(1 \times \gamma_i^{-1})\psi_i|D_j$ is C^r and tends to the identity.

It follows that $\psi|(-1-h,1+h)^n \to R^n$ is a C^r-chart for $\mathcal{G}$, which tends to the identity in the C^r-topology as δ tends to zero. This completes the proof of Lemma 3.1.

Theorem 3.7. Let $\mathcal{F}$ be a C^r-foliation of M of codimension k and let $(I, \{\phi_i\},\{K_i\})$ be a neighbourhood scheme for $\mathcal{F}$. Then the sets $N(I,\{\phi_i\}, \{K_i\}, \{\delta_i\})$ form a basis for the neighbourhoods of $\mathcal{F}$ in the fine C^r-topology on the space of C^r-foliations of M of codimension k.

(The theorem and its proof, slightly modified, are also valid for the compact C^r-topology.)

Proof Let $(J,\{\psi_j\}, \{L_j\})$ be a neighbourhood scheme for $\mathcal{F}$ and let $\varepsilon_j > 0$ for $j \in J$. We must find $\{\delta_i\}$ such that

$$N(I,\{\phi_i\},\{K_i\},\{\delta_i\}) \subset N(J,\{\psi_j\}, \{L_j\},\{\varepsilon_j\}).$$

Given $j \in J$, let $I(j) = \{i \in I: L_j \cap K_i \neq \emptyset\}$. Then $I(j)$ is a finite subset of I, since $\{K_i\}$ is locally finite and L_j is compact. Let $\psi_j L_j \subset L_j' \subset \operatorname{im}\psi_j$, where L_j' is a cube, slightly larger than and concentric with $\psi_j L_j$. We now apply Lemma 3.1 according to the following dictionary:

Role played in Lemma 3.1 by	is played here by
$\mathcal{F}$	$\psi_j \mathcal{F}$
$\mathcal{G}$	$\psi_j \mathcal{G}$
$[-1,1]^n$	$\psi_j L_j$
$[-1-2b,1+2b]^n$	L'_j
U_i	$\psi_j(\text{dom } \phi_i)$ $(i \in I(j))$
ϕ_i	$\phi_i \psi_j^{-1}$ $(i \in I(j))$
K_i	$L'_j \cap \psi_j K_i$ $(i \in I(j))$
ε	ε_j.

Lemma 3.1 gives us a certain $\delta(j) > 0$ for this situation. We choose δ_i so that if $i \in I(j)$, then $\delta_i < \delta(j)$. This is possible since K_i is compact and $\{L_j\}$ is locally finite.

Let $\mathcal{G}$ be a foliation in $N(I,\{\phi_i\}, \{K_i\},\{\delta_i\})$. We have admissible charts $\phi'_i : U'_i \to R^n$ for $\mathcal{G}$, such that for $|\alpha| \leq r$, $|D^\alpha(\phi'_i \phi_i^{-1}) - id)| < \delta_i$ on $\phi_i K_i$. Then $|D^\alpha((\phi'_i \psi_j^{-1}) (\psi_j \phi_i^{-1}) - id) < \delta_i$ on $\phi_i(K_i \cap \psi_j^{-1} L'_j)$. By Lemma 3.1, there is a C^r-chart ψ''_j for $\psi_j \mathcal{G}$ on a neighbourhood $\psi_j U'_j$ of $\psi_j L_j$, such that $|D^\alpha(\psi''_j - id)| < \varepsilon_j$ on $\psi_j U'_j$ and $\psi''_j \psi_j U'_j$ is an open cube. Let $\psi'_j = \psi''_j \psi_j$. Then $'_j$ is a chart for $\mathcal{G}$ on U'_j, $\psi'_j U'_j$ is an open cube and $L_j \subset U'_j$. We have

$$|D (\psi'_j \psi_j^{-1} - id)| < \varepsilon_j \text{ on } \psi_j L_j.$$

This proves the theorem.

<u>Proposition 3.8</u> Let $\mathcal{F}$ be a C^r-foliation of M of codimension k. Let I be an indexing set and let $\phi_i: U_i \to R^n$ be a family of distinguished charts for $\mathcal{F}$. Let $K_i \subset U_i$ be compact and suppose that $\phi_i K_i$ is a cube for each i. Suppose also that $M = \bigcup_{i \varepsilon I} \text{int } K_i$.

Then we obtain a neighbourhood of $\mathcal{F}$ in the fine C^r-topology by specifying a family $\{\delta_i\}_{i \varepsilon I}$ of positive numbers. We take all C^r-foliations $\mathcal{G}$ of M which have distinguished projections $g_i: \text{int } K_i \to R^k$ such that for $|\alpha| \leq r$, $|D^\alpha(g_i \phi_i^{-1} - \Pi_2)| < \delta_i$, on ϕ_i (int K_i). A neighbourhood in the compact C^r-topology is obtained by insisting on such inequalities for only a finite subset of I.

<u>Proof</u>. Suppose $\mathcal{G}$ is near to $\mathcal{F}$. Then $\mathcal{G}$ has distinguished charts $\psi_i: U_i' \to R^n$, where U_i' is a neighbourhood of K_i and $\psi_i \phi_i^{-1}$ is C^r-near to the identity on $\phi_i K_i$. Hence $\mathcal{G}$ has distinguished projections $g_i = \Pi_2 \psi_i$ such that $g_i \phi_i^{-1}$ is C^r-near to Π_2 on ϕ_i(int K_i).

Conversely, suppose $|D^\alpha(g_i \phi_i^{-1} - \Pi_2)| < \delta_i$ on ϕ_i(int K_i) for $|\alpha| \leq r$. Let ψ_i: int $K_i \to R^n$ be defined by $(\Pi_1 \phi_i, g_i)$. Then $\psi_i \phi_i^{-1} : \phi_i(\text{int } K_i) \to R^n$ is clearly a diffeomorphism onto its image, and it is C^r-near to the identity.

Let $\{L_i\}$ be a shrinking of $\{K_i\}$, such that $\phi_i L_i$ is a cube concentric with $\phi_i K_i$ and such that $M = \bigcup_{i \varepsilon I}$ int L_i. Under the conditions above, $\mathcal{G} \in N(I, \{\phi_i\}, \{L_i\}, \{\epsilon_i\})$ for $\{\delta_i\}$ sufficiently small.

<u>Corollary 3.9</u>. Let $\mathcal{F}$ be a C^r-foliation of M of codimension k. Let f: N $\to$ M be a C^r-map transverse to $\mathcal{F}$. Let $U \subset N$ be an open subset with compact closure. If $\mathcal{G}$ is near to $\mathcal{F}$ in the compact C^r-topology, then $f|U$ is transverse to $\mathcal{G}$, and $(f|U)^*\mathcal{G}$ is near to $(f|U)^*\mathcal{F}$ in the compact C^r-topology.

Proof. This follows easily from the preceding proposition.

§4. The axioms

Here we prove the axioms mentioned in the introduction. Axiom 1 is trivial for both the compact and fine C^r-topologies on the space of C^r-foliations. We can of course assume that ϕ is close to the identity in the compact C^r-topology in order to deduce that $\phi\mathcal{F}$ is near $\mathcal{F}$ in the compact C^r-topology.

To prove Axiom 2 (or the stronger version Axiom 2' in the case of the fine C^r-topology), we note first that there is no loss of generality in assuming that h is defined on $R \times D(1-2\varepsilon)$, where $\varepsilon > 0$ is arbitrarily small. To see this, define a distinguished chart on a neighbourhood of $h(\{0\} \times \overline{D(1-\varepsilon)})$. For appropriate coordinates on this chart and for t small and $t \geq 0$, $h(t,y) = (\alpha(t,y),y)$. Then α can be extended to $(-\infty,\varepsilon) \times D(1-2\varepsilon)$.

We now have $h^{-1}\mathcal{F}$, which is the standard foliation of $R \times D(1-2\varepsilon)$ by horizontal lines. If $\mathcal{G}$ is near $\mathcal{F}$ then we can apply Corollary 3.9 to the foliation $h^{-1}\mathcal{G}$, which is near the standard foliation $h^{-1}\mathcal{F}$ in the compact C^r-topology. We obtain a chart $\psi:(1-\varepsilon, 1+\varepsilon) \times D(1-3\varepsilon) \to R^{k+1}$ for $h^{-1}G$ such that $\psi(t,y)=(t,g(t,y))$, $g(0,y) = y$ and ψ is C^r-near the identity on $[-1,1] \times D(3/4)$. We define $k(t,y) = h\psi(t,y)$.

This completes the proof of Axiom 2 for the compact C^r-topology. Clearly the same proof works for Axiom 2' and the fine C^r-topology.

§5. Remark on the definition

It is important to take K_i as a cube in the definition of a topology on the space of C^r-foliations. Suppose for example K is the standard closed solid torus in R^3, consisting of all points within a distance of 1 from the circle $z = 0$, $x^2 + y^2 = 9$. Let U be an ε-neighbourhood of K. Then if $\psi: U \to R^3$ is near the identity, and $\mathcal{F}$ is the standard foliation of R^3 by horizontal planes, $\psi^{-1}\mathcal{F}$ will inevitably contain annular leaves.

However in our topology, there are foliations of U arbitrarily near $\mathcal{H}U$, with each leaf simply connected.

§6 The C^{∞}-case.

Proposition. The fine/compact C^{∞}-topology on the space of foliations, as defined above, is equal to the fine/compact C^{∞}-topology on the space of sections of the Grassmannian bundle (as explained in the introduction).

Proof. Clearly if $\mathcal{G}$ is near $\mathcal{F}$ in the fine/compact C^r-topology, then the associated section $\sigma\mathcal{G}$ of the Grassmannian bundle is near to $\sigma\mathcal{F}$ in the fine/compact C^{r-1} topology on the space of sections.

Conversely suppose that $\mathcal{F}$ is given and that $\mathcal{G}$ is such that $\sigma\mathcal{G}$ is near $\sigma\mathcal{F}$ in the compact/fine C^r-topology. In order to prove that $\mathcal{G}$ is near $\mathcal{F}$ in the compact/fine C^r-topology, we may examine what the hypotheses mean on a coordinate chart. So suppose that $\mathcal{F}$ is a foliation of $R^n = R^{n-k} \times R^k$ by planes of the form $R^{n-k} \times \{y\}$. The Grassmannian of (n-k) planes in R^n can be coordinatized near to $R^{n-k} \times 0$ by associating to a plane of dimension (n-k) a linear map $\lambda: R^{n-k} \to R^k$, such that the graph of λ is the given plane.

Therefore a foliation $\mathcal{G}$ of $(-1-h,1+h)^n$ has $\sigma\mathcal{G}$ near to $\sigma\mathcal{F}$ in the uniform C^r-topology, then at each point $(x,y) \in (-1-h,1+h)^n$, the tangent plane to $\mathcal{G}$ corresponds to a linear map

$$\lambda(x,y): R^{n-k} \to R^k$$

and λ is C^r-near to zero.

Let $x \in (-1-2h,1+2h)^{n-k}$, $y \in (-1-h,1+h)^k$ and $0 \le t \le 1$. Consider the differential equation for $\rho(x,y,t)$,

$$\frac{d\rho}{dt} = \lambda(tx,\rho)x$$

with initial condition $\rho(x,y,0) = y$. For λ small, it is clear that a solution exists. It is well-known that this solution is unique and is a C^r-function of x,y,t. For fixed x and y, the curve $(tx,\rho(x,y,t))$ is tangent to $\mathcal{G}$ and therefore lies in a fixed leaf of $\mathcal{G}$. Let

$$\phi(x,y) = (x,\rho(x,y,1)).$$

Then ϕ is C^r and is C^r-near to the identity (since if $\lambda = 0$ then $\phi = \mathrm{id}$). Clearly $\phi^{-1}: (-1-h,1+h)^n \to R^n$ is a chart for $\mathcal{G}$ which is C^r-near to the identity. The proposition follows.

Corollary of the proof.

The C^r-topology on the space of sections of the Grassmannian bundle is finer than our C^r-topology on the space of C^{r+1}-foliations and coarser than our C^{r+1}-topology.

BIBLIOGRAPHY

1. M.W. Hirsch, "Stability of compact leaves of foliations" Dynamical Systems (Proc. Symp. University of Bahia Salvador 1971) pp 135-153. Academic Press New York 1973.

2. S. Kobayashi and K. Nomizu, "Foundations of Differential Geometry, Vol. 1", Interscience (1963).

3. G. Reeb, "Sur les espaces fibrés et les variétés feuilletées" Actualités Scientifiques et Industrielles 1183. Paris (1952) Herman.

Mathematics Institute
University of Warwick
COVENTRY CV4 7AL
ENGLAND

STABILITY OF COMPACT FOLIATIONS

by

D.B.A. Epstein and H. Rosenberg*

§1 Introduction

Let M be a connected n-manifold without boundary, which is foliated by $\mathcal{F}$, a foliation in which all leaves are compact. Such a foliation is called a compact foliation. Let $q:M \to Q$ be the map which identifies each leaf to a point, and let Q have the quotient topology. By an example due to Sullivan [6], Q is not necessarily Hausdorff. We say that $\mathcal{F}$ is a compact Hausdorff foliation if Q is Hausdorff.

Compact Hausdorff foliations enjoy many nice properties (see Epstein [1]) and there are many equivalent ways of stating this condition. In the case of a C^r-foliation $r \geq 1$, we obtain complete local information by the following theorem which is proved in [1].

Thereom 1.1. Let $\mathcal{F}$ be a compact Hausdorff foliation which is C^r $(r \geq 1)$. Then there is a "generic leaf" L_o with the property that there is an open dense subset of M, where the leaves all have trivial holonomy and are all diffeomorphic to L_o. Given a leaf L, we can describe a neighbourhood of L, together with the foliation on the neighbourhood, as follows. There is a finite subgroup H of O(k), where k is the codimension of $\mathcal{F}$.

---

* This paper is the result of research carried out during the Foliations Symposium at the University of Warwick in 1975/76. The Symposium was supported by the Science Research Council.

H acts freely on L_o on the right and $L_o/H \cong L$. Let D^k be the unit k-disk. We foliate $L_o \times D^k$ with leaves of the form $L_o \times \{pt\}$. This foliation is preserved by the diagonal action of H, defined by $h(x,y) = (xh^{-1},hy)$ for $h \in H$, $x \in L_o$ and $y \in D^k$. So we have a foliation induced on $\mathcal{U} = L_o \times_H D^k$. The leaf corresponding to $y = 0 \in D^k$ is L_o/H. This theorem then states that there is a C^r-embedding $\phi : \mathcal{U} \to M$, which preserves leaves and is the given identification of L_o/H with L on the central leaf corresponding to $y = 0$.

A special case of this situation occurs when $q:M \to Q$ is a C^r-fibre bundle. Then the leaves of $\mathcal{F}$ are the fibres of q and Q is a manifold. Under these circumstances Rosenberg and Langevin [4] used the ideas of Thurston's stability theorem [7] to prove that if $H^1(L_o;R) = 0$ then the foliation $\mathcal{F}$ is stable for small perturbations. That is to say, if $\mathcal{G}$ is a foliation which is near to $\mathcal{F}$ in some suitable topology, then there is a diffeomorphism of M throwing $\mathcal{G}$ onto $\mathcal{F}$.

Hamilton has recently proved a similar stability theorem for compact Hausdorff foliations of a compact manifold [8] . His methods use elliptic operator theory and the Nash-Moser implicit function technique. In this paper we plan to generalize Hamilton's results in two ways. Firstly we extend the results from the C^∞-case to the C^r-case ($r \geq 1$). Secondly we prove the theorem without the assumption that M is compact.

The following is our main result.

Main Theorem Let $r \geq 1$ and let $\mathcal{F}$ be a compact Hausdorff C^r-foliation of a manifold M. Let L_o be the generic leaf and suppose that $H^1(L_o;R) = 0$. Let N be a neighbourhood of the identity in the fine C^r-topology on the space of C^r-diffeomorphisms of M. Then, if $\mathcal{G}$ is sufficiently near $\mathcal{F}$ in the fine C^r-topology (see Epstein [2] for the definition of this topology), there is a C^r-diffeomorphism $\phi \in N$ such that $\phi \mathcal{G} = \mathcal{F}$.

§2 Approximating differentiable actions

There are very general and powerful results concerning the approximation of differentiable actions of compact Lie groups on manifolds (see for example Palais [5] and Grove and Karcher [3]). In the case of a linear action of a compact group on R^k, we obtain a simple formula (which is probably well-known). We give the proof explicitly because we need the result in a rather special form which is not available in the literature.

Let V be a finite dimensional real vector space and let G be a compact Lie group. Let $\alpha : G \times V \to V$ be a continuous linear action of G on V (so that α is analytic). Let K be a compact subset of V.

Theorem 2.1 Let W be an open neighbourhood of K and let $\beta : G \times W \to W$ be a C^r-action of G on W. Then there exists a C^r-map $f: W \to V$ such that $\alpha(g, fw) = f(\beta(g,w))$ for $g \in G$ and $w \in W$. Moreover:

i) if $x \in W$ and $\beta|G \times \{x\} = \alpha|G \times \{x\}$, then $f(x) = x$;

ii) $f|K$ depends continuously on $\beta|G \times K$ (all function spaces topologized with the C^r-topology);

iii) if $\beta|G \times K$ is C^r-near to $\alpha|G \times K$, then $f|K$ is C^r-near to the identity;

iv) if β is defined on V, and is linear, then f is linear. We have the following immediate corollary.

Corollary 2.2 The isomorphism classes of linear representations of a compact Lie group form a discrete space. (This corollary is of course very well-known).

Proof of Theorem 2.1. We define

$$f(v) = \int_G \alpha(g,\beta(g^{-1},v))\, dg \text{ for } v \in W.$$

Then for $v \in W$ and $h \in G$,

$$\alpha(h,fv) = \int_G \alpha(hg,\beta(g^{-1},v))dg \text{ since } \alpha \text{ is linear}$$

$$= \int_G \alpha(g,\beta(g^{-1}h,v))dg \text{ since the measure is invariant}$$

$$= f(\beta(h,v)).$$

All the results stated follow easily.

Remark. Actually we only use derivatives along V, not along G.

§3 Proof of the main theorem

Since M is foliated by a compact Hausdorff foliation, each point of the quotient space Q has a neighbourhood which is the quotient of a k-disk by a finite group acting orthogonally. It follows that Q is locally compact. Since the projection map $\pi : M \to Q$ is open, Q has a countable basis. Therefore we can write Q first as an increasing union of compact sets, and then in the form

$$Q = \bigcup_{i=1}^{\infty} \text{int } K_i, \text{ where } K_i \text{ is compact, and if } K_i \cap K_j \neq \emptyset$$

then $|i-j| \leq 1$.

We then cover M with an open covering $\{\mathcal{U}_r\}_{r=1}^{\infty}$ with the following properties:

<u>3.1.1</u>. For each r there is a finite subgroup H_r of the orthogonal group O(k), and a free action of H_r on L_o, the generic leaf (see Theorem 1.1). U_r is diffeomorphic to $L_o \times_{H_r} \text{int } D^k$ and this diffeomorphism extends to a diffeomorphism of $\bar{U}_r$ with $L_o \times_{H_r} D^k$, both diffeomorphisms respecting the foliation.

<u>3.1.2</u>. For each r, there exists i such that $\bar{U}_r \subset \text{int } K_i$.

<u>3.1.3</u>. Given i, the set of indices r such that $\pi\bar{U}_r \cap K_i \neq \emptyset$ is finite.

Let $D_i \subset \text{int } D^k$ be a concentric disk, such that if V_i is the subspace of $U_i \cong L_o \times_{H_i} \text{int } D^k$ corresponding to $L_o \times_{H_i} \text{int } D_i$, then $\{V_i\}$ is an open covering of M. Let E_i be a concentric disk such that

$$D_i \subset \text{int } E_i \subset E_i \subset \text{int } D^k.$$

Let $\phi_i \colon R \to R$ be a C^∞-monotonic function such that $\phi_i(\|x\|) = 1$ for x in a neighbourhood of D_i and $\phi_i(\|x\|) = 0$ for x in a neighbourhood of $R^k \setminus \text{int } E_i$.

Let $L_i \cong L_o/H_i$ be the central leaf of U_i. We fix a basepoint $\ell_o \in L_o$ (independent of i) and we define $\ell_i \in L_i$ to be the image of this point under the map $L_o \to L_o/H_i \cong L_i$. Let $\gamma_i \colon L_i \to [0,1]$ be a bump function which is 1 near to ℓ_i and zero outside a small ball S_i containing ℓ_i. For i = 1,2,3,..., let $p_i \colon L_o \times \text{int } D^k \to U_i$ be the covering map associated to the above situation. The group of covering translations is H_i, and we have a well-defined epiomorphism $\theta_i \colon \Pi_1(L_i, \ell_i) \to H_i$.

Now let $\mathcal{G}$ be a foliation of M and let $\mathcal{G}|U_i$ be near $\mathcal{F}|U_i$ in the compact C^r-topology (see Epstein [2]). Then $(p_i)^* \mathcal{G} = \mathcal{G}_i$ is a foliation of $L_o \times \text{int } D^k$ which is near to the product foliation in the compact C^r-topology (Corollary 3.9 of [2]). According to the calculation made by Rosenberg and Langevin [4], if $H^1(L_o;R) = 0$, we may assume that for each i, there is an open neighbourhood N_i of $L_o \times E_i$ in $L_o \times \text{int } D^k$, such that each leaf of $\mathcal{G}_i$ meeting N_i is diffeomorphic to L_o and meets each disk of the form $\{\ell\} \times \text{int } D^k$ $(\ell \in L_o)$ exactly once. We may then assume without loss of generality that N_i is saturated for the foliation $\mathcal{G}_i$. Furthermore, since $\mathcal{G}_i$ is invariant under the group H_i of covering translations, we may assume that N_i is invariant under H_i.

Let M_i be the image in U_i of N_i and let $P_i = M_i \cap \{\ell_i\} \times \text{int } D^k$. We obtain a holonomy homomorphism from $\mathcal{G}_i$.

$$\Pi_1(L_i, \ell_i) \to \text{Diff } P_i$$

by following a path on L_i as closely as we can while staying on the same leaf of $\mathcal{G}_i$. Clearly this homomorphism factors through θ_i in the form

$$\Pi_1(L_i, \ell_i) \xrightarrow{\theta_i} H_i \xrightarrow{\beta_i} \text{Diff } P_i.$$

By Corollary 3.9 of [2], we may assume that

$$\beta_i | H_i \times E_i : H_i \times E_i \to P_i$$

is C^r-near to the given orthogonal action of H_i. (We may think of P_i as a subset of int D^k by lifting to the disk through ℓ_o).

From Theorem 2.1, we obtain a map $f_i : P_i \to R^k$, such that $f_i | E_i$ is C^r-near to the inclusion.

We now proceed to construct an isotopy of M which throws $\mathcal{G}$ onto $\mathcal{F}$ on V_i. Let

$$\psi(\ell,x) = \phi_i (\|x\|) \gamma_i \ell \qquad \text{for} \quad \ell \in S_i,\ x \in D^k.$$

Let

$$F_t(\ell,x) = (\ell,\ (1 - t\psi(\ell,x))x + t\psi(\ell,x)f_i(x))$$

for $\ell \in S_i$, $x \in E_i$, $0 \le t \le 1$.

Then F_t has the following properties.

3.2.1). F_t is an isotopy of $S_i \times E_i$ into $S_i \times D^k$, if f_i is sufficiently close to the identity.

2). F is C^r-close to the constant isotopy if f_i is C^r-close to the identity.

3). $F_o = \text{id}$.

4). $F_1 | \{\ell_i\} \times D_i = f_i | D_i$.

5). F_t is the identity near the boundary of $S_i \times E_i$.

6). If $f_i x = x$, then $F_t(\ell,x) = (\ell,x)$ for all $0 \le t \le 1$.

Now we can think of $S_i \times E_i$ as a tubular neighbourhood of S_i in M. By 3.2.5), the isotopy F_t can be extended to all of M as the identity outside this neighbourhood. Applying F_1 to $\mathcal{G}$ we obtain by Theorem 2.1, a foliation where the holonomy on D_i is equal to the holonomy homomorphism of $\mathcal{F}$

$$\Pi_1(L_i,\ell_i) \to H_i \to O(k) \subset \text{Diff } D_i.$$

For each point v of $M_i = p_i N_i$, we get another point v' as follows. Let $v = p_i(\ell,x)$ where $(\ell,x) \in N_i$. Let τ be a path in L_i from $p_i(\ell,0)$ to ℓ_i. The holonomy of $F_1\mathcal{G}$ along τ followed by the holonomy of $\mathcal{F}$ along $(-\tau)$ gives us $v' = p_i(\ell,x)$. Since the holonomy groups of $F_1\mathcal{G}$ and $\mathcal{F}$ agree at ℓ_i, it is easy to see that v' is independent of the choice of τ. By making appropriate choices for τ, we see from the results of Epstein [2], that the map $v \to v'$ is a C^r map of M_i into U_i which is near the identity in the compact C^r-topology, if $\mathcal{G}|U_i$ is near $\mathcal{F}|U_i$ in the compact C^r-topology. Using ϕ_i as in the construction of F_t, and the map $v \to v'$ instead of f_i, we can construct an isotopy G_t of M such that:

3.3.1) G_t is the identity outside $p_i(L_o \times E_i)$.

2) G is C^r-close to the identity, if $\mathcal{G}|U_i$ is near to $\mathcal{F}|U_i$ in the compact C^r-topology.

3) $G_o = id$.

4) The foliations $G_1\ F_1\mathcal{G}$ and $\mathcal{F}$ agree on V_i.

3.4. Note that if Z is a set on which $\mathcal{F}$ and $\mathcal{G}$ agree, and which is saturated for both $\mathcal{F}$ and $\mathcal{G}$, then by 2.1i), f_i is the identity on points corresponding to points of Z. Hence $F_t|Z = id$ by 3.2.6). Finally $G_t|Z = id$ by the statement for G_t, which is analogous to 3.2.6).

Now we proceed to isotope $\mathcal{G}$ to $\mathcal{F}$ by applying the above construction to another of the open sets U_j, this time replacing $\mathcal{G}$ by $G_1F_1\mathcal{G}$. By Axiom 1 of [2], $G_1F_1\mathcal{G}$ is C^r-near to $\mathcal{G}$.

We continue in this way until all the U_i's are dealt with. According to 3.4), successive steps will not spoil what has already been accomplished. The only problem, since we are applying a countable number of isotopies one at a time to $\mathcal{G}$, is that their cumulative effect might be to move $\mathcal{G}$ far from $\mathcal{F}$. To counter this objection, we recall 3.1.2 and 3.1.3. We first perform the isotopies for the finite set of U_i's meeting $\pi^{-1}K_5$, then the isotopies for the U_i's meeting $\pi^{-1}K_{10}$, then for $\pi^{-1}K_{15}$, $\pi^{-1}K_{20}$, $\pi^{-1}K_{25}$. etc. Finally we construct isotopies in order to fix up U_i's meeting $\pi^{-1}K_1$, $\pi^{-1}K_2$, $\pi^{-1}K_3$, $\pi^{-1}K_4$, $\pi^{-1}K_6$ etc.

It is clear that above the construction of the isotopy for a U_i meeting K_j will work if $\mathcal{G}$ is near enough to $\mathcal{F}$ in the compact C^r-topology over $\bigcup_{p=j-10}^{j+10}$ int K_p. Hence the entire construction goes through if $\mathcal{G}$ is near enough to $\mathcal{F}$ in the fine C^r-topology.

This completes the proof of the theorem.

§4 Remarks 1) If $H^1(L_o;R) \neq 0$, then the foliation $\mathcal{F}$ is not stable. This is shown in [4].

2) The above results lead one to the following question. Let M be a manifold foliated with all leaves compact. Then there is a "generic leaf" L_o. Suppose that $H^1(L_o;R) = 0$. Is it true that the foliation is Hausdorff. (See [1] for further background). In the counterexamples to this statement when L_o is a circle, the fact that compact Hausdorff foliations, with L_o as a generic leaf, are not stable, plays a crucial role (see [6]).

REFERENCES

[1] D.B.A. Epstein, "Foliations with all leaves compact", Ann. de l'Inst. Fourier, 26 (1976) pp.265-282.

[2] D.B.A. Epstein, "A topology for the space of foliations", these Proceedings.

[3] K. Grove and H. Karcher, "How to conjugate C^1-close group actions", Math. Z. 132 (1973) pp. 11-20.

[4] R. Langevin and H. Rosenberg, "On Stability of Compact Leaves and Fibrations", (to appear).

[5] R. S. Palais, "Equivalence of nearly differentiable actions of a compact group", B.A.M.S. 67 (1961) pp.362-364.

[6] D. Sullivan, "A counterexample to the periodic orbit conjecture", Pub. Math. de l'I.H.E.S. 46 (1976).

[7] W. Thurston, "A generalisation of the Reeb stability theorem", Topology, 13 (1974) pp.347-352.

[8] R. S. Hamilton, "Deformation theory of foliations", (available from Cornell University in mimeographed form).

Mathematics Institute,
University of Warwick,
COVENTRY, ENGLAND
Dept. de Mathématiques, Université de Paris XI
91 Orsay, FRANCE.

SECONDARY OPERATIONS IN K-THEORY AND THE GENERALIZED VECTOR FIELD PROBLEM

by S. Feder and W. Iberkleid

1. INTRODUCTION

It is the purpose of this paper to present a method for computing secondary characteristic classes in K-theory of (unstable) vector bundles of the form

$$E \otimes \xi \to X \times P^m$$

The method works best for torsion spaces X. To the best of our knowledge this is the first such method. As an application we obtain a result about the generalized vector field problem. Let ξ_n denote the canonical line bundle over the real proyective space P^n. The generalized vector field problem is the determination of the maximal number of independent sections of the bundle $k\xi_n$. By our use of secondary operations in K-theory we obtain:

THEOREM 1 If $\binom{k-1}{n} \equiv 1(2)$ then $2k\xi_{2n}$ does not have $2k-2n+2\nu_2(2k)$ sections if $k \equiv 1(2)$ or $k \equiv 0(2)$ and $\nu_2(k) \equiv n(2)$ and it does not have $2k-2n=2\nu_2(2k)+2$ sections otherwise.

This is very similar to the result of Theorem 1.1(i) of [3] and in half the cases improves it. The computations necessary to obtain this result are, however, much simpler in our case.

One hopes that other theories, mainly MU^* and BP^* used in a similar fashion will lead to further results which should give general necessary conditions for existence of sections of $k\xi_n$. Work in this direction is in progress.

2. Secondary characteristic classes of a vector bundle.

Given a vector bundle $p:E \to B$ and a cohomology theory h^* one has an exact sequence

$$h^*(E) \xrightarrow{i^*} h^*(B) \xrightarrow{p^*} h^*(SE) \xrightarrow{\delta} h^{*+1}(E) \longrightarrow \cdots$$

Where $h^*(E)$ is cohomology with compact support and $i:B \to E$ denotes the zero-section. SE denotes the sphere bundle associated to E. If φ is any natural cohomology operation we can define a secondary operation on $\mathrm{Im}\delta \cap \mathrm{Ker}\varphi$ with values in $p^*h^*(B)/p^*\varphi h^*(B)$. Indeed, let $v \in h^{*+1}(E)$ be such that $\varphi v = 0$. Let $u \in h^*(SE)$ with $\delta u = v$. There is then a well defined class

$$S_\varphi(v) \in p^*h^*(B)/p^*\varphi h^*(B).$$

Since $\varphi v = 0$ and $\varphi\delta = \delta\varphi$ we have $\delta\varphi(u) = 0$ and $\varphi(u) \in p^*h^*(B)$. Define

$$S_\varphi(v) = \varphi(u) \bmod p^*\varphi h^*(B).$$

This is well defined, for if $\delta u' = v$ then $u - u' \in p^*h^*(B)$ and $\varphi(u') = \varphi(u) \bmod p^*\varphi h^*(B)$.

Moreover we have the following

LEMMA 2. If the bundle $p:E \to B$ admitts a non-zero section then $S_\varphi(v) = 0$ whenever it is defined.

PROOF. Let $r:B \to SE$ be such a section and let $\alpha \in h^*(B)$ be such that $p^*\alpha = \varphi(u)$.

Since $\alpha = r^*p^*\alpha = r^*\varphi(u) = \varphi(r^*u)$ we have

$$\varphi(u) = p^*\varphi(r^*u) \in p^*h^*(B).$$

REMARK. Suppose a section exists and a choice of u such that $\delta u = v$ has been made. Let φ_1, φ_2 be two operations such that $\varphi_1(v) = \varphi_2(v) = 0$. Then there is a unique element $\alpha(=r^*u)$ in $h^*(B)$ such that

$$\varphi_i(u) = \varphi_i(p^*\alpha) \qquad i = 1,2.$$

If the bundle E were orientable for the cohomology theory

h^* then $h^*(E) \cong h^*(X)$ (with a possible shift in grading) and $h^*(E)$ is generated over $h^*(X)$ by an element $\Delta \in h^*(E)$ called the Thom class. The image $i^*(\Delta) = \chi(E)$ is called the Euler class of E. If $\chi(E) = 0$ the long exact sequence of the fibration decomposes into short exact sequences:

$$0 \longrightarrow h^r(B) \longrightarrow h^r(SE) \longrightarrow h^{r+1}(E) \longrightarrow 0$$

There always is some class $U \in h^*(SE)$ such that $\delta U = \Delta$ and if $\varphi(\Delta) = 0$ we define a secondary characteristic class $c_\varphi(E) = p^{*-1} s_\varphi(\Delta) \in h^*(B)$. Such classes must, of course, vanish in the presence of a section in view of Lemma 2.

The difficulty in computing secondary characteristic classes in order to get obstructions to section problems lies in the fact that one rarely has a hold on the operations in $h^*(SE)$ since the above mentioned exact sequence is usually the only source of information about $h^*(SE)$.

In what follows we shall apply equivariant K-theory to get further information, which in the case of the generalized vector field problem leads to strong results.

3. Equivariant K-theory.

Given a compact group G one has a cohomology theory $K_G^*(X)$ for G-spaces. If X is a free G-space we have $K_G^*(X) \cong K^*(X/G)$ the usual K-theory of the orbit space X/G. If X is a trivial G-space one has $K_G^*(X) \cong K^*(X) \otimes R(G)$ where $R(G)$

is the complex representation ring of the group G. The theory K_G^* has all the pleasant properties of cohomology theories and all complex G-bundles are orientable with respect to this theory. Moreover, if $H \subset G$ is a subgroup we have a restriction homomorphism $K_G(X) \to K_H(X)$ which is natural in the category of G-spaces. For our purposes we shall use the groups $Z_2 \subset S^1$. Their representation rings are: $R(S^1) = Z[t,t^{-1}]$ - the polynomial ring on the standard representation $t: S^1 \to U(1)$ and its inverse t^{-1}; $R(Z_2) = Z[t]/(t^2-1)$ where t stands for the restriction of $t \in R(S^1)$ to the subgroup $Z_2 \subset S^1$.

Let us illustrate the use of equivariant K-theory to obtain a fact which will be of use further on. Denote by t the trivial bundle (of complex dimention 1) with nontrivial action over P^{2m}. Then for the bundle mt we have the following exact sequence

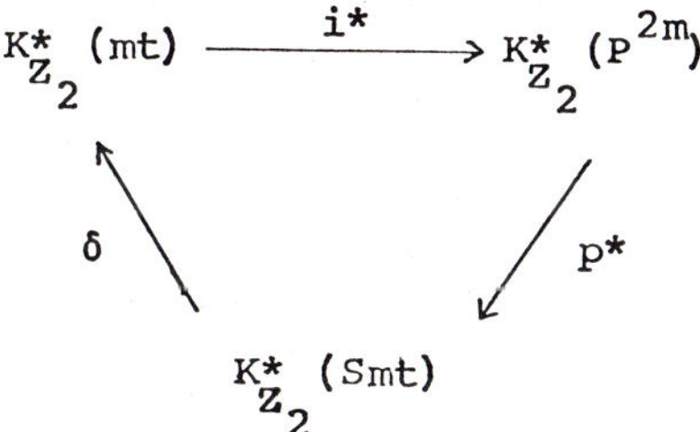

The groups $K^*_{Z_2}(mt)$ are taken with compact support and by the Thom isomorphism theorem we have

$$K^*_{Z_2}(mt) \cong K^*_{Z_2}(P^{2m}) \cong K^*(P^{2m}) \otimes R(Z_2)$$

The restriction i^* composed with the Thom isomorphism

becomes multiplication by the Euler class of the bundle, which for a line bundle L is simply $1-L$, and so for mt is $(1-t)^m$. We obtain

$$0 \to K^1(P^{2n} \times P^{2m-1}) \to K^0(P^{2m}) \otimes R(Z_2) \xrightarrow{(1-t)^m} K^0(P^{2m}) \otimes R(Z_2) \to K^0(P^{2n} \times P^{2m-1}) \to 0$$

This exact sequence (note that $Smt/Z_2 \approx P^{2n} \times P^{2m-1}$) yields a complete description of $K^*(P^{2n} \times P^{2m-1})$. Note that $\psi^p y = y$ for all $y \in K^0(P^{2n} \times P^{2m-1})$ and p odd. Here ψ^p is the Adams operation. This is a typical example of the use of equivariant K-theory to compute ordinary K-groups.

For further information on equivariant K-theory we refer the reader to [1].

4. Secondary operations in K-theory

Let $E \to X$ be a complex vector bundle and let ξ denote the Hopf bundle over P^{2m-1}. The product bundle $E \hat{\otimes}_R \xi$ has a complex structure given by that of E and can be thought of as $E \otimes \eta$ where η is the complexification of ξ. If $E \otimes \eta$ has a non-zero section then the sequence

$$0 \longrightarrow K^{-1}(X \times P^{2m-1}) \xrightarrow{P^*} K^{-1}(S(E \otimes \eta)) \xrightarrow{\delta} K^0(E \otimes \eta) \longrightarrow 0$$

should split with respect to all operations ψ^P. We now note that all groups in this sequence can be thought of as equivariant:

$$X \times P^{2m-1} \approx S(mt)/Z_2; \quad S(E \otimes \eta) \approx S(mt) \times_X S(Et)/Z_2$$
$$E \otimes \eta \approx S(mt) \times_X Et/Z_2$$

Here t should be thought of as the line bundle $X \times C$ with the action of Z_2 given by $(x,z) \rightarrow (x,-z)$. The now equivariant groups can be fitted into other exact sequences and we get:

$$
\begin{array}{ccccccc}
 & & K^{-1}_{Z_2}(X) & & & & K^{0}_{Z_2}(Et) \\
 & & \downarrow & & & & \downarrow \\
 & \longrightarrow & K^{-1}_{Z_2}(Smt) & \xrightarrow{P^*} & K^{-1}_{Z_2}(Smt \times SEt) & \xrightarrow{\delta} & K^{0}_{Z_2}(Smt \times Et) \\
 & & \downarrow \delta & & \downarrow \delta_1 & & \downarrow \delta \\
K^{0}_{Z_2}(mt \times Et) & \longrightarrow & K^{0}_{Z_2}(mt) & \xrightarrow{P^*} & K^{0}_{Z_2}(mt \times SEt) & \xrightarrow{\delta} & K^{1}_{Z_2}(mt \times Et) \\
\downarrow & & \downarrow & & \downarrow & & \\
K^{0}_{Z_2}(Et) & \longrightarrow & K^{0}_{Z_2}(X) & \longrightarrow & K^{0}_{Z_2}(SEt) & \longrightarrow & \qquad (*)
\end{array}
$$

This diagram is commutative and, since all boundary homomorphisms are induced by maps of spaces, all homomorphisms commute with the operations ψ^P. Note that, by the Thom isomorphism theorem all four groups in the lower left corner are isomorphic. Taking these isomorphisms into account we get the diagram:

$$
\begin{array}{ccc}
K^0(X) \otimes R(Z_2) & \xrightarrow{\lambda_{-1}^{Et}} & K^0(X) \otimes R(Z_2) \\
\downarrow \lambda_{-1}^{mt} & & \downarrow \lambda_{-1}^{mt} \\
K^0(X) \otimes R(Z_2) & \xrightarrow{\lambda_{-1}^{Et}} & K^0(X) \otimes R(Z_2)
\end{array}
$$

where $\lambda_{-1}mt$ and $\lambda_{-1}Et$ are the K-theoretical (equivariant) Euler classes of mt over X and Et over X respectively. Once these are computed we look for an alement $U \in K_{Z_2}(mt)$ such that $(\lambda_{-1}mt)\cdot U \neq 0$ is "divisible" by $\lambda_{-1}Et$. For such an element U we have $p^*U = \delta_1 u$ for some $u \in K^{-1}_{Z_2}(Smt \times SEt) \cong K^{-1}(S(E\otimes\eta))$. One now carries on the computations with the element U -this amounts to functional operations and the indeterminacy usually increases. Note that $\delta u \in K^{0}_{Z_2}(Smt \times Et) \approx K^{0}(E\otimes\eta)$ is a multiple of the thom class Δ, say $a\cdot\Delta$, with $a \in K^{0}(X \times P^{2m-1})$. If $\psi^{p}a = a$ we can introduce the operation

$$\varphi_p = \psi^{p} - \rho_p(Et)$$

Where $\rho_p(Et)$ is the "canibalistic" class introduced by Bott [2]. We have $\varphi_p(a\cdot\Delta) = 0$ and $\delta u = a\cdot\Delta$ and so the secondary operation $S_{\varphi_p}(a\cdot\Delta)$ is well defined. We now prove the following proposition.

<u>PROPOSITION 3</u>. If ψ^{p} acts as identity on $K^{0}(X)$ then $S\varphi_p$ is well defined on all elements δu for which $\delta_1 u = p^*U$ for some $U \in K^{0}_{Z_2}(mt)$.

<u>PROOF</u>. Consider the diagram

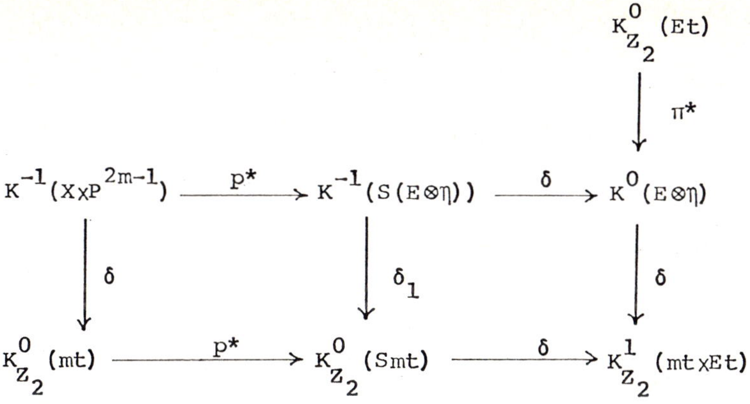

Let $U \in K^{0}_{Z_2}(mt)$ and $u \in K^{-1}(S(E\otimes\eta))$ be such that $p^{*}U = \delta_{1}u$. Let $v = \delta u \in K^{0}(E\otimes\eta)$. Since the bottom row is exact $\delta p^{*}U = 0$ and so $\delta\delta_{1}u = 0$. This means that there is an element $w \in K^{0}_{Z_2}(Et)$ such that $\pi^{*}w = v$. But since ψ^{p} is the identity on $K^{0}(X)$ it is also the identity on $K^{0}_{Z_2}(X)$ and

$$\psi^{p}w = \rho_{p}(Et)\cdot w$$

Therefore $\varphi_{p}(v) = \pi^{*}\varphi_{p}(w) = 0$.

REMARK. Note that even if ψ^{p} acts as identity on $K^{0}(X)$ it is not generally true that it does so on $K^{0}(X\times P^{2m-1})$. Using the above proposition we will be able to evaluate $\varphi_{p}(U)$. Note that if $p^{*}\delta: K^{-1}_{Z_2}(Smt) \to K^{0}_{Z_2}(mt\otimes SEt)$ is a monomorphism the fact that our operation is "functional" will cause no loss of information. In order to evaluate $\delta_{p}(U)$ which, by Proposition 3, is well defined whenever ψ^{p} is the identity on $K^{0}(X)$ we note that, via the Thom isomorphism,

$$U = A\cdot\Delta(mt)$$

for some $A \in K^0_{Z_2}(X)$. Here $\Delta(mt)$ is the Thom class of mt. If we define $\varphi_p(1) = \rho_p(mt) - \rho_p(Et)$ we get

$$\varphi_p(U) = \varphi_p(1)\cdot U$$

In general, for any element $y \in K^0_{Z_2}(mt)$ we have $\varphi_p(y) = \varphi_p(1)\cdot y$.

5. Application to the generalized vector field problem.

We use the standard conversion of the many-section problem into a one-section problem. Namely, if $E \to X$ is a vector bundle which has m linearly independent sections then $E\otimes\xi \to X\times P^{m-1}$ where ξ is the Hopf bundle over P^{m-1}, has one section. In the meta-stable situation the implication can be reversed so that the two problems become equivalent. We are interested in the case when X is the real projective space and E is a multiple of the Hopf bundle. For technical reasons we shall take $X = P^{2n}$ and $E = k\eta-(k-n-1)$ - a complex vector bundle, where η is the complexification of the real Hopf bundle. Note that E is still a stable bundle. This "desuspension" of $2k\xi$ is necessary to make the method work. The direct application of the method to the bundle $2k\xi$ gives no results whatsoever, unless $2k \leq 2n+2$.

If E has $2m$ sections then $E\otimes_{\mathbb{R}}\xi = E\otimes_{\mathbb{C}}\eta$ over $P^{2n}\times P^{2m-1}$ has one section. We now have the situation of the previous section with $X = P^{2m}$. Consider the diagram

$$\begin{array}{ccc}
K^0(P^{2n})\otimes R(Z_2) & \xrightarrow{\lambda_{-1}Et} & K^0(P^{2n})\otimes R(Z_2) \\
\Big\downarrow \lambda_{-1}mt & & \Big\downarrow \lambda_{-1}mt \\
K^0(P^{2n})\otimes R(Z_2) & \xrightarrow{\lambda_{-1}Et} & K^0(P^{2n})\otimes R(Z_2)
\end{array}$$

The Euler class $\lambda_{-1}mt = (1-t)^m$ we have seen in section 3. In order to compute $\lambda_{-1}Et$ we note that this class is the restriction of the Euler class $\lambda_{-1}Et \in K^0_{S^1}(P^{2n})$ where t is thought of as the standard S^1-module. In $K^0_{S^1}(P^{2n})$ we have

$$\lambda_{-1}Et = \frac{(1-\eta t)^k}{(1-t)^{k-n-1}}$$

and the division is unique. We now set $1-\eta = X$, $1-t = T$. We have the relations $X^2 = 2X$ and $X^{n+1} = 0$.

$$\lambda_{-1}Et = \frac{(-\eta t)^k}{(1-t)^{k-n-1}} = \frac{(T+Xt)^k}{T^{k-n-1}} = \sum_{i=0}^{n} \binom{k}{i} t^i X^i T^{n+1-i}$$

Restricting now to $K^0_{Z_2}(P^{2n})$, which amounts to introducing the relation $t^2 = 1$, i.e. $T^2 = 2T$ we have

$$\lambda_{-1}Et = T^{n+1} + XT^n \sum_{1}^{n} \binom{k}{i}(-1)^i = T^{n+1} + \epsilon_{k,n} XT^n$$

where $\epsilon_{k,n} = \sum_{1}^{n} (-1)^i \binom{k}{i} = -1 + (-1)^{n-1}\binom{k-1}{n-1}$.

Note that $XT^n = X^nT$ is an element of order 2 and so $\epsilon_{k,n}$ is a mod2 number: $\epsilon_{k,n} = 1+\binom{k-1}{n-1}$. The element $A = T^{n-m+1}+\epsilon_{k,n}XT^{n-m}\in K^0_{Z_2}(P^{2n})$ corresponds under the Thom isomorphism to a desired element $U\in K^0_{Z_2}(mt)$ for which $p^*U=\delta_1 u$ for some $u\in K^{-1}(S(E\otimes\eta))$.

In order to evaluate $\varphi_p(U)$ we must calculate

$$\varphi_p(1) = \rho_p(mt) - \rho_p(Et)$$

Recall that ρ_p is multiplicative and for a line bundle L, $\rho_p(L) = 1+L+\ldots+L^{p-1}$. We have

$$\rho_3(mt) = (1+t+t^2)^m$$

and

$$\rho_3(Et) = \frac{(1+\eta t+t^2)^k}{(1+t+t^2)^{k-n-1}}$$

In order to carry out the second division we first compute $\rho_3(Et)\in K^0_{S^1}(P^{2n})$ and then, by restriction, find the desired element in $K^0_{Z_2}(P^{2n})$. We find, after dividing and then letting $t^2=1$

$$\rho_3(Et) = \sum \binom{k}{i}(-t)^i X^i(3-T)^{n+1-i}$$

This can further be simplified to

$$\rho_3(Et) = 3^{n+1}+\frac{1-3^{n+1}}{2}T+\frac{1-3^k}{2}\left[1-\frac{3^{k-n-1}+1}{2}T\right]3^{-(k-n-1)}X.$$

We also have

$$\rho_3(mt) = 3^m + \frac{1-3^m}{2} T,$$

and a rather cumbersome expression

$$\varphi_3(1) = (3^m - 3^{n+1}) + \frac{3^{n+1}-3^m}{2} T - \frac{1-3^k}{2}\left[1 - \frac{3^{k-n-1}+1}{2} T\right] 3^{-k+n+1} X.$$

Note, however, that

$$\varphi_3(1)\cdot T = \frac{1-3^k}{2} TX$$

In order to simplyfy the computations in the next section we also compute $\varphi_{-1}(1)$, the operation associated to ψ^{-1}, the complex conjugation on vector bundles. We find

$$\rho_{-1}(mt) = (-1)^m t^m \ , \ \rho_{-1}(Et) = (-1)^{n+1} t^{n+1} \eta^k$$

and

$$\varphi_{-1}(1) = (-1)^m t^m \left[1 + (-1)^{m+n} t^{n+m+1} \eta^k\right]$$

Using the fact that $Tt = -T$ we find

$$\varphi_{-1}(1)\cdot T = T(1-\eta^k)$$

When $m \equiv n(2)$ and $k \equiv 0(2)$ we have

$$\varphi_{-1}(1) = \pm(2-T) ; \varphi_{-1}(1)\cdot T = 0$$

6. PROOF OF THEOREM 1

Since we are looking for sections of $2k\xi$ over P^{2n}, the first obstruction to get below the dimention $2n$ is the Stiefel-Whitney class $w_{2n} = \binom{2k}{2n} X^{2n}$.

If this is zero, i.e. $\binom{2k}{2n} \equiv \binom{k}{n} \equiv 0(2)$ then $\binom{k-1}{n-1} \equiv \binom{k-1}{n}(2)$. Under the assumption $\binom{k-1}{n} \equiv 1(2)$ we have $\epsilon_{k,n} = 0$ and the class $A \in K^0_{Z_2}(mt) \cong K^0_{Z_2}(P^{2n})$ becomes

$$A = 2^{n-m}T$$

We have $\varphi_3(A) = 2^{n-m} \cdot \varphi_3(T) = 2^{n-m} \cdot \frac{1-3^k}{2} TX$

We also have $\varphi_{-1}(A) = 0$ if $k \equiv 0(2)$ and $m \equiv n(2)$. Now by the remark after Lemma 2, if $\varphi_3(A)$ is in the indeterminacy i.e. $\varphi_3(A) = \varphi_3(B)$ with $B \in \delta K^{-1}_{Z_2}(Smt) = \mathrm{Ker}(T^m : K^0_{Z_2}(P^{2n}) \to K^0_{Z_2}(P^{2n}))$ then $\varphi_{-1}(B) = 0$.

Suppose $B = a+bX+(\alpha+\beta X)T$. We get

$$\varphi_{-1}(B) = \pm(2-T)(a+bX) = 0 \quad \Rightarrow a=b=0.$$

Since $B \in \mathrm{Ker}(T^m)$ we get, finally, $B = \beta XT$ with $\beta XT \cdot 2^m = 0$ i.e. $\underline{\beta = 2^{n-m} \cdot c}$. But

$$\varphi_3(B) = \varphi_3(2^{n-m}cXT) = 2^{n-m+1} \cdot c \, \frac{1-3^k}{2} XT.$$

This means that in order that $\varphi_3(B) = \varphi_3(A)$ (actually modulo the image of multiplication by T^{n+1}) we must have $\varphi_3(A) = 0$ (then $\varphi_3(B) = 0$ as well) i.e.

$$m \leq \nu_2 \frac{(1-3^k)}{2} = \begin{cases} \nu_2(k)+1 & \text{if} \quad \nu_2(k)>0 \\ 0 & \text{if} \quad \nu_2(k)=0 \end{cases}$$

This gives the result of Theorem 1 for $k \equiv 0(2)$ and $m \equiv n(2)$. The same result is valid without these assumptions and can be deduced, even if it is a bit móre complicated, using just the operation φ_3.

BIBLIOGRAPHY

1.- Michael Atiyah, K-theory, Benjamin 1967

2.- Raul Bott, Lectures on K-theory (Mimeographed) 1963

3.- Donald Davis, Generalized homology and the generalized vector field problem. Quart. J. Math. Oxford (2), 25(1974), 169-93.

CENTRO DE INVESTIGACION Y ESTUDIOS AVANZADOS DEL IPN.

ON HOLOMORPHIC SOLUTIONS OF CERTAIN KIND OF PFAFFIAN SYSTEMS WITH SINGULARITIES.*

par

R. GÉRARD

In this lecture I will give some of the results obtained in collaboration with Y. Sibuya during his stay in Strasbourg.

The initial aim of our work was the study of linear connexion with irregular singularities in the several variables case. In particular the problem of reduction of this connection and the behavior of the horizontal sections near the singular set. But to do this we first have to solve some problems concerning non linear pfaffian systems with singularities. The results that we have obtained are themself interesting both for people working on singular linear connection in several variables and for people studying the foliation defined by a pfaffian system near the singular set.

For simplicity reasons we will restrict ourself to the case of two variables.

Consider a pfaffian system of the form

$$(1) \qquad dy_i = \frac{f_i^1(x_1,x_2,y_1,y_2\dots y_n)dx_1}{x_1^{p_1}} + \frac{f_i^2(x_1,x_2,y_1,y_2,\dots y_n)dx_2}{x_2^{p_2}}$$

$i = 1,2,\dots n$,

where :

---

* Lecture given in July 1976 at the

" III Escola Latino Americana de Matematica"

Rio de Janeiro.

1) for each $i = 1,2,\ldots n$, f_i^1 and f_i^2 are holomorphic functions near the origin of $\mathbb{C}^2(x) \times \mathbb{C}^n(y)$.

2) p_1 and p_2 are non negative integers.

For the system (1) we shall use the condensed notation :

(1)
$$dy = \frac{f^1(x,y)\, dx_1}{x_1^{p_1}} + \frac{f^2(x,y)\, dx_2}{x_2^{p_2}} .$$

Moreover we shall assume :

1) $f^i(0,0) = 0$ for $i = 1,2$.

2) that the system (1) is completely integrable.

A <u>formal solution</u> of (1) near the origin is by definition a formal power serie

$$\varphi = \sum_{r_1+r_2>0} \varphi_{r_1 r_2} x_1^{r_1} x_2^{r_2} \qquad \varphi_{r_1 r_2} \in \mathbb{C}^n$$

who satisfies formally the system (1). This means that we have formally

$$d\varphi = \frac{f^1(x,\varphi)\, dx_1}{x_1^{p_1}} + \frac{f^2(x,\varphi)\, dx_2}{x_2^{p_2}} .$$

The questions for which we are going to give some answers are the following :

1) <u>Under what conditions does the system (1) has formal solutions</u> ?

2) <u>Under what conditions is a formal solution convergent</u> ?

3) <u>Under what conditions are the formal or convergent solutions unique</u> ?

§ 1. - <u>The case</u> $p_1 = p_2 = 0$.

In this case the system (1) has no singularities and the classical Frobenius theorem gives us the answers to the questions 1,2,3 .

§ 2. - The case $p_1 = p_2 = 1$.

The system (1) can be put in the following form :

$$(1')\quad \begin{cases} x_1 \dfrac{\partial y}{\partial x_1} = f_o^1(x_1,x_2) + A^1(x_1,x_2)y + R^1(x_1,x_2,y) \\ x_2 \dfrac{\partial y}{\partial x_2} = f_o^2(x_1,x_2) + A^2(x_1,x_2)y + R^2(x_1,x_2,y) \end{cases}$$

where for $i = 1,2$:

$A^i(x_1,x_2)$ is an $n \times n$ - matrix which is holomorphic at the origin of C^2.

$R^i(x_1,x_2,y)$ is a holomorphic vector of order greater or egal to two in y.

And we have :

THEOREM 1. - Every formal solution of (1) is convergent.

THEOREM 2. - If one of the matrices $A^i(0,0)$ has no non negative entire eigenvalue then the system (1) has one and only one formal solution.

Idea of the proof of theorem 1.

Assume that $u = \sum_{r_1+r_2>o} a_{r_1r_2} x_1^{r_1} x_2^{r_2}$ is a formal solution of the system (1).

Write u in the form

$$u = \sum_{r_1=o}^{+\infty} a_{r_1}(x_2)x_1^{r_1}$$

where

$$a_{r_1}(x_2) = \sum_{r_2=o}^{+\infty} a_{r_1r_2} x_2^{r_2}.$$

The formal serie u is a formal solution of the ordinary differential system

$$x_2 \frac{dy}{dx_2} = f^2(x_1,x_2,y)$$

where x_1 is considered as a parameter.

This means that we have formally

$$x_2\left[\sum_{r_1=0}^{+\infty} \frac{d}{dx_2}(a_{r_1}(x_2))x_1^{r_1}\right] = f^2(x_1,x_2,\sum_{r_1=0}^{+\infty} a_{r_1}(x_2)x_1^{r_1})$$

in particular

$$x_2 \frac{d}{dx_2}(a_o(x_2)) = f^2(0,x_2,a_o(x_2)) .$$

And $a_o(x_2)$ is a formal solution of an ordinary differential system of the form

$$x \frac{dy}{dx} = f(x,y)$$

where f is holomorphic at the origin and satisfies $f(0,0) = 0$.

But as it is well known in the theory of ordinary differential systems in the complex domain this implies that $a_o(x_2)$ is convergent in a neighborhood of the origin.

Setting $y = v + a_o(x_2)$, we find that the formal power serie

$$\sum_{m=1}^{+\infty} a_m(x_2)x_1^m$$

is a formal solution of a system of the form

$$x_2 \frac{dv}{dx_2} = h_o(x_1,x_2) + B(x_1,x_2)v + O(v^2) .$$

By identification we have for all $m = 1,2,\ldots$

$$x_2 \frac{d}{dx_2} a_m(x_2) = B(0,x_2)a_m(x_2) + Q_m(x_2)$$

where Q_m is determined when all the $a_p(x_2)$ are known for all $p < m$.

This means that each $a_m(x_2)$ is a formal solution of an ordinary linear system of the form

$$x_2 \frac{dy}{dx_2} = B(0,x_2)y + Q_m(x_2)$$

and this implies that it does exist a neighborhood of the origin in $C(x_2)$ in which all the power series $a_m(x_2)$ are convergent.

We have proved that the coefficients $a_{r_1}(x_2)$ of the formal solution

$$u = \sum_{r_1=0}^{+\infty} a_{r_1}(x_2)x_1^{r_1}$$

of our original pfaffian system are all convergent, but now the convergence of this power serie in x_1 follows from

LEMMA 1. - Consider a differential system including a parameter x_2

(E) $$x_1 \frac{dy}{dx_1} = f(x_1,x_2,y)$$

where f is holomorphic near the origin of $C^2(x_1,x_2) \times C^n(y)$ then every formal solution

$$\sum_{m=0}^{+\infty} a_m(x_2)x_1^m$$

of (E) where

1) for all m, a_m is holomorphic in $\{x_2 \in C | \ |x_2| < \delta\}$

2) $a_o(0) = 0$

is convergent.

The proof of this lemma is very technical and make use of the integral equation associated to the differential equation.

Remark : The result of this lemma is not true if the $a_m(x_2)$ are only formal in fact consider the equation

$$\frac{du}{dx_1} = (1+x_2)u$$

the solution are given by

$$u = C(x_2)\exp[(1+x_2)x_1]$$

where $C(x_2)$ can be an arbitrary formal power serie.

But we have the following result.

PROPOSITION 1. - *If the equation (E) has a formal solution of the form*

$$\sum_{m=0}^{+\infty} a_m(x_2)x_1^m$$

where the a_m's are formal power series in x_2 and $a_0(0) = 0$ then (E) admits a formal solution of the same form where the $a_m(x_2)$ are convergent in a neighborhood of the origin of $\mathbb{C}$.

The proof of this result makes use of a theorem due to Artin concerning the convergence of formal solutions of ordinary equations.

Idea of the proof of theorem 2.

The proof of theorem 2 is easy and follows by putting in the pfaffian system (1) a formal power serie

$$u = \sum_{r_1+r_2>0} a_{r_1 r_2} x_1^{r_1} x_2^{r_2}$$

and proceeding by identification.

§ 3. The case $p_1 = 1$ and $p_2 > 1$.

Now we have to study pfaffian systems of the form

$$(1)\qquad \begin{cases} x_1 \dfrac{\partial y}{\partial x_1} = f_0^1(x_1,x_2) + A^1(x_1,x_2)y + R^1(x_1,x_2,y) \\ x_2^{p_2} \dfrac{\partial y}{\partial x_2} = f_0^2(x_1,x_2) + A^2(x_1,x_2)y + R^2(x_1,x_2,y) \end{cases}$$

where $p_2 > 1$ and A^1 and A^2 are $n \times n$-matrices and R^1 and R^2 are

$n \times 1$ - matrices which are of order greater or egal to two in y. We have

THEOREM 3. - _If_ $A^1(0,0)$ _has no non negative entire eigenvalue or if_ $A^2(0,0)$ _is regular then_ (1) _has one and only one formal solution_.

THEOREM 4. - _If_ $A^1(0,0)$ _has no non negative entire eigenvalue then the system_ _(1) has one and only one holomorphic solution which is zero at the origin_.

The proof of theorem 3 is easy by the classical procedure of identification.

Proof of theorem 4.

The hypothesis on $A^1(0,0)$ implies that the system (1) has a unique formal solution $\hat{\varphi}$ of the form

$$\hat{\varphi} = \sum_{r_1+r_2>0} \varphi_{r_1 r_2} x_1^{r_1} x_2^{r_2} .$$

We have to prove that this formal solution is convergent.

If we write this formal solution in the form :

$$\hat{\varphi} = \sum_{r_1=0}^{+\infty} \varphi_{r_1}(x_2) x_1^{r_1}$$

we have formally

$$x_1 \frac{\partial \hat{\varphi}}{\partial x_1} = f^1(x_1, x_2, \hat{\varphi})$$

$$= f_0^1(x_1,x_2) + A^2(x_1,x_2)\hat{\varphi} + O((\hat{\varphi})^2)$$

and the coefficients $\varphi_{r_1}(x_2)$ are given by the equations :

$$f^1(0,x_2,\varphi_0(x_2)) = 0$$

$$\left[\frac{\partial f^1}{\partial y}(0,x_2,\varphi_0(x_2)) - I\right]\varphi_1(x_2) = H_1(x_2)$$

$$\left[\frac{\partial f^1}{\partial y}(0,x_2,\varphi_0(x_2)) - 2I\right]\varphi_2(x_2) = H_2(x_2)$$

- -
- -

$$\left[\frac{\partial f^1}{\partial y}(0,x_2,\varphi_o(x_2)) - r_1 I\right]\varphi_{r_2}(x_2) = H_{r_1}(x_2)$$

- -

- -

where $H_{r_1}(x_2)$ is known when we know $\varphi_r(x_2)$ for all $r < r_1$.

The implicit function theorem gives you $\varphi_o(x_2)$ as a convergent power serie and moreover it does exist a disk centered at the origin of the x_2-plane such that for all $r_1 > 0$ the matrix

$$\frac{\partial f^1}{\partial y}(0,x_2,\varphi_o(x_2)) - r_1 I$$

is invertible. Then by induction on r_1 it does exist a neighborhood of the origin in the x_2-plane in which all the power series $\varphi_{r_1}(x_2)$ converge.

Now we have a formal serie

$$\sum_{r_1=o}^{+\infty} \varphi_{r_1}(x_2)x_1^{r_1}$$

with holomorphic coefficients which is solution of

$$x_1 \frac{\partial y}{\partial x_1} = f^1(x_1,x_2,y) .$$

By lemma 1 this power serie is convergent and the theorem 4 is proved.

§ 4. The case $p_1 > 1$, $p_2 > 1$.

We are considering now a completely integrable system of the form

$$(1) \qquad dy = \frac{f^1(x_1,x_2,y)}{x_1^{p_1}} dx_1 + \frac{f^2(x_1,x_2,y)}{x_2^{p_2}} dx_2$$

where $p_1 > 1$ and $p_2 > 1$.

Whe have

THEOREM 5. - *If both matrices*

$$A^i(0,0) = \frac{\partial f^i}{\partial y}(0,0) \quad \text{for} \quad i = 1,2$$

<u>are regular then the system</u> (1) <u>has one and only one holomorphic solution</u> φ <u>near the origin and such that</u> $\varphi(0,0) = 0$.

<u>Proof</u>. - Assume that f^1 and f^2 are holomorphic in $U^1(x_1) \times U^2(x_2) \times V(y)$ where

$$U^i(x_i) = \{x_i \in C \mid 0 \le |x_i| < r\}$$

$$V(y) = \{y \in C^n \mid 0 \le |y_i| < \rho_i\} .$$

We have also $f^i(0,0,0) = 0$ for $i = 1,2$.

The system (1) can also be written in the form

$$(1_1) \qquad x_1^{p_1} \frac{\partial y}{\partial x_1} = f^1(x_1,x_2,y)$$

$$(1_2) \qquad x_2^{p_2} \frac{\partial y}{\partial x_2} = f^2(x_1,x_2,y) .$$

Consider the system

$$(1_1) \qquad x_1^{p_1} \frac{dy}{dx_1} = f^1(x_1,x_2,y)$$

where x_2 is considered as a parameter.

From a theorem due to Y. Sibuya it follows that it does exist :

1) a covering $S_1^1, S_2^1, \ldots, S_{N_1}^1$ of $U^1 - \{0\}$ where each S_h^1 is a sector of opening greater then $\frac{\pi}{p_1 - 1}$

2) for each $h = 1,2\ldots N_1$, a solution $\varphi_h(x_1,x_2)$ of (1_1) holomorphic and defined in $S_h^1 \times U^2$ such that φ_h has an asymptotic expansion of the form

$$\sum_{k=0}^{+\infty} \varphi_{h,k}(x_2) x_1^k$$

valid in S_h^1 uniformely in U^2 where moreover each $\varphi_{h,k}$ is holomorphic in

U^2 and $\varphi_{h,o}(0) = 0$.

Now we are going to prove that for each $h = 1,2,\dots N_1$ we have

$$x_2^{P_2} \frac{\partial \varphi_h}{\partial x_2} = f^2(x_1,x_2,\varphi_h(x_1,x_2))$$

on $S_h^1 \times U^2$. We have

$$x_1^{P_1} \frac{\partial}{\partial x_1}(x_2^{P_2} \frac{\partial \varphi_h}{\partial x_2}) = x_2^{P_2} \frac{\partial}{\partial x_2}(x_1^{P_1} \frac{\partial \varphi_h}{\partial x_1})$$

on $S_h^1 \times U^2$ for all $h = 1,2,\dots N_1$, and using the equation (1_1).

$$x_1^{P_1} \frac{\partial}{\partial x_1}(x_2^{P_2} \frac{\partial \varphi_h}{\partial x_2}) = x_2^{P_2} \frac{\partial f^1}{\partial x_2}(x_1,x_2,\varphi_h) + \frac{\partial f^1}{\partial y}(x_1,x_2,\varphi)x_2^{P_2} \frac{\partial \varphi_h}{\partial x_2}$$

but using the complete integrability condition you will find easily that :

$$x_1^{P_1} \frac{\partial}{\partial x_1}(x_2^{P_2} \frac{\partial \varphi_h}{\partial x_2} - f^2(x_1,x_2,\varphi_h)) = \frac{\partial f^1}{\partial y}(x_1,x_2,\varphi_h)\left[x_2^{P_2} \frac{\partial \varphi}{\partial x_2} - f^2(x_1,x_2,\varphi_h)\right]$$

ou $S_h^1 \times U^2$ for all $h = 1,2,\dots N_1$.

This means that the function

$$v_h(x_1,x_2) = x_2^{P_2+1} \frac{\partial \varphi_h}{\partial x_2} - f^2(x_1,x_2,\varphi_h)$$

is on $S_h^1 \times U^2$ a solution of the linear system

$$\text{(s)} \qquad x_1^{P_1} \frac{du}{dx_1} = \frac{\partial f^1}{\partial y}(x_1,x_2,\varphi_h(x_1,x_2))\, u$$

where

$$\frac{\partial f^1}{\partial y}(x_1,x_2,\varphi_h(x_1,x_2)) = A^1(x_1,x_2) + O(\varphi_h)$$

with $A^1(0,0)$ regular.

For each h , v_h has an asymptotic expansion in x_1 on S_h^1 uniformely valid in U^2 and this implies that v_h is bounded near the origin.

But the opening of the sector S_h^1 being greater then $\frac{\pi}{P_1-1}$ the linear system (s) has no other bounded solution then the trivial solution.

This implies that we have for all $h = 1,2,\dots N_1$, $v_h = 0$ on $S_h^1 \times U^2$ and it follows that for all h , $\varphi_h(x_1,x_2)$ is a solution of

$$x_2^{p_2} \frac{\partial y}{\partial x_2} = f^2(x_1,x_2,y) \quad \text{on} \quad S_h^1 \times U^2 .$$

In the same way as before it also does exist

1) a covering $S_1^2, S_2^2, \dots S_{N_2}^2$ of $U^2 - \{0\}$ where each S_h^1 is a sector of opening greater then $\frac{\pi}{p_2 - 1}$

2) for each $k = 1,2,\dots N_2$, a solution $\psi_k(x_1,x_2)$ of (1_2) holomorphic and defined in $S_k^2 \times U^1$ such that

$$\psi_k \underset{u.U^1}{\overset{S_k^2}{\sim}} \sum_{i=0}^{+\infty} \psi_{k_i}(x_1) x_2^i$$

with ψ_{k_i} holomorphic in U^1 and $\psi_{ko}(0) = 0$.

And moreover each ψ_k is also a solution of (1_1) on $S_k^2 \times U^1$.

Another result due also to Y. Sibuya and concerning ordinary differential systems with parameters implies then that we have for all h and k .

$$\varphi_h(x_1,x_2) = \psi_k(x_1,x_2) \quad \text{on} \quad S_h^1 \times S_k^2 .$$

Now consider φ_1 which is holomorphic in $S_h^1 \times U^2$, for all $k = 1,2,\dots N_2$

$$\varphi_1 = \psi_k \quad \text{on} \quad S_1^1 \times S_k^2 .$$

But ψ_k is holomorphic in $U^1 \times S_k^2$ and so is the analytic continuation of φ_1 to $U^1 \times S_k^2$. This being true for all k we have then

$$\psi_h = \psi_k \quad \text{on} \quad U^1 \times S_h^2 \cap S_k^2$$

and it does exist a function ψ holomorphic on $U^1 \times U^2 - \{0\}$ such that for all k

$$\psi | U^1 \times S_k^2 = \psi_k .$$

But for all k , ψ_k admits an asymptotic expansion in $U^1 \times S_k^2$ so ψ_k is bounded near the origin of $C(x_2)$ and ψ can be extented to a holomorphic function in $U^1 \times U^2$.

And this function ψ is the holomorphic solution of our given theorem.

The unicity is easy to prove, by proving unicity of formal power series solutions of our system.

§ 5. Application to linear connexions.

To avoid the complications coming from the abstract notion of linear connexions, we will restrict ourself to linear pfaffian system which gives the expression of our connexion in a basis.

Consider a completely integrable linear pfaffian system of the form

$$(1) \qquad dy = \left(\frac{A(x_1,x_2)}{x_1^{p_1}} dx_1 + \frac{B(x_1,x_2)}{x_2^{p_2}} dx_2\right) y$$

where A and B are $n \times n$ - matrices which are holomorphic near the origin of C^2 . Assume $p_1 > 1$ and $p_2 > 1$ it is easy to see what we have to change in the other cases. Then we have.

THEOREM 6. (Formal reduction theorem). - *If you have*

$$A(0,0) = \begin{pmatrix} A_o^1 & 0 \\ 0 & A_o^2 \end{pmatrix} \quad B(0,0) = \begin{pmatrix} B_o^1 & 0 \\ 0 & B_o^2 \end{pmatrix}$$

where one of the pairs of matrices (A_o^1, A_2^2) , $B_o^1, B_o^2)$ *have no common eigenvalues then it does exist a formal transformation*

$$T = \begin{pmatrix} I & T^{12} \\ T^{21} & I \end{pmatrix}$$

which put the system (1) *by setting* $y = Tz$ *in the form*

$$dz = \left(\frac{\begin{pmatrix} \mathcal{A}^1 & 0 \\ 0 & \mathcal{A}^2 \end{pmatrix}}{x_1^{p_1}} dx_1 + \frac{\begin{pmatrix} \mathcal{B}^1 & 0 \\ 0 & \mathcal{B}^2 \end{pmatrix}}{x_2^{p_2}} dx_2 \right) z$$

with

$$\mathcal{A}^i(0,0) = A_o^i \qquad \mathcal{B}^i(0,0) = B_o^i \qquad i = 1,2 \, .$$

<u>Idea of the proof.</u>

Write
$$A = \begin{pmatrix} A^{11} & A^{12} \\ A^{21} & A^{22} \end{pmatrix} \qquad B = \begin{pmatrix} B^{11} & B^{12} \\ B^{21} & B^{22} \end{pmatrix}$$

and try by identification to find T. It is easy to see that T^{12} and T^{21} are given by the following systems.

$$\begin{cases} x_1^{p_1} \dfrac{\partial T^{12}}{\partial x_1} = A^{12} + A^{11}T^{12} - T^{12}A^{22} - T^{12}A^{21}T^{12} \\[2ex] x_2^{p_2} \dfrac{\partial T^{12}}{\partial x_2} = B^{12} + B^{11}T^{12} - T^{12}B^{22} - T^{12}B^{21}T^{12} \end{cases}$$

$$\begin{cases} x_1^{p_1} \dfrac{\partial T^{21}}{\partial x_1} = A^{21} + A^{22}T^{21} - T^{21}A^{11} - T^{21}A^{12}T^{21} \\[2ex] x_2^{p_2} \dfrac{\partial T^{21}}{\partial x_2} = B^{21} + B^{22}T^{21} - T^{21}B^{11} - T^{21}B^{12}T^{21} \, . \end{cases}$$

Proceeding by formal identification you can proof that it does exist formal solution for this systems.

THEOREM 7. - <u>If both pairs of matrices</u> (A_o^1, A_2^2) <u>and</u> (B_o^1, B_o^2) <u>have no common eigenvalues then the formal transformation</u> T <u>is convergent</u>.

<u>Proof</u> : Convergence of T^{12} for example. Order arbitrary the elements of the matrix T^{12} and put them together to make a one column matrix $\mathcal{T}^{12}$. Then this matrix is a formal solution of a pfaffian system of the form

$$x_1^{p_1} \frac{\partial y}{\partial x_1} = f_o^1(x_1,x_2) + C_1(x_1,x_2)z + f_2^1(x_1,x_2,z)$$

$$x_2^{p_2} \frac{\partial y}{\partial x_2} = f_o^2(x_1,x_2) + C_2(x_1,x_2)z + f_2^2(x_1,x_2,z) \text{ .}$$

The hypothesis on the pairs $(A_o^1,A_o^2),(B_o^1,B_o^2)$ imply then that both matrices $C_1(0,0)$ and $C_2(0,0)$ are regular and by theorem 5 our formal solution is convergent.

As an application let us solve a pfaffian system of the form

$$(1)\begin{cases} x_1^{p_1} \dfrac{\partial y}{\partial x_1} = A^1(x_1,x_2)y \\ x_2^{p_2} \dfrac{\partial y}{\partial x_2} = A^2(x_1,x_2)y \end{cases}$$

in the case where $A^1(0,0)$ and $A^2(0,0)$ have both district eigenvalues. The complete integrability condition implies that $[A^1(0,0),A^2(0,0)] = 0$ and it is possible by constant transformation to put them in the diagonal form. But then theorem 7 says that it does exist a convergent transformation $T(x_1,x_2)$ which put the hole system in the diagonal form

$$\begin{cases} x_1^{p_1} \dfrac{\partial z}{\partial x_1} = \mathcal{A}^1(x_1,x_2)y \\ x_2^{p_2} \dfrac{\partial z}{\partial x_2} = \mathcal{A}^2(x_1,x_2)y \end{cases}$$

where

$$\mathcal{A}^1(x_1,x_2) = \begin{pmatrix} a_1^1(x_1,x_2) & & 0 \\ & \ddots & \\ 0 & & a_n^1(x_1,x_2) \end{pmatrix}$$

$$\mathcal{A}^2(x_1,x_2) = \begin{pmatrix} a_1^2(x_1,x_2) & & 0 \\ & \ddots & \\ 0 & & a_n^2(x_1,x_2) \end{pmatrix}$$

This means that we have now to solve n-scalar systems

$$\begin{cases} x_1^{p_1} \dfrac{\partial z_i}{\partial x_1} = a_i^1(x_1,x_2) z_i \\ x_2^{p_2} \dfrac{\partial z_i}{\partial x_2} = a_i^2(x_1,x_2) z_i \end{cases}$$

which are also completely integrable. But this can be done very easily. At the end we find that the system (1) has a fondamental matrix of solutions of the form

$$U(x_1,x_2) x_1^{\Lambda_1} x_2^{\Lambda_2} \exp P_1(\frac{1}{x_1}) \exp P_2(\frac{1}{x_2})$$

where : U is a $n \times n$-matrix holomorphic at the origin, Λ_1 and Λ_2 are constant diagonales matrices, P_1 and P_2 are polynomials in $\frac{1}{x_1}$ and $\frac{1}{x_2}$ with matrix coefficients.

INSTITUT DE RECHERCHE MATHEMATIQUE AVANCEE
Laboratoire Associé au C.N.R.S.
7, rue René Descartes
67084 - STRASBOURG Cedex

SCATTERING PROBLEMS IN DIFFERENTIAL GEOMETRY

by Herman Gluck and David Singer

We study here the scattering of geodesic fields on a Riemannian manifold, and describe an obstruction whose vanishing is necessary and sufficient for deforming the metric so as to deflect one geodesic field into another. This is analogous to the inverse scattering problems of physics and to the design of lenses with preassigned distortion.

As applications of these techniques, we state without proof the following.

THEOREM A. *Every smooth manifold of dimension ≥ 2 can be given a Riemannian metric with a nontriangulable cut locus.*

THEOREM B. *There exists a smooth convex closed surface of revolution in three-space having nontriangulable cut loci with respect to a nonempty open set of points.*

These two theorems may be viewed as evidence that a topological space C can appear as a cut locus on a manifold in the Riemannian sense if and only if it can appear there in a corresponding topological sense.

Announcements of these results have appeared in [2] and [7]. Proofs of Theorems A and B, as well as alternative proofs of the scattering theorems, can be found in [3] and [8]. We thank Professors Eugenio Calabi, Emil Grosswald, Richard Hamilton, Albert Heins, Jürgen Moser, Albert Nijenhuis, Oscar Rothaus, Robert Strichartz and David Tischler for many insights gained in conversation with them. We also thank the National Science Foundation for support.

TABLE OF CONTENTS

1. The scattering theorems
2. Two handy lemmas
3. Proof of necessity for theorem 1
4. Proof of sufficiency for theorem 1
5. Integrable geodesic fields
6. Relative scattering theorems
7. Surfaces of revolution
8. Geodesics on surfaces of revolution
9. Deflection of geodesics by a bump
10. The inverse scattering problem
11. Proof of theorem 10.1

1. THE SCATTERING THEOREMS

Consider a Riemannian manifold N and a compact separating hypersurface M. Two geodesic fields G and G' are given on N (or at least on a neighborhood of M), with both crossing M transversely. This is shown as "BEFORE" in the figure on the next page, while as "AFTER" we show the metric on N deformed in a neighborhood U of M so as to produce a field G" of geodesics connecting G below M to G' above. The region U, with its deformed metric, can be thought of as a "lens" which "focuses" G into G'.

As shown in the "AFTER" figure, the geodesic of G which crosses M at x has been connected by a geodesic of G" to the geodesic of G' which crosses M at $\phi(x)$, thus defining a map $\phi: M \to M$. We say that G has been connected to G' according to ϕ.

PROBLEM: Given G, G' and ϕ, when is this possible?

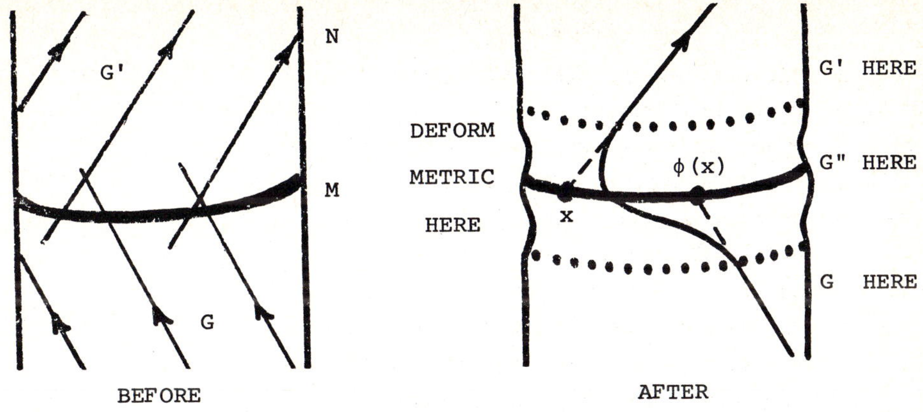

FIGURE 1

Let V denote the unit vector field tangent to G and ω the dual one-form on N defined by $\omega(U) = U \cdot V$. Similarly define V' and ω' relative to G', orienting V' so that G' crosses M from the same side as does G.

THEOREM 1. <u>Necessary and sufficient conditions for being able to deform the metric on N near M so as to produce a field of geodesics connecting G to G' according to ϕ are:</u>

(a) <u>The map ϕ: M $\to$ M is a diffeomorphism, concordant to the identity; and</u>

(b) <u>The one-form on M, $\omega|_M - \phi^*(\omega'|_M)$ is exact.</u>

We use the phrase "near M" to mean "within an arbitrarily small neighborhood of M".

In studying these deformation problems, the special case of "integrable" geodesic fields occurs repeatedly and naturally. A field G of geodesics on N^n is <u>integrable</u> if the orthogonal (n-1)plane distribution is integrable in the usual sense of being tangent to a foliation (which is then a "Riemannian foliation" [6]). In this case

$d\omega = 0$, so that ω determines a cohomology class $[\omega]$ in $H^1(N;R)$. If $i: M \to N$ is the inclusion, we may also consider the restriction $i^*[\omega] = [\omega|_M]$ in $H^1(M;R)$.

Then Theorem 1 simplifies to

THEOREM 2. *Necessary and sufficient conditions for being able to deform the metric on N near M so as to produce a field of geodesics connecting the integrable geodesic fields G to G' according to ϕ are:*

(a) *The map ϕ is a diffeomorphism, concordant to the identity; and*

(b) $[\omega|_M] = [\omega'|_M]$ *in* $H^1(M;R)$.

A simple illustration of Theorem 2: let $N = S^1 \times R^1$ and $M = S^1 \times 0$. Let G_α be the rising family of helices making an angle α with the vertical, $-\pi/2 < \alpha < \pi/2$. Then we easily conclude from this theorem:

(1) Any rising field of geodesics on a neighborhood of $S^1 \times 0$ can be connected to some G_α by deformation of metric near $S^1 \times 0$.

(2) But G_α can *not* be so connected to G_β if $\alpha \neq \beta$.

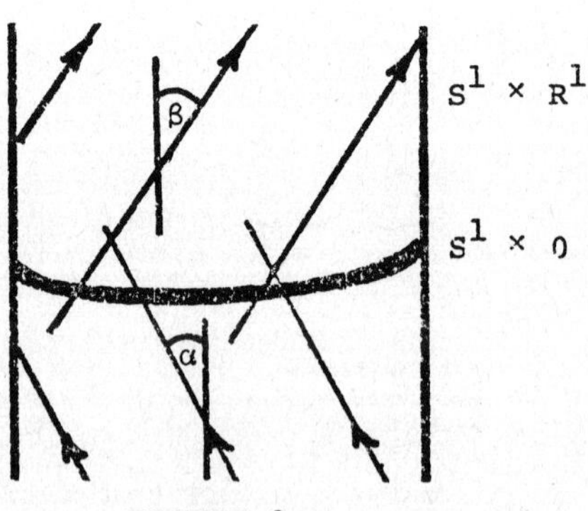

FIGURE 2

Note that in Theorem 2, conditions (a) and (b) are now disconnected, so that if the required deformation exists for one ϕ concordant to the identity, then it exists for any such ϕ. If we weaken Theorem 2 by relinquishing the right to choose ϕ, then we can strengthen it in another direction.

THEOREM 3. *If condition (b) of Theorem 2 is satisfied, then for some ϕ the deformed metric on N can be chosen pointwise conformal to the original metric.*

Recall that in optics, pointwise conformality is the mathematical counterpart of an isotropic lens.

Theorems 1, 2 and 3 admit relative versions, in which the separating hypersurface M is only required to be closed, not compact. In that case the geodesic fields G and G' agree outside some compact subset of M, in a neighborhood of which the metric is to be deformed so as to connect G to G'. What happens here is that conditions (a) and (b), suitably relativized, are still necessary but no longer sufficient. A third condition appears, which is easiest to appreciate once we have gone through the proof in the absolute case. So we delay this material until section 6.

The equivariant case, dealing with scattering problems on a surface of revolution, is taken up in sections 7 through 11.

2. TWO HANDY LEMMAS

We will make repeated use of the following simple facts. The first tells how to recognize when a field of curves on a Riemannian manifold consists of geodesics.

LEMMA 2.1. *Let G be a field of curves on a Riemannian manifold N^n and let V, $\{g_t\}$ and ω be the corresponding unit tangent vector field, flow and dual one-form. Let σ be the (n-1) plane distribution on N orthogonal to G. Then the following conditions are equivalent:*

(1) *The curves of G are geodesics.*

(2) *The one-form ω is invariant under the flow $\{g_t\}$, or in other words, $L_V\omega = 0$, where L_V denotes Lie differentiation with respect to V.*

(3) *The orthogonal distribution σ is invariant under the flow.*

Conditions (2) and (3) are easily seen to be equivalent, and it is a standard result of differential geometry that they are a consequence of condition (1). It is a straightforward exercise to check that such a proof is reversible.

LEMMA 2.2. *Let G be a field of geodesics on a Riemannian manifold N, and let V, $\{g_t\}$ and ω be as above. Let C: $[0,1] \to N$ be a smooth curve in N, and t: $[0,1] \to R^1$ a smooth function such that the curve $C'(s) = g_{t(s)}(C(s))$ is defined. Then*

$$\int_{C'} \omega = \int_C \omega + t(1) - t(0).$$

In particular, if t(0) = t(1) (a natural condition if C is a closed curve), then $\int_{C'} \omega = \int_C \omega$.

Using the notation $\frac{dC}{ds}$ for the tangent vector $C_*(\frac{\partial}{\partial s})$ to the curve C, and likewise for C', we have by the "Chain Rule"

$$\frac{dC'}{ds} = g_{t(s)_*}(\frac{dC}{ds}) + \frac{dt}{ds} V_{C'(s)} . \tag{2.3}$$

Hence

$$\int_{C'} \omega = \int_0^1 \omega(\frac{dC'}{ds})\, ds = \int_0^1 \omega(g_{t(s)_*} \frac{dC}{ds})\, ds + \int_0^1 \omega(\frac{dt}{ds} V_{C'(s)})\, ds$$

$$= \int_0^1 \omega(\frac{dC}{ds})\, ds + \int_0^1 \frac{dt}{ds}\, ds ,$$

where the first integral has been simplified by applying Lemma 2.1, and the second simplified because $\omega(V) = 1$.

Therefore

$$\int_{C'} \omega = \int_C \omega + t(1) - t(0) ,$$

proving the lemma.

3. PROOF OF NECESSITY FOR THEOREM 1

We start with a Riemannian manifold N and a compact hypersurface M which separates it into components A and A'. On N we have geodesic fields G and G', both crossing M transversely. Furthermore, we are given a map $\phi: M \to M$.

We assume that the metric on N can be deformed inside an arbitrarily small neighborhood U of M, and a geodesic field G" with respect to the new metric found which agrees with G on A - U, agrees with G' on A' - U, and connects G to G' according to ϕ. We want to verify conditions (a) and (b) of Theorem 1.

First we must show that ϕ is a diffeomorphism of M, concordant to the identity. Recall that ϕ is <u>concordant</u> to the identity if there exists a diffeomorphism $F: M \times [0,1] \to M \times [0,1]$ such that $F(x,0) = (x,0)$ and $F(x,1) = (\phi(x),1)$.

Let $\{g_t\}$ and $\{g'_t\}$ be the flows associated with the unit vector fields V and V' tangent to G and G'. By compactness of M, there is a number $\varepsilon > 0$ such that $g_t(x)$ and $g'_t(x)$ are defined for all x in M and all t in $[-\varepsilon,\varepsilon]$. The map $x \mapsto g_{-\varepsilon}(x)$ pushes M down along the geodesics of G to the submanifold M_- of N, while the map $x \mapsto g'_{\varepsilon}(x)$ pushes M up along the geodesics of G' to M_+. Since the neighborhood U of M in N is arbitrarily small, we may assume that it lies in the region between M_- and M_+.

Notice that if we push M down to M_- along the geodesics of G, then up to M_+ along the geodesics of G" and finally back down to M along G', then the composite map of $M \to M$ is just ϕ. This makes it intuitively clear that ϕ is a diffeomorphism of M, concordant to the identity. The verification is a routine exercise using the fundamental technical lemma about concordance found in Munkres [5, p. 59] .

To check condition (b), we must show that the one-form $\omega|_M - \phi^*(\omega'|_M)$ is exact, and we do this by showing that its integral over any closed curve C in M vanishes. Equivalently, we must show that $\int_C \omega = \int_{C'} \omega'$, where $C' = \phi(C)$.

Let C_- be the curve on M_- obtained by pushing C down to M_- along the geodesics of G. Push C_- up along G" to get the curve C_+ on M_+, and push this down along G' to get C' on M again.

Now,

$$\int_C \omega = \int_{C_-} \omega \qquad \text{by Lemma 2.2, pushing down along G}$$

$$= \int_{C_-} \omega'' \qquad \text{because } \omega = \omega'' \text{ there}$$

$$= \int_{C_+} \omega'' \qquad \text{pushing up along } G''$$

$$= \int_{C_+} \omega' \qquad \text{because } \omega'' = \omega' \text{ there}$$

$$= \int_{C'} \omega' \qquad \text{pushing down along } G' \quad ,$$

completing the proof of necessity for Theorem 1.

4. PROOF OF SUFFICIENCY FOR THEOREM 1

We start as before with the Riemannian manifold N and the compact hypersurface M separating it into components A and A'. On N we have geodesic fields G and G', both crossing M transversely. This time we are given a diffeomorphism $\phi: M \to M$ which is concordant to the identity, such that the one-form $\omega|_M - \phi^*(\omega'|_M)$ is exact. We must modify the metric within a preassigned neighborhood U of M and exhibit a field G" of geodesics in this new metric which connects G to G' according to ϕ.

The proof begins by using the concordance between ϕ and 1_M to connect G and G' by a family G" of curves which will be geodesics in the yet-to-be-chosen metric on N. Then, using the exactness of $\omega|_M - \phi^*(\omega'|_M)$, we interpolate between ω and ω' a one-form ω'' and define a tangent vector field V" to G" so that $\omega''(V'') = 1$. Using this data, we define the new metric on N and check that the curves of G" become geodesics.

STEP 1. Construction of G".

Let $\{g_t\}$ and $\{g'_t\}$ be the flows associated with the unit vector fields V and V' tangent to G and G'. As in the proof of necessity, we invoke the compactness of M to find a number $\varepsilon > 0$ such that $g_t(x)$ and $g'_t(x)$ are defined for all x in M and all t in $[-\varepsilon,\varepsilon]$.

Using these flows, we define two tubular neighborhoods of M in N,

$$H \text{ and } H' : M \times [-\varepsilon,\varepsilon] \to N$$

by

$$H(x,t) = g_t(x) \qquad \text{and} \qquad H'(x,t) = g'_t(x).$$

We choose ε small enough so that both these neighborhoods lie within the preassigned neighborhood U of M.

Then, using Munkres' lemma [5, p.59], H and H' can be pieced together to yield

$$\overline{H} : M \times [-\varepsilon,\varepsilon] \to N$$

such that

$$\overline{H}(x,t) = \begin{cases} H(x,t) & \text{for } -\varepsilon \leqq t \leqq -\varepsilon/2 \\ H'(x,t) & \text{for } \varepsilon/2 \leqq t \leqq \varepsilon . \end{cases}$$

Since ϕ is concordant to the identity, we can use Munkres' lemma once more to produce a diffeomorphism

$$F : M \times [-\varepsilon,\varepsilon] \to M \times [-\varepsilon,\varepsilon]$$

such that

$$F(x,t) = \begin{cases} (x,t) & \text{for } -\varepsilon \leqq t \leqq -\varepsilon/2 \\ (\phi(x),t) & \text{for } \varepsilon/2 \leqq t \leqq \varepsilon . \end{cases}$$

Next we extend ϕ to a diffeomorphism

$$\Phi : \text{Image of } H \to \text{Image of } H'$$

by

$$\Phi(g_t(x)) = g'_t(\phi(x)) ,$$

that is,

$$\Phi = H' \circ (\phi \times 1) \circ H^{-1} .$$

Thus Φ takes an arc of the geodesic of G through x to an arc of the geodesic of G' through $\phi(x)$.

Finally we define a diffeomorphism

$$\Psi : \text{Image of } H \to \text{Image of } \overline{H}$$

by

$$\Psi = \bar{H} \circ F \circ H^{-1} .$$

Then Ψ = identity on $H(M \times [-\varepsilon,-\varepsilon/2])$ and $\Psi = \Phi$ on $H(M \times [\varepsilon/2,\varepsilon])$.

So Ψ takes an arc of the geodesic of G through x to a curve which begins coincident with that arc and ends coincident with an arc of the geodesic of G' through $\phi(x)$. In this way, Ψ provides the family G" of curves connecting G to G' via ϕ.

STEP 2. Construction of ω''.

We know by hypothesis that $\omega|_M - \phi^*(\omega'|_M)$ is exact, so let $f: M \to R$ be a smooth function such that $\omega|_M - \phi^*(\omega'|_M) = df$. Now extend f to a function $f: \text{Im } H \to R$ by defining $f(H(x,t)) = f(x)$; in other words, f is constant along the geodesics of G. Then we have

(4.1) $\omega - \Phi^*(\omega')$ is exact on Im H and equals df there.

To see this, let C be a curve in Im H, running from H(x,s) to H(y,t). Then

$$\int_C \omega - \Phi^*(\omega') = \int_C \omega - \int_{\Phi(C)} \omega' .$$

Push C along the geodesics of G to the curve C_o on M, running from x to y. Then

$$\int_C \omega = \int_{C_o} \omega + t - s$$

by Lemma 2.2, and similarly

$$\int_{\Phi(C)} \omega' = \int_{\phi(C_o)} \omega' + t - s ,$$

by pushing $\Phi(C)$ along the geodesics of G' to $\phi(C_o)$ on M. Therefore,

$$\int_C \omega - \Phi^*(\omega') = \int_C \omega - \int_{\Phi(C)} \omega' = \int_{C_o} \omega - \int_{\phi(C_o)} \omega'$$

$$= \int_{C_o} \omega|_M - \phi^*(\omega'|_M) = f(y) - f(x)$$

$$= f(H(y,t)) - f(H(x,s)) .$$

It follows that $\omega - \Phi^*(\omega') = df$ on Im H.

Next, let ρ: Im H $\to$ R be a smooth function depending only on the t coordinate, which vanishes for $-\varepsilon \leqq t \leqq -\varepsilon/2$, equals 1 for $\varepsilon/2 \leqq t \leqq \varepsilon$, and is monotonically increasing inbetween.

Now define a one-form $\overline{\omega}$ on Im H by the equation

$$\overline{\omega} = \omega - d(\rho f).$$

Note that for $-\varepsilon \leqq t \leqq -\varepsilon/2$, $\rho = 0$ and hence $\overline{\omega} = \omega$. Similarly, for $\varepsilon/2 \leqq t \leqq \varepsilon$, $\rho = 1$ and so $\overline{\omega} = \omega - df = \Phi^*(\omega')$. We assert

(4.2) <u>$\overline{\omega}(V) > 0$, and hence $\overline{\omega}$ is nondegenerate.</u>

Now

$$\overline{\omega}(V) = \omega(V) - d(\rho f)(V) = \omega(V) - V(\rho f)$$
$$= \omega(V) - \rho\, V(f) - f\, V(\rho) .$$

But $\omega(V) - \Phi^*\omega'(V) = df(V) = V(f)$, and using this value for $V(f)$, we get

$$\overline{\omega}(V) = (1 - \rho)\, \omega(V) + \rho\, \Phi^*\omega'(V) - f\, V(\rho) .$$

But $\omega(V) = 1$ and $\Phi^*\omega'(V) = \omega'(\Phi_* V) = \omega'(V') = 1$, so

(4.3) $$\overline{\omega}(V) = 1 - f \frac{d\rho}{dt} ,$$

where we have written $\frac{d\rho}{dt}$ in place of $V(\rho)$. Since $\frac{d\rho}{dt} > 0$, (4.2) would certainly follow if $f \leqq 0$. Since Im H is compact, we can simply add a constant to f to make it a negative function, without altering the relation $\omega - \Phi^*(\omega') = df$. We do this, and (4.2) follows.

<u>NOTE.</u> It will <u>not</u> be possible to do this in the proof of the relative deformation theorems, and a third condition, involving an appropriate inequality for f, will reveal itself.

Finally we define the one-form ω'' on Im $\overline{H}$ by

$$\omega'' = (\Psi^{-1})^* \overline{\omega} .$$

Note that $\omega'' = \omega$ on $H(M \times [-\varepsilon,-\varepsilon/2])$ because $\overline{\omega} = \omega$ there and Ψ^{-1} is the identity there. Also, $\omega'' = \omega'$ on $H'(M \times [\varepsilon/2,\varepsilon])$ because $\overline{\omega} = \Phi^*\omega'$ on $H(M \times [\varepsilon/2,\varepsilon])$ and $\Psi = \Phi$ there. Since $\overline{\omega}$ is nondegenerate, so is ω''.

STEP 3. Construction of the new metric.

So far we have the family G" of curves and the nondegenerate one-form ω''.

Since Ψ: Im $H \to$ Im $\overline{H}$ takes curves of G to curves of G", the vector field Ψ_*V is tangent to G". Moreover,

$$\omega''(\Psi_*V) = (\Psi^{-1})^* \overline{\omega} (\Psi_*V) = \overline{\omega}(V) > 0,$$

by (4.2). Hence multiplying Ψ_*V by a smooth positive function, we get a tangent vector field V" to G" such that

$$\omega''(V'') = 1.$$

It is easy to check that V" = V on $H(M \times [-\varepsilon,-\varepsilon/2])$ and that V" = V' on $H'(M \times [\varepsilon/2,\varepsilon])$.

Now we define a Riemannian metric on Im $\overline{H}$ by stipulating:

(1) V" is to be a unit vector field.

(2) ker ω'' is to be perpendicular to V".

(3) The original metric is to be used on ker ω'' .

Since V" = V and $\omega'' = \omega$ on $H(M \times [-\varepsilon,-\varepsilon/2])$, the new metric agrees with the old metric there. Similarly, V" = V' and $\omega'' = \omega'$ on $H'(M \times [\varepsilon/2,\varepsilon])$, so the new and old metrics agree there. Hence the new metric can be extended via the old metric over the rest of N.

STEP 4. Verification that G" is a field of geodesics in the new metric.

Invoking Lemma 2.1, we will do this by showing that the Lie derivative $L_{V''}\,\omega'' = 0$. Now

(4.4) $\qquad L_{V''}\,\omega'' \;=\; V'' \lrcorner\, d\omega'' \;+\; d(V'' \lrcorner\, \omega''),$

where $\lrcorner$ indicates left interior multiplication [9, p. 102]. Since $V'' \lrcorner\, \omega'' = \omega''(V'') = 1$, the second term on the right of (4.4) is certainly zero. Next observe that

$$d\omega'' \;=\; (\Psi^{-1})^*\, d\bar{\omega} \;=\; (\Psi^{-1})^*\, d\omega\ .$$

Therefore

$$\begin{aligned} V'' \lrcorner\, d\omega'' &= V'' \lrcorner\, (\Psi^{-1})^*\, d\omega \\ &= k\ \Psi_* V \lrcorner\, (\Psi^{-1})^*\, d\omega \qquad \left(\begin{array}{l}\text{because } V'' \text{ is some}\\ \text{multiple } k \text{ of } \Psi_* V\end{array}\right) \\ &= k\ (\Psi^{-1})^*\ (V \lrcorner\, d\omega)\ . \end{aligned}$$

However,

$$L_V(\omega) \;=\; V \lrcorner\, d\omega \;+\; d(V \lrcorner\, \omega)\ ,$$

and the left side must vanish by Lemma 2.1, while $d(V \lrcorner\, \omega) = d(1) = 0$. It follows that $V \lrcorner\, d\omega = 0$, and inserting this above, we get $V'' \lrcorner\, d\omega'' = 0$. But then by (4.4) $L_{V''}\,\omega'' = 0$, so the curves of G" are indeed geodesics in the new metric, completing the proof of Theorem 1.

5. INTEGRABLE GEODESIC FIELDS

In this section we prove Theorems 2 and 3. Let G be a field of curves (not necessarily geodesics) on the Riemannian manifold N^n, V a unit vector field tangent to G and ω the dual one-form. By the Frobenius theorem, the (n-1) plane distribution orthogonal to G will be tangent to a foliation if and only if $d\omega \wedge \omega = 0$. If this condition holds, then in fact $d\omega = \omega \wedge L_V\omega$.

If, in addition, G happens to be a field of geodesics, then by Lemma 2.1, $L_V\omega = 0$. But then $d\omega = 0$, and as remarked in section 1, an integrable field of geodesics determines in this fashion a cohomology class $[\omega]$ in $H^1(N;R)$, and by restriction to the hypersurface M, a cohomology class $[\omega|_M]$ in $H^1(M;R)$.

Note that a geodesic field streaming out orthogonally from a submanifold (of any dimension) of N will be integrable. In particular, the geodesics streaming out from a point form an integrable field, and hence the relevance of this notion to the study of the cut locus.

Let us check that Theorem 2 follows immediately from Theorem 1. If ω and ω' are closed forms, then $\omega|_M - \phi^*(\omega'|_M)$ is exact if and only if $[\omega|_M] = \phi^*[\omega'|_M]$ in $H^1(M;R)$. But if ϕ is concordant to the identity, then it is certainly homotopic to the identity, in which case $\phi^* = 1$ on the cohomology level. Hence condition (b) of Theorem 1 reduces to $[\omega|_M] = [\omega'|_M]$, that is, condition (b) of Theorem 2.

We turn now to Theorem 3, in which it is given that the closed one-forms ω and ω', corresponding to the integrable geodesic fields G and G', satisfy $[\omega|_M] = [\omega'|_M]$ in $H^1(M;R)$. Without loss of generality, we can assume that N is a tubular neighborhood of M. In that case, the inclusion $M \hookrightarrow N$ is a homotopy equivalence, hence induces an isomorphism $H^1(N;R) \to H^1(M;R)$. From this we conclude that $[\omega] = [\omega']$ in $H^1(N;R)$, and so there is a smooth function $f: N \to R$ such that $\omega - \omega' = df$.

As in Step 2 of the proof of Theorem 1 (but this time more directly) we obtain a nondegenerate one-form

$$\omega'' = \omega - d(\rho f),$$

which coincides with ω somewhat below M and with ω' somewhat above. Notice that the field G" of curves has yet to be constructed.

Now let V" be the vector field orthogonal in the old metric to the (n-1) plane $\ker \omega''$ at each point of N, satisfying $\omega''(V'') = 1$, and oriented so that it coincides with V somewhat below M and with V' somewhat above. Inbetween, however, it may not be a unit vector field. There is a unique new metric on N, pointwise conformal to the old one, making V" a unit vector field everywhere. This new metric coincides with the old one outside a neighborhood of M.

Finally, let G" be the field of curves tangent to V". Clearly, G" coincides with G somewhat below M and with G' somewhat above. But notice that we have lost control over which curves of G and G' are connected by G".

It remains to check that the curves of G" are geodesics in the new metric. We do this by showing that $L_{V''}\omega'' = 0$, again by using formula (4.4):

$$L_{V''}\,\omega'' = V'' \lrcorner\, d\omega'' + d(V'' \lrcorner\, \omega'').$$

But this time ω'' is a closed form (because ω is), so we see directly that the first term on the right vanishes. And the second term vanishes as before, since $V'' \lrcorner\, \omega'' = 1$. So the curves of G" are indeed geodesics, completing the proof of Theorem 3.

6. RELATIVE SCATTERING THEOREMS

We turn now to formulating and proving relative versions of Theorems 1, 2 and 3. In them, the separating hypersurface M is only required to be closed, not necessarily compact. On M we are given K, a compact submanifold-with-boundary, of the same dimension as M and smoothly embedded. The geodesic fields G and G' on N cross M transversely and agree outside K; that is, if x is a point of $\overline{M - K}$, then the geodesics of G and G' crossing M at x coincide. The map $\phi: M \to M$ is the identity on $\overline{M - K}$.

THEOREM 1'. <u>Necessary and sufficient conditions for being able to deform the metric on N in an arbitrarily small neighborhood of K so as to produce a field of geodesics connecting G to G' according to ϕ are:</u>

(a) <u>The map $\phi: M \to M$ is a diffeomorphism, concordant to the identity relative to $\overline{M - K}$;</u>

(b) <u>The one-form on M, $\omega|_M - \phi^*(\omega'|_M)$ is exact rel $\overline{M - K}$; and</u>

(c) <u>For any curve C on M beginning in $\overline{M - K}$,</u>

$$\int_C \omega|_M - \phi^*(\omega'|_M) \leqq 0 .$$

<u>If the words "arbitrarily small neighborhood" are replaced by "neighborhood of width 2ε", then the integral in (c) is only required to be less than 2ε.</u>

COMMENTS. (1) We explain the language of the theorem. To obtain a small neighborhood of K in N, first choose a small compact neighborhood W of K on M. Then there is an $\varepsilon > 0$ such that $g_t(x)$ and $g'_t(x)$ are defined for all x in W and all t in $[-\varepsilon,\varepsilon]$. Using these flows, we define tubular neighborhoods H and H': $W \times [-\varepsilon,\varepsilon] \to N$ as in the absolute case, and mean $H(W \times [-\varepsilon,0]) \cup H'(W \times [0,\varepsilon])$ as a neighborhood of K of width 2ε in the statement of the theorem. Furthermore, a deformation of metric on this neighborhood will be understood to restrict to the identity on the geodesic arcs of G = G' passing through points of $\overline{W - K}$.

A concordance $F: M \times [0,1] \to M \times [0,1]$ between the identity and ϕ is said to be <u>relative to</u> $\overline{M - K}$ if F is the identity on $\overline{M - K} \times [0,1]$. A one-form on M which vanishes on $\overline{M - K}$ is said to be exact <u>rel</u> $\overline{M - K}$ if it is the differential of a function on M which vanishes on $\overline{M - K}$.

(2) Notice that if M is a compact manifold and we choose $K = M$, then Theorem 1' reduces to Theorem 1, and in particular, condition (c) disappears.

PROOF OF NECESSITY. With only obvious changes, the argument given in Section 3 that conditions (a) and (b) are necessary in the absolute case works here also. So we turn to condition (c).

First, by condition (b), there is a smooth function $f: M \to R$, vanishing on $\overline{M - K}$, such that $df = \omega|_M - \phi^*(\omega'|_M)$. Thus if C is any curve on M, beginning at any point of $\overline{M - K}$ and ending at x, then

$$f(x) = \int_C \omega|_M - \phi^*(\omega'|_M) \quad . \tag{6.1}$$

Next, let G_x denote the geodesic of G passing through $x \in M$, $G'_{\phi(x)}$ that of G' passing through $\phi(x)$, and G''_x that of G" connecting these two. Assume the metric on N has been deformed within a tubular neighborhood of W of width 2ε. The lower face of this neighborhood is $W_- = g_{-\varepsilon}(W)$ and the upper face is $W_+ = g'_\varepsilon(W)$. Let $L(x)$ denote the length of the arc of G''_x within this neighborhood, measured in the new metric.

Now connect the points x_o in $\overline{W - K}$ and x in W by a curve C on W. Then C can be pushed down along the geodesics of G to the curve C_- on W_-, then up along G" to C_+ on W_+, and then down along G' to $C' = \phi(C)$ back on W.

By Lemma 2.2, we have

$$\int_{C_+} \omega'' = \int_{C_-} \omega'' + L(x) - L(x_o) \quad .$$

Furthermore, by the already familiar application of Lemma 2.2,

$$\int_{C_+} \omega'' = \int_{C_+} \omega' = \int_{C'} \omega' = \int_C \phi^*(\omega'|_M) \quad ,$$

and

$$\int_{C_-} \omega'' = \int_{C_-} \omega = \int_C \omega|_M .$$

Hence

$$L(x) - L(x_o) = - \int_C \omega|_M - \phi^*(\omega'|_M) = - f(x) + f(x_o).$$

But $L(x_o) = 2\varepsilon$ and $f(x_o) = 0$, so we get

(6.2) $$L(x) = 2\varepsilon - f(x).$$

Now we certainly have $L(x) > 0$. So if the required deformation of metric takes place within a tubular neighborhood of K of width 2ε, we must have $f(x) < 2\varepsilon$. And if it can take place within an arbitrarily small neighborhood of K, we must have $f(x) \leqq 0$. So condition (c) is necessary.

PROOF OF SUFFICIENCY. Again with only obvious changes, the arguments given in Section 4 for the absolute case will carry us as far as equation (4.3):

$$\overline{\omega}(V) = 1 - f \frac{d\rho}{dt} .$$

It is still true that f is bounded from above on the compact neighborhood $H(W \times [-\varepsilon, \varepsilon])$, but we are no longer free to add a constant to f to make it negative. For this would destroy the agreement between G" and the common value of G and G' outside of K.

Since $\rho: [-\varepsilon, \varepsilon] \to [0,1]$, we can arrange that

$$0 \leqq \frac{d\rho}{dt} \leqq \frac{1}{2\varepsilon} + \delta$$

for any preassigned $\delta > 0$, and still keep ρ constant on neighborhoods of $-\varepsilon$ and ε. By condition (c), $f < 2\varepsilon$, and then for sufficiently small δ we get $\overline{\omega}(V) > 0$. If $f \leqq 0$, we get this immediately, no matter the value of ε.

The rest of the proof continues as in the absolute case, and we have Theorem 1'.

COMMENT. Think again of the tubular neighborhood of K, together with its deformed metric, as the "lens" which "focuses" G into G'. In that context we think of f(x) as the "divergent power" required of the lens, and measured along any curve from the "edge" of the lens to x. Condition (c) may then be paraphrased as saying: "the divergent power of a lens is less than the width at its edge".

In the integrable case, Theorem 1' simplifies to the following relative version of Theorem 2.

THEOREM 2'. <u>Necessary and sufficient conditions for being able to deform the metric on N in an arbitrarily small neighborhood of K so as to produce a field of geodesics connecting the integrable geodesic fields G to G' according to ϕ are:</u>

(a) <u>Same as in Theorem 1';</u>

(b) <u>$[\omega|_M - \omega'|_M] = 0$ in $H^1(M, \overline{M-K}; R)$</u>;

(c) <u>Same as in Theorem 1'</u>.

<u>If we use a neighborhood of K of width 2ε, then the integral in (c) is only required to be less than 2ε.</u>

Condition (b) of Theorem 1' reduces in the integrable case to $[\omega|_M - \phi^*(\omega'|_M)] = 0$ in $H^1(M, \overline{M-K}; R)$. But if ϕ is concordant to the identity rel $\overline{M-K}$, then it is certainly homotopic to the identity rel $\overline{M-K}$, and this implies that $[\phi^*(\omega'|_M) - \omega'|_M] = 0$ in $H^1(M, \overline{M-K}; R)$. Adding, we get condition (b) of Theorem 2'.

Theorem 3 relativizes to

THEOREM 3'. <u>If conditions (b) and (c) are satisfied in Theorem 2', then for some ϕ the deformed metric on N can be chosen pointwise conformal to the original metric.</u>

There is nothing new in the proof.

7. SURFACES OF REVOLUTION

We turn now to the scattering of geodesics on a surface of revolution, restricting ourselves to deformations of metric which remain within this class. In a sense, this is the equivariant version of the theory already described. We begin with some general remarks about surfaces of revolution.

Let $z = f(r)$ be an even C^∞ function. If we rotate its graph about the z-axis in three-space, we obtain a C^∞ surface of revolution. The Euclidean metric on three-space induces a Riemannian metric on the surface; in a neighborhood of the "center" $(0,0,f(0))$ the metric can be described in polar coordinates (via projection to the x-y plane) by

$$ds^2 = (1 + f'(r)^2)\, dr^2 + r^2\, d\theta^2 \quad . \tag{7.1}$$

Such surfaces of revolution have a peculiar feature which limits their generality. The circle $r = r_o$ on the surface has circumference $2\pi r_o$. To get to that circle from the center, one must travel a distance

$$\int_0^{r_o} \sqrt{1 + f'(r)^2}\; dr \geq r_o$$

on the surface.

Yet one can imagine a surface possessing rotational symmetry, on which one can travel from the center to a circle of circumference $2\pi r_o$ along a path of length less than r_o. For example, the formula

$$ds^2 = \cos^2 r\, dr^2 + r^2\, d\theta^2$$

defines a C^∞ Riemannian metric on the open disc $r < \pi/2$ in the plane. Let $r_o = \pi/4 \approx .785$. On such a surface, one can travel from the center to the circle of circumference $2\pi r_o$ along a path of length

$$\int_0^{r_o} \cos r\, dr = \sin r_o = \sqrt{2}/2 \approx .707.$$

Surfaces of this sort cannot be isometrically embedded as surfaces of revolution in three-space, in spite of their obvious rotational symmetry. In polar coordinates, their metrics look like

$$ds^2 = E(r)\, dr^2 + r^2\, d\theta^2 , \tag{7.2}$$

where in contrast to (7.1), we allow $E(r) < 1$. Such surfaces cannot be avoided; they arise naturally in what follows.

When does formula (7.2) define a smooth Riemannian metric? If $E(r)$ is a strictly positive C^∞ function, defined for $0 < r < a \leq \infty$, then (7.2) obviously defines a C^∞ Riemannian metric on the punctured disc $0 < r < a$. We need to know when this metric is C^∞ at the origin. The answer is given by

LEMMA 7.3. _Necessary and sufficient conditions for the metric (7.2) to be C^∞ at the origin are:_

(1) $\lim_{r\to 0} E(r) = 1$;

(2) _The function E, when extended over the interval $(-a,a)$ by defining $E(0) = 1$ and $E(-r) = E(r)$, is still C^∞. In other words, $E(r)$ should be a C^∞ even function on $(-a,a)$ with $E(0) = 1$._

The C^∞ character of this metric at the origin can be decided by switching from polar to rectangular coordinates. The proof of the lemma is straightforward and omitted.

Is any further generality possible? If a two-dimensional Riemannian manifold has rotational symmetry about a point p, then in geodesic polar coordinates about p its metric has the form

$$ds^2 = dR^2 + G(R)\, d\Theta^2 . \tag{7.4}$$

It is another technical exercise to show that on some neighborhood of p a change of coordinates can be found which transforms (7.4) to (7.2), and it is in this spirit that we assert the generality of the latter. Note, however, that the representation (7.4) is often valid over a larger portion of the surface than is (7.2).

8. GEODESICS ON SURFACES OF REVOLUTION

Let M be a surface of revolution with center p, and let (r,θ) be polar coordinates on a neighborhood of p, in terms of which the metric is given by (7.2). A geodesic running through this neighborhood and given parametrically in terms of arc length by $r = r(s)$ and $\theta = \theta(s)$ must satisfy the differential equations [10, p. 134]:

$$\ddot{r} = \frac{1}{2E(r)} \left(- E'(r)\, \dot{r}^2 + 2r\, \dot{\theta}^2 \right) \tag{8.1}$$

$$\ddot{\theta} = - \frac{2}{r}\, \dot{r}\, \dot{\theta} . \tag{8.2}$$

Here the dot indicates d/ds and the prime indicates d/dr.

These equations can be integrated explicitly to give the (nonparametric) equations of a geodesic. Indeed, (8.2) integrates immediately to

$$\dot{\theta} = c\, r^{-2} , \tag{8.3}$$

where c is a constant along the given geodesic. And this, together with (7.2), easily leads to

(8.4) $$\theta = \theta_o + \int_{r_o}^{r} \pm \frac{c\sqrt{E(r)}}{r\sqrt{r^2-c^2}}\, dr ,$$

where (r_o, θ_o) are the coordinates of a fixed point on the geodesic.

CAUTION. Along half of each geodesic, the plus sign above must be used; along the other half, the minus sign.

Notice that setting $c = 0$ in (8.4) yields $\theta = \theta_o$, which is the equation of one of the geodesics through p (which we call meridians). In general, the geometric meaning of the constant c is revealed by the following lemma.

LEMMA 8.5. The absolute value of the constant c above is the radius (in the r-θ plane) of the smallest circle touched by the given geodesic.

As noticed above, this is true for the geodesics passing through p. Otherwise, if a geodesic has a point (r_o, θ_o) of closest approach to p, then at that point we have $dr/ds = 0$. Substituting this into (7.2), we get $1 = r^2 \dot{\theta}^2$ at (r_o, θ_o). Comparing this with (8.3), we get $|c| = r_o$, as claimed.

The fact that a geodesic not passing through p must have a point of closest approach to p follows easily from the convergence of the improper integral in (8.4) at the lower limit $r_o = |c| \neq 0$.

We close this section with Clairaut's Theorem.

THEOREM 8.6 (Clairaut). If the geodesic $r = r(s)$, $\theta = \theta(s)$ makes an angle $\phi(s)$ with the meridian, then

$$r \sin\phi = c ,$$

where c is the same constant as above.

We see easily that $\sin \phi = \frac{r\, d\theta}{ds} = r\,\dot{\theta}$. But then

$$r \sin \phi = r^2\,\dot{\theta} = c$$

by (8.3), as claimed.

REMARK. In general, a surface of revolution may possess a "neck", that is, a closed geodesic of the form r = constant, though this cannot occur in a neighborhood within which the metric can be represented by (7.2). On such surfaces, Theorem 8.6 remains valid, while Lemma 8.5 becomes false, in the sense that a geodesic can spiral in to a smallest circle (the neck) without actually touching it.

9. DEFLECTION OF GEODESICS BY A BUMP

Let M be a surface of revolution with center p, and let (r,θ), $r < a$, be polar coordinates on a neighborhood of p, in terms of which the Riemannian metric is given by (7.2):

$$ds^2 = E(r)\, dr^2 + r^2\, d\theta^2 .$$

Suppose now that this metric is altered in a smaller neighborhood but is still rotationally symmetric about p:

$$ds_1^{\,2} = E_1(r)\, dr^2 + r^2\, d\theta^2 , \tag{9.1}$$

with $E_1(r) = E(r)$ for $b < r < a$. We want to measure the effect of this change on the geodesics, and will do so by defining and calculating the "deflection" they undergo.

Let γ be a geodesic with respect to the original metric, and let (r_o, θ_o) be a point of γ somewhere in the annular region $b < r < a$ where the two metrics agree. If γ does not pass into the inner region $r < b$ (the "bump") where the metrics differ, then it is obviously unaffected by the change and remains a geodesic in the new metric. But if it does go into the bump, then in general it will no longer be a geodesic in the new metric.

Let γ_1 be the geodesic in the new metric which begins coincident with γ in the region $b < r < a$. As γ_1 enters the bump, it alters course from γ. When it comes out of the bump, as it must by symmetry, it will again be a geodesic in both metrics; in fact, a "rotated version" of the corresponding portion of γ.

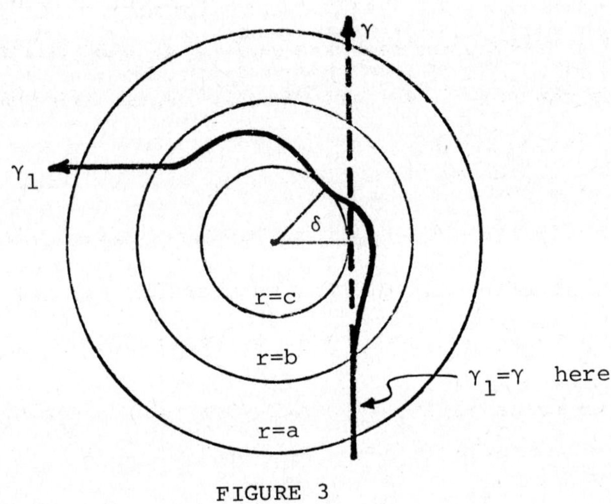

FIGURE 3

The constant c for the geodesic γ can be computed from either (8.3) or (8.6) at the point (r_o, θ_o). The same computation, with the same result, will be valid for γ_1, since the two coincide in a neighborhood of this point. So we have

(9.2) <u>The radius in the r-θ plane of the smallest circle touched by the geodesic γ, is the same for γ1.</u>

But the points on this circle at which γ and γ_1 touch are different, and if they are denoted by (c,θ) and (c,θ_1), then the quantity $\theta_1 - \theta$ can be used to measure the <u>deflection</u> due to the change of metric. It follows from the rotational symmetry that all geodesics in the original metric having the same value of c will undergo the same amount of deflection due to the bump. So $\theta_1 - \theta$ is a function of c, and we write

(9.3) $$\delta(c) = \theta_1 - \theta ,$$

thus defining the <u>deflection function</u> δ.

REMARK. If the geodesic γ is given by the parametric equations $r = r(s)$ and $\theta = \theta(s)$, then the same equations will describe γ_1 before it enters the bump. When it finally leaves the bump, γ_1 can be described by the equations

$$r = r(s) \qquad \text{and} \qquad \theta = \theta(s) + 2\,\delta(c) .$$

This follows immediately from the symmetry of a geodesic about its point of closest approach to p. So the quantity $2\,\delta(c)$ measures the total deflection due to the change of metric.

We turn next to an explicit formula for $\delta(c)$. Applying (8.4) to γ, we get

$$\theta = \theta_o - \int_{r_o}^{c} \frac{c\sqrt{E(r)}}{r\sqrt{r^2-c^2}}\, dr.$$

The minus sign is used because on this portion of γ, $d\theta/dr < 0$. Similarly for γ_1 we have

$$\theta = \theta_o - \int_{r_o}^{c} \frac{c\sqrt{E_1(r)}}{r\sqrt{r^2-c^2}}\, dr.$$

Subtracting these two formulas, we get

$$\delta(c) = \int_{c}^{r_o} \frac{c\, g(r)}{r\sqrt{r^2-c^2}}\, dr ,$$

where

$$g(r) = \sqrt{E_1(r)} - \sqrt{E(r)}$$

represents the change in metric. Note that $g(r) = 0$ for $b \leq r < a$, so that we may think of g as being defined on $(0,\infty)$ with $g(r) = 0$ for $r \geqq b$. With that understanding, we rewrite the formula for $\delta(c)$ as

$$(9.4) \qquad \delta(c) = \int_c^\infty \frac{c\, g(r)}{r\sqrt{r^2-c^2}}\, dr \quad .$$

In this formula, $c > 0$, and since $\delta(c) = 0$ for $c \geqq b$, we may also think of δ as being defined on $(0,\infty)$. Note that the integral in (9.4) is improper but convergent at its lower limit.

We close this section by recording the differentiability properties of the functions $g(r)$ and $\delta(c)$.

(9.5) <u>The change of metric function $g(r) = \sqrt{E_1(r)} - \sqrt{E(r)}$ satisfies</u>

(1) <u>$\lim_{r\to 0} g(r) = 0$; and</u>

(2) <u>If one extends g over $(-\infty,\infty)$ by defining $g(0) = 0$ and $g(-r) = g(r)$, then g is of class C^∞.</u>

<u>Briefly, g is a C^∞ even function with $g(0) = 0$ and support in $[-b,b]$.</u>

This follows immediately from Lemma 7.3, since $E(r)$ and $E_1(r)$ are strictly positive C^∞ even functions taking the value 1 at $r = 0$ and agreeing for $|r| \geqq b$.

(9.6) <u>The deflection function $\delta(c)$, initially defined for $c > 0$, extends to a C^∞ odd function with support in $[-b,b]$.</u>

We omit the simple geometric argument, since this result reappears as part (4) of Theorem 10.1.

10. THE INVERSE SCATTERING PROBLEM

Having just computed the amount by which a geodesic is deflected due to a change in metric, we turn to the inverse problem: given a preassigned deflection function, is there a change in metric which will produce it? The answer is yes. We formulate the results in this section.

Let $C_0^\infty(0,\infty)$ denote the set of C^∞ functions $f\colon (0,\infty) \to R$ with bounded support. That is, for each such f, there is a number b_f such that $f(x) = 0$ for $x \geq b_f$. Let E_0 denote the set of C^∞ even functions $f\colon R \to R$ with bounded support, and O_0 the set of C^∞ odd functions $f\colon R \to R$ with bounded support. Both E_0 and O_0 may be regarded as subsetsof $C_0^\infty(0,\infty)$, since such functions are determined by their values on $(0,\infty)$. Finally let $E_{00} = \{f \in E_0 \colon f(0) = 0\}$.

We want to analyze the transformation $T\colon g \mapsto \delta$ defined by formula (9.4):

$$\delta(c) = \int_c^\infty \frac{c\, g(r)}{r\sqrt{r^2-c^2}}\, dr , \qquad c > 0 .$$

Its behavior is exposed by the following theorem.

THEOREM 10.1. <u>The transformation T has the following properties:</u>

(1) <u>$T\colon C_0^\infty(0,\infty) \to C_0^\infty(0,\infty)$.</u>

(2) <u>T is a bijection, with inverse $T^{-1}\colon \delta \mapsto g$ given by</u>

$$g(r) = -\frac{2}{\pi}\int_r^\infty \frac{c\,\delta'(c)}{\sqrt{c^2-r^2}}\, dc , \qquad r > 0. \tag{10.2}$$

(3) <u>T preserves supports, in the sense that $g(r) = 0$ for $r \geq b$ if and only if $\delta(c) = 0$ for $c \geq b$.</u>

(4) <u>$T(E_{00}) = O_0$, and hence $T^{-1}(O_0) = E_{00}$.</u>

(5) <u>$T|E_{00}\colon E_{00} \to O_0$ is continuous if E_{00} is given the C^{2k+2}</u>

topology and O_o the C^k topology, $0 \leq k \leq \infty$. Similarly, $T^{-1}|O_o: O_o \to E_{oo}$ is continuous if O_o is given the C^{2k+1} topology and E_{oo} the C^k topology, $0 \leq k \leq \infty$.

The proof of this theorem will be given in the next section. It is clear that this result contains the solution to the inverse scattering problem for geodesics on a surface of revolution, as follows.

THEOREM 10.3. Let M be a surface of revolution with center p, in a neighborhood of which the Riemannian metric is given in polar coordinates by

$$ds^2 = E(r)\, dr^2 + r^2\, d\theta^2 , \qquad r < a .$$

Let $\delta(c)$ be a preassigned deflection function; that is, a C^∞ odd function with support in $[-b,b]$, $b < a$. If δ is sufficiently small in the C^1 topology, then there exists a perturbation of metric,

$$ds_1^2 = E_1(r)\, dr^2 + r^2\, d\theta^2 , \qquad r < a ,$$

with $E_1(r) = E(r)$ for $b \leq r < a$, for which δ is the corresponding deflection function.

By Theorem 10.1, there is a C^∞ even function $g(r)$, vanishing at 0 and supported in $[-b,b]$, for which $T(g) = \delta$. If we define

$$E_1(r) = (\sqrt{E(r)} + g(r))^2 ,$$

then E_1 will be a C^∞ even function on $(-a,a)$ which agrees with E outside $(-b,b)$ and such that $E_1(0) = 1$. If δ is sufficiently small in the C^1 topology, then g will be small in the C^0 topology, and so we can arrange that $E_1 > 0$ on $(-a,a)$. But then by Lemma 7.3, $ds_1^2 = E_1(r)\, dr^2 + r^2\, d\theta^2$ defines a C^∞ Riemannian metric on the given neighborhood of p in M, agreeing with the old metric outside the ball $r < b$. Clearly δ is the corresponding deflection function, and the theorem follows.

11. PROOF OF THEOREM 10.1

PROOF OF (1). If in formula (9.4) we make the substitution $r = cx$, we get

$$(11.1) \qquad \delta(c) = \int_1^\infty \frac{g(cx)}{x\sqrt{x^2-1}}\, dx \,, \qquad c > 0.$$

Note that if $g(r) = 0$ for $r \geqq b$, then the upper limit of this integral may be replaced by $x = b/c$, so is really finite. At the lower limit, $x = 1$, the integral is improper but convergent as usual.

From (11.1) we see that if $g\colon (0,\infty) \to R$ is continuous, then so is $\delta\colon (0,\infty) \to R$. Furthermore, as often as g is continuously differentiable, we may differentiate (11.1) with respect to c and get

$$(11.2) \qquad \delta^{(n)}(c) = \int_1^\infty \frac{x^n\, g^{(n)}(cx)}{x\sqrt{x^2-1}}\, dx.$$

In particular, if g is of class C^∞, so is δ. It is clear from (9.4) that if $g(r) = 0$ for $r \geqq b$, then $\delta(c) = 0$ for $c \geqq b$. This proves (1) and half of (3).

PROOF OF (2). Given $g \in C_0^\infty(0,\infty)$, we already know from (1) that $\delta = T(g)$ lies in $C_0^\infty(0,\infty)$, so that the integral

$$\int_r^\infty \frac{\delta(c)}{\sqrt{c^2\;r^2}}\, dc \,, \qquad r > 0,$$

converges. We evaluate it by Fubini's theorem, as follows.

$$\int_r^\infty \frac{\delta(c)}{\sqrt{c^2-r^2}}\, dc = \int_r^\infty \frac{1}{\sqrt{c^2-r^2}} \left[\int_c^\infty \frac{c\, g(s)}{s\sqrt{s^2-c^2}}\, ds\right] dc$$

$$= \int_r^\infty \left[\int_r^s \frac{c\, dc}{\sqrt{c^2-r^2}\,\sqrt{s^2-c^2}}\right] \frac{g(s)}{s}\, ds \,.$$

Now the inner integral just above has the value $\pi/2$, as is easily checked. Hence

$$(11.3) \qquad \int_r^\infty \frac{\delta(c)}{\sqrt{c^2-r^2}}\, dc = \frac{\pi}{2} \int_r^\infty \frac{g(s)}{s}\, ds .$$

If we now define

$$\tilde{g}(r) = \int_r^\infty \frac{g(s)}{s}\, ds , \qquad r > 0,$$

then on the one hand we have from this very definition

$$(11.4) \qquad \tilde{g}'(r) = -\frac{g(r)}{r} ,$$

while on the other we have from (11.3)

$$\tilde{g}(r) = \frac{2}{\pi} \int_r^\infty \frac{\delta(c)}{\sqrt{c^2-r^2}}\, dc .$$

But we can differentiate such a formula directly, just as was done in obtaining (11.2). That is, first letting $c = rx$, we get

$$\tilde{g}(r) = \frac{2}{\pi} \int_1^\infty \frac{\delta(rx)}{\sqrt{x^2-1}}\, dx ,$$

and therefore

$$\tilde{g}'(r) = \frac{2}{\pi} \int_1^\infty \frac{x\, \delta'(rx)}{\sqrt{x^2-1}}\, dx = \frac{2}{\pi} \int_r^\infty \frac{c}{r} \frac{\delta'(c)}{\sqrt{c^2-r^2}}\, dc.$$

But then by (11.4)

$$(11.5) \qquad g(r) = -r\, \tilde{g}'(r) = -\frac{2}{\pi} \int_r^\infty \frac{c\, \delta'(c)}{\sqrt{c^2-r^2}}\, dc .$$

Thus if we define the transformation $S: \delta \mapsto g$ by (11.5), then we can conclude, just as we did for T, that $S: C_o^\infty(0,\infty) \to C_o^\infty(0,\infty)$ and interpret the calculation leading to (11.5) as showing that

$$S \circ T = \text{identity}.$$

To show that $T \circ S = \text{identity}$, we first notice that $S(\delta)$ can also be given by the formula

$$(11.6) \qquad S(\delta)(r) = \frac{2}{\pi} \int_r^\infty \frac{r^2 [\delta(c) - c\, \delta'(c)]}{c^2 \sqrt{c^2-r^2}}\, dc .$$

Comparing with (11.5), we must show that

$$\int_r^\infty \frac{r^2\,[\delta(c) - c\,\delta'(c)]}{c^2\sqrt{c^2-r^2}} + \frac{c\,\delta'(c)}{\sqrt{c^2-r^2}} \quad dc \quad = \quad 0.$$

But the integrand is the derivative of

$$\frac{\sqrt{c^2-r^2}\quad \delta(c)}{c} ,$$

which vanishes at $c = r$ and at $c = \infty$. So (11.6) follows from the Fundamental Theorem of Calculus, the fact that the integrals are improper notwithstanding.

To check now that $T \circ S$ = identity, we compute:

$$(TS\,\delta)(c) = \int_c^\infty \frac{c\,(S\delta)(r)}{r\sqrt{r^2-c^2}}\,ds$$

$$= \int_c^\infty \frac{c}{r\sqrt{r^2-c^2}} \left[\frac{2}{\pi}\int_r^\infty \frac{r^2\,[\delta(s) - s\,\delta'(s)]}{s^2\sqrt{s^2-r^2}}\,ds\right] dr,$$

by (11.6). Then by Fubini, this equals

$$\frac{2}{\pi}\int_c^\infty \left[\int_c^s \frac{r\,dr}{\sqrt{r^2-c^2}\sqrt{s^2-r^2}}\right] c\ \frac{\delta(s) - s\,\delta'(s)}{s^2}\ ds \quad .$$

As noted earlier, the inner integral has the value $\pi/2$, and we get

$$(TS\,\delta)(c) = c\int_c^\infty \frac{\delta(s) - s\,\delta'(s)}{s^2}\,ds = c\int_c^\infty -\left[\frac{\delta(s)}{s}\right]'\,ds$$

$$= \delta(c) \ .$$

Thus $T\colon C_o^\infty(0,\infty) \to C_o^\infty(0,\infty)$ is a bijection with inverse S, proving (2).

PROOF OF (3). As for the remaining half of (3), it is clear from (10.2) that if $\delta(c) = 0$ for $c \geqq b$, then also $g(r) = 0$ for $r \geqq b$.

PROOF OF (4). Turning to (4), we must show that $T(E_{oo}) \subset O_o$ and $S(O_o) \subset E_{oo}$. Suppose first that $\delta \in O_o$. Then we must show that the function g defined by (10.2):

$$g(r) = -\frac{2}{\pi} \int_r^\infty \frac{c\,\delta'(c)}{\sqrt{c^2-r^2}}\,dc\,, \qquad r > 0,$$

is the restriction to $(0,\infty)$ of a C^∞ even function which vanishes at 0.

Now δ odd implies δ' even, and this in turn implies the existence of a C^∞ function $F: R \to R$ such that $\delta'(c) = F(c^2)$. Then

$$(11.7) \qquad g(r) = -\frac{1}{\pi} \int_{r^2}^\infty \frac{F(y)}{\sqrt{y - r^2}}\,dy\,, \qquad r > 0\,,$$

via the substitution $y = c^2$.

Consider the function $G: R \to R$ defined by

$$(11.8) \qquad G(x) = -\frac{1}{\pi} \int_x^\infty \frac{F(y)}{\sqrt{y - x}}\,dy\,.$$

Letting $y - x = z$, we get

$$(11.9) \qquad G(x) = -\frac{1}{\pi} \int_0^\infty \frac{F(x+z)}{\sqrt{z}}\,dz.$$

From this it is apparent that we can differentiate under the integral sign and conclude that G is also of class C^∞.

But then $g(r) = G(r^2)$, showing that g is C^∞ and even. For this argument, which is much simpler than our original one, we are indebted to Jürgen Moser.

To continue,

$$g(0) = -\frac{2}{\pi} \int_0^\infty \frac{c\,\delta'(c)}{\sqrt{c^2-0^2}}\,dc = -\frac{2}{\pi}\,\delta(c)\Big|_0^\infty = 0.$$

Thus $\delta \in O_o$ implies $S(\delta) \in E_{oo}$.

To complete the proof of (4), suppose that $g \in E_{oo}$. We must show that the function δ defined by (9.4):

$$\delta(c) = \int_c^\infty \frac{c\, g(r)}{r\sqrt{r^2-c^2}}\, dr, \qquad c > 0,$$

is the restriction to $(0,\infty)$ of a C^∞ odd function. We rewrite this as

$$\frac{\delta(c)}{c} = \frac{1}{2}\int_c^\infty \frac{g(r)/r^2}{\sqrt{r^2-c^2}}\, dr^2 .$$

Now $g \in E_{oo}$ implies $g(r)/r^2 \in E_o$, and the preceding argument then shows that $\delta(c)/c \in E_o$. Hence $\delta \in O_o$, and (4) is established.

PROOF OF (5). We will show first that $T^{-1}|O_o\colon O_o \to E_{oo}$ is continuous if O_o is given the C^{2k+1} topology and E_{oo} the C^k topology for any k, $0 \leq k \leq \infty$. The passage from $\delta \in O_o$ via T^{-1} to $g \in E_{oo}$ can be broken into several steps, which are summarized in the following diagram.

ELEMENT:	δ		δ'		F		G		g
	$\cap$		$\cap$		$\cap$		$\cap$		$\cap$
SPACE:	O_o	$\xrightarrow{S_1}$	E_o	$\xrightarrow{S_2}$	$C_o^\infty[0,\infty)$	$\xrightarrow{S_3}$	$C_o^\infty[0,\infty)$	$\xrightarrow{S_4}$	E_o
TOPOLOGY:	C^{2k+1}		C^{2k}		C^k		C^k		C^k

ACTION OF $T^{-1}|O_o$

This decomposition of $T^{-1}|O_o$ was used in the proof of (4), just above. In the diagram, $C_o^\infty[0,\infty)$ denotes the set of C^∞ functions $F\colon [0,\infty) \to R$ with bounded support.

The map S_1 above is just differentiation, and is clearly continuous from the C^{2k+1} topology to the C^{2k} topology. The map S_3 is defined by (11.9), and by repeated differentiation under the integral sign there, we easily see that S_3 is continuous from the C^k topology to the C^k topology. This leaves the maps S_2 and S_4, which are inverses of each other.

The map S_4 starts with a function $G \in C_o^\infty[0,\infty)$ and associates with it the C^∞ even function g defined by $g(t) = G(t^2)$. Repeated differentiation of this formula makes it obvious that S_4 is continuous from the C^k topology to the C^k topology.

But its inverse, S_2, is not and that is where the loss of derivatives occurs. Start with a C^∞ even function f (in our application, $f = \delta'$) having bounded support, and define a function $F: [0,\infty) \to R$ by $F(t) = f(\sqrt{t})$. It is a straightforward exercise to check that F is indeed C^∞ at the origin, and hence an element of $C_o^\infty[0,\infty)$.

LEMMA 11.10. Under these conditions,

$$|F^{(k)}(t)| \leq \frac{k!}{(2k)!} \max |f^{(2k)}| \quad , \qquad k \geq 1 .$$

COMMENTS. (1) It is amusing to contrast this inequality with the obvious equality:

$$F^{(k)}(0) = \frac{k!}{(2k)!} f^{(2k)}(0) \quad .$$

(2) The above lemma shows that if f is close to 0 in the C^{2k} topology, then F is close to 0 in the C^k topology. Hence the map S_2 is continuous, as asserted.

PROOF. Consider Taylor's formula with remainder for f:

$$f(x) = f(0) + f'(0)x + \ldots + \frac{f^{(n)}(0)}{n!} x^n + \frac{1}{n!} \int_0^x (x-\xi)^n f^{(n+1)}(\xi) d\xi \ .$$

Since f is an even function, its odd order derivatives vanish at the origin. Take the above formula for $n = 2k-1$ and insert in it the definition $F(t) = f(\sqrt{t})$, getting

$$(11.11)\quad F(t) = f(0) + \frac{f''(0)}{2}\, t + \dots + \frac{f^{(2k-2)}(0)}{(2k-2)!}\, t^{k-1}$$

$$+ \frac{1}{(2k-1)!} \int_0^{\sqrt{t}} (\sqrt{t}-\xi)^{2k-1} f^{(2k)}(\xi)\, d\xi \, .$$

If we make the substitution $\xi = \sqrt{\tau}$, the remainder term becomes

$$(11.12)\qquad \frac{1}{(2k-1)!} \int_0^t (\sqrt{t}-\sqrt{\tau})^{2k-1} f^{(2k)}(\sqrt{\tau})\, \frac{d\tau}{2\sqrt{\tau}} \, .$$

Now differentiate (11.11) k times with respect to t, using the remainder term in the form (11.12). All the terms on the right hand side of (11.11) drop out, with the exception of the remainder term, yielding

$$F^{(k)}(t) = \frac{1}{(2k-1)!} \frac{d^k}{dt^k} \int_0^t (\sqrt{t}-\sqrt{\tau})^{2k-1} f^{(2k)}(\sqrt{\tau})\, \frac{d\tau}{2\sqrt{\tau}} \, .$$

In differentiating this integral with respect to t, the term obtained by evaluating the integrand at $\tau = t$ always drops out. Therefore

$$F^{(k)}(t) = \frac{1}{(2k-1)!} \int_0^t \frac{d^k}{dt^k} (\sqrt{t}-\sqrt{\tau})^{2k-1} f^{(2k)}(\sqrt{\tau})\, \frac{d\tau}{2\sqrt{\tau}} \, .$$

Hence

$$|F^{(k)}(t)| \leqq \frac{\max |f^{(2k)}|}{(2k-1)!} \int_0^t \left| \frac{d^k}{dt^k} (\sqrt{t}-\sqrt{\tau})^{2k-1} \right| \frac{d\tau}{2\sqrt{\tau}} \, .$$

But

$$\frac{d^k}{dt^k} (\sqrt{t}-\sqrt{\tau})^{2k-1} = \frac{(2k)!}{2^{2k}\, k!} \; \frac{(1-\frac{\tau}{t})^{k-1}}{\sqrt{t}} \, ,$$

so the integrand above is positive for $0 \leqq \tau \leqq t$, and the absolute value signs may be dropped. Then the operator d^k/dt^k can be moved back outside the integral sign. Finally we change back to the variable of integration ξ, and the result is

$$|F^{(k)}(t)| \leqq \frac{\max |f^{(2k)}|}{(2k-1)!} \frac{d^k}{dt^k} \int_0^{\sqrt{t}} (\sqrt{t}-\xi)^{2k-1}\, d\xi \, .$$

But the value of this integral is just $t^k/2k$, and hence its kth derivative is $k!/2k$. Inserting this, we get the lemma.

This completes the argument that $T^{-1}|O_o : O_o \to E_{oo}$ is continuous if the domain has the C^{2k+1} topology and the range the C^k topology.

The argument that $T|E_{oo} : E_{oo} \to O_o$ is continuous if E_{oo} is given the C^{2k+2} topology and O_o the C^k topology is the same, and is summarized by the following diagram.

ELEMENT:	g	$g(r)/r^2$	F	G	$\delta(c)/c$	δ
	$\cap$	$\cap$	$\cap$	$\cap$	$\cap$	$\cap$
SPACE:	E_{oo} $\longrightarrow$	E_o $\longrightarrow$	$C_o^\infty[0,\infty)$ $\longrightarrow$	$C_o^\infty[0,\infty)$ $\longrightarrow$	E_o $\longrightarrow$	O_o
TOPOLOGY:	C^{2k+2}	C^{2k}	C^k	C^k	C^k	C^k

ACTION OF $T|E_{oo}$

Here $F(t) = \dfrac{g(\sqrt{t})}{t}$ and $\delta(c)/c = G(c^2)$.

This completes the proof of Theorem 10.1.

COMMENT. The transformation T given by (9.4) is essentially an Abel integral, and this is seen most clearly in the form (11.8). See [1] and [4] for details.

REFERENCES

1. N. H. Abel, Résolution d'un problème de mécanique, J. Reine Angew. Math. 1(1826), 13-18.

2. H. Gluck and D. Singer, Deformations of geodesic fields, Bull. Amer. Math. Soc. 82(1976), 571-574.

3. __________, Scattering of geodesic fields, I, to appear.

4. J. B. Keller, Inverse problems, Amer. Math. Monthly 83(1976), 107-118.

5. J. R. Munkres, ELEMENTARY DIFFERENTIAL TOPOLOGY, Annals of Math. Studies 54, Princeton U. Press, revised ed. (1966).

6. B. L. Reinhart, Foliated manifolds with bundle-like metrics, Annals of Math. 69(1959), 119-132.

7. D. Singer and H. Gluck, The existence of nontriangulable cut loci, Bull. Amer. Math. Soc. 82(1976), 599-602.

8. __________, Scattering of geodesic fields, II, to appear.

9. S. Sternberg, LECTURES ON DIFFERENTIAL GEOMETRY, Prentice-Hall (1964).

10. D. J. Struik, LECTURES ON CLASSICAL DIFFERENTIAL GEOMETRY, Addison-Wesley (1950).

UNIVERSITY OF PENNSYLVANIA

CASE WESTERN RESERVE UNIVERSITY

THE CONNECTED COMPONENTS OF MORSE-SMALE VECTOR FIELDS ON TWO MANIFOLDS

by

C. Gutierrez and W. de Melo

Let Σ denote the set of Morse-Smale vector fields on a compact, two dimensional manifold M. Recall that a vector field X is Morse-Smale if it has finitely many hyperbolic critical elements (singularities and closed orbits), the ω-limit set of any orbit is a critical element and there is no saddle connection. Peixoto proved in [7] that Σ is open in $\mathfrak{X}^r(M^2)$, $r \geq 1$, and it is also dense if M^2 is orientable. Using Pugh's closing lemma it follows that Σ is also dense in $\mathfrak{X}^1(M^2)$ for any two dimensional compact manifold M. Furthermore those vector fields are structurally stable [6], [5], and in [9] Peixoto gave a classification of the topological equivalence classes in Σ. It is clear that a topological equivalence class in Σ may have infinitely many connected components. Here we give a classification of $\pi_0(\Sigma)$, the set of connected components of Σ.

In [8] Peixoto stated a theorem which relates the connected components of Σ with the isotopy classes of homeomorphisms of M. More precisely, two vector fields X and Y are in the same connected component of Σ iff there is a homeomorphism h, isotopic to the identity, sending oriented orbits of X onto oriented orbits of Y. This result does not establish a bijection between the connected components of a topological equivalence class in Σ and the isotopy classes of homeomorphisms. In fact, the topological equivalence class of the north pole-south pole vector field on S^2 is connected whereas there are two isotopy classes of homeomorphisms of S^2.

In our classification of $\pi_0(\Sigma)$ we caracterize the connected components of a given vector field in terms of some geometric properties of the vector field itself. Furthermore we give an explicit method to construct representatives for all connected components of Σ. Peixoto's result discussed above follows from our theorem.

In section 1 we classify the connected components of polar Morse-Smale vector fields, i.e., gradient-like vector fields in Σ having only one sink and one source. In section 2 we describe the connected components of gradient-like vector fields and in section 3 we study the connected components of Morse-Smale vector fields with closed orbits. To simplify the exposition we deal only with orientable manifolds although the results extend, with minor modifications, to non-orientable manifolds.

We finish by stating two problems:

Problem 1. Compute the fundamental group of each connected component of Morse-Smale vector fields on M^2.

Problem 2. Describe the connected components of Morse-Smale diffeomorphisms of M^2.

§1 Polar Morse-Smale vector fields

Let $X \in \mathfrak{X}^r(M)$, $r \geq 1$, be a polar Morse-Smale vector field. The union of the stable manifold of a saddle of X and the source is a simple closed curve in M called a stable cycle of X. If M is a k-torus and X is a polar Morse-Smale vector field on M then X has $2k$ stable cycles which intersect each other at the source. Since X is Morse-Smale, the complement of the union of the stable cycles is the stable manifold of the sink. Therefore if we cut M through the stable cycles we get a polygon with $4k$ edges whose vertices correspond to the source. Conversely, let us start with a regular polygon in the plane with $4k$ edges. Consider a vector field which has sources in the vertices, a saddle in the middle of each edge and one sink whose stable manifold is the disc bounded by the polygon. If we identify the edges in order to get the k-torus, then this vector field induces a polar Morse-Smale vector field on the k-torus. Each pair of identified edges gives rise to a stable cycle. With the same method we construct representatives of every equivalence class of polar Morse-Smale vector fields on any non-orientable manifold.

Let $X, Y \in \mathfrak{X}^r(M)$ be Morse-Smale vector fields. We say that X and Y are isotopically equivalent if they are in the same connected component of Σ. This means that there is a continuous mapping $F: [0,1] \to \mathfrak{X}^r(M)$ such that $F_0 = X$, $F_1 = Y$ and F_t is a Morse-Smale vector field for any $t \in [0,1]$. F is called an isotopical equivalence.

Theorem 1.1 - Let $X, Y \in \mathfrak{X}^r(M)$ be polar Morse-Smale vector fields.

Then X and Y are isotopically equivalent if and only if there is a bijection between the stable cycles of X and the stable cycles of Y such that the corresponding cycles are homotopic.

Before proving the above theorem we state the second theorem which gives the classification of the connected components.

Consider the collection of all subsets of $2p$ elements of the fundamental group $\pi_1(M)$. We say that two such subsets A and B are equivalent, $A \sim B$, if for any $\alpha \in A$ either $\alpha \in B$ or $-\alpha \in B$. Denote by $\pi_1^{(2p)}$ the set of equivalence classes. Let $X \in \mathfrak{X}^r(M)$, $r \geq 1$, be a polar Morse-Smale vector field and M be a p-torus. Then the $2p$ stable cycles of X represent an element of $\pi_1^{(2p)}$ which we denote by $\pi(X)$.

Proposition 1: Let $X, Y \in \mathfrak{X}^r(M)$ be polar Morse-Smale vector fields. Then

(i) If h is an automorphism of $\pi_1(M)$, there exists a Morse-Smale vector field $Z \in \mathfrak{X}^r(M)$ which is topologically equivalent to X and such that $\pi(Z) = h(\pi(X))$.

(ii) X and Y are topologically equivalent if and only if there is an automorphism h of $\pi_1(M)$ with $h(\pi(X)) = \pi(Y)$.

Proof. The first part of the proposition follows from a theorem of Nielsen [4]. In fact it states that there exists a diffeomorphism $H: M \to M$ such that the induced automorphism on $\pi_1(M)$ is $H^* = h$. Then it follows that $Z = H_*X$ satisfies the above conditions. Here $H_*X = dH \circ X \circ H^{-1}$.
Now we prove (ii). If $H: M \to M$ is a topological equivalence between X and Y, then $H^*: \pi_1(M) \to \pi_1(M)$ takes $\pi(X)$ onto $\pi(Y)$. Conversely, if there is an automorphism $h: \pi_1(M) \to \pi_1(M)$ then by (i) there exists a vector field Z which is topologically equivalent to X and such that $\pi(Z) = \pi(Y)$. From theorem 1.1 it follows that Z is topologically equivalent to Y. This proves the proposition.
Let $X^1,\dots,X^N \in \mathfrak{X}^r(M)$ be representatives of the different equivalence classes of polar Morse-Smale vector fields (see [3] and [9]. Let

A_i $i = 1,2,\ldots,N$, be the set of all elements $h(\pi(X^i)) \in \pi_1^{(2p)}$ where $h: \pi_1(M) \to \pi_1(M)$ is an automorphism. It follows from theorem 1.1 and proposition 1 that $A_i \cap A_j = \emptyset$ if $i \neq j$. Then we have

<u>Theorem 1.2</u>. The mapping π: {C^r polar Morse-Smale vector fields} $\to$ $\to A = \bigcup_{i=1}^{N} A_i$ induces a bijection between the connected components of polar Morse-Smale vector fields and A.

Let us consider, as an example, the torus T^2. There is only one equivalence class of polar Morse-Smale vector field on T^2. Let $\tilde{A}$ be the set of 2×2 matrices with integers entries and determinant ±1. In $\tilde{A}$ consider the equivalence relation $\sim$ which identifies two matrices B and C if the columns of B are equal to either the columns of C or of $-C$. Since the automorphisms of $\pi_1(T^2)$ are defined by the elements of $\tilde{A}$, it follows that $A = \tilde{A}/\sim$.

It remains to prove theorem 1.1. For that we need some lemmas.

<u>Lemma 1.1</u>. Let $X, Y \in \mathfrak{X}^r(M)$ be Morse-Smale vector fields and p be a singularity of X and Y of the same index of stability. Then X and Y are isotopically equivalents to vector fields $\tilde{X}$ and $\tilde{Y}$ such that, $\tilde{X} = \tilde{Y}$ in a neighborhood of p, $\tilde{X} = X$ and $\tilde{Y} = Y$ outside a neighborhood of p.

<u>Proof</u>. We construct an isotopical equivalence which does not modify the vector field outside a neighborhood of p. Therefore, we can assume that $X \in \mathfrak{X}^r(R^2)$ and $0 = p$. Let $L = DX_0$.
We consider first the case where 0 is a sink of X. Let S_t, $t \in (0,2]$, be a family of circles, each of them bounding a convex disc V_t which contains 0 such that the circles are disjoints, transversal to the vector fields X and L and $V_s - \{0\} = \bigcup_{t \leq s} S_t$. Since the discs V_t are convex, it follows that the circles S_t are also

transversal to the linear vector field $A = \begin{pmatrix} -1 & 0 \\ 0 & -1 \end{pmatrix}$.

Let $\varphi: \mathbb{R}^2 \to [0,1]$ be a C^∞ function such that $\varphi = 0$ outside V_2 and $\varphi = 1$ in V_1. If $R = A-X$ we have that the vector field $F_s = X + s\varphi R$ is transversal to S_t for all $t \in (0,2]$ and for all $s \in [0,1]$. Thus F_s is an isotopical equivalence between X and the vector field $\tilde{X} = F_1$ which coincides with A in a neighborhood of 0. Similarly, Y is isotopically equivalent to a vector field $\tilde{Y}$ which coincides with A in a neighborhood of 0.

Suppose now that 0 is a saddle of X. Let g be a diffeomorphism with support in a small neighborhood of the origin sending the local stable and unstable manifolds into the horizontal and vertical axis, respectively, and such that there exists an isotopy g_t with $g_0 =$ identity and $g_1 = g$. Let $X^s = (g_s)_* X$, then X^1 is isotopically equivalent to X and $L = DX^1(0) = \begin{pmatrix} \lambda & 0 \\ 0 & \mu \end{pmatrix}$ with $\lambda < 0$ and $\mu > 0$.

We are going to construct an isotopical equivalence F_s such that $F_o = X_1$ and F_1 is a vector field which coincides with L in a neighborhood of 0.

Let V_2 be a neighborhood of the origin whose boundary is a union of four segments transversal to X_1 and four segments of orbits of X_1. Let $\varphi: \mathbb{R}^2 \to [0,1]$ be a C^∞ function such that $\varphi = 0$ outside V_2 and $\varphi = 1$ in a neighborhood V_1 of 0. It is easy to see that 0 is the only singularity of $F_s = X_1 + s\varphi(X_1 - L)$ in V_2 for all $s \in [0,1]$. Thus from Poincaré-Bendixon theorem it follows that any orbit of F_s which enters V_2 either goes to 0 or leaves V_2 through one of the transversal segments to F_s and then goes to a sink. Therefore F_s is Morse-Smale for all $s \in [0,1]$.

With the same argument we can construct an isotopical equivalence between F_1 and a vector field $\tilde{X}$ which coincide with the vector field $\begin{pmatrix} -1 & 0 \\ 0 & 1 \end{pmatrix}$ in a neighborhood of 0.

Lemma 1.2. Let $X, Y \in \mathfrak{X}^r(M)$ be Morse-Smale vector fields, $B \subset M$ be an open ball and S a curve transversal to X such that $B \subset \bigcup_{t \in I} X_t(S)$, where I is a compact interval. Assume that B is contained in the stable manifold of an attractor or in the unstable manifold of a repellor of X.

Let U be an open set containing the closure $\overline{B}$ of B. If $Y = X$ in $U-B$ then Y is isotopically equivalent to a vector field $\tilde{Y}$ which coincides with X in U and with Y in $M-U$.

Proof. First of all by a small perturbation of Y, with support in B, we get a vector field Y^1 which is in general position with the fibration $S_t = X_t(S)$. This implies that Y^1 is either transversal to S_t or the set of tangencies is a union of circles. In each of those circles there are finitely many points where the tangency is of cubic type and in the other points the tangency is parabolic. Y^1 is isotopically equivalent to Y since it is near Y. If Y^1 is transversal to S_t then it is clear that Y^1 is isotopically equivalent to a vector field $\tilde{Y}$ which coincides with X in B and with Y in $M-B$. If Y^1 is not transversal to S_t there is a circle of tangencies C which bounds a disc D such that Y^1 is transversal to S^t in the interior of D. Then we can get an isotopical equivalence between Y^1 and a vector field Y^2 such that $Y^2 = Y^1$ outside a small neighborhood of D and Y^2 is tangent to S_t in the interior of D.
It is clear that Y^2 is isotopically equivalent to a vector field $\tilde{Y}^2$ which is transversal to S_t in a neighborhood of D and coincides with Y^2 outside this neighborhood.

By using this construction several times we can eliminate all circles of tangencies and we finally get a vector field Y^n isotopically equivalent to Y, which is transversal to S_t. Therefore Y^n is isotopically equivalent to a vector field Y satisfying the condition of this lemma.

Lemma 1.3. Let $X \in \mathfrak{X}^r(M)$ be a Morse-Smale vector field, $S_1 \subset M$ be a circle transversal to X contained in the stable manifold of an attractor or in the unstable manifold of a repellor of X. If $h: S_1 \to S_1$ is an orientation preserving diffeomorphism, then there is a vector field Y isotopically equivalent to X satisfying the following properties:

1) $Y = X$ in $M - \bigcup_{t\in[-1,0]} X_t S_1$;

2) The Poincaré mapping of Y from $X_{-1}S_1$ to S_1 is given by $h\circ X_1$.

3) If h is the identity in an interval $I \subset S_1$ then $Y = X$ in the union of the orbits of X through points of I.

Proof. By using the method of suspension of diffeomorphisms we get a vector field Y topologically equivalent to X satisfying the properties 1), 2) and 3). It remains to prove that X and Y are isotopically equivalent. Certainly Y is isotopically equivalent to a vector field Y^1 with the same properties and such that Y^1 has an orbit transversal to the circles S_t, $t \in [-1, 0]$. By a small perturbation of Y^1 we get a vector field Y^2 which is in general position with respect to the family of circles S_t, $t \in [-1, 0]$. Since Y_1 has an orbit transversal to all these circles any circle of tangency bounds a disc. Therefore we can use the arguments of lemma 1.2 to construct an isotopical equivalence between Y^2 and X. Hence X and Y are isotopically equivalent.

Lemma 1.4: Let $X, \tilde{X} \in \mathfrak{X}^r(M)$ be Morse-Smale vector fields. Assume that X and $\tilde{X}$ coincide in a C^∞, two-dimensional submanifold with boundary $U \subset M$. Suppose X and $\tilde{X}$ have trajectories

γ and $\tilde{\gamma}$, respectively, satisfying the properties:

(i) γ and $\tilde{\gamma}$ are transversal to the boundary of U and $\gamma \cap \partial U =$ $= \tilde{X} \cap \partial U = \{a,b\}$.

(ii) The α and ω-limit sets of γ and $\tilde{\gamma}$ are contained in the interior of U.

(iii) The arcs of trajectory $ab \subset \gamma$ and $\tilde{ab} \subset \tilde{\gamma}$, joining a and b, are homotopic, relatively to $\{a,b\}$, in the complement of the interior of U.

Then X is isotopically equivalent to a vector field Y which coincides with $\tilde{X}$ in U and in a neighborhood of $\tilde{\gamma}$.

<u>Proof</u>. It follows from [1] that there exists an isotopy of M taking ab on $\tilde{ab}$ which does not move a and b. Using the same argument we can construct a C^{∞} diffeotopy F_t, $0 \leq t \leq 1$, such that F_o is the identity, $F_1(ab) = F_1(\tilde{ab})$ and $F_t(p) = p$ for all $p \in U$. In fact, since any diffeomorphism of M close to the identity is diffeotopic to the identity we can assume that ab intersects transversally $\tilde{ab}$ (except at a and b) in finitely many points. In [2] it is proved the existence of subarcs mn and $\tilde{mn}$ of ab and $\tilde{ab}$, respectively, such that $mn \cup \tilde{mn}$ bounds a closed disc D in the complement of the interior of U with $D \cap ab = mn$ and $D \cap \tilde{ab} =$ $= \tilde{mn}$. Thus, there is a diffeotopy between the identity of M and a diffeomorphism f such that: i) f is the identity is U and outside a neighborhood of D; ii) the number of points in the intersection of f(ab) and $\tilde{ab}$ is either two or strictly smaller than the number of points in ab $\cap$ $\tilde{ab}$ as in the picture below.

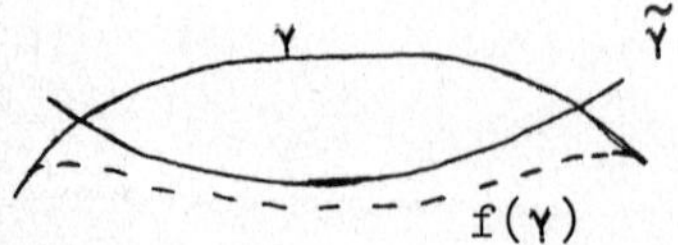

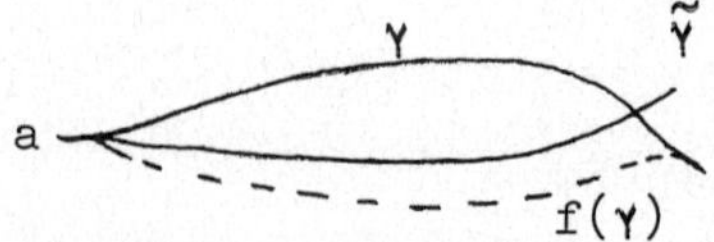

Therefore we can assume that $ab \cap \widetilde{ab} = \{a,b\}$. Since $ab \cup \widetilde{ab}$ bounds a disc we can get as before a diffeotopy F_t with the desired properties. Thus $(F_1)_*X$ is an isotopical equivalence between X and the vector field $(F_1)_*X$ which coincides with X in U and has $\widetilde{\gamma}$ as a trajectory. Now it is easy to see that this vector field is isotopically equivalent to a vector field which coincides with $\widetilde{X}$ in a neighborhood of $\widetilde{\gamma}$.

Proof of theorem 1.1. From lemma 1.1 we can assume that X and $\widetilde{X}$ have the same source p, the same attractor q and that they coincide in two disjoint discs D_1 containing p and D_2 containing q, whose boundaries are transversal to X. Let α and $\widetilde{\alpha}$ be homotopic stable cycles of X and $\widetilde{X}$ respectively. From lemma 1.3 we can assume that α and $\widetilde{\alpha}$ intersect the boundary of D_1 at the same points. By lemma 1.4, $\widetilde{X}$ is isotopically equivalent to a vector field X^1 which coincides with X in $D_1 \cup D_2$ and such that α is a stable cycle of X^1. Let β be another stable cycle of X and $\widetilde{\beta}$ be the stable cycle of X^1 homotopic to β. Since α is a stable cycle of both vector fields it follows from the arguments of lemma 1.4 that β is homotopic to $\widetilde{\beta}$ in the complement of α. Thus, using again lemmas 1.3 and 1.4, we construct an isotopical equivalence between X^1 and a vector field X^2 which has α and β as stable cycles. Therefore $\widetilde{X}$ is isotopically equivalent to a vector field Y which has the same stable cycles as X. From lemmas 1.1 and 1.3 it follows that Y is isotopically equivalent to a vector field Y^1 which coincides with X in $D_1 \cup D_2$ and in a neighborhood of the stable cycles. Since the complement of those neighborhoods is contained in a disc in the stable manifold of the sink, it follows that Y^1 is isotopically equivalent to X.

§2 Gradient-like Morse-Smale vector fields

We consider in this section Morse-Smale vector fields without closed orbits defined on a compact oriented two manifold.

The lemma below is well known.

Lemma 2.1. Let $X \in \mathfrak{X}^r(M)$, $r \geq 1$, be a gradient-like Morse-Smale vector field. Then there exist circles C_1 and C_2 transversal to X, bounding disjoint discs D_1 and D_2 respectively, such that D_1 contains all sources of X and D_2 contains all sinks of X.

Let C_1 and C_2 be circles as above. We say that C_i is a circle of type i. A point $p \in C_1$ is of type 1 (type 2), with respect to the pair (C_1,C_2), if it belongs to the unstable (stable) manifold of a saddle contained in $D_1(D_2)$ respectively. A point $p \in C_1$ is of type 3 if it belongs to the stable manifold of a saddle in $M-D_1 \cup D_2$. The stable cycle of a saddle $s \in M-D_1 \cup D_2$ is a simple closed curve which is the union of the stable manifold of s and an interval contained in the interior of D_1. We consider the circle C_i oriented as the boundary of D_i while D_i has the orientation induced by that of M.

If $X \in \mathfrak{X}^r(M^2)$ is a Morse-Smale vector field without closed orbits we denote by $\tilde{\pi}(X)$ the set of all 6-tuples $(C_1, C_2, F, S, T, \Gamma)$ where C_1 and C_2 are circles of type 1 and 2, F, S, T are the set of points of type 1, 2, 3 resp. and Γ is the set of stable cycles of X with respect to the pair (C_1,C_2).

Let $X, Y \in \mathfrak{X}^r(M)$ be gradient like Morse-Smale vector fields. We say that $(C_1,C_2,F,S,T,\Gamma) \in \tilde{\pi}(X)$ is equivalent to $(\tilde{C}_1,\tilde{C}_2,\tilde{F},\tilde{S},\tilde{T},\tilde{\Gamma}) \in \tilde{\pi}(Y)$ if there is an orientation preserving diffeomorphism $h: C_1 \to \tilde{C}_1$, satisfying the following properties:

1) $h(F) = \tilde{F}$, $h(S) = \tilde{S}$ and $h(T) = \tilde{T}$;

2) if $a, b \in C_1$ belong to the stable (unstable) manifold of the same saddle of X then $h(a)$ and $h(b)$ belong to the stable (unstable) manifold of the same saddle of Y;

3) if $a \in T$ then the stable cycle of X containing a is homotopic to the stable cycle of Y containing $h(a)$.

Denote by $\pi(X)$ the set of equivalence classes of $\tilde{\pi}(X)$.

Lemma 2.2. If $X \in \mathfrak{X}^r(M^2)$ is a gradient-like Morse-Smale vector field then $\pi(X)$ is a finite set.

Proof. Let (C_1,C_2,F,S,T,Γ) and $(\tilde{C}_1,\tilde{C}_2,\tilde{F},\tilde{S},\tilde{T},\tilde{\Gamma})$ be elements of $\tilde{\pi}(X)$. Denote by D_i and $\tilde{D}_i$ the discs bounded by C_i and $\tilde{C}_i$ respectively. We claim that if D_i and $\tilde{D}_i$ contain the same saddles of X then (C_1,C_2,F,S,T,Γ) is equivalent to $(\tilde{C}_1,\tilde{C}_2,\tilde{F},\tilde{S},\tilde{T},\tilde{\Gamma})$. In fact, since D_1 and $\tilde{D}_1$ contain the same saddles there is a real number $\lambda > 0$ such that $X_{-\lambda}(D_1) \subset \tilde{D}_1$. Let $h\colon X_{-\lambda}(C_1) \to \tilde{C}_1$ be the Poincaré mapping. Consider the diffeomorphism $h = \tilde{h} \circ X_{-\lambda}\colon C_1 \to \tilde{C}_1$. It is clear that h is orientation preserving. Since D_2 and $\tilde{D}_2$ contains the same saddles of X it follows that h defines an equivalence between (C_1,C_2,F,S,T,Γ) and $(\tilde{C}_1,\tilde{C}_2,\tilde{F},\tilde{S},\tilde{T},\tilde{\Gamma})$. Since X has finitely many saddles it follows that $\pi(X)$ is finite and the lemma is proved.

Theorem 2.1. Let $X, Y \in \mathfrak{X}^r(M)$ be gradient-like Morse-Smale vector fields. Then X and Y are isotopically equivalent if and only if $\pi(X) = \pi(Y)$.

Proof. If $\pi(X) = \pi(Y)$ there exist $(C_1,C_2,F,S,T,\Gamma) \in \tilde{\pi}(X)$ equivalent to some $(\tilde{C}_1,\tilde{C}_2,\tilde{F},\tilde{S},\tilde{T},\tilde{\Gamma}) \in \tilde{\pi}(Y)$. By an isotopy of the manifold we may assume that $\tilde{C}_1 = C_1$ and $\tilde{C}_2 = C_2$. Let $h\colon C_1 \to C_1$ be the diffeomorphism defining the equivalence between (C_1,C_2,F,S,T,Γ) and $(C_1,C_2,\tilde{F},\tilde{S},\tilde{T},\tilde{\Gamma})$. Since h is orientation preserving we can use the

same argument as in lemma 1.3 to get a vector field Y^1, isotopically equivalent to Y, such that the identity of C_1 is an equivalence between $(C_1,C_2,F,S,T,\Gamma) \in \tilde{\pi}(X)$ and $(C_1,C_2,F,S,T,\tilde{\Gamma}) \in \tilde{\pi}(Y^1)$. Let D_1 be the disc whose boundary is C_1. Using lemmas 1.4 and 1.1 several times we get a vector field Y^2, isotopically equivalent to Y^1, such that Y^2 coincides with X in D_1 and $(C_1,C_2,F,S,T,\Gamma) \in \tilde{\pi}(Y^2)$. By lemma 1.4 Y^2 is isotopically equivalent to a vector field Y^3 whose stable cycles coincide with those of X and such that $(C_1,C_2,F,S,T,\Gamma) \in \tilde{\pi}(Y^3)$. It is easy to see that Y^3 is isotopically equivalent to X.
The converse is easy.

Now we will describe all connected components of gradient like Morse-Smale vector fields on M^2. We start with vector fields on S^2.

Let $D \subset \mathbb{R}^2$ be the closed unit disc. A <u>cordal system</u> in D is a finite set of disjoint segments in D, called cords, each one joining exactly two points of the boundary, ∂D, of D. Two cordal systems are equivalent if there is an orientation preserving diffeomorphism $f: \partial D \to \partial D$ such that $a, b \in \partial D$ belong to the same cord of the first system if and only if $f(a)$ and $f(b)$ belong to the same cord of the other system. Let $\mathcal{C}^1, \mathcal{C}^2, \mathcal{C}^3, \ldots$ be the equivalence classes of cordal systems. For each $i = 1,2,\ldots$ we choose a cordal system in $\mathcal{C}^i$ and a vector field X^i on D satisfying the properties: i) X^i is transversal to ∂D; ii) X^i has one and only one saddle in each cord and this cord is contained in the unstable manifold of the saddle; iii) X^i has one and only one source in each connected component of the complement of the cords; iv) the α-limit set of any point in ∂D is either a source or a saddle. In the picture below we sketch the vector fields associated to the equivalence classes of cordal systems with one, two and three cords.

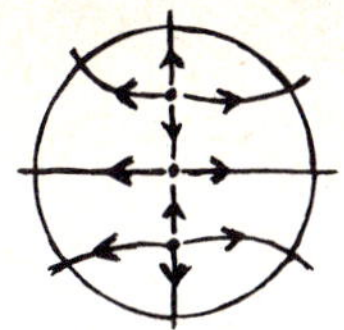 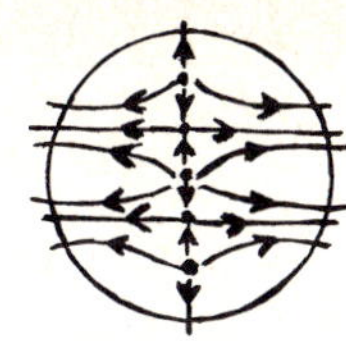 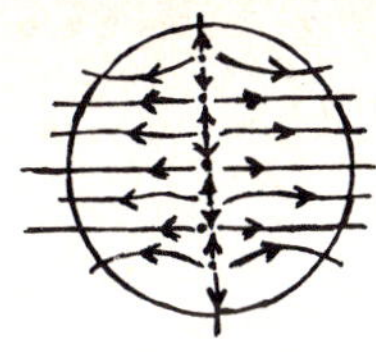 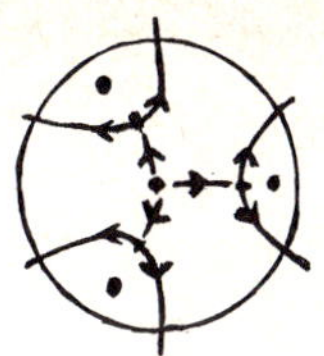

Consider two equivalence classes of cordal systems C^i, C^j and the corresponding vector fields X^i and X^j. Let H_{ij} be the set of orientation preserving diffeomorphisms $h: \partial D \to \partial D$ such that $h(p)$ is in the stable manifold of a sink of $-X^j$ whenever p is in the unstable manifold of a saddle of X^i. We say that $h_1, h_2 \in H_{ij}$ are equivalent if there is an orientation preserving diffeomorphism $h: \partial D \to \partial D$ satisfying the following properties: a) if $a \in \partial D$ is in the unstable manifold of a saddle of X^i then $h(a)$ is in the unstable manifold of the same saddle; b) if $b \in \partial D$ is in the stable manifold of a saddle of $-X^j$ then $h(h_1(b)) = h_2(b)$.

Denote by $\tilde{H}_{ij}$ the set of equivalence classes in H_{ij}. Clearly $\tilde{H}_{ij}$ is a finite set. If $h \in H_{ij}$ we can construct a gradient-like vector field on S^2 by taking two copies of D, one of them with the vector field X^i and the other with the vector field $-X^j$, and identifying the boundaries via h. Let us denote this vector field by $X(i,j,h)$. From theorem 2.1 it follows that $X(i,j,h_1)$ is isotopically equivalent to $X(i,j,h_2)$ whenever h_1 is equivalent to h_2. Thus we have defined a mapping from $\tilde{H}_{ij}$ into the set of connected components of Morse-Smale vector fields. From lemma 2.1 any connected component of gradient-like Morse-Smale vector field on S^2 can be obtainned in this way. Therefore there is a surjective mapping from $\cup \tilde{H}_{ij}$ onto the set of connected components of gradient-like Morse-Smale vector fields on S^2. This mapping together with theorem 2.1 provides the classification of those connected components. We observe that the above mapping is finite to one but it is not one to one as the example below shows.

Example. Consider the vector field X on S^2 whose orbit structure on the discs D_1 and D_2 are as in the picture. The Poincaré mapping from C_1 to C_2 is the identity.

Here, C_1, the boundary of D_1 is a circle of type 1. The points of type 1 in C_1 are F_1, $F_1' \in W^u(s_1)$ and $F_2, F_2' \in W^u(s_2)$. The points of type 2 are S_1, $S_1' \in W^s(s_3)$ and $S_2, S_2' \in W^s(s_4)$. We may consider another circle of type 1, $\tilde{C}_1$, which is described by broken lines in the picture. The points of type 1 in $\tilde{C}_1$ are $\tilde{F}_1, \tilde{F}_1' \in$ $\in W^u(s_2)$ and $\tilde{F}_2, \tilde{F}_2' \in W^u(s_3)$. The points of type 2 in $\tilde{C}_1$ are $\tilde{S}_1, \tilde{S}_1' \in W^s(s_1)$ and $\tilde{S}_2, \tilde{S}_2' \in W^s(s_4)$. It is easy to see that there is no diffeomorphism $h: C_1 \to \tilde{C}_1$ with $h(F) = \tilde{F}$ and $h(S) = \tilde{S}$.

Remark. A topological equivalence class of a Morse-Smale vector field on S^2 may have one or two connected components. In fact, the equivalence class of the polar vector field is connected. On the other hand the vector fields X and Y sketched in the picture below are topologically equivalent but belong to different connected components.

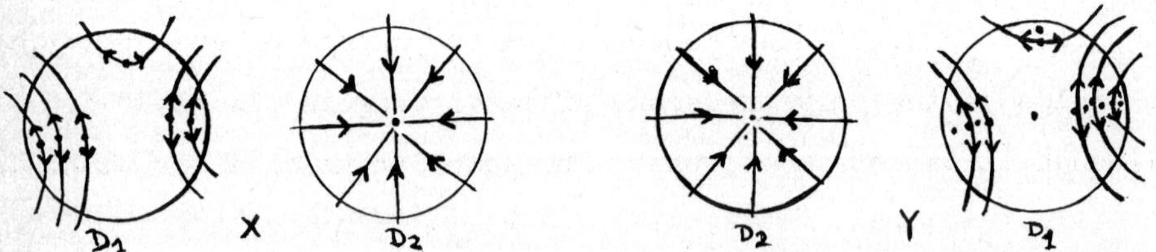

We describe now the connected components of gradient like-Morse-Smale vector fields on an orientable two manifold M. We start with vector fields having only one sink.
Let $\mathcal{B}_1$, $\mathcal{B}_2$,... be the connected components of polar Morse-Smale vector fields on M. For each $j = 1,2,...$ we choose a representative

$Y^j \in \mathfrak{B}_j$ and for each equivalence class of cordal system $\mathfrak{C}^i$ we pick a vector field X^i on D representing this class. Let C^j be a circle transversal to Y^j and enclosing a disc D^j which contains the source of Y^j. Denote by $\mathfrak{H}_{ij}$ the set of diffeomorphisms $h: \partial D \to C^j$ such that if $a \in \partial D$ is in the unstable manifold of a saddle of X^i then $h(a)$ is in the stable manifold of a sink of Y^j. We say that $h_1, h_2 \in \mathfrak{H}_{ij}$ are equivalent if there is an orientation preserving diffeomorphism $h: C^j \to C^j$ satisfying the following properties: i) if $p \in C^j$ is in the stable manifold of a saddle of Y^j then $h(p)$ is in the stable manifold of the same saddle; ii) if $q \in \partial D$ is in the unstable manifold of a saddle of X^i then $h(h_1(q)) = h_2(q)$. Let $\tilde{\mathfrak{H}}_{ij}$ be the set of equivalence classes in $\mathfrak{H}_{ij}$. Clearly $\tilde{\mathfrak{H}}_{ij}$ is a finite set. If $h \in \mathfrak{H}_{ij}$ we can construct, by glueing D and M-D with h, a vector field Z on M such that: a) C^j is a circle of type 1 for Z and the stable cycles of Z coincide with those of Y^j; b) $q \in C^j$ is a point of type 1 if and only if $h^{-1}(q)$ is in the unstable manifold of a saddle of X^i. From theorem 2.1 it follows that if $h_1, h_2 \in \mathfrak{H}_{ij}$ are equivalent then the corresponding vector fields $Z^1, Z^2 \in \mathfrak{X}^\infty(M)$ are isotopically equivalent. Thus we have defined a mapping from $\cup\tilde{\mathfrak{H}}_{ij}$ into the set of connected components of Morse-Smale vector fields. By lemma 2.1 this mapping is onto the set of connected components of gradient-like vector fields having only one sink. This mapping is clearly finite to one although it is not one to one. In fact, let us consider the vector field X on the torus T^2 as in the picture below.

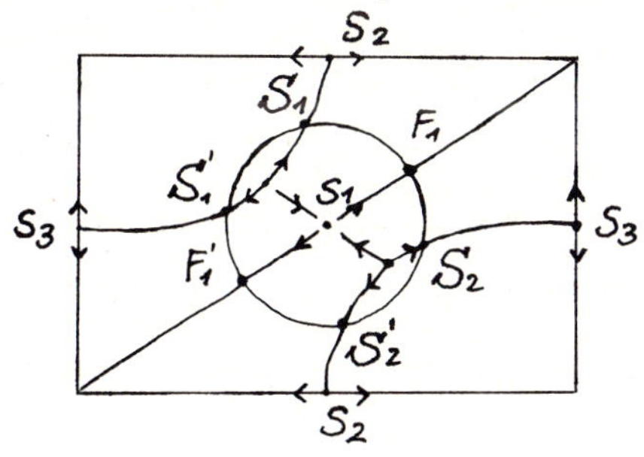

Here $\pi(X)$ contains (C_1,C_2,F,S,T,Γ) where $F = F_1,F_1' \in W^u(s_1)$, $S = \{S_1,S_1',S_2,S_2'\}$ and $\Gamma = \Gamma_1, \Gamma_2$. The stable cycle Γ_1, which contains the saddle s_2, is represented in $\pi_1(T^2) = \mathbb{Z} \times \mathbb{Z}$ by $(0,1)$ and Γ_2, the stable cycle of the saddle s_3, is represented by $(1,0)$. We may consider another circle of type 1, $\tilde{C}_1$, as in the follwing picture:

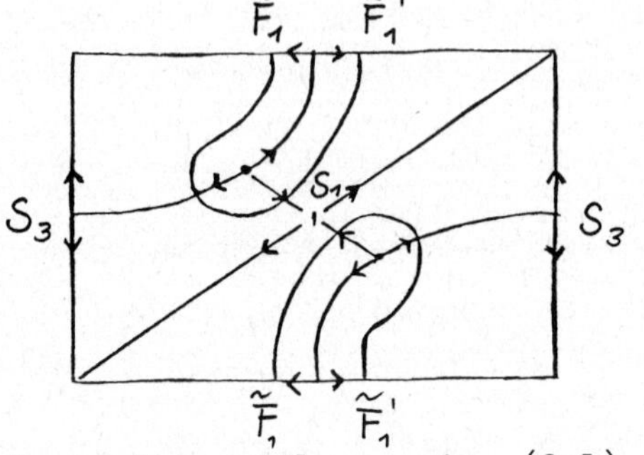

Here the stable cycle of the saddle s_1 is $(0,1)$ and the stable cycle of the saddle s_3 is $(1,1)$. Therefore $(\tilde{C}_1,\tilde{C}_2,\tilde{F},\tilde{S},\tilde{T},\tilde{\Gamma}) \in \tilde{\pi}(X)$ is not equivalent to $(C_1,C_2,F,S,T,\Gamma) \in \tilde{\pi}(X)$.

To describe the connected components of gradient like Morse-Smale vector fields on M we start with the connected components $\mathfrak{s}_1,\mathfrak{s}_2,\ldots$ of vector fields with only one sink as discussed above. In each $\mathfrak{s}_j$ we choose a representative $Z^j \in \mathfrak{X}^r(M^2)$. Let C_1^j and C_2^j be circles of type 1 and 2, respectively, for the vector field Z^j. Let X^i be a vector field on the disc D representating the equivalence class of cordal systems $\mathfrak{C}_i$. Let $h\colon \partial D \to C_1^j$ be a diffeomorphism such that if $a \in \partial D$ is in the unstable manifold of saddle of X^i then $h(a)$ is in the stable manifold of a sink of $-Z^j$. Using h to glue D and $M - D_1^j$ we get a gradient-like Morse-Smale vector field on M^2 such that: a) C_1^j and C_2^j are circles of type 1 and 2 for Y^2; b) the stable cycles of Y^j coincide with those of $-Z^j$; c) the points of type 2 and type 3 of Y^j are the points of type 2 and 3 of $-Z^j$; d) $p \in C_i^j$ is a point of type 1 for Y^j if and only if $h^{-1}(q)$ is in the unstable manifold of a saddle of X^i. By lemma 2.1 we can describe all the connected components of gradiente like Morse-Smale vector fields in this way. Now we consider the set $\mathcal{H}_{ij}$ of diffeomorphisms as above and define an equivalence relation on $\mathcal{H}_{ij}$ as before. Then the set $\widetilde{\mathcal{H}}_{ij}$ of equivalence classes is finite and there is a mapping from $\cup\widetilde{\mathcal{H}}_{ij}$ into the set of connected components. This mapping is finite to one and onto. This mapping and theorem 2.1 provide a classification of the connected componentes of gradiente-like Morse-Smale vector fields on M^2.

§3 Morse-Smale vector fields with closed orbits.

We consider here the set Σ^r, $r \geq 1$, of Morse-Smale vector fields on M. Let $X \in \Sigma^r$ and Λ be the set of closed orbits of X. Let $M_1, M_2, \ldots, M_\ell$, be the connected components of $M - \Lambda$. Certainly each hole of M_i, $i = 1, \ldots, \ell$, acts as an attractor or a reppelor for the restriction of X to M_i. If we assume that each hole of M_i is a singularity of X then X/M_i will have the phase space of a Morse-Smale vector field without closed orbits. Under this assumption we can define, as before, $\widetilde{\pi}(X/M_i)$. Let $\pi^*(X/M_i)$ denote the set of 8-tuples $(C_1, C_2, F, S, T, \Gamma, P, Q)$ where $(C_1, C_2, F, S, T, \Gamma) \in \widetilde{\pi}(X/M_i)$ and P(resp. Q) is the set of maximal intervals of $C_1 - F$ (resp. $C_1 - T$) such that $I \in P (I \in Q)$ if $\alpha(I)$, the set of α-limit points of orbits through points of I, is a closed orbit (resp. $\omega(I)$, of ω-limit points of orbits through I, is a closed orbit).

Let $X, Y \in \Sigma^r$ have the same set of closed orbits Λ. Let M_i, $i = 1, 2, \ldots, \ell$ be the connected components of $M - \Lambda$. We say that $(C_1, C_2, F, S, T, \Gamma, Q) \in \pi^*(X/M_i)$ is equivalent to $(\widetilde{C}_1, \widetilde{C}_2, \widetilde{F}, \widetilde{S}, \widetilde{T}, \Gamma, \widetilde{P}, \widetilde{Q}) \in \pi^*(Y/M_i)$ if there is an orientation preserving diffeomorphism $h: C_1 \to \widetilde{C}_1$ which gives an equivalence between $(C_1, C_2, F, S, T, \Gamma) \in \widetilde{\pi}(X/M_i)$ and $(\widetilde{C}_1, \widetilde{C}_2, \widetilde{F}, \widetilde{S}, \widetilde{T}, \Gamma) \in \widetilde{\pi}(Y/M_i)$ and moreover for each $I \in P$, $J \in Q$ we have $h(I) \in \widetilde{P}$, $h(J) \in \widetilde{Q}$, $\alpha(h(I)) = \alpha(I)$ and $\omega(h(J) = \omega(J)$. We denote by $\pi(X/M_i)$ the equivalence class of elements of $\pi^*(X/M_i)$.

Finally we write $\pi(X) = (\pi(X/M_1),\dots, \pi(X/M_\ell))$. We note that $\pi(X) = \pi(Y)$ for $X, Y \in \Sigma^r$ makes sence only when the set of closed orbits of X and Y coincide. Let $X, Y \in \Sigma^r$. We say that $\pi(X)$ is equivalent to $\pi(Y)$ if there is a vector field $Z \in \Sigma^r$, isotopically equivalent to Y, such that $\pi(X) = \pi(Z)$. Using the same arguments as in the proof of theorem 2.1 we get the following:

Theorem 3.1. $X, Y \in \Sigma^r$ are isotopically equivalent if and only if $\pi(X)$ is equivalent to $\pi(Y)$.

Now we will describe the connected components of Morse-Smale vector fields on $\mathfrak{X}^r(M)$. This description will be given by recurrence on the number of closed orbits and on the genus of the manifold M.

Let $X \in \mathfrak{X}^\infty(M)$ be a Morse-Smale vector field. Let p be an attracting (respec. a repelling) singularity of X and C_1, C_2 be circles transverse to X enclosing discs D_1, D_2, respectively such that $p \in D_1 \subset D_2$ and D_2 is contained in the stable manifold of p (respec. unstable manifold of p). Let M^* (respec. M_*) be the closure of $M-D_1$ and $X^* \in \mathfrak{X}^\infty(M^*)$ (respec. $X_* \in \mathfrak{X}^\infty(M_*)$) a vector field which coincides with X in $M-D_2$ and such that its phase portrait (the phase portrait of $-X$) in the closure of $D_2 - D_1$ is as in the picture below.

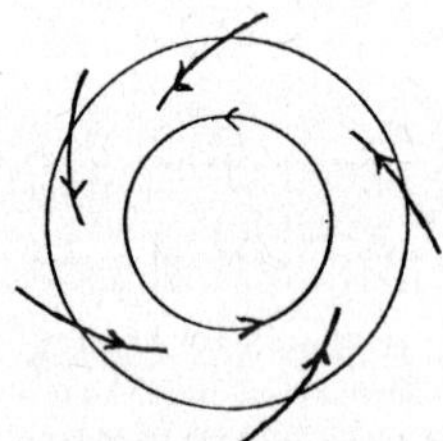

Let T^k denote the k-torus and $\pi^1_{jk}(T^{k+j})$ denote, for, $k,j=0,1,2,\dots,$ the set of elements α of the fundamental group $\pi^1(T^{k+j})$ such that

(i) α has a representative which is a simple closed curve (see [10])

(ii) If γ is a simple closed curve which represents $\alpha \in \pi^1_{jk}(T^{k+j})$ then $T^{k+j} - \gamma$ has two connected components so that one of them has Euler characteristic j.

Let $\alpha \in \pi^1_{jk}(T^{k+j})$. If $X^* \in \mathfrak{X}^\infty((T^k)^*)$ and $Y^* \in \mathfrak{X}^\infty((T^j)^*)$ are obtained as above and moreover C_k and C_j are the boundary of $(T^k)^*$ and $(T^j)^*$, respectively, then we can construct a C^∞ vector field, say $X^* \textcircled{\alpha} Y^*$, defined on the $k+j$ torus $(T^k)^* \textcircled{\alpha} (T^j)^*$ obtained by identifying C_k with C_j in such a way $C_k = C_j$ becomes a representative of α. Similarly $X_* \textcircled{\alpha} Y_*$ and $(T^k)_* \textcircled{\alpha} (T^j)_*$ can be defined.

Let $\pi^1_1(T^{k+1})$ be the set of elements α of the fundamental group $\pi^1(T^{k+1})$ such that α has a representative γ which is a simple closed curve and $T^{k+1} - \gamma$ is connected.

Let $\alpha \in \pi^1_1(T^{k+1})$. If $X \in \mathfrak{X}^\infty(T^k)$ is a Morse-Smale vector field with two attracting (resp. repelling) singularities, in a similar way as we have constructed X^*(resp. X_*) above, we can obtain a C^∞ vector field $\tilde{X}$ defined on a manifold $\tilde{M}$ whose boundary is* formed by two closed orbits of $\tilde{X}$, say C_1 and C_2. $X^{**}(\alpha)$ (respec. $X_{**}(\alpha)$) will denote the vector field defined on the $k+1$ torus $M^{**}(\alpha)$ (resp. $M_{**}(\alpha)$) obtained by identifying C_1 with C_2 in such a way $C_1 = C_2$ becomes a representative of α.

Let us assume that we have describe all the connected components of Morse-Smale vector fields on the sphere, on the torus,..., on the k-torus and moreover that we have described all the connected components of Morse-Smale vector fields on the (k+1)-torus with at most λ attracting closed orbits and θ repelling closed orbits. Let $X^1(j, \lambda_j, \theta_j),\ldots, X^N(j, \lambda_j, \theta_j)$, be representative of all the connected components defined on the j-torus, with λ_j attracting closed orbits and θ_j repelling closed orbits. Given two representatives $X = X^{s_1}(A, \lambda_A, \theta_A)$ and $Y = X^{s_2}(B, \lambda_B, \theta_B)$ having attracting singularities with $A+B = k+1$, $\lambda_A + \lambda_B = \lambda$ and $\theta_A+\theta_B=\theta$ and given $\alpha \in \pi^1_{A,B}(T^{k+1})$, there exists $X^s(k+1, \lambda+1, \theta)$ in the same connected component as that of $X^* \; ⓐ \; Y^*$. In the same way, given a representative $X^s(k, \lambda, \theta)$ with at least two attracting singularities and given $\alpha \in \pi^1_1(T^{k+1})$, there exist $X^s(k+1, \lambda+1, \theta)$ in the same connected component as that of $[X^s(k, \lambda, \theta]^{**}(\alpha)$.

Certainly any representative $X^s(k+1, \lambda+1, \theta)$ can be obtained in this form. Moreover given $X^s(k+1, \lambda+1, \theta)$, there exist only a finite number of vector fields in the sequence of representatives which can be used as above in order to obtain $X^s(k+1, \lambda+1, \theta)$. A similar statement can be done for any representative $X^s(k+1, \lambda, \theta+1)$.

REFERENCES

[1] R. Baer, Isotopie von Kurven auf orientierbaren, geschlossenen Flächen und ihr Zusammenhang mit der topologrochen Deformation der Flächen, J. reine augew. Math. 159(1928).

[2] D.B.A. Epstein, Curves on 2-manifolds and isotopies, Acta Math. 115 (1966).

[3] G.Fleitas, Classification of gradient like flows in dimension two and three, to appear in Bol. Soc. Bras. Mat.

[3] J. Nilsen, Untersuchungen zur Topologie der geschlossenen zweiseitigen Flächen, Acta math. 58, (1932) 87-167.

[5] J. Palis, S.Smale, Structural Stability Theorems, Global Analysis, Proc.Symp. in Pure Math., XIV, AMS (1970).

[6] M.C. Peixoto - M.M. Peixoto, Structural stability in the plane with enlarged boundary conditions. An.Acad.Bras.Ci. 31(1959) 135-160.

[7] M.M. Peixoto, Structural stability on two-dimensional manifolds, Topology 1 (1962) 101-120.

[8] M.M. Peixoto, Structural Stability on two-dimensional manifolds, Bol.Soc.Mat. Mexicana (1960) 188-189.

[9] M.M. Peixoto, On the Classification of Flows on 2-Manifolds Proc. Symp.Dyn.Systems. Salvador. Ac. Press (1973) 389, 419.

[10] B.L. Reinhart, Algorithm for Jordan curves on compact surfaces, Ann. of Math (2) 75 (1962), 209-222.

Instituto de Matemática Pura e Aplicada
Rio de Janeiro, RJ
Brasil.

FEUILLETAGES EN CYLINDRES

Gilbert HECTOR.

Un feuilletage de codimension 1 sur une variété compacte, est un <u>*feuilletage en cylindres*</u> *s'il possède des feuilles non compactes, toutes homéomorphes au cylindre* $S^1 \times \mathbb{R}$. *Nous nous proposons dans ce travail de :*

i) caractériser les variétés qui possèdent un feuilletage en cylindres de classe C^2 *;*

ii) classifier ces feuilletages à conjugaison topologique près.

INTRODUCTION.-

Du point de vue des résultats, le présent travail est analogue à celui de H. Rosenberg et R. Roussarie sur les feuilletages par plans (i.e. feuilletages dont les feuilles non compactes sont des plans (cf. [12], [13], [14]). Par contre, les méthodes utilisées sont très différentes, les difficultés à surmonter n'étant pas du tout les mêmes dans les deux cas. Ainsi notre étude est basée sur les trois observations suivantes dont deux au moins sont triviales pour les feuilletages par plans (*) :

i) un feuilletage en cylindres de classe C^2 sans feuille compacte est sans holonomie;

ii) toute feuille compacte d'un feuilletage en cylindres de classe C^2 est un tore ou une bouteille de Klein;

iii) tout feuilletage en cylindres de classe C^2 est presque sans holonomie.

(*) En outre, toute feuille compacte d'un feuilletage en plans de classe C^2 est un tore d'après la remarque 4 de [9].

Au paragraphe 1, on dresse un "catalogue" de feuilletages en cylindres, les paragraphes suivants ayant pour but de montrer que ce catalogue est exhaustif. Au § 2 [resp. § 4], on établit (i) [resp. (ii)], ce qui permet de classifier les feuilletages en cylindres de classe C^2 sans feuille compacte [resp. avec feuille compacte] au § 3 [resp. § 5] (la propriété (iii) est démontrée au § 5).

Ce travail a profité (à son origine) de l'ambiance stimulante du Symposium 1976 sur les feuilletages de l'Université de WARWICK et surtout de plusieurs conversations avec H. Rosenberg. L'intérêt manifesté par Alcides L. Neto de l'IMPA de RIO de JANEIRO a également permis de surmonter quelques difficultés.

1 - CATALOGUE DE FEUILLETAGES EN CYLINDRES.-

On note I l'intervalle [0,1] et on appelle <u>tronc de cylindre</u> le produit $S^1 \times I$. On fera abondamment usage dans la suite des quatre fibrés en troncs de cylindre au-dessus de S^1. Ces fibrés sont déterminés par le fait que l'homéomorphisme de recollement obtenu en coupant le long d'une fibre, conserve ou non l'orientation de $S^1 \times I$ et échange ou non les deux cercles du bord de $S^1 \times I$. Leur espace total est l'une des variétés suivantes :

i) le produit $T^2 \times I$ [resp. $K^2 \times I$] où T^2 et K^2 désignent respectivement le tore et la bouteille de Klein ;

ii) le fibré non trivial T [resp. K] de fibre I au-dessus de T^2 [resp. K^2].

On vérifie aisément que le bord des deux dernières est connexe, homéomorphe à T^2.

a) Suspensions de la fibration usuelle en cercles de T^2 *et* K^2.

Si L est l'une des deux surfaces T^2 ou K^2, on note $p : L \to \mathbb{S}^1$ la fibration usuelle en cercles "méridiens" de L.

Soit $\Phi(L)$ le groupe des homéomorphismes ψ de L tels que (*) :

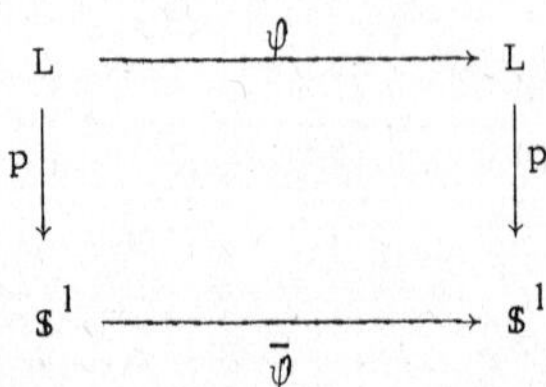

i) ψ est compatible avec la fibration p ;

ii) l'homéomorphisme $\bar{\psi} : \mathbb{S}^1 \to \mathbb{S}^1$ induit par ψ, conserve l'orientation de $\mathbb{S}^1$.

On désigne par M_ψ la variété obtenue en recollant les deux bords de $L \times I$ à l'aide de ψ et F_ψ le feuilletage de M_ψ induit par la fibration en troncs de cylindre de $L \times I$.

Définition 1.- On dira que (M_ψ, F_ψ) *est une suspension de la fibration en cercles de* L.

Pour toute suspension (M_ψ, F_ψ), l'homéomorphisme ψ est isotope à un homéomorphisme linéaire dont la matrice est de la forme $\begin{bmatrix} \varepsilon & n \\ 0 & 1 \end{bmatrix}$, où $\varepsilon = \pm 1$ suivant que ψ conserve ou non l'orientation des fibres de L et n est un entier relatif. Le couple (ε, n) caractérise la variété M_ψ.

(*) La condition (ii) permet de simplifier certains énoncés ultérieurs ; d'ailleurs les feuilletages que l'on obtiendrait à l'aide de "suspensions" ne vérifiant pas cette propriété, possèdent des feuilles compactes et donc se retrouvent sous la rubrique §(1,c).

Par ailleurs, toute feuille non compacte de F_ψ est évidemment un cylindre sans holonomie et F_ψ est sans feuille compacte si et seulement si le nombre de rotation $\rho(\bar{\psi})$ de $\bar{\psi}$ est irrationnel. Enfin si ψ est de classe C^2 et si $\rho(\bar{\psi})$ est irrationnel, toutes les feuilles de F_ψ sont partout denses et $\rho(\bar{\psi})$ caractérise la classe de conjugaison topologique de F_ψ d'après le théorème de Denjoy (cf. [2]).

b) *Modèles de feuilletages en cylindres.*

Soit $V = \{T^2 \times I, K^2 \times I, T, K\}$. Par analogie avec la classification introduite dans [4], on pose la définition suivante :

Définition 2.- *Un feuilletage en cylindres* (M,F) *est un modèle (de feuilletages en cylindres) de type (1) ou (2) si* $M \in V$, F *est presque sans holonomie, tangent à* ∂M *et si la condition correspondante est satisfaite :*

1) la restriction $\overset{\circ}{F}$ *de* F *à* $\overset{\circ}{M}$ *est une fibration en cylindres de* $\overset{\circ}{M}$ *sur* S^1 *;*

2) F *est transverse à la fibration de fibre* I *de* M *et toutes ses feuilles sont propres.*

Un modèle de type (1) est encore dit élémentaire.

A conjugaison topologique près, il y a exactement six modèles élémentaires. En effet, si (M,F) est un tel modèle, on peut, par les méthodes habituelles, relever chaque composante connexe L de ∂M dans un voisinage tubulaire du bord, transversalement à F et la variété obtenue en coupant le long de ces sous-variétés tansverses est un fibré en troncs de cylindre . D'autre part, l'holonomie de toute feuille compacte L est cyclique (puisque $\overset{\circ}{F}$ est une fibration) et donc il est facile de voir que tous les germes de feuilletages correspondants sont topologiquement conjugués. Ceci implique immédiatement que :

i) il y a exactement un modèle élémentaire sur T ou K ;

ii) il y a deux modèles élémentaires sur $T^2 \times I$ ou $K^2 \times I$, (suivant que les feuilles cylindriques de $\overset{o}{F}$ spiralent le long de deux composantes du bord dans le même sens ou en sens contraire).

Par exemple, les deux modèles sur $T^2 \times I$ peuvent être obtenus en faisant le produit par S^1 des feuilletages du tronc de cylindre représentés ci-dessous :

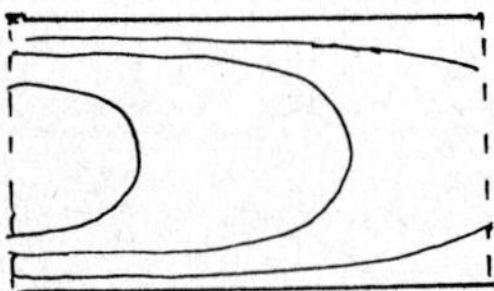
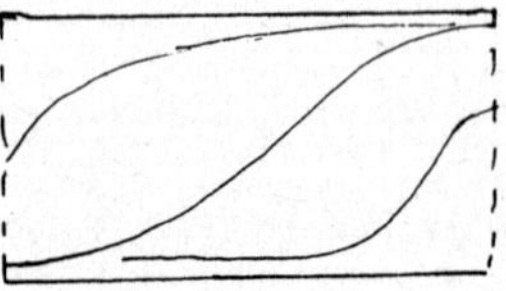

Par contre, il existe beaucoup de modèles de type (2). On peut cependant les classifier à conjugaison topologique près (au moins théoriquement) en remarquant que, d'après le théorème 2 de [4], tout modèle de type (2) se decompose en une famille au plus dénombrable de modèles élémentaires.

c) *Recollement de modèles.*

Enfin, on peut fabriquer de nouveaux feuilletages en cylindres en recollant à l'aide d'un homéomorphisme f :

i) soit les deux composantes du bord de $T^2 \times I$ [resp. $K^2 \times I$], la variété obtenue étant fibrée sur S^1 de fibre T^2 [resp. K^2] ;

ii) soit deux composantes connexes du bord de deux exemplaires de variétés appartenant à V ; la variété obtenue appartient à V si elle est à bord, elle est de la forme $M_1 \cup_f M_2$, avec M_i homéomorphe à T ou K, si elle est fermée.

2 - LE LEMME FONDAMENTAL.-

Au point de départ de cette étude, il y a évidemment le problème de savoir s'il existe ou non des minimaux exceptionnels dans les feuilletages en cylindres.

Lemme fondamental.- *Toutes les feuilles d'un feuilletage en cylindres* (M,F) *de classe* C^2 *sans feuille compacte sont partout denses et sans holonomie.*

En particulier, F *ne possède pas de minimal exceptionnel.*

La démonstration de ce lemme repose sur le fait qu'un minimal est un espace de Baire (cf. [1]) et que le pseudo-groupe d'holonomie (global) d'un feuilletage (de codimension 1) sur une variété compacte est dénombrable. Nous commençons par rappeler brièvement la construction de ce pseudo-groupe.

Soient (M,F) une variété feuilletée et $\Omega = \{\omega_1,\ldots,\omega_p\}$ un recouvrement ouvert distingué fini de M tel que, si $\omega_i \cap \omega_j \neq \emptyset$, toute plaque de ω_i rencontre au plus une plaque de ω_j, l'intersection $\omega_i \cap \omega_j$ définissant un homéomorphisme local

$$h_{ji} : X_i \longrightarrow X_j ,$$

où X_i est une sous-variété transverse à F représentant l'espace des plaques de ω_i. L'ensemble fini $\{h_{ji}\}$ engendre un pseudo-groupe dénombrable Γ d'homéomorphismes locaux de $X = \coprod_i X_i$, que l'on appelle le pseudo-groupe d'holonomie de F (réduit à X). Et si on note G_x le groupe des germes en $x \in X$ des éléments du sous-pseudo-groupe d'isotropie Γ_x de Γ en x, la représentation d'holonomie Ψ_x, au point x, de la feuille F_x passant par x, est un homomorphisme surjectif de $\pi_1(F_x,x)$ sur G_x.

Le premier lemme ci-dessous est immédiat (et sans doute connu) :

Lemme 1.- *Pour tout feuilletage* (M,F), *la réunion* H *des feuilles à holonomie non triviale est maigre au sens de Baire* (cf. [1]).

En effet, si $\partial\text{Fix}(g) = \text{Fix}(g) - \overset{\circ}{\overline{\text{Fix}(g)}}$ pour $g \in \Gamma$, on a $H \cap X = \bigcup_{g\in\Gamma} \partial\text{Fix}(g)$, donc $H \cap X$ est maigre dans X.

Pour $g \in \Gamma$, on note D_g le domaine de définition (que l'on peut toujours supposer connexe) de g, $\text{Fix}(g) \subset D_g$ l'ensemble de ses points fixes et $[g_x]$ son germe en $x \in \text{Fix}(g)$. Alors si F n'a pas de cycle évanouissant (cf. [10]) l'égalité $\Psi_x^{-1}([g_x]) = \{0\}$ pour un point $x \in \text{Fix}(g)$ implique $\Psi_y^{-1}([g_y]) = \{0\}$ pour tout $y \in \text{Fix}(g)$.

En particulier, si (M,F) est un feuilletage en cylindres de classe C^2 sans feuille compacte, on sait que F ne possède pas de cycle évanouissant d'après le théorème de Novikov (cf. [10]). Dans ce cas, on peut donc définit $\hat{\Gamma} \subset \Gamma$ comme étant l'ensemble des éléments g de Γ tels que $\Psi_x^{-1}([g_x]) \neq \{0\}$ pour tout $x \in \text{Fix}(g)$.

Lemme 2.- *Soient* (M,F) *un feuilletage en cylindres de classe* C^2 *sans feuille compacte,* M *un minimal de* F *et* M_X *sa trace sur* X.

S'il existe $g \in \hat{\Gamma}$ *tel que* $\text{Fix}(g) \cap M_X$ *est d'intérieur non vide (dans* M_X*), alors* $M = M$ *et* F *est sans holonomie.*

Démonstration.- Soit $g \in \hat{\Gamma}$ tel que $\text{int}(\text{Fix}(g) \cap M_X)$, intérieur de $\text{Fix}(g) \cap M_X$, soit non vide. Pour tout $x \in \text{int}(\text{Fix}(g) \cap M_X)$, l'holonomie correspondant à un élément non nul σ de $\pi_1(F_x,x)$ appartenant à $\Psi_x^{-1}([g_x])$ est non contractante. Donc puisque $\pi_1(F_x,x)$ est isomorphe à $\mathbb{Z}$, ceci implique que l'holonomie de F_x n'est pas contractante (i.e. ne possède aucun germe de contraction).

Or toute feuille F de F contenu dans M est dense dans M et donc coupe $\text{int}(\text{Fix}(g) \cap M_X)$. Par suite M n'est pas un minimal exceptionnel d'après le théorème de Sacksteder (cf. [15]),autrement dit $\mathcal{M} = M$ et l'holonomie de toute feuille F_x , $x \in \mathrm{int}(\mathrm{Fix}(g) \cap \mathcal{M}_X)$, est triviale. Bref $\mathcal{F}$ est sans holonomie.

Démonstration du lemme fondamental.- Soit $\mathcal{M}$ un ensemble minimal de $\mathcal{F}$.

a) Si $\mathcal{M} = M$, il existe $F \in \mathcal{F}$ à holonomie triviale d'après le lemme 1 et donc $\mathcal{F}$ est sans holonomie d'après le lemme 2.

b) Il nous reste donc simplement à montrer que $\mathcal{F}$ ne possède pas de minimal exceptionnel.

Or si $\mathcal{M}$ est un minimal exceptionnel de $\mathcal{F}$, on sait que $\mathrm{Fix}(g) \cap \mathcal{M}_X$ est rare dans $\mathcal{M}_X$ pour tout $g \in \hat{\Gamma}$ d'après le lemme 2. De plus si $\mathcal{H}$ désigne la réunion des feuilles à holonomie non triviale, on a :

$$\mathcal{H} \cap \mathcal{M}_X = \bigcup_{g \in \hat{\Gamma}} (\mathrm{Fix}(g) \cap \mathcal{M}_X) \ ;$$

donc $\mathcal{H} \cap \mathcal{M}_X$ est maigre dans l'espace de Baire $\mathcal{M}_X$ et il existe une feuille F dans $\mathcal{M}$ à holonomie triviale. Mais ceci est à nouveau impossible d'après le lemme 2. D'où le lemme.

Le lemme fondamental appelle quelques commentaires et remarques :

Remarque 1.- Le lemme fondamental n'est plus vrai en classe C^1 ; en effet le produit par S^1 du feuilletage de Denjoy sur T^2 est un feuilletage en cylindres de T^3 ayant un minimal exceptionnel.

Par contre le lemme reste valable en classe C^2 pour les minimaux locaux (i.e. minimaux pour le feuilletage induit dans un ouvert saturé).Il faut utiliser dans ce cas la version "généralisée" du théorème de Sacksteder de [4].

Remarque 2.- Il ne faudrait pas vouloir déduire de ce lemme que tout minimal exceptionnel de classe C^2 contient une feuille sans holonomie ; il est facile de construire un contre-exemple dans $V_3 \times S^1$ (où V_3 est la surface de genre 3) avec un feuilletage de classe C^∞ transverse à S^1.

Remarque 3.- On ne peut espérer améliorer le lemme en se restreignant à des conditions portant sur l'holonomie des feuilles (au lieu de leur groupe fondamental). En effet, le théorème 11 de [5] montre que l'holonomie de toute feuille exceptionnelle d'un feuilletage analytique est cyclique.

Remarque 4.- H. Rosenberg a suggéré d'étendre ce lemme aux feuilletages de variétés de dimension supérieure par des feuilles de la forme $T^k \times \mathbb{R}^\ell$. Ce sera possible grâce au théorème de N. Kopell, à condition de trouver une interprétation "à la Novikov" des cycles évanouissants dans cette situation particulière.

3 - FEUILLETAGES EN CYLINDRES SANS FEUILLE COMPACTE.-

Le lemme fondamental permet de classifier les feuilletages en cylindres sans feuille compacte.

Théorème 1.- *Tout feuilletage en cylindres, sans feuille compacte, de classe* C^2 *est une suspension de la fibration en cercles de* T^2 *ou* K^2.

La démonstration du théorème requiert un résultat préliminaire :

Lemme 3.- *Dans les conditions du théorème 1, il existe une sous-variété* L *de* M*, difféomorphe à* T^2 *ou* K^2, *transverse à* F *et sur laquelle* F *induit la fibration usuelle en cercles.*

Démonstration.- D'après le lemme fondamental, on sait que F est sans holonomie et que toutes ses feuilles sont partout denses. D'autre part, F ne possède pas de cycle évanouissant et donc on peut construire aisément deux plongements A et B du tronc de cylindre $S^1 \times I$ dans M tels que :

i) A est transverse à F et F induit sur A la fibration usuelle en cercles, chaque fibre de cette fibration représentant un générateur du groupe fondamental de la feuille correspondante ;

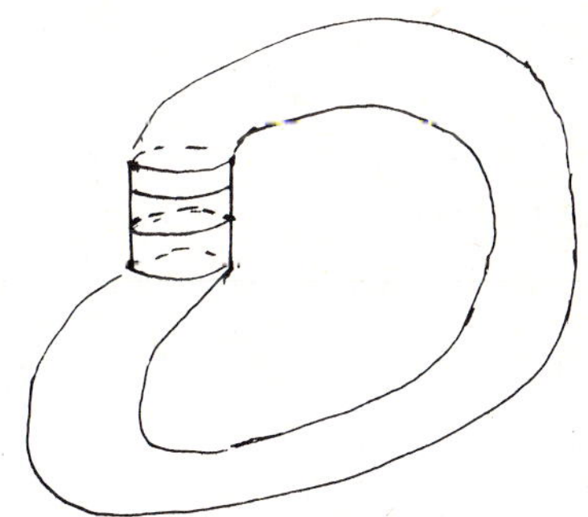

ii) B est contenu dans une feuille de F et $A \cap B = \partial A = \partial B$.

La variété topologique $L_o = A \cup B$ est homéomorphe à T^2 ou K^2 et si F est transversalement orientable, on peut lisser L_o par les méthodes habituelles pour la rendre transverse à F. On obtient ainsi une sous-variété L transverse à F et fibrée en cercles par F.

Si F n'est pas transversalement orientable, on lisse, après passage à un revêtement à deux feuillets, de manière compatible avec la projection. Dans tous les cas, on obtient la variété L annoncée.

Démonstration du théorème 1.-

a) Tout d'abord, on peut supposer que la variété L construite au lemme 3 ne sépare pas M. En effet, dans le cas contraire, soit $\hat{M}$ une des deux composantes obtenues en coupant M le long de L. La variété $\hat{M}$ est fibrée en troncs de cylindre par F et son bord est connexe, donc $\hat{M}$ est difféomorphe à T ou K. Or il est clair qu'il existe une sous-variété $\hat{L}$ de T [resp. K], difféomorphe à T^2 [resp. K^2], transverse à la fibration en troncs de cylindre de $\hat{M}$ et ne séparant pas $\hat{M}$; elle ne séparera donc pas non plus M.

b) La variété $\tilde{M}$ obtenue en coupant M le long de L (qui ne sépare pas M) est fibrée en troncs de cylindres par F et son bord n'est pas connexe. Donc $\tilde{M}$ est difféomorphe à $T^2 \times I$ ou $K^2 \times I$ et F est une suspension.

4 - GERMES DE FEUILLETAGES EN CYLINDRES.-

Pour caractériser les feuilles compactes des feuilletages en cylindres et classifier les germes de feuilletages correspondants, on utilise la classification générale des germes de feuilletages de [5].

a) Soient L une variété compacte et G le germe au voisinage de L d'un feuilletage F de codimension 1 de $L \times [0,1)$, transverse au facteur $[0,1)$ et admettant $L = L \times \{0\}$ comme seule feuille compacte (germe attractant). Pour tout $\varepsilon > 0$, on peut supposer que $L \times \{\varepsilon\}$ est en position générale par rapport à F. Le feuilletage F_ε induit par F dans $L \times [0,\varepsilon]$ est une réalisation de G qui est définie par une représentation d'image P_ε du groupe fondamental $\pi_1(L)$ de L dans le pseudo-groupe des homéomorphismes locaux de $[0,\varepsilon]$, définis au voisinage de 0 et ayant 0 comme point fixe. Enfin, le groupe G des germes en 0 des éléments de P_ε (groupe d'holonomie de L) caractérise le germe G d'après le théorème 2.7 de [3].

Par ailleurs, pour tout $\varepsilon > 0$, la relation d'équivalence ρ_ε associée à l'action de P_ε sur $(0,\varepsilon]$ possède un ensemble minimal non compact M_ε, contenant 0 dans son adhérence (dans $[0,\varepsilon]$). Et si F est de classe C^2, on sait d'après [5], que G vérifie l'une des deux conditions suivantes :

(I) il existe $\varepsilon > 0$ tel que M_η est une trajectoire propre monotone de limite 0 pour tout $\eta \leq \varepsilon$;

(II) $M_\varepsilon = (0,\varepsilon]$ pour tout ε ;

Suivant les cas, le germe G est dit de type (I) ou (II); il est dit presque sans holonomie s'il possède une réalisation presque sans holonomie.

<u>*Germes de feuilletages contractants au voisinage de*</u> T^2 <u>*ou*</u> K^2.-

1) Si $L = T^2$, tout germe attractant G de classe C^2 au voisinage de L est presque sans holonomie, de type (I) ou (II) suivant que le groupe d'holonomie de L est isomorphe à $\mathbb{Z}$ ou $\mathbb{Z}^2$ (cf. [8]). Tous les germes de type (I) sont topologiquement conjugués. De plus, pour toute réalisation F_ε de G, on peut supposer que $L \times \{\varepsilon\}$ est transverse à F, de sorte que les feuilles non compactes de F_ε sont des demi-cylindres $\mathbb{R} \times \mathbb{R}^+$ si G est de type (I) et des demi-plans $\mathbb{R} \times \mathbb{R}^+$ si G est de type (II).

2) Si $L = K^2$, on vérifie de même que tous les germes attractants de classe C^2 au voisinage de L sont presque sans holonomie de type (I). Ils sont tous topologiquement conjugués et les feuilles non compactes des réalisations F_ε d'un tel germe sont des demi-cylindres.

b) Venons-en maintenant aux germes de feuilletages en cylindres et soit (M,F) un feuilletage en cylindres sur une variété à bord $\partial M \neq \emptyset$, sans feuille compacte intérieure (i.e. tel que F_x est un cylindre, pour tout $x \in \overset{\circ}{M}$). Pour toute composante connexe L de ∂M, le germe G de F au voisinage de L est attractant et on a les caractérisations suivantes :

<u>Lemme</u> 4.- *Si G est de type (I) de classe C^2, la feuille L est difféomorphe à T^2 ou K^2 et G est l'un des deux germes de type (I) décrits plus haut.*

<u>*Démonstration*</u>.- Choisissons $\varepsilon > 0$ tel que, avec les notations de (a), M_ε soit une trajectoire propre monotone de P_ε. Comme $L \times \{\varepsilon\}$ est une position générale par rapport à F, on peut supposer que la feuille F_ε de la réalisation F_ε de G, correspondant à la trajectoire M_ε, soit transverse à $L \times \{\varepsilon\}$. Par suite F_ε est homéomorphe à $[F_\varepsilon \cap L \times \{\varepsilon\}] \times R^+$

et donc F_ε est un demi-plan ou un demi-cylindre.

En outre, si $P_\varepsilon(x)$ désigne le sous-pseudo-groupe d'isotropie de P_ε en un point $x \in [0,\varepsilon]$, il est facile de voir que :

i) $P_\varepsilon(x) = P_\varepsilon(y)$ pour tout $(x,y) \in M_\varepsilon \times M_\varepsilon$;

ii) $P_\varepsilon/P_\varepsilon(x)$, $x \in M_\varepsilon$, est un pseudo-groupe à un générateur.

D'où par passage aux germes, on obtient un homomorphisme $\Psi : \pi_1(L) \to \mathbb{Z}$ dont le noyau est représenté par les lacets de L qui se relèvent en lacets aux divers points de M_ε , autrement dit : $\text{Ker}\ \Psi = \pi_1(F_\varepsilon) = \mathbb{Z}^\alpha$, avec $\alpha \in \{0,1\}$.

La suite exacte $0 \longrightarrow \mathbb{Z}^\alpha \longrightarrow \pi_1(L) \longrightarrow \mathbb{Z} \longrightarrow 0$ implique que $L = T^2$ ou K^2 et donc G est l'un des germes décrits en (a). D'où le lemme.

<u>Lemme</u> 5.- *Si* G *est de type (II) de classe* C^2, G *est presque sans holonomie,* L *est difféomorphe à* T^2 *et* G *est encore l'un des germes décrits en (a).*

<u>*Démonstration*</u>.- Si G est de type (II), on déduit de la remarque 1 (à la suite du lemme fondamental) que G est presque sans holonomie. Mais il est alors bien connu que pour toute réalisation F_ε de G, les éléments de P_ε sont des contractions ou des dilatations et que le groupe d'holonomie G de L est isomorphe à $\mathbb{Z}^p$, $p \in \mathbb{N}$.

En procédant comme au lemme précédent, on obtient la suite exacte suivante, où $\alpha \in \{0,1\}$:

$$0 \longrightarrow \mathbb{Z}^\alpha \longrightarrow \pi_1(L) \longrightarrow \mathbb{Z}^p \longrightarrow 0 \quad .$$

On en conclut à nouveau que L est difféomorphe à T^2 ou K^2 ; donc finalement T^2 puisque G est de type (II) (cf (a)). D'où le lemme.

<u>Remarque</u> 5.- En fait nous montrerons au paragraphe 5 que tout germe de feuilletage en cylindres est de type (I).

5 - FEUILLETAGES EN CYLINDRES - CAS GENERAL.-

Nous commençons par étudier les feuilletages en cylindres sans feuille compacte intérieure.

<u>Lemme</u> 6.- *Soit* (M,F) *un feuilletage en cylindres de classe* C^2 *sans feuille compacte intérieure. Si* $\partial M \neq \emptyset$ *, la restriction* $\overset{o}{F}$ *de* F *à* $\overset{o}{M}$ *est une fibration de* $\overset{o}{M}$ *sur* S^1.

<u>Démonstration</u>.- Comme $\partial M \neq \emptyset$, toute feuille non compacte F de F contient (au moins), une composante connexe L de ∂M dans son adhérence. Par suite l'holonomie de F est triviale :

- d'après le remarque 1 si le germe G de F en L est de type (II),

- d'après la description des germes de feuilletages en cylindres s'il est de type (I).

Alors (M,F) est un modèle de feuilletages presque sans holonomie de type (1) au sens de [4]. Comme tel, il est transversalement orientable et de plus d'après le théorème 2 de [4] on sait que si $\overset{o}{F}$ est le feuilletage induit par F dans $\overset{o}{M}$, on a l'une des deux situations suivantes :

a) $\overset{o}{F}$ est une fibration de $\overset{o}{M}$ sur S^1 ;

b) toutes les feuilles de $\overset{o}{F}$ sont difféomorphes, partout denses et $\overset{o}{F}$ est topologiquement conjugué à un feuilletage défini par une forme fermée $\overset{o}{\omega}$ (qui ne s'étend pas à ∂M).

Pour terminer la démonstration du lemme, il nous suffit donc de montrer que le cas (b) est impossible. Or dans ce cas, le théorème de fibration de Tischler (cf. [16]), perfectionné à la manière de Joubert-Moussu (cf. [7]), et étendu aux feuilletages presque sans holonomie (cf. [6] et [4]) montre qu'il existe une fibration $\hat{F}$ de $\overset{o}{M}$ sur S^1 de fibre $\hat{F}$ telle que la feuille type F de $\overset{o}{F}$ soit un revêtement galoisien de $\hat{F}$ de groupe $\mathbb{Z}^p$, $p \in \mathbb{N}$. Comme F est un cylindre, ceci implique que $p = 0$ et donc $\hat{F}$ est également un cylindre. Bien plus en reprenant la démonstration du théorème de Joubert-Moussu on voit que $\overset{o}{F}$ et $\hat{F}$ sont topologiquement conjugués, donc $\overset{o}{F}$ est une fibration, ce qui démontre le lemme.

<u>Théorème</u> 2.- *Tout feuilletage en cylindres* (M,F) *de classe* C^2 *sans feuille compacte intérieure est un modèle élémentaire si* $\partial M \neq \emptyset$.

<u>Démonstration</u>.- D'après le lemme 6, le germe de F au voisinage de toute composante connexe L de ∂M est de type (I). Par suite, un cylindre $S^1 \times \mathbb{R}$ ayant deux bouts, ∂M a au plus deux composantes connexes qui sont difféomorphes à T^2 ou K^2 d'après le lemme 4. Par les méthodes habituelles, on peut relever chacune de ces composantes dans un voisinage tubulaire de ∂M, transversalement à F et la variété obtenue en coupant le long de ces sous-variétés transversales est un fibré en troncs de cylindre difféomorphe à M. Le théorème s'ensuit aisément.

Nous en arrivons enfin au cas général :

<u>Théorème</u> 3.- *Pour tout feuilletage en cylindres* (M,F) *de classe* C^2 *ayant au moins une feuille compacte,* F *est presque sans holonomie et* M *s'obtient en recollant en nombre fini de modèles.*

<u>Démonstration</u>.- On sait que pour tout feuilletage sur une variété compacte M, la réunion C des feuilles compactes est un fermé de M. Alors pour toute composante connexe W de $M - C$, le feuilletage F_W induit

par un feuilletage en cylindres F dans la variété $\hat{W}$, obtenue en coupant $\bar{W}$ le long des feuilles compactes intérieures, est p.s.h. d'après le lemme 6. Donc F lui-même est presque sans holonomie et la fin du théorème est simplement l'extension au cas non transversalement orientable du théorème de structure des feuilletages p.s.h. de [4] (théorème 1).

Comme on sait classifier les modèles (cf. paragraphe § (1,b)), le théorème 3 permettra de classifier les feuilletages en cylindres ayant au moins une feuille compacte. Enfin, des théorèmes 1 et 3, on peut déduire les quelques corollaires suivants :

Théorème 4.- *Une variété compacte* M *de dimension 3 possède un feuilletage en cylindres de classe* C^2 *si et seulement si elle vérifie l'une des trois propriétés ci-dessous :*

i) M *appartient à* $V = \{T^2 \times I, K^2 \times I, T, K\}$;

ii) M *est fibrée sur* S^1 *de fibre* T^2 *ou* K^2 ;

iii) $M = M_1 \cup_f M_2$, *où* M_i *est difféomorphe à* T *ou* K.

Corollaire 1.- *Pour tout feuilletage en cylindres* F *de classe* C^2, *toutes les feuilles de* F *sont simutanément propres ou partout denses.*

Corollaire 2.- *Toute feuille cylindrique d'un feuilletage en cylindres de classe* C^2 *a une croissance linéaire (au sens de* [4]*).*

Corollaire 3.- *Si la variété compacte* M *possède un feuilletage en cylindres de classe* C^2, *alors* $\pi_1(M)$ *est résoluble.*

Corollaire 4.- *Si la variété compacte* M *possède un feuilletage en cylindres analytique, alors* $H_1(M,\mathbb{R}) \neq 0$.

Pour ce dernier corollaire il suffit de remarquer que tous les germes de feuilletages en cylindres au voisinage d'une feuille compacte étant contractants cycliques, les arguments de Plante-Thurston dans [11] s'appliquent sans condition de croissance.

EN GUISE DE CONCLUSION.-

Bien sûr, connaissant la classification des feuilletages par plans et des feuilletages en cylindres, il serait tentant de vouloir classifier les feuilletages par cylindres <u>et</u> plans.

Pour ce faire, il semble que l'on puisse procéder de deux manières assez différentes :

a) Montrer que si $(M,\mathcal{F})$ est un feuilletage par cylindres et plans de classe C^2, alors $\pi_1(M)$ est résoluble.

En effet, d'après une communication orale, S. Goodman et J. Plante auraient complètement classifié les feuilletages des 3-variétés à groupe fondamental résoluble qui n'ont pas de composante de Reeb.

b) Déterminer une famille de "modèles" de feuilletages par cylindres et plans.

Dans cette direction, on se heurte aux deux problèmes préliminaires suivants :

Problème 1 : Montrer qu'un feuilletage par cylindres et plans de classe C^2 ne possède pas de minimal exceptionnel.

Problème 2 : Montrer que si $(M,\mathcal{F})$ est un feuilletage par cylindres et plans de classe C^2 sur une variété compacte à bord, sans feuille compacte intérieure, toutes les feuilles non compactes de $\mathcal{F}$ sont simultanément ou des cylindres ou des plans.

Remarquons que notre lemme fondamental ne permet pas de résoudre le problème 1 et que sur les variétés fermées, il existe des feuilletages par cylindres et plans de classe C^2 à holonomie non triviale dont toutes les feuilles sont partout denses (avec mélange de cylindres et plans).

REFERENCES.

[1] N. BOURBAKI - *Topologie Générale, chap. IX : Utilisation des nombres réels en topologie générale.*

[2] A. DENJOY - *Sur les courbes définies par les équations différentielles à la surface du tore.* J. de Math. Pures et Appl., 9-11 (1932), 333-375.

[3] A. HAEFLIGER *Variétés feuilletées.* Ann. Scuola Norm. Sup. Pisa, 16 (1964), 367-397.

[4] G. HECTOR - *Croissance des feuilletages presque sans holonomie.* A paraître dans Lecture Notes, School of Topology, PUC-RJ, 1976.

[5] G. HECTOR - *Classification cohomologique des germes de feuilletages (preprint).*

[6] H. IMANISHI - *Structure of codimension one foliations which are almost without holonomy.* J. of Math. Kyoto Univ., 16-1 (1976), 93-99.

[7] G. JOUBERT et R. MOUSSU - *Feuilletages sans holonomie d'une variété fermée.* C.R. Acad. Sc. Paris, 270 (1970), 507-509.

[8] N. KOPELL - *Commuting diffeomorphisms.* Proc. of Symp. in Pure Math., XIV, 165-184.

[9] R. MOUSSU et R. ROUSSARIE - *Relations de conjugaison et de cobordisme entre certains feuilletages.* Publ. Math. IHES, 43 (1974), 143-168.

[10] S.P. NOVIKOV - *Topology of foliations.* Trudy Mosk. Mat. Obshch., 14 (1965), 513-583. A.M.S. Translations, (1967), 268-304.

[11] J. PLANTE et W. THURSTON - *Polynomial growth in holonomy groups of foliations (preprint).*

[12] H. ROSENBERG - *Foliations by planes.*
Topology, 7 (1968), 131-138.

[13] H. ROSENBERG et R. ROUSSARIE - *Reeb foliations.*
Ann. of Math., 91 (1970), 1-24.

[14] H. ROSENBERG et R. ROUSSARIE - *Topological equivalence of Reeb foliations.*
Topology, 9 (1970), 231-242.

[15] R. SACKSTEDER - *Foliations and pseudo-groups.*
Amer. J. of Math., 87 (1965), 79-102.

[16] D. TISCHLER - *On fibering certain foliated manifolds over* S^1.
Topology, 9 (1970), 153-154.

Gilbert HECTOR
Université des Sciences et Techniques de Lille I
U.E.R. de Mathématiques Pures et Appliquées
B.P. 36
59650 - VILLENEUVE D'ASCQ

MESURE DE LEBESGUE ET NOMBRE DE ROTATION

Michael Robert HERMAN

Centre de Mathématiques de l'Ecole Polytechnique

Nous nous proposons de généraliser un théorème d'Arnold et de montrer qu'en un certain sens les difféomorphismes du cercle de classe C^r $(r \geq 3)$ qui sont C^{r-2}-conjugués à des rotations ont de "la mesure de Lebesgue" sur les chemins C^1 pour lesquels le nombre de rotation varie.

Soit $[a,b] \to f_t \in \mathrm{Diff}_+^r(T^1)$ un chemin de classe C^1 de difféomorphismes de T^1 préservant l'orientation de classe C^r, $r \geq 3$, et m la mesure de Lebesgue.

Posons $M(f_t) = m\ \{ t \in [a,b] \mid f_t$ est C^{r-2} conjugué à translation irrationnelle$\}$.

En 6, nous montrerons que, si $\rho(f_a) \neq \rho(f_b)$ = nombre de rotation de f_b, et si $r \geq 3$, alors $M(f_t) > 0$.

En 7, nous montrerons que, si $[0,1] \to f_t \in \mathrm{Diff}_+^r(T^1)$ "tend dans la C^3-topologie" vers le chemin standard, $\alpha \in [0,1] \to R_\alpha$ avec $R_\alpha(x) = x + \alpha$ (x et α sont pris modulo 1), alors $M(f_t) \to 1$.

En fait, ce deuxième résultat persiste sur T^n une fois définie en 2 la fonction vectorielle $\rho(f)$ pour f homéomorphisme de T^n dans T^n homotope à l'identité

($\rho(f)$ est une fonction de T^n dans T^n non nécessairement constante comme cela se produit sur T^1).

L'idée est très simple : la fonction "nombre de rotation" est lipschitzienne de rapport 1 aux points R_α, (rotation de nombre de rotation α) et même (d'après P. Brunovský [2]) est "dérivable" pour α irrationnelle. Ensuite on conjugue à des rotations par des difféomorphismes de classe C^1 en utilisant [4] et [6].

En fait, il est plus commode de travailler dans le revêtement $\mathbb{Z}$-cyclique de $\mathrm{Diff}^r_+(T^1)$, $D^r(T^1)$, car ce groupe est canoniquement un sous-groupe de $\mathrm{Diff}^r_+(\mathbb{R})$.

<u>Plan</u>

1. Notations.
2. Nombre de rotation.
3. Les problèmes et exemples.
4. Quelques lemmes.
5. Dérivabilité de la fonction ρ en R_α, $\alpha \in \mathbb{R} - \mathbb{Q}$.
6. Théorème des chemins.
7. Cas de T^n : généralisation d'un théorème d'Arnold en C^r.
8. Caractère localement lipschitzien de la fonction ρ sur des sous-espaces.

1. NOTATIONS

$m = dx$ = mesure de Lebesgue de $\mathbb{R}^n$ et désigne aussi la mesure de Haar de $T^n = \mathbb{R}^n/\mathbb{Z}^n$. Si $A \subset \mathbb{R}^n$ est m-mesurable, $m(A)$ désigne la mesure de Lebesgue de A.

1.1 Pour $r = 0$, ou $r \geq 1$ $r \in \mathbb{R}$, ou $r = +\infty$ ou ω, (ce qu'on écrit $0 \leq r \leq \omega$) on a le groupe

$$D^r(T^n) = \{ f = Id + \varphi \in \mathrm{Diff}^r(\mathbb{R}^n) \mid \varphi \in C^r(T^n, \mathbb{R}^n) \}$$

avec

$$C^r(T^n, \mathbb{R}^n) = (C^r(T^n))^n = \{ \varphi \in C^r(\mathbb{R}^n, \mathbb{R}^n) \mid \varphi \text{ est } \mathbb{Z}^n\text{-périodique} \}$$

(pour $r = 0$, $D^r(T^n)$ est un groupe d'homéomorphismes).

On a les translations (ou rotations) $R_\alpha : x \to x + \alpha$. R_α est une translation ergodique (sur T^n) si et seulement si les composantes de α sont irrationnelles et rationnellement indépendantes.

Pour $r = 0, 1, 2, \ldots$ $D^r(T^n)$ est un groupe topologique pour la C^r-topologie et $C = \{R_p \mid p \in \mathbb{Z}^n\}$ est le centre de $D^r(T^n)$.

$D^r(T^n)/C$ est isomorphe à $\mathrm{Diff}^r_o(T^n)$ qui est le groupe des difféomorphismes de classe C^r qui sont homotopes comme applications continues de T^n dans T^n à l'identité.

Si $r \geq 1$ est entier (ou non) $D^r(T^n)$ est un groupe qui est homéomorphe à un ouvert de $C^r(T^n, \mathbb{R}^n)$.

$D^r_+(T^n)$ est le sous-groupe composante connexe de l'Id de $D^r(T^n)$ pour la C^r-topologie. Si $r \geq 1$, $D^r_+(T^n)$ est homéomorphe à un ouvert connexe de $C^r(T^n, \mathbb{R}^n)$.

Pour $n = 1, 2, 3$, $D^r_+(T^n) = D^r(T^n)$, et si $n \geq 6$, pour tout $r \geq 0$, $\pi_o(D^r(T^n))$ n'est pas un groupe de type fini (d'après Hatcher et Wagoner).

On pose, si $x = (x_1, x_2, \ldots, x_n) \in \mathbb{R}^n$, $|x| = \sup_i |x_i|$; et si $\varphi : \mathbb{R}^n \to \mathbb{R}^n$ est une fonction bornée $\mathbb{Z}^n$-périodique

$$|\varphi|_o = \sup_{x \in \mathbb{R}^n} |\varphi(x)| .$$

On appellera "C^r-conjugaison" la conjugaison dans le groupe $D^r(T^n)$ et comme $D^r(T^n)$ est le revêtement $\mathbb{Z}^n$-cyclique de $\mathrm{Diff}_o^r(T^n)$, c'est la conjugaison dans le groupe $\mathrm{Diff}_o^r(T^n)$ en passant (mod $\mathbb{Z}^n$).

Si $f : \mathbb{R}^n \to \mathbb{R}^n$ et $k \in \mathbb{N}$, on notera f^k l'itérée k-ième de f. Si $f \in D^1(T^n)$, Df est la dérivée de f.

1.2 <u>Définition</u> : <u>Soient</u> $-\infty < a < b < +\infty$, n <u>et</u> p <u>des entiers et</u> $0 \leq r \leq \omega$.

1) $t \in [a,b]^p \to f_t = \mathrm{Id} + \varphi_t \in D^r(T^n)$ <u>est continue si</u> $t \to \varphi_t$ <u>est continue de</u> $[a,b]^p$ <u>dans</u> $C^r(T^n, \mathbb{R}^n)$. <u>Si</u> $p = 1$, $t \to f_t$ <u>s'appelle un chemin</u>.

2) $t \to f_t$ <u>est de classe</u> C^k, $k \geq 0$, <u>si</u> $t \to \varphi_t \in C^r(T^n, \mathbb{R}^n)$ <u>est de classe</u> C^k <u>et on pose</u> $k \geq 1$,

$$\frac{\partial f_t(x)}{\partial t} = \frac{\partial \varphi_t(x)}{\partial t} \in \mathcal{L}(\mathbb{R}^p, C^r(T^n, \mathbb{R}^n)).$$

1.3 <u>Proposition</u> : <u>Soient</u> $h_1 \in D^{r+1}(T^n)$, $h_2 \in D^r(T^n)$ <u>et</u> $[a,b]^p \ni t \to f_t \in D^r(T^n)$ <u>une application de classe</u> C^1, <u>alors</u> $t \in [a,b]^p \to h_1 \circ f_t \circ h_2 \in D^r(T^n)$ <u>est de classe</u> C^1, <u>et</u>

$$\frac{\partial}{\partial t}(h_1 \circ f_t \circ h_2) = (Dh_1 \circ f_t \circ h_2) . \left(\frac{\partial f_t}{\partial t} \circ h_2\right) .$$

<u>Démonstration</u> : Elémentaire ; voir [3].

<u>Remarque</u> : Noter le phénomène classique de la perte de dérivabilité.

2. NOMBRE DE ROTATION

2.1 Soit $f = Id + \varphi$ avec $\varphi \in C^{o}(T^{n}, \mathbb{R}^{n})$ (f n'est pas nécessairement un homéomorphisme).

On a par récurrence sur l'entier k

$$f^{k} = Id + \sum_{i=0}^{k-1} \varphi \circ f^{i} .$$

2.2 On pose

$$\rho(f) = \limsup_{k \to +\infty} \frac{1}{k} \sum_{i=0}^{k-1} \varphi \circ f^{i} \equiv \limsup_{k \to +\infty} \frac{f^{k} - Id}{k}$$

(lim sup signifie lim sup de chaque composante). (On peut alternativement prendre la lim inf.)

2.3 $\rho(f)$ est une fonction de $\mathbb{R}^{n}$ dans $\mathbb{R}^{n}$, $\mathbb{Z}^{n}$-périodique en général non constante (voir [4]).

2.4 On a les propriétés suivantes :

a) $\rho(R_{p} \circ f) = p + \rho(f)$, si $p \in \mathbb{Z}^{n}$;

b) $\rho(R_{\alpha}) = \alpha \in \mathbb{R}^{n}$ (i.e. $\rho(R_{\alpha})$ est constante $= \alpha$) ;

c) Si $f = h^{-1} \circ R_{\alpha} \circ h$ avec $h \in D^{o}(T^{n})$ alors on a $\rho(f) = \alpha \in \mathbb{R}^{n}$, (et même si $k \to +\infty$, $(f^{k} - Id)/k$ converge uniformément vers $\alpha \in \mathbb{R}^{n}$) ;

d) si $g = h^{-1} \circ f \circ h$ avec $h \in D^{o}(T^{n})$, alors on a $\rho(g) = \rho(f) \circ h$ (écrire $h \circ g^{k} = f^{k} \circ h$ si $k \in \mathbb{N}$, et remplacer $h = Id + \varphi$).

2.5 Cas de T^{1}

Si $f = Id + \varphi \in D^{o}(T^{1})$ (ou plus généralement si $f = Id + \varphi$ est continue monotone non décroissante, avec φ $\mathbb{Z}$-périodique), alors on a $\rho(f) \in \mathbb{R}$, et

$$\left| \frac{f^k - Id}{k} - \rho(f) \right| < \frac{1}{k} .$$

Il en résulte que, lorsque $k \to +\infty$, $\frac{f^k - Id}{k}$ converge uniformément vers le nombre $\rho(f)$; la fonction $\rho : D^o(T^1) \to \mathbb{R}$ est continue et constitue un invariant de C^o-conjugaison à la rotation $R_{\rho(f)}$. Par 2.4 a), ρ passe au quotient et définit un invariant continu de conjugaison sur $Diff^o_+(T^1)$ à valeurs dans T^1 que l'on note encore ρ. On désigne aussi par ρ la restriction de $\rho : D^o(T^1) \to \mathbb{R}$ à un sous-espace de $D^o(T^1)$.

On a aussi pour tout entier k, $f^k - Id - k\rho(f)$ s'annule (i.e. a un zéro). Il suit la propriété : soient $p/q \in \mathbb{Q}$, $p \in \mathbb{Z}$, $q \in \mathbb{N}$, $q \geq 1$, $(p,q) = 1$, et $f \in D^o(T^1)$. $\rho(f) = p/q$ si et seulement si $f^q - R_p$ s'annule $(f^q - R_p \in C^o(T^1))$.

Il en résulte que (mod 1) f a un point périodique d'ordre minimal q, et donc pour $1 \leq i < q$, f^i (mod 1) est sans point fixe.

Il en résulte aussi que $\rho(f) \in \mathbb{R} - \mathbb{Q}$ si et seulement si (mod 1) f est sans point périodique.

3. LES PROBLEMES ET EXEMPLES

3.1 Soit $\alpha \in \mathbb{R} - \mathbb{Q}$ et $\alpha = a_o + 1/(a_1 + 1/a_2 + \ldots)$ son développement en fraction continue.

Définition : α satisfait à une condition A, si on a

$$\lim_{B \to +\infty} \limsup_{N \to +\infty} \left(\sum_{a_i \geq B,\, 1 \leq i \leq N} \operatorname{Log}(1 + a_i) \Big/ \sum_{1 \leq i \leq N} \operatorname{Log}(1 + a_i) \right) = 0 .$$

Soit A l'ensemble des $\alpha \in \mathbb{R} - \mathbb{Q}$ qui satisfont à une condition A.

alors $\mathbb{R} - A$ est la mesure de Lebesgue nulle et, si $\alpha \in A$, pour tout $\epsilon > 0$, il existe $C_\epsilon > 0$, tel que, pour tout $p/q \in \mathbb{Q}$, on ait

$$(*) \qquad \left| \alpha - \frac{p}{q} \right| \geq C_\epsilon / q^{2+\epsilon} .$$

Observons que ceci ne caractérise pas $\alpha \in A$, par exemple le nombre $e \notin A$ mais satisfait à la condition $(*)$.

3.2 Nous avons montré en [4], [5] et [6] le :

<u>Théorème</u> : <u>Si</u> $3 \leq r \leq \omega$ <u>et</u> $f \in D^r(T^1)$ <u>avec</u> $\rho(f) = \alpha \in A$, <u>alors</u> f <u>est</u> $C^{r-1-\beta}$ $(\forall\ \beta > 0)$-<u>conjugué à</u> R_α <u>(avec la convention que</u> $\infty - 2 = \infty$ <u>et</u> $\omega - 2 = \omega$).

3.3 On montre (voir [4]) que, si $0 \leq r \leq \omega$, pour tout $p/q \in \mathbb{Q}$, le fermé $F^r_{p/q} = \{f \in D^r(T^1) \mid \rho(f) = p/q\}$ est connexe et d'intérieur non vide dans $D^r(T^1)$. De plus, $0^r_{p/q} = \{f \in D^r(T^1) \mid f^q = R_p\} \equiv \{g^{-1} \circ R_{p/q} \circ g \mid g \in D^r(T^1)\}$ est un fermé contenu dans la frontière de $\mathrm{Int}(F^r_{p/q})$ dans $D^r(T^1)$. Comme $\mathrm{Int}(F^\omega_{p/q}) \neq \emptyset$ (et est même connexe), on peut construire dans $\mathrm{Int}(F^\omega_{p/q})$ des chemins analytiques non constants $[a,b] \in t \to f_t$ vérifiant $\frac{\partial f_t}{\partial t} > 0$, mais tels que f_t ne soit C^0-conjugué à la rotation $R_{p/q}$ pour aucun $t \in [a,b]$ puisque $(f_t)^q \neq R_p$.

De plus, l'ouvert $U^r = \bigcup_{p/q \in \mathbb{Q}} \mathrm{Int}(F^r_{p/q})$ est dense dans $D^r(T^1)$, et si $r \geq 1$, U^r contient l'ouvert dense des difféomorphismes structurellement stables (voir [1]).

Si $t \in [a,b] \to f_t \in D^0(T^1)$ est un chemin continu avec $\rho(f_a) \neq \rho(f_b)$, on doit s'attendre à ce qu'en général la fonction $t \in [a,b] \to \rho(f_t)$ reste constante sur un intervalle d'intérieur non vide chaque fois que $\rho(f_t) \in \mathbb{Q}$ (voir [2]).

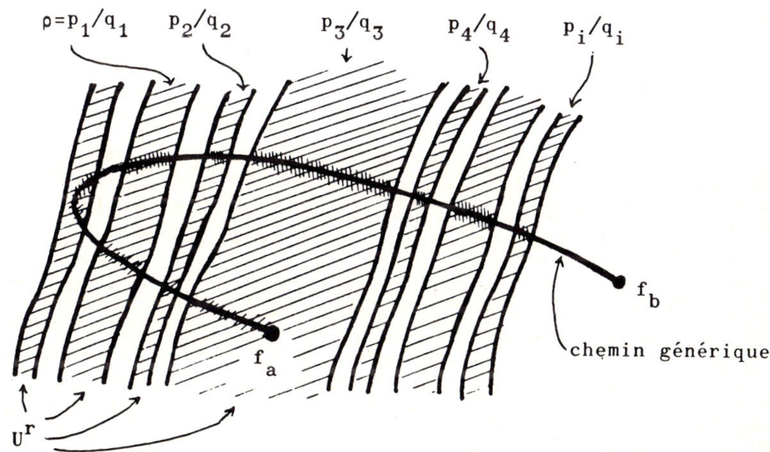

Nous allons rappeler des exemples explicites.

3.4 Etude d'un chemin (voir [4] pour les démonstrations).

Soient $f \in D^o(T^1)$ et le chemin C^ω, $t \in \mathbb{R} \to R_t \circ f \in D^o(T^1)$.

Posons $\rho(t) = \rho(R_t \circ f)$. On a les propriétés suivantes :

a) $\rho(t+1) = \rho(t) + 1$;

b) ρ est continue et donc prend toutes les valeurs réelles ;

c) ρ est monotone non décroissante ;

d) si $\rho(t_o) \in \mathbb{R} - \mathbb{Q}$, alors ρ est strictement croissante en t_o ;

e) si f a la propriété A_o (i.e. pour tout $p \in \mathbb{Z}$ et $q \in \mathbb{Z} - \{0\}$ et tout $t \in \mathbb{R}$, on a $(R_t \circ f)^q \neq R_p$) alors pour tout $p/q \in \mathbb{Q}$, $\rho^{-1}(p/q)$ est un intervalle d'intérieur non vide.

(Si $p/q \in \mathbb{Q}$, $h^{-1}(p/q)$ = un seul point, est équivalent à $(R_t \circ f)^q = R_p$).

Si f satisfait la propriété A_o, alors $\mathbb{R} - \mathrm{Int}\rho^{-1}(\mathbb{Q})$ est un fermé parfait totalement discontinu et (mod 1) c'est un Cantor.

3.4.1 Remarque : Soit $f = Id + \varphi \in D^o(T^1)$ avec φ non constante, alors, pour tout $t \in [\mathrm{Min}\,\varphi, \mathrm{Max}\,\varphi]$, $\rho(R_{-t} \circ f) = 0$.

3.5 Définition : $t \in \mathbb{R} \to f_t \in D^o(T^1)$ est un chemin positif si on a :

a) $f_{t+1} = 1 + f_t$ pour tout $t \in \mathbb{R}$;

b) et si pour tout $x \in \mathbb{R}$ et tout $t_1 < t_2$, on a l'inégalité $f_{t_1}(x) < f_{t_2}(x)$.

Exemples :

1) $t \to R_t \circ f \in D^o(T^1)$.

2) $t \to f_t \in D^o(T^1)$, chemin C^1 vérifiant a) et tel que, pour tout t, $\dfrac{\partial f_t}{\partial t} > 0$.

Remarque : Pour les chemins positifs, on a les mêmes propriétés que pour le chemin $t \to R_t \circ f$ (soit 3.4 a) à e)).

3.6 Exemples de f satisfaisant 3.4 e) (voir [4]).

a) $f = Id + \varphi \in D^{\omega}(T^1)$ avec φ une fonction entière non constante ; par exemple, si $0 < |a| < 1/2\pi$,

$$f_a(x) = x + a \sin 2\pi x .$$

b) Nous avons montré en [4] que, pour tout $0 \leq r \leq \omega$, l'ensemble des $f \in D^r(T^1)$ qui satisfont la propriété A_o est un résiduel dense dans $D^r(T^1)$ pour la C^r-topologie (en fait c'est une intersection dénombrable d'ouverts denses).

3.7 Description d'un exemple.

3.7.1 Fixons $0 < |a| < 1/2\pi$ et soit $f_a(x) = x + a \sin 2\pi x$, et posons $\rho_a(t) = \rho(R_t \circ f_a)$. On a $\rho_a(0) = 0$, $\rho_a(1) = 1$, et $\rho_a : [0,1] \to [0,1]$ a les propriétés de 3.4.

$K_a = [0,1] - \mathrm{Int}\, \rho_a^{-1}(\mathbb{Q})$ est un Cantor et, pour tout $p/q \in [0,1] \cap \mathbb{Q}$, $\rho_a^{-1}(p/q)$ est un intervalle d'intérieur non vide.

Notons que ρ_a n'est pas de classe C^1, puisqu'elle est constante sur chaque composante connexe de l'ouvert dense $\mathrm{Int}\, \rho_a^{-1}(\mathbb{Q})$ et donc de dérivée nulle sur cet ouvert dense : si ρ_a était C^1, alors ρ_a serait constante, absurde.

Bien que $\rho_a(K_a) = [0,1]$, il se pourrait que l'on ait $m(K_a) = 0$ et ρ_a serait alors une fonction singulière (i.e. de dérivée m-presque partout nulle : une fonction de Lebesgue).

V.I. Arnold a montré que, si $|a| \to 0$, $m(K_a) \to m([0,1]) = 1$ (voir [1]).

Nous montrons qu'en fait, pour tout $|a| < 1/2\pi$, on a $m(K_a) > 0$ et même ρ_a est absolument continue sur $[0,1]$.

3.7.2 Lignes de niveau $\rho = C^{te}$.

On considère dans le plan (t,a) le rectangle $0 \leq t \leq 1$, $0 \leq a \leq 1/2\pi$ et la famille à 2 paramètres $(t,a) \to R_t \circ f_a$.

On peut tracer les lignes de niveau ρ = constante.

$\rho(R_t \circ f_a) = 0$ si et seulement si $R_t \circ f_a$ a un point fixe soit $|t| \leq |a|$.
Pour chaque $p/q \in \mathbb{Q}$, $F_{p/q} = \{(t,a) \mid \rho(R_t \circ f_a) = p/q\}$ est un fermé d'intérieur non vide ; et, si $a_o \neq 0$, la frontière de $F_{p/q}$ est coupée en deux points par la droite $a = a_o$, qui sont ainsi des graphes de fonctions continues de a.

Par 3.4 d), si $\alpha \in \mathbb{R} - \mathbb{Q}$, l'ensemble $F_\alpha = \{(t,a) \mid \rho(R_t \circ f_a) = \alpha\}$ est le graphe d'une fonction continue de a.

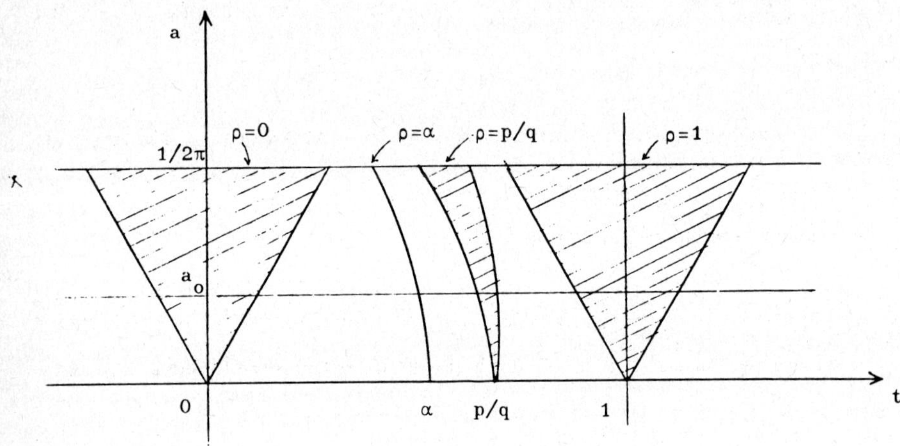

3.8 Les fonctions et problèmes.

(tous les ensembles considérés sont m-mesurables, voir [4]).

3.8.1 Soit $f \in D^o(T^1)$; et posons

$$M_1(f) = m\{t \in [0,1] \mid \rho(R_t \circ f) \in \mathbb{R} - \mathbb{Q}\} .$$

(i.e. la mesure de Lebesgue de l'ensemble considéré).

On montre que $M_1 : D^o(T^1) \to [0,1]$ est semi-continue supérieurement. Evidemment, pour tout $\lambda \in \mathbb{R}$, $M_1(f) = M_1(R_\lambda \circ f)$. Si f n'est pas une rotation, par 3.4.1, $M_1(f) < 1$.

$M_1(f)$ n'est pas continue dans la C^0-topologie au point $f = Id$, $(M_1(Id) = 1)$.

Exemple : Considérons $q \in \mathbb{N}$, et $0 < a < 1/2\pi$ fixé et posons :

$$f_q(x) = x + \frac{1}{q} a \sin 2\pi qx \quad .$$

Si $q \to +\infty$, $f_q \to Id$ dans la C^0-topologie, mais

$$1 - M_1(f_q) = 1 - M_1(f_1) \geq a = m\{t \in [0,1] \mid \rho(R_t \circ f_1) = 0\} \ .$$

3.8.2 On suppose $\beta > 0$ (petit), $2n + 2\beta \leq r \leq \omega$ et $f \in D^r(T^n)$, posons $\beta' > \beta$; disons $\beta' = \frac{3}{2}\beta$ et $M(f) = m\{t \in [0,1]^n \mid R_t \circ f$ est $C^{r-n-\beta'}$-conjugué à une translation ergodique$\}$.

Théorème : Soit n fixé ; si $f \to Id$ dans $C^{2n+2\beta}$-topologie, alors $M(f) \to 1$ (i.e. $M : D^r(T^n) \to [0,1]$ est continue au point $f = Id$).

La démonstration se trouve au § 7.4.

3.8.3 Soient $3 \leq r \leq \omega$ et $t \in [a,b] \to f_t \in D^r(T^1)$ un chemin continu ; posons $M(f_t) = m\{t \in [a,b] \mid R_t \circ f$ est C^{r-2}-conjugué à une rotation irrationnelle$\}$.

Problème : Si $\rho(f_a) \neq \rho(f_b)$, est-ce que $M(f_t) > 0$?

Il faut nécessairement faire une hypothèse de dérivabilité sur le paramétrage ainsi que le montre l'exemple suivant.

Exemple : On reprend l'exemple de 3.7.1.

Soit $G \subset]0,1[$ un Cantor de mesure nulle et $\psi : [0,1] \to [0,1]$ un homéomorphisme croissant tel que $\psi(G) = K_a$.

Soit le chemin $t \in [0,1] \to R_{\psi(t)} \circ f_a \in D^\omega(T^1)$, continu pour la C^ω-topologie (mais non absolument continue). On a :

$$0 = m(G) = m\{t \in [0,1] \mid \rho(R_{\psi(t)} \circ f) \in \mathbb{R} - \mathbb{Q}\} .$$

4. QUELQUES LEMMES

On suppose dans la suite que $-\infty < a < b < +\infty$.

4.1 LEMME. Soit $f = Id + \alpha + \varphi$ avec $\alpha \in \mathbb{R}^n$ et $\varphi \in C^o(T^n, \mathbb{R}^n)$; on a

$$|\rho(f) - \rho(R_\alpha)|_o \leq |\varphi|_o = |f - R_\alpha|_o .$$

Démonstration : $\rho(f) = \alpha + \limsup_{k \to +\infty} \frac{1}{k} \sum_{i=0}^{k-1} \varphi \circ f^i$, et $\rho(R_\alpha) = \alpha$,

donc $$|\rho(f) - \rho(R_\alpha)|_o \leq \frac{1}{n}\, n\, |\varphi|_o = |\varphi|_o = |f - R_\alpha|_o . \quad \square$$

4.2 Remarque : Le lemme 4.1 dit que ρ est lipschitzienne en R_α (de rapport de Lipschitz égal à 1).

4.3 LEMME. Soit $t \in [a,b]^n \to f_t = Id + \alpha + \varphi_t$ tel que $t \to \varphi_t \in C^o(T^n, \mathbb{R}^n)$ soit de classe C^1 et que, pour un $t_o \in [a,b]^n$, $\varphi_{t_o}(x) \equiv 0$. Alors, si $t_o + \Delta t \in [a,b]^n$, on a l'inégalité :

$$|\rho(f_{t_o + \Delta t}) - \rho(f_{t_o})| \leq C\, |\Delta t| ,$$

avec $$C = \sup_{x \in \mathbb{R}^n,\, t \in [a,b]^n} \left| \frac{\partial f_t}{\partial t}(x) \right| .$$

Démonstration : Comme $f_{t_o} = R_\alpha$, on applique 4.1 en utilisant

$$|\varphi_{t_o + \Delta t}(x) - \varphi_{t_o}(x)| = |\varphi_{t_o + \Delta t}(x)| \leq C\, |\Delta t|$$

qui résulte de la formule de la moyenne. $\square$

4.4 LEMME. Soit $t \in [a,b] \to f_t \in D^o(T^1)$ un chemin de classe C^1. On suppose que $f_{t_o} = h^{-1} \circ R_\alpha \circ h$ pour un $t_o \in [a,b]$ et $h \in D^1(T^1)$. Alors on a l'inégalité :

$$\left| \frac{\rho(f_{t_o + \Delta t}) - \rho(f_{t_o})}{\Delta t} \right| \leq C ,$$

avec $$C = |Dh|_0 \sup_{x \in \mathbb{R},\, t \in [a,b]} \left| \frac{\partial f_t(x)}{\partial t} \right| .$$

Démonstration : Si $g_t = h \circ f_t \circ h^{-1}$ avec $g_{t_0} = R_\alpha$, l'invariance par conjugaison de ρ entraîne :

$$\left| \frac{\rho(f_{t_0+\Delta t}) - \rho(f_{t_0})}{\Delta t} \right| = \left| \frac{\rho(g_{t_0+\Delta t}) - \rho(g_{t_0})}{\Delta t} \right| ,$$

l'inégalité cherchée résulte alors de 1.3 et 4.4. □

4.5 Remarque : Soient $K \subset [a,b]^n$ un compact, et $t \in [a,b]^n \to f_t \in D^0(T^n)$ une application de classe C^1. On suppose que, pour tout $t \in K$, f_t est C^0-conjugué à une translation et que, de plus, pour un $t_0 \in K$, f_{t_0} est C^1-conjugué à R_α. Soit $f_{t_0} = h^{-1} \circ R_\alpha \circ h$. Alors, pour tout $t_0 + \Delta t \in K$, on a l'inégalité :

$$|\rho(f_{t_0+\Delta t}) - \rho(f_{t_0})| \le C |\Delta t| , \quad \text{avec } C = |Dh|_0 \sup_{x \in \mathbb{R}^n,\, t \in [a,b]^n} \left| \frac{\partial f_t(x)}{\partial t} \right| .$$

5. DERIVABILITE DE LA FONCTION ρ en R_α pour $\alpha \in \mathbb{R} - \mathbb{Q}$

5.1 Rappel : Soit $w : [0,1] \to [0,1]$ un module de continuité.

$C^w(T^1) = \{\varphi \in C^0(T^1) \mid \sup_{x \neq y,\, |x-y| \le 1} \left| \frac{\varphi(x) - \varphi(y)}{w(|x-y|)} \right| = |\varphi|_w < +\infty\}$ est une algèbre de Banach pour la norme $|\ |_{C^w} = |\ |_0 + |\ |_w$.

5.2 Proposition : Soient $\alpha \in \mathbb{R} - \mathbb{Q}$ et w un module de continuité, il existe une suite de nombres réels positifs $(C_n(\alpha))_{n \in \mathbb{N}}$ avec $\lim_{n \to +\infty} C_n(\alpha) = 0$ (la suite ne dépend que de w et de α), telle que, pour tout $\varphi \in C^w(T^1)$, on ait, si $n \ge 1$:

$$\left| \frac{1}{n} \sum_{i=0}^{n-1} \varphi \circ R_{i\alpha} - \int_{T^1} \varphi(x)dx \right|_0 \le |\varphi|_w C_n(\alpha) \;^{\dagger}.$$

† FIxons $\beta > 0$ et C petit ; si $K_C = \{\alpha \in [0,1] \mid \forall p/q \in \mathbb{Q},\ |\alpha - p/q| \ge C/q^{2+\beta}\}$ on peut choisir la suite $C_n(\alpha)$ indépendante de $\alpha \in K_C$ (et dépendant seulement de C et β).

Pour la démonstration, voir [7, p. 146].

Le théorème de 5.3 b) est de P. Brunovský (voir [2]).

5.3 Théorème : Soit $\alpha \in \mathbb{R} - \mathbb{Q}$.

a) ρ: $D^1(T^1) \to \mathbb{R}$ est dérivable en R_α et l'application dérivée est :

$$\varphi \in C^1(T^1) \longrightarrow \int_{T^1} \varphi(x)dx \quad \in \quad \mathbb{R}$$

(ρ n'est pas dérivable en $R_{p/q}$, $p/q \in \mathbb{Q}$).

b) Si $t \in [a,b] \to f_t(x) = x + \alpha + \varphi_t(x) \in D^o(T^1)$ est un chemin de classe C^1 et si $\varphi_o(x) \equiv 0$, alors $t \to \rho(f_t)$ est dérivable pour $t = 0$, et on a :

$$\frac{\partial}{\partial t} \rho(f_t)_{t=0} = \int_0^1 \frac{\partial f_t}{\partial t}(x)dx \quad .$$

Démonstration de a) : On pose $|\Delta\varphi|_{C^1} = |\Delta\varphi|_o + |D\Delta\varphi|_o$. Il faut montrer que, pour tout $\epsilon > 0$, il existe $\eta > 0$ tel que, si $|\Delta\varphi|_{C^1} \leq \eta$, on ait

$$|\rho(f) - \alpha - \int_{T^1} \Delta\varphi(x)dx|_o \leq \epsilon\, |\Delta\varphi|_{C^1}$$

où $f = Id + \alpha + \Delta\varphi$.

Par 5.2, il existe $n_o \geq 1$ tel que, pour $n \geq n_o$, on ait

$$|\frac{1}{n} \sum_{i=0}^{n-1} \Delta\varphi \circ R_{i\alpha} - \int_{T^1} \Delta\varphi(x)dx|_o \leq |\Delta\varphi|_{C^1}\, \epsilon \, .$$

On conclut, si $|\Delta\varphi|_o$ est assez petit, que :

$$|\frac{1}{n_o} \sum_{i=0}^{n_o-1} \Delta\varphi \circ f^i(x) - \int_{T^1} \Delta\varphi(x)dx|_o \leq 2\epsilon\, |\Delta\varphi|_{C^1}$$

(car $|\Delta\varphi \circ f^i(x) - \Delta\varphi \circ R_{i\alpha}|_o \leq |D\Delta\varphi|_o\, |f^i - R_{i\alpha}|_o$ par la formule de la moyenne et pour i fixé, $f \to f^i$ est continue pour la C^o-topologie).

Si k est un entier :

$$\left| \frac{1}{n_o k} \sum_{i=0}^{n_o k-1} \Delta\varphi \circ f^i - \int_{T^1} \Delta\varphi(x)dx \right|_o \leq 2|\Delta\varphi|_{C^1}\, \epsilon .$$

Un raisonnement élémentaire de division $(n = n_o k + r,\ r < n_o)$ donne finalement qu'il existe un entier n_1 et $\eta > 0$ tel que, pour $n \geq n_1$ et $|\Delta\varphi|_{C^1} \leq \eta$, on ait

$$\left| \frac{1}{n} \sum_{i=0}^{n-1} \Delta\varphi \circ f^i - \int_{T^1} \Delta\varphi(x)dx \right|_o \leq 3|\Delta\varphi|_{C^1}\, \epsilon ,$$

et donc, par passage à la limite,

$$\left| \rho(f) - \rho(R_\alpha) - \int_{T^1} \Delta\varphi(x)dx \right|_o \leq 3|\Delta\varphi|_{C^1}\, \epsilon . \quad \square$$

5.4 <u>Remarque</u> : Dans la démonstration de 5.3 a), on peut remplacer la topologie C^1 par une topologie C^o plus module de continuité (en utilisant 5.2) et 5.3 b) résulte alors immédiatement de cette remarque.

5.5 <u>Corollaire</u>. <u>Soit</u> $f \in D^1(T^1)$ <u>de la forme</u> $f = h^{-1} \circ R_\alpha \circ h$, <u>avec</u> $\alpha \in \mathbb{R} - \mathbb{Q}$ <u>et</u> $h \in D^1(T^1)$. <u>Alors</u> $\rho : D^1(T^1) \to \mathbb{R}$ <u>est dérivable en</u> f, <u>de dérivée</u>

$$\Delta\varphi \in C^1(T^1) \longrightarrow \int_{T^1} (Dh \circ f \circ h^{-1}) \,.\, \Delta\varphi \circ h^{-1}\, dm \in \mathbb{R} .$$

<u>Démonstration</u> : Cela résulte de 5.4 et de ce que l'application $f \to h \circ f \circ h^{-1}$ de $D^1(T^1)$ dans $D^o(T^1)$ est dérivable (voir [3]). $\square$

5.6 <u>Corollaire</u> : <u>Soient</u> $t \in [a,b] \to f_t \in D^o(T^1)$ <u>un chemin de classe</u> C^1, <u>et</u> $t_o \in [a,b]$ <u>tel que</u> $f_{t_o} = h^{-1} \circ R_\alpha \circ h$, <u>avec</u> $\alpha \in \mathbb{R} - \mathbb{Q}$ <u>et</u> $h \in D^1(T^1)$. <u>Alors</u> $t \in [a,b] \to \rho(f_t)$ <u>est dérivable en</u> t_o, <u>de dérivée en</u> t_o,

$$\left.\frac{\partial \rho(f_t)}{\partial t}\right|_{t=t_o} = \int_{T^1} Dh(f_{t_o}(h^{-1}(x))) \,.\, \frac{\partial f_{t_o}}{\partial t}(h^{-1}(x))dx \in \mathbb{R} .$$

Démonstration : Cela résulte de 1.3 et de 5.3 b). □

5.7 On pourrait peut être penser que, si $t \in [a,b] \to f_t \in D^{\omega}(T^1)$ est un chemin C^{ω}, alors $t \to \rho(f_t)$ est une fonction lipschitzienne. †

Contre-exemple (dû à Pliss, voir aussi [1]) : Soit $X_t(x) = t + \sin^2 2\pi x \in C^{\omega}(T^1)$ un champ de vecteur C^{ω} dépendant de façon C^{ω} de $t \in \mathbb{R}$. Soit $t \to f_t$ le chemin analytique de difféomorphisme de classe C^{ω} obtenu en prenant pour f_t le difféomorphisme au temps $\xi = 1$ dans le groupe à un paramètre $h_t(\xi)$ engendré par X_t (i.e. $f_t = \exp(\xi X_t)|_{\xi=1}$). Si $t \in [-1,0]$, $\rho(f_t) = 0$, et, si $t > 0$, $\rho(f_t)$ dépend de façon analytique de t. En effet X_t est alors C^{ω}-conjugué au champ de vecteur constant $= \rho(f_t)$, (i.e. il existe $g_t \in D^{\omega}(T^1)$ tel que :

$$Dg_t(g_t^{-1}(x)) \cdot X_t(g_t^{-1}(x)) = \rho(f_t)) .$$

On a finalement pour $t > 0$

$$\rho(f_t) = 1/\int_0^1 (1/X_t(u))du ,$$

et, par un calcul élémentaire de primitive $\lim_{t \to 0_+} \rho(f_t)/t = +\infty$, et donc $t \to \rho(f_t)$ n'est pas lipschitzienne au voisinage de $t = 0$.

---

† Soit $K \subset [0,1]$ un Cantor, $m(K) > 0$, et χ sa fonction caractéristique. Soit $h(x) = \frac{1}{m(K)} \int_0^x \chi(t)dt$. Alors $0 = h(0)$, $h(1) = 1$, h est monotone non décroissante et lipschitzienne et, sur chaque composante connexe de $[0,1]-K$, h est constante. Cet exemple se produit dans certains des contre-exemples construits par Denjoy.

6. THEOREME DES CHEMINS (réponse au problème 3.8.3)

Soient $3 \leq r \leq \omega$, $t \in [a,b] \to f_t \in D^r(T^1)$ de classe C^1 (en fait, $t \to f_t \in D^r(T^1)$ continue, et $t \to f_t \in D^0(T^1)$ C^1 suffit) ; posons $M(f_t) = m\,\{t \in [a,b] \mid f_t$ est C^{r-2}-conjugué à une translation irrationnelle$\}$ (i.e. la mesure de Lebesgue de l'ensemble considéré).

et

$$\rho(t) = \rho(f_t) .$$

6.1 Théorème : Pour tout chemin comme ci-dessus, avec $\rho(f_a) \neq \rho(f_b)$, alors $M(f_t) > 0$; et si $N \subset [a,b]$ est un borélien de m-mesure nulle, alors $m(\rho(N)) = 0$.

6.1.1 Corollaire : Si $t \in [a,b] \to \rho(t)$ est une fonction à variation bornée, alors ρ est une fonction absolument continue.

6.1.2 Exemples :

1) Si $t \in [0,1] \to f_t \in D^r(T^1)$ un chemin de classe C^1 positif, alors $t \in [0,1] \to \rho(f_t)$ est monotone non décroissante donc à variation bornée et il résulte que $t \to \rho(f_t)$ est absolument continue.

2) Si $f \in D^3(T^1)$, alors $t \in [0,1] \to \rho(R_t \circ f)$ est absolument continue.

3) Génériquement $t \to \rho(f_t)$ est à variation bornée (voir [2]).

Notations : Posons $\rho([a,b]) = J$, qui est un intervalle d'intérieur non vide, puisque $\rho(f_a) \neq \rho(f_b)$ et

$$C = \sup_{x \in \mathbb{R},\, t \in [a,b]} \left| \frac{\partial f_t(x)}{\partial t} \right| .$$

6.2 Proposition : Il existe une suite croissante de compacts de $[a,b]$ $D_1 \subset D_2 \subset \ldots, \bigcup_k D_k = D$, telle que $t \in D$ si et seulement si $f_t = h_t^{-1} \circ R_{\rho(f_t)} \circ h_t$ où h_t est un difféomorphisme de classe C^1, de plus, si $t \in D_k$, on a $|Dh_t|_0 \leq k$.

Démonstration : Soit $t \in [a,b] \to \mathfrak{n}(t) = \sup_{n \in \mathbb{Z}} |Df_t^n|_0 \in \overline{\mathbb{R}} = \mathbb{R} \cup \{+\infty\}$. $\mathfrak{n}(t)$ est semi-continue inférieurement, et $D_k = \{t \in [a,b] \mid \mathfrak{n}(t) \leq k\}$ est donc compact par la semi-continuité inférieure de $\mathfrak{n}$; la proposition résulte alors de [4]. □

Remarques :

a) Le fait que, si f_t est C^1-conjugué à $R_{\rho(f_t)}$, alors $\mathfrak{n}(t) < +\infty$ est immédiat.

b) Si $f \in D^1(T^1)$ est C^1-conjugué à la rotation R_α, alors, si $n \to +\infty$, $\frac{1}{n} \sum_{i=0}^{n-1} f^i - i\alpha$ converge dans la C^1-topologie vers $h \in D^1(T^1)$ et on a $f = h^{-1} \circ R_\alpha \circ h$. En particulier, si $n \to +\infty$, $\frac{1}{n} \sum_{i=0}^{n-1} Df^i \to Dh$.

Par conséquent, si $f \in D^1(T^1)$ est C^1-conjugué à une rotation, $\sup_{n \in \mathbb{Z}} |Df^n|_0 = k < +\infty$ implique que $|Dh|_0 \leq k$ est aussi immédiat.

c) Dans l'exemple 3.7.1, $\{t \in K_a \mid \mathfrak{n}(t) = \infty\}$ est un G_δ dense dans K_a (voir [4]).

6.3 Démonstration de 6.1 : Par 6.2 et 4.4, $\rho|_{D_k} : D_k \to J$ est lipschitzienne de rapport $\leq kC$. Par 3.2, $\rho(\bigcup_k D_k) \supset A \cap J$, et donc $\lim_{k \to +\infty} m(\rho(D_k)) = m(J \cap A) = m(J) > 0$.

On peut donc trouver un compact $B \subset [a,b]$ et un entier k tels que :

$$m(\rho(B)) > 0, \quad \rho(B) \subset A, \quad \text{et} \quad B \subset D_k .$$

Par 3.2, $M(f_t) \geq m(B)$.

Puisque $B \subset D_k$, $\rho|_B : B \to J$ est donc lipschitzienne de rapport $\leq kC$. Le fait que $M(f_t) > 0$ résulte du lemme bien connu suivant (puisqu'on a $m(B) \geq \frac{1}{kC} m(\rho(B)) > 0$).

6.4 Lemme. Soit $K \subset \mathbb{R}^n$ un compact et $f : K \to \mathbb{R}^n$ une fonction lipschitzienne de rapport $\leq \ell$ (i.e. pour tout x et y dans K, $|f(x) - f(y)| \leq \ell\, |x - y|$) alors $m(f(K)) \leq \ell^n\, m(K)$.

6.5 Fin de la démonstration de 6.1 : Le fait que $m(\rho(N)) = 0$ résulte de ce que

$$m(\rho(N - D)) \leq m(J - A) = 0 ,$$

et de ce que

$$m(\rho(N \cap D_k)) = 0$$

pour tout k, puisque $\rho|_{D_k}$ est lipschitzienne. Ceci démontre le théorème 6.1. □

7. CAS DE T^n : GENERALISATION D'UN THEOREME D'ARNOLD EN C^r

Nous allons démontrer le théorème 3.8.2. On suppose que $n \geq 1$ est un entier fixé. (Toutes les constantes dépendent de n et nous n'indiquerons pas cette dépendance).

On fixe $\beta > 0$ petit ($\beta \leq 10^{-100}$ disons) et $\beta' > \beta$; disons $\beta' = \frac{3}{2}\beta$.

7.1 Théorème : Soit $f \in D^r(T^n)$ avec $r \geq 2n + 2\beta$; si $f \to \mathrm{Id}$ dans la $C^{2n+2\beta}$-topologie alors

$M(f) = m\,\{t \in [0,1]^n \,|\, R_t \circ f$ est $C^{r-n-\beta'}$-conjugué à une translation ergodique$\} \to 1$,

(i.e. la mesure de Lebesgue de l'ensemble considéré tend vers 1 si $f \to \mathrm{Id}$).

7.2 Construction de Cantors.

Soit $0 < C$ donné. On pose :

$$K_C = \{\alpha \in [0,1]^n \,|\, \|\langle k,\alpha\rangle\| \geq C/|k|^{n+\beta} \quad \text{si } k \in \mathbb{Z}^n - \{0\}\ \}$$

avec les notations si $\alpha = (\alpha_1, \ldots, \alpha_n)$, $k = (k_1, \ldots, k_n)$,

$$\langle k, \alpha\rangle = \sum_i k_i\, \alpha_i , \quad |k| = \sup_i |k_i|$$

et, si $x \in \mathbb{R}$, $\qquad \|x\| = \operatorname{Inf}_{p \in \mathbb{Z}} |x - p|$.

K_C est un compact de $[0,1]^n$, et si $C \to 0$, $m(K_C) \to 1$.

Si C est assez petit, quitte à soustraire un ensemble dénombrable, K_C est un Cantor.

Si $n = 1$, on remplace K_C par $K'_C \subset A$, K'_C compact, tel que $K'_C \subset K_C$ et que, si $C \to 0$, $m(K'_C) \to 1$ (c'est possible puisque $m(A \cap [0,1]) = 1$).

7.3 Rappel ($n \geq 1$ est fixé).

Nous avons démontré dans [4] le théorème suivant (la continuité en α est immédiate car à chaque étape de la démonstration tout dépend continuement de $\alpha \in K_C$).

Ce théorème est dû essentiellement à V.I. Arnold et J. Moser.

Théorème : Soit $C > 0$; il existe $\mu_0(C) > 0$, ($\mu_0(C) \to 0$ si $C \to 0$) et $L(C) > 0$ ($L(C) \to +\infty$ si $C \to 0$) tels que, si $2n + 2\beta \leq r \leq \omega$ et $f \in D^r(T^n)$ avec $|f - \mathrm{Id}|_{C^{2n+2\beta}} = \mu \leq \mu_0$, alors il existe des applications continues, qui dépendent de f, $\alpha \in K_C \to \lambda_f(\alpha) \in \mathbb{R}^n$ et $\alpha \in K_C \to h_f(\alpha) \in D^{r-n-\beta'}(T^n, 0)$ telles que

$$R_{\lambda_f(\alpha)} \circ f = h_f^{-1}(\alpha) \circ R_\alpha \circ h_f(\alpha)$$

et que, si $\alpha \in K_C$:

$$|\lambda_f(\alpha) - \alpha| \leq \mu\, L(C), \quad (|h_f(\alpha) - \mathrm{Id}|_0 + |Dh_f(\alpha) - \mathrm{Id}|_0) \leq \mu\, L(C).$$

(On pose $D^r(T^n, 0) = \{g \in D^r(T^n) \,|\, g(0) = 0\}$ et puisque R_α est sur T^n une translation ergodique, ceci détermine uniquement le difféomorphisme $h_f(\alpha)$.)

7.4 Démonstration de 7.1 : Soient $0 < \epsilon < 1$ et C tel que $m(K_C) \geq 1 - \frac{\epsilon}{2}$, ce qui détermine $L(C)$. On choisit $\mu > 0$ ($\mu \leq \mu_0$) tel que $\mu\, L(C) \leq (1 + \frac{\epsilon}{2})^{1/n} - 1$ alors, si $|f - \mathrm{Id}|_{C^{2n+2\beta}} \leq \mu$, on a

$$|Dh_f(\alpha)|_0 \leq (1 + \tfrac{\epsilon}{2})^{1/n} .$$

Soit $D_C = \lambda(K_C)$; D_C est un compact de $\mathbb{R}^n$ puisque $\alpha \to \lambda(\alpha)$ est continue.
Par 2.4 c), si $t \in D_C$ alors

$$\lim_{k\to+\infty} \frac{(R_t \circ f)^k - Id}{k} = \rho(R_t \circ f) \in \mathbb{R}^n .$$

Posons $\rho(t) = \rho(R_t \circ f)$, si $\alpha \in K_C$, on a $\rho \circ \lambda(\alpha) = \alpha$ (i.e. $\rho \circ \lambda = Id|_{K_C}$).
$t \in D_C \to \rho(t)$ est continue et elle est même lipschitzienne de rapport $\leq (1+\frac{\epsilon}{2})^{1/n}$ par 4.5.

Par le lemme 6.4, on a

$$m(K_C) \equiv m(\rho(D_C)) \leq (1+\frac{\epsilon}{2})\, m(D_C) ,$$

et donc $m(D_C) \gtrsim 1-\epsilon$. Or, par 7.3, $M(f) \gtrsim m(D_C)$.
Comme pour tout $\epsilon > 0$, on a déterminé $\mu > 0$ tel que, si $|f - Id|_{C^{2n+2\beta}} \leq \mu$, alors $M(f) \geq 1-\epsilon$, le théorème est démontré. □

7.5 Cas $n = 1$.

Le théorème 7.3 s'énonce ainsi si $n = 1$ (voir 7.2 pour la définition de K'_C).

Théorème ; Soit $C > 0$, il existe $\mu_o(C) > 0$ et $L(C)$ tels que, si $|f - R_{\rho(f)}|_{C^{2+2\beta}} = \mu \leq \mu_o(C)$ et $\rho(f) = \alpha \in K'_C$, alors il existe $h_f \in D^{r-1-\beta'}(T^1, 0)$ tel que $f = h_f^{-1} \circ R_{\rho(f)} \circ h_f$ avec l'inégalité $|h_f - Id|_{C^1} \leq \mu\, L(C)$.
De plus l'application $f \to h_f$ est continue.

7.6 Par le même raisonnement qu'en 7.4, en utilisant 7.5, on a le :

Théorème : Soient $3 \leq r \leq \omega$, et $t \in [0,1] \to f_t \in D^r(T^1)$ un chemin positif de classe C^1 ; si f_t est "suffisamment" proche de $t \in [0,1] \to R_t$, alors $M(f_t)$ est voisin de 1. (Suffisamment proche veut dire : pour tout $t \in [0,1]$, $|f_t - R_t|_{C^{2+2\beta}}$ est assez petit, et, pour tout $t \in [0,1]$ et $x \in \mathbb{R}$, $|\partial f_t(x)/\partial t - 1|$ est assez petit.)

8. CARACTERE LOCALEMENT LIPSCHITZIEN DE LA FONCTION ρ SUR DES SOUS-ESPACES

8.1 Soient $C > 0$ petit, K_C' défini comme en 7.2.

Soit $$G_C \equiv F^{3+3\beta'}_{K_C'} = \{ f \in D^{3+3\beta'}(T^1) \mid \rho(f) \in K_C' \} .$$

G_C est un fermé dans $D^{3+3\beta'}(T^1)$.

8.2 Proposition : $\rho: G_C \to \mathbb{R}$ est localement lipschitzienne pour la $C^{3+3\beta'}$-topologie.

Démonstration : Si $f \in G_C$, par 3.2, f est $C^{2+5\beta'/2}$-conjugué à $R_{\rho(f)}$ par un unique h_f, (unique si $h_f(0) = 0$), (i.e. $f = h_f^{-1} \circ R_{\rho(f)} \circ h_f$). Soit $f_o \in G_C$, par 7.5, il existe un ouvert U de $D^{3+3\beta'}(T^1)$, avec $f_o \in U$, et un entier $k \geq 1$, tels que, pour tout $f \in G_C \cap U$, on ait $1/k \leq Dh_f \leq k$.

En utilisant 4.2, par conjugaison, on a, pour tout f_1 et f_2 dans $U \cap G_C$, l'inégalité :

$$|\rho(f_1) - \rho(f_2)| \leq k|f_1 - f_2|_o . \quad \square$$

8.3 Proposition : L'application $f \in G_C \to h_f \in D^1(T^1,0)$ est lipschitzienne en R_α si $\alpha \in K_C'$.

Démonstration : On a, par 7.5 et 4.2, si $|f - R_{\rho(f)}|_{C^{2+2\beta}} \leq \mu_o$,

$$|h_f - h_{R_\alpha}|_{C^1} = |h_f - Id|_{C^1} \leq L(C)|f - R_{\rho(f)}|_{C^{2+2\beta}}$$

$$\leq L(C)|f - R_\alpha|_{C^{2+2\beta}} + L(C)|\rho(f) - \rho(R_\alpha)|$$

$$|h_f - h_{R_\alpha}|_{C^1} \leq L(C)\,|f - R_\alpha|_{C^{2+2\beta}} + L(C)\,|f - R_\alpha|_0 \,. \square$$

8.4 Remarques :

a) Nous avons utilisé 7.5 en 8.2 et 8.3 pour avoir une uniformité de l'application $f \to h_f$ au voisinage de R_α pour $\alpha \in K'_C$.

b) Quitte à remplacer $3+3\beta'$ par $4+4\beta'$, $f \subset \Gamma^{4+4\beta'}_{K'_C} \to h_f \in D^1(T^1,0)$ est une application localement lipschitzienne.

BIBLIOGRAPHIE

[1] V.I. ARNOLD, Small denominators I, Transl. Amer. Soc. 2nd series, vol. 46, p. 213-284.

[2] P. BRUNOVSKÝ, Generic properties of the rotation number of one-parameter diffeomorphisms of the circle, Czech. Math. J. 24 (99), 1974, 74-90.

[3] J. DIEUDONNÉ, Fondement de l'analyse moderne, t. I, Gauthier-Villars, Paris, 1963 (voir ch. VIII, 12, ex. 8).

[4] M.R. HERMAN, Sur la conjugaison différentiable des difféomorphismes du cercle à des rotations, (à paraître).

[5] M.R. HERMAN, Conjugaison C^∞ des difféomorphismes du cercle dont le nombre de rotation satisfait à une condition arithmétique, C.R. Acad. Sc. Paris, t. 282, (1976), 503-506.

[6] M.R. HERMAN, Conjugaison C^∞ des difféomorphismes du cercle pour presque tout nombre de rotation, C.R. Acad. Sc. Paris, t. 283 (1976), 579-582.

[7] L. KUIPERS and H. NIEDERREITER, Uniform distribution of sequences, Interscience, Wiley, New-York, 1974.

Centre de Mathématiques
de l'Ecole Polytechnique

Route de Saclay - 91128 Palaiseau cedex -
France

THE GODBILLON-VEY INVARIANT OF FOLIATIONS BY PLANES OF T^3

by

Michael Robert HERMAN

In [5] Harold Rosenberg and William Thurston asked among others the following question:

Question: Is the Godbillon Vey invariant of a C^2-foliation by planes of T^3 equal to zero?

We propose to show the answer is yes.

Let $T^n = \mathbb{R}^n/\mathbb{Z}^n$, $R_\alpha : x \to x+\alpha$, and m be the Haar measure of T^n.

1. If $\mathfrak{F}$ is a C^r transversely oriented foliation of T^3 with all the leaves diffeomorphic to $\mathbb{R}^2$, then $\mathfrak{F}$ is called a foliation by planes (any compact connected 3-manifold with a foliation by planes is diffeomorphic to T^3; see [4]).

H. Rosenberg and R. Roussarie in [4] showed if $r \geq 2$ ($r = 1,0$ should work also) that there exists a C^r-diffeomorphism H of T^3 such that $\mathfrak{F}' = H_*(\mathfrak{F})$ is a foliation transverse to a C^∞-fibration $T^1 \to T^3 \to T^2$. $\mathfrak{F}'$ is therefore a suspension foliation of a representation of $\pi_1(T^2) = \mathbb{Z} \oplus \mathbb{Z} \to \mathrm{Diff}^r_+(T^1)$. Let f and g be generators of the representation (so that $f \circ g = g \circ f$). As $\mathfrak{F}'$ is a foliation by planes, f and g are commuting diffeomorphisms with irrational and rationally independant rotation numbers $\rho(f) = \alpha$ and $\rho(g) = \beta$ (α and β are in T^1).

Let us recal rapidly how one obtains the suspension foliation: Let $V = \mathbb{R}^2 \times T^1$. Define $F(x_1,x_2,\theta) = (x_1+1,x_2,f(\theta))$ and $G(x_1,x_2,\theta) = (x_1,x_2+1,g(\theta))$. F and G generate a fixed point free action of

$A \cong \mathbb{Z} \oplus \mathbb{Z}$ on V. This action of A leaves invariant the foliation of V, $\{\theta = \text{constant}\}$, and V/A (diffeomorphic to T^3) has the quotient foliation (called the suspension foliation).

2. By Denjoy's theorem, f is C^0-conjugate to R_α, $f = h^{-1} \circ R_\alpha \circ h$.

As α is irrational the centralizer of R_α in $\mathrm{Homeo}_+(T^1)$ is the standard group of rotation T^1. Therefore h conjugates g to the rotation R_β. One has that $H(x_1, x_2, \theta) = (x_1, x_2, h(\theta))$ is an equivariant homeomorphism from (V, A) to $(V, \mathbb{Z}^2)$ where $\mathbb{Z}^2$ is the action on V defined by $f = R_\alpha$ and $g = R_\beta$. One concludes (see [4]) that H conjugates the suspension foliation on V/A to a standard foliation by planes of T^3 defined by a constant one form (i.e. a linear foliation of T^3).

Even if the foliation $\mathcal{F}$ is C^∞ the conjugating homeomorphism is not necessarily absolutely continuous: there exists a Borel $\mathcal{F}$-invariant set B (i.e. B is a union of leaves of $\mathcal{F}$) with $m(B) = 0$ but $m(H(B)) = 1$. If the conjugating homeomorphism H is singular then any conjugating homeomorphism is also singular with respect to m (see the following Lemma).

Let us indicate how this follows from [2]. We constructed $f \in \mathrm{Diff}^\infty_+(T^1)$, $f = h^{-1} \circ R_\alpha \circ h$, α irrational, such that h sends a Borel set of Haar measure 0 to a set of Haar measure 1. Furthermore the C^∞ centralizer of f, $\mathrm{Cent}^\infty(f)$, has the power of the continuum. One can therefore find $g \in \mathrm{Diff}^\infty_+(T^1)$ commuting with f such that $\rho(g) = \beta$ is irrational and rationnaly independent of α for we have the injective continuous group homomorphism $0 \to \mathrm{Cent}^\infty(f) \xrightarrow{\rho} T^1$. (One can show that $m(\rho(\mathrm{Cent}^\infty(f)) = 0)$).

The foliation obtained by suspension of f and g has the desired properties by the following elementary lemma:

Lemma - *Let $\mathfrak{F}$ be a foliation by hyperplanes of T^n defined by a constant one form of T^n. Let H_1 be a homeomorphism of T^n that leaves $\mathfrak{F}$ invariant. If A is an $\mathfrak{F}$-invariant Borel set of T^n with $m(A) = 0$, then $m(H_1(A)) = 0$.*

The fact that H is either singular or absolutely continuous is related to the ergodic properties of the foliation.

Let us recall a definition given by Denjoy and Birkhoff:

Definition - *Let M be a paracompact Lipschitz manifold and $\mathfrak{F}$ a Lipschitz foliation of M. $\mathfrak{F}$ is said to be ergodic with respect to a Lebesgue measure m on M if for every Borel $\mathfrak{F}$-invariant set B we have $m(B) = 0$ or $m(M-B) = 0$.*

Of course the definition depends only the class of sets of Lebesgue measure 0, which is intrinsically defined on a Lipschitz manifold.

Caution: The $\mathfrak{F}$-saturation of a set of m-measure 0 on M is not necessarily m-measurable, and the $\mathfrak{F}$-saturation of a Borel set of M is not necessarily Borel but only Souslinian.

In [2] we proved a theorem that implies the following

Theorem - *Let $A \subset \mathrm{Diff}^2_+(T^1)$ be a countable subgroup such that there exists $f \in A$ with $\rho(f)$ irrational. Then a foliation of a connected manifold obtained by suspension of the group A is ergodic with respect to a Lebesgue measure.*

In particular any C^2-foliation of T^3 by planes is m-ergodic.

Furstenberg (see [2]) gave an example of a C^ω codimension 2 foliation of T^3 that is minimal but not ergodic with respect to the Haar measure.

We have shown (see [2] and [3]) that if $r \geq 3$ and α or β satisfies a certain arithmetical condition (in fact belongs to a certain set of Haar measure one), then one can choose the conjugating homeomorphism H of the Foliation $\mathfrak{F}$ of class C^r to a standard linear foliation of T^3 to be a diffeomorphism of class C^{r-2} (with $+\infty-2 = +\infty$ and $\omega = \omega-2$).

Remark: In case the conjugatinghomeomorphism H is singular with respect to m then the foliation $\mathfrak{F}$ is not defined by a closed one form of class C^1 (for otherwise $\mathfrak{F}$ would be C^1-conjugate to linear foliation of T^3).

3. Giving a representation of $\pi_1(T^2) = \mathbb{Z} \oplus \mathbb{Z} \to \mathrm{Diff}^2_+(T^1)$ with generators f and g such that $\rho(f) = \alpha$ and $\rho(g) = \beta$ are irrational and rationnally independent is equivalent, by suspension to giving a C^2-foliation $\mathfrak{F}$ of T^3 by planes.

We can associate to the representation the Godbillon-Vey invariant of $\mathfrak{F}$ $\mathcal{G}v(f,g) \in \mathbb{R}$. By 2. the fact that the Godbillon-Vey invariant of $\mathfrak{F}$ is zero does not "immediately" follow from the fact $\mathfrak{F}$ is topologically conjugate to a foliation defined by a constant one form for "usually" the conjugating map is not C^2.

Let us recal how one calculates the Godbillon-Vey invariant $\mathcal{G}v(\mathfrak{F})$ of a transversely oriented codimension 1, C^3-foliation $\mathfrak{F}$ of M^n.

Let w be a non singular one form defining $\mathfrak{F}$. By integrability we have $w \wedge dw = 0$. Let η be a one form such that $dw = \eta \wedge w$.

$\eta \wedge d\eta$ is a closed 3 form and defines a cohomology class $[\eta \wedge d\eta] \in H^3(M,\mathbb{R})$ that depends only on the foliation $\mathfrak{F}$ of M^n.

If M^3 is 3 dimensional compact connected orientable manifold by integration we have the Godbillon-Vey number of $\mathfrak{F}$:

$$H^3(M,R) \xrightarrow{\cong} \mathbb{R}$$

$$[\eta \wedge d\eta] \longrightarrow \int_M \eta \wedge d\eta.$$

4. Let $D^r(T^1) = \{f \in \mathrm{Diff}^r_+(\mathbb{R}^1) \mid f\text{-}Id \in C^r(T^1)\}$ with $C^r(T^1) =$ $= \{\varphi \in C^r(\mathbb{R}) \mid \varphi$ is $\mathbb{Z}$ periodic$\}$. Let $R_\alpha : x \to x+\alpha$ be the translation of rotation number α. $D^r(T^1)$ is the universal covering group of $\mathrm{Diff}^r_+(T^1)$, for $\mathrm{Diff}^r_+(T^1) = D^r(T^1)/C$ where $C = \{R_p \mid p \in \mathbb{Z}\}$ is the center of $D^r(T^1)$.

If $f \in \mathrm{Diff}^r_+(T^1)$, $D^r f \in C^r(T^1)$ is the rth derivative of f.

One shows (see [2]) that if $f,g \in \mathrm{Diff}^r_+(T^1)$ commute and if $\tilde{f} \in D^r(T^1)$ (resp. $\tilde{g}$) is any covering of f (resp. g), then $\tilde{f}$ and $\tilde{g}$ commute.

Let $\omega(g,f) = \int_{T^1} (\mathrm{Log}\, Df(\theta))\cdot(D\, \mathrm{Log}\, D(g\circ f)(\theta))d\theta$ for f and g in $D^2(T^1)$. We shall leave out the T^1 in the integral formulas.

ω is a continuous 2-cochain on $D^2(T^1)$ (i.e. $\omega : D^2(T^1)^2 \to \mathbb{R}$).

We suppose that $D^2(T^1)$ acts trivially on $\mathbb{R}$.

Proposition: ω is a 2-cocycle.

Proof: We have to verify that for every f,g,h in $D^2(T^1)$ we have

$$0 = d\omega(f,g,h) = \omega(g,h) - \omega(f\circ g,h) + \omega(f,g\circ h) - \omega(f,g).$$

Consider

$$\text{(i)} \quad \omega(g,h) = \int (\mathrm{Log}\, Dh)\cdot D\, \mathrm{Log}\, D(g\circ h)dm,$$

then by integration by parts (i.e. $\int \psi\cdot D\varphi\, dm = -\int \varphi\cdot D\psi\, dm$),

$$\text{(i)} = -\int (\mathrm{Log}\, D(g\circ h))\cdot D\, \mathrm{Log}\, Dh\, dm = -\int (D\, \mathrm{Log}\, Dh)\cdot[(\mathrm{Log}\, Dg)\circ h + \mathrm{Log}\, Dh]dm,$$

$$\text{(i)} = -\int (\mathrm{Log}(Dg)\circ h)\cdot D\, \mathrm{Log}\, Dh\, dm.$$

$$\text{(ii)} \quad \omega(f\circ g,h) = \int (\mathrm{Log}\, Dh)\cdot D\, \mathrm{Log}\, D(f\circ g\circ h)\, dm.$$

$$\text{(iii)} \quad \omega(f,g\circ h) = \int (\mathrm{Log}\, D(g\circ h))\cdot D\, \mathrm{Log}\, D(f\circ g\circ h)\, dm =$$

$$= \int [(\mathrm{Log}\, Dg)\circ h + \mathrm{Log}\, Dh]\cdot D\, \mathrm{Log}\, D(f\circ g\circ h)\, dm.$$

We have

$$\text{(iii)}-\text{(ii)} = \int (\mathrm{Log}\, Dg)\circ h \cdot D[\mathrm{Log}\, D(f\circ g)\circ h + \mathrm{Log}\, Dh]dm,$$

so by the formula of change of variables (i.e. $\int (\varphi\circ h)\cdot D(\psi\circ h)dm = \int \varphi\cdot D\psi\, dm$) we have

$$(iii)-(ii) = \int (\mathrm{Log}\, Dg)\cdot D\, \mathrm{Log}\, D(f\circ g)dm + \int ((\mathrm{Log}\, Dg)\circ h)\cdot D\, \mathrm{Log}\, Dh\, dm.$$

$$(iv) \quad \omega(f,g) = \int (\mathrm{Log}\, Dg)\cdot \mathrm{Log}\, D(f\circ g)dm.$$

We finally have $(i)-(ii)+(iii)-(iv) = 0$. ∎

ω defines a cohomology class in $H^2(K(D^2(T^1),1),\mathbb{R})$, and even in the continuous group cohomology H_c^2 of $D^2(T^1)$ with the trivial action on $\mathbb{R}$.

<u>Proposition</u> - <u>Given</u> f,g <u>in</u> $D^2(T^1) = K$ <u>with</u> $f\circ g = g\circ f$ <u>then the</u> 2-<u>chain</u> $C_{(f,g)} = (f,g) - (g,f) \in \mathbb{Z}[K\times K]$ <u>is a</u> 2-<u>cycle</u>.

<u>Proof</u>: $C_{(f,g)}$ is closed for we have

$$\partial C_{(f,g)} = (f) + (g) - (f\circ g) - (f) - (g) + (g\circ f) = 0 \in \mathbb{Z}[K].$$ ∎

We have the following formula of Thurston proved in [1].

<u>Theorem</u> - $\omega(C_{(f,g)}) = \omega(f,g) - \omega(g,f) = \mathcal{G}v(f,g) \in \mathbb{R}$.

<u>Sketch of proof</u>: We use the notations of 1, f and g are in $\mathrm{Diff}_+^3(T^1)$.

Let $w = \varphi(x_1,x_2,\theta)d\theta$ $(\varphi > 0)$ be a one form on V that defines the foliation $\{\theta = \text{constant}\}$.

Let us suppose that w is invariant under the group action A (i.e. w is A-invariant). We have the conditions:

a) $$\varphi(x_1,x_2,\theta) = \varphi(x_1+1,x_2,f(\theta))Df(\theta) \Leftrightarrow \psi(x_1,x_2,\theta) = \psi(x_1+1,x_2,f(\theta)) + \mathrm{Log}\, Df(\theta),$$

b) $$\varphi(x_1,x_2,\theta) = \varphi(x_1,x_2+1,g(\theta))Dg(\theta) \Leftrightarrow \psi(x_1,x_2,\theta) = \psi(x_1,x_2+1,g(\theta)) + \mathrm{Log}\, Dg(\theta),$$

with $\psi = \mathrm{Log}\, \varphi$.

If w is A-invariant then we have

$$dw = \eta \wedge w \quad \text{with} \quad \eta = \frac{1}{\varphi}\frac{\partial\varphi}{\partial x_1}\, dx_1 + \frac{1}{\varphi}\frac{d\varphi}{\partial x_2}\, dx_2 = \frac{\partial\psi}{\partial x_1}\, dx_1 + \frac{\partial\psi}{\partial x_2}\, dx_2 .$$

One verifies by a) and b) that η is A-invariant.

It results that $\eta \wedge d\eta$ is A-invariant so we have

$$(*) \quad \mathcal{G}v(f,g) = \int_\Delta \eta\wedge d\eta = \int_\Delta \left(\frac{\partial\psi}{\partial x_2}\frac{\partial^2\psi}{\partial x_1\partial\theta} - \frac{\partial\psi}{\partial x_1}\frac{\partial^2\psi}{\partial x_2\partial\theta}\right) dx_1\otimes dx_2\otimes d\theta ,$$

Δ being the fundamental domain of A: $\Delta = [0,1] \times [0,1] \times T^1$.

Everything reduces to choosing ψ C^2 verifying a) and b) on neighborhood of Δ.

For $(x_1,x_2,\theta) \in \Delta$ let

$$\psi_1(x_1,x_2,\theta) = \begin{cases} x_1 \text{ Log } Df^{-1}(\theta) + x_2 \ (\text{Log } Dg^{-1})\circ(f^{-1}(\theta)) & \text{if} \quad x_2 \leq x_1, \\ x_1 \ (\text{Log } Df^{-1})\circ(g^{-1}(\theta)) + x_2 \text{ Log } Dg^{-1}(\theta) & \text{if} \quad x_1 \leq x_2. \end{cases}$$

ψ_1 is continuous (for f and g commute so we have $(\text{Log } Df^{-1})\circ g^{-1} + \text{Log } Dg^{-1} = (\text{Log } Dg^{-1})\circ f^{-1} + \text{Log } Df^{-1}$). ψ_1 satisfies a) and b) on Δ, and an easy calculation of (*) (integrate θ first) gives the result+.

But unfortunately ψ_1 is not C^2 ! Let $\gamma(t)$ be a C^∞ function from $[0,1]$ to $[0,1]$ with $\gamma(t) = 0$ if $0 \leq t \leq C$ and $\gamma(t) = 1$ if $t \geq 1-C$ $(C > 0)$.

Define $\psi_2(x_1,x_2,\theta)$ by the same formula as for ψ_1 but replacing x_1 by $\gamma(x_1)$ and x_2 by $\gamma(x_2)$.

Again the calculation of (*) gives the result but still ψ_2 is not C^2 on the diagonal $\{(x_1,x_2,\theta) \in \Delta \mid x_1 = x_2\}$.

Let $0 < 4\varepsilon < C$ and $0 \leq \delta_\varepsilon(x_1,x_2) \leq 1$ be a C^∞ function $[0,1]^2$ with the conditions $(\|x\| = (x_1^2+x_2^2)^{1/2})$:

+ One gets the formula f replaced by f^{-1} and g by g^{-1}. We will see in 5 that it is the same formula (that is also clear by group cohomology argument).

- $\delta_\varepsilon(x_1,x_2) = 1$ if $|x_1-x_2| \geq \varepsilon$, or $\|x\| \leq \frac{3\varepsilon}{2}$ or $\|x-(1,1)\| \leq \frac{3\varepsilon}{2}$;
- $\delta_\varepsilon(x_1,x_2) = \gamma\left(\frac{|x_1-x_2|}{\varepsilon}\right)$ if $x_1+x_2 \geq 4\varepsilon$ and $x_1+x_2 \leq 2-4\varepsilon$;
- $\delta_\varepsilon(x_1,x_2)$ is C^∞ on $[0,1]^2$ and $0 \leq \delta_\varepsilon(x_1,x_2) \leq 1$.

In words δ_ε is equal to 1 outside a neighborhood of the diagonal, in a neighborhood of $(0,0)$ and $(1,1)$ and in a neighborhood of part of the diagonal δ_ε is 0.

$\psi_\varepsilon = \delta_\varepsilon \cdot \psi_2$ is of class C^2. If we calculate (*) with ψ_ε we get that $\omega(C_{(f,g)}) = c\, \mathcal{G}v(f,g)$ with c a universal constant. If $\varepsilon \to 0$ one proves the proposition. ■

5. <u>Proposition</u> - <u>We have the following equality</u>

$$\mathcal{G}v(f,g) = -\int_a^{a+1} [(\mathrm{Log}\, Dg)\cdot D\, \mathrm{Log}\, Df^{-1} + (\mathrm{Log}\, Dg^{-1})\cdot D\, \mathrm{Log}\, Df]\, dm,$$

with $a \in \mathbb{R}$.

<u>Proof</u>: We have

$$\omega(f,g) = \int (\mathrm{Log}\, Dg)\cdot D\, \mathrm{Log}\, D(f\circ g)dm =$$

$$= \int (\mathrm{Log}\, Dg)\cdot D\, [\mathrm{Log}\, Df)\circ g + \mathrm{Log}\, Dg]\, dm =$$

$$= \int (\mathrm{Log}\, Dg)\cdot D\, \mathrm{Log}(Df)\circ g\, dm = \int (\mathrm{Log}\, Dg)\circ g^{-1}\cdot (D\, \mathrm{Log}\, Df)\, dm =$$

$$= -\int (\mathrm{Log}\, Dg^{-1})\cdot D\, \mathrm{Log}\, Df\, dm.$$

By integration by parts we have:

$$\omega(g,f) = -\int (\mathrm{Log}\, Df^{-1})\cdot D\, \mathrm{Log}\, Dg\, dm = \int (\mathrm{Log}\, Dg)\cdot D\, \mathrm{Log}\, Df^{-1}\, dm.$$

The proposition follows from the fact that $\mathcal{G}v(f,g) = \omega(f,g) - \omega(g,f)$. ■

We shall write

$$\mathcal{G}v(f,g) = \int_a^{a+1} d\, \mathcal{G}v(f,g) \quad \text{with } a \in \mathbb{R} \text{ and}$$

$$d\mathcal{G}v(f,g) = -[(\text{Log } Dg)\circ D \text{ Log } Df^{-1} + (\text{Log } Dg^{-1})\circ D \text{ Log } Df]dm.$$

6.1 Let G be a group, and $\chi: \mathbb{Z}^2 \to G$ a representation with $\chi(m,p) = f^m g^p$.

To the representation we associate the 2-cycle

$$C_{(f,g)} = (f,g) - (g,f) \in \mathbb{Z}[G\times G].$$

Let B_2 the subgroup of $\mathbb{Z}[G\times G]$ generated by the boundaries of 3-chains (i.e. the group of 2-boundaries).

Let us recall that

$$\partial(f,g,h) = (g,h) - (f\cdot g,h) + (f,g\cdot h) - (f,g) \in \mathbb{Z}[G\times G];$$

and if ω is a 2-cocycle and $c \in B_2$ then $\omega(c) = 0$.

6.2 Proposition - Let f and g be in G with $fg = gf$ then

$$C_{(f^n,g)} - n\, C_{(f,g)} \in B_2.$$

Proof: Let us consider the homomorphism $\varphi: \mathbb{Z}^2 \to \mathbb{Z}^2$ defined by $\varphi(m,p) = (nm,p)$ and the homomorphism $\chi_1 = \chi\circ\varphi: \mathbb{Z}^2 \to G$.

To χ_1 associate the 2-cycle $C_{(f^n,g)}$.

$\varphi_*: H_2(T^2,\mathbb{Z}) \to H_2(T^2,\mathbb{Z})$ is the multiplication by $\deg\varphi = n$, and $T^2 = K(\mathbb{Z}^2,1)$. Therefore $C_{(f^n,g)} - n\, C_{(f,g)}$ is equal to 0 in $H_2(K(G,1),\mathbb{Z})$.

Second proof: Let us give a more constructive proof of 6.2 by induction on n. For $n=1$ 6.2 is true. Suppose 6.2 is true for n and let us prove 6.2 for $n+1$.

Writting $\partial(f^n,g,f) + \partial(f,g,f^n)$ we have

$$C_{(g,f)} + C_{(f^n,gf)} = C_{(f^n,g)} + C_{(f^n g,f)} \quad (\text{mod } B_2).$$

Writing $\partial(f,gf^n,f)$ we conclude that

$$C_{(gf^n,f)} = C_{(gf^{n+1},f)} = C_{(g,f)} \quad (\text{mod } B_2).$$

Writing $\partial(g,f,f^n) + \partial(f^n,f,g)$ we deduce

$$c_{(f^{n+1},g)} = c_{(f,g)} + c_{(f^n,fg)} \qquad (\text{mod } B_2),$$

and, by induction, it follows that

$$c_{(f^{n+1},g)} = (n+1)\, c_{(f,g)} \qquad (\text{mod } B_2).$$

■

6.3 Corollary - Let f and g in $D^2(T^1)$ with $f\circ g = g\circ f$ then $\mathcal{G}v(f^n,g^p) = np\,\mathcal{G}v(f,g)$ if n and p are in $\mathbb{Z}$.

7. The following theorem contains a result of G. Wallet (see [7]) and provides a different proof of this result.

Theorem - If f and g are in $D^2(T^1)$ with $f\circ g = g\circ f$ then $\mathcal{G}v(f,g) = 0$.

Proof: By 6.3 we have if n is an integer, $n \geq 1$, $\mathcal{G}v(f,g) = \frac{1}{n}\mathcal{G}v(f,g^n)$.

By the formula of 5

$$\mathcal{G}v(f,g) = \frac{1}{n}\mathcal{G}v(f,g^n) =$$

(1)

$$= -\int \left[\left(\frac{\text{Log } Dg^n}{n}\right)\cdot D \text{ Log } Df^{-1} + \left(\frac{\text{Log } Dg^{-n}}{n}\right)\cdot D \text{ Log } Df\right] dm.$$

A) Case $\rho(g) = \alpha \in \mathbb{R}-\mathbb{Q}$ ($\rho(g)$ is the rotation number of g).

In 1932 A. Denjoy proved that as $n \to +\infty$, then $\frac{\text{Log } Dg^{\pm n}}{n}$ converges uniformly to 0 (g is in $D^1(T^1)$ with $\rho(g) \in \mathbb{R}-\mathbb{Q}$ is all that is needed). Let us recall the argument (see [2] for more details),

$$\frac{\text{Log } Dg^n}{n} = \frac{1}{n}\sum_{i=0}^{n-1} (\text{Log } Dg)\circ g^i,$$

and as $\rho(g) = \alpha \in \mathbb{R}-\mathbb{Q}$ g is therefore uniquely ergodic: as $\text{Log } Dg \in C^0(T^1)$ if $n \to +\infty$, $\frac{\text{Log } Dg^n}{n}$ converges uniformly to a constant $a \in \mathbb{R}$. The constant a is equal to 0 because $\int_0^1 Dg^n(x)dx = 1$.

By (1) we conclude

$$\mathcal{G}v(f,g) = \lim_{n\to+\infty} \frac{1}{n} \mathcal{G}v(f,g^n) = 0.$$

B) Case $\rho(f) = \frac{p_1}{q_1} \in \mathbb{Q}$ and $\rho(g) = \frac{p_2}{q_2} \in \mathbb{Q}$, p_1 and p_2 are integers, and q_1 and q_2 are strictly positive integers.

By the formula

$$\mathcal{G}v(R_{-p_1} \circ f^{q_1}, R_{-p_2} \circ f^{q_2}) = q_1 q_2 \, \mathcal{G}v(f,g),$$

is enough to prove that

$$\mathcal{G}v(f,g) = 0$$

if f and g have fixed points on $\mathbb{R}$, (f and g are in $D^2(T^1)$ and $f \circ g = g \circ f$), and we suppose this is the case is what follows and we suppose also that $f \neq \mathrm{Id}$ and $g \neq \mathrm{Id}$.

a) The immediate case

Let us suppose that at every fixed point a of g, $Dg(a) = 1$.

Let us remark that for every positive interger n, $\frac{1}{n}$ Log Dg^n and $\frac{1}{n}$ Log Dg^{-n} are bounded in norm by $|\mathrm{Log}\, Dg|_{C^0}$. We have as $n \to +\infty$, $\frac{1}{n}$ Log Dg^n and $\frac{1}{n}$ Log Dg^{-n} converge simply to 0 (since Log $Dg^{\pm n}$ converge simply to 0 as $n \to +\infty$; and convergence implies Cesaro convergence). It follows by dominated convergence that

$$\left(\frac{\mathrm{Log}\, Dg^n}{n}\right) \cdot D \,\mathrm{Log}\, Df^{-1} + \left(\frac{\mathrm{Log}\, Dg^{-n}}{n}\right) \cdot D \,\mathrm{Log}\, Df \to 0 \quad \text{in} \quad L^1(T^1, dx)$$

as $n \to +\infty$.

By (1) we conclude that

$$\mathcal{G}v(f,g) = \lim_{n\to+\infty} \frac{1}{n} \mathcal{G}v(f,g^n) = 0.$$

b) The general case

Let $[a,b]$ with $a < b$, be an interval with $g(a) = a$, $g(b)=b$ and g is without fixed any point on $]a,b[$, and a is the attracting fixed point of g. Then by dominated convergence we conclude that

$$\int_a^b d\mathcal{G}v(f,g) = -\mathrm{Log}\, Dg(a)(\mathrm{Log}\, Df^{-1}(b) - \mathrm{Log}\, Df^{-1}(a)) + \tag{2}$$

$$+ \mathrm{Log}\, Dg(b)(\mathrm{Log}\, Df(b) - \mathrm{Log}\, Df(a)).$$

We need the two following lemmas:

Lemma 1 - Let $Id \neq f$ and $g \neq Id$ be in $D^2(T)$ with $f\circ g = g\circ f$, f and g have fixed points. Let $a < b$ with $g(a) = a$, $g(b) = b$ g is with out any fixed point on $]a,b[$. Then $f(a) = a$, $f(b) = b$ and either f is without any fixed point an $]a,b[$, or $f|_{[a,b]} = Id_{[a,b]}$.

Proof: Let c be a fixed point of f, $c \in]a,b[$. As g leaves invariant the fixed points of f, (since $f\circ g = g\circ f$), we conclude that $f(a) = a$, $f(b) = b$; but c is a fixed point f and $c \in]a,b[$ and this is contrary to [6] if we suppose that $f|_{[a,b]} \neq Id$. The lemma follows by changing the roles of f and g. ∎

Lemma 2 - Let f and g as in Lemma 1 then there exists $k \in \mathbb{R}$ such that

$$k\, \mathrm{Log}\, Dg(a) = \mathrm{Log}\, Df(a) \quad \text{and} \quad k\, \mathrm{Log}\, Dg(b) = \mathrm{Log}\, Df(b).$$

Proof: We can suppose that $\mathrm{Log}\, Dg(a) \neq 0$ and $\mathrm{Log}\, Dg(b) \neq 0$ for if $\mathrm{Log}\, Dg(a) = 0$ then we have $\mathrm{Log}\, Df(a) = 0$ and then the lemma is immediate. We apply then [6, Lemma 3 p.172]. ∎

End of the proof of the theorem

By Lemma 1 and 2 the formula (2) becomes

$$\int_a^b d\mathcal{G}v(f,g) = -k(\mathrm{Log}\, Dg(a))^2 + k(\mathrm{Log}\, Dg(b))^2, \tag{3}$$

(this is independent from the fact that a is an attracting fixed point of g).

Let us consider $b_o < b_1$ $g(b_o) = b_o$, $g(b_1) = b_1$, g has only hyperbolic fixed points in $]b_o,b_1[$, the number is either finite or infinite.

We consider 2 cases

(i) $b_1 = b_o + 1$,

(ii) $b_1 < b_o + 1$ and $Dg(b_o) = 1 = Dg(b_1)$.

By (3) and Lemma 2 there exists $k \in \mathbb{R}$, such that for every interval $[a,b] \subset [b_o,b_1]$, with $g(a) = a$, $g(b) = b$ and g is without any fixed point on $]a,b[$, then

$$\int_a^b d\mathcal{G}v(f,g) = k(\mathrm{Log}\, Dg(b))^2 - k(\mathrm{Log}\, Dg(a))^2;$$

hence in both cases (i) and (ii)

$$\int_{b_o}^{b_1} d\mathcal{G}v(f,g) = k \sum_{\text{Intervals } [a,b]} (\mathrm{Log}\, Dg(b))^2 - (\mathrm{Log}\, Dg(a))^2 = 0,$$

and the theorem easily follows. ∎

Remarks: 1) The proof of case A) and of case B) a) works if f and g are C^1 plus absolutely continuous, and case B) b) works if f and g are C^1 plus Lipschitz.

2) I believe there is a proof of the general case B) without using [6].

3) If $a < b$ and G is the group of C^2-increasing diffeomorphisms of $[a,b]$ with derivative equal to 1 at a and b, then

$$\omega(g,f) = \int_a^b (\mathrm{Log}\, Df)\cdot D\, \mathrm{Log}\, D(g\circ f)\, dm$$

is 2-cocycle on G, by the same proof as in 4, (notice that one can integrate by parts without constants for $\mathrm{Log}\, Df(a) = \mathrm{Log}\, Df(b) = 0$).

Finally let us ask the following problem:

Problem - *Given* $r \geq 2$ *and* f *and* g *in* $\mathrm{Diff}_{+}^{r}(T^1)$ *with* $f \circ g = g \circ f$ *and* $\rho(f) \in T^1 - \mathbb{Q}/\mathbb{Z}$. *Can one approximate in the* C^r*-topology* f *and* g *by* f_i *and* g_i *with* $f_i \circ g_i = g_i \circ f_i$ *and for every* i, f_i *is* C^r*-conjugate to a rotation?*

Bibliography

[1] R. Bott - On some formulas for the characteristic classes of groups-actions, to appear in Springer Lecture Notes as proceeding of the conference at Rio 1976 on foliations.

[2] M.R. Herman - Sur la conjugaison différentiable des difféomorphismes du cercle à des rotations, to appear.

[3] M.R. Herman - Conjugaison C^∞ des difféomorphismes du cercle pour presque tout nombre de rotations, C.R. Aca.Sci.Paris, t.283, p.579-582.

[4] H. Rosenberg and R. Roussarie - Topological equivalence of Reeb foliations, Topology, (9), 1970, p. 231-242.

[5] H. Rosenberg and W. Thurston - Some remarks on foliations in Dynamical Systems, edited by M.M. Peixoto, Academic Press, New York, 1973, p. 463-478.

[6] N. Kopell - Commuting diffeomorphisms, Global Analysis, Symp. Pure Math., Vol. XIV, A.M.S., 1970 p.165-184.

[7] G. Wallet - Nullité de l'invariant de Godbillon-Vey d'un tore, to appear in C.R. Acad. Sci. Paris.

Centre de Mathématiques
École Polytechnique
91128 PALAISEAU Cedex
FRANCE

ON THE CONSTRUCTION OF THE TRACE IN SERRE DUALITY

by M. Herrera *

Consider a compact complex manifold X of dimension n and complex hypersurfaces $Y_1,\dots,Y_n$ in X such that $Y = \cap(Y_i : 1 \leq i \leq n)$ is a finite set.

There is a commutative diagram (cf. CARRELL [1], VERDIER [11])

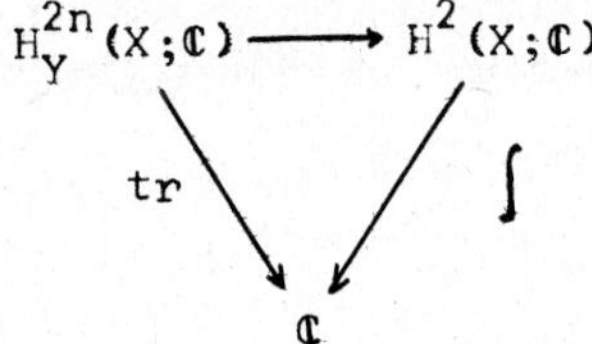

where the upper terms are the cohomology groups of X with supports in Y and with closed supports, and where the vertical arrows are given, respectively, by Grothendieck's residue (HARTSHORNE [8], EL ZEIN [4]) and by integration.

The object of this lecture is to construct this diagram in the chain level, in the case Y is a complete intersection of arbitrary codimension. To this purpose, we represent local cohomology clases of X by semi-meromorphic forms with poles in the union $\cup(Y_i : 1 \leq i \leq n)$, and define the trace by means of the residual currents introduced in [2] and [3].

The method still works for X non-compact, using cohomology with compact supports. With minor modifications, it can be adapted to describe in the chain level the Poincare duality isomorphism $H_Y^{\cdot}(X;\mathbb{C}) \to$ $\to H_{2n-\cdot}(Y;\mathbb{C})$ (cf. [9]). We refer to [2] and [3] for the definitions and notations not explicitly given here.

* Supported by a Guggenheim Fellowship, 1976-77.

1. Preliminaries. Let $\tilde{Y}$ be a hypersurface in the paracompact complex manifold X, $\dim_{\mathbb{C}} X = n$, $W = X - \tilde{Y}$ and $j:W \to X$ the immersion. We use the following notations:

$\mathbb{C}_W$ is the constant sheaf on W of stalk $\mathbb{C}$.

$\mathcal{E}^\cdot$ is the complex of sheaves of smooth differential forms on X with coefficients in $\mathbb{C}$.

$\Omega^p(*\tilde{Y})$ and $\mathcal{E}^p(*\tilde{Y})$ are, respectively, the sheaves of meromorphic and semi-meromorphic p-forms on X with poles on $\tilde{Y}$; locally, the sections of $\mathcal{E}^p(*\tilde{Y})$ are quotients of smooth p-forms on X by equations of $\tilde{Y}$.

$\Omega^\cdot(*\tilde{Y}) = \sum(\Omega^p(*\tilde{Y}): 0 \leq p \leq n)$ and

$\mathcal{E}^\cdot(*\tilde{Y}) = \sum(\mathcal{E}^p(*\tilde{Y}): 0 \leq p \leq 2n)$ are then complexes, with the usual exterior differentiation d.

$\mathcal{H}(\Omega^\cdot(*\tilde{Y}))$ and $\mathcal{H}(\mathcal{E}^\cdot(*\tilde{Y}))$ denote the cohomology sheaves of these complexes.

$\underset{\sim}{R}j_*(\mathbb{C}_W)$ is the sheaf on X generated by the presheaf $U \to H^\cdot(W \cap U;\mathbb{C})$, U open in X.

Let $\mathcal{E}^\cdot_W = \mathcal{E}^\cdot|W$. Then

$$\Omega^\cdot(*\tilde{Y}) \subset \mathcal{E}^\cdot(*\tilde{Y}) \subset j_* \mathcal{E}^\cdot_W;$$

these inclusions induce homomorphisms of the associated homology sheaves.

1.1 Lemma: $\mathcal{H}(\Omega^\cdot(*\tilde{Y})) \simeq \mathcal{H}(\mathcal{E}^\cdot(*\tilde{Y})) \simeq \underset{\sim}{R}j_*(\mathbb{C}_W)$.

D) The first isomorphism is proved by hand, using the exactness properties of $\bar{\partial}$. The second isomorphism follows then from the result of Grothendieck ([7], th.2), which assures that

$$\mathcal{H}(\Omega^\cdot(*\tilde{Y})) \xrightarrow{\sim} \mathcal{H}(j_*\mathcal{E}^\cdot_W),$$

and from $\mathcal{H}(j_*\mathcal{E}^\cdot_W) \simeq \underset{\sim}{R}j_*(\mathbb{C}_W)$.

One deduces as usual

1.2 Corollary: $\underline{\underline{H}}(X;\Omega^{\cdot}(*\ \tilde{Y})) \simeq H\Gamma(X,E^{\cdot}(*\ \tilde{Y})) \simeq H^{\cdot}(W;\mathbb{C})$.

2. Construction of the local cohomology

Let $Y_0,\ldots,Y_s$ be hypersurfaces in X, $Y = \cap(Y_i : i \in \Delta)$, $\Delta = \{0,\ldots,s\}$; then $\mathcal{U} = \{U_i = X - Y_i : i \in \Delta\}$ is an open covering of $U = X - Y$.

Denote by Δ^t the set of t-simplexes $S = (i_0,\ldots,i_{t-1}) \subset \Delta$; for each $S \in \Delta^t$, set $Y_S = \cup(Y_i : i \in S)$.

Let $\check{C}^{t,q}(*)$ be the sheaf of alternated t-cochains on $\mathcal{U}$ with values on $\{E^q(*\ Y_S) : S \in \Delta^t\}$; on an open set $V \subset X$ we have:

$$\Gamma(V,\check{C}^{t,q}(*)) = \prod_{S\in\Delta^t} \Gamma(V,E^q(*\ Y_S)).$$

Represent by

$$\lambda = (\lambda_S \in \Gamma(V,E^q(*\ Y_S)) : S \in \Delta^t)$$

an alternate cochain in $\Gamma(V,\check{C}^{t,q}(*))$. There are differentials

$$d: \begin{cases} \check{C}^{t,q}(*) \to \check{C}^{t,q+1} \\ \lambda = (\lambda_S : S \in \Delta^t) \to (d\lambda_S : S \in \Delta^t), \end{cases}$$

(2.1)

$$\delta: \begin{cases} \check{C}^{t,q}(*) \to \check{C}^{t+1,q}(*) \\ \lambda \to \delta\lambda, \end{cases}$$

where for each $T = (i_0,\ldots,i_{t+1}) \in \Delta^{t+1}$:

$$(\delta\lambda)_T = \sum_{k=0}^{t+1} (-1)^k \lambda_{T(\check{k})},$$

$$T(\check{k}) = (i_0,\ldots,\check{i}_k,\ldots,i_{t+1}).$$

We consider also $\check{C}^m(*) = \sum(\check{C}^{t,q}(*):t+q = m)$ and the total complex

$$\check{C}^{\cdot}(*) = \sum(\check{C}^m(*):0 \leq m \leq 2n+s),$$

with total differential $D = \delta + (-1)^t d$ on $\check{C}^{t,\cdot}(*)$.

There is an injection $i:E^{\cdot} \to C^{\cdot}(*)$ given by $E^m \to C^{0,m}(*)$, $\omega \to (\omega_i : i \in \Delta)$, $\omega_i = \omega(i \in \Delta)$.

We define $Q(*)$ as the quotient complex:

$$0 \to E^{\cdot} \to C^{\cdot}(*) \to Q^{\cdot}(*) \to 0. \tag{2.2}$$

2.3 Theorem: There is a canonical diagram

$$\begin{array}{ccccccccc}
\ldots \to H^{2n-1}\Gamma(X,E^{\cdot}) & \to & H^{2n-1}\Gamma(X,\check{C}^{\cdot}(*)) & \to & H^{2n-1}\Gamma(X,\check{Q}^{\cdot}(*)) & \to & H^{2n}\Gamma(X,E^{\cdot}) \to \ldots \\
I\downarrow \approx & (c) & I\downarrow \approx & (c) & \sigma\downarrow \approx & (a) & I\downarrow \approx \quad (c) \\
\ldots \to H^{2n-1}(X;\mathbb{C}) & \to & H^{2n-1}(U;\mathbb{C}) & \to & H_Y^{2n}(X;\mathbb{C}) & \to & H^{2n}(X;\mathbb{C}) \to \ldots
\end{array}$$

where the upper eact sequence is deduced from (2.2) and the bottom is the long exact sequence of local cohomology; (c) squares commute and (a) squares anticommute.

D) I is the classical De Rham isomorphism. To define I, let $\check{C}^{\cdot}(E_U^m)$ be the Čech resolution of E_U^m(= The sheaf of smooth m-differential forms on U) associated to the covering $\mathfrak{U}$. Construct the double complex $\check{C}^{\cdot,\cdot} = \check{C}^{\cdot}(E_U^{\cdot})$ with differentials as in (2.1) and the corresponding total complex $\check{C}^{\cdot}$. There is an homomorphism

$$\mu : \check{C}^{\cdot}(*) \to h_{*}\check{C}^{\cdot} \qquad (2.4)$$

where $h:U \to X$ is the immersión, given by the restrictions

$$\Gamma(V, E^{m}(*\ Y_S)) \to \Gamma(V - Y_S, E_U^{m}),$$

(V open in X).

In cohomology μ induces:

$$H\Gamma(X, \check{C}^{\cdot}(*)) \overset{\simeq}{\to} H\Gamma(X, h_{*}\check{C}^{\cdot}) \overset{\simeq}{\to} H^{\cdot}(U;C).$$

In fact, one deduces the 2nd. isomorphism by classical Čech arguments ([6], II. 5.3), since both $\check{C}^{\cdot}(*)$ and $\check{C}^{\cdot}$ are fine sheaves. The 1st. isomorphism is obtained considering the $E_1^{t,q}$ terms of the first spectral sequences of the complexes

$$\Gamma(X, \check{C}^{\cdot}(*)) \to h_{*}\Gamma(X, \check{C}^{\cdot}) = \Gamma(U, \check{C}^{\cdot}),$$

which are

$$\prod_{S \in \Delta^t} H^q\Gamma(X, E^{\cdot}(*\ Y_S)) \overset{\mu'}{\to} \prod_{S \in \Delta^t} H^q\Gamma(X - Y_S; E_U^{\cdot})$$

$$\simeq \prod_{S \in \Delta^t} H^q(X - Y_S; \mathbb{C});$$

μ' is here an isomorphism by Corollary 1.2. So, the $E_1^{t,q}$ terms of both complexes being isomorphic $\Rightarrow$ μ is an isomorphism.

One constructs easily σ, with the asserted commutativity properties.

3. Definition of the trace

3.1 Suppose now that X is compact, $s = p-1$, and that

$$\operatorname{codim}_{\mathbb{C}} Y_0 \cap \dots \cap Y_i = n - i - 1,$$

$$0 \leq i \leq p-1,$$

so that all intersections in the family $F = \{Y_0,\dots,Y_{p-1}\}$ are complete.

In these conditions, we recall the definitions (cf. [2] or [3]) of the $\mathbb{C}$-homomorphisms *residue* R_F and *residue-principal value* RP_F:

$$R_F : \Gamma(X, E^{2n-i-1}(* \bigcup_{j=0}^{i} Y_j)) \to \mathbb{C},$$

$$RP_F : \Gamma(X, E^{2n-i}(* \bigcup_{j=0}^{i} Y_j)) \to \mathbb{C}, \tag{3.2}$$

$$0 \leq i \leq p-1.$$

Let $\varphi_0,\dots,\varphi_i$ be holomorphic equations of $Y_0,\dots,Y_i$ on an open set $V \subset X$, and let

$$\omega' \in \Gamma_c(V, E^{2n-i-1}(* \bigcup_{j=0}^{i} Y_j))$$

$$\omega'' \in \Gamma_c(V, E^{2n-i}(* \bigcup_{j=0}^{i} Y_j));$$

then

$$R_F(\omega') = \lim_{\delta\to 0} \int_{T^i_{\underset{\sim}{\delta}}} \omega' ,$$

$$RP_F(\omega'') = \lim_{\delta\to 0} \int_{D^i_{\underset{\sim}{\delta}}} \omega'' , \tag{3.3}$$

where the tubes

$$T^i_{\underset{\sim}{\delta}} = (|\varphi_0| = \delta_0,\ldots,|\varphi_i| = \delta_i),$$

$$D^i_{\underset{\sim}{\delta}} = (|\varphi_0| = \delta_0,\ldots,|\varphi_{i-1}| = \delta_{i-1},|\varphi_i| > \delta_i)$$

are oriented conveniently and the vector $\underset{\sim}{\delta} = (\delta_0,\ldots,\delta_i)$ tends to zero such that $\delta_j/\delta^q_{j+1} \to 0$ for all $q > 0$, $j = 0,\ldots,i-1$. The local definitions (3.3) are then patched together to give the global R_F and RP_F. If $s = 0$, one retroves the operators defined in [10]. In this case we abreviate RP_F by P_F.

3.4 Consider

$$\lambda \in \Gamma(X,C^{2n-1}(*)),$$

$$\lambda = \sum(\lambda^{t,j}:t+j = 2n-1,\ 0 \leq t \leq p-1);$$

we define the trace $tr(\lambda)$ of λ as

$$tr(\lambda) = R_F(\lambda^{p-1,2n-p}).$$

It is clear that

a) $tr(\lambda)$ depends only of the class of λ in $\Gamma(X,\Omega^{2n-1}(*))$, and

b) $tr(D\lambda) = 0$.

In fact, if $\lambda = \sum(\lambda^{t,j}:t+j = 2n-2)$, then

$$tr(D\lambda) = R_F(\delta\lambda^{p-2,2n-p}) = 0,$$

by property 1.7.6(2) of [3].

As a consequence, one can define

$$\mathrm{tr}: \begin{cases} H^{2n-1}\Gamma(X,\mathcal{Q}^{\cdot}(*)) \to \mathbb{C} \\ [\lambda] \qquad\qquad \to \mathrm{tr}(\lambda) \end{cases} .$$

3.5 <u>Theorem</u>. The following diagram commutes

$$\begin{array}{ccc} H^{2n}_{Y}(X;\mathbb{C}) & \longrightarrow & H^{2n}(X;\mathbb{C}) \\ \sigma \uparrow \simeq & & I \uparrow \simeq \\ H^{2n-1}\Gamma(X,\mathcal{Q}^{\cdot}(*)) & \xrightarrow{\partial} & H^{2n}\Gamma(X,\mathcal{E}^{\cdot}_{X}) \\ & {\scriptstyle (-1)^{a(p)}\mathrm{tr}} \searrow \quad \swarrow {\scriptstyle \int} & \\ & \mathbb{C} & \end{array} \qquad)$$

where $a(p) = 1 + p(p-1)/2$ and ∂ is the conexion homomorphism deduced from (2.2).

D) Take $\lambda = \sum(\lambda^{t,j} : t+j = 2n-1) \in Z\Gamma(X,\mathcal{Q}^{2n-1}(*))$. Then

$$\begin{aligned} d\lambda^{t,j} &= (-1)^{t-1}\delta\lambda^{t-1,j+1}, \quad 0 < t \leq p-1, \\ d\lambda^{0,2n-1} &\in \mathrm{Im}(i:\Gamma(X,\mathcal{E}^{2n}) \to \Gamma(X,\check{\mathcal{E}}^{0,2n})). \end{aligned} \qquad (3.6)$$

By [3], 1.7.5(1), one has

$$\begin{aligned} \mathrm{tr}[\lambda] &= R_{F}(\lambda^{p-1,2n-p}_{\{0,\dots,p-1\}}) = (-1)^{p}\, RP_{F}(d\lambda^{p-1,2n-p}_{\{0,\dots,p-1\}}) \\ &= RP_{F}(\delta\lambda^{p-2,2n-p+1}_{\{0,\dots,p-1\}}), \end{aligned}$$

where

$$\delta\lambda^{p-2,2n-p+1}_{\{0,\dots,p-1\}} = \sum_k (-1)^k \lambda^{p-2,2n-p+1}_{\{0,\dots,\check{k},\dots,p-1\}}.$$

By [3], 1.7.6 (2), the contribution to RP_F of all terms in this sum is zero, with the exception of the last term, which gives ([3], 1.7.7 (3)):

$$\mathrm{tr}[\lambda] = (-1)^{p-1} RP_F(\lambda^{p-2,2n-p+1}_{\{0,\dots,p-2\}})$$

$$= (-1)^{p-1} R_F(\lambda^{p-2,2n-p+1}_{\{0,\dots,p-2\}})$$

By iteration

$$\mathrm{tr}(\lambda) = (-1)^{\frac{p(p-1)}{2}} R_F(\lambda^{0,2n-1}_{\{0\}})$$

$$= (-1)^{a(p)} P_F(d\lambda^{0,2n-1}_{\{0\}}) = (-1)^{a(p)} P_F(i(\omega)),$$

where $a(p) = 1 + p(p-1)/2$ and $\omega \in \Gamma(X, E^{2n})$, by (3.6).

Then

$$\mathrm{tr}[\lambda] = (-1)^{a(p)} P_F(i(\omega)) = (-1)^{a(p)} \int_X \omega,$$

since ω is regular on X.

It is clear from the construction that the class $[\omega] \in H^{2n}(X, E^{\cdot}_X)$ verifies $\partial[\lambda] = [\omega]$, which gives the theorem.

3.7 <u>Remarks</u>. a) Suppose $p = n$, so that Y is a finite set in X, and take $\lambda \in \Gamma(X, \check{C}^{n-1,n}(*))$. Then

$$\mathrm{tr}[\lambda] = R_F[\lambda] = \sum_{y \in Y} R_{F,y}[\lambda],$$

where $R_{F,y}[\lambda]$ is the punctual residue of λ at $y \in Y$([2] or [3]). If λ is meromorphic one has $d\lambda = 0$, so that $D\lambda = 0$ in $\Gamma(X, C^{2n-1}(*))$.

Hence, the cohomology class of λ in

$$H^{2n-1}\Gamma(X, \Omega^{\cdot}(*)) \simeq H^{2n}_{Y}(X;\mathbb{C})$$

comes from $H^{2n-1}\Gamma(X, \check{C}^{\cdot}(*))$, and its image in $H^{2n}\Gamma(X, E^{\cdot}) \simeq H^{2n}(X;\mathbb{C})$ is zero. It follows that for a meromorphic $\lambda \in \Gamma(X, \check{C}^{n-1,n}(*))$ one has

$$\operatorname{tr}[\lambda] = \sum_{y \in Y} R_{F,y}[\lambda] = 0,$$

which is a result of Griffiths [5].

In fact, this result is still true when the intersection Y of the hypersurfaces $Y_0, \ldots, Y_{n-1}$ in X (compact) has codimensión less than n([2], [3]).

b) Let f be a divisor on X: on some open covering $\mathcal{V}$ of X, f is given by meromorphic functions $f_V (V \in \mathcal{V})$ such that $f_{V'}/f_{V''}$ is holomorphic and inversible on $V' \cap V''$, for V' and $V'' \in \mathcal{V}$. The singular set of f is a hypersurface $\tilde{Y}$ of X.

It is clear that the family $\{\frac{df_V}{f_V} : V \in \mathcal{V}\}$ of meromorphic 1-forms defines a cycle $c(f) \in \Gamma(X, \Omega^1(*))$, where $\Omega^{\cdot}(*)$ is the complex defined in (2.2) with poles in the single surface $\tilde{Y}$. By 2.3, c(f) represents a cohomology class $[c(f)] \in H^2_{\tilde{Y}}(X;\mathbb{C})$. Its image in $H^2(X;\mathbb{C})$ is the Chern class(over $\mathbb{C}$) of f.

On the other hand, one can see that the current $R[c(f)]$ is the integration current on the cycle z(f) of f, which is the Poincaré dual of $[c(f)]$. So, c(f) and $R_{[c(f)]}$ represent respectively the Chern class of f and its dual.

REFERENCES

[1] CARRELL, J.B.:A remark on the Grothendick residue map, preprint, 1975.

[2] COLEFF, N. and HERRERA, M.: Fibering of residual currents. Summer Institute of Several Complex Variables, 1975, Williamstown, U.S.A.

[3] COLEFF, N. and HERRERA, M.: Les courants résiduels associés a une forme méromorphe; preprint, Instituto Argentino de Matemática, Viamonte 1636, B.Aires.

[4] EL ZEIN, F.: Complexe dualisant et applications, thèse 1976, PARIS VII.

[5] GRIFFITHS, P.: Variations on a theorem of Abel, to be published in Inventiones Math. 1976.

[6] GODEMENT, R.: Topologie algébrique et théorie des faisceaux. Hermann, Paris, 1958 (Act. scient. et ind. 1252; Publ. Inst. Math. Univ. Strasbourg 13).

[7] GROTHENDICK, A.: On the De Rham cohomology of algebraic Varieties; Publ. Math. I.M.E.S. n. 29.

[8] HARTSHORNE, R.: Residues and duality, Lecture notes in mathematics 20, Springer, New York.

[9] HERRERA, M.: Résidus multiples sur les espaces complexes. Exp. aux Journées Complexes de Metz, 1972. I.R.M.A., Université Louis Pasteur, Strasbourg.

[10] HERRERA, M. and LIEBERMAN, D.: Residues and principal values on complex spaces. Math. Annalen 194, 259-294 (1971).

[11] VERDIER, J.L.: Base change for twisted inverse image of coherent sheaves, Internat. Colloq., TaTa Inst. Fund. Res., Bombay (1968); Oxford Univers. Press, London, 1969.

Universidad de Buenos Aires
Argentina

Localization theories for groups and homotopy types

Peter Hilton

1. Introduction

Let C be a category and let P be a family of (rational) prime numbers. Let us suppose that we have a full subcategory C_P of C and that we agree to call the objects of C_P the P-*local* objects of C. We will say we have a P-*localization theory* in C if we may associate with each object X of C an object X_P of C_P and a morphism $e = e_P : X \to X_P$ of C such that, given any morphism $f : X \to Y$ in C with Y in C_P, there exists a unique morphism $\bar{f} : X_P \to Y$ in C_P with $\bar{f}e = f$.

(1.1)
$$\begin{array}{ccc} X & \xrightarrow{e} & X_P \\ {\scriptstyle f}\downarrow & \swarrow {\scriptstyle \bar{f}} & \\ Y & & \end{array}$$

This is equivalent to asking that C_P be *coreflexive* in C ; that is, the embedding functor $C_P \subseteq C$ should have a left-adjoint, left-inverse $L : C \to C_P$ (so that $LX = X_P$). If we have a P-localization theory for all families P, we say that we have a *localization theory* in C. We write g_P for Lg, for any morphism g of C ; we call e (or any morphism equivalent to e) the P-*localizing* morphism.

In this article we discuss localization theories in four categories: the category Ab of abelian groups; the category N of nilpotent groups; the homotopy category H_1 of one-connected spaces of the based homotopy type of CW-complexes; and the homotopy category NH of nilpotent spaces [4]. We concentrate on certain very specific aspects of these localization theories and take the opportunity, at the end of the article, to correct a false statement made in [3] and repeated in [4].

The material of the next three sections is essentially contained in the monograph [4] by Mislin, Roitberg and the author. The material of the final section will appear shortly, in extended form, in a paper by the same trio of authors [5].

2. The category Ab of abelian groups

We define an abelian group B to be P-local if it admits unique division by all primes q outside P. It is then very elementary to construct, for a given abelian group A, the P-localizing homomorphism $e : A \to A_P$ and to verify the universal property (1.1). Namely, we start with the ring Z_P of P-local integers, that is, the subring of the rational field $\mathbb{Q}$ consisting of rational numbers expressible as fractions $\frac{m}{n}$, where no member of P divides n; and then we set

$$A_P = A \otimes Z_P \ ; \ ea = a \otimes 1 \ , \ a \in A \ .$$

In this localization theory we have the following crucial properties, all easily proved.

2.1 The homomorphism $\varphi : A \to B$ is an isomorphism if and only if $\varphi_p : A_p \to B_p$ is an isomorphism for all primes p.

2.2 Let A, B be finitely-generated. Then $A \cong B$ if and only if $A_p \cong B_p$ for all primes p.

2.3 Let A be finitely-generated. Then A is the pullback of the collection $\{A_p\}$ over the rationalization A_o.

2.4 (Detection principle) The homomorphism $\varphi : A \to B$ P-localizes A if and only if B is P-local and φ is P-bijective. (Proposition I.1.9 of [4]).

We now comment on and clarify these four assertions. First, if P consists of a single prime p, we write A_p, φ_p instead of $A_{(p)}$, $\varphi_{(p)}$. Second, the conclusion of 2.2 is stronger than that of 2.1 since, in 2.2, no assumption is made that the collection of 'local' isomorphisms $A_p \cong B_p$ is derived from a 'global' homomorphism $A \to B$. On the other hand, the conclusion of 2.2 is false if we allow B not to be finitely-generated; indeed with $A = Z$ there are infinitely many mutually non-isomorphic (and, of course, non-finitely-generated) groups B with $A_p \cong B_p$ for all primes p. Third, the meaning of 2.3 is the following. The rationalization A_o of A is the localization of A at the empty family of primes. For each p, there is a canonical rationalizing homomorphism $r_p : A_p \to A_o$, so that it is evident what we should mean by a collection of elements $\{a_p\}$, $a_p \in A_p$, agreeing over A_o - we mean that $r_p(a_p)$ is independent of p. Then 2.3 asserts that, given a collection of elements $\{a_p\}$, $a_p \in A_p$, agreeing over A_o, there exists a unique element $a \in A$ such that $e_p a = a_p$. It is important to remark that it is only the existence of a which requires that A be finitely generated; the uniqueness of a follows easily from 2.4. Fourth, the significance of 2.4 is that it replaces the definition of P-localization, which requires in principle the study of all homomorphisms from A to P-local groups, by a criterion which involves only a study of the homomorphism φ itself. We say that a homomorphism $\varphi : A \to B$ is P-*injective* if the kernel of φ consists of elements of finite orders prime to the members of P ; that it is P-*surjective* if, given any $b \in B$, there exists a positive integer n admitting no member of P as factor such that $nb \in$ image φ ; and that it is P-*bijective* if it is P-injective and P-surjective.

3. *The category N of nilpotent groups*

Our definition of a P-*local* nilpotent group H differs from that given in Section 2 only insofar as we use multiplicative notation in N and additive notation in Ab ; thus we ask that H have unique q^{th} roots for all primes q outside P . It is then true, but far from immediate, that N admits a localization theory; see, for example, [1] for details. Moreover, this theory does *generalize* the theory described in Section 2 in the sense that, if the abelian group A is regarded as a nilpotent group then its P-localization, as a nilpotent group, coincides with its P-localization as an abelian group. Indeed, more is true; if N_c is the full subcategory of N consisting of groups of nilpotency class $\leqslant c$ (so that $Ab = N_1$), then N_{c+1} admits a localization theory generalizing that of N_c , in the sense described above. Thus the existence of a localization theory in N is demonstrated by an argument by induction on c .

We turn now to the analogies of the four properties of the localization theory in Ab . With regard to the localization theory in N we have:

3.1 The homomorphism $\varphi : G \to H$ is an isomorphism if and only if $\varphi_p : G_p \to H_p$ is an isomorphism for all primes p . (Theorem I.3.12 of [4]).

3.2 There exist finitely-generated nilpotent groups G , H such that $G_p \cong H_p$ for all primes p but $G \not\cong H$. (p. 32 of [4]).

3.3 Let G be finitely-generated. Then G is the pullback of the collection $\{G_p\}$ over the rationalization G_0 . (Theorem I.3.6 of [4]).

3.4 (*Detection principle*) The homomorphism $\varphi : G \to H$ P-localizes G if and only if H is P-local and φ is P-bijective. (Fundamental Theorem, p. 7 of [4]).

We again comment. Plainly we may say that, in generalizing from

Ab to N , the analogues of 2.1, 2.3, 2.4 remain valid, whereas the analogue of 2.2 is false. The first example of two groups G , H (belonging to N_2) satisfying 3.2 was given by Milnor and subsequently Mislin gave a systematic method of generating examples; see [4,6]. Arising from 3.2 comes the concept, due to Mislin, of the genus of a finitely-generated nilpotent group G . This is the set of isomorphism classes of finitely-generated nilpotent groups H such that $G_p \cong H_p$ for all primes p . This set is finite; moreover if the commutator subgroup of G is finite the genus has the structure of a (finite) abelian group [2]. As to 3.3, we again really have an existence and uniqueness statement, and only the existence statement requires that G be finitely-generated. In 3.4 we have the notion of a P-bijective homomorphism; the definition of P-injective, P-surjective, P-bijective homomorphism in N is simply obtained from that in Ab by replacing additive by multiplicative notation.*

4. The homotopy categories H_1 and NH

The localization theory in Ab may be applied in order to develop a localization theory in the homotopy category H_1 of one-connected spaces of the based homotopy type of CW-complexes. We may then generalize this theory using the localization theory in N , to the homotopy category NH of nilpotent spaces. We recall that a (based) space X is said to be nilpotent if it is of the based homotopy type of a CW-complex, and if moreover $\pi_1 X$ is nilpotent and operates nilpotently on the higher homotopy groups of X . The importance of the category NH stems from the facts that it contains H_1 ; it contains all connected Lie groups; it contains the classifying spaces of all nilpotent Lie groups; and, most important from the technical point of view, if X is nilpotent and Y is a compact polyhedron, then every component of the function space X^Y is nilpotent.

We say that a space X in NH is P-local if all its homotopy groups are P-local; it turns out that this is equivalent to requiring that all its homology groups be P-local. It is then not difficult to

---

* We could have said that φ , in Ab , is P-surjective if the cokernel of φ consists of elements of finite orders prime to the members of P . This would not have generalized so smoothly to N ; but it does actually turn out to characterize P-surjections in N , provided 'cokernel' is given its categorical meaning.

construct the P-localization in H_1, using the P-localization in Ab and the cellular structure of the objects of H_1. Basic to this construction is that of a P-<u>local sphere</u> S^n_P, $n \geq 2$. We may either simply regard S^n_P as the Moore space having Z_P as its only (reduced) nonvanishing homology group, in dimension n; or we may construct S^n_P by the <u>telescope</u> construction. In the latter construction we take a sequence of maps

$$S^n \xrightarrow{h_1} S^n \xrightarrow{h_2} S^n \xrightarrow{h_3} \cdots,$$

where each prime outside P occurs infinitely often as the degree of a map among $h_1, h_2, h_3, \ldots$. We then construct S^n_P as a union of mapping cylinders of the maps $h_1, h_2, h_3, \ldots$.

When the space X is nilpotent but not one-connected, the construction of the P-localization X_P is far less direct and involves a principal refinement of the Postnikov tower of X; see [4] for details. However, it does turn out that there exists a localization theory in NH, generalizing that in H_1, and we may again turn to the analogues of properties 2.1 - 2.4. The situation is the following:

4.1 A morphism $f : X \to Y$ in NH is an equivalence if and only if $f_p : X_p \to Y_p$ is an equivalence for all primes p. (This follows from 3.1 and the Whitehead Theorem).

4.2 There exist spaces X, Y of finite type in H_1 such that $X_p \simeq Y_p$ for all primes p but $X \not\simeq Y$. (See, e.g., Example III.1.3 of [4]).

4.3 (<u>Hasse principle</u>) Let X be nilpotent of finite type and let W be a compact connected polyhedron. Then $[W,X]$ is the pullback of the collection $\{[W,X_p]\}$ over $[W,X_0]$. (Theorem II.5.1 of [4]).

4.4 (<u>Detection principle</u>) The morphism $f : X \to Y$ in NH P-localizes X if and only if Y is P-local and $\pi_n f : \pi_n X \to \pi_n Y$ is P-bijective for all $n \geq 1$ (or, equivalently, Y is P-local and $H_n f : H_n X \to H_n Y$ is P-bijective for all $n \geq 1$). (Theorem II.3B of [4], together with 3.4).

Once more we would wish to comment on these assertions. 4.1 is, of course, the straight generalization of 3.1 and is proved via the Whitehead theorem. As to 4.2, we have emphasized that examples are to be found not merely in NH - as they must be, since they are present in N - but even in H_1. Thus the phenomenon recorded in 4.2 is not to be thought of as a 'non-abelian' phenomenon. The requirement that X, Y be of <u>finite type</u> (that is, that their homotopy groups - or, equivalently, their homology groups - all be finitely-generated) is the natural generalization of the finite-generation condition imposed in 2.2; however, we may even obtain examples of 4.2 in which X, Y are smooth, closed manifolds. In fact let us consider principal S^3-bundles over S^7. Such bundles are classified by the group $\pi_6(S^3)$ which is cyclic of order 12 generated by the Blakers-Massey element ω which measures the non-commutativity of quaternionic multiplication. Let us write E_k for the total space of the bundle classified by $k\omega$, $0 \leq k \leq 11$. Then E_1 is the symplectic group $Sp(2)$ and it may be shown that

$$(E_1)_p \simeq (E_5)_p \text{ , for all primes } p \text{ ; } E_1 \not\simeq E_5 \text{ .}$$ (See III.2 of [4]).

Thus, once again, the notion of genus arises; and the genus of a compact one-connected polyhedron is again finite (it is surely true that the genus of a compact nilpotent polyhedron is finite). It is a fascinating problem to study genus-invariant properties (for example, it turns out that the property of being a Hopf manifold with globally finitely- generated homology groups is genus-invariant, and this result has led to the discovery, in fairly systematic fashion, of new Hopf manifolds* by Zabrodsky and others). It is also a fascinating problem to develop effective algebraic invariants capable of distinguishing between spaces of the same genus; so much of the powerful apparatus of modern algebraic topology - for example, the Steenrod operations - is genus-invariant.

We come now to the Hasse principle 4.3 which forms the starting point of our discussion in the next section. If W, X are based spaces we write $[W,X]$ for the set of based homotopy classes of maps of W into X. Note that in the statement of the Hasse principle we do not require W to be nilpotent. Note also that if $W = S^n$ then 4.3 is an immediate consequence of 3.3; and thus 3.3 plays a key role in the proof of 4.3. We cannot eliminate the role of W in the enunciation of the principle since the homotopy category does not, in general, admit pullbacks. Moreover it is not sufficient to require that W be of finite type, although it is easy to deduce from the Hasse principle that the conclusion of 4.3 remains true if W is quasi-finite, that is, if W is nilpotent with globally finitely-generated homology groups. An example due to Mislin shows that such a polyhedron W need not have the homotopy type of a compact polyhedron.

---

*A Hopf manifold is a connected manifold M admitting a continuous multiplication with 2-sided unity.

With regard to the detection-principle 4.4 we remark that this is a consequence of the following more basic result. We say that $f : X \to Y$ is a P-equivalence, where $X, Y \in NH$, if $f_P : X_P \to Y_P$ is a homotopy equivalence. Then f is a P-equivalence if and only if $\pi_n f : \pi_n X \to \pi_n Y$ is P-bijective for all $n \geq 1$ (or, equivalently, $H_n f : H_n X \to H_n Y$ is P-bijective for all $n \geq 1$). Of course, the equipotency of the roles of homotopy and homology in the study of nilpotent spaces is by no means a triviality - they are certainly not equivalent for general connected polyhedra - and accounts for the tractability of nilpotent spaces and their amenability to study by the established techniques of algebraic topology.

5. The Hasse principle and related properties of the homotopy sets [W,X]

A key element in the proof of the Hasse principle 4.3 is the following. Let $W = V \cup e^n$, $n \geq 2$, let $g : W \to X$, $\bar{g} = g|V$. There is then a sequence

$$(5.1) \qquad \pi_1(X^V, \bar{g}) \xrightarrow{\varphi=\varphi(g)} \pi_n X \longrightarrow [W,X]_g \longrightarrow [V,X]_{\bar{g}}$$

Here $[W,X]_g$ is the homotopy set of maps $W \to X$, based at g : $[V,X]_{\bar{g}}$ is defined similarly; and (5.1) is exact in the sense that

(i) coker φ operates faithfully on the element $g \in [W,X]$;

(ii) if $h : W \to X$, then $\bar{h} = \bar{g} \Leftrightarrow h = g^{\alpha}$, $\alpha \in$ coker φ .

It is important to stress that the image of φ depends on the choice of $g \in [W,X]$. Thus, as we vary g, we obtain various groups coker $\varphi(g)$ embedded (by exactness) in the set $[W,X]$. Notice that the set $[W,X]$ has a preferred element 0, the class of the constant map.

We may use (5.1) to provide an inductive proof of the Hasse principle; thus we observe, as a group-theoretical fact, that 4.3 holds if W is a wedge of circles and are thus able to prove it for a general compact connected polyhedron W by assuming it true for V and applying (5.1). The same technique enables us to prove the following collection of propositions [5]. In all these propositions W is a compact connected polyhedron and X is nilpotent of finite type. We say that a family of primes is cofinite if its complement is finite.

Proposition 5.1. Let $S \subset T$ be families of primes. Then $[W,X_T] \to [W,X_S]$ is finite-to-one.

Proposition 5.2. Let $f : W \to X_0$. Then there exists a cofinite family of primes P such that f lifts uniquely (up to homotopy) into X_P.

Since the proofs of these two propositions are similar, we will be content to sketch the proof of Proposition 5.2. We will assume the group-theoretical facts established, so that we have merely to carry out the inductive step, based on (5.1). We first prove the existence of a cofinite family Q such that f lifts into X_Q, and will handle the uniqueness assertion later. Thus let $W = V \cup e^n$ and let $\bar{f} = f|V$ lift into $X_{\bar{Q}}$, as $\bar{g} : V \to X_{\bar{Q}}$. If $a : S^{n-1} \to V$ is the attaching map for e^n, then $\bar{g}a$ represents an element in the kernel of $\pi_{n-1}X_{\bar{Q}} \to \pi_{n-1}X_o$. This kernel is finite, so that there is a cofinite Q^* in $\bar{Q}$ such that, if $\bar{g}$ 'lowers' to $\bar{g}^* : V \to X_{Q^*}$, then $\bar{g}^*a = 0$. Thus $\bar{g}^*$ extends to $g^*: W \to X_{Q^*}$. From (5.1) we now infer

$$\text{(5.2)}\qquad \begin{array}{ccccc} \pi_n X_{Q^*} & \longrightarrow & [W,X_{Q^*}]_{g^*} & \longrightarrow & [V,X_{Q^*}]_{\bar{g}^*} \\ \downarrow & & \downarrow & & \downarrow \\ \pi_n X_o & \longrightarrow & [W,X_o]_{f^*} & \longrightarrow & [V,X_o]_{\bar{f}} \end{array},$$

where f^* is obtained by lowering g^*; of course, $f^*|V = \bar{f}$. Thus $f = f^{*\alpha}$, $\alpha \in \pi_n X_o$. For some k, $k\alpha$ is in the image of $\pi_n X_{Q^*}$. Thus there exists a cofinite Q contained in Q^* such that α is the image of some element β, say, in $\pi_n X_Q$. Set $g = g'^{\beta}$, where $g' : W \to X_Q$ is obtained by lowering g^*. Then g lifts f into X_Q.

We now handle the uniqueness: to do so we strengthen the proposition by claiming that _there exists a cofinite family_ P _such that_ f _lifts uniquely into_ X_S _for all_ $S \subseteq P$. Thus we assume inductively that $\bar{f}$ lifts uniquely into X_T for $T \subseteq \bar{Q}$, and let f lift into X_R, with R and $\bar{Q}$ cofinite. Take $Q^* = R \cap \bar{Q}$ and let f lift to $g^* : W \to X_{Q^*}$. From (5.1) we infer

$$\text{(5.3)}\qquad \begin{array}{ccccc} \pi_1(X_{Q^*},g^*) & \xrightarrow{\varphi^*} & \pi_n X_{Q^*} & \twoheadrightarrow & \operatorname{coker} \varphi^* \\ \downarrow & & \downarrow & & \downarrow e \\ \pi_1(X_o,f) & \xrightarrow{\psi} & \pi_n X_o & \twoheadrightarrow & \operatorname{coker} \psi \end{array}$$

Now we show that $\operatorname{coker} \varphi^*$ is a finitely-generated Z_{Q^*}-module and e in (5.3) is the rationalizing map. Thus the kernel of e is finite. It follows, as before, that we may find a cofinite $P \subseteq Q^*$ such that, for all $S \subseteq P$, we have

$$\text{(5.4)}\qquad \begin{array}{ccccccccc} \pi_1(X_S,g) & \xrightarrow{\varphi} & \pi_n X_S & \twoheadrightarrow & \operatorname{coker} \varphi & \rightarrowtail & [W,X_S]_g & \longrightarrow & [V,X_S]_g \\ \downarrow & & \downarrow & & \downarrow e & & \downarrow & & \downarrow \\ \pi_1(X_o,f) & \xrightarrow{\psi} & \pi_n X_o & \twoheadrightarrow & \operatorname{coker} \psi & \rightarrowtail & [W,X_o]_f & \longrightarrow & [V,X_o]_f \end{array},$$

where e is injective and g is obtained by lowering g^*. We claim that g is the unique lift of f into X_S. For if h also lifts f, then $\bar{g}$ and $\bar{h}$ both lift $\bar{f}$ so that $\bar{h} = \bar{g}$. Thus $h = g^{\alpha}$, $\alpha \in \operatorname{coker} \varphi$, whence $f = f^{e\alpha}$. This implies that $e\alpha$ is the neutral element of $\operatorname{coker} \psi$, so that α is the neutral element of $\operatorname{coker} \varphi$, whence $h = g$.

As an immediate consequence of Proposition 5.2, in its strengthened form, we have

Corollary 5.3. *There exists a cofinite family of primes* P *such that* $[W,X_S] \to [W,X_0]$ *is weakly injective (i.e., the counterimage of* 0 *is* 0) *for all* $S \subseteq P$.

In [4] the authors incorrectly asserted a strengthening of Corollary 5.3, in which 'weakly injective' was replaced by 'injective'. Doubt was cast on this stronger claim by J.F. Adams in a letter to one of the authors (P.H.); moreover, Adams proposed a counterexample. In [5] the authors show that Adams' scepticism was justified and that his proposed counterexample did, indeed, disprove the stronger claim. Before describing the counterexample, we point out that Corollary 5.3 does, in fact, have the following consequences [5].

Theorem 5.4. *There exists a cofinite family of primes* P *such that* $[W,X_P] \to [W,X_0]$ *is injective, provided that one of the following conditions holds*:

(i) W *is a 1-connected co-H-space*;

(ii) W *is a suspension space*;

(iii) X *is a rational H-space (i.e.,* X_0 *is an H-space*).

The rationale behind this theorem is that, under any of the three stated assumptions, there is homogeneity in the counterimages of elements of $[W,X_0]$ under the map $[W,X_P] \to [W,X_0]$, so that weak injectivity implies injectivity. The Adams counterexample displays the absence of such homogeneity (without the supplementary assumptions) in very strong form. We will sketch the proof of (iii).

Suppose $\dim W \leq n$. We may then, without real loss of generality, suppose that X has vanishing homotopy groups in dimensions $\geq n+1$; we may also (see Theorem II.4.1 of [4]) suppose that $(X \times X)^{n+1}$ is a finite complex. Consider the map $X \times X \to X_0 \times X_0 \to X_0$, composing rationalization with the multiplication on X_0, and let u be its restriction to $(X \times X)^{n+1}$. There is then, by Proposition 5.2, a cofinite

P_1 such that u and $u|(X \vee X)^{n+1}$ lift uniquely to X_S for all $S \subseteq P_1$. Let v_1 be the lift of u; then v_1 extends uniquely to $v_2 : X \times X \to X_S$, inducing $v : X_S \times X_S \to X_S$. It is plain that v is an H-structure on X_S making $X_S \to X_O$ an H-map. Let P_2 be cofinite such that $[W,X_S] \to [W,X_O]$ is weakly injective for all $S \subseteq P_2$ and let $P = P_1 \cap P_2$, $S \subseteq P$. Then P is cofinite and $[W,X_S] \to [W,X_O]$ is a weakly injective homomorphism of loops and hence injective.

We now describe explicitly the Adams counterexample.

We take $X = S^3 \vee S^3$, $W = S^2 \times S^3$. Reverting to (5.1) we regard W as $V \cup e^5$, where $V = S^2 \vee S^3$; and we choose a positive integer m and define $g : W \to X$ to be the map which projects W onto S^3 and then sends S^3 to the second target S^3 in X by a map of degree m. We will use the notation m for g or $\bar{g}$, so that (5.1) becomes

$$(5.2) \qquad \pi_1(S^3 \vee S^{3^{S^2 \vee S^3}}, m) \xrightarrow{\varphi = \varphi(m)} \pi_5(S^3 \vee S^3) \longrightarrow [S^2 \times S^3, S^3 \vee S^3]$$

It turns out that for any element ξ of $\pi_1(S^3 \vee S^{3^{S^2 \vee S^3}}, m)$, $\varphi(2\xi)$ is a multiple of mw, where $w \in \pi_5(S^3 \vee S^3)$ is the universal Whitehead product $[\iota_1, \iota_2]$, while there exists ξ_0 with $\varphi(\xi_0) = mw$. Thus if m is odd coker $\varphi = \mathbb{Z}/m \oplus T$, where T is a finite 2-group. It follows that coker φ is annihilated by rationalization. However, given <u>any</u> cofinite P, we may choose m to be a P-number so that the $\mathbb{Z}/m$-summand in coker φ is preserved under P-localization. Since coker φ embeds in $[S^2 \times S^3, S^3 \vee S^3]$ it is clear that $[S^2 \times S^3, S^3_P \vee S^3_P] \to [S^2 \times S^3, S^3_O \vee S^3_O]$ is not injective for any cofinite P; indeed, a slight strengthening of the argument shows that it is not injective for any <u>non-empty</u> P whatsoever!

Let us be explicit about the inhomogeneity responsible for the fact that, in our example, $[W,X_P] \to [W,X_O]$ cannot be injective for any cofinite P. Given P and m odd, then $[W,X_P]$ contains the group coker φ_P where $\varphi = \varphi(m)$ is as in (5.2); and coker φ_P contains the summand $\mathbb{Z}/m_P$ where m_P is the P-part of m. Thus $[W,X_P]$ contains finite subgroups of arbitrarily large order (as we vary m) and all these subgroups are annihilated under $[W,X_P] \to [W,X_O]$, since coker φ_O is zero. Here we are to understand by 'annihilated' that the elements of such a subgroup are all mapped to the same element of $[W,X_O]$ (not necessarily, of course, the 'zero' element, since we are not dealing with a <u>group</u>-homomorphism). That is to say, counterimages of elements of $[W,X_O]$ are arbitrarily large sets, differing in size one from another.

Bibliography

[1] Peter Hilton, Localization and cohomology of nilpotent groups, Math. Zeits. 132 (1973), 263-286.

[2] Peter Hilton and Guido Mislin, On the genus of a nilpotent group with finite commutator subgroup, Math. Zeits. 146 (1976), 201-211.

[3] Peter Hilton, Guido Mislin and Joseph Roitberg, Homotopical localization, Proc. Lond. Math. Soc. 3, XXVI (1973), 693-706.

[4] Peter Hilton, Guido Mislin and Joseph Roitberg, Localization of Nilpotent Groups and Spaces. Mathematics Studies 15, North Holland (1975), pp. X+156.

[5] Peter Hilton, Guido Mislin and Joseph Roitberg, On maps of finite complexes into nilpotent spaces of finite type: a correction to 'Homotopical Localization'. Proc. London Math. Soc. (1977) (to appear).

[6] Guido Mislin, Nilpotent groups with finite commutator subgroups, Springer Lecture Notes 418 (1974), 103-120.

Case Western Reserve University, Cleveland, USA
Battelle Research Center, Seattle, USA
Mathematical Institute, University of Warwick, England

On Hyperbolic Attractors of Codimension One[1)]

by

Heinrich Kollmer

Let M be a differentiable manifold of dimension n and let $f \in \mathrm{Diff}(M)$.

Definition - A compact subset $\Lambda \subset \Omega(f)$ (= non-wandering points of f) is called a hyperbolic attractor of f if

1) There is a compact neighborhood $N \supset \Lambda$ with $f(N) \subset \mathrm{int}(N)$ and $\bigcap_{i\geq 0} f^i(N) = \Lambda$.
2) The periodic points $\mathrm{per}(f|\Lambda)$ are dense in Λ
3) $f|\Lambda: \Lambda \to \Lambda$ has a hyperbolic structure
4) There exists a transitive point in Λ

Hyperbolicity of Λ means that there is a splitting of the tangent bundle over Λ: $T_\Lambda M = E^s + E^u$ into a contracting (stable) bundle E^s and an expanding (unstable) bundle E^u, both of which are invariant under df (See [4]). s and u denote the fibre dimension of E^s and E^u respectively. s is called the codimension of Λ. Hyperbolic attractors come up in the theory of Axiom A diffeomorphisms [4].
In [2] Plante proves that Λ determines a non-trivial homology class of $H_{n-1}(M,R)$, if $\Lambda \subset M$ is a hyperbolic attractor of codimension one and if the bundles E^s and E^u are both oriented.
In [1] it is shown that simply connected 3-manifolds do not admit

---

1) This paper was written while the author was supported by the Deutsche Forschungsgemeinschaft.

expanding attractors [5] of codimension one whose stable foliation is C^1. This result is generalized by the following:

<u>Theorem A</u> - If M is a simply connected, compact differentiable manifold with $\dim M \geq 3$, then M does not admit a hyperbolic attractor of codimension one.

This theorem constrasts with Plykin's example in [3] where he describes a construction of a hyperbolic attractor of codimension one on S^2. The difference between $\dim M = 2$ and $\dim M = n \geq 3$ is caused by the fundamental group of $S^1 = S^{2-1}$ and S^{n-1}. Theorem A is a consequence of the following result:

<u>Theorem B</u> - Let M be an oriented, differentiable manifold of dimension n and let $f \in \mathrm{Diff}(M)$. If Λ is a hyperbolic attractor of codimension one, then there exists a neighborhood $N \supset \Lambda$, which is a compact manifold with boundary (not necessarily smooth) with the following properties:

1) $f(N) \subset \mathrm{int}\, N$
2) $\bigcap_{i \geq 0} f^i(N) = \Lambda$
3) the boundary of N, ∂N, is a disjoint union of $(n-1)$-spheres
4) $f: N \to N$ is isotopic to a homeomorphism $g: N \to N$.

The assumption that M is oriented can probably be dropped with some additional work using a double cover argument.

<u>Sketch of the Proof of Theorem B:</u>

I. Construct a neighborhood N of Λ with the following properties:

a) N is a compact C^0-manifold with boundary.
b) $f(N) \subset \mathrm{int}\, N$.
c) If H is a connected component of $M - N$ then $f^r(H) \supset H$ for some $r \in \mathbb{N}$.

d) N is foliated by $W^s(\Lambda)|N$, the restriction of the stable manifolds restricted to N, where a leaf is considered to be connected.

e) Let l be a leaf of $W^s(\Lambda)|N$ and let I be a connected component of $l \cap \partial N$. I is a single point if and only if I contains an end point of l.

f) Each leaf of $W^s(\Lambda)|N$ intersects Λ.

II. For $r \in \mathbb{N}$ define $\Delta_r = \text{clos}\,(N-f^r(N))$. Δ_r is foliated by $W^s(\Lambda)|\Delta_r$. Let D_r be the union of leaves of $W^s(\Lambda)|\Delta_r$, which intersect $f^r(\partial N)$ in exactly two points. We prove that $\Delta_r = D_r \cup E_r$ and $D_r \cap E_r = \emptyset$.

III. Here we analyze E_r. Let F_r be a connected component of E_r, where r is picked in such a way that $f^r(F_r \cap \partial N) = F_r \cap f^r(\partial N)$. F_r is a compact n-manifold with boundary foliated by $W^s(\Lambda)|F_r$. The leaves are compact intervals. ∂F_r is connected and it is the union of three (n-1)-dimensional manifolds with boundary: $F_r = G_r \cup H_r \cup K_r$ where $G_r = \partial F_r \cap \partial f^r(N)$, H_r is the union of leaves of $W^s(\Lambda)|F_r$ which intersect ∂N, and $K_r = f^r(\partial N \cap F^r)$.
Put $H'_r = G_r \cap H_r$ and $K'_r = G_r \cap K_r$. Both H'_r and K'_r are non-empty compact (n-2)-manifolds and $H'_r \cap K'_r = \emptyset$. $H'_r \cup K'_r$ is the boundary of G_r in ∂F_r.
If l is a leaf of $W^s(\Lambda)|F_r$, then the two end points of l are in G_r, and every point of G_r is such an end point. Let l be as above and let L and L' be as the two unstable manifolds closest to the end points of l, measured along l in $f^r(N)$. L and L' are independent of l. Define $p_r: G_r \to L \cup L'$ to be the map, which assigns to $x \in G_r$ the closest point of $L \cup L'$ measured as above. p_r is a homeomorphism onto its image.

The same construction can be done for ir $(i \in \mathbb{N})$ instead of r. L and L' are independent of i and one gets the inclusions

$$p_r(G_r) \subset p_{2r}(G_{2r}) \subset p_{3r}(G_{3r}) \subset \dots \subset L \cup L'$$

and the equalities

$$p_r(H'_r) = p_{ir}(H_{ir}) \quad \text{for all} \quad i \in \mathbb{N}.$$

f induces a sequence of continuous expanding injections $(f_i)_{i\in\mathbb{N}}: p(H'_r) \to \bigcup_{i\in\mathbb{N}} p_{ir}(G_{ir})$ with $f_i(p_r(H'_r)) = p_{ir}(K'_{ir})$. There exists a constant $\epsilon > 1$ such that the minimal expansion of f_i is greater than ϵ. From this we deduce that the compact (n-2)manifolds $p_r(H'_r)$ and $p_{ir}(K_{ir})$ do not link in $L \cup L'$ (L and L' are "copies" of $\mathbb{R}^{n-1}$) and that $L \cap p_r(H'_r)$ and $L' \cap p_r(H'_r)$ consist of one connected component. From this and the fact that M is oriented we conclude that N can be modified such that $p_r(H'_r)$ consists of two (n-2)-spheres and that F_r is homeomorphic to $S^{n-2} \times [0,1] \times [0,1]$ where the sets $(s,t) \times [0,1]$ are the "stable leaves" of F_r and $S^{n-2} \times [0,1] \times \{0\}$ and $S^{n-2} \times [0,1] \times \{1\}$ are the sets H_r and K_r respectively.

IV. To describe D_r we make a similar construction. Let C_r be a connected component of E_r. $C_r \cap f^r(N)$ can be projected into the closest unstable manifold L (measured along the stable manifolds). The closure of its image, C'_r, is a compact (n-2)-manifold with boundary ∂ in L. ∂ is a disjoint union of sets of the form $p_r(H_r)$ (as above). $(f^r)^{-1}$ induces a series of contracting injections on C'_r. And similarly as in III one obtains that C'_r is an open (n-1)-disk, D^o_{n-1}. From this one concludes that C_r is homeomorphic to $D^o_{n-1} \times I$ where $\{x\} \times I$ is a "stable leaf".

V. Putting together D_r and F_r one gets Δ_r as a union of sets homeomorphic to $S^{n-1} \times I$, where $S^{n-1} \times \{0\}$, $S^{n-1} \times \{1\}$ are the connected components of ∂N and $\partial f(N)$ respectively. Now it is also clear that we can construct the required isotopy. This ends the proof of Theorem B.

Sketch of the Proof of Theorem A:

$\pi_1(M) = e$ and the fact that ∂N is a union of spheres implies that $\pi_1(N) = e$, if $n > 2$ (Seifert - van Kampen - Theorem). Therefore, the stable and unstable foliations of N are oriented. According to Plante [2], $H_{n-1}(M,\mathbb{R}) \neq 0$. On the other hand, $\pi_1(M) = e$ implies $H_{n-1}(M,\mathbb{R}) = 0$. This then proves Theorem A.

BIBLIOGRAPHY

[1] Kollmer, H. "On the Existence of Expanding Attractors of Codimension One". Thesis, Northwestern University, (1975).

[2] Plante, J. "The Homology Class of an Expanded Invariant Manifold". Dynamical Systems, Warwick 1974. Springer Lecture Notes in Mathematics, 468, pp. 251-257.

[3] Plykin, "Sources and Sinks of A Diffeomorphisms". Math. Sb. 94 (136), (1974), pp. 243-264). (Russian)

[4] Smale, S. "Differentiable Dynamical Systems". BAMS 73 (1967), pp. 747-817.

[5] Williams, R.F. "Expanding Attractors". Institut des Hautes Etudes Scientifiques. Publ. Math. 43 (1974), pp. 169-703.

Mathematisches Institut der Universität
D8520 Erlangen, West Germany

On Liénard's Equation

by

A. Lins, W. de Melo and C. C. Pugh

Introduction - In the International Congress of Mathematics held in Paris in 1900, Hilbert made a list of 23 problems. The second part of Hilbert's 16th problem is still an open and difficult question: to find a bound for the number of isolated closed orbits for a polynomial vector field of degree d on the plane. The solution of this problem is the key step in the classification, modulo topological equivalence, of the phase portrait of an open and dense set of polynomial vector fields of degree d on the plane (Morse-Smale vector fields).

Here we are interested in a particular case of the above problem. Consider the equation

$$(*)\quad \begin{cases} \dot{x} = y-f(x) \\ \dot{y} = -x \end{cases}$$

where f is a polynomial of degree d. This is called Liénard's equation. The problem is again to find a bound for the number of limit cycles for those equations, depending only on the degree d.

This equation plays an important role in the theory of non-linear electrical circuits. In fact, Van der Pol [7], was the first to consider this type of equation, when studying vacuum tube oscilators. Liénard [5] proved that if f is a continuous odd function, which has a unique positive root at $x = a$ and is monotone increasing for $x \geq a$, then $(*)$ has a unique limit cycle. In

[6] it is proved that if f is an odd polynomial of degree five, then (*) has at most two limit cycles. As far as we know there is no result in this direction whithout symmetry assumptions on f.

In §1 we study, using Poincaré compatification, the behavior at infinity of the orbits of (*). As a consequence we get a complete description of the phase space of (*), modulo the number of closed orbits. We prove also that if $f = p+q$. Where p is even, q is odd and has a unique root at 0, then (*) has no closed orbits.

In §2 we prove that if f is any polynomial of degree three then (*) has at most one limit cycle. In fact we give a complete classification of the phase space of the cubic Liénard's equation, in terms of some explicit algebraic conditions on the coefficient of f.

Using a method due to Poincaré we show in §3 that if $d = 2n+1$ or $2n+2$, then for any $0 \leq k \leq n$ there exists a Liénard's equation of degree d with exactly k closed orbits. This motivates the following.

<u>Conjecture</u>. If f has degree $2n+1$ or $2n+2$, then (*) has at most n limit cycles.

We are grateful to J. Palis, S.Smale, F.Takens for many stimulating conversations.

§1 - <u>The phase space of polynomial Liénard's equation</u>

Consider the equation (*) with $f(x) = a_d x^d + \ldots a_1 x$. Denote by X^f the vector field associated to (*), namely, $X^f(x,y) = (y-f(x), -x)$. We make some elementary remarks about the phase space of X^f.

(1) The origin is the only singularity of X^f. This singularity is an attractor if $a_1 > 0$ and a repellor if $a_1 < 0$.

(2) The vertical axis, $E^V = \{(0,y) | y \in \mathbb{R}\}$ is the set of points where X^f is horizontal and the graph of f is the set of points where X^f is vertical.

(3) Let d be odd and $a_d > 0$. It follows easily from (2) that the positive orbit of any point $p \in E^V - \{0\}$ must intersect the graph of f and that the positive orbit of a point $q \in \text{graph}(f)$ must intersect the vertical axis E^V, as in the figure 1. The same is true for the negative orbit, if $a_d < 0$.

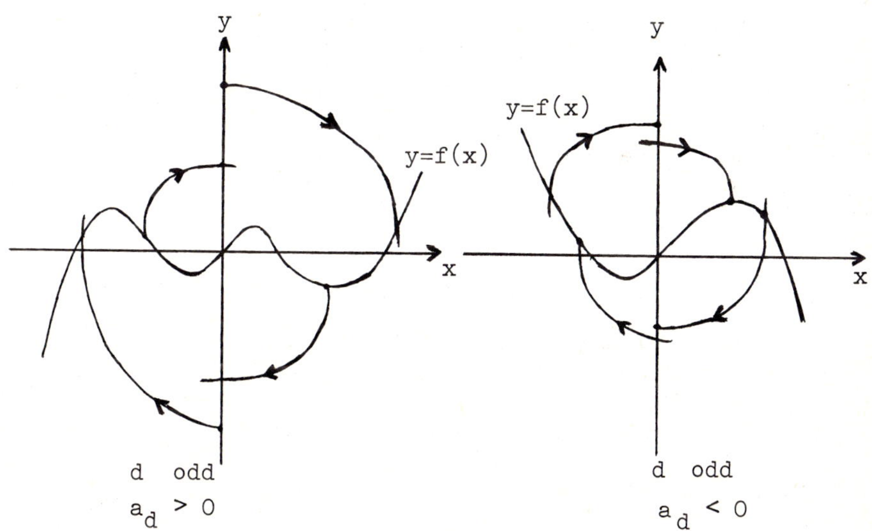

FIGURE 1

Therefore the positive orbit $\mathcal{O}^+(p)$ of any point $p \in E^V_+ = \{(0,y) | y > 0\}$ must intersect E^V_-. Thus we can define a Poincaré mapping $P: E^V_+ \to E^V_+$, where $P(p)$ is the first intersection of $\theta^+(p)$ with E^V_+. Since any closed orbit of X^f must enclose the origin, it follows that there are as many closed orbits of X^f, as fixed points of P. If $a_d < 0$, we can define similarly a Poincaré mapping $P_-: E^V_+ \to E^V_+$, where $P_-(p)$ is the first point of intersection of the negative orbit of p, with E^V_+.

(4) Let d be odd and $a_d > 0$. If $p \in E^V_- = \{(0,y) | y < 0\}$, then the negative orbit of p, either intersects E^V_+, or goes to infinity and remains below the graph of f. There exists at least one point whose negative orbit goes to infinity. In fact let $\varepsilon > 0$ and consider the curve $y = f(x)-\varepsilon$. There exists $a > 0$, such that for $x \geq a$ the vector $X^f(x,f(x)-\varepsilon) = (\varepsilon,-x)$ is transversal to the curve $y = f(x)-\varepsilon$ and points to the interior of the region bounded by this curve and the graph of f. Since the same thing occurs for points in the graph of f, it follows that there exists an orbit which remains inside this region for negative time. The first point where this orbit intersects E^V_-, satisfies the above property. Using the Poincaré compactification and the blowing-up method we are going to prove the existence of a unique point $p_- \in E^V_-$ whose negative orbit goes to infinity, is assymptotic to the graph of f and such that the negative orbit through any point in E^V_-, above p_-, intersects the positive vertical axis. Similarly there exists a unique point $p^+ \in E^V_+$ with the same properties (see fig. 2)

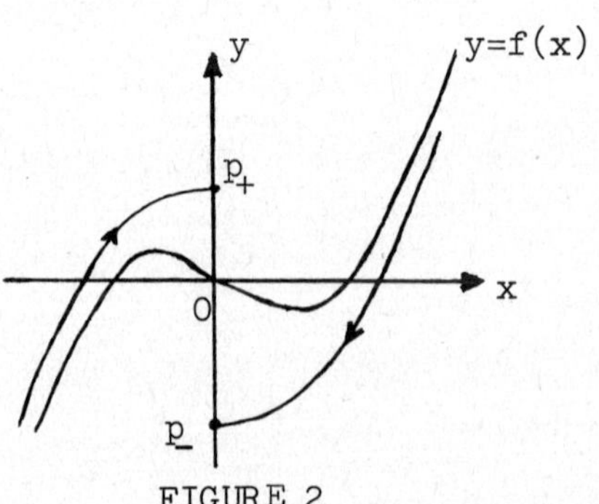

FIGURE 2

Thus the intersection of a closed orbit with the vertical axis, is a pair of points in the segment (p_-, p_+). If $a_d < 0$ we can make similar statements with respect to positive orbits.

(5) Let d be even and $a_d > 0$. If $p \in E_+^v$, then both the positive and negative orbits of p intersect the graph of f and therefore they intersect the negative vertical axis. If $p \in E_-^v$ then the negative orbit of p either intersects E_+^v or goes to infinity and remains below the graph of f. The same is true for positive orbits. In fact we are going to prove the existence of a unique point $p_+ \in E_-^v$ (resp. $p_- \in E_-^v$) such that the positive (resp. negative) orbit of p_+ (resp. p_-) goes to infinity and is assymptotic to the graph of f. If $p \in E_-^v$ is above p_+(resp. p_-) then the positive (resp. negative) orbit of p intersects E_+^v.

Assume $p_- \neq p_+$. Let p_m be the point in $\{p_-, p_+\}$ closest to the origin and let $\tilde{p}_m \in E_+^v$ be the first point where the orbit (positive or negative) of p_m intersects E_+^v. Then any closed orbit of X^f must intersect E^v in a pair of points in the interval $(p_m, \tilde{p}_m)$. Furthermore we can define a Poincaré mapping in the interval $(0, \tilde{p}_m)$. If $p_- = p_+$, then the Poincaré mapping is defined in E_+^v and the closed orbits must intersect E^v above p_+. The pictures below ilustrate the situations we have discussed.

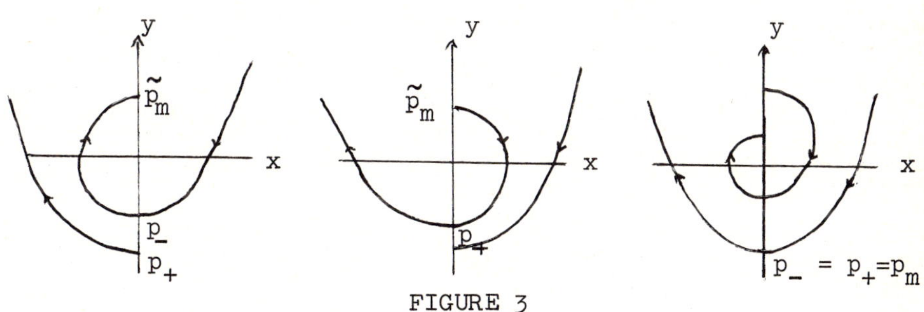

FIGURE 3

If d is even and $a_d < 0$ we have similar results.

Now we discuss the Poincaré compactification [2]. Consider the sphere $S^2 = \{y \in \mathbb{R}^3 \mid \sum_{i=1}^{3} y_i^2 = 1\}$, let $p_N = (0,0,1)$ be the north pole of S^2 and $T_{p_N}S^2$ be the plane $\{y \in \mathbb{R}^3 | y_3 = 1\}$. Let $\pi^+ : T_{p_N}S^2 \to S^2$ and $\pi^- : T_{p_N}S^2 \to S^2$ be the central projections, i.e., $\pi^+(p)$ (resp. $\pi^-(p)$) is the intersection of the live joinning p to the origin with the northern (resp. southern) hemisphere of S^2. Let X be a polynomial vector field of degree d on the plane and let $\varphi: S^2 \to \mathbb{R}$ be defined by $\varphi(y) = y_3^{d-1}$. Then the vector fields $\varphi.(\pi^+)_*X = \varphi \cdot d\pi^+(X \circ (\pi^+)^{-1})$ and $\varphi \cdot (\pi^-)_*X$, extend to an analytic vector field $\pi(X)$, on S^2. The equator is invariant by the flow of $\pi(X)$ and a neighborhood of the equator corresponds to a neighborhood of infinity in $\mathbb{R}^2$.

In orther to find nice analytic expressions for $\pi(X)$ it is convenient to use the following coordinate systems on S^2. Let $U_i^+ = \{y \in S^2 \mid y_i > 0\}$ and $U_i^- = \{y \in S^2 \mid y_i < 0\}$. Let $\varphi_i^+ : U_i^+ \to \mathbb{R}^2$ be given by $\varphi_i^+(y) = (\frac{y_j}{y_i}\ \frac{y_k}{y_i})$ for $j < k$ and $j,k \neq i$. We define $\varphi_i^- : U_i^- \to \mathbb{R}^2$ by the same expression. Consider the vector field $\pi(X^f)$, where $X^f(x,y) = (y-f(x),-x)$. Then $(\varphi_1^+)_* (\pi(X^f))$ is given by:

$$\begin{cases} \dot{u} = \dfrac{1}{\Delta^{d-1}} [-v^{d-1} - u^2v^{d-1} + uv^d\ f(1/v)] \\ \\ \dot{v} = \dfrac{1}{\Delta^{d-1}} [-uv^d + v^{d+1}\ f(1/v)] \end{cases}$$

where $\Delta = \sqrt{1 + u^2 + v^2}$. Here the points of the equator are represented by $v = 0$, the points of the northern hemisphere by $v>0$ and the point $(0,0)$ corresponds to the point $(1,0,0) \in S^2$. $(\varphi_2^+)_*(\pi(X^f))$ is given by

$$\begin{cases} \dot{u} = \dfrac{1}{\Delta^{d-1}} [v^{d-1} + u^2 v^{d-1} - v^d f(u/v)] \\ \dot{v} \quad \dfrac{1}{\Delta^{d-1}} [u\, v^d] \end{cases}$$

As before the northern hemisphere corresponds to $v > 0$ and $(0,0) = \varphi_2^+(0,1,0)$.

The vector fields $(\varphi_i^-)_*(\pi(X^f))$ have the same expressions as $(\varphi_i^+)_*(\pi(X^f))$, multiplied by $(-1)^{d-1}$, but in this case the northern hemisphere corresponds to $v < 0$.

It follows from the above expressions that $\pi(X)$ has four singularities in the equator: $p_1 = (1,0,0)$, $p_2 = (-1,0,0)$, $q_1 = (0,1,0)$ and $q_2 = (0,-1,0)$. It is clear that p_1 and p_2 are hyperbolic singularities. If $a_d > 0$ (resp. $a_d < 0$), then p_1 is a repellor (resp. an attractor) and p_2 is a repellor (resp. attractor) if d is odd and an attractor (resp. repellor) if d is even. The theorem below gives the behavior of $\pi(X^f)$ in a neighborhood of the equator.

<u>Theorem 1</u> - Let $\pi(X^f)$ be as above. Then the orbit structure of $\pi(X^f)$ in a neighborhood of the equator is given by one of the pictures below:

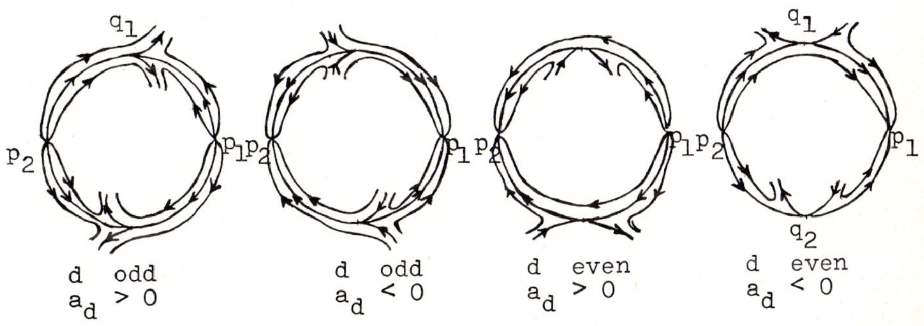

FIGURE 4

Here X^f is topologically equivalent to a saddle in a neighborhood of q_1 and in a neighborhood of q_2 and their separatrices are tangent to the equator.

<u>Proof</u> - It suffices to study the singularities q_1 and q_2, since we already know that p_1 and p_2 are hyperbolic singularities. We are going to study the singularity q_1 when d is odd and $a_d > 0$. The other cases are similar.

For each $k > 0$ consider the region $A_k = \{(x,y) \in \mathbb{R}^2 | f(x) - k \leq y \leq f(x)$ and $x \geq a\}$. Denote by $A_{k,1}$ and $A_{k,2}$ the part of the boundary of A_k containned in the curve $y = f(x)$ and $y=f(x)-k$ respectively. If a is large enough then X^f is transversal to $A_{k,1}$ and $A_{k,2}$ for every $k \geq 1$. At points of $A_{k,1}$ and $A_{k,2}$, X^f points to the interior of A_k. Therefore the positive orbit of a point in $A_{k,1} \cup A_{k,2}$ enters the region A_k and intersects the line $x = a$. It is easy to see that the negative orbit of a point $p \in A_{k,1}$ intersects the line $\Delta - = \{(x,-x) | x \in \mathbb{R},\ x < 0\}$. Choosing a large enough, the vector $X^f(x,x)$ will be transversal to the diagonal Δ for any $x \geq a$. We set $b = f(a)-a$ and choose a so that $b > 1$. Consider the region $B_b = \{x,y) | x \leq y \leq f(x)-b\}$. Let $A_b^+ = \pi^+(A_b)$ and $B_b^+ = \pi^+(B_b)$, as in the picture below

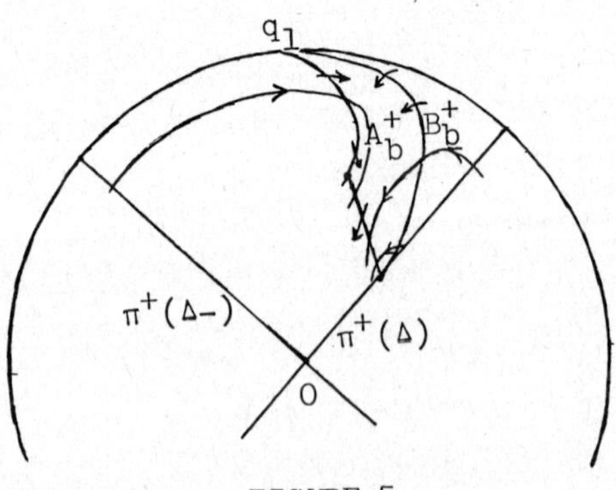

FIGURE 5

Thus the orbit of $\pi(X^f)$ through any point in the interior of B_b^+ enters B_b^+ through $\pi^+(\Delta)$ and leaves B_b^+ through the boundary of A_b^+. Hence the negative orbit through a point in $A_{b,1}$ (resp. $A_{b,2}$) intersects $\pi^+(\Delta_-)$ (resp. $\pi^+(\Delta_+)$). Thus, in orther to get a complete picture of the orbits around q_1, it remains to prove the existence of a unique orbit in A_b^+, whose α-limit set is q_1. We are going to use the directional blowing-up construction [1]. Let $\varphi: \mathbb{R}^2 \to \mathbb{R}^2$, $\varphi(u',v') = (u,v)$, where $u = u'$ and $v = u'v'$. Then $\varphi^{-1}(0,0) = \{(0,v')|v' \in \mathbb{R}\}$ and φ^{-1} is a diffeomorphism outside the axis $u = 0$. Let $\tilde{Y}^1 = (\varphi^{-1})_* Y$, where $Y = (\varphi_1^+)_* \pi(X^f)$. The vector field $\tilde{Y}^1$ has the following expression:

$$\begin{cases} \dot{u}' = \dfrac{1}{\Delta^{d-1}} [(u')^{d-1}(v')^{d-1} + (u')^{d+1}(v')^{d-1} - (u')^d(v')^d f(1/v')] \\[2ex] \dot{v}' = \dfrac{1}{\Delta^{d-1}} [(u')^{d-1}(v')^{d-1} f(1/v') - (u')^{d-2}(v')^d] \end{cases}$$

Consider the vector field $Y^1 = (\Delta^{d-1}/(u')^{d-2})\tilde{Y}^1$. Clearly Y^1 has the same orbits as $(\varphi^{-1})^* Y$ in the half planes $u' > 0$ and $u' < 0$.

The orbits have the same orientation in $u' > 0$ and the opposite orientation in $u' < 0$.

Let $\psi: \mathbb{R}^2 \to \mathbb{R}^2$, $\psi(u,v)=(u',v')$, where $u' = uv^{d-1}$ and $v' = v$. Then $\psi^{-1}(0,0) = \{(u,0);\ u \in \mathbb{R}\}$ and ψ^{-1} is a diffeomorphism outside the axis $v' = 0$. Let $\tilde{Y}^2 = (\psi^{-1})_* Y^1$. The vector field $Y^2 = \frac{1}{v^{d-1}} \tilde{Y}^2$, whose expression is given below, have the same orbits as $\tilde{Y}^2$ in the half planes $v>0$ and $v<0$. The orbits have the same orientation (see fig. 6).

$$Y^2 \begin{cases} \dot{u} = d(u - u^2 v^d f(1/v)) + u^3 v^{2d-2} \\[2ex] \dot{v} = -v + uv^{d+1} f(1/v) \end{cases}$$

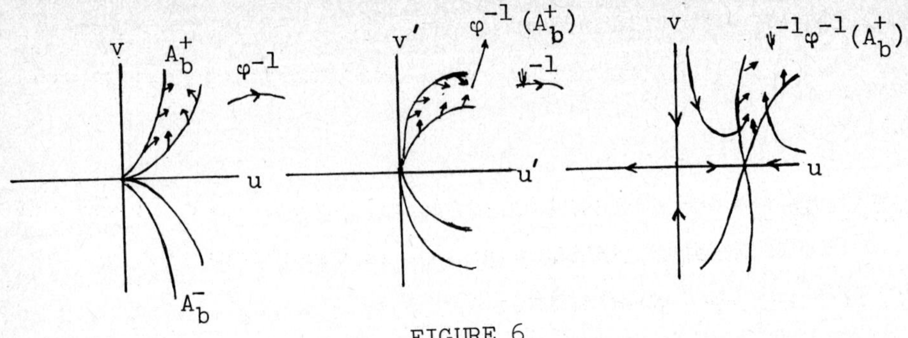

FIGURE 6

Y^2 has two singularities, namely $(0,0)$ and $(1/a_d,0)$. The derivative of Y^2 at $(0,0)$ is $\begin{pmatrix} d & 0 \\ 0 & -1 \end{pmatrix}$ and therefore $(0,0)$ is a saddle point. The derivative of Y^2 at $(1/a_d,0)$ is $\begin{pmatrix} -d & -da_{d-1}/a_d^2 \\ 0 & 0 \end{pmatrix}$, therefore $(1/a_d,0)$ is a partially hyperbolic singularity [3]. Since $(1/a_d,0)$ is normally attracting and there is no orbit inside $A_b^+ \cup A_b^-$ whose ω-limit is q_1, it follows that the center manifold is unique. This proves the existence and unicity of the separatrices.

<u>Corollary</u> - Let f be an even pollynomial. Then if $a_d > 0$ (resp. $a_d < 0$), there is a disc D in the northern hemisphere of S^2, containing the north pole, whose boundary is formed by separatrices of q_1 (resp. q_2). Any orbit inside D is compact. If γ is an orbit outside D, then the α-limit set of γ is p_1(resp. p_3) and the ω-limit set of γ is p_2(resp. p_1).

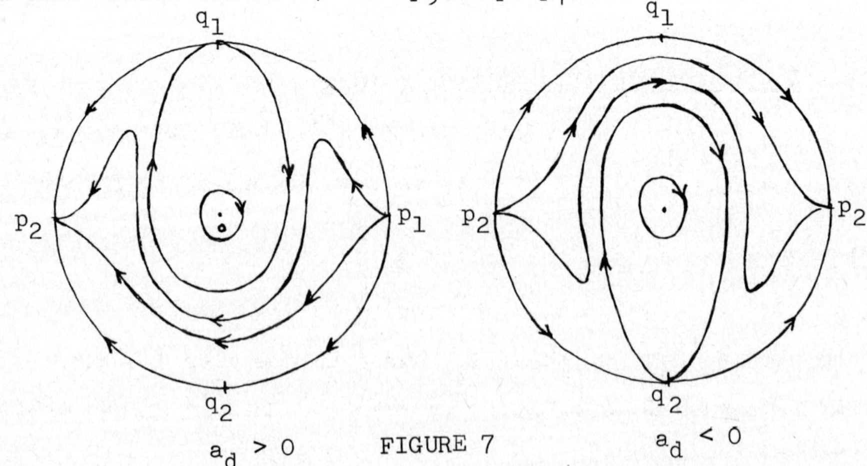

FIGURE 7

Proof - Let $\psi: \mathbb{R}^2 \to \mathbb{R}^2$ be given by $\psi(x,y) = (-x,y)$. Then $\psi_*(X^f) = -X^f$, because f is even. Now the proof follows from theorem 1.

Proposition 1 - Let $f(x) = e(x) + \Theta(x)$, where $e(x)$ is even and $\Theta(x)$ is odd. If 0 is the unique root of $\Theta(x)$, then X^f has no closed orbits.

Proof - Consider the vector field $Y(x,y) = (y-e(x),-x)$. If the coefficient of the leading term of $e(x)$ is positive, then, by the corollary, every orbit of Y intersects the negative vertical axis in a unique point. Let $V: \mathbb{R}^2 \to \mathbb{R}$ be defined as follows: if $p \in \mathbb{R}^2-\{0\}$, then $V(p)$ is the intersection of the orbit of p with the negative vertical axis and $V(0) = 0$. Clearly V is C^∞ in $\mathbb{R}^2-\{0\}$ and 0 is the unique minimum of V. The level curves of V are the orbits of Y. Since V is a first integral of Y, $\frac{\partial V}{\partial y} = xk(x,y)$, $\frac{\partial V}{\partial y} = (y-e(x))k(x,y)$, where $k(x,y) > 0$ if $(x,y) \neq (0,0)$. If $\Theta(x) > 0$ for $x > 0$, then $\dot{V}(x,y) = (y-f(x))\frac{\partial V}{\partial x} - x\frac{\partial V}{\partial y} = -k(x,y)x\Theta(x)$ is not positive for $x \neq 0$ and $\dot{V}(x,y) = 0$ if and only if $x=0$. Hence there is no closed orbit of X^f, since V is strictly decreasing along orbits of X^f. This proves the proposition.

§2 - Liénard's equation of degree three

Consider the vector field $X^f(x,y) = (y-f(x),-x)$, where $f(x) = a_3x^3+a_2x^2+a_1x$. In this section we prove the following:

Theorem 2 - We have the following possibilities:

a) If $a_1 a_3 > 0$, then X^f has no closed orbits

b) If $a_1 \cdot a_3 < 0$, then X^f has a unique closed orbit.

c) If $a_1 = 0$ and $a_3 \neq 0$, then X^f has no closed orbits. The origin is a weak attractor for $a_3 > 0$ and a weak repellor for $a_3 < 0$.

d) If $a_3 = 0$ and $a_1 \neq 0$, then X^f has no closed orbits. The origin is a hyperbolic attractor for $a_1 > 0$ and a hyperbolic repellor for $a_1 < 0$.

e) If $a_1 = a_3 = 0$, then the origin is a center.

We sketch below the orbit structure of the vector field $\pi(X^f)$.

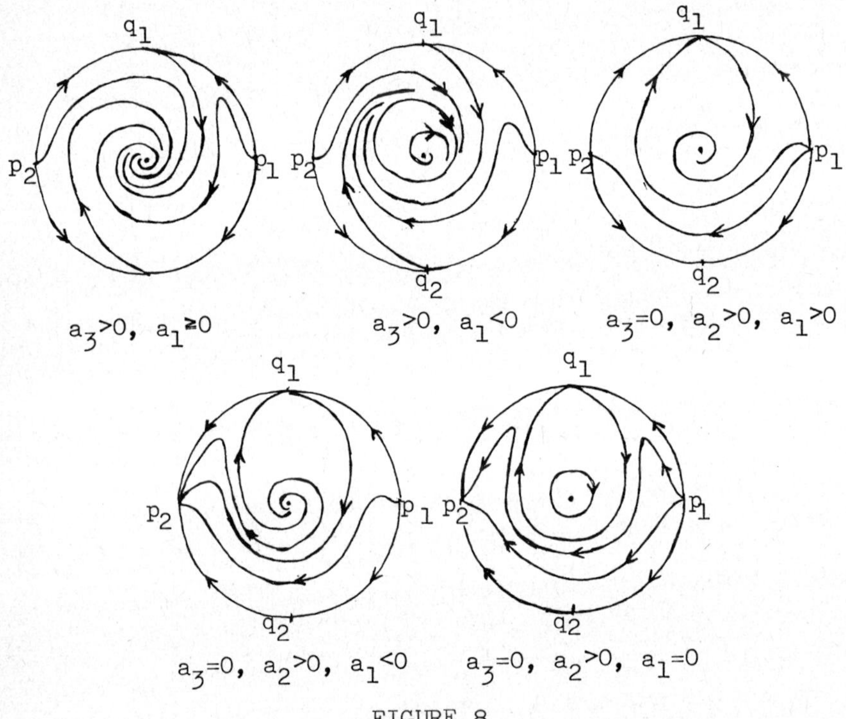

FIGURE 8

<u>Remarks</u>: 1 - When $a_3 \neq 0$ and a_1 changes sign, we have a Hopf bifurcation: a closed orbit colapses in the origin.

2 - When $a_1 \neq 0$ and a_3 changes sign, we have a bifurcation at infinity: a closed orbit goes to infinity.

<u>Proof of theorem 2</u> - a), c), d) and e), follow from proposition 1.

We prove b) when $a_3 > 0$, $a_1 < 0$. The other case is similar. We divide the proof in three steps.

(1) - Every closed orbit of X^f must intersect the lines $x = \pm\sqrt{-\frac{a_1}{a_3}}$.

(2) - X^f has at least one and at most two closed orbits, one of them is hyperbolic.

(3) - X^f has exactly one closed orbit.

Proof of (1) - Consider the function $V(x,y) = e^{-2a_2y}(y-a_2x^2+1/2a_2)$ if $a_2 \neq 0$ and $V(x,y) = x^2+y^2$ if $a_2 = 0$ (V is a first integral of the vector field $(x,y) \to (y-a_2x^2,-x)$. It follows that

$$\dot{V}(x,y) = DV(x,y) \cdot X^f(x,y) = 2a_2e^{-2a_2y}x^2(a_3x^2 + a_1) .$$

Therefore $\dot{V}$ does not change sign inside the strip bounded by the lines $x = \pm\sqrt{-\frac{a_1}{a_3}}$. Since $V(-x,y) = V(x,y)$, it is easy to see that there exists a closed level curve of V which is tangent to both lines $x = \sqrt{-\frac{a_1}{a_3}}$ and $x = -\sqrt{-\frac{a_1}{a_3}}$ and encloses a disc D containning the origin. Since V does not change sign inside D, no closed orbit of X^f can intersect D, which proves (1).

Proof of (2) - If follows from theorem 1 and Poincaré-Bendixson theorem, that X^f has at least one closed orbit.

Let $X = (P,Q)$ be a vector field in the plane and $\gamma(t) = (x(t),y(t))$ be a periodic orbit of X, of period T. We use the following criterion: (*) If $\int_\gamma \operatorname{div}(X)dt = \int_0^T(\frac{\partial P}{\partial x}+\frac{\partial Q}{\partial y})(x(t),y(t))dt < 0$ (resp. > 0), then γ is a hyperbolic attracting (resp. repelling) closed orbit.

Let $X^f(x,y) = (y-f(x),-x)$, where $f(x)= a_3x^3 + a_2x^2 + a_1x$,

$a_3 > 0$, $a_1 < 0$. Let γ be the innermost closed orbit. Since the origin is a repellor, it follows that γ cannot be a repellor. Hence $I = \int_\gamma \operatorname{div}(X^f)dt = \int_\gamma -(3a_3x^2 + 2a_2x + a_1)dt \leq 0$. Suppose that $\tilde{\gamma}$ is another closed orbit of X^f. The idea is to show that $\tilde{I} = \int_{\tilde{\gamma}} \operatorname{div}(X^f)dt < I \leq 0$. From this fact, it follows that every closed orbit which contains γ in its interior is attracting and this implies that there exist at most two closed orbits.

Observe first that $\int_\gamma x dt = \int_\gamma (-\dot{y})dt = 0$. Hence $I = \int_\gamma -(3a_3x^2 + a_1)dt$ and similarly $\tilde{I} = \int_{\tilde{\gamma}} -(3a_3x^2 + a_1)dt$. In orther to compare I and $\tilde{I}$ we divide the integrals in four parts. Let $p_i = (x_i, y_i)$ and $\tilde{p}_i = (x_i, y_i)$, $i = 0,1,2,3$ be the points of intersection of γ and $\tilde{\gamma}$, respectively, with the lines $x = \pm\sqrt{-\frac{a_1}{3a_3}}$, as in the picture below.

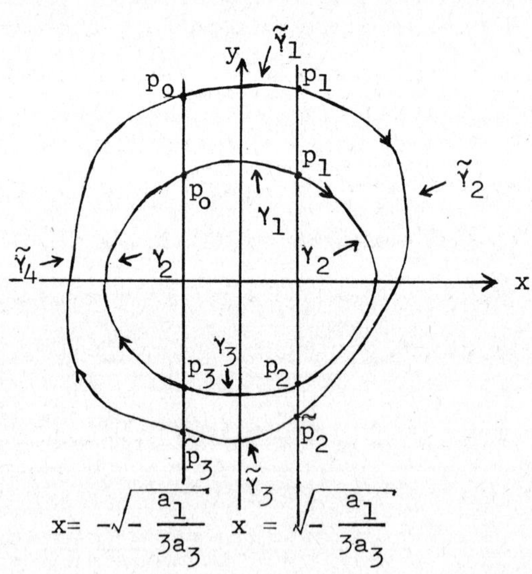

FIGURE 9

Let γ_i (resp $\tilde{\gamma}_i$) $i = 1,2,3,4$, be the segment of γ (resp. $\tilde{\gamma}$) between p_{i-1} and p_i (resp. $\tilde{p}_{i-1}$ and $\tilde{p}_i$), where we put $p_4 = p_o$ (resp. $\tilde{p}_4 = \tilde{p}_o$). Let $I_i = \int_{\gamma_i} -(3a_3x^2 + a_1)dt$ and $\tilde{I}_i = \int_{\tilde{\gamma}_i} -(3a_3x^2 + a_1)dt$. Consider first the integrals I_1 and $\tilde{I}_1$. Since the curves γ and $\tilde{\gamma}$ do not intersect the curve $y = f(x)$ in the strip bounded by the lines $x = \pm\sqrt{-\dfrac{a_1}{3a_3}}$, we consider γ_1 and $\tilde{\gamma}_1$ parametrized by the variable x. Let $(x,y_1(x))$, $(x,\tilde{y}_1(x))$ be such parametrizations. We can write

$$I_1 = \int_{\gamma_1} -(3a_3x^2 + a_1)dt = \int_{x_o}^{x_1} \frac{-(3a_3x^2 + a_1)}{y_1(x) - f(x)}dx$$

and similarly $\tilde{I}_1 = \displaystyle\int_{x_o}^{x_1} - \frac{(3a_3x^2 + a_1)}{\tilde{y}_1(x) - f(x)}\,dx$, where $x_o = -\sqrt{-\dfrac{a_1}{3a_3}}$ and $x_1 = \sqrt{-\dfrac{a_1}{3a_3}}$.

Since $-(3a_3x^2 + a_1) > 0$ and $\tilde{y}_1(x) - f(x) > y_1(x) - f(x) > 0$ for $x_o < x < x_1$, it follows that $\tilde{I}_1 < I_1$. Similarly one can prove that $\tilde{I}_3 < I_3$.

Let us consider the integrals I_2 and $\tilde{I}_2$. Since X^f is horizontal only in the points of the axis $x = 0$, we can consider γ_2 and $\tilde{\gamma}_2$ parametrized by the variable y. Let $(x_1(y),y)$, $(\tilde{x}_1(y),y)$ be such parametrizations. We have

$$I_2 = \int_{\gamma_2} -(3a_3x^2 + a_1)dt = \int_{y_1}^{y_2} \frac{3a_3(x_1(y))^2 + a_1}{x_1(y)}\,dy =$$

$$= -\int_{y_2}^{y_1}(3a_3\, x_1(y) + \frac{a_1}{x_1(y)})dy < 0$$

and similarly

$$\tilde{I}_2 = - \int_{\tilde{y}_2}^{\tilde{y}_1} (3a_3 \tilde{x}_1(y)) + \frac{a_1}{\tilde{x}_1(y)}) \, dy < 0, \quad \text{where} \quad y_i, \ \tilde{y}_i \quad i = 1,2$$

are as in the figure 9. Since $\tilde{x}_1(y) > x_1(y)$ for $y \in (y_2, y_1)$, we have $3a_3 \tilde{x}_1(y) + \frac{a_1}{\tilde{x}_1(y)} > 3a_3 x_1(y) + \frac{a_1}{x_1(y)}$ for $y \in (y_2, y_1)$

therefore

$$\tilde{I}_2 = - \int_{\tilde{y}_2}^{\tilde{y}_1} (3a_3 \tilde{x}_1(y) + \frac{a_1}{\tilde{x}_1(y)}) dy < - \int_{y_2}^{y_1} (3a_3 \tilde{x}_1(y) + \frac{a_1}{\tilde{x}_1(y)} dy <$$

$$< - \int_{y_2}^{y_1} (3a_3 x_1(y) + \frac{a_1}{x_1(y)}) \, dy = I_2 .$$

Similarly one can prove that $\tilde{I}_4 < I_4$, which proves (2).

Proof of (3) - Suppose by contradiction that X^f has two closed orbits. Let γ be the innermost and $\tilde{\gamma}$ be the outermost. Let $D \subset \mathbb{R}^2$ be the disc whose boundary is γ and A be the annulus whose boundary is $\gamma \cup \tilde{\gamma}$. Since the origin is a repellor and $\tilde{\gamma}$ is an attractor, γ attracts the orbits containned in D and repels the orbits containned in A. It follows that there exist closed curves $\mathcal{C}_1$, $\mathcal{C}_2$, such that $\mathcal{C}_1 \subset D$, $\mathcal{C}_2 \subset A$, X^f is transversal to both $\mathcal{C}_1$, $\mathcal{C}_2$ and points to the exterior of $\mathcal{C}_1$ and $\mathcal{C}_2$ (see the picture below).

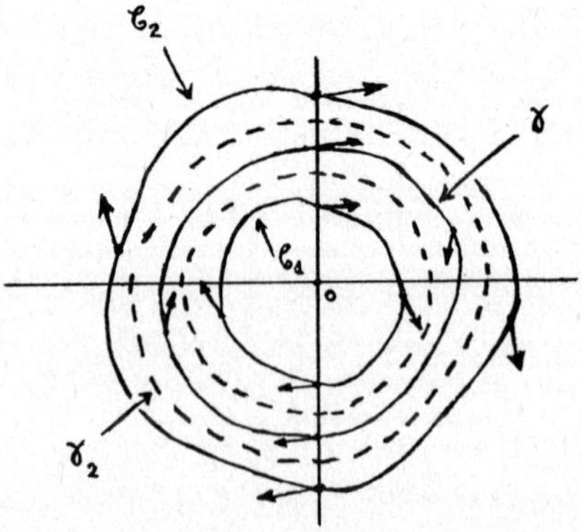

The idea is to perturb $f(x)$ by a small term of the form ϵx, in orther to get a new vector field $X^{\bar{f}}(x,y) = (y-f(x)-\epsilon x,-x)$ which has at least three closed orbits, which contradicts (2). Since $\tilde{\gamma}$ is hyperbolic, C_1 and C_2 are compact, it is possible to choose $\epsilon > 0$ such that $X^{\bar{f}}$ has at least one chosed orbit $\bar{\gamma}$ outside C_2 and X^f points to the exterior of C_1 and C_2. Since $\begin{vmatrix} y-f(x)-\epsilon x & -x \\ y-f(x) & -x \end{vmatrix} = \epsilon x^2$, $X^{\bar{f}}$ points to the interior of γ, except in the two points $\gamma \cap \{(x,y) | x = 0\}$. Therefore by Poincaré-Bendixson theorem, $X^{\bar{f}}$ has at least one closed orbit γ_1 between C_1 and γ and at least one closed orbit γ_2 between γ and C_2. Hence $X^{\bar{f}}$ has at least three closed orbits, which proves the theorem.

<u>Remarks</u> 1 - With the same argument as above, one can prove that the equation

$$\begin{cases} \dot{x} = y-f(x)+\alpha x^2 \\ \dot{y} = -\beta x \end{cases}$$

has exactly one closed orbit, if we impose the following hypothesis:

a) $\beta > 0$

b) f is odd and $\lim\limits_{x\to\infty} (f(x)-\alpha x^2) = \infty$

c) $f'(x) = 0$ has exactly one positive root a, and $f'(0) < 0$.

d) $\dfrac{f'(x_1)}{x_1} > \dfrac{f'(x_2)}{x_2}$ if $x_1 > x_2 \geq a$.

<u>Example</u> - The equation $\begin{cases} \dot{x} = y-(x^5 + ax^3 + bx^2 + cx) \\ \dot{y} = -\beta x \end{cases}$

has exactly one closed orbit if $c < 0$ and $\beta > 0$.

2 - This remark is interesting for the computation of the limit cycle by a computer. Consider the equation $(*)\begin{cases} \dot{x} = y - f(x) \\ \dot{y} = -\beta x \end{cases}$ where

$\beta > 0$ and $f(x) = a_3x^3 + a_2x^2 + a_1x$, $a_2 \geq 0$.

Argument (1) of the proof of the theorem 2, proves that the limit cycle γ of (*) intersects the line $x = \sqrt{-\frac{a_1}{a_3}}$. It is easy to see that γ intersects this line in the segment defined by

$$-\frac{a_1a_2}{a_3} - \varepsilon < y < -\frac{a_1a_2}{a_3}, \quad \text{where} \quad \varepsilon = \frac{\alpha\beta}{2(\alpha a_2 + |a_1|)}, \quad \alpha = \sqrt{-\frac{a_1}{a_3}}.$$

If $a_2 < 0$ γ intersects the line $x = -\sqrt{-\frac{a_1}{a_3}}$ in the segment

$$\left(-\frac{a_1a_2}{a_3} < y < -\frac{a_1a_2}{a_3} + \varepsilon\right) \quad \text{where} \quad \varepsilon = \frac{\alpha\beta}{2(\alpha|a_2| + |a_1|)}.$$

§3 Liénard's equation near the center $\begin{cases} \dot{x} = y \\ \dot{y} = x \end{cases}$

In this section we prove the following:

Theorem 3. Let $d = 2n+1$ or $2n+2$. Given an integer $0 \leq k \leq n$ there exists a polynomial $f(x) = a_dx^d + \ldots + a_1x$ such that the equation

$$\dot{x} = y - f(x)$$
$$\dot{y} = -x$$

has exactly k closed orbits.

In the proof of this theorem we use a method due to Poincaré (see [4] p. 221). We recall this methods in the case of Liénard's equation. Let $x(y_o,t,\varepsilon)$, $y(y_o,t,\varepsilon)$ be the solution, through the point $(0,y_o)$, of the equation $(*)\begin{cases} \dot{x} = y - \varepsilon g(x) \\ \dot{y} = -x \end{cases}$ where $g(x) = b_dx^d + \ldots + b_1x$, ε and b_d are positive. Since the trajectory through $(0,y_o)$ comes back to the positive vertical axis, there exists an analytic function $t(y_o,\varepsilon)$ such that $x(y_o,t(y_o,\varepsilon),\varepsilon) = 0$ and $t(y_o,0)=2\pi$. We can write

$$x(y_o,t,\epsilon) = y_o \sin t - \epsilon \int_0^t \cos\,(t-\tau)\; g(x(y_o,\tau,\epsilon))d\tau$$

$$y(y_o,t,\epsilon) = y_o \cos t + \epsilon \int_0^t \sin\,(t-\tau)\; g(x(y_o,\tau,\epsilon))d\epsilon$$

Since $y(y_o,2\pi,0) - y_o = 0$ there exists an analytic function $F(y_o,\epsilon)$ such that

$$\epsilon F(y_o,\epsilon) = y(y_o,t(y_o,\epsilon),\epsilon) - y_o =$$

$$= y_o(\text{oos}\; t-1) + \epsilon \int_0^t \sin\,(t-\tau)\; g(x(y_o,\tau,\epsilon))d\tau$$

Then the orbit of (*) through $(0,y_o)$ is closed if and only if y_o is a root of the equation $F(y_o,\epsilon) = 0$. Let us consider the polynomial

$$F(y_o,0) = \frac{\partial y}{\partial \epsilon}(y_o,0) = -\int_0^{2\pi} (\sin\tau)\; g(y_o \sin\tau)d\tau$$

Let $y_1(\epsilon)$ be a root of the equation $F(y_o,\epsilon) = 0$ and suppose $y_1(\epsilon)$ converges to y_1 as $\epsilon \to 0$. Then y_1 is a root of $F(y_o,0) = 0$. Conversely, suppose y_1 is a root of $F(y_o,0)$ and $\frac{\partial F}{\partial y_o}(y_1,0) \neq 0$. Then, by the implicit function theorem, there exists a unique function $y_1(\epsilon)$ such that $y_1(0) = y_1$ and $F(y_1(\epsilon),\epsilon) = 0$. This is Poincaré's method.

<u>Lemma</u>. Let d be odd and $g'(0) \neq 0$. Suppose that the polynomial $F(y_o,0)$ has k positive roots which are simple. Then if ϵ is small enough the equation (*) has exactly k closed orbits.

<u>Proof</u> - Let $y_1,\ldots,y_k$ be the roots of $F(y_o,0)$. From Poincaré's method it follows that (*) has at least k closed orbits given by the roots $y_1(\epsilon),\ldots,y_k(\epsilon)$ of $F(y_o,\epsilon)$. In order to prove that for ϵ small (*) has exactly k closed orbits it is enough to

show the existence of numbers $M_2 > M_1 > 0$ such that for $y_o < M_1$, or $y_o > M_2$ there is no closed orbits of (*) through $(0,y_o)$. In fact, if this not the case, there would exist a sequence $\bar{y}(\epsilon_n) \to \bar{y}$ with $\epsilon_n \to 0$ such that $F(\bar{y}(\epsilon_n),\epsilon_n) = 0$ and $\bar{y}_n(\epsilon_n) \neq y_i(\epsilon_n)$ for $i = 1,\dots,k$. This is a contradiction since y must be a root of $F(y_o,0)$. Let us prove the existence of M_1. Choose M_1 so that $g(x) \neq 0$ is $0 < |x| < M_1$. Consider the function $V_\epsilon(x,y) = \frac{1}{2\epsilon}(x^2+y^2)$. Then $\dot{V}_\epsilon = -xg(x)$. If $g'(0) > 0$ then $\dot{V}_\epsilon$ is negative in the strip $-M_1 < x < M_1$. Thus the positive orbit through a point $(0,y_o)$, with $y_o < M_1$, remains in this strip and converges to the origin. If $g'(0) < 0$ the same is true for the negative orbit.

Now we show the existence of M_2. We assume $b_d > 0$. Consider the rectangle $R_\beta = \{(x,y),\ |x| \le 1$ and $|y| \le \beta\}$. Let $a \le 0 \le b$ be the roots of g such that all the other roots are containned in the interval (a,b). Let K be the supremum of $|xg(x)|$ on the interval $[a,b]$. Choose $\beta > 0$ so that $\beta > |a|$, b and $g(\beta) > 2K(b-a)$. Let $M_2 > 0$ be such that the disc of center 0 and radius M_2 contains the rectangle R_β. Let $\epsilon_o > 0$ be so small that if $\epsilon \le \epsilon_o$ the segment of orbit of $(*)_\epsilon$ between $(0, M_2)$ and the point where this orbit intersects the positive vertical axis for the first time does not intersect the rectangle R_β. We also take ϵ_o so small that $|\epsilon_o . g(x)| < \frac{1}{2}$ for all $x \in [a,b]$. Let $y_o > M_2$ and $P(y_o)$ be the point where the positive orbit through $(0,y_o)$ intersects the positive vertical axis. We claim that $V(0,P(y_o)) > V(0,y_o)$ and therefore the orbit through $(0, y_o)$ is not closed. In fact $V(0, P(y_o)) - V(0,y_o) = \int_0^T \dot{V}\,dt$ where T is the time between $(0,y_o)$ and $(0,P(y_o))$. We may write

$$\int_0^T \dot{V}dt = \int_0^{T_1} \dot{V}dt + \int_{T_1}^{T_2} \dot{V}dt + \int_{T_2}^{T_3} \dot{V}dt + \int_{T_3}^{T_4} \dot{V}dt + \int_{T_4}^{T} \dot{V}dt =$$

$$= I_1 + I_2 + I_3 + I_4 + I_5 , \quad \text{where } T_1 < T_2 < T_3 < T_4$$

are such that $x(y_o, T_1, \epsilon) = x(y_o, T_2, \epsilon) = b$, $x(y_o, T_3, \epsilon) =$ $= x(y_o, T_4, \epsilon) = a$. Since, $\dot{x} \neq 0$ in the intervals $[0,T_1]$, $[T_2,T_3]$ and $[T_4,T]$ we have

$$|I_1| = |\int_o^{T_1} -xg(x)dt| = |\int_o^b \frac{-xg(x)}{y - \epsilon g(x)} dx| \le \int_o^b \frac{k}{1-\frac{1}{2}} dx = 2kb$$

$$|I_3| = |\int_{T_2}^{T_3} - xg(x)dt| = |\int_b^a \frac{-xg(x)}{y-\epsilon g(x)} dx| \le |\int_b^a \frac{k}{1-\frac{1}{2}} dx| = 2kb$$

$$|I_5| = |\int_{T_4}^{T} -xg(x)dt| = |\int_a^o \frac{-xg(x)}{y-\epsilon g(x)} dx| \le -2ka$$

Thus $|I_1 + I_3 + I_5| \le 4k(b-a)$. On the other hand, since $\dot{y} \neq 0$ on the interval $[T_1,T_2]$ we have

$$I_2 = \int_{T_1}^{T_2} -xg(x)dt = \int_{y_1}^{y_2} g(x)dy \ge \int_1^{-1} g(x)dy \ge \int_1^{-1} g(\beta)dy = -2g(\beta)$$

Hence $y_1 = y(y_o,T_1,\epsilon)$ and $y_2 = y(y_o,T_2,\epsilon)$.

Since I_4 is also negative and $g(\beta) > 2k(b-a)$ it follows that $\int_o^T \dot{V}dt < 0$. This proves the lemma.

Proof of Theorem 3. 1) Assume $d = 2n+1$. Consider a polynomial $g(x) = b_d x^d + \ldots + b_1 x$. It is easy to see that $F(y_o,0)$ is an odd polynomial of degree $2n+1$ and we can choose the coefficient of g so that $F(y_o,0)$ has $2k+1$ real roots which are simple. Then the theorem follows from the above lemma.

2) Assume $d = 2n+2$. From 1) it follows that there exists a a polynomial $f_o(x) = a_{2n+1}x^{2n+1} + \ldots + a_1 x$ such that the equation

$$\dot{x} = y-f_o(x)$$
$$\dot{y} = -x$$

has exactly k closed orbits which are hyperbolic. We claim that if δ is small enough and $f_\delta(x) = \delta x^{2n+2} + f_o(x)$ then the vector field $X^\delta(x,y) = (y-f_\delta(x),-x)$ has k closed orbits. In fact, let $k > 0$ and $A_\delta\{(x,y) \in R^2;\ f_o(x) - k \leq y \leq f_\delta(x),\ x \geq a\}$. It is easy to see that given $\delta_1 > 0$, then the vector field X^δ is transversal to the curves $y=f_\delta(x)$ and $y=f_o(x)-k$ for $x \geq a$ and any $\delta \leq \delta_1$. Furthermore X^δ points to the interior of A_δ. For any $\delta < \delta_1$ the orbit of X^δ through a point P of the segment $\{(a,y),\ f_o(a) - k \leq y \leq f_{\delta_1}(a)$ intersects the positive vertical axis in a point $T_\delta(P)$. Since this mapping is continuous there exists $M > 0$ such that $T_\delta(P) < M$ for any $p \in \{(a,y);\ f_o(a) - k \leq y \leq f_{\delta_1}(a)\}$ and $\delta \leq \delta_1$. Therefore all the closed orbits of X^δ intersect the positive vertical axis in a point of the compact inverval $[0,M]$. It follow that if δ is small then the vector field X^δ will have exactly k hyperbolic closed orbits.

REFERENCES

[1] Dumortier, F. - Singularities of Vector Fields on the Plane, J. Diff. Eq., 23, 53-106 (1977).

[2] Gonzales, E.A.V. - Generic properties of polynomials vector fields at infinity, Trans. of A.M.S., 143, 1969.

[3] Kelley, A. - The stable, center-stable, center, center unstable and unstable manifolds. Appendix C in "Tranversal mappings and flows" by R Abraham and J.Robbin, Benjamin, New York, 1967.

[4] Lefschetz, Differential Equations: Geometric Theory, Interscience Publishers, New York, 1957.

[5] Liénard, A. - Étude des oscillations entretenues, Revue Génerale de l'Électricité 23: 901-912, 946-954 (1928)

[6] Ryckov, G.S. - The maximal number of limit cycles of the system $\dot{y} = -x, \quad \dot{x} = y - \sum_{i=0}^{2} a_i x^{2i+1}$ is equal to two, Differential' 11 (1975), 390-391, 400.

[7] Van der Pol, B. On oscillation hysteresis in a triode generator with two degrees of freedom. Phil. Mag (6) 43, 700-719 (1922).

Instituto de Matemática Pura e Aplicada
Rio de Janeiro - Brazil

University of California, Berkeley

On the Weierstrass Preparation Theorem

S. Lojasiewicz

In [3] we have given a proof of the Malgrange-Mather differentiable preparation theorem ([1] and [2]). In this note we will show how the method used in [3] works in the simple cases of the complex analytic or formal one, giving thus some proof of the classical Weierstrass preparation theorem in these cases.

Let $A_{xy..}$ denote $\mathcal{O}_{xy..}$ or $\hat{\mathcal{O}}_{xy..}$ i.e. respectively the algebra of germs at 0 of holomorphic functions or of complex formal power series, in variables $x = (x_1, x_2, \ldots)$, $y = (y_1, y_2, \ldots), \ldots$ The preparation theorem says that if an $f \in A_{xt}$ satisfies

(r_n) $\quad f(0,t) = ht^n$ with some unit $h \in A_t$,

then for each $g \in A_{xt}$ the following formula (of division by f) holds:

$$g(x,t) = \chi(x,t)\, f(x,t) + \sum_0^{n-1} \alpha_i(x) t^i$$

with some $\chi \in A_{xt}$ and $\alpha_i \in A_x$, $i = 0, \ldots, n-1$.

First we repeat here an elegant trick of Malgrange which reduce the theorem to the division by the "generic polynomial" $Q(c,t) = t^n + c_{n-1}t^{n-1} + \ldots + c_o$.

Given an f satisfying (r_n), assume we can divide by Q. Thus we have $f(x,t) = \chi(c,x,t)\, Q(c,t) + \sum_0^{n-1} \alpha_i(c,t) t^i$ with some χ and α_i. Then we obtain $\chi(0) \neq 0$ and $\alpha_i(0) = 0$ (put $c = 0$, $x = 0$ and compare the Taylor polynomials of degree n), and next $\frac{\partial \alpha_j}{\partial c_j}(0) = -\chi(0) \neq 0$ and $\frac{\partial \alpha_i}{\partial \gamma_j}(0) = 0$ for $i < j$). Hence by the implicit function theorem there is a $c \in A_x^n$ such that $\alpha_i(c(x),x) = 0$ and $c(0) = 0$. Therefore $f(x,t) = \chi(c(x),x,t)\, Q(c(x),t)$, and, given any $g \in A_{xt}$, as we have $g(x,t) = \tilde{\chi}(c,x,t)\,Q(c,t) + \sum_0^{n-1} \tilde{\alpha}_i(c,x)t^i$ with some $\tilde{\chi}$ and $\tilde{\alpha}_i$, it follows

$$g(x,t) = \frac{\tilde{\chi}(c(x),x,t)}{\chi(c(x),x,t)}\, f(x,t) + \sum_0^{n-1} \tilde{\alpha}_i(c(x),x)t^i \ .$$

Thus it is sufficient to divide by $P(c,t) = t^n - c_1 t^{n-1} + \ldots + (-1)^n c_n$.

We need now some simple observations.

Let $\varphi \in A_x{}^n$, $x = (x_1,\ldots,x_n)$, be such that $\varphi(0) = 0$ and $\operatorname{jac}\varphi = \det \frac{\partial\varphi_i}{\partial x_j} \neq 0$ (as an element of A_x); then, (for $F \in A_y$, $y=(y_1,\ldots,y_n)$), $F \circ \varphi = 0$ implies $F = 0$ (since $F \circ \varphi = 0$ implies, by differentiation and Cramer formulas, $\frac{\partial F}{\partial x_j} \circ \varphi = 0$, and therefore by induction $D^pF \circ F = 0$ and $D^pF(0) = 0$ for every p); by the same way, in the case of $\hat{\mathcal{O}}$, $F_v \circ \varphi \to 0$ implies $F_v \to 0$ in the Krull topologies.

Put $\sigma = (\sigma_1,\ldots,\sigma_n)$ where $\sigma_i \in A_\gamma$, $(\gamma = (\gamma_1,\ldots,\gamma_n))$, are the elementary symmetric functions. Observe that $\operatorname{jac}\sigma = \prod_{i<j} (\gamma_i - \gamma_j) \neq 0$ and $P(\sigma(\gamma), t) = (t-\gamma_1)\ldots(t-\gamma_n)$. Next $\sigma(V)$ is a neighborhood of 0 when V is (the roots of $P(c,z) = 0$ are small if the coefficients c_i are).

An $F \in A_{xu}$ is symmetric in respect to x iff $F \circ \pi_\alpha = F$ for each permutation $\alpha = (\alpha_1,\ldots,\alpha_n)$ where $\pi_\alpha = (x_{\alpha_1},\ldots,x_{\alpha_n}, u)$. The symmetric function theorem holds:

If $F \in A_{xu}$ is symmetric in respect to x, then $F(x,u) = \varphi(\sigma(x),u)$ with some $\varphi \in A_{yu}$.

This follows from the symmetric function theorem for polynomials, since we have $F = \sum_0^\infty F_v$ with F_v homogeneous and symmetric polynomial of degree v, so that $F_\nu(x,u) = \varphi_\nu(\sigma(x),u)$ with some polynomial φ_ν ; now $\sum_0^\infty \varphi_\nu$ is convergent: uniformely in a neighborhood of 0 in the case of $\mathfrak{G}$ (as $\sum_0^\infty \varphi_\nu(\sigma(x),u)$ is uniformely convergent in a neighborhood V of 0 , and $(\sigma,u)(V)$ is also a neighborhood of 0), and in the Krull topology in the case of $\hat{\mathfrak{G}}$, since $\varphi_\nu \to 0$ (as $F_\nu \to 0$ and $\mathrm{jac}(\sigma,u) \neq 0$).

Now we have unique division by $(t-\gamma_1)\ldots(t-\gamma_n)$: to each $f \in A_{\gamma x t}$ we have $f(\gamma,x,t) = \chi(\gamma,x,t)\,(t-\gamma_1)\ldots(t-\gamma_n) + \sum_0^{n-1} \alpha_i(\gamma,x)t^i$ with unique $\chi \in A_{\gamma x t}$ and $\alpha_i \in A_{\gamma x}$.

For uniqueness, assume $f = 0$ and put $t = \gamma_j$, so $\sum_0^{n-1} \alpha_i(\gamma,x)\gamma_j^i = 0$ which implies $\alpha_i = 0$ and then $\chi = 0$. For division (i.e. existence) one does it step by step dividing by each $t-\gamma_j$ following the observation that, given any $f \in A_{\gamma x t}$, we have $f(\gamma,x,t) = g(\gamma,x,t)(t-\gamma_j) + f(\gamma,x,\gamma_j)$ with some $g \in A_{\gamma x t}$.

To divide by P take any $f \in A_{cxt}$. We can divide by $(t-\gamma_1)\ldots(t-\gamma_n)$ so that $f(\sigma(\gamma),x,t) = \tilde{\chi}(\gamma,x,t)(t-\gamma_1)\ldots(t-\gamma_n) + \sum_0^{n-1} \tilde{\alpha}_i(\gamma,x)t^i$ with some $\tilde{\chi}$ and $\tilde{\alpha}_i$, and the uniqueness of this division gives easily the symmetry of $\tilde{\chi}$ and $\tilde{\alpha}_i$ in respect to γ , so that $\tilde{\chi}(\gamma,x,t) = \chi(\sigma(\gamma),x,t)$ and $\tilde{\alpha}_i(\gamma,x) = \alpha_i(\sigma(\gamma),x)$ with some $\chi \in A_{cxt}$ and $\alpha_i \in A_x$. Thus $(f-\chi P-\sum_0^{n-1} \alpha_i t^i) \circ (\sigma,x,t) = 0$ which follows $f - \chi P - \sum_0^{n-1} \alpha_i t^i = 0$.

References

[1] B. Malgrange, Ideals of differentiable functions, Oxford, 1966.

[2] J. Mather, Stability of C^∞ mappings: I. The division theorem, Annals of Math. 87 (1968), 89-104.

[3] S. Lojasiewicz, Whitney fields and the Malgrange-Mather preparation theorem, Liverpool Singularities-Symposium I (1971), 106-115.

University of Krakow

REDUCTION OF SEMILINEAR PARABOLIC EQUATIONS TO FINITE DIMENSIONAL C^1 FLOWS

by

Ricardo Mañé

Introduction

Let E be a Banach space. A C^r semiflow on E is a C^r map $\varphi: (0,+\infty)\times E \to E$ satisfying:

a) $\lim\limits_{t\to 0^+} \varphi(t,u) = \qquad\qquad \forall\, u \in E$

b) $\varphi(t+s,u) = \varphi(t,\varphi(s,u)) \qquad \forall\, u \in E,\ \ t,\ s > 0$

Roughly speaking, semiflows play the same role in the theory of semilinear parabolic equations that flows in the theory of ordinary differential equations. For instance consider the following simple example: let $f: \mathbb{R}^{2m} \times [0,\ell] \to \mathbb{R}$ be a C^r map $(r \geq 1)$. Consider the problem of finding functions $u: [0,+\infty)\times[0,\ell] \to \mathbb{R}$ satisfying:

I) u is continuous, $u(t,\cdot): [0,\ell] \to \mathbb{R}$ is C^{2m} for all $t > 0$ and $u(\cdot,x): (0,+\infty) \to \mathbb{R}$ is C^1 for all $x \in [0,\ell]$.

II) Denoting u_t the derivative of u respect to the first variable and $u^{(j)}$ the j-th derivative in the second variable, u satisfies:

$$u_t = -(-1)^m\, u^{(2m)} + f(u^{(2m-1)},\dots,u^{(1)},\ u,\ x)$$

III) $u^{(j)}(t,0) = u^{(j)}(t,\ell) = 0$ for all $t > 0$, $0 \leq j \leq m-1$.

Suppose that $\{\|f'(s)\| \mid s \in \mathbb{R}^{2m}\times[0,\ell]\} < +\infty$ and let E be the Sobolev space $H^{2m-1,p}([0,\ell]) \cap H_0^{m,p}([0,\ell])$, $p \geq 2$. Then there exists [1] a C^r semiflow φ on E such that if $v \in E$ the function $u: [0,+\infty)\times[0,\ell] \to \mathbb{R}$ defined by $u(t,x) = \varphi(t,v)(x)$ for $t > 0$ and $u(0,x) = v(x)$ is the unique function satisfying conditions (I), (II), (III) and $u(0,\cdot)= v$.

In this note we shall prove that in certain cases the semiflow generated by a semilinear parabolic equation satisfies the following property: there exists a finite dimensional C^1 submanifold $V \subset E$, invariant under φ, and a continuous retraction $\pi: E \to V$ such that x is asymptotic with $\pi(x)$ (i.e. $\lim_{t\to+\infty} \|\varphi(t,x)-\varphi(t,\pi(x))\| = 0$) for all $x \in E$. For example, this theorem can be applied to the semiflow generated by the P.D.E. described above if the derivative $D_1 f$ of f respect to its first real variable is small. More precisely if $|D_1 f(s)| \le \frac{1}{7}(\frac{\pi}{\ell})^{1/2m}$ for all $s \in \mathbb{R}^{2m}\times[0,\ell]$.

General background and statement of the theorem.

We shall consider differential equations of the form:

$$\dot{u} = -Au + f(u) \tag{1}$$

where A satisfies the following hypothesis:

(H1) A is a sectorial operator of a Banach space X^0 i.e. is densely defined, closed and there exists $K > 0$ such that for all $\lambda \in \mathbb{C}$ with $\mathrm{Re}(\lambda) \le 0$ the resolvent $(\lambda-A)^{-1}$ is defined and satisfies:

$$\|(\lambda-A)^{-1}\| \le \frac{K}{|\lambda| + 1}$$

Under this assumption fractional powers A^α, $\alpha \in \mathbb{R}$ are defined [1] and for $\alpha > 0$ A^α is closed with dense domain. Let X^α be the domain of A^α endowed with the graph norm $\|\cdot\|_\alpha = \|A^\alpha(\cdot)\|_0$ where $\|\cdot\|_0$ is the norm of X^0. If $0 \le \alpha < 1$, $V \subset X^\alpha$ is open and

$f: V \to X^{o}$ is a C^{1} map then (1) has local solutions, i.e. for all $a \in V$ there exists $\delta > 0$ and a C^{1} map $u: (0,\delta) \to V$ such that:

$$u((0,\delta)) \subset X^{1}$$

$$\lim_{t\to 0^{+}} u(t) = a$$

$$\dot{u}(t) = -Au(t) + f(u(t)) \qquad \forall\, t \in (0,\delta)$$

Suppose that f satisfies the stronger assumption:

(H2) f is a C^{r} map $f: X^{\alpha} \to X^{o}$ and is C^{1} bounded i.e. $\sup\{\|f'(x)\| \; x \in X^{\alpha}\} < +\infty$ where $f'(x)$ denotes the derivatives $f'(x): X^{\alpha} \to X^{o}$ of f at x.

Under this hypothesis there exists a C^{r} semiflow φ on X^{α} such that:

a) $\varphi((0,+\infty) \times X^{\alpha} \subset X^{1}$

b) $\dfrac{\partial\varphi}{\partial t}(t,u) = -A\varphi(t,u) + f(\varphi(t,u))$

Assume that A and f also satisfy the following properties:

(H3) The exists a Hilbert space Y^{o} such that X^{o} is continuously embedded as a dense subspace of Y^{o} and A extends to a sectorial normal operator $\tilde{A}$ of Y^{o} with compact inverse. Moreover if Y^{α} is the domain of $\tilde{A}^{\alpha}$ endowed with the norm $|\cdot|_{\alpha} = |\tilde{A}^{\alpha}(\cdot)|_{o}$ where $|\cdot|_{o}$ is the norm of Y^{o}, then for all $\alpha_{1} > 0$ there exists $\alpha_{2} > 0$ such that $Y^{\alpha_{2}} \subset X^{\alpha_{1}}$ and the inclusion is continuous.

(H4) Let $\{\mu_{n} | \; n \in \mathbb{Z}^{+}$ be the sequence of eigenvalues of $\tilde{A}$ and let $\lambda_{n} = \mathrm{Re}(\mu_{n})$. Suppose $\lambda_{1} < \lambda_{2} < \dots$. There exists $0 < \gamma\, \alpha$

$k_1 > 0$, $k_2 > 0$ such that:

$$\limsup \frac{\lambda_{n+1} - |\mu_n|}{|\mu_n|^{\gamma}} > 0$$

$$|f'(u)v|_0 \le k_1 |v|_\gamma + k_2 |v|_\beta \quad \forall\, u,v \in X^\alpha$$

$$k_1 < \frac{1}{2e+1} \limsup \frac{\lambda_{n+1} - |\mu_n|}{|\mu_n|^{\gamma}}$$

Before stating the theorem let us see an example of a semilinear parabolic P.D.E. that generates a differential equation of type (1) satisfying hypothesis (H1) - (H4). Let $\Omega \subset \mathbb{R}^n$ be a bounded domain with C^∞ boundary. Consider the equation:

$$u_t = -\sum_{|\alpha| \le 2m} a_\alpha(x) D^\alpha u + F(u^{(2m-1)}, \dots, u, x)$$

with boundary condition:

$$u^{(j)}(t,x) = 0 \quad \forall\, t > 0, \quad x \in \partial\Omega, \quad j \le m-1$$

where $u^{(j)}$ denotes the j-th derivative of $u: [0,+\infty) \times \Omega \to \mathbb{R}$ respect to the second variable, u_t denotes the derivative respect to the first variable, $L = \sum_{|\alpha| \le 2m} a_\alpha(x) D^\alpha$ is a self adjoint uniformly elliptic operator with $a_\alpha \in C^\infty(\overline{\Omega}, \mathbb{R})$ and $F: \mathbb{R}^{\ell_{2m-1}} \times \dots \times \mathbb{R}^{\ell_0} \times \Omega \to \mathbb{R}$ is a C^r map. Take $X^0 = L^p(\Omega)$ and let A be the sectorial operator of X^0 with domain $H^{2m,p}(\Omega) \cap H_0^{m,p}(\Omega)$ defined by L [1]. If p is large and we take $0 < \alpha < 1$ near 1, the space X^α is continuously imbedded in $C^{2m-1}(\overline{\Omega})$, therefore the map $f: X^\alpha \to X^0$ defined by:

$$f(u)(x) = F(u^{(2m-1)}(x), \dots, u(x), x)$$

is C^r. If the first partial derivatives of F respect to the first

2m-1 variables is uniformely bounded, then f is C^1 bounded and with these definitions of A, f and X^o, equation (1) satisfies hypothesis (H1), (H2). Defining $Y^o = L^2(\Omega)$ and $\tilde{A}$ as the linear operator with domain $H^{2m,2}(\Omega) \cap H_o^{m,2}(\Omega)$ defined by L, hypothesis (H3) is also satisfied. In order that the nonlinear term f satisfy hypothesis (H4) we have to require the condition $2m \geq n$. Then the sequence $\lambda_1 \leq \lambda_2 \leq \ldots$ of eigenvalues of $\tilde{A}$ satisfies ([3] Theorem 14.6):

$$\limsup_{n\to+\infty} \frac{\lambda_{n+1} - \lambda_n}{\lambda_n^{(2m-n)/n}} > 0$$

Denoting $D_j f$ the derivative of f respect to the j-th variable it follows that if for all $p \in \mathbb{R}^{\ell_{2m-1}} \times \ldots \times \Omega$, $1 \leq j \leq 2m-1$.

$$|D_j f(p) < \frac{1}{2e+1} \limsup_{n\to+\infty} \frac{\lambda_{n+1} - \lambda_n}{\lambda_n (2m-n)/n}$$

then taking $\gamma = \frac{2m-n}{n}$ hypothesis (H4) is satisfied.

In the statement and proof of the next theorem we shall denote $X^\alpha = E$, $\|\cdot\|_\alpha = \|\cdot\|$, and E_n^c, Y_n^c will be the subspace of Y^o associated to the spectral subsets $\{\mu_1, \ldots, \mu_n\}$, $\{\mu_j \mid j \geq n+1\}$. By (H3) we have $E_n^c \subset X^\alpha$. Hence $X^\alpha = E_n^c \oplus (Y_n^s \cap X^\alpha)$. Let $E_n^s = Y_n^s \cap X^\alpha$.

<u>Theorem</u> - There exist $n \in \mathbb{Z}^+$, a C^1 map $h: E_n^c \to E_n^c$ and a continuous retraction $\pi: E \to V$, where $V = \text{graph}(h)$, satisfying:

a) $\pi \circ \varphi_t = \varphi_t \circ \pi \quad \forall\ t > 0$

b) For all $x \in E$, $W^{ss}(x) \overset{\text{def}}{=} \pi^{-1}(\pi(x))$ is a C^r submanifold and the map

$$x \to E_x^{ss} \overset{\text{def}}{=} T_x W^{ss}(x)$$

is continuous and $E = E_x^{ss} \oplus E_n^c \quad \forall\, x \in E.$

c) There exist $K_o > 0$, $\sigma > 0$ satisfying:

$$\|\varphi(t,x_1) - \varphi(t,x_2)\| \le K_o\, e^{-\sigma t}\|x_1 - x_2\| \quad \forall\, x_1 \in E,\ x_2 \in W^{ss}(x_1),\ t>0$$

$$\|\varphi_t'(x)/E_x^{ss}\| \cdot \|\varphi_t'(\varphi(t,x))/T_{\varphi(t,x)} V\| \le K_o\, e^{-\sigma t} \quad \forall\, x \in V,\quad t > 0$$

d) $\sup\{\|h(x)\|,\ \|h'(x)\| \mid h \in E_n^c\} < +\infty$

The smoothness of V cannot be improved without strenghtning condition (H4). To obtain a C^m invariant manifold a condition of type

$$\limsup_{n\to+\infty} \frac{\lambda_{n+1} - m|\mu|}{|\mu|^{\gamma}} > 0$$

for some $0 \le \gamma < \alpha$ must be required. Unfortunately selfadjoint elliptic operators, the main source of examples of linear operator satisfying our assumptions, don't satisfy this property.

Proof of the theorem - Let $M(2 \times 2)$ be the space of real 2×2 matrices endowed with the norm of the supremum of the entries and the order relation $M_1 \le M_2$ if all the entries of M_1 are less or equal than the corresponding entries of M_2. Let Σ_δ be the space of bounded maps $F\colon [0,\delta] \to M(2\times 2)$ with the norm $\|F\| = \sup\{\|F(s)\| \mid s \in [0,\delta]\}$ and the order relation induced by the order defined on $M(2\times 2)$. Given $n \in \mathbb{Z}^+$ let $\pi_n^c\colon Y^o \to E_n^c$, $\pi_n^s\colon Y^o \to Y_n^s$ be the orthogonal projections. It is easy to prove $E_n^c \subset \bigcap_{\beta>0} Y^\beta$, hence for all $\beta \ge 0$, $Y^\beta = E_n^c \oplus (Y^\beta \cap Y_n^s)$ and E_n^c is orthogonal to $Y^\beta \cap Y_n^s$ in the Hilbert structure of Y^β. Let $0 \le \gamma < \alpha$ be the constant given by (H4). Observe that by (H4) $\varphi_t'(x)$ extends to a bounded map of Y^γ for all $x \in E$, $t > 0$. Define:

$$\theta_n(t) = \sup_{x\in E} \begin{pmatrix} \|\pi_n^s \ \varphi_t'(x)\pi_n^s\| & \|\pi_n^s \ \varphi_t'(x)\pi_n^c\| \\ \|\pi_n^c \ \varphi_t'(x)\pi_n^s\| & \|\pi_n^c \ \varphi_t'(x)\pi_n^c\| \end{pmatrix}$$

$$r_n(t) = \sup_{x\in E} \begin{pmatrix} \|\pi_n^s \ f'(x)\pi_n^s\| & \|\pi_n^s \ f'(x)\pi_n^c\| \\ \|\pi_n^c \ f'(x)\pi_n^s\| & \|\pi_n^c \ f'(x)\pi_n^c\| \end{pmatrix}$$

$$\beta_n(t) = \begin{pmatrix} \exp(-\lambda_{n+1} \ t) & 0 \\ 0 & \exp(-\lambda_1 \ t) \end{pmatrix}$$

By (5) θ_n and r_n are well defined. We want exponential estimates for θ_n. For that we shall use the following abstract version of Gronwall's inequality.

Lemma I - Let Σ be a complete metric space endowed with a reflexive, transitive and closed order relation. Let $F: \Sigma \circlearrowleft$ be an order preserving map such that for some $0 < \sigma < 1$ satisfies $d(F(x_1), F(x_2)) \le \sigma d(x_1,x_2)$ for all $x_1, x_2 \in \Sigma$. Then $x \le F(x)$ and $F(y) \le y$ imply $x \le y$.

Proof - Let $\Sigma(y) = \{x \in \Sigma \mid z = y\}$. $\Sigma(y)$ is closed and $F(\Sigma(y)) \subset \subset \Sigma(y)$. Hence the fixed point x_o of F is in $\Sigma(y)$. The condition $x \le F(x)$ implies $x \le F^n(x) \ \forall n \in \mathbb{Z}^+$. Therefore

$$x \le \lim_{n\to+\infty} F^n(x) = x_o \le y.$$

Lemma II - For all $\tilde{k}_1 > k_1$ there exists $\delta > 0$, $K > 0$ and $n_o > 0$ such that if:

$$\gamma_n(t) = \begin{matrix} \exp(-\lambda_{n+1} \ t) + \tilde{k}_1 \ t^{1-\gamma}, & \tilde{k}_1 \ t^{1-\gamma} \\ \tilde{k}_1 \ t^{1-\gamma} & , \quad K \end{matrix}$$

then $\theta_n(t) \le \gamma_n(t)$ for all $n \ge n_o$, $0 < t < \delta$.

Proof - Given $\tilde{k}_1 > k_1$ by (H4) there exists $n_o > 0$ such that

$$r_n(t) \le \begin{pmatrix} \tilde{k}_1 & \tilde{k}_1 \\ \\ \tilde{k}_1 & \tilde{k}_o \end{pmatrix}$$

with $K_o = k_1+k_2$. Let $e^{-\tilde{A}t}$ be the semigroup generated by $\tilde{A}$ [1]. This semigroup satisfies the following inequalities:

$$|e^{-\tilde{A}t} v|_\gamma \le e^{-\lambda_{n+1} t} |v|_\gamma \qquad \forall\, v \in Y_n^s \cap Y^\gamma$$

$$|e^{-\tilde{A}t} v|_\gamma \le t^{-\gamma} |v|_\gamma \qquad \forall\, v \in Y$$

and $\varphi_t'(x)$ satisfies:

$$\varphi_t'(x) = e^{-\tilde{A}t} + \int_0^t e^{-\tilde{A}(t-s)} f'(\varphi(s,x))\, \varphi_s'(x)ds$$

Hence:

$$\theta_n(t) \le \beta_n(t) + \int_0^t (t-s)^{-\alpha} r_n(s)\, \theta_n(s)ds \tag{1}$$

for all $n \in \mathbb{Z}$, $t > 0$. Take $K > K_o$. It is easy to see that there exists $\delta_o > 0$ such that

$$\beta_n(t) + \int_0^t (t-s)^{-\alpha} r_n(s)\, \gamma_n(s)ds \le \gamma_n(t) \tag{2}$$

for all $n \ge n_o$, $0 < t < \delta_o$. Now take $0 < \delta < \delta_o$ and define $\Sigma = \Sigma_\delta$ and $F_n\colon \Sigma \circlearrowleft$: by

$$F_n(\theta)(t) = \beta_n(t) + \int_0^t (t-s)^{-\alpha} r_n(s)\, \theta(s)ds$$

If δ is small enough F_n is a contraction for all $n \in \mathbb{Z}$. By Lemma I, (1) and (2), it follows that $\theta_n \le \gamma_n$ in Σ for all $n \ge n_o$.

For the next step of the proof we shall need the definition of angular contraction of a splitting of a Banach space. Consider a

Banach space F with a splitting $F = F_1 \oplus F_2$. Let $\pi_i: F \to F_i$ $i = 1, 2$ be the projections associated with the splitting and denote by $\mathcal{L}_\varepsilon(F_i,F_j)$ $i = 1,2,$ $j = 1,2$ the set of bounded linear maps $L: F_i \to F_j$ with $\|F\| \leq \varepsilon$.

Definition - A set of bounded linear maps $\{L_\xi \mid \xi \in \mathfrak{a}\}$ of F is a set of angular contractions of the splitting $F = F_1 \oplus F_2$ if there exists $\varepsilon > 0$ such that:

a) For all $T \in \mathcal{L}_\varepsilon(F_1,F_2)$ and $\xi \in \mathfrak{a}$ there exists a unique $T' \in \mathcal{L}_\varepsilon(F_1,F_2)$ such that:

$$L_\xi \text{ graph}(T) = \text{grap}(T')$$

b) For all $T \in \mathcal{L}_\varepsilon(F_2,F_1)$ and $\xi \in \mathfrak{a}$ there exists a unique $T' \in$ $\in \mathcal{L}_\varepsilon(F_2,F_1)$ such that $L_\xi^{-1}(\text{graph}(t)) = \text{graph}(T')$

c) Define the graph transforms $\Phi_1^\xi: \mathcal{L}_\varepsilon(F_1,F_2)\hookleftarrow$, $\Phi_2^\xi: \mathcal{L}_\varepsilon(F_2,F_1)\hookleftarrow$ by the properties:

$$L_\xi \text{ graph}(T) = \text{graph}(\Phi_1^\xi(T))$$

$$L_\xi^{-1} \text{ graph}(T) = \text{graph}(\Phi_2^\xi(T))$$

Then there exists $0 < \sigma < 1$ such that for all $\xi \in \mathfrak{a}$:

$$\|\Phi_1^\xi(T') - \Phi_1^\xi(T'')\| \leq k\|T' - T''\| \qquad \forall\, T', T'' \in \mathcal{L}_\varepsilon(F_1,F_2)$$

$$\|\Phi_2^\xi(T') - \Phi_2^\xi(T'')\| \leq k\|T' - T''\| \qquad \forall\, T', T'' \in \mathcal{L}_\varepsilon(F_2,F_1)$$

We shall use the following criteria to recognize sets of angular contractions:

Lemma III - If there exist $a, b, c \in \mathbb{R}$ satisfying for all $\xi \in \mathfrak{a}$:

$$\|\pi_2 L_\xi/F_2\| \leq a \tag{3}$$

$$\|\pi_1 L_\xi/F_2\| \leq b \tag{4}$$

$$\|\pi_2 L_\xi/F_1\| \leq b \tag{5}$$

$(\pi_1 L_\xi/F_1)^{-1}$ exists and $\|(\pi_1 L_\xi/F_1)^{-1}\| \leq c$ and a, b, c satisfy for some $\varepsilon > 0$:

$$cb\varepsilon < 1 \tag{6}$$

$$cb\varepsilon^2 - (1 - ac)\varepsilon + bc < 0 \tag{7}$$

$$\frac{ac + b^2 c^2}{(1 - cb\varepsilon)^2} < 1 \tag{8}$$

then $\{L_\xi \mid \xi \in \mathfrak{a}\}$ is a set of angular contractions of $F = F_1 \oplus F_2$.

<u>Corollary</u> - There exists $n_1 > 0$ such that for all $n \geq n_1$ there exists $\rho > 0$ such that $\{\varphi_\rho'(x) \mid x \in E\}$ is a set of angular contractions of the splitting $Y^\gamma = E_n^c \oplus (Y^\gamma \cap Y_n^s)$.

<u>Proof</u> - Take $\tilde{k}_1$ satisfying:

$$k_1 < \tilde{k}_1 < \frac{1}{2e+1} \limsup \frac{\lambda_{n+1} - |\mu_n|}{|\mu_n|^\gamma}$$

If $n \in \mathbb{Z}^+$ let $\rho = (|\mu_n| + \tilde{k}_1)^{-1}$. Define:

$$a = \exp(-\lambda_{n+1}\rho) + \tilde{k}_1 \rho^{1-\gamma}$$

$$b = \tilde{k}_1 \rho^{1-\gamma}$$

$$c = \exp(|\mu_n| + \tilde{k}_1 + 1)\rho \qquad \varepsilon = 1$$

Using (H4) it follows that when n is large enough a, b and c satisfy condition (6), (7) and (8) of Lemma III. Moreover Lemma II implies that conditions (3) and (4) are also satisfied. Finally (5) will follow from the next Lemma

Lemma IV - Given $\tilde{k}_1 > k_1$ there exist $k_o > 0$, $n_1 \in \mathbb{Z}^+$ such that:

$$\|(\pi_n^c \varphi_t'(x)/E_n^c)^{-1}\| \leq \exp((|\mu_n| + k_o)t) \tag{9}$$

for all $x \in E$, $0 \leq t \leq (|\mu_n| + \tilde{k}_1)^{-1}$, $n \geq n_1$.

Proof - Define:

$$U_n(x,t) = \pi_n^c \varphi_t'(x)/E_n^c$$

$$A_n = \tilde{A}/E_n^c$$

$$B_n^c(x,t) = \pi_n^c f'(x)/E_n^c$$

$$B_n^s(x,t) = \pi_n^c f'(x)/(Y^\gamma \cap Y_n^s)$$

$$V_n(x,t) = \pi_n^s \varphi_t'(x)/E_n^c$$

Then:

$$\dot{U} = -(A_n - B_n^c)U_n + B_n^s V_n$$

If $P_n(x,t) = (U_n(x,t))^{-1}$

$$\dot{P}_n = P_n(A_n - B_n^c) - P_n B_n^s V_n P_n \tag{10}$$

and if $u_n(x,t) = \|P_n(x,t)\|$ and n is large:

$$\dot{u}_n \leq (|\mu_n| + \tilde{k}_1)u_n + \tilde{k}_1^2 t^{1-y} u_n^2$$

Let $v_n = u_n \exp(-(|\mu_n| + \tilde{k}_1)t)$. Then:

$$\dot{v} \leq \tilde{k}_1^2 \, t^{1-\gamma} \exp((|\mu_n| + \tilde{k}_1)t)v_n^2$$

If:

$$f(t) = \int_0^t \tilde{k}_1^2 \, s^{1-\gamma} \exp((|\mu_n| + \tilde{k}_1)s)ds$$

it follows that:

$$v_n(t) \leq \frac{1}{1 - f(t)}$$

for all t such that $f(s) < 1$ for $s \in [0,t]$. Since $\delta < 1$ we have:

$$f(t) \leq \frac{e\, k_1^2}{1 - \gamma} t^2$$

If n is large enough

$$1 - \exp(-t) \geq \frac{e\, \tilde{k}_1^2}{1 - \gamma} t^2$$

for all $0 \leq t \leq (|\mu_n| + \tilde{k}_1)^{-1}$. Hence $f(t) \leq 1 - \exp(-t)$ and then $v_n(x,t) \leq \exp \, t$ and the Lemma is proved.

Using Corollary I we can improve estimate (9):

<u>Lemma V</u> - There exists $k > 0$ satisfying

$$\|(\pi_n^c \, \varphi_t'(x)/E_n^c)^{-1}\| \leq \exp(|\mu_n| + k)t$$

for all $n \geq n_1$, $x \in E$, $t > 0$.

<u>Proof</u> - If $n \geq n_1$ by Corollary I we have that:

$$V_n(x,t) \; P_n(x,t)\| \leq 1$$

for all $x \in E$, $t > 0$, $n \geq n_1$. By (10):

$$\dot{u}_n = (|\mu_n| + 2\tilde{k}_1 + 1)u_n$$

Taking $k = 2\tilde{k}_1 + 1$ the Lemma follows.

Using Corollary I and the fixed point theorem for contractions it is not difficult to prove the following property:

Lemma VI - If $n \geq n_1$ and $\mathcal{L}(Y^\gamma \cap Y_n^s, E_n^c)$ is the Banach space of bounded linear operators from $Y^\gamma \cap Y_n^s$ in E_n^c there exists a bounded continuous map $\Phi_n : E \to \mathcal{L}(Y^\gamma \cap Y_n^s, E_n^c)$ such that $L_n(x) =$ $=$ graph $\Phi_n(x)$ satisfies:

$$L_n(x) = (\varphi_t'(x))^{-1} L_n(\varphi(t,x)) \tag{11}$$

$$L_n(x) \oplus E_n^c = Y^\gamma \tag{12}$$

for all $x \in E$.

Lemma VII - For $n \geq n_1$ $\tilde{L}_n(x) = \{v + \Phi_n(x)v \mid v \in E_n^s\}$ satisfies the following properties:

a) $\tilde{L}(x) \oplus E_n^c = E \quad \forall\, x \in E$

b) $\sup\{\|\Phi_n(x)\|/\|v\| \mid x \in E,\ v \in E_n^s\} < +\infty$

c) $\tilde{L}(x) = (\varphi_t'(x))^{-1}\, \tilde{L}(\varphi(t,x)) \quad \forall\, x \in E,\ t > 0$

d) There exist $K_n > 0$, $C > 0$ such that for all $x \in E$, $t > 0$:

$$\|\varphi_t'(x)/\tilde{L}_n(x)\| \leq K_n \exp((-\lambda_{n+1} + C)t)$$

e) Let $\tilde{\pi}_n^c(x) : E \to E_n^c$, $\tilde{\pi}_n^s(x) : E \to \tilde{L}(x)$ be the projections associated with the splitting ( ). There exists $K > 0$ such that:

$$\|\tilde{\pi}_n^c(\varphi(t,x))\ \varphi_t'(x)/E_n^c\| \le K\ \exp((|\mu_n| + 2k_1)t) \quad \forall\ x \in E(\ )$$

<u>Proof</u> - Recalling that $E_n^c \subset E$ property (a) follows from (12). Since the inclusion $E \to Y^\gamma$ is bounded, (b) is implied by Lemma VI. (a) implies that $\tilde{L}(x) = L(x) \cap E$. Hence (c) follows from (11). To prove (d) take $\alpha_1 > 0$ such that $Y^{\alpha_1} \hookrightarrow E$ and the inclusion is bounded. Such α_1 exists by hypothesis (H4) of the theorem. Let m_1, m_2 be the norms of the inclusions $Y^{\alpha_1} \hookrightarrow E$ and $X^{\alpha_1} \hookrightarrow Y^{\alpha_1}$. Using some well known properties of semigroups [1] we have for $v \in E_n^s$:

$$\|e^{-At}\ v\| \le m_1 |e^{-\tilde{A}t}\ v|_{\alpha_1} |e^{-\tilde{A}(t-t_o)}\ e^{-\tilde{A}t_o}\ c|_{\alpha_1} \le$$

$$\le m_1\ \exp(-\lambda_{n+1}\ t) m_2\ t^{-\alpha_1}\ \|v\|$$

Then for some $M_n > 0$ we obtain:

$$\|e^{-At}/E_n^s\| \le M_n\ \exp\ (-\lambda_{n+1}\ t) \tag{13}$$

for all $t > 0$. By a similar argument we can assume

$$\|e^{-At}\ v\| \le M_n\ t^{-\alpha}\ \exp(-\lambda_{n+1}\ t) \tag{14}$$

for all $x \in E$, $t > 0$. Now define:

$$U^s(x,t) = \pi_n^s\ \varphi_t'(x)$$

$$U^c(x,t) = \pi_n^c\ \varphi_t'(x)$$

$$B^s(x,t) = \pi_n^s\ f'(\varphi(t,x))/E_n^s$$

$$B^c(x,t) = \pi_n^s\ f'(\varphi(t,x))/E_n^c$$

$$M = \sup\{\|B^s(x,t)\|,\ \|B^c(x,t)\|\ \mid\ x \in E,\ \ t > 0\}.$$

Moreover by VII (b) there exists $\delta > 0$ satisfying:

$$\|U^c(x,t)/\tilde{L}_n(x)\| \leq \delta \|U^s(x,t)/\tilde{L}_n(x)\| \tag{15}$$

$$\|\varphi_t^!(x)/\tilde{L}_n(x)\| \leq \delta \|\pi_n^s \, \varphi_t^!(x)/\tilde{L}_n(x)\| \tag{16}$$

for all $x \in E$, $t > 0$. Consider the equation:

$$\pi_n^s \, \varphi_t^!(x) = e^{-At} \, \pi_n^s + \int_0^t e^{-A(t-w)} \, (B^c(x,w) \, U^c(x,w) + B^s(x,t) \, U^s(x,w))dw$$

By (13), (14) and (15) this equation implies:

$$\|\pi_n^s \, \varphi_t^!(x)/\tilde{L}_n(x)\| \leq M_n \, \exp(-\lambda_{n+1} \, t) \, \|\pi_n^s/\tilde{L}_n(x)\| +$$
$$+ M_n \, M(1+\delta) \int_0^t (t-w)^{-\alpha} \, \exp(-\lambda_{n+1} \, 1) \, \|\pi_n^s \, \varphi_w^!(x)/\tilde{L}_n(x)\| ds$$

and this inequality together with (16) proves (d). To prove (e) define the cones $V_\delta(x) = \{v \in Y^\gamma \mid |\hat{\pi}_n^s(x)v| \leq \delta |\hat{\pi}_n^c(x)v|_\gamma\}$ where $\hat{\pi}_n^s(x) \colon Y^\gamma \to Y^\gamma \cap Y_n^s$, $\hat{\pi}_n^c(x) \colon Y^\gamma \to E_n^c$ are the projections associated to the splitting $\tilde{L}(x) \oplus E_n^c = E$. By Corollary I we can take δ such that

$$\varphi_t^!(x) \, V_\delta(x) \subset V_\delta(\varphi(t,x)) \tag{17}$$

$$L(x) \cap V_{2\delta}(x) = \{0\} \tag{18}$$

for all $x \in E$, $t > 0$.

By (17):

$$\varphi_t^!(x) E_n^c \subset V_\delta(\varphi(t,x)) \qquad \forall \, x \in E, \quad t > 0$$

Then by (18) there exists $K > 0$ satisfying:

$$\|\hat{\pi}_n^c(x)v\| \leq K \|\pi_n^c \, v\|$$

for all $x \in E$, $t > 0$, $v \in \varphi_t'(x)\, E_n^c$. Therefore by Lemma V:

$$\|\hat{\pi}_n^c(\varphi(t,x))\, \varphi_t'(x)/E_n^c\| \leq K \exp((|\mu_n| + 2k_1^{+1})t) \tag{19}$$

Since E_n^c is finite dimensional and $\hat{\pi}_n^c(\cdot)/E = \tilde{\pi}_n^c(\cdot)$ (19) implies property (e).

Proof of the theorem - Take $n \geq n_1$. Define $F_t: E_n^c \times \mathcal{L}_{\epsilon_1}(E_n^c, E_n^s) \to$
$\to E_n^c \times \mathcal{L}_{\epsilon_2}(E_n^c, E_n^s)$, where $\mathcal{L}_\epsilon(E_n^c, E_n^s)$ denotes the set of bounded linear maps of E_n^c in E_n^s with norm $\leq \epsilon$, as $F_t(x,L) = (\varphi(t,x), L_t)$ where L_t satisfies:

$$\varphi_t'(x) \text{ graph } L = \text{graph } L_t$$

By Lemma VII there exist $\epsilon_2 > \epsilon_1 > 0$ such that F_t is well defined for all $t > 0$ and for some $K > 0$, $\lambda > 0$ we have:

$$\|F_t(x,L') - F_t(x,L'')\| \leq K \exp(-\lambda t)\, \|L' - L''\|$$

for all $t > 0$, $L', L'' \in \mathcal{L}_{\epsilon_1}(E_n^c, E_n^s)$, $x \in E$. Using the fixed point theorem for contractions we can find a map $\Phi: E_n^c \to \mathcal{L}_{\epsilon_1}(E_n^c, E_n^s)$ satisfying:

$$F_t(x, \Phi(x)) = (\varphi(t,x), \Phi(\varphi(t,x))) \tag{20}$$

$$\|F_t(x,L) - \Phi(\varphi(t,x))\| \leq Ke^{-\lambda t}\, \|L\| \tag{21}$$

for all $t > 0$, $(x,L) \in E_n^c \times \mathcal{L}_{\epsilon_1}(E_n^c, E_n^s)$. By Lemma VII we can aply the methods in [2] to find a family of C^r submanifolds $\{W^{ss}(x) \mid x \in E\}$ and constants $K_1 > 0$, $\lambda_1 > 0$ satisfying:

$$x \in W^{ss}(x)$$

$$\tilde{L}(x) = T_x W^{ss}(x)$$

$$\|\varphi(t,y) - \varphi(t,x)\| \le K_1 e^{-\lambda_1 t}\|x-y\|$$

for all $x \in E$, $t > 0$, $y \in E^{ss}(x)$. Let Σ_ε be the space of C^1 maps $h: E_n^c \to E_n^s$ such that $\sup\{\|h(x)\| \,|\, x \in E_n^c\} < +\infty$, $\sup\{\|h'(x)\| \,|\, x \in E_n^c < +\infty$.

For $h \in \Sigma_\varepsilon$ let $\|h\|_o = \sup\{\|h(x)\| \mid x \in E_n^c\}$. Define $\psi_t: \Sigma_{\varepsilon_1} \to \Sigma_{\varepsilon_2}$ by:

$$\varphi_t(\text{graph}(h)) = \text{graph}(\psi_t(h))$$

By Lemma VII there exist $\varepsilon_2 > \varepsilon_1 > 0$ such that ψ_t is well defined and using the foliation $\{W^{ss}(x) \mid x \in E\}$ it follows that for some $K_2 > 0$, $\lambda_2 > 0$:

$$\|\psi_t(h_1) - \psi_t(h_2)\|_o \le K_2 \exp(-\lambda_2\ t)\ \|h_1-h_2\|_o \tag{22}$$

for all $t > 0$, $h_1, h_2 \in \Sigma_{\varepsilon_1}$. Take $h_o \in \Sigma_{\varepsilon_1}$. By (22) there exists $C > 0$ such that $\|\psi_t(h_o)\|_o \le C$ for all $t > 0$. Therefore

$$\begin{aligned}\|\psi_{t_1}(h_o) - \psi_{t_1}(h_o)\|_o &= \|\psi_{t_2}(h_o) - \psi_{t_2}(\psi_{t_2-t_1}(h_o))\| \le \\ &\le 2KC\ \exp(-\lambda t_2)\end{aligned} \tag{23}$$

for all $t_2 > t_1 > 0$. Moreover by the definition of ψ_t and (21):

$$\|(\psi_t(h_o))'(x) - \Phi(\psi_t(h_o)(x))\| \le K\varepsilon_1\ \exp(-\lambda t)$$

Hence:

$$\begin{aligned}\|(\psi_{t_2}(h_o))'(x) - (\psi_{t_1}(h_o))'(x)\| &\le 2K\varepsilon_1\ \exp(-\lambda t_1) + \\ &+ \|\Phi(\psi_{t_2}(h_o)(x)) - \Phi(\psi_{t_1}(h_o)(x))\|\end{aligned} \tag{24}$$

By (23) when $t \to +\infty$ ψ_1 converges uniformely to a bounded continuous map $h: E_n^c \to E_n^s$. By (24) $\|(\psi_{t_2}(h_o))'(x) - (\psi_{t_1}(h_o))'(x)\|$ converges to 0 uniformly on compact sets. Therefore $h \in \Sigma_{\varepsilon_1}$ and $\psi_t(h) = h$ for all $t > 0$. Let $V = \text{graph}(h)$ and $\pi: E \to V$ a retraction defined by

$$\pi(x) = W^{ss}(x) \cap V$$

Using the properties of the manifolds $W^{ss}(x)$ and Lemma VII it is not difficult to show that h and π satisfy properties (a) - (d) of the theorem.

References

[1] A. FRIEDMAN - Partial Differential Equations. Holt, Rinehart and Winston (1969).

[2] M. HIRSCH; C.PUGH; M. SHUB - Invariant Manifolds (to appear)

[3] S. AGMON - Lectures on Elliptic Boundary value Problems. D.Van Nostrand (1965).

Instituto de Matemática Pura e Aplicada
Rio de Janeiro, RJ - Brasil

AXIOM A FOR ENDOMORPHISMS

by

Ricardo Mañé

Let M be a compact C^∞ Riemann manifold without boundary and let $End^r(M)$ be the space of C^r endomorphisms of M endowed with the C^r topology. Following the diffeomorphisms theory let us say that f is C^r stable if there exists a neighborhood $\mathcal{U}$ of f in $End^r(M)$ such that for all $g \in \mathcal{U}$ there exists a homeomorphism $h: M$ satisfying $fh = hg$. Denote by $\Omega(f)$ the set of non wandering points of f [1] and by $S(f)$ the set of singularities of f i.e. the points x of M where Tf/T_xM is not injective. In [2] the following conjecture was stated:

<u>Conjecture</u> - If f is C^r stable $\Omega(f) \cap S(f) = \phi$.

A motivation for this conjecture is that $Per(f) \cap S(f) = \phi$ when f is C^r stable. To see this suppose that there exists $x \in Per(f) \cap$ $\cap\ S(f)$. Take g_1 C^r near to g and such that $x \in Per(g_1)$ and g_1/U is not injective for all neighborhood U of x. Take g_2 C^r near to g satisfying $S(g_2) \cap Per(g_2) = \phi$ [1]. If f is C^r stable there exists a homeomorphism $h: M \hookleftarrow$ satisfying $hg_1 = g_2h$. Then $h(Per(g_1)) = Per(g_2)$ and $h(x) \in S(g_2)$ because g_2/U is not injective for all neighborhood U of $h(x)$. Therefore $h(x) \in S(g_2) \cap$ $\cap\ Per(g_2)$.

In this paper we shall prove that $S(f) \cap \Omega(f) = \phi$ when f is absolutely C^r stable. Absolute stability is defined as follows:

<u>Definition</u> - $f \in End^r(M)$ is C^r absolutely stable if there exists a

neighborhood $\mathcal{U}$ of f in $\mathrm{End}^r(M)$ and a constant $K > 0$ such that for all $g \in \mathcal{U}$ there exists a homeomorphis $h: M \circlearrowleft$ satisfying $fh = hg$ and $d_o(h,I) \le K d_o(f,g)$ where $d_o(\cdot,\cdot)$ is defined by $d_o(f_1,f_2) = \sup\{d(f_1(x), f_2(x)) \mid x \in M\}$ $d(\cdot,\cdot)$ being the Riemannian metric on M and I is the identity map.

The property $\Omega(f) \cap S(f) = \phi$ for absolutely stable endomorphism is part of the next theorem. For its statement we need the following definition:

Definition - We say that $f \in \mathrm{End}^r(M)$ satisfies Axiom A if there exists a continuous splitting $TM/\Omega(f) = E^s \oplus E^u$ and constants $K > 0$, $0 < \lambda < 1$ satisfying:

a) $(Tf)E^s \subset E^s$

$(Tf)E^u \subset E^u$

b) $\|(Tf)^n/E^s_x\| \le K\lambda^n \qquad \forall\, x \in \Omega(f), \quad n \ge 0$

$\|(Tf)^n\, v\| \ge K\lambda^{-n}\|v\| \qquad \forall\, x \in \Omega(f), \quad v \in E^u_x,\ n \ge 0$

c) $\mathrm{Per}(f)$ is dense in $\Omega(f)$

d) If $x_1, x_2 \in \Omega(f)$ and $f(x_1) = f(x_2)$ then $E^s_x = \{0\}$

e) $S(f) \cap \Omega(f) = \phi$

Theorem - C^r absolutely stable endomorphisms satisfy Axiom A.

Obviously Axiom A alone it is not sufficient to imply stability. What is missing is a property playing the role of the strong transversality condition in diffeomorphisms theory. For research and conjectures in that direction see [7] and [8]. Axiom A implies local stability in the following sense: if g is C^r near to f there exists an injective continuous map $h: \Omega(f) \to M$ satisfying $gh = hf/\Omega(f)$. This property suggest the possibility of proving an Ω-stability theorem for Axiom A endomorphism (obviously requiring a no cycles conditions). However such a result would be rather unsatisfying since there exist examples of Ω-stable endomorphisms with singularities in the non wandering set [7].

For diffeomorphisms the theorem above was proved by Franks [5] and Guckenheimer [6] in the case $r=1$ and in [3] in the general case. Here we shall use an adaptation of [3] plus a couple of steps concerning parts (d) and (e) of our definition of Axiom A. The methods in [3] are specially suited for endomorphisms since they don't involve closing lemmas (unknown for endomorphisms even in the C^1 topology) and work in the C^r topology (observe that the C^1 topology is not useful to study stability of endomorphisms because C^1 stable endomorphisms cannot have singularities).

To sketch the proof of the theorem we need first some definitions and a Lemma. If $\Lambda \subset M$ is a compact subset of M let $\Gamma^b(\Lambda)$ be the space of bounded sections of TM/Λ with the norm $\|\eta\| = \sup\{\|\eta(x)\| \,|\, x \in M\}$ and let $\Gamma^o(\Lambda)$ be the closed subspace of continuous sections. If $f \in E^r(M)$ and $f(\Lambda) = \Lambda$ let T_fM/Λ be the vector bundle on Λ consisting of couples (p,v) with $p \in \Lambda$, $v \in T_pM$. Let $\Gamma_f^b(\Lambda)$, $\Gamma_f^o(\Lambda)$ be the corresponding spaces of bounded and continuous sections. Let $L_f\colon \Gamma^b(\Lambda) \to \Gamma_f^b(\Lambda)$ the linear map defined by:

$$L_f(\eta) = Tf \circ \eta - \eta \circ f$$

<u>Definition</u> - If $f \in \mathrm{End}^r(M)$ and $\Lambda \subset M$ is a compact subset such that $f(\Lambda) = \Lambda$ we say that Λ is a prehyperbolic set for f if there exist a splitting $TM/\Lambda = E^s \oplus E^u$ and constants $K>0$, $0 < \lambda < 1$ satisfying:

a) $(Tf)^{-1} E^s \subset E^s$

$(Tf)E^u = E^u$

b) $\|(Tf)^n/E^s_x\| \leq K\lambda^n \qquad \forall\, x \in \Omega(f), \quad n > 0$

$\|(Tf)^n\, v\| \geq K\lambda^{-n}\|v\| \qquad \forall\, x \in \Omega(f), \quad n > 0, \quad v \in E^u_x$

c) If $x_1, x_2 \in \Lambda$ and $f(x_1) = f(x_2) = x$ then $E^s_x = \{0\}$

Lemma A - Λ Is a prehyperbolic set for f if and only if $L_f\colon \Gamma^o(\Lambda) \to \Gamma^o_f(\Lambda)$ is an isomorphism.

This Lemma will be proved in the next section. The theorem is proved through the following steps:

Step I - $L_f\colon \Gamma^o(M) \to \Gamma^o_f(M)$ is surjective

Step II - If Λ is a minimal set for f then $L_f\colon \Gamma^o(\Lambda) \to \Gamma^o_f(\Lambda)$ is surjective. Therefore by Lemma A and the previous stpe Λ is prehyperbolic.

Step III - Let $\mathfrak{F}$ be the family of all prehyperbolic sets of f and let Σ be the closure of $\cup\{\Lambda \mid \Lambda \in \mathfrak{F}\}$. Then Σ is prehyperbolic and can be characterized as the union of the orbits $\{x_n \mid n \in \mathbb{Z}\} \subset M$ of f (i.e. sequences satisfying $f(x_n) = x_{n+1}$) such that if a sequence $\{v_n \mid n \in \mathbb{Z}^+\} \subset TM$ satisfies $v_n \in T_{x_n}M$, $(Tf)v_n = v_{n+1}$ and $\sup\{\|v_n\| \mid n \in \mathbb{Z}\} < +\infty$ then $\|v_n\| = 0$ for all n.

Step IV - If $x \in M$ let $\omega(x)$ be the set of ω-limit points of x and let $\overline{L}^+(f)$ be the closure of $L^+(f) = \{\omega(x) \mid x \in M\}$. Then $\overline{L}^+(f) \subset \Sigma$.

Step V - $\operatorname{Per}(f)$ is dense in $\overline{L}^+(f)$.

Step VI - $\overline{L}^+(f) \cap S(f) = \phi$.

Step VII - $\overline{L}^+(f) = \Omega(f)$.

Steps I and II are proved following the case of diffeomorphisms in [6] and [3]. To prove III observe that $\mathfrak{F}$ is nonempty by II and that L_f: $\Gamma^o(\Sigma) \to \Gamma^o_f(\Sigma)$ is one to one because $\cup\{\Lambda \in \mathfrak{F}\}$ is dense in Σ. By I it follows that is an isomorphism and by Lemma A this implies that Σ is prehyperbolic. The proof of Step IV is an adaptation of the similar case for diffeomorphisms [3]. For the proof of the next two steps we need the following Lemma:

<u>Lemma B</u> - Suppose that $f \in End^r(M)$ and that $\bar{L}^+(f)$ is prehyperbolic. Then given $x \in \bar{L}^+(f)$ and a neighborhood $\mathcal{U}$ of f there exists a neighborhood U of x and $g \in \mathcal{U}$ such that:

a) $g(x) = f(x)$ for all $x \in U$

b) $x \in Per(g)$

<u>Remark</u> - Without the hyperbolicity hypothesis this property is unknown even assuming that f is a diffeomorphism, that $Per(f)$ is dense in $\bar{L}^+(f)$ and $r=1$.

To prove V take $x \in \bar{L}^+(f)$ and $g \in End^r(M)$ nearby f such that $x \in Per(g)$. Let $h: M \circlearrowleft$ be a homeomorphism satisfying $hg = fh$. Then $h(x) \in Per(f)$. Moreover we can suppose that $d_o(h,I)$ is small. Hence $h(x)$ is nearby x.

To prove VI suppose that $x \in \bar{L}^+(f) \cap S(f)$. Let $g \in End^r(M)$ nearby f coinciding with f in a neighborhood of d (therefore $x \in S(g)$) and such that $x \in Per(g)$. Hence $x \in Per(g) \cap S(g)$ contradicting the stability of g.

Finally to prove Step VII take a descomposition

$\overline{L}^+(f) = \bigcup_{j=1} L_j$ in disjoint compact sets satisfying:

a) $f(L_i) = L_i \qquad i = 1,\dots,k$

b) For all $x \in M$ there exists L_i such that $\omega(x) \subset L_i$

c) For all L_i there exists $p \in L_i$ such that $\omega(p) = L_i$

The existence of this descomposition is proved following the spectral descomposition theorem for diffeomorphisms [10]. If $x \in M$ define its α-limit set $\alpha(x)$ as follows: $y \in \alpha(x)$ if there exists a sequence $\{n_k\}$ of positive integers and $y_k \in f^{-n_k}(x)$ such that $\lim_{k\to+\infty} y_k = y$. Following once more the diffeomorphism theory let us say that f has a cycle if there exist $L_{j_1},\dots,L_{j_m}$ and $x_i \in M$ $i = 1,\dots,m$ such that $\omega(x_i) = l_{j_i}$ and $\alpha(x_i) \cap L_{j+1} \neq \phi$ for $i = 1,\dots,m-1$ and $\omega(x_m) = L_{j_m}$, $\alpha(x_m) \cap L_{j_1} \neq \phi$. If $\Omega(f) \neq \overline{L}^+(f)$ it is easy to see that there exists a cycle and using the method in [9] we find g nearby f satisfying $g/\Omega(f) = f$ and $Per(g) \neq Per(f)$ thus contradicting the stability of f.

<u>Proof of Lemma A</u> - Suppose that $L_f\colon \Gamma^o(\Lambda) \to \Gamma^o_f(\Lambda)$ is surjective. We claim that $L_f\colon \Gamma^b(\Lambda) \to \Gamma^b_f(\Lambda)$ is one to one. If it is not there exists $\xi \in \Gamma^b(\Lambda)$ satisfying $L_f(\xi) = 0$ and using the method in [4] theorem A there exists a sequence $\{\xi_n\} \subset \Gamma^o(\Lambda)$ such that $\|L_f(\xi_n\| \leq \frac{1}{n}\|\xi_n\|$ contradicting our hypothesis on L_f. Moreover $L_f\colon \Gamma^b(\Lambda) \to \Gamma^p_f(\Lambda)$ is surjective because if $\xi\ \Gamma^b_f(\Lambda)$ there exists a generalized sequence $\{\xi_\alpha \mid \alpha \in \mathcal{G}\} \subset \Gamma^o_f(\Lambda)$ such that for all $x \in \Lambda$ $\lim \xi_\alpha(x) = \xi(x)$ and $\|\xi_\alpha\| \leq 2\|\xi\|$, $\alpha \in \mathcal{G}$. Let $\eta_\alpha = L_f^{-1}(\xi_\alpha)$. Since $\{\|\eta_\alpha\| \mid \alpha \in \mathcal{G}\}$ is bounded there exists a subsequence $\{\eta_{\overline{\alpha}} \mid \overline{\alpha} \in \mathcal{G}\}$ such that for some $\eta \in \Gamma^b(\Lambda)$ $\lim \eta_{\overline{\alpha}}(x) = \eta(x)$ for all $x \in \Lambda$. Then

$L_f(\eta) = \xi$.

For $x \in \Lambda$ define E^s_x as the set of $v \in T_xM$ such that $\{\|(Tf)^n v\| \mid n \in \mathbb{Z}^+\}$ is bounded and E^u_x as the set of $v \in T_xM$ such that if $\{x_{-n} \mid n \in \mathbb{Z}^+\} \subset \Lambda$ is a sequence satisfying $f(x_{-n})=f(x_{-n+1})$ there exists a sequence $\{v_{-n} \mid n \in \mathbb{Z}^+\}$, $v_{-n} \in T_{x_{-n}}M$, $\{\|v_{-n}\| \mid n \in \mathbb{Z}^+\} < +\infty$ such that $(Tf)v_{-n} = (Tf)v_{-n+1}$. Using that $L_f: \Gamma^b(\Lambda) \to \Gamma^b_f(\Lambda)$ is an isomorphism it follows that $E^s_x \oplus E^u_x = T_xM$ for all $x \in M$. Suppose $x_1, x_2 \in \Lambda$ $x_1 \neq x_2$ and $f(x_1) = f(x_2) = x$. We claim that $E^s_x = \{0\}$. To prove this take $v \in E^s_x$. Define $\xi \in \Gamma^b_f(\Lambda)$ by $\xi(x_1) = v$, $\xi(x_2)=v$ $\xi(x) = 0$ if $x \neq x_1, x_2$. Let $\eta = L_f^{-1}(\xi)$. Then:

$$(Tf)\ \eta(x_1) = v + \eta(x) \tag{1}$$

$$(Tf)\ \eta(x_2) = 2v + \eta(x) \tag{2}$$

$$(Tf)^n(\eta(y_i)) = \eta(x_i) \tag{3}$$

if $f^n(y_i) = x_i$, $i = 1, 2$, and:

$$(Tf)^n\ \eta(x) = \eta(f^n(x)) \tag{4}$$

for all $n \in \mathbb{Z}^+$. By (4) $\eta(x) \in E^s_x$. Therefore $\eta(x) + v$ and $\eta(x) + 2v$ are also in E^s_x. Hence, by (1) and (2), $\eta(x_i) \in E^s_{x_i}$, $i = 1, 2$. But (3) implies $\eta(x_i) \in E^u_{x_i}$, $i = 1, 2$. Then $\eta(x_i) = 0$, $i = 1, 2$. Then $v = 0$. It remains to prove that the maps $x \to E^s_x$, $x \to E^u_x$ define continuous subbundles satisfying condition (a) of the definition of prehyperbolic set. These properties follow from the property $E^s_x \oplus E^u_x = \{0\}$ for all $x \in \Lambda$, using the methods in [4].

Proof of Lemma B - Take a neighborhood U of x_0 and a C^∞ splitting $TM/(f^{-1}(U) \cup f^2(U)) = \tilde{E}^s \oplus \tilde{E}^u$ such that

$\tilde{E}^s/(f^{-1}(U) \cap \bar{L}^+(f))$ and $\tilde{E}^u/(f^2(U) \cap \bar{L}^+(f))$ are ε_o-near to $E^s/(f^{-1}(U) \cap \bar{L}^+(f))$ and $E^u/(f^2(U) \cap \bar{L}^+(f))$. Let $\xi_1,\ldots,\xi_m$ be C^∞ vectorfields with support in $f^2(U)$ such that $\{\xi_1(x),\ldots,\xi_m(x)\}$ is a basis for $\tilde{E}^u_x$ for all $x \in f^2(U)$ and let $\xi_{m+1},\ldots,\xi_{m+k}$ be C^∞ vectorfields with support in $f^{-1}(U)$ such that $\{\xi_{m+1}(x),\ldots,\xi_{m+k}(x)\}$ is a basis for $\tilde{E}^s_x$ for all $x \in f^{-1}(U)$. Let $n_o = m+k$. If $(s_1,\ldots,s_m, s_{m+1},\ldots,s_{n_o}) \in \mathbb{R}^{n_o}$ define $\xi_s = \Sigma\, s_i \xi_i$. For $n \in \mathbb{Z}^+$, $g \in \mathrm{End}^r(M)$ define $F(g,n)\colon \mathbb{R}^{n_o} \to M$ by

$$F(g,n)(s) = (\exp \xi_s \circ g)^n(x_o)$$

Using the fact that Λ is prehyperbolic it is not difficult to prove the following estimate:

Lemma I - If U and ε_o are small enough there exist a neighborhood $\mathcal{U}_1$ of f in $\mathrm{End}^r(M)$, a neighborhood $U' \subset U$ of x_o and constants $k > 0$, $c > 0$, $r > 0$ such that:

$$\|F(g,n)'(s)v\| \geq k\|v\|$$

if $g \in \mathcal{U}_1$, $n \in \mathbb{Z}^+$, $\|s\| \leq r$, $v \in \mathbb{R}^{n_o}$, $F(g,n)(s) \in U'$ and $d(F(g,j)(s), \bar{L}^+(f)) \leq c$ for all $0 \leq j \leq n$.

The prehyperbolicity of $\bar{L}^+(f)$ implies the following expansivity property: there exists $\delta > 0$ such that if $x \in \Lambda$ and $y \in M$ satisfy $d(f^n(x), f^n(y)) \leq \delta$ for all $n \in \mathbb{Z}$ then $x = y$.

From this property follows the next Lemma:

Lemma II - For all $\delta > 0$ the exist $\varepsilon(\delta) > 0$ and a neighborhood $\mathcal{U}(\delta)$ of f such that if $g_1, g_2 \in \mathcal{U}(\delta)$, $x \in \Lambda$, $y \in M$ and $d(g_1^n(x), g_2^n(x)) \leq \varepsilon(\delta)$ and $d(g_2^j(x), L^+(f)) \leq \varepsilon(\delta)$ for all $0 \leq j \leq n$ then $d(g_1^j(x), g_2^j(x)) \leq \delta$ for all $0 \leq j \leq n$.

To prove Lemma B take $\bar{g} \in \mathcal{U}_1$ ($\mathcal{U}_1$ given by Lemma I),

$n, \in \mathbb{Z}^+$ and $r_1 > 0$ satisfying:

1) $\bar{g}^n(x_o) \in U'$

2) $\bar{g}(x) = f(x)$ for all $x \in U$

3) $\exp \xi_s \circ g \in \mathcal{U} \cap \mathcal{U}(c)$ for all s with $\|s\| \le r_1$, where $\mathcal{U}$ is the neighborhood given by Lemma II and c the constant given by Lemma I.

4) $d(\bar{g}^n(x_o), x_o) \le kr_1$, k given by Lemma I

5) $d(\bar{g}^n(x_o), x_o) \le \epsilon(\frac{c}{2})$, $d(\bar{g}^j(x_o), \bar{L}^+(f)) \le \epsilon(\frac{c}{2})$, $\epsilon(\cdot)$ as in Lemma II.

Let λ: $[0, \delta_1]$ be a geodesic arc parametrized by arc lenght with $\lambda(0) = \bar{g}^n(x)$, $\lambda(\delta_1) = x$. Let S be the set of $s \in \mathbb{R}^{n_o}$ such that $\|s\| \le r_1$ and:

$$d((\exp \xi_s \circ \bar{g})^j(x_o), \bar{g}^j(x_o)) \le c$$

if $0 \le j \le n$. Define $\tau = \sup\{t | \lambda([0,t]) \subset F(\bar{g},n)S\}$. If $\tau = \delta_1$ the Lemma is proved taking $g = \exp \xi_s \circ \bar{g}$ where $s = F(\bar{g},n)^{-1}(x)$. Suppose $\tau < \delta_1$. Then $s_1 = F(\bar{g},n)^{-1} \lambda(\tau) \in \partial S$. But by Lemma I the lenght of the arc μ: $[0,\tau] \to \mathbb{R}^{n_o}$ defined by $\mu = F(g,n)^{-1} \circ \lambda$ is $\le \frac{1}{k_1} d(\bar{g}^n(x_o), x_o) \le r_1$. Hence $\|s_1\| \le r_1$ and since $s_1 \in \partial S$ we must have:

$$d((\exp \xi_{s_1} \circ \bar{g})^{m_1}(x_o), \bar{g}^{m_1}(x_o)) = c \tag{1}$$

for some $0 < m_1 \le n$. On the other side observe that $(\exp \xi_s \circ \bar{g})^n(x_o)$ is in the arc λ joining $\bar{g}^n(x)$ and x, hence $d((\exp \xi_{s_1} \circ \bar{g})^n (x_o), x_o) \le$ $\le d(\bar{g}^n(x), x) \le \epsilon(\frac{c}{2})$. By Lemma II it follows that

$$d((\exp \xi_{s_1} \circ \bar{g})^m (x_o), x_o) \le c/2$$

for all $0 \le m \le n$ contradicting (1).

References

[1] M. SHUB - Endomorphisms of compact manifolds. Amer.J.Math., 91 (1969) 175-199.

[2] R. MAÑÉ & C. PUGH - Stability of endomorphisms, Dynamical Systems, Warwick 1974. Lectures Notes in Mathematics 468, Springer--Verlag.

[3] R. MAÑÉ - On infinitesimal and absolute stability of diffeomorphisms. Dynamical Systems, Warwick 1974. Lectures Notes in Mathematics 468, Springer-Verlag.

[4] R. MAÑÉ - Quasi-Anosov diffeomorphisms and hyperbolic manifolds. (to appear Trans.Amer.Math.Soc.).

[5] J. FRANKS - Absolutely structural stable diffeomorphisms. Proc. Amer.Math.Soc., 37(1973) 293-296.

[6] J. GUCKENHEIMER - Absolutely Ω-stable diffeomorphism. Topology 11(1972) 195-197.

[7] J. FRANKS - Structural Stability of smooth contracting endomorphisms. Asterisque 31(1976) 141-187.

[8] M.V. JACOBSON - On smooth mappings of the circle to itself. Math. Sbornik 85(1971) 163-188.

[9] J. PALIS - A note on Ω-stability. Global Analysis. Proc.Symp. Pure Math. (1970) 14.

[10] S. SMALE - Differentiable Dynamical Systems. Bull.Amer.Math.Soc. 73(1967) 747-817.

Instituto de Matemática Pura e Aplicada
Rio de Janeiro, RJ

Characterizations of AS diffeomorphisms

by

Ricardo Mañé

The purpose of this note is to prove some simple characterizations of AS diffeomorphisms i.e. diffeomorphisms satisfying Axiom A and the strong transversality condition [1]. We shall use the concept of quasi-Anosov isomorphism of a vector bundle: we say that an isomorphism Φ of a continuous finite dimensional finslered vector bundle F on a compact space K is quasi-Anosov if for all $0 \neq v \in F$ the set $\{\|\Phi^n v\| \mid n \in \mathbb{Z}\}$ is unbounded. In [3] and [6] these isomorphisms were studied and its properties applied to diffeomorphisms f of a compact manifold without boundary such that Tf is a quasi-Anosov isomorphism of TM. Here we shall consider the case when the cotangent derivative T^*f of f is a quasi-Anosov isomorphism of the cotangent bundle T^*M. Recall that T^*f is defined as an isomorphism of T^*M covering f^{-1} and such that $\langle (T^*f)v,w\rangle = \langle v,(Tf)w\rangle$ for all $x \in M$, $v \in T^*_{f(x)}M$, $w \in T_xM$. We shall prove that T^*f is quasi-Anosov if and only if f is an AS diffeomorphism. Using this property is quite easy to deduce another characterization involving only properties of Tf (property (b) in the statement of the theorem) and finally we shall give a new and shorter proof of the theorem [2] stating that f is AS if and only if it is infinitesimally stable, i.e. if $I-f_\#: \Gamma^0(TM)\hookleftarrow$ is surjective, where $\Gamma^0(TM)$ denotes the space of C^0 sections of TM and $f_\#: \Gamma^0(TM)\hookleftarrow$ the linear map defined by $f_\#(\xi) = Tf\circ\xi\circ f^{-1}$.

<u>Theorem</u> - Let f be a C^1 diffeomorphism of a smooth compact

boundaryless manifold M. Then the following properties are equivalent:

a) f is an AS diffeomorphism

b) Every $v \in TM$ can be decomposed as $v = v^+ + v^-$ satisfying:

$$\liminf_{n\to+\infty} \|(Tf)^n v^+\| = \liminf_{n\to+\infty} \|(Tf)^{-n} v^-\| = 0 \qquad (1)$$

c) The cotangent derivative T^*f is a quasi-Anosov isomorphism of the cotangent bundle T^*M

d) f is infinitesimally stable

<u>Proof</u> - (a) $\Rightarrow$ (d) is proved in [1]

(d) $\Rightarrow$ (c). Endow $\Gamma^0(TM)$ with the Banach norm $\|\eta\| = \sup\{\|\eta(x)\| \mid x \in M\}$. Let $\Gamma^0(TM)'$ the dual space of $\Gamma^0(TM)$ endowed with the norm $\|\xi\| = \sup\{|\langle\xi,\eta\rangle| \mid \eta \in \Gamma^0(TM),\ \|\eta\| = 1\}$ and let $f'_{\#}: \Gamma^0(TM)' \circlearrowleft$ the adjoint of $f_{\#}$. Suppose that for some $x \in M$ and $v \in T^*_xM$ we have $\sup\{\|(T^*f)^n v\| \mid n \in \mathbb{Z}\} < +\infty$. Define for $n \in \mathbb{Z}^+$, $\varphi_n \in \Gamma^0(TM)'$ by:

$$\langle\varphi_n,\eta\rangle = \sum_{-n}^{n} \langle (T^*f)^j v,\ \eta(f^{-j}(x))\rangle$$

Then:

$$\|\varphi_n\| = \sum_{-n}^{n} \|(T^*f)^j v\|$$

and:

$$\langle (I-f'_{\#})\varphi_n,\ \eta\rangle = \langle\varphi_n,\eta\rangle - \langle\varphi_n,\ f'_{\#}\eta\rangle =$$

$$= \langle (T^*f)^{-n} v,\ \eta(f^n(x))\rangle - \langle (T^*f)^{n+1} v,\ \eta(f^{-(n+1)}(x))\rangle$$

Therefore:

$$\|(I-f'_{\#})\varphi_n\| = \|(T^*f)^{n+1} v\| + \|(T^*f)^{-n} v\|$$

Since $I-f_{\#}$ is surjective it follows that $I-f'_{\#}$ is injective and has closed range. Then there exists $k > 0$ such that $\|(I-f'_{\#})\varphi\| \geq$ $\geq k\|\varphi\|$ for all $\varphi \in \Gamma^0(TM)'$. Then:

$$\|(T^*f)^{n+1} v\| + \|(T^*f)^{-n} v\| \geq \sum_{-n}^{n} \|(T^*f)^j v\| \tag{2}$$

and since $\sup\{\|(T^*f)^n v\| \mid n \in \mathbb{Z}\} < +\infty$ the series $\sum_{-\infty}^{+\infty} \|(T^*f)^j v\|$ converges and therefore

$$\lim_{n\to\pm\infty} \|(T^*f)^n v\| = 0$$

This and (2) imply that $\sum_{-\infty}^{+\infty} \|(T^*f)^j v\| = 0$ hence $v = 0$.

(c) $\Rightarrow$ (a) For $x \in M$ define E^s_x (resp. $\tilde{E}^s_x$) as the set of vectors $v \in T_xM$ $(v \in T^*_xM)$ such that $\{\|(Tf)^n v\| \; n \in \mathbb{Z}^+\}$ $(\{\|(T^*f)^n v\| \mid n \in \mathbb{Z}^+\}$ is bounded and E^u_x (resp. $\tilde{E}^u_x$) as the set of $v \in T^*_xM$ such that $\{\|(Tf)^{-n} v\| \mid n \in \mathbb{Z}^+\}$ $(\{(T^*f)^{-n} v\| \mid n \in \mathbb{Z}^+\})$ is bounded. Let $L^+(f)$ be the set of ω-limit points of f and $\overline{L}^+(f)$ its closure. By [3], [6] $\overline{L}^+(f)$ is a hyperbolic set for f. Defining the stable and unstable manifolds of x as:

$$W^s(x) = \{y \in M \mid \lim_{n\to+\infty} d(f^n(x), f^n(y)) = 0\}$$

$$W^u(x) = \{y \in M \mid \lim_{n\to+\infty} d(f^{-n}(x), f^{-n}(y)) = 0\}$$

follows from the hyperbolicity of $\overline{L}^+(f)$ that

$$T_x W^s(y_1) = E^s_x$$

$$T_x W^u(y_2) = E^u_x$$

for all $y_1, y_2 \in \overline{L}^+(f)$ and $x \in W^s(y_1) \cap W^u(y_2)$. Hence if we prove that:

$$E_x^s + E_x^u = T_xM \tag{3}$$

for all $x \in M$ it will follow that stable and unstable manifolds of points in $\overline{L}^+(f)$ intersect transversaly and therefore by [4] this implies that $\overline{L}^+(f)$ coincides with the set of non wandering points of f and then f is an AS diffeomorphism. Defining

$$(E_x^s)_o = \{v \in T_x^*M \mid \langle v,w\rangle = 0 \quad \text{for all} \quad w \in E_x^s\}$$

$$(E_x^u)^o = \{v \in T_x^*M \mid \langle v,w\rangle = 0 \quad \text{for all} \quad w \in E_x^u\}$$

(3) is equivalent to:

$$(E_x^s)_o \cap (E_x^u)_o = \{0\} \tag{4}$$

for all $x \in M$. Since T^*f is quasi-Anosov we know that $\tilde{E}_x^s \cap \tilde{E}_x^u =$ $= \{0\}$ for $x \in M$, hence to prove (4) it is sufficient to show:

$$(E_x^s)^o \subset \tilde{E}_x^u \tag{5}$$

$$(E_x^u)^o \subset \tilde{E}_x^s \tag{6}$$

for all $x \in M$. We shall prove only (5) ((6) follows in a similar way). If $x \in M$, using the hyperbolicity of $\overline{L}^+(f)$ it is easy to show (for details see [3]) that for all $n \in \mathbb{Z}^+$ there exists an splitting

$$E^s_{f^n(x)} \oplus \hat{E}_n^u = T_{f^n(x)}M \tag{7}$$

satisfying:

I) If $\lim_{k\to+\infty} f^{n_k}(x) = y \in L^+(f)$ then $\lim_{k\to+\infty} \hat{E}^u_{n_k} = E^u_y$

II) $(Tf)\,\hat{E}^u_n = \hat{E}^u_{n+1}$ for all $n \in \mathbb{Z}^+$

III) There exists $K > 0$ such that if $\pi_n : T_{f^n(x)}M \to \hat{E}^u_n$ is the projection associated with the splitting (7) then $\|\pi_n\| \le K$ for all $n \in \mathbb{Z}^+$.

From these properties follows that:

$$\lim_{n\to+\infty} \|(Tf)^{-n} / \hat{E}^u_n\| = 0 \tag{8}$$

Then if $w \in (E^s_y)^o$ we have for all $n \in \mathbb{Z}^+$ and $v \in T_{f^n(x)}M$:

$$|\langle (T^*f)^{-n}v, w\rangle| = |\langle w, (Tf)^{-n}v\rangle| = |\langle w, \pi_o(Tf)^{-n}\rangle| =$$

$$|\langle w, (Tf)^{-n}\pi_n v\rangle| \le \|w\|\,\|(Tf)^{-n} / \hat{E}^u_n\|\,\|K\|\,\|v\|$$

Hence $\|(T^*f)^{-n}v\| \le K\|(Tf)^{-n} / \hat{E}^u_n\|$. By (8) this implies $v \in \tilde{E}^u_x$.

(a) ⇒ (b) Let $W^s(x)$, $W^u(x)$ as before. Since f satisfies Axiom A, for all $x \in M$ there exist y_1, y_2 in the set $\Omega(f)$ of non wandering points of f [1] such that $x \in W^s(y_1) \cap W^u(y_2)$ [5] and since f satisfies the strong transversality condition

$$T_x W^s(y_1) + T_x W^u(y_2) = T_x M$$

By the hyperbolicity of $\Omega(f)$ we have:

$$\lim_{n\to+\infty} \|(Tf)^n / T_x W^s(y_1)\| = \lim_{n\to+\infty} \|(Tf)^{-n} / T_x W^u(y_2)\| = 0$$

thus proving (b).

(b) ⇒ (c) Suppose that $x \in M$ and $w \in T^*_x M$ satisfies $\sup\{\|(T^*f)^n v\| \mid n \in \mathbb{Z}\} < K$. Take $v \in T_x M$ such that $\langle v, w\rangle = \|v\|$. Let $v = v^+ + v^-$ where v^+, v^- satisfy (1). Then:

$$|\langle w, v^+\rangle| = |\langle w, (Tf)^{-n} (Tf)^n v^+\rangle| = |\langle (T^*f)^{-n} w, (Tf)^n v^+\rangle| \leq$$

$$\leq K\|(Tf)^n v^+\|$$

Hence $|\langle w, v^+\rangle| = 0$. In a similar way it follows that $|\langle w, v^-\rangle| = 0$ hence $\langle w, v\rangle = \|v\| = 0$.

REFERENCES

[1] - J. Robbin - A structural stability theorem, Ann. of Math. 94 (1971) 447-493.

[2] - R. Mañé - On infinitesimal and absolute stability of diffeomorphisms Dynamical Systems, Warwick 1974, Lecture Notes 468, Springer-Verlag.

[3] - R. Mañé - Hyperbolic manifolds and quasi-Anosov diffeomorphisms (to appear Trans. Amer. Math. Soc.).

[4] - S. Newhouse - Hyperbolic limit sets. Trans. Amer. Math. Soc. 167 (1972) 125-150.

[5] - M. Hirsch, J. Palis, C. Pugh, M. Shub - Neighborhoods of hyperbolic sets, Inventiones Math. 9 (1970) 121-134.

[6] - J. Selgrade - Isolated invariant sets for flows on vector boundles, Trans. Amer. Math. Soc. 203 (1973) 359-390.

Ricardo Mañé

IMPA

Rio de Janeiro, Brasil

Structural Stability of Integrable Differential Forms

by

Airton S. de Medeiros*

§1. Preliminaries

Let ω be a differential p-form defined on a manifold M. A point $x \in M$ such that $\omega(x) = 0$ is called a singularity of ω. The set of singular points of ω will be denoted by Sing(ω).

Definition 1.1 - Let ω be a differential p-form defined on a manifold M. We say that ω is integrable, if for each $x \in M - \mathrm{Sing}(\omega)$ there exists a neighborhood $V(x)$ and an integrable system of p 1-forms $\eta_1,\ldots,\eta_p$ (i.e., $\eta_1 \wedge\ldots\wedge \eta_p \wedge d\eta_j = 0$, $j=1,\ldots,p$) defined on V such that $\omega = \eta_1\wedge\ldots\wedge \eta_p$ on V.

Note: If p=1 this definition is equivalent to the fact that $\omega \wedge d\omega = 0$. on M.

We shall denote by $\mathcal{J}_r^p(M)$ (resp. $\mathcal{J}_r^p(0)$) the set of integrable p-forms defined on M (resp. defined on a neighborhood of $0 \in R^n$) and we consider in $\mathcal{J}_r^p(M)$ (resp. $\mathcal{J}_r^p(0)$) the C^1-topology. In the case p=1 we shall write $\mathcal{J}_r(M)$ (resp. $\mathcal{J}_r(0)$) rather then $\mathcal{J}_r^1(M)$ (resp. $\mathcal{J}_r^1(0)$).

A form $\omega \in \mathcal{J}_r^p(M)$ induces a codimension p foliation on $M - \mathrm{Sing}(\omega)$, which will be denoted by $\mathcal{F}(\omega)$. This foliation is given

* This paper is essentially the author's doctoral dissertation prepared under the supervision of Professor C. Camacho at IMPA, Rio de Janeiro. This work was supported by CNPq and UFRN.

by the distribution of (n-p)-planes $T_x = \{v \in TM_x \mid i(v)\,\omega(x) = 0\}$ which is integrable by the theorem of Frobenius. Where $i(v)\,\omega(x)$ is the (p-1)-form defined by

$$i(v)\omega(x).(v_1,\dots,v_{p-1}) = \omega(x)(v,v_1,\dots,v_{p-1}) \qquad \forall\ v_1,\dots,v_{p-1} \in TM_x.$$

If $\omega \in \mathcal{J}_r(0)$ we denote by ω'_0 the linear part of ω at 0. If $\omega = \sum_{i=1}^{n} \omega_i dx_i$, the vector field grad $\omega = (\omega_1,\dots,\omega_n)$ is called the gradient of ω.

Now let ω be a p-form on M^{p+1} and let Ω be a volume form on M. We associate to ω the vector field $\vec{\omega}$ in the following way:

$\vec{\omega}$ is the only vector field on M such that $\omega = i(\vec{\omega})\Omega$. When $M = \mathbb{R}^{p+1}$ we always consider $\Omega = dx_1 \wedge \dots \wedge dx_{p+1}$.

<u>Definition 1.2</u> - Let M and M' be differentiable manifolds and let $\omega \in \mathcal{J}^p_r(M)$ and $\omega' \in \mathcal{J}^p_r(M')$. We say that ω and ω' are topologically equivalent if there exists a homeomorphism $h: M \to M'$ satisfying

(a) $h(\text{Sing}(\omega)) = \text{Sing}(\omega')$

(b) $h: M-\text{Sing}(\omega) \to M'-\text{Sing}(\omega')$ is a topological equivalence between the foliations $\mathcal{F}(\omega)$ and $\mathcal{F}(\omega')$ i.e. the image under h of a leaf of $\mathcal{F}(\omega)$ is a leaf of $\mathcal{F}(\omega')$.

<u>Definition 1.3</u> - Let $\omega \in \mathcal{J}^p_r(M)$. We say that ω is C^1-estructurally stable, if there exists a neighborhood $N(\omega) \subset \mathcal{J}^p_r(M)$ consisting of forms topologically equivalent to ω.

<u>Definition 1.4</u> - Let $\omega \in \mathcal{J}^p_r(0)$ such that $\omega(0) = 0$. We say that ω is locally structurally stable at 0, if for any neighborhood $V(0) \subset \mathbb{R}^n$, there exists a neighborhood $N(\omega) \subset \mathcal{J}^p_r(0)$ such that if $\pi \in N$ then there exists $0' \in V$ such that $\pi(0') = 0$

furthermore, there are neighborhoods $U(0)$ and $U'(0')$ such that $\omega|U$ and $\pi|U'$ are topologically equivalent.

Definition 1.5 - Le $\omega \in \mathcal{I}_r^p(M)$ and let F be a leaf of $\mathcal{F}(\omega)$. We define the limit set of F to be the set $\lim(F)$ of all points in M, that are limits of sequences that diverge in F. The limit set of ω, $\lim(\omega)$, is the union of all $\lim(F)$ for all leaves of $\mathcal{F}(\omega)$.

Note: Let ω, π be integrable p-forms and let h be a topological equivalence between ω and π, then

$$h(\lim(F)) = \lim(h(F)) \quad \text{and} \quad h(\lim(\omega)) = \lim(\pi).$$

§2. Local Structural Stability

The problem of structural stability of integrable forms was posed by Kupka in [2]. He studied necessary local conditions around the singularities for the global C^1-structural stability of integrable 1-forms defined on a compact manifold. In this section we study the problem from another point of view. We shall prove the following theorems:

Theorem A - Let $\omega = \sum_1^n \omega_i \, dx_i$ $(n \geq 3)$ be in $\mathcal{I}_r(0)$ $(r \geq 2)$, $\omega(0) = 0$ and let $A = (a_{ij})$ where $a_{ij} = \dfrac{\partial \omega_i}{\partial x_j}(0)$.

If ω satisfies one of the conditions below, then, ω is locally C^1-structurally stable at 0.

(i) A is symmetric, has rank n, the number of positive eigenvalues is different of two, as well as the number of negative eigenvalues.

(ii) A is not symmetric, and, if $a_{ij} \neq a_{ji}$ the submatrix

$$A_{ij} = \begin{pmatrix} a_{ii} & a_{ij} \\ a_{ji} & a_{jj} \end{pmatrix}$$

has non-vanishing determinant.

Theorem A' - Let ω be in $\mathcal{J}^p_r(0)$ $(r \geq 2)$, $\omega(0) = 0$ and $d\omega(0) \neq 0$. If there exists a decomposition $\mathbb{R}^n = \mathbb{R}^{p+1} \oplus \mathbb{R}^{n-p-1}$ such that ${}^{\perp}(\omega|\mathbb{R}^{p+1})$ is hyperbolic. Then ω is C^1-stable at $0 \in \mathbb{R}^n$. This condition is necessary for the C^{r-2}-stability if $r > 2$.

Note: In the case $p=1$ Theorem A' is just a restatement of part two of Theorem A.

Before we prove Theorems A and A' we recall some results about integrable forms, which will be used in the proof of these theorems and in Remark 2.2 below.

Lemma 2.1 - Let $\omega \in \mathcal{J}_r(0)$ be such that $\omega(0) = 0$. Then ω'_o is integrable.

Proof: See [4], page 192.

Lemma 2.2 - Let $\eta = \Sigma\, T_i\, dx_i$ be an integrable 1-form in $\mathbb{R}^n$ with linear coefficients. Then, one of the conditions below must occur:

(1) η is exact

(2) η has rank two.

Proof: See [4], page 142.

Remark 2.1: The "Fundamental Lemma" which we shall prove later in this section contains a proof of Lemma 2.2. This proof implies the following facts.

If $d\eta = \sum_{k<\ell} \alpha_{k\ell}\, dx_k \wedge dx_\ell$ and if $\alpha_{ij} \neq 0$ then we can reduce η to the canonical form

(c) $\eta = L_i(x_i,x_j)\, dx_i + L_j(x_i,x_j)\, dx_j$ where $L_m(x_i,x_j) = T_m(0,\dots,0,x_i,0,\dots,x_j,0,\dots,0)$, $m = i,j$.

If $(L_i,L_j): \mathbb{R}^2 \to \mathbb{R}^2$ is not an isomorphism, we can reduce η to the canonical form.

(d) $\eta = x_1\, dx_2$.

Definition 2.1 - Let $\omega \in \mathcal{J}_r(0)$ $(n \geq 3)$, $\omega(0) = 0$, be such that $\omega_0' = df$ and f is a positive or negative definite quadratic form, then we say that 0 is a central point of ω.

Reeb (see [4], pag. 145) proved the following

Theorem 2.1 - Let $\omega \in \mathcal{J}_r(0)$. If 0 is a central point of ω, then the integral manifolds of ω in a neighborhood of 0, are homeomorphic to the sphere S^{n-1}.

Remark 2.2: The conditions of Theorem A are also necessary in the case of forms with linear coefficients.

In fact let $\omega \in \mathcal{J}_r(0)$ be a 1-form with linear coefficients, which fails to satisfy conditions (i) and (ii) of Theorem A. Then we must distinguish the following cases:

1st case - A is symmetric and rank of A is less then n.

In this case $d\omega = 0$ and so $\omega = dg$, where g is a degenerate quadratic form. We define $\tilde{\omega} = d\tilde{g}$ where $\tilde{g}$ is a non-degenerate quadratic form C^∞-close to g. It is clear that ω and $\tilde{\omega}$ are not equivalent.

2nd case - A is symmetric, has rank n and the number of positive eigenvalues of A is equal to two.

As before we have $\omega = dg$, but now g is a non-degenerate quadratic form. So, under a linear change of variables we can reduce ω to the canonical form $\omega = xdx + ydy - (z_1\, dz_1 + \ldots + z_\ell\, dz_\ell)$, $\ell = n-2$.

Let $W = \{(x,y,z_1,\ldots,z_\ell) \in \mathbb{R}^n \mid x^2+y^2-(z_1^2+\ldots+z_\ell^2) > 0\}$ and let

$$\eta(x,y,z_1,\ldots,z_\ell) = \begin{cases} \dfrac{\varepsilon\, r \exp(-\frac{1}{r})}{x^2+y^2}(-ydx+xdy) & \text{if } (x,y,z_1,\ldots,z_\ell) \in W \\ 0 & \text{otherwise} \end{cases}$$

where $r = x^2+y^2 - (z_1^2+\ldots+z_\ell^2)$.

It is easy to see that η is a C^∞ 1-form which is C^∞-close to zero, and that $\tilde{\omega} = \omega+\eta$ is integrable. Then ω and $\tilde{\omega}$ are not equivalent because $\lim(\tilde{\omega}) \supset \partial W = \{(x,y,z_1,\ldots,z_\ell) \mid r = 0\}$ and $\lim(\omega) = \{0\}$.

Note: The case where the number of negative eigenvalues is equal to two reduces to the 2nd case by considering the form $-\omega$.

3rd case - A is not symmetric, $\operatorname{rank}(A) > 1$ and there exists $a_{ij} \neq a_{ji}$ such that $\det A_{ij} = 0$.

Since A is not symmetric, ω is not exact. By Lemma 2.2 and Remark 2.1 we can write $\omega = L_1(x_1,x_2)dx_1 + L_2(x_1,x_2)dx_2$ where $L_1(x_1,x_2) = a_{ii}x_1 + a_{ij}x_2$ and $L_2(x_1,x_2) = a_{ji}x_1 + a_{jj}x_2$, hence the matrix of ${}^{\perp}(\omega|x_1x_2)$ is

$$\begin{pmatrix} a_{ji} & a_{jj} \\ -a_{ii} & -a_{ij} \end{pmatrix}$$

which has zero as an eigenvalue, so $\bar{}(\omega|x_1\ x_2)$ is not hyperbolic. If we choose $(\tilde{L}_2,-\tilde{L}_1)$ C^1-close to $(L_2,-L_1)$ not equivalent to it, the 1-form $\tilde{\omega} = \tilde{L}_1\, dx_1 + \tilde{L}_2\, dx_2$ is not equivalent to ω.

4th case - A is not symmetric and has rank one.

By Remark 2.1 we can write $\omega = x_1\, dx_2$, then we consider the form $\tilde{\omega} = x_1\, dx_2 + \epsilon\, x_2\, dx_1$. It is obvious that ω and $\tilde{\omega}$ are not equivalent.

Proof of Theorem A: - (i) A is symmetric, has rank n, the number of positive eigenvalues of A is different of two, as well as the number of negative eigenvalues.

We shall show that under these hypothesis, any form $\tilde{\omega}$ in a neighborhood of ω in $\mathcal{J}_r(0)$ is equivalent to ω_o'. Since A is symmetric $\omega_o' = df$ and f is a non-degenerate quadratic form. By means of a change of variables we can write $\omega_o' = x_1\, dx_1 + \ldots + x_k dx_k - (y_1 dy_1 + \ldots + y_\ell dy_\ell)$ where $k+\ell = n$ and $k,\ell \neq 2$.

Let V be a neighborhood of $0 \in \mathbb{R}^n$. We observe that grad ω_o' has 0 as a hyperbolic singularity, hence, there exists a neighborhood $\mathfrak{n}(\text{grad}\ \omega_o')$ in $\mathfrak{X}_1^r(\mathbb{R}^n)$ such that each $X \in \mathfrak{n}$, has exactly one singularity on V.

There is a neighborhood $N_1(\omega)$ in $\mathcal{J}_r(0)$, such that $\tilde{\omega} \in N_1(\omega)$ implies grad $\tilde{\omega} \in \mathfrak{n}$. This shows that each form $\tilde{\omega} \in N_1$ has exactly one singularity $0' \in V$. We may assume without loss of generality that $0' = 0$.

Now we shall construct the topological equivalence between $\tilde{\omega}$ and ω_o'.

1st case - $\ell = 0$ (or $k = 0$). This case is an immediate consequence of Theorem 2.1. Actually, we can choose $N(\omega) \subset N_1(\omega)$ such that 0 is a central point of all $\omega \in N(\omega)$ and such that the integral manifolds of $\tilde{\omega}$ are transverse to the rays coming from 0. So each ray intercepts each leaf of $\mathcal{F}(\tilde{\omega})$ exactly once, in a neighborhood $W(0)$. Analogously there exists a neighborhood $U(0)$, such that the leaves of $\mathcal{F}(\omega)$ in U are spheres and intercept each ray exactly once.

We define the topological equivalence $h: U \to W$ as follows

(1) We choose rays $r' \subset U$ and $r'' \subset W$ and a homeomorphism $f: r' \to r''$ such that $f(0) = 0$.

(2) Given $x \in U-0$ there exists exactly one ray r_x and a leaf F_x of $\mathcal{F}(\omega)$ such that $x = r_x \cap F_x$. Let $x' = r' \cap F_x$ and let $\tilde{F}_{f(x')}$ be the leaf of $\mathcal{F}(\tilde{\omega})$ by $f(x')$, then we define $h(x) = r_x \cap \tilde{F}_{f(x')}$.

(3) $h(0) = 0$.

It is immediate to check that h is a homeomorphism.

<u>2nd case</u> - $\ell = 1$ (or $k = 1$). We can write $\omega_o' = x_1 dx_1 + \ldots + x_{n-1} dx_{n-1} - y\, dy$, of course we assume that $n \geq 4$.

We shall denote by $\mathbb{R}^{n-1}$ the subspace $y = 0$, and by $\mathbb{R}$ the subspace $x_i = 0$, $i = 1,\ldots,n-1$. We observe that 0 is a central point of $\omega_o'|\mathbb{R}^{n-1}$, so there exists $N_2 \subset N_1$ such that 0 is a central point of $\tilde{\omega}|\mathbb{R}^{n-1}$ for all $\tilde{\omega} \in N_2$. In view of the first case above if U and W are sufficiently small neighborhoods of 0, we can define a topological equivalence $h_o: U \cap \mathbb{R}^{n-1} \to W \cap \mathbb{R}^{n-1}$, between $\omega_o'|\mathbb{R}^{n-1}$ and $\tilde{\omega}|\mathbb{R}^{n-1}$ that preserves the rays.

If $\lambda \in \mathbb{R}^{n-1}$ and $|\lambda| = 1$, we denote by E_λ the ray in the direction of λ, and we denote by P_λ the plane generated by E_λ and $\mathbb{R}$. Le $\tilde{X}_\lambda = {}^{\perp}(\tilde{\omega}|P_\lambda)$, we obtain, in this way, a continuous family

$(\tilde{X}_\lambda)$ of vector fields, all in a neighborhood of $X = {}^{\perp}(\omega_0'|P_\lambda)$.

The vector field X is hyperbolic and $(\tilde{X}_\lambda)$ is a continuous family, hence, if we choose N_2 sufficiently small, there are neighborhoods U and W of 0, such that X and $\tilde{X}_\lambda$ are equivalent on $U \cap P_\lambda$ and $W \cap P_\lambda$, respectively.

It is not difficult to define a continuous family (h_λ) of topological equivalences $h_\lambda\colon P_\lambda \cap U \to P_\lambda \cap W$, between X and $\tilde{X}_\lambda$, satisfying

(1) $h_\lambda|E_\lambda = h_0|E_\lambda$

(2) $h_\lambda|\mathbb{R} = h_1$ where $h_1\colon \mathbb{R} \to \mathbb{R}$ is a homeomorphism, satisfying $h_1(0) = 0$.

Actually this construction is done by using the vector field $\operatorname{grad} \omega_0'|P_\lambda = (x,-y)$ as follows:

Let C_λ (resp. $\tilde{C}_\lambda$) be the cone obtained as the union of the stable and unstable manifolds of ${}^{\perp}(\omega_0'|P_\lambda)$ (resp. ${}^{\perp}(\tilde{\omega}|P_\lambda)$). Then we define a continuous family (h_λ') of homeomorphisms between C_λ and $\tilde{C}_\lambda$, satisfying $h_\lambda'(0) = 0$. We define $h_\lambda|E_\lambda = h_0|E_\lambda$, $h_\lambda|\mathbb{R} = h_1$ and $h_\lambda|C_\lambda = h_\lambda'$, in order to extend h_λ defined as above, we take x in the complement of the set where h is already defined. Let γ_x be the orbit of $\operatorname{grad} \omega_0'|P_\lambda$ passing through x, this orbit intersects C_λ in a unique point $\hat{x}$. Let x' be the intersection of the orbit of X through x with the axis generated by λ. We define $h_\lambda(x)$ to be the only point of intersection of $\gamma_{h_\lambda'(\hat{x})}$ and the orbit of $\tilde{X}_\lambda$ passing through $h|E_\lambda(x')$.

It is clear that the family (h_λ) satisfies properties (1) and (2). The fact that h_λ is a topological equivalence between X and $\tilde{X}_\lambda$ is simple to prove.

Now we define $h\colon U \to V$ as follows

(1) $h|\mathbb{R} = h_1$

(2) If $x \in U - \mathbb{R}$ there exists a unique plane P_λ such that $x \in P_\lambda$. Then we define $h(x) = h_\lambda(x)$.

The fact that h is a homeomorphism follows from the fact that (h_λ) is a continuous family and that $h_\lambda|\mathbb{R} = h_1$ for all λ.

Now let $\tilde{E}^s_\lambda$ and $\tilde{E}^u_\lambda$ denote respectively the stable and unstable manifolds of the vector field $^\perp(\tilde{\omega}|P_\lambda)$. We denote by $(\tilde{E}^u_\lambda)^+$ (resp. $(\tilde{E}^u_\lambda)^-$) the intersection of $\tilde{E}^u_\lambda$ with the half-plane that contains the positive part of the axis $\mathbb{R}$ (resp. the negative part). Analogously we define $(\tilde{E}^s_\lambda)^\pm$.

To show that h is an equivalence between $\mathfrak{F}(\tilde{\omega})$ and $\mathfrak{F}(\omega'_o)$, we need the following lemma which is easy to prove:

<u>Lemma 2.3</u> - If we denote by p a generic point of $\mathbb{R}$ and by $\tilde{F}_\ell$ a generic leaf of $\mathfrak{F}(\tilde{\omega}|\mathbb{R}^\ell)$, we have that $\Sigma_p = \bigcup_\lambda \tilde{F}^\lambda_p$, $\Sigma_\ell = \bigcup_{x \in \tilde{F}_\ell} \tilde{F}_x$, $\tilde{C}^+ = \bigcup_\lambda (\tilde{E}^u_\lambda)^+ \cup (\tilde{E}^s_\lambda)^+$ and $\tilde{C}^- = \bigcup_\lambda (\tilde{E}^u_\lambda)^- \cup (\tilde{E}^s_\lambda)^-$ are the leaves of $\mathfrak{F}(\tilde{\omega})$, where

(a) $\tilde{F}^\lambda_p$ is the leaf of $\mathfrak{F}(\tilde{\omega}|P_\lambda)$ that contains p.

(b) $\tilde{F}_x$ is the leaf of $\mathfrak{F}(\tilde{\omega}|P_\lambda)$ that contains x.

The lemma above caracterizes the leaves of $\mathfrak{F}(\tilde{\omega})$ in terms of saturated of "certain" sets, by the orbits of the $\tilde{X}_\lambda$'s. Such caracterization trivially holds for the leaves of $\mathfrak{F}(\omega'_o)$.

Now we observe that the elements envolved in these caracterizations are invariant under h, which implies that h is an equivalence as desired.

<u>3rd case</u> - $\ell \geq 3$ and $k \geq 3$. We have $\omega'_o = x_1\,dx_1 + \ldots + x_k\,dx_k - (y_1 dy_1 + \ldots + y_\ell dy_\ell)$.

We denote by $\mathbb{R}^k$ the subspace $y_j = 0$, $j = 1,\ldots,\ell$ and by

$\mathbb{R}^\ell$ the subspace $x_i = 0$, $i = 1,\dots,k$. We denote by S^{k-1} and $S^{\ell-1}$ the unitary spheres of $\mathbb{R}^k$ and $\mathbb{R}^\ell$ respectively.

Given $\lambda \in S^{\ell-1}$ let P_λ be the subspace generated by $\mathbb{R}^k$ and λ, and let $\omega'_\lambda = \omega'_o|P_\lambda$, we observe that

$$\omega'_\lambda = x_1\, dx_1 + \dots + x_k\, dx_k - \lambda d\lambda$$

and that

$$\omega'_o|\mathbb{R}^\ell = -(y_1\, dy_1 + \dots + y_\ell\, dy_\ell)$$

hence 0 is a central point of $\omega|\mathbb{R}^\ell$. As in the first case we define a topological equivalence $H_1\colon U \cap \mathbb{R}^\ell \to W \cap \mathbb{R}^\ell$ between $\mathcal{F}(\tilde{\omega}|\mathbb{R}^\ell)$ and $\mathcal{F}(\omega'_o|\mathbb{R}^\ell)$, in such a way that the rays of $\mathbb{R}^\ell$ are preserved by H_1.

Since $\omega'_o|\mathbb{R}^k = x_1\, dx_1 + \dots + x_k\, dx_k$ we can define a topological equivalence $H_o\colon U \cap \mathbb{R}^k \to W \cap \mathbb{R}^k$, between $\omega'_o|R^k$ and $\tilde{\omega}|\mathbb{R}^k$, in such a way that the rays of $\mathbb{R}^k$ are preserved by H_o.

We observe that for each $\lambda \in S^{\ell-1}$ it is possible to define a topological equivalence H_λ between $\omega'_o|P_\lambda$ and $\tilde{\omega}|P_\lambda$ by applying the process of the second case, where

1) $h_o = H_o$

2) $h_1 = H_1|E_\lambda$

It is clear that the construction of H_λ depends continuously on the homeomorphism h_1. So we obtain a continuous family (H_λ) of topological equivalences between $\omega'_o|P_\lambda$ and $\tilde{\omega}|P_\lambda$. Then we define $H\colon U \to W$ by $H(x) = H_\lambda(x)$, where λ is such that $x \in P_\lambda$.

It is clear that H is well defined and that it is a homeomorphism.

By considerations similiar to those done before in the 2nd case, the fact that H is an equivalence is contained in

Lemma 2.4 - If we denote by $\tilde{F}_k$ and $\tilde{F}_\ell$, generic leaves of $\mathcal{F}(\tilde{\omega}|\mathbb{R}^k)$ and $\mathcal{F}(\tilde{\omega}|\mathbb{R}^\ell)$ respectivelly, we have that

$$\Sigma_k = \bigcup_\lambda \tilde{F}_k^\lambda, \quad \Sigma_\ell = \bigcup_{x \in \tilde{F}_\ell} \tilde{F}_x \quad \text{and} \quad \tilde{C} = \bigcup_\lambda \tilde{C}_\lambda^+$$

are the leaves of $\mathcal{F}(\tilde{\omega})$ where

a) $\tilde{F}_k^\lambda$ is the leaf of $\mathcal{F}(\tilde{\omega}|P_\lambda)$ containing $\tilde{F}_k$.

b) $\tilde{F}_x$ is the leaf of $\mathcal{F}(\tilde{\omega}|P_\lambda)$ containing x.

c) $\tilde{C}_\lambda^+$ is the leaf of $\mathcal{F}(\tilde{\omega}|P_\lambda)$ obtained in Lemma 2.3.

(ii) A is not symmetric and if $a_{ij} \neq a_{ji}$, $\det(A_{ij}) \neq 0$. Of course we may suppose $i=1$ and $j=2$. The fact that A is not symmetric implies that $d\omega(0) \neq 0$. We are going to prove the

Fundamental Lemma - Let $\omega = \sum_1^n \omega_i \, dx_i$ be an integrable 1-form such that $d\omega(0) \neq 0$. Then, by means of a change of variables we can reduce ω (in a neighborhood of 0) to the form

$$\omega = \omega_i(0,\ldots,0,x_i,0,\ldots,0,x_j,0,\ldots,0)\, dx_i + $$
$$+ \omega_j(0,\ldots,0,x_i,0,\ldots,0,x_j,0,\ldots,0)\, dx_j.$$

Proof: We shall devide the proof into three other lemmas.

Lemma 2.5 - Let ω be as above, then there are forms η_1 and η_2 such that $d\omega = \eta_1 \wedge \eta_2$, in a neighborhood of 0.

Proof: Let $d\omega = \sum_{i<j} \alpha_{ij}\, dx_i \wedge dx_j$, since $d\omega(0) \neq 0$, there exist i and j such that $\alpha_{ij}(0) \neq 0$. We may suppose without loss of generality that $i = 1$ and $j = 2$.

We define $\Omega_1 = i(e_1)d\omega$ and $\Omega_2 = i(e_2)d\omega$ where e_k denotes the vector $(0,\ldots,0,1,0,\ldots,0)$, 1 in the k-th coordenate. We are going to show that $\Omega_1 \wedge \Omega_2 = \alpha_{12}\, d\omega$. The integrability condition gives $d\omega \wedge d\omega = 0$, so $0 = i(e_1)(d\omega \wedge d\omega) = 2(i(e_1)d\omega) \wedge d\omega \Rightarrow \Omega_1 \wedge d\omega = 0$, and $0 = i(e_2)(\Omega_1 \wedge d\omega) = (i(e_2)\Omega_1) \wedge d\omega - \Omega_1 \wedge (i(e_2)d\omega)$, but $i(e_2)\Omega_1 = \alpha_{12}$ and we have then $\alpha_{12} d\omega = \Omega_1 \wedge \Omega_2$. If we define $\eta_1 = \Omega_1$ and $\eta_2 = \frac{1}{\alpha_{12}} \Omega_2$ we get $d\omega = \eta_1 \wedge \eta_2$ as desired.

<u>Lemma 2.6</u> - Let ω be as in the previous lemma. The equation

$$\text{(I)} \qquad d\omega(x)\cdot(X,v) = 0 \qquad \forall X \in \mathbb{R}^n$$

defines in a neighborhood of 0, a distribution τ of (n-2) planes which is integrable.

<u>Proof</u>: We want to show that $\{\eta_1, \eta_2\}$ is a local representation of the distribution τ.

Let τ_x be the plane of τ at x and let $X \in \tau_x$ we have

$$0 = i(X)(d\omega(X)) = i(X)(\eta_1(x) \wedge \eta_2(x)) =$$
$$= (i(X)\eta_1(x))\eta_2(x) - \eta_1(x)(i(X)\eta_2(x))$$

but $\eta_1 \wedge \eta_2 \neq 0$ so $\eta_1(x)\cdot X = 0$ and $\eta_2(x)\cdot X = 0$. Conversselly if $\eta_1(x)\cdot X = 0$ and $\eta_2(x)\cdot X = 0$ the same equalities above shows that $i(X)(d\omega(x)) = 0$.

The system $\{\eta_1, \eta_2\}$ is integrable, for the relation $d\omega = \eta_1 \wedge \eta_2$ gives immediately $\eta_1 \wedge \eta_2 \wedge d\eta_1 = \eta_1 \wedge \eta_2 \wedge d\eta_2 = 0$. This completes the proof of the lemma.

Lemma 2.7 - Let $\mathcal{G}(\omega)$ be the codimension two foliation associated to τ. Then in a neighborhood of $0 \in \mathbb{R}^n$, we have:

(1) $\mathcal{G}(\omega)$ is transverse to the x_1x_2-plane.

(2) The leaves of $\mathcal{G}(\omega)$ are integral manifolds of ω.

(3) $\mathrm{Sing}(\omega)$ is saturated by the leaves of $\mathcal{G}_\omega$.

Proof: (1) It is obvious since $i(e_1)d\omega \neq 0$ and $i(e_2)d\omega \neq 0$.

(2) We know that if X is tangent to a leaf of $\mathcal{G}(\omega)$ at x, then $i(X)(d\omega(x)) = 0$. On the other hand we have

$0 = i(X)(\omega \wedge d\omega)(x) = (i(X)(\omega(x))d\omega(x)$ and since $d\omega \neq 0$ we must have $\omega(x).X = 0$ which completes the proof of (2).

(3) Let $0' \in \mathrm{Sing}(\omega)$ and let G' be the leaf of $\mathcal{G}(\omega)$ through $0'$, we want to show that $G' \subset \mathrm{Sing}(\omega)$. Let $p \in G'$ and let X be a vector field tangent to $\mathcal{G}(\omega)$ such that the orbit $\gamma(t)$ of X through $0'$ contains p. The Lie derivative of ω with respect to X is given by the formula $L_X(\omega) = i(X)(d\omega) + d(i(X)\omega)$. The Lemma 2.6 implies $i(X)d\omega = 0$, so, $L_X\omega = 0$, then $\omega(\gamma(t)) = 0 \quad \forall t \Rightarrow \gamma(t) \subset \mathrm{Sing}(\omega)$ which completes the proof.

Proof of the Fundamental Lemma: Let V be a neighborhood of 0, where the conclusions of the lemmas proved above, hold and let $f: V \to V$ be a difeomorphism such that

(a) $f = \mathrm{Id}$ in the x_1x_2-plane

(b) f transforms the leaves of $\mathcal{G}(\omega)$ into the subspaces $x_1 = \text{cte}$, $x_2 = \text{cte}$.

We shall see that

$$f^*\omega = \omega_1(x_1,x_2,0,\ldots,0)dx_1 + \omega_2(x_1,x_2,0,\ldots,0)dx_2 .$$

We observe that $\mathrm{grad}(f^*\omega)$ is parallel to the x_1x_2-plane since $\mathrm{grad}(f^*\omega)$ is normal to the subspaces $x_1 = \mathrm{cte}$, $x_2 = \mathrm{cte}$, then we have $f^*\omega = F_1\, dx_1 + F_2\, dx_2$. On the other hand

$$d(f^*\omega) = f^*(d\omega) = f^*(\eta_1 \wedge \eta_2) = (f^*\eta_1) \wedge (f^*\eta_2).$$

The same remark made above applies to the forms $f^*\eta_1$ and $f^*\eta_2$, so $d\omega = G\, dx_1 \wedge dx_2$. Then we have

$$G\, dx_1 \wedge dx_2 = d\omega = d(F_1 dx_1 + F_2 dx_2) = \left(\frac{\partial F_2}{\partial x_1} - \frac{\partial F_1}{\partial x_2}\right) dx_1 \wedge dx_2 +$$

$$+ \sum_{i=3}^{n} \frac{\partial F_1}{\partial x_i} dx_2 \wedge dx_i + \sum_{j=3}^{n} \frac{\partial F_2}{\partial x_j} dx_j \wedge dx_2$$

and so $\dfrac{\partial F_1}{\partial x_k} = \dfrac{\partial F_2}{\partial x_k} = 0 \quad \forall\, k \geq 3$, then F_1 and F_2 depend only on x_1 and x_2, and since $f = \mathrm{Id}$ in the x_1x_2-plane it follows that $F_i(x_1,x_2) = \omega_i(x_1,x_2,0,\ldots,0)$, $i=1,2$.

Remark: It is easy to see that the change of variables whose existence is asserted in the Fundamental Lemma may be chosen to have the particular form $f(x_1,\ldots,x_n) = (f_1(x_1,\ldots,x_n),\ f_2(x_1,\ldots,x_n),\ x_3,\ldots,x_n))$ where (f_1,f_2) is the identity in the x_1x_2-plane. Under this assumption f is unique.

Now we can show without difficulty that if ω satisfies the condition (ii) of Theorem A, then ω is locally C^1-stable at 0.

By the fundamental lemma there exists a neighborhood U of 0 such that ω and any form $\tilde{\omega}$ C^1-close to ω are respectively equivalent to the forms

$$\omega^* = \omega_1(x_1,x_2,0,\ldots,0)dx_1 + \omega_2(x_1,x_2,0,\ldots,0)dx_2$$

$$\tilde{\omega}^* = \tilde{\omega}_1(x_1,x_2,0,\ldots,0)dx_1 + \tilde{\omega}_2(x_1,x_2,0,\ldots,0)dx_2 .$$

Then $\mathcal{F}(\omega)$ and $\mathcal{F}(\tilde{\omega})$ are equivalent if and only if $\mathcal{F}(\omega^*)$ and $\mathcal{F}(\tilde{\omega}^*)$ are equivalent and these last two foliations are equivalent if and only if $\omega|x_1x_2$ and $\tilde{\omega}|x_1x_2$ are equivalent.

Since $\det A_{12} \neq 0$ and $B=\begin{pmatrix} a_{21} & a_{22} \\ -a_{11} & -a_{12} \end{pmatrix}$ is the matrix of the linear part of ${}^{\perp}(\omega|x_1x_2)$ at 0, it follows that 0 is a hyperbolic singularity of ${}^{\perp}(\omega|x_1x_2)$. Then there exists $\tilde{0} \in \mathrm{Sing}(\tilde{\omega}|x_1x_2) \cap U$ such that $\omega|x_1x_2$ and $\tilde{\omega}|x_1x_2$ are equivalent in neighborhoods of 0 and $\tilde{0}$ respectively which completes the proof of Theorem A.

<u>Remark 2.3</u>: The hypothesis $r \geq 2$ of Theorem A, was used only in the proof of part (ii), for, the forms η_1 and η_2 have the same class of differentiability of $d\omega$.

<u>Remark 2.4</u>: The construction of the topological equivalence h in the 2nd case of part (i), in the case $k=2$, fails on the definition of h_o, for, since $k=2$ the foliation $\mathcal{F}(\omega_o'|R^2)$, which is a center, is not structurally stable. However the definition of h in the set $U = \{(x_1,x_2,y_1,\ldots,y_\ell)|x_1^2+x_2^2 - (y_1^2+\ldots+y_\ell^2) \leq 0\}$ is independent of the existence of h_o. Hence $\mathcal{F}(\omega_o'|U)$ and $\mathcal{F}(\tilde{\omega}|h(U))$ are equivalent. Note that $h(U)$ is the set of all points whose leaf does not intersect the x_1x_2-plane.

If we consider instead of $\mathcal{J}_r(0)$ the set $\mathcal{J}_A(0)$ of the analytic integrable 1-forms, then the case analyzed above $(k=2)$ is also C^1-structurally stable. To see this we observe that if $\tilde{\omega}$, C^1-close to ω, is analytic then $\mathcal{F}(\tilde{\omega}|x_1x_2)$ is a center and hence we can define h_o.

Remark 2.5: Let $\omega \in \mathcal{J}_r(0)$ such that $\omega(0) = 0$. If ω does not satisfy the conditions of Theorem A, then one of the three cases below must occur:

(a) $d\omega'_o \neq 0$ and there exist i and j such that $a_{ij} \neq a_{ji}$, but, $\det A_{ij} = 0$.

(b) $\omega'_o = df$, f is non-degenerate quadratic form, but the index or the coindex of f is equal to two.

(c) $\omega'_o = df$ and f is a degenerate quadratic form.

In case (a), using the Fundamental Lemma, it is easy to see that ω is not C^{r-2}-structurally stable. The cases (b) and (c) are open problems. We believe that (b) and (c) imply in non-stability.

Remark 2.6: If $\omega \in \mathcal{J}_r(0)$ and $\omega'_o \equiv 0$, then ω is not C^1-stable at 0. To see this let φ be a C^∞-function that vanishes identically in a sufficiently small neighborhood of 0, and that is equal to one outside of some neighborhood of 0. The form $\tilde{\omega} = \varphi\omega$ is C^1-close to ω and is integrable. Since $\tilde{\omega}$ vanishes identically in a neighborhood of 0, it is not stable there.

The perturbation above is not C^2-small and actually there exist integrable 1-forms such that $\omega'_o \equiv 0$ that are C^2-stable, as in the case of the form $\omega = ayzdx + bxzdy + cxydz$ $(a \neq b \neq c \neq a)$.

This phenomena is due to the fact that the vector field $Y = \mathrm{curl}(\omega)$, varies continuously in class C^{r-1}, when ω varies in class C^r. In the example above $Y = ((c-b)x, (a-c)y, (b-a)z)$ which has the origin as a hyperbolic saddle. The leaves of $\mathcal{F}(\omega)$ coincide with the orbits of a hyperbolic linear action of $\mathbb{R}^2$. (See [1],pg.38). This example was indicated to me by Alcides Lins Neto.

Proof of Theorem A' - It is clear that all we need here is a version of the Fundamental Lemma for integrable p-forms and actually we have the following generalization.

Proposition A (Fundamental Lemma for p-forms) - Let $\omega \in \mathcal{J}^p_r(0)$, $r\geq 2$, be such that $d\omega(0) \neq 0$. Then there exists a decomposition $\mathbb{R}^n = \mathbb{R}^{p+1} \oplus \mathbb{R}^{n-p-1}$ and a C^{r-1} change of variables f, such that $f^*\omega = I^*(\omega)$ where $I: \mathbb{R}^n \to \mathbb{R}^n$ is the projection of $\mathbb{R}^n$ into $\mathbb{R}^{p+1}$.

Proof: Let ω be as above, and let $d\omega =$
$= \sum_{1\leq k_1<\ldots<k_{p+1}\leq n} \alpha_{k_1\ldots k_{p+1}} dx_{k_1}\wedge\ldots\wedge dx_{k_{p+1}}$. We suppose that $\alpha_{1,\ldots p+1}(0) \neq 0$ and let V be a neighborhood of 0 where $\alpha_{1,\ldots,p+1} \neq 0$. From now on we shall consider ω restricted to V. Before proving the theorem we are going to prove

Proposition 2.1 - Under the above hypothesis there exist $p+1$ 1-forms $\eta_1,\ldots,\eta_{p+1}$ such that $d\omega = \eta_1\wedge\ldots\wedge\eta_{p+1}$.

The proof of this proposition will be devided into three lemmas:

Lemma 2.8 - Let $\{e_1,\ldots e_n\}$ be the standard basis of $\mathbb{R}^n$. Under the above hypothesis we have that

$$(i(e_{i_\ell})\ldots i(e_{i_1})d\omega) \wedge (i(e_{j_i})\ldots i(e_{j_m})d\omega) = 0$$

since $\ell+m = p-1$.

Proof: We have to show that the above equality holds, only for those points $x \in V - \mathrm{Sing}(\omega)$, for, since $d\omega \neq 0$, the closure of $V\text{-}\mathrm{Sing}(\omega)$ in V is all of V. Now let $x \in V\text{-}\mathrm{Sing}(\omega)$ and let $U \subset V$ be a neighborhood of x such that $\omega = \pi_1 \wedge\ldots\wedge \pi_p$ in U, then we have:

(i) $d\omega \wedge \pi_j = 0 \quad j = 1,\dots,p.$

(ii) $(i(e_{i_k})\dots i(e_{i_1})d\omega) \wedge (\pi_{j_1} \wedge\dots\wedge \pi_{j_{k+1}}) = 0$

$(i(e_{i_k})\dots i(e_{i_1})d\omega) \wedge (\pi_{j_1} \wedge\dots\wedge \pi_{j_k} \wedge d\pi_{j_s}) = 0.$

Proof:

(i) It is obvious that $\omega \wedge \pi_j = 0 \quad j = 1,\dots,p,$ on the other hand the integrability condition says that $\omega \wedge d\pi_j = 0$ $j = 1,\dots,p$ hence $d\omega \wedge \pi_j = 0 \quad j = 1,\dots,p.$

(ii) We use induction on k. Since $d\omega \wedge \pi_j = 0$ it follows that $(i(e_{i_1})d\omega) \wedge \pi_j = [(-1)^p \pi_j(e_{i_1})]d\omega$ so

$(i(e_{i_1})d\omega) \wedge (\pi_{j_1} \wedge \pi_{j_2}) = 0$ and $(i(e_{i_1})d\omega) \wedge (\pi_{j_1} \wedge d\pi_{j_s}) = 0,$ which proves the assertion for $k=1$. Now we suppose that (ii) is true for k. Taking the product $i(e_{i_{k+1}})$ of both sides we get

$$(1) \quad (i(e_{i_{k+1}})\dots i(e_{i_1})d\omega) \wedge (\pi_{j_1} \wedge\dots\wedge \pi_{j_{k+1}}) =$$
$$= (-1)^{p-k-1} (i(e_{i_k})\dots i(e_{i_1})d\omega) \wedge (\sum_{\ell=1}^{k} (-1)^{\ell} \pi_{j_1} \wedge\dots\wedge \pi_{j_{\ell-1}} \wedge$$
$$\wedge (i(e_{i_{k+1}})\pi_{j_\ell}) \wedge \pi_{j_{\ell+1}} \wedge\dots\wedge \pi_{j_{k+1}}).$$

$$(2) \quad (i(e_{i_{k+1}})\dots i(e_{i_1})d\omega) \wedge (\pi_{j_i} \wedge\dots\wedge \pi_{j_k} \wedge d\pi_{j_s}) =$$
$$= (-1)^{p-k-1}(i(e_{i_k})\dots i(e_{i_1})d\omega) \wedge$$
$$\wedge (\sum_{\ell=1}^{k} (-1)^{\ell} \pi_{j_1} \wedge\dots\wedge \pi_{j_{\ell-1}} \wedge (i(e_{i_{k+1}})\pi_{j_\ell}) \wedge \pi_{j_{\ell+1}} \wedge\dots\wedge \pi_{j_k} \wedge d\pi_{j_s} +$$
$$+ \pi_{j_i} \wedge\dots\wedge \pi_{j_k} \wedge (i(e_{i_{k+1}})d\pi_{j_s})).$$

Since in the second member of (1) each summand has as a factor of $(i(e_{i_k})\dots i(e_{i_1})d\omega)$ a product of k π_j's when we multiply them by $\pi_{j_{k+2}}$ we get zero by the induction hypothesis which proves the first statement of (ii). In the second member of (2) the term which is the sumation will vanish when multiplied by $\pi_{j_{k+2}}$ by the induction hypothesis. The missing term will vanish whem multiplied by $\pi_{j_{k+2}}$ because of the first statement of (ii).

Now to finish the proof of Lema 2.8 we observe that each term in the development of $(i(e_{j_1})\dots i(e_{j_m})d\omega)$ in terms of the π_j's, has either a product of $(p-m-1) = \ell$ π_j's and one $d\pi_j$ or a product of $p-1-m+1 = \ell+1$ π_j's.

Lemma 2.9 - With the same notation of Lemma 2.8, we have:

$$(i(e_{i_{p-\ell}})\dots i(e_{i_1})d\omega) \wedge (i(e_{i_{p-\ell+1}})\dots i(e_{i_p})d\omega) =$$

$= (-1)^{\epsilon_\ell}\Delta$, where $1 \le \ell \le p$, $\Delta = (i(e_{i_p})\dots i(e_{i_1})d\omega) \wedge d\omega$ and $\epsilon_\ell = (\ell-1) + \frac{1}{2}(\ell+1)(\ell+2)$.

Proof: We use induction on ℓ. From Lemma 2.8 we have that

$$(i(e_{i_{p-1}})\dots i(e_{i_1})d\omega) \wedge d\omega = 0$$ taking the product by $i(e_{i_p})$

we get $\Delta = (-1)(i(e_{p-1})\dots i(e_{i_1})d\omega) \wedge (i(e_{i_p})d\omega)$ which proves the case $\ell = 1$. Now suppose the lemma is true for some $1 \le \ell \le p$, from Lemma 2.8 we have $(i(e_{i_{p-\ell-1}})\dots i(e_{i_1})d\omega \wedge (i(e_{i_{p-\ell+1}})\dots i(e_{i_p})d\omega) = 0$ multiplying by $i(e_{i_{p-\ell}})$ we get $(i(e_{i_{p-\ell}})\dots i(e_{i_1})d\omega)\wedge(i(e_{i_{p-\ell+1}})\dots i(e_{i_p})d\omega =$ $= (-1)^{\ell+3}(i(e_{i_{p-\ell-1}})\dots i(e_{i_1})d\omega) \wedge (i(e_{i_{p-\ell}})\dots i(e_{i_p})d\omega)$ then by the induction hypothesis we have

$$\Delta = (-1)^{\epsilon_\ell+\ell+3} (i(e_{i_{p-\ell-1}})\dots i(e_{i_1})d\omega) \wedge(i(e_{i_{p-\ell}})\dots i(e_{i_p})d\omega)$$

and it is clear that $\epsilon_\ell + \ell + 3 = \epsilon_{\ell+1}$.

<u>Lemma 2.10</u> - The notation being as in the previous lemma we have that $\Delta = 0$.

<u>Proof</u>: Lemma 2.9 implies that $\Delta = (-1)^{\epsilon_p} d\omega\wedge(i(e_{i_1})\dots i(e_{i_p})d\omega) =$
$= (-1)^{\epsilon_p+p+1} (i(e_{i_1})\dots i(e_{i_p})d\omega)\wedge d\omega =$
$= (-1)^{\epsilon_p+p+1+\frac{1}{2}p(p-1)}\Delta$ but $\epsilon_p + p + 1 + \frac{1}{2}p(p-1)$ is an odd integer hence $\Delta = -\Delta$ and then $\Delta = 0$.

Now we can finish the proof of Proposition 2.1. If $1\le i\le p+1$ we set $\omega_i = i(e_1)\dots i(e_{i-1})i(e_{i+1})\dots i(e_{p+1})d\omega$. Lemma 2.10 implies that $\omega_i \wedge d\omega = 0$ and since $\{\omega_1,\dots,\omega_{p+1}\}$ is a set of linearly independent 1-forms, for $\omega_1 \wedge\dots\wedge \omega_{p+1} (e_1,\dots,e_{p+1}) = (\alpha_{12\dots p+1})^{p+1} \neq$ $\neq 0$, it follows easyly that $(\alpha_{12\dots p+1})^p\, d\omega = \omega_1\wedge\dots\wedge \omega_{p+1}$, then we define $\eta_1 = (\alpha_{12\dots)p+1})^{-p}\, \omega_1$, $\eta_i = \omega_i$ if $2 \le i \le p+1$.

To finish the proof of the Proposition A we proceed as in the proof of the Fundamental Lemma for 1-forms. For this reason we shall only indicate the various steps omitting the details.

First we observe that the system $(\eta_1,\dots,\eta_{p+1})$ is integrable and we denote by $\mathcal{G}(\omega)$ the codimension $p+1$ foliation defined by this system. Then we have:

(a) $\mathcal{G}(\omega)$ is transverse to the linear subspace spanned by $\{e_1,\dots,e_{p+1}\}$.

(b) The leaves of $\mathcal{G}(\omega)$ are integral manifolds of ω.

(c) $\mathrm{Sing}(\omega)$ is saturated by the leaves of $\mathcal{G}(\omega)$.

Now let $U \subset V$ be a neighborhood of 0 saturated by the leaves of $\mathcal{G}(\omega)$, then we define $f\colon W \to U$ in the following way:

Let $P\colon U \to \mathbb{R}^{p+1}$ be the projection of U into $\mathbb{R}^{p+1}$ along the leaves of $\mathcal{G}(\omega)$. If $x = (x_1,\ldots,x_{p+1}, x_{p+2},\ldots,x_n) \in U$ we define $g(x) = (P(x),x_{p+2},\ldots,x_n)$ and we set $f = g^{-1}$.

Then by using the properties (a), (b) and (c) we show that $f^*(\omega) = I^*(\omega)$ where $I: \mathbb{R}^n \to \mathbb{R}^n$ is the projection of $\mathbb{R}^n$ into $\mathbb{R}^{p+1}$.

Now that we have the fundamental lemma for p-forms the proof of Theorem A' is completed in the same way we did for the proof of part (ii) of Theorem A.

§3. Global Structure of Integrable Differential Forms

In this paragraph M indicates a connected compact orientable manifold. Let $\omega \in \mathcal{J}_r(M)$ satisfy the conditions of Theorem A at each singular point, i.e. if $p \in \mathrm{Sing}(\omega)$ and if $f: B_1(0) \to U(p) \subset \subset M$ is a local chart around p, the form $f^*\omega$ in $\mathcal{J}_r(0)$ satisfies the conditions of the Theorem. It is not difficult to prove that this property does not depend on the local chart f. The set of all such forms will be denoted by $S_r(M)$.

If $\omega \in S_r(M)$ the connected components of $\mathrm{Sing}(\omega)$ are either isolated points or codimension two compact submanifolds of M. This is an immediate consequence of the Fundamental Lemma. These connected components will be called isolated singularities or codimension two singularities.

<u>Lemma 3.1</u> - Let Γ be a codimension two singularity of $\omega \in S_r(M)$ and let $p \in \Gamma$. Let G be a Riemannian metric on M and let Σ be a transverse section to Γ through p. The vector field ${}^{\perp}(\omega \mid \Sigma)_G$ has a hyperbolic singularity at p, and the type of this singularity (saddle, source, etc) depends only on the connected

component Γ.

Proof: By the Fundamental Lemma there exists a codimension two foliation $\mathcal{G}(\omega)$ defined in a neighborhood U of p such that the leaves of $\mathcal{G}(\omega)$ are integral manifolds of ω and $U \cap \Gamma$ is a leaf of $\mathcal{G}(\omega)$. Let $q \in U \cap \Gamma$ and let Σ' be a transverse section to Γ, through q. The singularity p of $^{\perp}(\omega|\Sigma)_G$ is of the same type of the singularity q of $^{\perp}(\omega|\Sigma')_G$ since these vector fields are differentiably equivalent by the projection of Σ into Σ' along the leaves of $\mathcal{G}(\omega)$. By the connectedness of Γ if follows that this type does not depend on p nor on Σ. If G' is annother Riemmanian metric, it is well known that $^{\perp}(\omega|\Sigma)_G$ and $^{\perp}(\omega|\Sigma)_{G'}$ have singularities of the same type. This completes the proof.

Definition 3.1 - Let Γ, p, G and Σ be as defined in Lemma 3.1. If p is a saddle point of $^{\perp}(\omega|\Sigma)_G$ we say that Γ is a saddle. In the other cases we say that Γ is an attractor.

Definition 3.3 - Let $\omega \in S_r(M)$ and let Γ be a codimension two singularity of ω. The index of Γ, $i(\Gamma)$ is $\chi(\Gamma)$ if Γ is an attractor and $-\chi(\Gamma)$ if Γ is a saddle where $\chi(\Gamma)$ denotes the Euler caracteristic of Γ.

Remark 3.1 - Let $\omega \in S_r(M)$ and let $W = M - \mathrm{Sing}(d\omega)$. We have on W a codimension two foliation $\mathcal{G}(\omega)$ given by Lemma 2.9, which satisfies

(1) The leaves of $\mathcal{G}(\omega)$ are integral manifolds of ω.

(2) The co-dimension two singularities of ω are leaves of $\mathcal{G}(\omega)$.

In the case $\dim M = 3$, $\mathcal{G}(\omega)$ is the foliation given by the vector field $Y = \mathrm{curl}\,\omega$, defined by $Y = \mathrm{grad}_G {}^{*}d\omega$ where $*$ is the star of Hoedge, and G is a Riemannian metric on M. The singularities of Y (which are not on W) are precisely the singularities of $d\omega$ and the codimension two singularities of ω are periodic orbits of Y.

Theorem B - Let $\omega \in S_r(M)$. Then the Euler caracteristic of M equals the sum of the indices of the singularities of ω.

Proof: Let $\Gamma_o, \ldots \Gamma_k$ be the codimension two singularities of ω.

For each $i = 0,\ldots,k$ let X_i be a Morse-Smale vector field on Γ_i. We take local charts on M around the singularities of the vector fields X_i and define a Riemannian metric on M, that coincides with the metric induced from $\mathbb{R}^n$ by the local charts in a neighborhood of the singularities of X_i. Let $X = \underset{G}{\text{grad}}\,\omega$. We take normal tubular neighborhoods $T\Gamma_i$ of Γ_i (i.e. the fibers are normal discs to Γ_i in the Riemannian metric G) and we take bump functions of class C^∞, $\varphi_i : T\Gamma_i \to \mathbb{R}$, such that $\varphi_i(p) = 1$ in a neighborhood of Γ_i, $\varphi_i(p) \geq 0$ and $\varphi_i(p) = 0$ outside some neighborhood of Γ_i. We extend the vector fields X_i to the $T\Gamma_i$ in the canonical way (by sliding the vector field normally to the fibers of $T\Gamma_i$), hence we get vector fields Y_i defined on $T\Gamma_i$, Let $\tilde{X}_i = \varphi_i Y_i$. Now we define

$$\tilde{X} = X + \tilde{X}_1 + \ldots + \tilde{X}_k \ .$$

The vector field $\tilde{X}$ satisfies:

(1) The singularities of $\tilde{X}$ are all isolated.

(2) $\tilde{X}$ is tangent to the Γ_i and $\tilde{X}|\Gamma_i = X_i$ $i = 1,\ldots,k$.

(3) The only singularities of $\tilde{X}$ which are not singularities of X, are those of the X_i's.

(4) The index of a singularity $p \in \Gamma_i$ with respect to $\tilde{X}$, is (± 1) times the index of p with respect to X_i, according to Γ_i is an attractor or a saddle respectively.

The proofs of (1) and (2) are immediate. As for (3) we observe that X and X_i are linearly dependent if and only if X or X_i is zero at that point, in fact X_i is tangent to the leaf Γ_i of the foliation $\mathcal{G}(\omega)$ while X is normal to this foliation. Of course if the tubular neighborhoods $T\Gamma_i$ are sufficiently "thin", the above conclusion holds. Hence in order to complete the proof of (3) it is sufficient to observe that X and X_i do not vanish simultaneously on $T\Gamma_i$.

To prove (4) we take the charts around the singularities of the X_i's, such that $(f^*\omega)'_0 = L_1(x_1,x_2)dx_1 + L_2(x_1,x_2)dx_2$, where f denotes a local chart around the singularity p of X_i, we observe that $\mathrm{Sing}(f^*\omega)$ is normal to the x_1x_2-plane. Let $e_1, e_2, E_1, \ldots, E_{n-2}$ be the standard basis of R^n where e_1, e_2 is a basis of the x_1x_2-plane. Since $\tilde{X}$ is tangent to Γ_i it follows that the matrix of $\tilde{X}'(p)$ has the form

$$\tilde{X}'(p) = \begin{pmatrix} \boxed{N} & 0 \\ 0 & \boxed{X_i'(p)} \end{pmatrix}$$

where N is the matrix of (L_1, L_2). Hence $\det \tilde{X}'(p) = \det N . \det X_i'(p)$, but $\det N$ is equal to the determinant of the matrix of $(L_2, -L_1)$ and $\dfrac{\det X_i'(p)}{|\det X_i'(p)|}$ is the index of p with respect to X_i. The above equality proves what we wanted.

Since the sum of the indexes of the vector field $\tilde{X}$ is the Euler caracteristic of M, the Theorem follows from properties (1), (2), (3) and (4).

<u>Proposition 3.1</u> - Let $\omega \in S_r(M)$ and let C_ω be the set of connected components of $\mathrm{Sing}(\omega)$. Then given $\varepsilon > 0$ there exists a neighborhood $N(\omega) \subset \mathcal{S}_r(M)$, such that for all $\pi \in N(\omega)$ there is a bijection $\Phi: C_\omega \to C_\pi$ such that for each $\Gamma \in C_\omega$, $\Phi(\Gamma)$ is ε-C^1-close to Γ and $\Phi(\Gamma)$ is an attractor if and only if Γ is an attractor.

<u>Proof</u>: By Theorem A we have to show only that the codimension two singularities of π, close to ω, are ε-C^1-close to the codimension two singularities of ω, and that the number of codimension two singularities of ω and π is the same.

If $\Gamma_0,\ldots,\Gamma_k$ are the codimension two singularities of ω, let $\varphi_i: T\Gamma_i \to \Gamma_i$ be tubular neighborhoods of the Γ_i $(i = 0,\ldots,K)$. If π is sufficiently close to ω the codimension two singularities of π are contained in $\bigcup_{i=0} T\Gamma_i$. Let $p \in \Gamma_i$ and let Σ be the fiber containing p. By the hyperbolicity of $^{\perp}(\omega|\Sigma)$, there exists a neighborhood $N_1(\omega)$, such that if $\pi \in N_1$, there exists exactly one singularity of π in Σ. Since p is locally structurally stable, if we consider N_1 sufficiently small, this singularity is not isolated and for all points q in neighborhood of p, there will exist exactly one singularity of π in the fiber containing q. By the compactness of Γ_i, there exists a neighborhood $N'(\omega)$ such that if $\pi \in N'$ there exists exactly one singularity of π at each fiber of $T\Gamma_i$, $\forall\, i = 0,\ldots,k$.

Hence in each $T\Gamma_i$ there exists exactly one codimension two singularity $\tilde{\Gamma}_i$ of π and $\varphi_i|\tilde{\Gamma}_i: \tilde{\Gamma}_i \to \Gamma_i$ is a diffeomorphism which is ε-C^1-close to the inclusion map of $\tilde{\pi}_i$ on M. To verify this last statement, we take a local chart $f: U(p) \to \mathbb{R}^n$ around $p \in \Gamma_i$.

Applying the Fundamental Lemma we see that $\Gamma(\omega)$ and $\Gamma(\pi)$ are obtained, respectively, as the inverse image of zero by the submersions

$P_\omega \circ f$ and $P_\pi \circ f$, where $P_\omega(x) = (\omega_1(x), \omega_2(x))$ and $P_\pi(x) =$ $= (\pi_1(x), \pi_2(x))$, it is clear that P_ω and P_π are C^1-close.

Remark 3.2: If M has dimension three and if Γ is an attractor of $\omega \in S_r(M)$, then there exists a torus T^2 (the boundary of a tubular neighborhood of Γ) that is transverse to $\mathcal{F}(\omega)$. To see this let U be a flow box of $\mathcal{G}(\omega)$ and let $p_1, p_2 \in U \cap \Gamma$. We recall that in this case $\mathcal{G}(\omega)$ is given by the vector field $Y = \text{curl } \omega$. Let Σ_1 and Σ_2 be transverse sections to Γ passing through p_1 and p_2 respectively. We had seen before that $Z_i = {}^{\perp}(\omega|\Sigma_i)$ has at p_i a hyperbolic singularity which is an attractor. In a neighborhood of p_1 in Σ_i where the Poincaré map of Y is defined, we take a transverse circle S_1 to Z_1. We may assume that Σ_1 is sent into Σ_2 by the diffeomorphis Y_1 induced at time $t = 1$, if we saturate S_1 by Y_t $0 \leq t \leq 1$, we get a cylinder C_1 which is trnasverse to $\mathcal{F}(\omega)$, for $\mathcal{F}(\omega)$ is invariant under Y_t. The boundary components of this cylinder are $S_1 \subset \Sigma_1$ and $S_2 \subset \Sigma_2$, where S_2 is transverse to Z_2 (for Y_1 is an equivalence between Z_1 and Z_2).

The construction of T^2 is completed if we construct a cylinder C_2 in U, transverse to $\mathcal{F}(\omega|U)$, with boundary components S_1 and S_2 in such a way that C_1 and C_2 glue in class C^1. The construction of C_2 is simple and we ommit it here.

Remark 3.3: Let Γ be a codimension two singularity of ω and let F be a leaf of $\mathcal{F}(\omega)$ such that some connected component of the intersection of F with a tubular neighborhood of Γ is a manifold with boundary equal to Γ. Then if γ is a closed curve on Γ which is homologous to zero on F, the holonomy of $\mathcal{G}(\omega)$ with respect to γ is trivial.

In fact let $C = S^1 \times R$ be a normal fence to F along γ. If the holonomy of γ is not trivial in any neighborhood of γ in C there exists a leaf of $\mathcal{G}(\omega) \cap C$ that is not a closed curve. Choose a point p on γ and Σ_p the vertical segment in C through p. Let t_1 be a point on Σ_p close to p and let $t_2 \neq t_1$ be the next point of intersection of the leaf α of $\mathcal{G}(\omega) \cap C$ through t_1, with Σ_p. Consider the closed curve $\bar{\alpha}$ obtained by completing the arc α from t_1 to t_2 with the segment $[t_1,t_2] \subset \Sigma_p$. Let η be the closed 1-form on F defined by $d\omega = \omega \wedge \eta$. Since γ is homologous to zero so is $\bar{\alpha}$, then $\int_{\bar{\alpha}} \eta = 0$ wich is an absurd since $\eta \equiv 0$ on α and $\eta|[t_1 t_2]$ is nowhere zero.

In dimension three if Γ is an attractor and all leaves of $\mathcal{F}(\omega)$ are simply connected then all the orbits of curl ω in a neighborhood of Γ are periodic, in this case the existence of the torus of Remark 3.2 is trivial.

Theorem C - Let $\omega \in S_r(M^3)$, where M^3 is an orientable manifold of dimension three. Suppose that

1. $Sing(\omega)$ is exactly one attractor
2. All leaves of $\mathcal{F}(\omega)$ are simply connected. Then $M^3 = S^3$ and ω is structurally stable.

Proof: Let Γ be the attractor of ω. We take a tubular neighborhood $T\Gamma$ (which is a solid torus) whose boundary T^2 is tranverse to $\mathcal{F}(\omega)$. The circle S^1 obtained by the intersection of a fiber of the tubular neighborhood with T^2 is tranverse to $\mathcal{F}(\omega)$.

Let F be a leaf of $\mathcal{F}(\omega)$, since all the leaves are simply connected and $Sing(\omega) = \Gamma$ it is easy to see that $\lim(\omega) = \Gamma$, hence $F \cap T^2$ is a finite number of closed curves, since F is simply connected it follows that $F \cap T^2$ is exactly one closed curve, and that this curve is not linked with Γ. Hence $F \cap S^1$

is exactly one point and the complement $T\Gamma^c$ of the interior of $T\Gamma$ is a fiber bundle over S^1 whose fiber is the disk D^2. Since M^3 is orientable it follows that $T\Gamma^c$ is a solid torus whose boundary is T^2.

We observe that the closed curves $F \cap T^2$, F being any leaf of $\mathcal{F}(\omega)$, are parallels with respect to $T\Gamma$ and meridians with respect to $T\Gamma^c$. Hence M^3 is obtained by pasting toghether two solid tori identifiying the boundaries by a map that sends parallels into meridians, it is well known that such a manifold is S^3.

Now we will show that ω is structurally stable.

Suppose that $T\Gamma$ is a normal tubular neighborhood of Γ with respect to some Riemannian metric of S^3, and let $T\Gamma^c$ and T^2 be as defined before. We shall denote by π the projection $\pi\colon T\Gamma \to \Gamma$, along the normal disks $D_x = \pi^{-1}(\pi(x))$.

The foliation $\mathcal{F}^c(\omega)$ induced by ω on $T\Gamma^c$ is equivalent to the trivial foliation of the solid torus, whose leaves are disks whose boundaries are the meridians of T^2. This property is still true for all foliations $\mathcal{F}^c(\tilde{\omega})$ induced in $T\Gamma^c$ by $\tilde{\omega} \in S_r(S^3)$ in a neighborhood N of ω. In fact by gluing two coppies of $T\Gamma^c$ identifying the boundaries by the identity map we obtain a foliation $\overline{\mathcal{F}}^c(\omega)$ of $S^2 \times S^1$ whose leaves are spheres S^2. By the second theorem of stability of Reeb (See [4], page 131) this foliation is C^0-stable. It is clear that $\overline{\mathcal{F}}^c(\omega)$ varies continuously in the C^0-topology when ω varies in the C^1-topology. Hence our statement is true.

Let F_o^c and $\tilde{F}_o^c$ be fixed leaves of $\mathcal{F}^c(\omega)$ and $\mathcal{F}^c(\tilde{\omega})$ respectively, and let $h_o\colon F_o^c \to \tilde{F}_o^c$ be a homemorphism.

Now we shall extend h_o to the interior of $T\Gamma^c$. First we extend the foliation of T^2 given by the closed curves $D_x \cap T^2$ boundary of the normal disk D_x $x \in T^2$, to the interior of $T\Gamma^c$, in such a way that each curve Δ_p $p \in T\Gamma^c$, of this foliation is closed and intersects each leaf of $\mathcal{F}^c(\omega)$ transversaly in exactly one point.

Now we fixe a point $x_o \in T^2$, and let $p \in T\Gamma^c$. We define $\bar{h}_o(p) = \Delta_{\hat{p}} \cap \tilde{F}^c_{\tilde{p}}$ where $\hat{p} = h_o(\Delta_p \cap F^c_o)$ and $\bar{p} = F^c_o \cap \Delta_{x_o}$. It is easy to see that $\bar{h}_o\colon T\Gamma^c \hookleftarrow$ is a homeomorphism that is a topological equivalence between $\mathcal{F}^c(\omega)$ and $\mathcal{F}^c(\tilde{\omega})$.

In order to extend $\bar{h}_o$ to the interior of $T\Gamma$ we proceed in an analogous way. First we extend h_o to the normal disk D_{x_o} whose boundary is Δ_{x_o}, in such a way that this extension is a topological equivalence between the vector fields $X_o = {}^{\perp}(\omega|D_{x_o})$ and $\tilde{X}_o = {}^{\perp}(\tilde{\omega}|D_{x_o})$.

Now we consider foliations τ of $T\Gamma - \Gamma$ and $\tilde{\tau}$ of $T\Gamma - \tilde{\Gamma}$ given respectively by tori of axis Γ and $\tilde{\Gamma}$ transverse to $\mathcal{F}(\omega)$ and $\mathcal{F}(\tilde{\omega})$ respectively. These foliations may be obtained, for example, by saturating T^2, respectively by the vector fields X and $\tilde{X}$ defined by $X(p) = {}^{\perp}(\omega|D_p)$ and $\tilde{X}(p) = {}^{\perp}(\tilde{\omega}|D_p)$. By joining to Γ (resp. to $\tilde{\Gamma}$) the intersections of τ with $\mathcal{F}(\omega)$ (resp. $\tilde{\tau}$ with $\mathcal{F}(\tilde{\omega})$) we obtain a foliation of $T\Gamma$ by closed curves ϕ_p (resp. $\tilde{\phi}_p$).

Let $q \in T\Gamma$ we define $\bar{h}_o(q) = D_{\hat{q}} \cap \tilde{\phi}_{\bar{q}}$ where $\hat{q} = h_o(D_q \cap \phi_o)$ and $\bar{q} = h_o(\phi_q \cap D_{x_o})$, where $\phi_o = F^c_o \cap T^2$. It is easy to see that $\bar{h}_o$ defined as above is a homeomorphism of S^3 that is a topological equivalence between $\mathcal{F}(\omega)$ and $\mathcal{F}(\tilde{\omega})$.

<u>Remark 3.6</u>: The same construction above shows that all forms $\tilde{\omega}$ in a neighborhood of ω are topologically equivalent to the form ω_o obtained as follows:

Let $i\colon S^n \to \mathbb{R}^{n+1}$ be the inclusion map of the sphere S^n into the euclidean space $\mathbb{R}^{n+1}$, and let $\Omega_o = -x_2 dx_1 + x_1\, dx_2$ in $\mathbb{R}^{n+1}$. We set $\omega_o = i^*\Omega_o$.

In the example below we shall describe a technic of

constructing integrable 1-forms in any compact manifold with a nice singular set.

This technic may be used also in the construction of integrable p-forms having nice singularities. Of course for big values of p we do not expect to obtain a general method of constructing examples, since this would require a good description of the singularities of maps whose range is p-dimensional.

<u>Example 3.1</u> - Let M be a compact manifold and let $\chi(M)$ denote the Euler characteristic of M, then:

(1) If $\chi(M)$ is even there exists $\omega \in S_r(M)$ such that $\mathrm{Sing}(\omega)$ has no isolated points.

(2) If $\chi(M)$ is odd there exists $\omega \in S_r(M)$ such that all singularities, but one, are of codimension two.

<u>Construction</u>: We use Theorems (1) and (2) of [3].

(1) <u>$\chi(M)$ is even</u>

Let $f: M \to \mathbb{R}^2$ be a generic map such that f has no cusps. Then the image Σ of the singular points of f into $\mathbb{R}^2$ is a finite number of closed curves and the intersecting points are double points.

Now we choose an orientation at each one of these curves (i.e. a normal C^∞ unit vector field along each curve). Of course this does not define a C^∞ vector field in the set Σ, for in the intersection points the two vectors do not agree. But if we consider around each intersection point a small neighborhood then in the complementary set of these neighborhoods in Σ we get a C^∞ vector field N. Now we define in each one of these neighborhoods a vector field that extends N as follows.

Let $V(p)$ be the neighborhood of the double point p. Then

$\Sigma \cap V - \{p\}$ has exactly four connected components each one of these components has an orientation induced by N and of course there are two consecutive components with the same orientation then we choose $V' \subset V$ such that $\overline{V}' \subset V$ and consider in V' a field of parallel lines that intersect both components. Now we orient this line-field compatibly with N and we extend it to V so that it extends N.

Now we extend this vector field to the plane and approximate it by a Morse-Smale vector field X in such a way that in a neighborhood of each singular point the 1-form $\overline{\omega}$ whose gradient is the orthogonal of X is not closed.

Then we set $\omega = f^*\overline{\omega}$. We claim that $(\omega, d\omega) \neq (0,0)$ in M.

In fact if $p \in M - f^{-1}(\Sigma)$ is such that $\omega_p = 0$ it means that $\overline{\omega}_{f(p)} = 0$ and $f(p) \notin \Sigma$ so $d\omega_p \neq 0$ since $d\overline{\omega}_{f(p)} \neq 0$. If $p \in f^{-1}(\Sigma)$ then $d\omega_p = 0$, but since X is transverse to Σ and $\omega . X = 0$ it follows that $\omega_p \neq 0$.

(2) <u>$\chi(M)$ is odd</u>

In this case we can take $f: M \to R^2$ a generic map that has exactly one cusp. Then we proceed as in the case above for the double points and in the neighborhood V' of the cusp point $p \in \mathbb{R}^2$ we introduce a point q of parabolic tangency of the vector field in V' with one of the components of the complement of p. Then we can construct the vector field as indicated in the figure bellow.

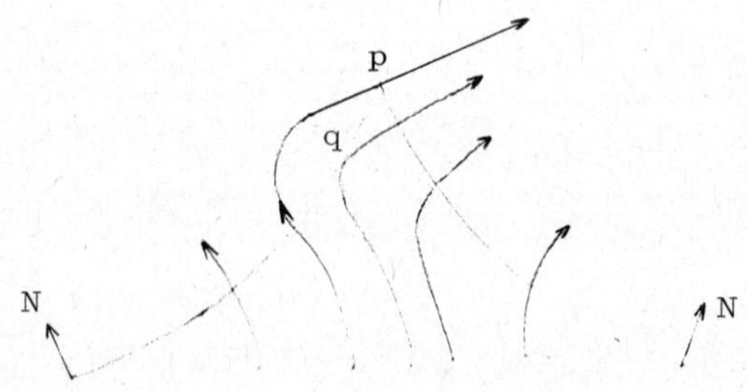

Of course $\omega q = 0$ and $d\omega q = 0$ but this is the only point with this property. The fact that ω satisfies condition (i) of Theorem A at q follows immediatly from the canonical form of f at q.

<u>Example 3.2</u> - Let M be a manifold with an Alexander decomposition $F: M \to \mathbb{C}$ (See [5]). Let Ω in $\mathbb{R}^2$ be the form $-x_2\,dx_1 + x_1\,dx_2$ and let $\omega = F^*\Omega$. The singularities of ω are all attractors. Let $P: \mathbb{C}^2 \to \mathbb{C}$ be the polynomial $P(z_o,z_1) = z_o^p - z_1^q$ and let $F: S^3 \to \mathbb{C}$ be given by $F = P|S^3$. If $p = q = n$ $F^*\Omega$ has exactly n attractors and it is easy to see that they are mutually linked. If p and q are relativelly prime, $\mathrm{Sing}(F^*\Omega)$ is exactly one attractor, that is, a closed curve with torus knot of type (p,q).

References

[1] C. Camacho - On $K^k \times Z^\ell$-actions, Dynamical Systems, Proceedings of Salvador Symposim (1971) 23-70.

[2] I. Kupka - The singularities of Integrable Structurally Stable Pfaffian Forms, Proc.Nat.Acad.Sci., USA 52 (1964).

[3] H. Levine - Elimination of cusps, Topology 3 (1964), 263-296.

[4] G. Reeb - Sur certaines propriétés topologiques des variétés feuilletées, Hermann, Paris (1952).

[5] H.E. Winkelnkemper - Manifolds as open books, Bull. A.M.S. 79 (1973).

Instituto de Matemática Pura e Aplicada
20.000 - Rio de Janeiro - Brasil

ACCESSIBILITY OF AN OPTIMUM

by

W. de Melo

In this paper we consider the problem of optimizing several functions simultaneously. We prove here that for almost all functions we can reach an optimum through a curve so that all the functions are strictly increasing along this curve. This was conjectured by Smale in [6]. In [9] Wan proved this result in the case of two functions.

Let M be a C^∞ compact manifold without boundary. We consider c functions $f_i: M \to \mathbb{R}$, $i = 1,\dots,c$, as the components of a C^∞ mapping $f: M \to \mathbb{R}^c$. A point $p \in M$ is called a local Pareto optimum of f if there is a neighborhood V of p such that no $q \in V$ satisfies: $f_i(q) \geq f_i(p)$ for all i and $f_j(q) > f_j(p)$ for some j. A differentiable curve $\alpha: \mathbb{R} \to M$ is called an admissible curve of f if $f_i \circ \alpha: \mathbb{R} \to \mathbb{R}$ is strictly increasing for all $i = 1,\dots,c$. We consider here the following question: given a point $q \in M$ is there an admissible curve $\alpha: [0,1] \to \mathbb{R}$ such that $\alpha(0) = q$ and $\alpha(1)$ is a local Pareto optimum?

It follows form [5] that, if $c \geq 2$ and the dimension of M is at least three, there exists a C^∞ mapping $f: M \to \mathbb{R}^c$ such that the above question has a negative answer.

Before stating our theorem we need some notations. Denote by θ_{op} the set of local optima of f. A point $p \in M$ is called a critical Pareto point of f if the image of the derivative of f at p, $\mathrm{Im}\, df_p$, does not intersect the positive orthant of $\mathbb{R}^c$, i.e., the the set of point of positive coordinative. We denote by θ or $\theta(f)$ the set of critical Pareto points of f. Clearly $\theta_{op} \subset \theta(f)$. In [4] it is proved that if $\dim M \geq c$ then, for generic mapping, $\theta(f)$ is a stratified set of M of dimension $c-1$. From now on we assume that $\dim M = m > \max\{c-1, 2c-4\}$.

<u>Theorem</u>. There exists a residual subset $G \subset C^\infty(M, \mathbb{R}^c)$ such that if $f \in G$ and $q \in M - \theta(f)$ then there is an admissible curve $\alpha: [0,1] \to \mathbb{R}$ such that $\alpha(0) = q$ and $\alpha(1) \in \theta_{op}$.

The idea of the proof is to construct a vector field X on M satisfying the properties: a) $X(p) = 0$ iff $p \in \theta(f)$; b) $df^i(p)\cdot X(p) > 0$ for all $p \in M - \theta(f)$ and $i = 1,\dots,c$; c) there is a dense subset of M such that if p is in this subset then the ω-limit set of the orbit of X through p is either an optimum or a "nice" point in the boundary of θ_{op}; d) if the ω-limit set of p is a "nice" point q in the boundary of θ_{op} then there exists an admissible curve from p to an optimum near q.

A vector field X satisfying properties a) and b) is called a generalized gradient of f. Of course such a vector field always exists. In order to construct a generalized gradient vector field satisfying properties c) and d) we have to impose some generic conditions on f.

Let G_0 be the set of C^∞ mapping $f: M \to \mathbb{R}^c$ such that $j^1 f: M \to J^1(M, \mathbb{R}^c)$ is transversal to the submanifolds $S_h(M, \mathbb{R}^c)$ [3].

Thus $S_h(f) = (j^1f)^{-1}(S_h(M, \mathbb{R}^c))$ is a submanifold of codimension $(m-c+h)h$ and coincides with the set of points where the rank of the derivative of f is $c-h$. It follows from Thom's transversality theorem that G_o is open and sense. Since $m > \max(c-1, 2c-4)$, the singular set of f, $S(f)$, coincides with $S_1(f)$. Hence $S(f)$ is a compact manifold of dimension $c-1$. We say that $p \in S(f)$ is a fold point if the kernel of the derivative of f is transversal to the tangent space of $S(f)$ at p. Denote by $S_{10}(f)$ the set of fold points of f. Let $G_1 \subset G_o$ be the set of mappings f such that $j^2f: M \to J^2(M, \mathbb{R}^c)$ is transversal to the submanifolds $S_{hk}(M, \mathbb{R}^c)$ [3]. Then G_1 is open and dense in $C^\infty(M, \mathbb{R}^c)$ and $S_{10}(f) =$ $= (j^2f)^{-1}(S_{10}(M, \mathbb{R}^c))$ is open and dense in $S(f)$. Furthermore $S(f) -$ $- S_{10}(f)$ is a union of submanifolds of codimension bigger or equal to one in $S(f)$. The proposition below, which is easy to prove, caracterizes a fold point in terms of the intrinsic second derivative. Denote by K the kernel of df_p. The intrinsic second derivative of f at p is a symmetric bilinear form $d^2f(p): K \times K \to \mathbb{R}^c/_{\mathrm{Imdf}_p}$ which is defined, in a coordinate system around p, as the restriction of the second derivative of f to $K \times K$.

<u>Proposition 1</u>. Let $f \in G_o$ and $p \in S(f)$. Then p is a fold point of f if and only if $d^2f(p)$ is a nondegenerate bilinear form.

We observe that since the rank of f at p is $c-1$, $\mathbb{R}^c/_{\mathrm{Imdf}_p}$ in isomorphic to $\mathbb{R}$. Furthermore, if p is a critical Pareto point of f then $\mathbb{R}^c/_{\mathrm{Imdf}_p}$ has a canonical positive ray and we can define the index of the bilinear form $d^2f(p)$.

It is clear that the critical Pareto set of f, $\theta(f)$, is a closed subset of $S(f)$. Consider the subset $\mathring{\theta}(f) \subset \theta(f)$ of points where the intersection of the image of the derivative of f with the closed positive cone of $\mathbb{R}^c$ is the origin. Clearly $\mathring{\theta}(f)$ is an open subset of $S(f)$. If a point p is in the boundary of $\mathring{\theta}(f)$ in $S(f)$ then $\mathrm{Im}\, df_p$ contains a coordinate axis of $\mathbb{R}^c$, say the i-th coordinate axis. Consider the projection $\pi_i: \mathbb{R}^c \to \mathbb{R}^{c-1}$ which drops the i-th coordinate. Then $\pi_i(\mathrm{Im}\, df_p)$ is a subspace of $\mathbb{R}^{c-1}$ of codimension 1. This means that p is a singular point of $\pi_i \circ f: M \to \mathbb{R}^{c-1}$ and in fact p is a critical Pareto point of $\pi_i \circ f$. It is clear that for an open and dense subset $G_1^1 \subset G_1$ the mappings $\pi_i \circ f$ satisfy the transversality conditions above and consequently $S(\pi_i \circ f)$ is a compact submanifold of dimension $c-2$. The boundary of $\mathring{\theta}(f)$ is containned in $\bigcup_{i=1}^{c} S(\pi_i \circ f)$ and $\theta(f) = \mathring{\theta}(f) \cup \bigcup_{i=1}^{c} \theta(\pi_i \circ f)$. Observe that $S(\pi_i \circ f) - S_{10}(\pi_i \circ f)$ is a union of submanifolds of codimension at least two in $S(f)$. To get a nice structure of $\theta(f)$ for generic mappings we have to continue the process and decompose $\theta(\pi_i \circ f)$ in a union of submanifolds. For each sequence of integers $I=(i_1,\ldots,i_s)$, with $0<i_1<i_2<\ldots<i_s\leq c$, consider the projection $\pi_I: \mathbb{R}^c \to \mathbb{R}^{c-s}$ which drops the coordinates with index in I. Let $G_2 \subset G_1$ be the set of mappings $f: M \to \mathbb{R}^c$ such that $\pi_I \circ f$ satisfies the transversality conditions above for all I. The following proposition is immediate:

<u>Proposition 2</u>. G_2 is open and dense in $C^\infty(M, \mathbb{R}^c)$. Furthermore, if $f \in G_2$ then $\theta(f) = \mathring{\theta}(f) \cup \bigcup_I \mathring{\theta}(\pi_I \circ f)$ where the union is over all sequences of integers $I = (i_1,\ldots,i_s)$. Each $\mathring{\theta}(\pi_I \circ f)$ is a submanifold of dimension $c-s-1$ where s is the number of elements in the sequence I.

We say that a critical Pareto point p is nondegenerate if it is either a fold point on a regular point for all mappings $\pi_I \circ f$. Otherwise it is called a degenerate point. Let $D(f) \subset \theta(f)$ be the set of degenerate points of f. The following proposition is easy to prove.

Proposition 3. Let $f \in G_2$ and $p \in \theta(f)$. Then the following is true: 1) p is a nondegenerate point if and only if for any sequence of integers I such that $p \in \theta(\pi_I \circ f)$, $d^2(\pi_I \circ f)$ is a nondegenerate bilinear form; 2) $D(f) \cap \mathring{\theta}(\pi_I \circ f)$ is a union of submanifolds of codimension ≥ 1 in $\mathring{\theta}(\pi_I \circ f)$.

Let $\mathring{\theta}_{op}(f) \subset \mathring{\theta}(f)$ be the set of points p such that $d^2 f_p$ is negative definite. Then $\mathring{\theta}_{op}$ is a submanifold of dimension $c-1$ and from [7] we get the following:

Proposition 4: If $p \in \mathring{\theta}_{op}(f)$ then p is a local Pareto optimum of f.

To get the set G in the statement of the theorem we need another generic condition. Let $G \subset G_2$ be the set of mappings $f: M \to \mathbb{R}^c$ satisfying the following multitransversality condition: if $p_1, \ldots, p_k \in M$ are such that $f(p_1) = f(p_2) = \ldots = f(p_k)$ then the subspaces $\mathrm{Imdf}_{p_1}, \ldots, \mathrm{Imdf}_{p_k}$ are in general position in $\mathbb{R}^c$. This implies that the codimension of $\bigcap_{i=1}^{k} \mathrm{Imdf}_{p_i}$ is equal to the sum of the codimensions of the subspaces Imdf_{p_i}. Therefore if $f \in G$ and $q \in \mathbb{R}^c$ then the number of points in $f^{-1}(q) \cap S(f)$ is at most c. It follows from the multijet Transversality Theorem that G is a residual subset of $C^\infty(M, \mathbb{R}^c)$ (see [1] p. 158).

Lemma 5. Let $f \in G$ and X be a generalized gradient vector field of f. The ω-limit set of an orbit of X is a unique point in $\theta(f)$.

Proof. Let γ be an orbit of X and $\omega(\gamma)$ denote the ω-limit set of γ. Since the components of f increase along γ it follows that $\omega(\gamma)$ cannot contain a regular point of X and therefore $\omega(\gamma) \subset \theta(f)$. Let $q \in \omega(\gamma)$. It is easy to see that $\omega(\gamma) \subset \bigcap_{i=1}^{c} (f^i)^{-1}(f^i(q))$ and since $\omega(\gamma)$ is connected and $\bigcap_{i=1}^{c} (f^i)^{-1}(f^i(q)) \cap S(f)$ is a finite set it follows that $\omega(\gamma) = q$. This proves the lemma.

Let $f \in G$ and consider the set $\mathring{\theta}_{op}(f)$ of points $p \in O(f)$ such that $d^2f(p)$ is negative definite. The boundary of $\mathring{\theta}_{op}(f)$ in $S(f)$, $\partial\mathring{\theta}_{op} = C\ell\, \mathring{\theta}_{op} - \mathring{\theta}_{op}$, is a union of submanifolds. Denote by θ^1_{op} the union of the submanifolds in $\partial\mathring{\theta}_{op}$ whose points are nondegenerate. Let X be a generalized gradient vector field of f. For $A \subset \theta(f)$ we define the stable set of A, $W^s(A)$, as the set of points in M whose ω-limit set is containned in A. From lemma 4 it follows that $W^s(A) = \bigcup_{p \in A} W^s(p)$.

<u>Lemma 6</u>. If $f \in G$ there exists a generalized gradient vector field X of f such that $W^s(\mathring{\theta}_{op} \cup \theta^1_{op})$ is dense in M.

<u>Proof</u>. Let L be the line through the origin of $\mathbb{R}^c$ containning the vector $(1,1,\ldots,1)$. Consider the orthogonal projection $\rho\colon \mathbb{R}^c \to L^{\perp}$, where $L^{\perp}$ is the orthogonal complement of L. If $p \in \theta(f)$ then $\rho(\mathrm{Im}\, df_p) = L^{\perp}$ and consequently there is a neighborhood U of $\theta(f)$ such that the restriction of $\rho \circ f$ to U is a submersion. Thus, if $p \in U$, $L_p = \{q \in U;\ (\rho\circ f)(q) = (\rho \circ f)(p)\}$ is a submanifold of U of dimension $m-c+1$ where m is the dimension of M. Consider the orthogonal projection $\tilde{\rho}\colon \mathbb{R}^c \to L \approx \mathbb{R}$. Let $\tilde{f}\colon U \to \mathbb{R}$ be defined by $\tilde{f} = \tilde{\rho}\circ f$. It is easy to see that if $p \in S(f) \cap U$ then the critical points of $\tilde{f}/L_p$ are the points of $L_p \cap S(f)$. Furthermore the index of $d(\tilde{f}/L_p)_p$ is equal to the index of d^2f_p. If $v \in (TL_p)_q$ is such that $d\tilde{f}(q)\, v > 0$ then $df^i(q)\cdot v > 0$ for all $i = 1,\ldots,c$. Let g be a riemannian metric on M. Define a vector field Y on U as follows: i) Y is tangent to the fibers of $\rho\circ f$; ii) the

restriction of Y to a fiber L_p, $p \in S(f)$, is the gradient of $\tilde{f}/L_p$ with respect to the metric g/L_p. It follows from the above remarks that $Y(p) = 0$ iff $p \in S(f) \cap U$ and $df^i(q) \cdot Y(q) > 0$ for all $i = 1,\dots,c$ whenever $Y(q) \neq 0$. However Y is not a generalized gradient vector field of f because it vanishes on $(S(f)-\theta(f)) \cap U$. We observe that if $p \in \theta(f)$ is a degenerate point then $W^s(p)$ is containned in the fiber L_p which has dimension $m-c+1$. Since $D(f)$, the set of degenerate points, is a union of submanifolds of dimension $\leq c-2$, $W^s(D(f))$ is containned in a union of submanifold of dimension $\leq m-1$. Furthermore, if $p \in \theta(f) - D(f)$ then p is a hyperbolic singularity of the restriction of Y to L_p and the index of this singularity is the same as the index of $d^2f(p)$. Thus if $d^2f(p)$ is not negative then the dimension of $W^s(p)$ is at most $m-c$. Consider the projection $\pi_I: \mathbb{R}^c \to \mathbb{R}^{c-s}$ where $I = (i_1,\dots,i_s)$ is a sequence of integers as before. Let L_I be the line $\pi_I(L)$ and let $\rho_I: \mathbb{R}^{c-s} \to L_I^{\perp}$ be the orthogonal projection where $L_I^{\perp}$ is the orthogonal complement of L_I. If $p \in \theta(\pi_I \circ f)$ then $\rho_I\ \pi_I(\mathrm{Im}df_p) = L_p^{\perp}$ and therefore there is a neighborhood U_I of $\theta(\pi_I \circ f)$ such that $\rho_I \circ \pi_I \circ f/U_I: U_I \to L_I^{\perp}$ is a submersion. Thus for any $p \in S(\pi_I \circ f) \cap U_I$ the fiber $L_{I,p} = \{q \in U_I;\ \rho_I \pi_I f(q) = \rho_I\ \pi_I\ f(p)\}$ is a submanifold of dimension $m-c+s$. It is clear that if J is a subsequence of I then $L_{J,p}$ is containned in $L_{I,p}$. Let $p \in \theta(\pi_i \circ f)$. For each $q \in L_{i,p}$ consider the cone $C_i(q) = \{v \in T(L_{i,p})_q;\ df^j(q) \cdot v > 0$ for $j = 1,\dots,c\}$. This defines a cone field on U^i. Let $V^i \subset S(f) \cap U^i$ be a compact neighborhood of $\theta(\pi_i \circ f)$. Let $W^i \subset V^i$ be the set of points where the cone C_i is empty. Clearly W^i is compact. Let Y^i be a vector field on V^i such that $Y^i(q) \in C_i(q)$ for all q, $Y^i(q) \neq 0$ if $q \in V^i - W^i$ and $Y^i(q) = 0$ whenever $q \in L_{i,p}$ for some $p \in W^i$. Let V be a neighborhood of $\theta(f)$ such that $V \subset U$ and $V \cap S(f) \subset \bigcup_{i=1}^{c} V_i$. We consider in V the vector field

$Z = Y + Y^1 + \ldots + Y^c$. It follows from the construction above that $Z(p) = 0$ if and only if $p \in \theta(f)$. Furthermore $df^j(p)\cdot Z(q) > 0$ for all $j = 1,\ldots,c$ and $q \in V - \theta(f)$. Now we extend Z to a vector field X on M so that X is a generalized gradient of X. Let $p \in \mathring{\theta}(\pi_I \circ f)$ where $I = (i_1,\ldots,i_s)$. Then X is tangent to the fiber $L_{I,p}$ whose dimension is $m-c+s$. Since $D(f) \cap S(\pi_I \circ f)$ is containned in a finite union of submanifolds of dimension at most $c-s-1$, it follows that $W^s(D(f) \cap \mathring{\theta}(\pi_I \circ f))$ is containned in a finite union of immersed submanifolds of dimension at most $m-1$. Therefore $M - W^s(D(f))$ is residual in M. Let $p \in \theta(f) - D(f)$ be such that $d^2f(p)$ is not negative definite. Since $p \notin D(f)$, p is a partially hyperbolic singularity whose center manifold is tangent to $S(f)$ at p and has dimension $c-1$. By [2], there exists a neighborhood V of p such that the stable set of $V \cap \theta(f)$ is containned in a submanifold of dimension $c-1+i$ where i is the index of $d^2f(p)$. Thus $M - W^s(V \cap \theta(f))$ is residual in M. This finishes the proof.

Let $p \in \mathring{\theta}(\pi_I \circ f) - D(f)$ where $I = (i_1,\ldots,i_s)$. Let $e_1,\ldots,e_c$ be the canonical basis of $\mathbb{R}^c$. Consider the following basis of $\mathbb{R}^c$: $E_1 = e_{i_1}$, $E_2 = e_{i_2},\ldots$, $E_s = e_{i_s}$, $E_{s+1} = e_{i_{s+1}},\ldots$, $E_{c-1} = e_{i_{c-1}}$ and $E_c = e_1 + \ldots + e_c$ where $i_{s+1},\ldots,i_c$ are such that $(i_1,\ldots,i_c)$ is a permutation of $(1,\ldots,c)$.

<u>Lemma 7</u>. Let $p \in \mathring{\theta}(\pi_I \circ f) - D(f)$ where $I = (i_1,\ldots,i_s)$. Suppose $d^2f(p)$ is negative definite on kern $df(p)$. Then there exists a coordinate system around p, vanishing at p, such that the expression of f in this coordinate system is given by:

$$f(u,x) = u_1E_1 + \ldots + u_{c-1}E_{c-1} + \left(\lambda(u) - \sum_{i=1}^{m-c+1} x_i^2\right) E_c + f(p)$$

where $\frac{\partial \lambda}{\partial u_i}(0,0) = 0$ for $i=1,\ldots,s$ and $\frac{\partial \lambda}{\partial u_j}(0,0) < 0$ for $j=s+1,\ldots,c-1$.

<u>Proof</u>. Since Imdf_p is transversal to E_c it follows from the implicit function theorem that there exists a coordinate system around p, vanishing at p, such that the expression of f in this coordinate system is given by

$$f(u,x) = u_1 E_1 + \ldots + u_{c-1} E_{c-1} + g(u,x) E_c + f(p)$$

The function $\beta(x) = g(0,x)$ is such that $d\beta(0) = 0$ and $d^2\beta(0)$ is negative definite. Thus β is right equivalent to $-\sum_{i=1}^{m-c+1} x_i^2$. Since p is a fold point it follows from [8] that g is the universal unfolding of β and, therefore, it is right equivalent, as an unfolding, to $\tilde{g}(u,x) = -\sum_{i=1}^{m-c+1} x_i^2$. Hence there is a coordinate system with the above properties.

<u>Lemma 8</u>. Let $p \in \overset{\circ}{\theta}(\pi_I \circ f) - D(f)$ be as in lemma 7. Let $\alpha: \mathbb{R} \to M$ be an integral curve of the generalized gradient vector field X of lemma 6 with $\lim_{t\to\infty} \alpha(t) = p$. If $\alpha(t)=(u(t),x(t))$ and $J \subset I$ is the smallest subsequence such that $\alpha(t) \subset L_{J,p}$ for all t sufficiently large, then there exists t_1 such that for $t \geq t_1$.

1) $u_j(t) \leq 0$ for $j = 1,\ldots,c-1$;

2) $\dfrac{\partial\lambda}{\partial u_j}(u(t)) > 0$ for all j with $i_j \in J$.

Proof. In the coordinates (u,x) of lemma 7 we have

$$S(f) = \{(u,0),\ u \in \mathbb{R}^{c-1}\}$$

$$\theta(f) = \{(u,0);\ \frac{\partial\lambda}{\partial u_i}(u) \leq 0,\quad i = 1,\ldots,c-1\}$$

$$L_{(u,0)} = \{(u,x);\quad x \in \mathbb{R}^{m-c+1}\}$$

$$L_{i_j,p} = \{(u,x);\ u_k = 0 \quad \text{if} \quad k \neq j\} \quad j = 1,\ldots,s$$

$$L_{I,p} = \{(u,x);\ u_{s+1} = \dots = u_{c-1} = 0\}$$

From the proof of lemma 6 we may write $X = Y + Y^1 + \dots + Y^c$ where Y is tangent to the fibers L_q and Y^i is tangent to the fibers $L_{i,q}$ for $i = 1,\dots,c$. Thus $du_j \cdot Y = 0$ and $Y^k = 0$ in a neighborhood of p whenever $k \notin I$. Since Y^{i_k} is tangent to the fibers $L_{i_k,q}$ it follows that $du_j \cdot Y^{i_k} = 0$ if $k \neq j$ and $du_j \cdot Y^{i_j} \geq 0$. Therefore u_j is increasing along, α. Hence $u_j(t) \leq 0$ for all $t \in \mathbb{R}$. By the construction of the vector field Y^{i_j} in lemma 6 it follows that $Y^{i_j}(u,x) = 0$ whenever $\frac{\partial\lambda}{\partial u_j}(u,x) \leq 0$. Therefore, for t sufficiently large, $\frac{\partial\lambda}{\partial u_j}(u(t),x(t)) > 0$ if $i_j \in J$ because otherwise $\alpha(t)$ whould be containned in $L_{\tilde{J},p}$ for $\tilde{J} = J - \{i_j\}$. This proves the lemma.

<u>Proof of the theorem</u>. Let $f \in G$ and $q \in M - \theta(f)$. Consider the set A of points $p \in M$ such that there exists an admissible curve from q to p. It is easy to see that A contains an open set. Therefore, by lemma 6, $A \cap W^s(\mathring{\theta}_{op} \cup \theta^1_{op})$ is not empty. Since the trajectories of the generalized gradient vector field X are admissible curves of f, it follows that there exists an admissible curve from q to a point p which belongs to either $\mathring{\theta}_{op}$ or θ^1_{op}. If $p \in \mathring{\theta}_{op}$ then, by proposition 4, p is an optimum and we are done. Suppose now that $p \in \theta^1_{op}$. Let $I = (i_1,\dots,i_s)$ be such that $p \in \mathring{\theta}(\pi_I \circ f) - D(f)$ and consider the coordinate system (u,x) of lemma 7. Let $J \subset I$ be the smallest subsequence of I such that the intersection of the admissible curve from q to p and a small neighborhood of p is containned in $L_{J,p}$. By lemma 8 there exists a point $p_o = (u^o,x^o) \in L_{J,p}$, near p, with the following properties: a) p_o belongs to the admissible curve from q to p; b) $u^o_i \leq 0$ for all $i = 1,\dots,c$; c) $\frac{\partial\lambda}{\partial u_j}(u^o) > 0$ for all j such

that $i_j \in J$. We are going to prove the existence of an in admissible curve from p_o to a point $p_1 \in \mathring{\theta}_{op}$ near p. Since p_o is near p and $p \notin D(f)$, there exists a differentiable curve $\alpha: \mathbb{R} \to \mathbb{R}^{c-1}$ with the following properties: i) $\alpha(0) = u^o$, $\alpha(1) = 0$; ii) $\alpha'(t) \neq 0$ $\forall\ t \in \mathbb{R}$; iii) $\alpha_j(t) = 0$ $\forall\ t \in \mathbb{R}$ if $i_j \notin J$; iv) $\alpha_j'(t) > 0$ for all $t \in [0,1]$ if $i_j \in J$. Then $\lambda(\alpha(t))$ is increasing in the interval $[0,1]$. Therefore there exists an increasing differentiable curve $\beta: \mathbb{R} \to \mathbb{R}$ with $\beta(0) = \lambda - \sum_{j=1}^{m-c+1} (x_j^o)^2$, $\beta(1) = 0$, $\beta'(t) > 0$ for $t \in [0,1]$ and $\beta(t) < \lambda(\alpha(t))$ for $t \in [0,1)$. Let $\gamma: \mathbb{R} \to \mathbb{R}^c$ be the curve defined by $\gamma(t)=\alpha_1(t)E_1+\ldots+\alpha_{c-1}(t)E_{c-1} + \beta(t)E_c + f(p)$. Then $\gamma(0) = f(p^o)$, $\gamma(1) = f(p)$ and $\gamma'(t)$ belongs to the positive cone of $\mathbb{R}^c$ for all $t \in \mathbb{R}$. Since p is in the closure of $\mathring{\theta}_{op}$ and the image of the singular set of f is $\{u_1E_1 +\ldots+ u_{c-1} E_{c-1} + \lambda(u) E_c + f(p);$ $u \in \mathbb{R}^{c-1}\}$, it follows that, by a small perturbation of γ, we can get a curve $\tilde{\gamma}: \mathbb{R} \to \mathbb{R}^c$, $\tilde{\gamma}(t) = \tilde{\alpha}_1(t)E_1+\ldots+ \tilde{\alpha}_{c-1}(t)E_{c-1} + \tilde{\beta}(t)E_c+f(p)$, with the following properties: a) $\tilde{\gamma}(0) = p^o$, $\tilde{\gamma}(1) \in f(\mathring{\theta}_{op})$; b) $\tilde{\gamma}'(t)$ belongs to the positive cone of $\mathbb{R}^c$ for all t; c) $\tilde{\beta}(t) < \lambda(\tilde{\alpha}(t))$ if $t < 1$. Now we consider a curve $\delta: [0,1] \to M$ defined by $\delta(t) =$ $= (\tilde{\alpha}(t), x(t)$ where $x(t)$ is such that $\sum_{i=1}^{m-c+1} (x_i(t))^2 - \lambda(\tilde{\alpha}(t)) -$ $- \tilde{\beta}(t)$. Clearly δ is an admissible curve with $\delta(0) = p^o$ and $\delta(1) \in \mathring{\theta}_{op}$. This proves the theorem.

REFERENCES

[1] M. Golubitsky and V. Guillemin, Stable Mappings and their Singularities, Graduate Texts in Mathematics 14, Springer Verlag, 1973.

[2] A. Kelley, The stable, center-stable, center, center unstable and unstable manifolds. Appendix C in "Transversal mappings and flows" by R Abraham and J. Robbin, Benjamin, New York, 1967.

[3] H.J. Levine, Singularities of Differentiable Mappings, Proc. of Liverpool Sing.Symp. I, Springer Lecture Notes, 192, 1971.

[4] W. de Melo, On the Structure of the Pareto Set, to appear in Atas da Soc.Bras. de Matemática.

[5] C.P. Simon and C. Titus, Characterization of Optima in Smooth Pareto Economic Systems, J. of Math. Economics 2 (1975).

[6] S.Smale, Global Analysis and Economics I, Pareto optimum and generalization of Morse Theory, Proc. Symp. Dyn. Systems at Salvador, Brazil. Ac. Press, New York (1973).

[7] S. Smale, Sufficient conditions for an optimum, Proc. Symp. Dyn. Systems at Warwick, Springer Lecture Notes, 468, Springer Verlag (1975).

[8] D.J.A. Trotman and E.C.Zeeman, Classification of elementary catastrophes of codimension ≤ 5, Warwick Lecture Notes, 1974.

[9] Y.H. Wan, Morse Theory for two functions, Topology 14, (1975).

Instituto de Matemática Pura e Aplicada
Rio de Janeiro, RJ - Brasil

THE SCATTERING PROBLEM FOR SOME PARTICLE SYSTEMS ON THE LINE

Jürgen Moser

Courant Institute of Mathematical Sciences, New York University

1. Introduction

We discuss the motion of n particles on the line under the influence of a potential $W(x)$, so that the differential equations take the form

$$\frac{d^2x_k}{dt^2} = -\frac{\partial W}{\partial x_k}, \qquad k = 1,2,\ldots,n,$$

where

$$W(x) = \sum_{1\leq k<\ell\leq n} U(x_k - x_\ell) .$$

We assume that $dU(x)/dx < 0$, so that the particles repel each other. Under an additional smallness condition on $U'(x)$ at infinity it is easy to show that asymptotically the particles behave like free ones, i.e.

$$\left.\begin{aligned} x_k(t) - \xi_k^+ t - \eta_k^+ &\to 0 \\ \dot{x}_k(t) - \xi_k^+ \qquad &\to 0 \end{aligned}\right\} \quad \text{for} \quad t \to +\infty, \quad k = 1,2,\ldots,n.$$

Here ξ_k^+, η_k^+ are called the asymptotic velocities and phases, respectively. Similarly one introduces ξ_k^-, η_k^- by

$$\left.\begin{aligned} x_k(t) - \xi_k^- t - \eta_k^- &\to 0 \\ \dot{x}_k(t) - \xi_k^- t \qquad &\to 0 \end{aligned}\right\} \quad \text{for} \quad t \to -\infty, \quad k = 1,2,\ldots,n.$$

The classical scattering problem consists in studying the scattering mapping

$$\sigma\colon (\xi^-,\eta^-) \to (\xi^+,\eta^+)$$

which relates the scattering data between the past and the future.

* This work was supported in part by the Office of Naval Research, Contract No. N00014-76-C-0301. Reproduction in whole or in part is permitted for any purpose of the U. S. Government.

Under rather general smoothness and smallness conditions on U one can show that this mapping σ is a diffeomorphism in an open domain which is canonical, i.e., satisfies

(a) $$\sum_{k=1}^{n} d\xi_k^- \wedge d\eta_k^- = \sum_{k=1}^{n} d\xi_k^+ \wedge d\eta_k^+$$

In general it is not possible, however, to give a precise description of this mapping, or even of its range or domain.

But recently some special potentials have been found for which the scattering mapping σ can be determined explicitly. To these potentials belong

$$U(x) = \frac{\lambda^2}{\sin h^2 \lambda x} \quad \text{for} \quad \lambda > 0$$

and its limit for $\lambda \to 0$

$$U(x) = \frac{1}{x^2}$$

In [6] it was proven that for these systems the asymptotic velocities are preserved up to their order. More precisely: If the particles are ordered according to $x_1 < x_2 < \ldots < x_n$ then $\xi_1^+ < \xi_2^+ < \ldots < \xi_n^+$ and

(b) $$\xi_{n-k+1}^+ = \xi_k^- \ .$$

The proof is based on the construction of n conserved quantities, of integrals, of the motion. This fact, combined with (a) shows that

$$d\left\{\sum_{k=1}^{n} (\eta_k^- - \eta_{n-k+1}^+)\, d\xi_k^-\right\} = 0 \ .$$

Hence, since the domain $\xi_1^- > \xi_2^- > \ldots > \xi_n^-$ of σ is simply connected one finds

$$\eta_{n-k+1}^+ - \eta_k^- = \frac{\partial S}{\partial \xi_k^-} = \delta_k$$

where S is a function of $\xi^- = (\xi_1^-, \ldots, \xi_n^-)$ only. The phase-shifts

$\delta_k = \partial S/\partial \xi_k^-$ are then independent of the phases $\eta_1^-, \ldots, \eta_n^-$. By an argument described in [7] for the determination of the phase shift of the Toda lattice, the function S can be shown to be of the form

$$S(\xi^-) = \sum_{k<\ell} s(\xi_k^- - \xi_\ell^-)$$

and therefore the phase shifts have the form

$$\text{(c)} \qquad \delta_k = - \sum_{j<k} s'(\xi_j^- - \xi_k^-) + \sum_{j>k} s'(\xi_j^- - \xi_k^-) \ .$$

This means, that the phase shifts are entirely determined by the pairwise interaction. Thus for these special systems the scattering map is reduced to the determination of $s(\xi)$, or equivalently to the computation of the phase shift of a two particle system which can be done by explicit integration.

Heuristically, (c) can be understood as follows: Since the phase shifts δ_k are independent of the phases, one can compute them by taking, say η_n, very large, so that the nth particle interacts with the first n-1 particles only after they have undergone their mutual interaction and already move like free particles. Now the nth particle interacts pairwise with each of the particles, experiencing a phase shift given by the two particle system interaction. These simple ideas allow the derivation of (c) by induction with respect to n. We will not carry out the details, since these results can also be found in [5].

It is most remarkable that for the potential $U(x) = x^{-2}$ the phase shifts δ_k vanish. This was mentioned in a footnote in p. 207 of [6] without proof. The above formula (c) shows that it suffices to verify this fact for n = 2, which we will do below. But this fact can also be deduced from some fairly explicitly representation of the solution given in [2]:

If $L(\xi,\eta) = L_{k\ell}(\xi,\eta)$ is the Hermitian matrix of the form

$$L_{kk} = \eta_k \ ; \quad L_{k\ell} = \frac{i}{\xi_k - \xi_\ell} \quad \text{for} \quad k \neq \ell$$

where $\xi_1 < \xi_2 < \ldots < \xi_n$, η_k real then the symmetric functions of the solutions $x_k = x_k(t)$ are given by

$$\sum_{k=1}^{n} x_k(1)^p = \operatorname{tr}(L(\xi,\eta + t\xi))^p \quad \text{for} \quad p = 1,2,\ldots,n \ .$$

In other words, the eigenvalues of $L(\xi,\eta + t\xi)$ agree, in appropriate order, with the solutions $x_k(t)$. We observe that for $U = x^{-2}$ the order of the particles is independent of t since the constant energy has the form

$$\frac{1}{2} \sum_{k=1}^{n} \dot{x}_k^2 + \sum_{k<\ell} (x_k - x_\ell)^{-2} \ ,$$

so that the distances $|x_k - x_\ell|$ are bounded away from zero.

For $\xi_1 < \xi_2 < \ldots < \xi_n$ the ordered eigenvalues of $L(\xi,\eta + t\xi)$ behave asymptotically for $t \to +\infty$ like

$$x_k(t) = t\xi_k + \eta_k + O(t^{-1})$$

and for $t \to -\infty$ like

$$x_k(t) = t\xi_{n-k+1} + \eta_{n-k+1} + O(t^{-1}) \ .$$

This proves

$$\xi_k^+ = \xi_k = \xi_{n-k+1}^- \ , \quad \eta_k^+ = \eta_k = \eta_{n-k+1}^-$$

i.e. $\delta_k = 0$.

Thus the scattering of this system is the same as for elastically reflecting mass points. In fact, the system

$$\frac{d^2 x_k}{dt^2} = -\varepsilon^2 \frac{\partial U}{\partial x_k} \ ; \quad U = \sum_{j<\ell} (x_j - x_\ell)^{-2}$$

has solutions, which for $\varepsilon \to 0$ tend to those of the elastically reflecting mass points.

2. An Inverse Problem

The above example raises the question for the most general potential yielding the phase shifts $\delta_k = 0$. More generally one may ask for the inverse problem: To determine the potential from the phase shifts δ_k. We will address ourselves to this question only for the case n = 2 and derive a definite result. Before formulating the result we observe that on account of the linear motion of the center of mass we have

$$\sum_{k=1}^{n} \xi_k^+ = \sum_{k=1}^{n} \xi_k^- , \qquad \sum_{k=1}^{n} \eta_k^+ = \sum_{k=1}^{n} \eta_k^-$$

and consequently

$$\sum_{k=1}^{n} \delta_k = 0 .$$

Hence, for n = 2, it suffices to know only <u>one</u> phase shift, say $\delta = \delta_2$, or $\delta = -\delta_1$, or $\delta = (\delta_2 - \delta_1)/2$, as a function of $\xi_2^- - \xi_1^-$. This function can be determined from the scalar equation for $x = x_1 - x_2$.

$$\text{(d)} \qquad \ddot{x} = \ddot{x}_1 - \ddot{x}_2 = \left(-\frac{\partial}{\partial x_1} + \frac{\partial}{\partial x_2}\right)(U(x_1 - x_2)) = -2\frac{\partial}{\partial x} U(x) .$$

For an even solution, i.e. $x(-t) = x(t)$, this phase shift is given by the asymptotic behavior

$$\text{(e)} \qquad x(t) \sim t x(+\infty) + \delta \quad \text{for} \quad t \to +\infty .$$

Indeed, for such an even solution one has

$$\eta_1^+ - \eta_2^+ = \eta_1^- - \eta_2^-$$

and therefore the relations

$$\delta_1 = \eta_2^+ - \eta_1^- , \qquad \delta_2 = \eta_1^+ - \eta_2^-$$

imply

$$\delta = \frac{1}{2}(\delta_2 - \delta_1) = \frac{1}{2}(\eta_1^+ - \eta_2^- - \eta_2^+ + \eta_1^-) = \eta_1^+ - \eta_2^+ .$$

Here we consider δ as a function of $x(+\infty) = \xi_1^+ - \xi_2^+ = \xi_2^- - \xi_1^-$.

Thus (e) contains the definition of δ for the system (d) with the potential U. In Section 3 formula (7), we will show that δ can be expressed by an integral operator acting on U, under appropriate assumptions on $U(x)$. We will require that $U(x)$ is defined in $A < x < \infty$, where A is possibly equal to $-\infty$; U is in $C'(A,\infty)$, $U' < 0$ and

$$\lim_{x \to A} U(x) = \infty , \qquad U(\infty) = 0 , \qquad \int_{A+1}^{\infty} U \, dx < \infty .$$

Under these hypotheses the phase shift $\delta = \delta(y)$ exists and depends continuously on $y = x(+\infty) > 0$.

To study the inverse problem we impose further conditions on $\delta = \delta(y)$: If $\delta \in C'(0,y1]$, $\delta' \leq 0$ and

$$\int_0^{y_1} y^\gamma \, |\delta'(y)| \, dy < \infty \quad \text{for some} \quad \gamma < \frac{1}{2}$$

then there exists a unique monotone decreasing potential $U(x)$ satisfying

$$\int^{\infty} U^\gamma \, dx < \infty$$

with the same constant γ giving rise to the phase shift δ. The inverse function $x = V(y)$ of $y = U(x)$ is given by an explicit formula, see (8).

However, if we admit the wider class of potentials satisfying only

$$\int^{\infty} U^\theta(x) \, dx < \infty \quad \text{for some} \quad \theta < 1 .$$

Then $U(x)$ is not any more uniquely determined by $\delta = \delta(y)$. In particular, we will show that the most general potential in this class with $\delta(y) \equiv 0$ is given by $U(x) = a/x^2$.

The precise results are formulated as a theorem in Section 5. The uniqueness proof is based on the Wiener-Hopf technique applied to an integral equation of Abel's type. In fact, our problem is closely related to a problem treated by N. Abel [1]: Consider a particle sliding up a hill under frictionless motion from an initial position P. The problem consists in determining the shape of the hill from the knowledge of the time $T = T(E)$ it takes the particle to return to P, if the initial energy E is allowed to vary. For a recent exposition and generalization of this problem see [4]. The differential geometrical problem discussed by H. Gluck in this conference is of similar character and related to an integral equation of Abel's type. Finally we mention the work of M. Adler [3] where the scatterings for other systems is discussed.

The author wishes to acknowledge interesting discussions on this topic with G. Galavotti, who suggested that the inverse square potential is uniquely characterized by zero phase shift.

3. Formulation of the Result

We consider a system of two mass points of equal mass on the line, governed by

$$\ddot{x}_k = -U_{x_k}, \qquad k = 1,2, \tag{1}$$

where $U = U(x_1 - x_2)$. Obviously, this problem reduces at once to a one particle system if one observes that the center of mass depends linearly on t and $x = x_1 - x_2$ satisfies

$$\ddot{x} = -2U_x(x) .$$

In the following we will require that U is continuously differentiable. Then the differential equations are to be understood as

$$\dot{x}^2 + 4U(x) = \text{const.} \tag{2}$$

We assume that we are dealing with repelling forces i.e. $U' < 0$ in $A < x < +\infty$ where A is possibly equal to $-\infty$. Furthermore we require

$$\lim_{x \to A} U(x) = +\infty \;, \quad U(+\infty) = 0 \;; \quad \int_{A+1}^{\infty} U\,dx < \infty \;. \tag{3}$$

Under these assumptions we can assert that all solutions of (2) exist for all real t and tend to infinity as $t \to \pm\infty$. Indeed, since $U(A) = \infty$ it follows from (2) that

$$4U(x(t)) \leq \text{const.}$$

is bounded from above and, since $U(A) = \infty$, that $x(t)$ is bounded away from A for all t for which the solution exists. Moreover, $x(t)$ is a convex function of t, since $\ddot{x} = -2U' > 0$. Since $\dot{x}^2 \leq$ const. and $x - A > p > 0$ for all t, for which the solution exists we conclude that $x(t)$ exists for all real t. For the behavior at infinity there are two possibilities: Either $x(t)$ tends to $+\infty$ for $t \to +\infty$ or $x(t)$ remains bounded for $t \geq 0$. In the latter case $x(t)$ is monotone decreasing and

$$A + p \leq x(t) \leq x(0) \quad \text{for} \quad t \geq 0$$

with some $p > 0$. Since $-U'(x)$ has a lower positive bound, say a, in this closed interval we conclude

$$\ddot{x} \geq 2a > 0 \quad \text{for} \quad t \geq 0 \;.$$

This implies, of course, that $x(t) \to +\infty$ and we have a contradiction. Thus $x(t) \to \infty$ for $t \to +\infty$ and also for $t \to -\infty$.

Any solution $x(t)$, being convex and tending to infinity as $t \to \pm\infty$, has a unique minimum, say $x(t_0)$. Replacing t by $t-t_0$ we can assume that this minimum is taken at $t = 0$, i.e.

$$\dot{x}(0) = 0$$

Then it is clear that $x(-t) = x(t)$ is symmetric in t. Next we study the asymptotic behavior of this solution for $t \to \pm\infty$. From (2) and the assumption $U(x) \to 0$ as $x \to \infty$ we conclude that $\dot{x}^2(t)$ has a limit, which we denote $\dot{x}^2(\infty)$, i.e.

$$\dot{x}^2 + 4U(x) = \dot{x}^2(\infty) \quad , \quad \text{where} \quad \dot{x}(\infty) > 0 \ .$$

Next we will show that

$$(4) \qquad x(t) - (\dot{x}(\infty)t + \delta) \to 0 \quad \text{for} \quad t \to \infty$$

where δ depends on $\dot{x}(\infty)$, in general. This is a consequence of the last assumption (3). The constant is called the phase shift of the solution. Since our solution is symmetric in t it is clear that

$$(5) \quad x(t) - (\dot{x}(-\infty)t + \delta) \to 0 \quad \text{also for} \quad t \to -\infty, \quad \dot{x}(-\infty) = -\dot{x}(\infty) \ ,$$

so that we do not have to distinguish between two phase shifts.

It is easy to prove (4) and at the same time derive a formula for δ, since one can solve (2) explicitly. For this purpose we introduce $x = V(y)$ in $0 < y < \infty$ as the inverse function of $y = 4U(x)$. Thus $V(y)$ is monotone decreasing and, by (3), tends to ∞ as $y \to 0$. Since $x(t)$ is monotone increasing for $t > 0$ we can consider t as a function of x, and also t as a function of y for $y > 0$. Then

$$\frac{dV}{dy} = \dot{x}\,\frac{dt}{dy}$$

and from (2) we obtain for $0 < y < y_0$

$$\frac{dt}{dy} = (y_0 - y)^{-1/2}\, V'(y) \text{ where } y_0 = 4U(x(0))$$

or

$$t(y) = -\int_y^{y_0} (y_0-s)^{-1/2}\, V'(s)\, ds.$$

From (2) we see that $\dot{x}^2(\infty) = y_0$; therefore the desired asymptotic formula (4) is equivalent to

$$t(y) - y_0^{-1/2}\,(V(y) - \delta) \to 0 \quad \text{as} \quad y \to 0.$$

We find

$$(6) \quad \sqrt{y_0}\, t(y) - V(y) = -V(y_0) - \int_y^{y_0} \left\{(1 - \frac{s}{y_0})^{-1/2} - 1\right\} V'(s)\, ds$$

and the asymptotic formula is established if we show that the integral on the right converges for $y \to 0$. Since

$$0 < (1 - \frac{s}{y_0})^{-1/2} - 1 \leq c\, \frac{s}{y_0} \quad \text{for} \quad 0 < \frac{s}{y_0} < \frac{1}{2}$$

it suffices to check that

$$\int_y^{y_0} s\, V'(s)\, ds = \int_y^{y_0} s\, dV$$

converge for $y \to 0$. This is an immediate consequence of the last assumption of (3). Thus $y \to 0$ yields in (6)

$$(7) \qquad \delta = V(y_0) + \int_0^{y_0} \left\{(1 - \frac{s}{y_0})^{-1/2} - 1\right\} dV(s)$$

where we consider δ as a function of $y_0 = \dot{x}^2(\infty)$. Clearly $\delta(y_0)$ is a continuous function of y_0 for $y_0 > 0$.

We consider some examples: (i) For the potential $U(x) = e^{-x}$ for which $A = -\infty$, one finds $V(y) = -\log y/4$ and

$$\begin{aligned} \delta(y) &= -\log \frac{y}{4} \int_0^y \left\{(1 - \frac{s}{y})^{-1/2} - 1\right\} \frac{ds}{s} \\ &= -\log \frac{y}{4} - \int_0^1 \left\{(1 - s)^{-1/2} - 1\right\} \frac{ds}{s} \\ &= -\log \frac{y}{4} - \log 4 = -\log y \end{aligned}$$

i.e.

$$\delta(y) = -\log y.$$

(ii) As second example, we mention

$$U(x) = \frac{1}{4} \frac{\lambda^2}{\sin h^2 \lambda x} \quad \text{for} \quad \lambda > 0 ,$$

with A = 0.

In this case we determine δ by computing the solution of

$$\dot{x}^2 + \frac{\lambda^2}{\sinh^2 \lambda x} = y \quad \text{where} \quad y = \dot{x}^2(\infty) .$$

With

$$\eta = \cosh \lambda x$$

one finds

$$\dot{\eta}^2 + \lambda^4 = \lambda^2 y(\eta^2 - 1)$$

$$\dot{\eta}^2 - \lambda^2 y \eta^2 = -\lambda^2(y + \lambda^2)$$

$$\eta = \sqrt{1 + \frac{\lambda^2}{y}} \cosh [\lambda \sqrt{y}\, t] .$$

If we compare the asymptotic behavior of this expression for $t \to \infty$ and of $\eta = \cosh \lambda x \sim e^{\lambda x}/2$ for $x \to \infty$ we find immediately

$$e^{\lambda\delta} = \sqrt{1 + \lambda^2/y^2}$$

or

$$\delta = \frac{1}{2\lambda} \log [1 + \frac{\lambda^2}{y^2}]$$

(iii) Letting $\lambda \to 0$ we obtain for

$$U(x) = \frac{a}{x^2}$$

the phase shift

$$\delta = 0.$$

We will see that this potential is, in some sense to be specified, the only potential with this property.

Of course, one can verify the fact that $\delta = 0$, directly, without referring to the potential $\lambda^2 \sinh^{-2} \lambda x$.

(iv) For the example

$$U(x) = \frac{1}{4} x^{-1/\alpha} \quad \text{or} \quad V = y^{-\alpha} \qquad (0 < \alpha < 1)$$

one finds

$$\delta(y) = \gamma(\alpha)\, y^{-\alpha}$$

where

$$\gamma(\alpha) = \sqrt{\pi}\, \frac{\Gamma(1-\alpha)}{\Gamma(\frac{1}{2}-\alpha)} .$$

Note that the right-hand side has poles at $\alpha = 1,2,\ldots$ and zeros at $\alpha = \frac{1}{2}, \frac{3}{2}, \frac{5}{2}, \ldots$ so that $\alpha = \frac{1}{2}$ is the only zero in

$$0 < \operatorname{Re} \alpha < 1 .$$

We turn to the inverse problem of reconstructing $U(x)$ if $\delta(y)$ is given. This requires inverting the linear integral equation (7) and, assuming $V(y)$ is monotone, taking the inverse function of $V(y)$. The integral equation is essentially of Abel's type and a solution can be found in explicit form: Assuming that $\delta(y)$ is given in the interval $0 < y \leq y_1$ we find a solution $V(y)$ of (7) in the same interval in the form

$$(8) \qquad V(y) = \frac{1}{\pi} \int_0^y (y-t)^{-1/2}\, t^{-1/2}\, \delta(t)\, dt .$$

In order to ensure the existence of this integral we have to impose growth conditions on $\delta(y)$ near $y = 0$. Moreover, it will suffice to assume $\delta(y)$ to be monotone decreasing so that $V(y)$ has the same property. However, this solution is not the only potential belonging to the phase shift $\delta(y)$ and the most general solution is given by the following theorem. This theorem could be proven under more general conditions but we are content with the assumptions sufficient in our context.

Theorem Let $\delta(y)$ be a continuously differentiable function in $0 < y \leq y_1$ with

(9) $$\int_0^{y_1} y^{\gamma} \, |d\delta(y)| < \infty \quad \text{for some} \quad \gamma < \frac{1}{2} .$$

Then formula (8) defines a continuously differentiable function $V(y)$ satisfying (7) and

(10) $$\int_0^{y_1} y^{\gamma} |dV(y)| < \infty .$$

Moreover, if $\delta'(y) < 0$ then $V'(y) < 0$ and

(11) $$V(y) \geq \delta(y) .$$

Thus, if (9) holds, $\delta'(0) < 0$ and $\delta(y) \to \infty$ for $y \to 0$ we obtain a $V(y)$ with the same properties and hence a monotone decreasing potential $U(x) > 0$ tending to zero for $x \to \infty$ and

$$\int^{\infty} U^{\gamma} \, dx < \infty .$$

The most general solution $V(y)$ of (7) which satisfies

$$\int_0^{y_1} y^{\theta} \, |dV(y)| < \infty \quad \text{for} \quad \text{some } 0 < \theta < 1$$

differs from (8) by $c/\sqrt{y}$, with a constant c. Therefore the most general potential $U(x)$ with the properties (3) and $\int^{\infty} U^{\theta} \, dx < \infty$ for which $\delta(y) = 0$ is $U(x) = a/x^2$.

4. Proof of the Theorem

The inverse formula: Replacing y by y/y_1 we can assume $y_1 = 1$ and write (8) in the form

(12) $$V(y) = \frac{1}{\pi} \int_0^1 (1-s)^{-1/2} \, s^{-1/2} \, \delta(sy) \, ds$$

by making the substitution $t = sy$. At first we restrict ourselves to $\delta \in C^1[0,1]$ so that (12) implies

$$V'(y) = \frac{1}{\pi}\int_0^1 (1-s)^{-1/2} s^{1/2} \delta'(sy)\, ds . \tag{13}$$

Inserting this expression into the right-hand side of (7) where we replace y_0 by y we obtain -- for the expression which should be $\delta(y)$ -- a sum $I + II + III$ of three terms where

$$I = V(y)$$

$$II = \int_0^1\int_0^1 (1-t)^{-1/2} \frac{y}{\pi} (1-s)^{-1/2} s^{1/2} \delta'(tsy)\, ds\, dt$$

$$III = -\frac{1}{\pi}\int_0^1\int_0^1 (1-s)^{-1/2} s^{-1/2} \frac{d}{dt}(\delta(tsy))\, dt\, ds$$

$$= -\frac{1}{\pi}\int_0^1 (1-s)^{-1/2} s^{-1/2} (\delta(sy) - \delta(0))\, ds$$

$$= -V(y) + \delta(0) .$$

To compute II we set $t = r/sy$ and obtain

$$II = \frac{1}{\pi}\iint_{0<r<sy<y} \left(1 - \frac{r}{sy}\right)^{-1/2} (1-s)^{-1/2} s^{-1/2} \delta'(r)\, ds\, dr$$

$$= \int_0^y \Phi\left(\frac{r}{y}\right) \delta'(r)\, dr$$

where

$$\Phi(\sigma) = \frac{1}{\pi}\int_\sigma^1 (s-\sigma)^{-1/2} (1-s)^{-1/2}\, ds \equiv 1.$$

The latter well known identity could be verified by complex integration. Thus we obtain the relation

$$I + II + III = V(y) + \int_0^y \delta'(r)\,dr - V(y) + \delta(0)$$

$$= \delta(y) - \delta(0) + \delta(0) = \delta(y) ,$$

which verifies our statement for $\delta \in C^1[0,1]$.

To obtain the general case we estimate this solution first for $\delta \in C^1[0,1]$. From (13) we find

$$|V'(y)| \leq \frac{1}{\pi} \int_0^1 (1-s)^{-1/2}\, s^{1/2}\, |\delta'(sy)|\, ds$$

and hence

$$\int_0^1 y^\gamma |V'(y)|\,dy \leq \frac{1}{\pi} \int_0^1 \int_0^1 (1-s)^{-1/2}\, s^{1/2} |\delta'(sy)|\, ds\, y^\gamma\, dy$$

$$\leq \frac{1}{\pi} \int_0^1 (1-s)^{-1/2}\, s^{(1/2)-\gamma-1} \left(\int_0^s t^\gamma |\delta'(t)|\,dt \right) ds$$

$$\leq c_1 \int_0^1 t^\gamma |\delta'(t)|\, dt$$

with

(14) $$c_1 = \frac{1}{\pi} B(\tfrac{1}{2}, \tfrac{1}{2} - \gamma) , \qquad 0 < \gamma < \frac{1}{2} .$$

Approximating an arbitrary function $\delta \in C^1(0,1]$ satisfying (9) in this norm we extend the definition of V to the general case of the theorem. Since the right-hand side of (7) is continuously dependent on V in the norm $\int_0^1 y^\gamma |\delta'(y)|\, dy$ we conclude the validity of (8) in the class of functions satisfying (9). The restriction $\gamma < 1/2$ is required for the finiteness of c_1 in (14).

The formula (13) makes the assertions that $\delta' < 0$ implies $V' < 0$ evident, since the kernel $(1-s)^{-1/2}\, s^{1/2}$ is positive. For the same reason one concludes from (12) and $\delta' < 0$ that

$$V(y) \geq \frac{1}{\pi} \int_0^1 (1-s)^{-1/2} s^{-1/2} \delta(y)\, ds = \delta(y)$$

proving (11).

It remains to prove the statement about the most general solution of (7) in $0 < y \leq 1$ satisfying

$$\int_0^1 y^\theta |dV| < \infty \quad \text{for some} \quad 0 < \theta < 1 .$$

We may assume $\delta = 0$ for $0 < y < 1$, so that

$$V(y) + \int_0^y \left\{\left(1 - \frac{s}{y}\right)^{-1/2} - 1\right\} dV(s) = 0 \quad \text{for} \quad 0 \leq y \leq 1.$$

For the proof we use the Wiener-Hopf technique: Setting

$$y = e^\eta$$

$$v(\eta) = \begin{cases} V(e^\eta) & \text{for } \eta \leq 0 \\ V(1) & \text{for } \eta > 0 \end{cases}$$

$$k(\eta) = \begin{cases} -1 & \text{for } \eta \leq 0 \\ (1-e^{-\eta})^{-1/2} - 1 & \text{for } \eta > 0 . \end{cases}$$

The above relation takes the form

$$\int_{-\infty}^{+\infty} k(\eta-\sigma)\, dv(\sigma) = -v(0) \quad \text{for } \eta \leq 0 . \tag{15}$$

It suffices to prove the

<u>Lemma 1</u> If $v(\eta) = v(0)$ for $\eta > 0$, v satisfy (15) and

$$\int_{-\infty}^{+\infty} e^{\theta\eta} |dv(\eta)| < \infty \tag{16}$$

then

$$v(y) = v(0)\, e^{-\eta/2} \quad \text{for} \quad \eta < 0 .$$

Indeed this lemma implies that $V(y) = V(1)\, y^{-1/2}$ for $0 < y \leq 1$ if $\delta = 0$ for $0 < y \leq 1$.

For the proof of the lemma we may assume $v(0) = 1$. We set

$$h(\eta) = \begin{cases} 0 & \text{for} \quad \eta \leq 0 \\ \displaystyle\int_{-\infty}^{+\infty} k(\eta-\sigma)\, dv(\sigma) & \text{for} \quad \eta > 0 \end{cases}$$

$$\Phi(\eta) = \begin{cases} 1 & \text{for} \quad \eta \leq 0 \\ 0 & \text{for} \quad \eta > 0 \end{cases}$$

so that

$$\int_{-\infty}^{+\infty} k(\eta-\sigma)\, dv(\sigma) = h(\eta) - \Phi(\eta) \quad \text{for all real} \quad \eta.$$

Indeed, for $\eta > 0$ this is evident and for $\eta \geq 0$ this corresponds precisely to (15) with $v(0) = 1$. To solve this equation we introduce the Laplace transforms

$$\hat{k}(z) = \int_{-\infty}^{+\infty} e^{\eta z}\, k(\eta)\, d\eta$$

$$(17) \qquad \hat{h}(z) = \int_{-\infty}^{+\infty} e^{\eta z}\, h(\eta)\, d\eta$$

$$\hat{w}(z) = \int_{-\infty}^{+\infty} e^{\eta z}\, dv(\eta) .$$

Using

$$\int_{-\infty}^{+\infty} e^{\eta z}\, \Phi(\eta)\, d\eta = \frac{1}{z} \quad \text{for} \quad \text{Re } z > 0$$

we find

$$\hat{k}(z)\ \hat{w}(z) = \hat{h}(z) - \frac{1}{z} \quad \text{for} \quad \theta < \text{Re } z < 1 . \tag{18}$$

To justify the domain of validity we prove

Lemma 2 Under the assumption

$$\int_{-\infty}^{+\infty} e^{\eta}\ |dv(\eta)| = \int_0^1 y\ |dV| < \infty$$

the function $\hat{h}(z)$ of (17) is analytic in $\text{Re } z < 1$ and satisfies

$$|\hat{h}(z)| \le c\ \frac{|\text{Re } z|^{1/2}}{1 - \text{Re } z} \quad \text{for} \quad \text{Re } z < 1 .$$

Furthermore,

$$|\hat{w}(z)| \le \int_0^1 y^{\theta}\,|dV| \qquad \text{for} \quad \text{Re } z \ge \theta .$$

Finally, the function $\hat{k}(z)$ of (17) is given by

$$\hat{k}(z) = B\left(\frac{1}{2}\ ,\ z\right) = \sqrt{\pi}\ \frac{\Gamma(\ - z)}{\Gamma(\frac{1}{2} - z)} \quad \text{in} \quad 0 < \text{Re } z < 1 .$$

We postpone the proof of these estimates to the end of this section. The right-hand side of (18) is (because of Lemma 2) meromorphic for $\text{Re } z < 1$ the left side meromorphic for $\text{Re } z > \theta$ so that both sides have an analytic continuation to a meromorphic function. In particular, we use that $\hat{k}(z)$ has as only zero in $0 < \text{Re } z < 1$ the point $z = 1/2$. Singling out this zero we multiply by (18) with $(z - 1/2)/\hat{k}(z)$ and find

$$\left(z - \frac{1}{2}\right)\ \hat{w}(z) = \frac{z - \frac{1}{2}}{\hat{k}(z)}\ \left(\hat{h}(z) - z^{-1}\right) . \tag{19}$$

Using the functional equation of the Γ-function we obtain the relations

$$\hat{k}\left(z - \frac{1}{2}\right)\ \hat{k}(z) = -\ \frac{\pi}{2}$$

$$\hat{k}(z) = \frac{z - \frac{1}{2}}{z} \hat{k}(z - 1)$$

with which we derive

$$(20) \qquad (z - \tfrac{1}{2}) \hat{k}^{-1}(z) = - \frac{z}{\pi} (z-1) \hat{k}(z - \tfrac{3}{2})$$

so that (19) takes the form

$$(19') \qquad \left(z - \tfrac{1}{2}\right) \hat{w}(z) = - \frac{z-1}{\pi} \hat{k}(z - \tfrac{3}{2}) \left(z\hat{h}(z) - 1\right) .$$

In this relation the left-hand side is analytic in $\operatorname{Re} z > \theta$ and the right-hand side for $\operatorname{Re} z < 1$. Indeed $\hat{k}(z - 3/2)$ is analytic for $\operatorname{Re} z < 3/2$ and can be estimated by

$$(21) \qquad |\hat{k}(z - \tfrac{3}{2})| \leq 1 \quad \text{for} \quad \operatorname{Re} z \leq 1 .$$

This follows from the trivial estimate

$$|B(\alpha,\beta)| = \int_0^1 |s^{\alpha-1}(1-s)^{\beta-1}| ds \leq B(\tfrac{1}{2},\tfrac{1}{2}) = \pi \quad \text{for} \quad \operatorname{Re} \alpha, \operatorname{Re} \beta \geq \tfrac{1}{2} .$$

Therefore the function $(z - 1/2) \hat{w}(z)$, at first defined in $\operatorname{Re} z > \theta$, has an analytic continuation to $\operatorname{Re} z < 1$ via (19), is therefore an entire function. By Lemma 2 and (21) the right-hand side of (19') is estimated by

$$c_1 |z|^{5/2} \quad \text{for} \quad \operatorname{Re} z < \theta'$$

for any $\theta' < 1$, and by Lemma 2 again by

$$c_2 |z| \quad \text{for} \quad \operatorname{Re} z \geq \theta .$$

Hence, taking $\theta < \theta' < 1$, it is a polynomial and by the last estimate even a linear function. Actually it is a constant, since for real positive $z \to +\infty$

$$|\hat{w}(z)| \to 0 .$$

Indeed, for given $\varepsilon > 0$ we can find $\rho = \rho(\varepsilon)$ such that

$$\int_{-\rho}^{0} e^{\eta} \, |dv(\eta)| < \frac{\varepsilon}{2} .$$

Hence for sufficiently large positive z

$$|\hat{w}(z)| \le \int_{-\infty}^{-\rho} |e^{\eta z} \, dv| + \frac{\varepsilon}{2} \le e^{-\rho(\mathrm{Re}\, z - 1)} \int_{-\infty}^{0} e^{\eta} \, |dv| + \frac{\varepsilon}{2} < \varepsilon .$$

Hence $(z - 1/2)\, \hat{w}(z)$ is a constant, or

$$\int_{-\infty}^{0} e^{\eta z} \, dv(\eta) = \frac{a}{z - \frac{1}{2}} .$$

Clearly, a possible solution of this relation is $v_0(\eta) = -2a\, e^{-\eta/2}$ for $\eta \le 0$, so that $q(\eta) = v(\eta) - v_0(\eta)$ satisfies

$$\int_{-\infty}^{0} e^{\eta z} \, dq = 0 \quad \text{for} \quad \mathrm{Re}\, z > 1$$

and

$$\int_{-\infty}^{0} e^{\eta} \, |dq| < \infty .$$

The uniqueness of the inverse of the Laplace transform finally gives $q(\eta) = 0$ for $\eta \le 0$ or

$$v(\eta) = -\, 2a\, e^{-\eta/2}$$

and because $v(0) = 1$

$$v(\eta) = e^{-\eta/2} .$$

This proves Lemma 1 and hence the last part of the theorem.

Proof of Lemma 2: First we estimate $h(\eta)$ where we can take $\eta > 0$. We will show that

$$|h(\eta)| \leq \begin{cases} \dfrac{c_2}{\sqrt{\eta}} & \text{for} \quad 0 < \eta \leq 1 \\ c_2 e^{-\eta} & \text{for} \quad \eta > 1 \end{cases}$$

With $y = e^{\eta} > 1$ we have

$$h(\eta) = \int_0^1 \left\{\left(1 - \frac{s}{y}\right)^{-1/2} - 1\right\} dV(s)$$

hence

$$|h(\eta)| \leq c_3 \int_0^{1/2} \frac{s}{y} |dV(s)| + \left\{\left(1 - \frac{1}{y}\right)^{-1/2} - 1\right\} \int_{1/2}^1 |dV(s)|$$

$$\leq c_3\, y^{-1} \int_0^{1/2} s|dV| + c_4 (y(y-1))^{-1/2}\, 2 \int_{1/2}^1 s|dV|$$

$$\leq c_5\, y^{-1/2}\, (y - 1)^{-1/2} \qquad \text{for } y > 1.$$

For $1 < y \leq e$ this can be estimated by

$$c_5 (y - 1)^{-1/2} \leq c_6\, \eta^{-1/2}$$

and for $y > e$ by $c_7\, y^{-1} = c_7\, e^{\eta}$ as we asserted.

Thus the Laplace transform $\hat{h}(z)$ of $h(\eta)$ can be estimated by

$$|\hat{h}(z)| \leq c_2 \left\{ \int_0^1 \frac{|e^{\eta z}|}{\sqrt{\eta}}\, d\eta + \int_1^{\infty} |e^{\eta(z-1)}|\, d\eta \right\}$$

$$= O\left(\frac{1}{1+\sqrt{p}}\right) + \frac{c_2}{1+p} \quad \text{for} \quad \operatorname{Re} z = -p < 0$$

and by

$$|\hat{h}(z)| \leq c_2 \int_0^1 \frac{d\eta}{\sqrt{\eta}} + \frac{c_2}{1 - \operatorname{Re} z} \text{ for } \quad 0 \leq \operatorname{Re} z < 1 \,.$$

We can combine these estimates to

$$|\hat{h}(z)| \leq c \frac{|\mathrm{Re}\ z|^{1/2}}{1 - \mathrm{Re}\ z} \quad \text{for} \quad \mathrm{Re}\ z < 1$$

as was claimed.

The estimate of $\hat{w}(z)$ follows trivially from

$$|\hat{w}(z)| \leq \int_{-\infty}^{0} e^{\eta\ \mathrm{Re}\ z} |dv(\eta)| \leq \int_{-\infty}^{0} e^{\eta\theta} |dv| \quad \text{for} \quad \mathrm{Re}\ z \geq \theta .$$

Finally, we identify $\hat{k}(z)$ and write

$$\hat{k}(z) = -\int_{-\infty}^{0} e^{\eta z}\, d\eta + \int_{0}^{\infty} \left\{ (1 - e^{-\eta})^{-1/2} - 1 \right\} e^{\eta z}\, d\eta$$

which is clearly analytic for $0 < \mathrm{Re}\ z < 1$. The first term being $-z^{-1}$ allows an analytic continuation to $\mathrm{Re}\ z < 1$ with a pole at $z = 0$. But for $\mathrm{Re}\ z < 0$ we can split off

$$\int_{0}^{\infty} (-1)\, e^{\eta z}\, d\eta = z^{-1}$$

and obtain for the analytic continuation of $\hat{k}(z)$

$$\hat{k}(z) = \int_{0}^{\infty} (1 - e^{-\eta})^{-1/2} e^{\eta z}\, d\eta = \int_{0}^{1} (1 - s)^{-1/2} s^{-z-1}\, ds .$$

Since the Beta function can be defined by

$$B(\alpha,\beta) = \int_{0}^{1} (1 - s)^{\alpha-1} s^{\beta-1}\, ds$$

we have

$$\hat{k}(z) = B(\tfrac{1}{2}, -z)$$

which proves Lemma 2.

References

[1] Abel, N. H., Résolution d'un problème de mecanique, Jour. Reine Angew. Math. 1, 13-18, 1826.

[2] Airault, H., McKean, H. P., and Moser, J., Rational and Elliptic Solutions of the Korteweg-deVries Equation and a Related Many-Body Problem, to appear in Comm. Pure Appl. Math., 1977.

[3] Adler, M., Some Finite Dimensional Integrable Systems, to appear in the Proceedings of Conference on Solitons, Tuscon, January 1976; also, under the same title, a dissertation at New York University, October 1976.

[4] Keller, J., Inverse Problems, Tha American Math. Monthly 83, 107-118, 1976.

[5] Kulish, P. P., Factorization of the classical and quantum S-matrix and conservation laws, preprint from the Institute of High Energy Physics, Serpukhov, 1975.

[6] Moser, J., Three Integrable Hamiltonian Systems Connected with Isospectral Deformations, Adv. Math. 16, 197-220, 1975.

[7] Moser, J., Finitely many mass points on the line under the influence of an exponential potential in an integrable system, Lecture Notes in Physics 38, Springer-Verlag, 1975.

PROOF OF A GENERALIZED FORM OF A FIXED POINT THEOREM DUE TO G. D. BIRKHOFF

J. Moser

Courant Institute of Mathematical Sciences, New York University

In his work on closed geodesics of manifolds W. Klingenberg [14, 15] applied a generalization of the Birkhoff fixed point theorem which is quite plausible but for which a proof has not been published. The statement is simply that in a punctured neighborhood of a fixed point a smooth symplectic mapping ϕ has generically infinitely many periodic points (i.e. fixed points of some iterate) provided that the linearized mapping $d\phi$ at the fixed point has at least one eigenvalue on the unit circle. The precise smoothness and nondegeneracy assumptions will be formulated below. The purpose of this paper is to make a complete proof for this statement available. The tools are all standard, and we do not claim any novelty in the approach.

Actually previous papers on this subject containing detailed proofs do not refer to symplectic mappings but Hamiltonian vector fields which depend periodically on the independent t-variable, in the neighborhood of a periodic orbit. Such a flow gives rise to a mapping to which the above theorem is applicable. Therefore our formulation covers that situation and is easier to prove. In Klingenberg's work the geodesic flow provides a Hamiltonian vector field and the above statement is applied to the Poincaré mapping of this flow near an elliptic periodic geodesic.

It is good to remember that the original fixed point theorem of Poincaré and Birkhoff of 1915 (see [1]) which asserts the existence of a fixed point of certain area-preserving annulus mappings is of global nature. It deals strictly with a two-dimensional mapping and

*This work was supported by the National Science Foundation, Grant No. NSF MCS 7403003-A02. Reproduction in whole or in part is permitted for any purpose of the U. S. Government.

no generalization to higher dimensions has been found (see also [7]). However, we are concerned with periodic points near a fixed point -- which is a local problem. This local problem can be generalized to higher dimensions, and the first announcement of such a result is [3], but the first proof, given in [4], is not quite correct in its technical details, and requires that all eigenvalues $\lambda_k, \lambda_k^{-1}$ $(k = 1,2,\dots,n)$ lie on the unit circle and satisfy

$$\prod_{k=1}^{n} \lambda_k^{j_k} \neq 1 \quad \text{for} \quad 0 < \sum_{k=1}^{n} |j_k| \leq \nu$$

where j_k are integers and $\nu = 16n+9$. This latter condition is rather artificial. It was improved to $\nu = 7$ by Harris [6], who showed that ν can be chosen independent of the dimension 2n of the space. But the natural condition is given by $\nu = 4$ and our proof requires just this (see Theorem 1 below).

In Section 2 we prove the above theorem in the stable case, i.e. when $|\lambda_k| = 1$, which is formulated as Theorem 2. (See also [4,5,6,7,8]). The general case is reduced to Theorem 2 by construction of the center manifold and a version of Darboux's lemma. For completeness these theorems are proven with the required smoothness assumptions, see Sections 4 and 5. In Section 6 we derive Birkhoff's normal form up to third order for symplectic mappings.

The proof in Section 2 which is the central part of this paper grew out of discussions with E. Zehnder and was presented by him in [12]. The author is grateful to Klingenberg for many discussions on this subject.

1. Statement

We consider a smooth canonical mapping ϕ in R^{2n} near an equilibrium point. It is our goal to prove the existence of infinitely many periodic points of ϕ, i.e. fixed points of some higher iterate ϕ^N $(N = 1,2,\dots)$ of ϕ, in any neighborhood of the given fixed point. If p is such a periodic point with $\phi^N(p) = p$ we call $\{\phi^k(p),\ k = 1,\dots,N-1\}$ the periodic orbit through p. We will require that the periodic orbits considered remain in a preassigned neighborhood of the given fixed point. Thus we are dealing with a local problem and we will express the relevant conditions in terms of the Taylor expansion of ϕ at the fixed point. It is clear, for example, that a necessary condition for the existence of such periodic orbits is that at least one of the eigenvalues of $d\phi$ at the fixed points is on the unit circle. Indeed, if none of these eigenvalues lies on the unit circle one calls the fixed point "hyperbolic" and it is easily shown that for every p the points $\phi^k(p)$ leave any neighborhood if k sufficiently large positive or negative, unless, of course, p is the fixed point itself.

To formulate the statement we observe that the eigenvalues $\lambda_1 \dots \lambda_{2n}$ of a linear canonical map, here $d\phi$, occur in pairs: If λ_k is an eigenvalue so is λ_k^{-1} and we may group the eigenvalues such that

$$\lambda_{k+n} = \lambda_k^{-1} \qquad (k = 1,2,\dots,n).$$

Then we have

Theorem 1 Let ϕ be a diffeomorphism in a neighborhood U of a fixed point O and assume ϕ to be canonical in U and in $C^4(U)$. If $\lambda_1, \lambda_2 \dots \lambda_m$ and $\lambda_1^{-1}, \dots \lambda_m^{-1}$ are all those eigenvalues of $d\phi|_0$ which lie on the unit circle we require

(i) $m \geq 1$

(ii) $\prod_{k=1}^{m} \lambda_k^{j_k} \neq 1$ for $1 \leq \sum_{k=1}^{m} |j_k| \leq 4$

where $j_1, j_2, \dots, j_m$ are any integers and if

(iii) The Taylor coefficients of ϕ up to order 3 satisfy a non-degeneracy condition (to be formulated) then ϕ possesses infinitely many periodic orbits in any neighborhood of 0.

<u>Remark</u>. Actually it suffices to assume $\phi \in C'''(U)$; our C^4 assumption is imposed merely to simplify the exposition somewhat (see the remark at the end of Section 4).

To illustrate this theorem we discuss it for n = 1, where ϕ represents an area-preserving mapping

$$\phi : \begin{pmatrix} x \\ y \end{pmatrix} \to \begin{pmatrix} f(x,y) \\ g(x,y) \end{pmatrix} \to \begin{pmatrix} ax + by + \hat{f} \\ cx + dy + \hat{g} \end{pmatrix}$$

with Jacobian $\equiv 1$, near the fixed point, say $x = y = 0$. In this case the eigenvalues of $d\phi|_0$ are λ, λ^{-1} and by assumption $|\lambda| = 1$, $\lambda^3 \neq 1$, $\lambda^4 \neq 1$. In this case, one can by a theorem of Birkhoff -- to be discussed in Section 6 -- introduce the coordinates x,y in such a way that ϕ takes the form

$$\begin{pmatrix} x \\ y \end{pmatrix} \to \begin{pmatrix} x \cos \Phi - y \sin \Phi \\ +x \sin \Phi + y \cos \Phi \end{pmatrix} + O\left((x^2 + y^2)^2\right) \tag{1}$$

where

$$\Phi = \alpha + \beta(x^2 + y^2) , \qquad \lambda = e^{i\alpha} \tag{2}$$

so that Φ is approximately a rotation y an angle which depends on the radius. In this case our condition (iii) means precisely that $\beta \neq 0$, which forces the mapping to be nonlinear. Indeed, if the mapping ϕ is linear and λ not a root of unity then ϕ has no periodic point, showing the necessity of some condition like (iii).

Also the condition (ii) is necessary. If $\lambda^2 + \lambda + 1 = 0$, i.e. a third root of unity then one can construct a canonical transformation which in complex coordinates $z = x + iy$ takes the form

$$\phi : z \to \lambda z + \bar{z}^2 + \ldots$$

One verifies that for this mapping the orbit through any point $z \neq 0$ leaves any neighborhood of the fixed point (see [2], p. 222). It also is easily seen that the statement is not true if the mapping is not required to be area preserving.

2. The Fixed Point Theorem for Higher Dimensions in the Stable Case

We consider a canonical map $\phi \in C^1$ near a fixed point of a form analogous to (1). Let $x \in R^m$, $y \in R^m$ and $x = y = 0$ be the fixed point, and assume that ϕ preserves the differential form

$$\omega = \sum_{k=1}^{m} dx_k \wedge dy_k \, . \tag{3}$$

The Euclidean norms in R^m, R^{2m} respectively will be denoted by $|x| = \left(\sum_1^m x_k^2\right)^{1/2}$, $(|x|^2 + |y|^2)^{1/2}$. We consider a mapping of the form

$$\begin{cases} x_k^{(1)} = x_k \cos \Phi_k - y_k \sin \Phi_k + f_k \\ y_k^{(1)} = x_k \sin \Phi_k + y_k \cos \Phi_k + f_{k+m} \\ \Phi_k = \alpha_k + \sum_{\ell=1}^{m} \beta_{k\ell}(x_\ell^2 + y_\ell^2) \end{cases} \tag{4}$$

The error terms $f_k = f_k(x,y)$ $(k = 1,2,\ldots 2m)$ are assumed to have vanishing derivatives up to order 3 at the origin, while otherwise f_k are in C^1. One could simply assume f_k to be in C''' but the following proof does not really require this, and this fine point will be useful later on. Thus with $r^2 = |x|^2 + |y|^2$ we can say that

$$r^{-3}\left\{|f(x,y)| + r\left(|D_x f| + |D_y f|\right)\right\} \tag{5}$$

tends to zero as $r \to 0$.

The following theorem is a modified version of a theorem due to Birkhoff and Lewis [4,5].

<u>Theorem 2</u> If $\phi \in C^1$ preserves the two form ω and the origin and is of the form (4) with a nonsingular matrix $(\beta_{k\ell})$ then every punctured neighborhood $0 < |x|^2 + |y|^2 < \varepsilon$ contains a periodic orbit.

For the proof we introduce polar coordinates by

(6) $$x_k = \sqrt{\varepsilon r_k} \cos \theta_k , \qquad y_k = \sqrt{\varepsilon r_k} \sin \theta_k$$

and observe that this transformation is singular if one or more of the r_k vanish. Therefore we restrict attention to a ball

$$B_\rho : \sum_{k=1}^{m} (r_k - \frac{1}{2m})^2 < \rho^2 \quad \text{with} \quad \rho < \frac{1}{2m}$$

in which $r_k > 1/2m - \rho > 0$. Moreover, on this ball we have

$$0 < \sum_1^m (x_k^2 + y_k^2) = \varepsilon \sum_{k=1}^{m} r_k < \varepsilon\left(\frac{1}{2} + m\rho\right) < \varepsilon$$

so that it lies in the considered punctured neighborhood. The mapping ϕ is defined in the domain $B_\rho \times T^m$, T^m denoting the torus defined by θ_k (mod 2π). The transformation (6) takes ω into

$$\omega = \sum dx_k \wedge dy_k = \frac{\varepsilon}{2} \sum dr_k \wedge d\theta_k$$

so that ϕ is also canonical in the new variables. Moreover, ϕ is what is called "exact canonical", i.e. for any closed curve Γ in $B_\rho \times T^m$ we have

$$\int_\Gamma \sum_{k=1}^{m} r_k \, d\theta_k = \int_{\phi(\Gamma)} \sum_1^m r_k \, d\theta_k$$

Indeed, we have $\varepsilon \, r_k \, d\theta_k = y_k \, dx_k - x_k \, dy_k$, and by Stokes' theorem

$$\varepsilon \int_\Gamma (r, d\theta) = \int_{\partial\Omega} (y, dx) - (x, dy) = 2 \int_\Omega dy \wedge dx = -2 \int_\Omega \omega$$

where Ω is a two-dimensional manifold in $|x|^2 + |y|^2 < \varepsilon$ with boundary $\partial\Omega = \Gamma$. Since our original mapping preserves ω, hence $\int_\Omega \omega$, the new one preserves $\int_\Gamma (r, d\theta)$ for closed curves Γ in $B_\rho \times T^m$. Note that we used in this argument that $|x|^2 + |y|^2 < \varepsilon$ is simply connected, which is, of course, not the case for $B_\rho \times T^m$. The concept "exact canonical" is indeed stronger than "canonical" as the example $r_k \to r_k + c_k$, $\theta_k \to \theta_k$ illustrates.

The new mapping which we call ϕ again is actually defined on the covering surface of $B_\rho \times T^m$ which is parametrized by r_k, θ_k. The mapping takes the form

$$(7) \qquad \begin{cases} r_k^{(1)} &= r_k + o_1(\varepsilon) \\ \theta_k^{(1)} &= \theta_k + \psi_k(r,\varepsilon) + o_1(\varepsilon) \\ \psi_k &= \alpha_k + \varepsilon \sum_{\ell=1}^{m} \beta_{k\ell} r_\ell \end{cases}$$

Here $o_1(\varepsilon)$ stands for a function $f = f(r,\theta,\varepsilon)$ of period 2π in the θ_k such that $\varepsilon^{-1}f$ and its first derivatives with respect to r, θ tend uniformly to zero as $\varepsilon \to 0$. If we use the notation $|f|_s$ for the maximum of all derivatives of order $\leq s$ in $B_\rho \times T^m$ we can express this by

$$\varepsilon^{-1}|f|_1 \to 0 \quad \text{as } \varepsilon \to 0 .$$

Functions, for which $\varepsilon^{-1}|f|_0 \to 0$, are symbolically denoted by $o(\varepsilon)$.

Theorem 3 Let ϕ be an exact canonical mapping in $B_\rho \times T^m$ of the form (7) with nonsingular $(\beta_{k\ell})$. Then for sufficiently small $\varepsilon > 0$ the domain $B_\rho \times T^m$ contains at least one perodic orbit.*

Clearly, this theorem implies Theorem 2.

We give the proof of Theorem 3 in several steps (see also [12]).

Lemma 1 Given a ρ' in $0 < \rho' < \rho$ and a positive constant c_1 we show that

* Here an orbit $(r^{(j)},\theta^{(j)})$, $j = 0,1,2,\dots,N$ is called periodic if ϕ: $(r^{(j)},\theta^{(j)}) \to (r^{(j+1)},\theta^{(j+1)})$ for $j = 0,1,\dots,N-1$ and $r_k^{(N)} = r_k^{(0)}$, $\theta_k^{(N)} = \theta_k^{(0)}$ (mod 2π).

$$\phi^j : B_{\rho'} \times T^m \to B_\rho \times T^m \quad \text{for} \quad j = 1,2,\ldots,N \tag{8}$$

if $\varepsilon N < c_1$ and $0 < \varepsilon < \varepsilon_0$ where ε_0 depends on ρ, ρ', c_1. Moreover,

$$\phi^j - \psi^j = \varepsilon^{-1} o_1(\varepsilon) \qquad \text{for} \quad j = 1,2,\ldots,N \tag{9}$$

where

$$\psi : (r,\theta) \to (r, \theta + \Psi(r,\varepsilon))$$

The purpose of this lemma is to show that iterates $\phi^j(\tau,\theta)$ do not run out of the domain of definition, if $j = O(\varepsilon^{-1})$. It will be important to have the estimates for the first derivatives of ϕ^j as they are given by (9).

Proof: Using the usual Euclidean norm in the r,θ space, we have

$$|r^{(j)} - r| < |r^{(1)} - r| + \ldots |r^{(j)} - r^{(j-1)}| < j\, o(\varepsilon)$$

and since $j < N < c_1/\varepsilon$ we can achieve that

$$|r^{(j)} - r| < c_1 \varepsilon^{-1} o(\varepsilon) < \rho - \rho'$$

if ε is small enough. This proves (8).

Also

$$|\theta^{(j)} - \theta - j\, \Psi(r,\varepsilon)| \leq |\theta^{(1)} - \theta - \Psi(r,\varepsilon)| + \ldots |\theta^{(j)} - \theta^{(j-1)} - \Psi(r,\varepsilon)|$$

$$\leq o(\varepsilon) + (o(\varepsilon) + |\Psi(r^{(1)},\varepsilon) - \Psi(r,\varepsilon)|) + \ldots$$

$$+ \left(o(\varepsilon) + |\Psi(r^{(j-1)},\varepsilon) - \Psi(r^{(j)},\varepsilon)|\right)$$

$$\leq N\, o(\varepsilon) + \|\beta\| \varepsilon \sum_{\ell=1}^{j-1} |r^{(\ell)} - r|$$

$$\leq N\, o(\varepsilon) + \varepsilon \sum_{\ell=1}^{j-1} \ell\, o(\varepsilon)$$

$$= N\, o(\varepsilon) + N^2 \varepsilon\, o(\varepsilon) \;\; = \;\; (1 + c_1) N\, o(\varepsilon) = \varepsilon^{-1} o(\varepsilon) \; .$$

Here $\|\beta\|$ is the norm of the matrix $\beta = (\beta_{k\ell})$. This proves $|\phi^j - \psi^j|_0 = o(\varepsilon)$ and to prove (9) it remains to estimate the Jacobians $d\phi^j$, $d\psi^j$.

We denote the Jacobian $d\phi$ at $r^{(j)}$, $\theta^{(j)}$ by F_j and that of $d\psi$ at $r^{(j)}$, $\theta^{(j)}$ by θ_j. But since the Jacobian G of $d\psi$ is independent of θ we have

$$G_j = G = \begin{pmatrix} I & 0 \\ \varepsilon\beta & I \end{pmatrix}$$

We note that $|G - I|_0 = O(\varepsilon)$ and

$$|F_j - I|_0 = |F_j - G|_0 + |G - I| = o(\varepsilon) + O(\varepsilon) = O(\varepsilon).$$

Therefore the product of N such mappings F_j or G can be estimated by

$$\left(1 + O(\varepsilon)\right)^N = e^{NO(\varepsilon)} = O(1) \; ,$$

i.e. is bounded, if $N\varepsilon$ is bounded. Therefore the Jacobians

$$|d\phi^j - d\psi^j|_0 = \left| \sum_{k=0}^{j-1} F_{j-1} \cdots F_{k+1} (F_k - G) \cdot G^{k-1} \right|_0 \leq N c_2\, o(\varepsilon) = \varepsilon^{-1}\, o(\varepsilon).$$

This proves the lemma.

Next we search for a fixed point of ψ^N in $B_{\rho'} \times T$. This requires us to find an $r \in B_{\rho'}$ such that $(2\pi)^{-1} N \Psi_k(r,\varepsilon)$ are integers for $k = 1,\ldots,m$, i.e.

$$(2\pi)^{-1} N\, \Psi_k(r,\varepsilon) \in Z^m \; . \tag{10}$$

<u>Lemma 2</u> If $0 < \rho'' < \rho' < \rho < 1/2m$ and

$$\varepsilon N \rho'' \geq 2\pi\sqrt{m}\, \|\beta^{-1}\|$$

and ε sufficiently small then there exists an $r^* \in B_{\rho''}$ satisfying

(10), i.e. the entire torus $r = r^*$ consists of fixed points of ψ^N.

Proof: The image of the ball $B_{\rho''}$ under the mapping $r \to \Psi(r,\varepsilon) = \alpha + \varepsilon\beta r$ contains a ball of radius

$$\varepsilon \,\|\beta^{-1}\|^{-1} \rho'' .$$

Indeed, if $s = \Psi(r,\varepsilon)$, $s' = \Psi(r',\varepsilon)$ then

$$|r - r'| \leq \varepsilon^{-1}\|\beta^{-1}\| \, |s - s'|$$

and taking r' as the center of $B_{\rho''}$ and s' as its image then $|s - s'| < \rho''\varepsilon\|\beta^{-1}\|^{-1}$ implies $|r - r'| < \rho''$ which proves the statement.

Thus the image of $B_{\rho''}$ under the map $r \to (2\pi)^{-1} N \,\Psi(r,\varepsilon)$ contains a ball of radius

$$R = (2\pi)^{-1} \,\|\beta^{-1}\|^{-1} \, N\varepsilon\rho''$$

and hence a cube of side length $m^{-1/2}R$. Thus, if $m^{-1/2}R \geq 1$ this cube certainly contains a lattice point of Z^m, which proves the lemma.

Lemma 3 Setting

$$c_0 = 2\pi\sqrt{m}\|\beta^{-1}\|/\rho'' , \qquad c_1 = c_0 + 1 ,$$

taking an integer N in

$$c_0 \leq \varepsilon N \leq c_1$$

and ε small enough there exists a torus $r = r(\theta,\varepsilon) = r^* + \varepsilon^{-1} o_1(\varepsilon)$ in $B_{\rho'} \times T^m$ which is mapped by ϕ^N "radially", that is in such a way that

$$\theta^{(N)} - \theta \in 2\pi \, Z^m .$$

Proof: By Lemma 1,

$$\theta^{(N)} - \theta = N \,\Psi(r,\varepsilon) + \varepsilon^{-1} o_1(\varepsilon)$$

and by the preceding lemma there exists an $r^* \in B_{\rho''}$ and a lattice point $h \in Z^m$ such that

$$(2\pi)^{-1} N\,\Psi(r^*,\varepsilon) = h \in Z^m$$

To prove our lemma we have to solve the equations

$$\theta^{(N)} - \theta = N\Psi(r,\varepsilon) + \varepsilon^{-1} o_1(\varepsilon) = 2\pi h$$

for r. We can apply the implicit function theorem: If we drop the error term $\varepsilon^{-1} o_1(\varepsilon) = o_1(1)$ we have the solution $r = r^*$, and since the Jacobian of $N\Psi$ has an inverse

$$\|(\varepsilon N \beta)^{-1}\| \leq c_0^{-1} \|\beta^{-1}\|$$

which is bounded, there exists a C^1-solution $r = r^* + \varepsilon^{-1} o_1(\varepsilon) \in B_{\rho'}$ as we wanted to show. At this point it was important to have an estimate for the first derivatives of ϕ^j for $j \leq N$, hence of $\theta^{(N)}(r,\theta,\varepsilon)$, independent of N.

To find the desired fixed point of ϕ^N we have to solve the equation

$$\theta^{(N)} - \theta = 2\pi h, \qquad h \in Z^m$$
$$r^{(N)} - r = 0 \quad .$$

So far we solved only the first half of the equation, which holds on an m-dimensional torus $\mathcal{T}: r = r(\theta,\varepsilon)$. Geometrically speaking, we have to find a point of intersection of $\mathcal{T} \cap \phi^N(\mathcal{T})$, and any point of this intersection is a fixed point of ϕ^N.

Generally two nearby m-dimensional tori in a 2m-dimensional space need not intersect, and it is here that the exact canonical character of comes into play.

In fact, on $r = r(\theta,\varepsilon)$ there exists a function $W(\theta,\varepsilon) \in C''$, of period 2π in $\theta_1,\dots,\theta_m$, such that

$$r_k^{(N)} - r_k = \frac{\partial W}{\partial r_k} \quad . \tag{11}$$

Indeed, since ϕ is an exact canonical mapping we have

$$\int \left\{ (r^{(N)}, d\theta^{(N)}) - (r, d\theta) \right\} = 0$$

for any closed curve on $r = r(\theta,\varepsilon)$. Since $\theta^{(N)} - \theta = 2\pi h$, hence $d\theta^{(N)} = d\theta$ we have

$$\int \left(r^{(N)} - r,\ d\theta \right) = 0$$

for any closed curve on the torus. Thus the integral

$$\int_0^\theta \left(r^{(N)} - r,\ d\theta \right)$$

is independent of the path and defines the desired function $W = W(\theta,\varepsilon)$.

To find a fixed point of ϕ^N we have to pick a point of $r = r(\theta,\varepsilon)$ with $r^{(N)} = r$, i.e. by (11) a critical point of W. For example, the maximum or minimum of W provides such points in $B_{\rho'} \times T^m$, and the corresponding orbit lies in $B_\rho \times T^m$. This proves Theorem 2.

3. Linear Normal Form

In the following it will be our aim to reduce the proof of Theorem 1 to Theorem 2. The procedure will be the following: We study a canonical map ϕ of the type described in Theorem 1, in particular, $d\phi$ may have eigenvalues which do not lie in the unit circle, for most points p the iterates $\phi^j(p)$ will escape at an exponential rate. This will ruin the estimates needed in the previous sections and therefore we construct an invariant submanifold on which such exponential escape does not occur. Restricting ourselves to this manifold, the so-called center manifold, we will be able to apply Theorem 2 to get the desired periodic points.

We begin with a study of $d\phi$ at the fixed point and bring it partially into a diagonal form.

The eigenvalues of a real linear canonical map -- like $d\phi$ -- occur in pairs λ, λ^{-1} and also in pairs $\lambda, \bar{\lambda}$. We denote the eigenspaces belonging to $|\lambda| > 1$, $|\lambda| < 1$, $|\lambda| = 1$ by E_+, E_-, E_0 respectively. Let $d\phi$ act on R^{2n} which is equipped with a symplectic bilinear form $[s_1,s_2] = -[s_2,s_1] \in R$ for $s_1,s_2 \in R^{2n}$ so that $[d\phi s_1, d\phi\, s_2] = [s_1,s_2]$. Moreover, $[\ ,\ ]$ is nondegenerate, i.e. $[s_1,s] = 0$ for all $s \in R^{2n}$ implies $s_1 = 0$.

<u>Lemma 4</u> Let E_α, E_β be the null spaces of $(d\phi - \alpha)^{2n}$, $(d\phi - \beta)^{2n}$ and $\alpha\beta \neq 1$. Then

$$[E_\alpha, E_\beta] = 0 .$$

<u>Proof</u>: Let E_α^k be the null space of $(d\phi - \alpha)^k$ so that

$$(0) = E_\alpha^0 \subset E_\alpha^1 \subset \ldots \subset E_\alpha^{2n} = E_\alpha .$$

It suffices to prove

$$[E_\alpha^k, E_\beta^\ell] = 0 \quad \text{if } \alpha\beta \neq 1$$

for all $k,\ell \geq 0$, which will be done by induction with respect to $k+\ell$.

For $k+\ell = 0$ the statement is trivial and we assume $[E_\alpha^\kappa, E_\beta^\lambda] = 0$ for $\kappa+\lambda < k+\ell$. Let

$$s_\alpha \in E_\alpha^k , \qquad\qquad s_\beta \in E_\beta^\ell$$

and set

$$s_\alpha' = (d\phi - \alpha)s_\alpha \in E_\alpha^{k-1} , \quad s_\beta' = (d\phi - \beta)s_\beta \in E_\beta^{\ell-1}$$

Thus

$$\alpha\, s_\alpha = d\phi\, s_\alpha - s_\alpha' , \qquad\qquad \beta\, s_\beta = d\phi\, s_\beta - s_\beta'$$

and therefore, since $d\phi$ is canonical,

$$\alpha\beta[s_\alpha,s_\beta] = [d\phi s_\alpha - s_\alpha', d\phi s_\beta - s_\beta'] = [s_\alpha,s_\beta] - [d\phi s_\alpha, s_\beta'] - [s_\alpha', d\phi s_\beta]$$

where we used the induction hypothesis. By the same reason and since $d\phi\ s_\alpha \in E_\alpha^k$, $d\phi\ s_\beta \in E_\beta^\ell$ the last two terms vanish, and we have

$$(\alpha\beta - 1)[s_\alpha, s_\beta] = 0$$

which proves the statement.

<u>Corollary 1</u> $[E_+, E_+] = 0$, $[E_-, E_-] = 0$, $[E_\pm, E_0] = 0$.

<u>Corollary 2</u> $[\ ,\ ]$ restricted to $E_+ + E_-$ or to E_0 is nondegenerate. In particular, $\dim E_+ = \dim E_-$ and $\dim E_0$ even.

Indeed, if $[s, E_+] = 0$, $[s, E_-] = 0$ and $s \in E_+ + E_-$ then, by Corollary 1, also $[s, R^{2n}] = 0$, i.e. $s = 0$. The same argument applies to E_0.

Set $\dim E_0 = 2m$, $\dim E_+ = \dim E_- = \ell$. Introduce a basis $\phi_1, \ldots, \phi_{2m}$ in E_0 such that

$$[\phi_k, \phi_j] = \delta_{k,j-m} \quad \text{for} \quad k \leq j \leq 2m \tag{12}$$

Then we can choose $\psi_1, \ldots, \psi_\ell$ as a basis in E_+ and $\psi_{\ell+1}, \ldots, \psi_{2\ell}$ as a basis in E_- so that by Corollary 1

$$[\psi_k, \psi_j] = 0 \quad \text{for} \quad k,j \leq \ell \text{ or } k,j \geq \ell+1$$

while, by Corollary 2, the matrix $[\psi_k, \psi_{j+\ell}] = R_{kj}$ for $k, \ell = 1,2,\ldots \ell$ is nonsingular. We replace the ψ_k by the basis ψ_k' defined by

$$\left.\begin{aligned} \psi_j &= \psi_j' \\ \psi_{j+\ell} &= \sum_{k=1}^{\ell} R_{kj}\ \psi_{k+\ell}' \end{aligned}\right\} \quad \text{for } j \leq \ell$$

so that

$$[\psi_k', \psi_j'] = \delta_{k,j-\ell} \quad \text{for} \quad 1 \leq k \leq j \leq 2\ell .$$

Thus if we introduce the variables $z_1, \ldots, z_{2n}$, $u_1, \ldots, u_\ell$, $v_1, \ldots v_\ell$ by

$$s = \sum_{k=1}^{2n} z_k \phi_k + \sum_{\ell=1}^{\ell} (u_j \psi_j + v_j \psi_{j+\ell})$$

Then $d\phi$ takes the block diagonal form

$$\text{(13)} \qquad \begin{cases} z \to Mz \\ u \to Pu \\ v \to Qv \end{cases}$$

and the bilinear form for $s, s' \in R^{2n}$ becomes

$$\text{(14)} \qquad \begin{aligned} [s,s'] &= \sum_{k=1}^{m} (z_k z'_{k+m} - z'_k z_{k+m}) + \sum_{j=1}^{\ell} (u_j v'_j - u'_j v_j) \\ &= (z, J_m z') + (w, J_\ell w') \end{aligned}$$

where w is a 2ℓ vector $(u_1, u_2, \ldots, u_\ell, v_1, \ldots, v_\ell)$ and J_m, J_ℓ are $2m$, 2ℓ matrices (respectively) of the form

$$J = \begin{pmatrix} 0 & I \\ -I & 0 \end{pmatrix}$$

In (13) the matrix M has eigenvalues on $|\lambda| = 1$, and P, Q have eigenvalues in $|\lambda| < 1$, $|\lambda| > 1$, respectively. Since the mapping preserves the form [ ] we conclude

$$(Pu, Qv') - (Pu', Qv) = (u, v') - (u', v)$$

hence

$$Q^T P = I \qquad \text{or} \qquad Q = (P^T)^{-1} .$$

Finally, we can always choose the norm in E_+ such that $|Pu| < |u|$, e.g. by replacing it by

$$\left(\sum_{j=0}^{\infty} |P^j u|^2 \right)^{1/2} .$$

Summarizing we can introduce variables z, u, v in such a way that $d\phi$ has the form (13) where M has eigenvalues on $|\lambda| = 1$ and

$$|P| < 1 , \qquad\qquad Q = (P^T)^{-1} ;$$

and the bilinear symplectic form is given by (14) i.e. the symplectic 2 form is given by

(18)
$$\begin{cases} \omega = \frac{1}{2}\,(dz,\ \wedge J_m\, dz) + \frac{1}{2}\,(dw,\ \wedge J_\ell\, dw) \\ \quad = \sum_{k=1}^{m} dx_k \wedge dy_k + \sum_{j=1}^{\ell} du_j \wedge dv_j \end{cases}$$

if we write $z_k = x_k$, $z_{k+m} = y_k$ for $k = 1,2,\ldots,m$.

If, moreover, the eigenvalues on the unit circle are distinct (hence $\neq \pm 1$) we can find a diagonal form for M: Set the distinct eigenvalues on the unit circle by $\lambda_k = e^{i\alpha_k}$, $\bar{\lambda}_k = e^{-i\alpha k}$ $(k = 1,2,\ldots,m)$ and the corresponding eigenvectors χ_k, $\bar{\chi}_k$, so that

$$[\chi_k,\chi_j] = 0\ ,\quad [\chi_k,\bar{\chi}_j] = \rho_k\,\delta_{k,j} \quad \text{for} \quad 1 \leq k \leq j \leq m,$$

where $\rho_k \neq 0$ is purely imaginary. Replacing χ_k by a multiple we can achieve that $\rho_k = \pm 2i$, and exchanging χ_k and $\bar{\chi}_k$, if necessary, we can achieve that $\rho_k = -2i$. Defining our real basis $\phi_1,\phi_2,\ldots\phi_{2m}$ by

$$\chi_k = \phi_k + i\phi_{k+m} \quad \text{for} \quad k = 1,2,\ldots,m$$

we find that (12) is satisfied and that M has the form

$$M = \begin{pmatrix} \cos\alpha & -\sin\alpha \\ +\sin\alpha & \cos\alpha \end{pmatrix} = e^{J_m A}$$

where $\alpha = \text{diag}\,(\alpha_1,\alpha_2,\ldots,\alpha_m)$ and $A = \text{diag}\,(\alpha_1,\ldots,\alpha_m, \alpha_1,\ldots,\alpha_m)$.

Thus, if $d\phi$ is a canonical map whose eigenvalues on the unit circle are assumed to be distinct then it can be represented in real variables $z = (x,y) = R^{2m}$, $\omega = (u,v) \in R^{2\ell}$ in the form

(15)
$$z \to e^{-J_m A} z\ , \qquad u \to Pu\ , \qquad v \to (P^T)^{-1} v$$

where the symplectic form is given by (14) and

$$|P| < 1 , \quad A = \text{diag}\,(\alpha_1,\ldots,\alpha_{2m}) \quad \text{with} \quad \alpha_{k+m} = \alpha_k$$

for $k = 1,2,\ldots,m$. This formula (15) gives the required normal form for $d\phi$ in canonical coordinates.

4. The Center Manifold

In our aim to reduce the mapping ϕ of Theorem 1 to that of Theorem 2 we construct a locally invariant manifold W of dimension $2m$ which is tangent to E_0. In this consideration the canonical character does not play a role.

Theorem 4 Let ϕ be a mapping of the form

$$\begin{cases} z^{(1)} = Mz + f(z,w) \\ w^{(1)} = Rw + g(z,w) \end{cases} \tag{16}$$

where $f,g \in C^4$ and f, g vanish with their first derivatives at the origin. Moreover, M has its eigenvalues on the unit circle, while R has no eigenvalue on the unit circle, M, R both nonsingular.

Then there exists an invariant manifold

$$W\colon \; w = \mu(z)$$

where $\mu \in C'''$ and $\mu(0) = 0$, $D_t\mu(0) = 0$, i.e. W is tangent to the hyperplane $w = 0$.

We apply this theorem to our canonical mapping with $d\phi$ given by (15), i.e.

$$R = \begin{pmatrix} P & 0 \\ 0 & (P^T)^{-1} \end{pmatrix}$$

and we obtain a submanifold $W \in C'''$. Restricting ϕ to W we obtain

$$\phi|_W\colon \; z^{(1)} = Mz + f(z,\mu(z))$$

which is a C'''-mapping with linear part given by $z \to Mz$.

For the proof of Theorem 4 one proceeds in steps and first constructs a manifold W_+ tangent to $E_+ + E_0$ and then a manifold W_- tangent to $E_- + E_0$. Then W is obtained as the intersection of these two. For the following proof we refer to [11] although the setup is slightly different. For references to A. Kelley's work see [11].

We change the notation completely and consider a map ϕ the form

$$
(17) \qquad \begin{cases} x^{(1)} = Ax + f(x,y) \\ y^{(1)} = By + g(x,y) \end{cases}
$$

near a fixed point. Here $x \in R^{\kappa}$, $y \in R^{\nu}$ and $f,g \in C^r$, $r \geq 2$, vanish with their first derivatives at the origin. Furthermore we assume

$$
(18) \qquad \|B^{-1}\| < 1 , \quad \|A\|^r \|B^{-1}\| < 1 .
$$

Theorem 5 The above map (17) possesses an invariant manifold

$$
W_-: y = u(x) \in C^{r-1}
$$

with $u(0) = 0$, $D_x u(0) = 0$, i.e. W_- is tangent to $y = 0$ at the origin.

Thus the mapping (17) restricted to W_- has only those eigenvalues in $|\lambda| \leq 1$ while those of $|\lambda| > 1$ are cut out. We apply this theorem to the mapping (16) with $w = (u,v)$

$$
R = \begin{pmatrix} P & 0 \\ 0 & Q \end{pmatrix} ; \quad \|P\| < 1 ; \quad \|Q^{-1}\| < 1 .
$$

Thus Theorem 5 assures the existence of

$$
W_-: v = \mu_-(z,u) \in C'''
$$

Applying the same argument to ϕ^{-1} we find a manifold

$$
W_+: u = \mu_+(z,v) \in C''' .
$$

Since μ_-, μ_+ vanish with their first derivative at the origin we can solve the last two equations to get

$$W = W_+ \cap W_- : \begin{pmatrix} u \\ v \end{pmatrix} = w = \mu(z) \in C'''$$

where $\mu(0) = 0$, $D_z\mu(0) = 0$ which proves Theorem 4.

Proof of Theorem 5: The manifolds are considered in a neighborhood of the origin only. If we replace x,y by $\varepsilon x, \varepsilon y$, A, B are unchanged and f,g replaced by $\varepsilon^{-1} f(\varepsilon x, \varepsilon y)$, $\varepsilon^{-1} g(\varepsilon x, \varepsilon y)$ in a fixed domain, say $|x| < 1$, $|y| < 1$, where $\varepsilon > 0$ is a small parameter. By choice of ε we can thus achieve that

$$|f|_r + |g|_r \tag{19}$$

is arbitrarily small.

Second, we can assume that $f(x,y)$ vanishes for $|x|$ near 1. Otherwise we multiply $f(x,y)$ by a smooth function $\zeta(x)$ which is equal to 1 for $|x| < \frac{1}{2}$ and equal to zero near the boundary. Thus $\zeta(x)\, f(x,y)$ has the desired property and (19) is multiplied by a fixed constant, i.e. can still be made arbitrarily small by choice of ε. We call $\zeta(x)\, f(x,y)$ again $f(x,y)$ and since this does not amount to a change for $|x| < \frac{1}{2}$ it suffices to find an invariant manifold $y = u(x)$ for the altered map. But since the new map is defined for all $x \in R^k$ we can and will look for a global invariant manifold $y = u(x)$, defined for all $x \in R^k$. In fact, this is the clue to the proof since the invariant manifold is locally not uniquely determined, while the global manifold is.

The invariance of the manifold $y = u(x)$ requires that for the image points we have $y^{(1)} = u(x^{(1)})$ or that

$$Bu + g(x,u) = u(Ax + f(x,u))$$

or equivalently

$$u = B^{-1}\Big\{u(Ax + f(x,u)) - g(x,u)\Big\} .$$

Denoting the right-hand side as a functional $F(u)$ we are looking for a fixed point $u = F(u)$ of the functional F. This can be found by the contraction principle, given $|B^{-1}| < 1$.

More precisely, we define the domain

$$D = \left\{u \in C^r,\ u(0) = 0,\ D_x u(0) = 0,\ |u|_r < 1\right\}$$

and show that F maps D into itself. Since $f,g \in C^r$ it is clear that $F(u) \in C^r$ and also $F(u) = 0$, $D_x F(u) = 0$ for $x = 0$ follows easily from the fact that f,g and the first derivatives vanish at the origin. It remains to estimate $|F(u)|_r$. Here we need to choose $|f|_r + |g|_r$ sufficiently small. Now we use for the norm $|f|_r$ a norm equivalent to the previously used one, but more adapted to this problem, namely, if $\xi^{(1)},\ldots,\xi^{(\rho)} \in R^K$, $\rho \le r$ we form

$$\left(\frac{\partial}{\partial\lambda_\rho} \cdots \frac{\partial}{\partial\lambda_1}\right) f(x + \lambda_1\xi^{(1)} + \ldots + \lambda_\rho\xi^{(\rho)})\Big|_{\lambda_1=0,\ldots,\lambda_\rho=0}$$

$$= \sum_{\beta_1,\ldots,\beta_\rho} \left(D_{x_{\beta_\rho}} D_{x_{\beta_{\rho-1}}} \cdots D_{x_{\beta_1}} f)\xi^{(1)}_{\beta_1}\xi^{(2)}_{\beta_2} \cdots \xi^{(\rho)}_{\beta_\rho}\right)$$

and take the maximum of the absolute value of this expression for all $x \in D$, $|\xi^{(1)}| \le 1, \ldots, |\xi^{(\rho)}| \le 1$, with $\rho \le r$ and call this norm $|f|_r$. It is a generalization of the familiar operator norm

$$\|A\| = \sup_{|\xi|\le 1} |A\xi|$$

To estimate $|F(u)|_r$ we abbreviate $D_{x_{\beta_\rho}}$ by D_{β_ρ} and observe that

$$D_{\beta_\rho} \cdots D_{\beta_1} u(Ax + f(x,u))$$

$$= \sum_{\gamma_1\cdots\gamma_\rho} (D_{\gamma_\rho} \cdots D_{\gamma_1} u(w))\, D_{\beta_1} w_{\gamma_1} \cdots D_{\beta_\rho} w_{\gamma_\rho} + O(\varepsilon)\ ,$$

where $w(x) = Ax + f(x,u(x))$, since all other terms contain a derivation of $f(x,u)$ which can be estimated by $O(\varepsilon)$.

Using our definition of the norm $|\ |_r$ we multiply the above relation by $\xi_{\beta_1}^{(1)} \dots \xi_{\beta_\rho}^{(\rho)}$ and estimate the expression by

$$|F(u)|_r \leq |B^{-1}|\left\{|u|_r \left\|\frac{\partial w}{\partial x}\right\|^r + O(\varepsilon)\right\}$$

$$= |B^{-1}|\ |u|_r\ \left(\|A\| + O(\varepsilon)\right)^r + O(\varepsilon)\ .$$

Since $|u|_r < 1$ and $\|A\|^r\|B^{-1}\| < 1$ one can choose ε so small that $\|F(u)\|_r < 1$, as we wanted to show.

The main observation is that F is a contraction in the maximum norm: For $u,v \in D$ we have

$$(20) \qquad |F(u) - F(v)|_0 \leq \theta|u - v|_0$$

with sum $\theta < 1$. Indeed

$$|g(x,u) - g(x,v)| < |D_y g|_0 \cdot |u - v|_0 = O(\varepsilon)|u - v|_0$$

and with $Ax + f(x,u) = x'$, $Ax + f(x,v) = x''$ we have

$$\begin{aligned} |u(x') - v(x'')| &= |u(x') - u(x'') + u(x'') - v(x'')| \\ &\leq |x' - x''| + |u - v|_0 \\ &\leq \left(|D_y f| + 1\right)|u - v|_0 = \left(1 + O(\varepsilon)\right)|u - v|_0 \end{aligned}$$

hence

$$|F(u) - F(v)|_0 \leq |B^{-1}|\left(1 + O(\varepsilon)\right)|u - v|_0\ .$$

Thus if we choose a θ in $|B^{-1}| < \theta < 1$ the above estimate (20) holds for sufficiently small ε.

Hence the sequence $u_0 = 0$,

$$u_{j+1} = F(u_j)$$

lies in D and converges uniformly to a solution u of our functional equation. Since all u_j lie in D (which is not closed in the $|\ |_0$-norm) we conclude that $D^{r-1}u_j$ are uniformly Lipshitz continuous, with Lipshitz constant 1. Since for any function $v \in C^r$ one has an estimate

$$|D^\rho v| \leq c|v|_0^{1-\rho/r} \, |D^r v|_0^{\rho/r} \quad \text{for} \quad \rho \leq r$$

we conclude that $D^\rho u_j$ converge uniformly for $\rho \leq r-1$ since

$$|D^\rho u_j - D^\rho u_\ell| \leq c|u_j - u_\ell|_0^{1-\rho/r}$$

The limit function $\lim_{j\to\infty} u_j = u$ lies in C^{r-1} and is the desired solution. This completes the proof of Theorem 5.

We remark that actually W_- and hence $W = W_- \cap W_+$ belongs to C^r. This requires somewhat more delicate estimates of the modulus of continuity as it was given by Fenichel [13]. We forego this refinement.

5. Darboux's Lemma

The restriction of ϕ to the center manifold preserves a closed differential form which, however, is not in the standard form as required for theorem 2. But by a theorem of Darboux locally any such two forms are equivalent under a diffeomorphism. In this section we prove this theorem with the required smoothness assumptions.

In our case

$$\omega = \frac{1}{2}\left(dz, J_m\, dz\right) + \frac{1}{2}\left(d\mu, J_\ell\, d\mu\right) = \frac{1}{2}\left(dz, A\, dz\right)$$

where $A = A(z)$ is an alternating matrix satisfying

$$A(0) = J_m\,, \qquad A \in C''$$

near $z = 0$, since $\mu \in C'''$.

Theorem 6 Given a closed differential form ω with the above properties and $\omega_0 = \frac{1}{2}(dz, A(0)\, dz)$ there exists a local diffeomorphism $\tau: z^* = f(z)$ with

$$f(0) = 0 \ , \quad D\, f(0) = I$$

and $f \in C''$. Moreover, at the origin f admits even 3 derivatives, i.e. has a Taylor expansion of order 3.

Note if $h(z) \in C''$ is a matrix function and $h(0) = 0$ then the function $f(z) = h(z) \cdot z$ has 3 derivatives at the origin. It is this phenomenon which occurs for the diffeomorphism τ.

If we subject our map to this transformation then $\phi|_W$ is mapped into a C" diffeomorphism admitting 3 derivations at the origin. Moreover ω is mapped into standard form and so we have reduced the mapping $\phi|_W$ to a canonical mapping as is required for Theorem 2. In the next section we will transform this mapping $\phi|_W$ into the form (4), so that Theorem 2 applies.

Proof: We shall need the Poincaré lemma for differential forms of degree 2, say

$$\alpha = \sum_{k,\ell=1}^{2m} a_{k\ell}(z)\, dz_k \wedge dz_\ell = (a(z)\, dz, dz)$$

with $a(z) \in C''$. If α is moreover closed there exists a one-form, say β satisfying

$$\beta = d\, \alpha$$

where β has coefficients in C", but at $z = 0$ they admit 3 derivatives.

Indeed, if we follow the standard construction of β by integrating along the rays tz we find

$$\beta = \sum b_k\, dz_k$$

with

$$b_k(z) = \int_0^1 \sum_j a_{kj}(tz) t z_j\, dt$$

where $a_{kj} \in C''$. From this formula the assertion is evident.

We prove our statement, using the arguments in [9], [10] and define the differential form

$$\omega_t = \frac{1}{2} (dz, A(tz)\, dz)$$

so that $\omega = \omega_1$, and ω_t is also a closed form. We determine a transformation $\tau_t: z \to f_t(z)$ such that

$$\tau_t^* \omega_t = \omega_0$$

by constructing a t-dependent vector field X_t which generates τ_t by

$$\frac{d}{dt} \tau_t = X_t(\tau_t) \,, \qquad \tau_0 = id \,.$$

Differentiating $\tau_t^* \omega_t = \omega_0$ we find

$$0 = \frac{d}{dt} (\tau_t^* \omega_t) = \tau_t^* \left\{ \dot{\omega}_t + d(\omega_t \lrcorner X_t) \right\}$$

where $\dot{\omega}_t$ is the form obtained by differentiating the coefficients $A(tz)$ with respect to t. Hence $\dot{\omega}_t$ is also closed and by Poincaré's lemma there exists a 1-form ψ such that $\dot{\omega}_t = d\psi$, Moreover $\psi \in C''$ and, by the above remark has 3 derivatives at the origin. Now we solve the above equation by setting

$$\omega_t \lrcorner X_t = - \psi$$

which has a unique solution, since ω_t is nondegenerate. Again $X_t \in C''$ and has 3 derivatives at the origin. Finally integrating the differential equations $d\tau_t/dt = X_t(\tau_t)$, $\tau_0 = id$, we find the transformation with the stated properties.

6. Birkhoff Normal Form

We summarize our results. The canonical map ϕ of Theorem 1 can be transformed by a transformation

$$\begin{cases} z \to \nu(z) \\ w \to w - \mu(z) \end{cases}$$

into the form, say, (16) where $g(z,0) = 0$. This expresses the fact that now $w = 0$ is the invariant manifold, because of our choice of $\mu(z) \in C'''$. The mapping $z \to \nu(z)$ was constructed according to Theorem 6 such that the restricted map

$$\phi|_W : z^{(1)} = Mz + f(z,0)$$

preserves the differential form

$$\frac{1}{2} (dz, J_m dz) ,$$

i.e. is canonical. Since $\nu \in C''$ but admits 3 derivatives at the origin the same is true for $\phi|_W$.

Therefore we consider a C^1-canonical mapping admitting a Taylor expansion of order 3 at the origin. We will subject it to a canonical transformation τ, which is real analytic, and bring it into the form (4) required in Theorem 2. The following result is a normal form due to G. D. Birkhoff, which we carry only to third order. Actually Birkhoff considered the corresponding question only for differential equations, autonomous or with periodic time dependence. The corresponding problem for the mapping can be treated in the same way but there is a minor technical difficulty.

<u>Theorem 7</u> Let $\phi \in C^1$ be a canonical mapping near the fixed point $z = 0$, 3 times differentiable at the origin and let $d\phi|_0$ have distinct eigenvalues $\lambda_k, \lambda_k^{-1}$, all on the unit circle and satisfying the condition

$$\prod_{k=1}^{m} \lambda_k^{j_k} \neq 1 \quad \text{for} \quad 1 \leq \sum |j_k| \leq 4 .$$

Then there exists a real analytic canonical transformation τ such that $\tau^{-1} \cdot \phi \cdot \tau$ has the form (4).

With this result theorem 1 is reduced to Theorem 2 since we succeeded in bringing $\phi|_W$, the restriction of ϕ to the center manifold, into the form for which the fixed point theorem of Section 2 applies, at least if $\beta = (\beta_{k\ell})$ is nonsingular. This is the condition alluded to under (iii) in Theorem 1. It is worth mentioning that this condition is invariant under diffeomorphisms in C''', as one easily verifies.

We turn to the proof of Theorem 7, which is actually a statement about formal power series aside from the remark that any formal canonical transformation agrees to finite order with a real analytic canonical transformation. To make this more precise we consider a formal canonical transformation τ for which the linear part $d\tau$ is the identity

$$\tau: z \to z^* = z + h(z) + \cdots \tag{21}$$

which is homogeneous of degree $\nu \geq 2$ in the coordinates $z_1, \ldots z_{2m}$.

<u>Lemma 5</u> For a formal canonical transformation of the form (21) there exists a homogeneous polynomial $H = H(z)$ such that

$$h(z) = J \frac{\partial}{\partial z} H(z) \quad . \tag{22}$$

Conversely, given such an $h(z)$ there exists a real analytic transformation whose Taylor expansion agrees with (21) to terms $\leq \nu$.

<u>Proof</u>: We make use of so-called generating functions to represent a canonical transformation. With the coordinates $x_k = z_k$, $y_k = z_{k+m}$, $x_k^* = z_k^*$, $y_k^* = z_{k+m}$ for $k = 1,2,\ldots,m$ we have on account of the canonical character

$$\sum dx_k^* \wedge dy_k^* = \sum dx_k \wedge dy_k$$

hence the one form

$$(x^* - x, dy^*) + (y - y^*, dx)$$

is closed. Using x, y^* as independent variables, which is possible since $d\tau$ at $z = 0$ is the identity we write the above 1-form as $dV(x,y^*)$ and hence

$$(23) \qquad x^* = x + \frac{\partial V}{\partial y^*}, \qquad\qquad y = y^* + \frac{\partial V}{\partial x}.$$

If τ is given by formal power series (21) then $V(x,y^*)$ is a formal power series beginning with terms of degree $\nu + 1$, if we neglect an irrelevant constant. By expansion one finds for the explicit representation (21)

$$x^* = x + \frac{\partial V^{(\nu+1)}}{\partial y} + \dots, \qquad y^* = y - \frac{\partial V^{(\nu+1)}}{\partial x} + \dots$$

if $V^{(\nu+1)}$ denotes the homogeneous terms of degree $\nu+1$ in V. Hence (22) is verified with $H = V^{(\nu+1)}$.

Conversely, given such a real homogeneous polynomial H we define a canonical mapping implicitly by (23) with $V = H(x,y^*)$, which gives rise to a real analytic canonical mapping of the desired form.

Proof of Theorem 7: We write ϕ in the form

$$\phi = \phi^{(1)} \circ \psi$$

where $\phi^{(1)}$ is the linear map $d\phi|_0$ and thus ψ has the identity as its Jacobian map at the origin and consequently can be represented in the form (21), (22) with a generating function H homogeneous of degree 3. We subject this mapping to a canonical transformation τ with the generating function V, also homogeneous of degree 3 by (23). The resulting mapping $\phi^* = \tau^{-1} \circ \phi \circ \tau$ can again be represented as $\phi^* = \phi^{(1)} \circ \psi^*$ with H replaced by

$$(24) \qquad H^* = H - V(Mz) + V(z)$$

where $\phi^{(1)}: z \to Mz$.

To prove this statement we compare the quadratic terms in the Taylor expansion of

$$\tau \circ \phi^{*} = \phi \circ \tau$$

and find readily with $H_z = \partial H/\partial z$,

$$M\, J\, H_z^{*} + J\, V_z(Mz) = M\, J(V_z + H_z)\ .$$

Now we use

$$\frac{\partial}{\partial z} V(Mz) = M^{T} V_z(Mz)$$

and the fact that M is symplectic: $M^{T} J\, M = J$ to conclude

$$\frac{\partial}{\partial z}\left\{H^{*}(z) + V(Mz) - V(z) - H(z)\right\} = 0\ .$$

Since all terms are homogeneous we obtain (24).

Next we show that V can be found so that $H^{*} = 0$, and V is uniquely determined by this condition. Thus ψ^{*} agrees with the identity to second degree and we can represent it again in the form (21) with a generating function H of fourth degree. Subjecting this mapping to another transformation (23) with generating function V homogeneous of degree 4 we replace H by H^{*} given by (24), by the same argument. It turns out that now H^{*} cannot be made $= 0$ by choice of V but only by a function satisfying

$$H^{*}(Mz) = H^{*}(z) \tag{25}$$

To prove these statements and characterize the resulting H^{*} given by (25) we introduce complex variables $\zeta_k = x_k + iy_k$, $\bar{\zeta}_k = x_k - ig_k$ and express V as a polynomial in $\zeta_k, \bar{\zeta}_k$:

$$V(z) = \sum V_{\rho\sigma}\, \zeta^{\rho}\, \bar{\zeta}^{\sigma} \quad \text{where} \quad \zeta^{\rho} = \prod_{k=1}^{m} \zeta_k^{\rho_k}\ .$$

Since $z \to Mz = e^{-JA}z$ goes over into $\zeta_k \to e^{\alpha_k}\zeta_k = \lambda_k \zeta_k$, $\bar{\zeta}_k \to \lambda_k^{-1}\bar{\zeta}_k$ we find

$$V(Mz) - V(z) = \sum (\lambda^{\rho-\sigma} - 1) V_{\rho\sigma} \zeta^{\rho} \bar{\zeta}^{\sigma} .$$

If $H_{\rho\sigma}$, $H^*_{\rho\sigma}$ are the coefficients of H, H^* in a similar expansion the equations (24), (25) take the form

$$\begin{cases} (\lambda^{\rho-\sigma} - 1) V_{\rho\sigma} = H_{\rho\sigma} - H^*_{\rho\sigma} \\ (\lambda^{\rho-\sigma} - 1) H^*_{\rho\sigma} = 0 \end{cases}$$

Hence $3 \leq \sum (\rho_k + \sigma_k) \leq 4$ hence $\sum |\rho_k - \sigma_k| \leq 4$ and so by assumption $\lambda^{\rho-\sigma} - 1 = 0$ only if $\rho = \sigma$. Hence the second equation implies $H^* = 0$ for $\rho \neq \sigma$ and therefore the first $(\lambda^{\rho-\sigma}-1)V_{\rho\sigma} = H_{\rho\sigma}$ for $\rho \neq \sigma$. For $\rho = \sigma$ we get $H_{\rho\rho} = H^*_{\rho\rho}$ and $V_{\rho\rho}$ remains arbitrary. Setting $V_{\rho\rho} = 0$ the solution is uniquely determined and defines a <u>real</u> homogeneous polynomial $V(z)$. This follows simply from the fact that the above equations in real coordinates z have the real form (24) and (25), and their solution is unique. Moreover, H^* is of a homogeneous polynomial in $\zeta_k \bar{\zeta}_k$; hence $H^*(z)$ of even degree. In particular $H^* = 0$ if its degree is 3 as was stated. For degree 4 it is a quadratic form in $\zeta_k \bar{\zeta}_k = x_k^2 + y_k^2 = r_k$ say

$$H^* = -\frac{1}{4} \sum_{k,\ell} \beta_{k\ell} r_k r_\ell$$

Thus the mapping ϕ has been transformed into one of the form $\phi = \phi^{(1)} \circ \psi$ where ψ has the form

$$z \to z + J \frac{\partial H^*}{\partial z} + \cdots$$

where

$$\frac{\partial H^*}{\partial z_k} = - b_k z_k \quad \text{with} \quad b_k = \sum_{\ell=1}^{m} \beta_{k\ell} r_\ell = b_{k+u}$$

and hence ϕ has the form

$$z \to e^{-JA}\{z - JBz + \cdots\} = e^{-J(A+B)} z + \cdots$$

where $B = \text{diag}(b_1, b_2, \ldots, b_{2m})$, $A = \text{diag}(\alpha_1, \ldots, \alpha_{2m})$ and J

commute with each other. This is the desired normal form to third order, proving Theorem 7. It is evident how to find a corresponding normal form to higher order, but that is not needed here.

References

[1] G. D. Birkhoff, Dynamical Systems, American Mathematical Society, 1966.

[2] Siegel, C. L., and Moser, J., Lectures on Celestial Mechanics, Springer-Verlag, 1971.

[3] Birkhoff, G. D., Une generalization à n-dimensions du dernier théorème de géométrie de Poincaré, Compt. Rend. Acad. Sci. 192, 1931, 196-198.

[4] Birkhoff, G. D., and Lewis, D. C., On the periodic motions near a given periodic motion of a dynamical system, Annali di Mat. 12, 1933, pp. 117-133.

[5] Lewis, D. C., Sulle oscillazioni periodiche d'une sistema dinamico, Atti Accad. Naz. Sincei Rend. Cl. Sci. Fis. Mat. 19, 1934, 234-237.

[6] Harris, T. C., Periodic solutions of arbitrary long periods in Hamiltonian systems, Jour. Diff. Eq. 4, 1968, 131-141.

[7] Arnold, V. I., Sur une properiété topologique des applications globalement canonique de la mécanique classique, Comp. Rend. Acad. Sci. 261, 1965, 3719-3722.

[8] Arnold, V. I., and Avez, A., Problèmes ergodiques de la mécanique classique, Gauthiers-Villars, 1967, Appendix 33.

[9] Weinstein, A., Symplectic structures on Banach manifolds, Bull. A.M.S., 1040-1041, 1969.

[10] Moser, J., On the volume elements on a manifold, Trans. Am. Math. Soc. 120, 286-294, 1965.

[11] Lanford III, O. E., Bifurcation of periodic solutions into invariant tori; the work of Ruelle and Takens, Lecture Notes

in Mathematics 322, 159-192, Springer-Verlag, 1972.

[12] Zehnder, E., Stability and Instability in Celestial Mechanics, Ensciquement du 3e Cycle de la Physique en Suisse Romande, Ecole Polytechnique Federale, Lausanne, 1975, esp. Appendix III.

[13] Fenichel, N., Persistence and smoothness of invariant manifolds for flows, Ind. Univ. Math. Jour. 21, 193-226, 1971.

[14] Klingenberg, W., Closed Geodesics on Riemannian manifolds, Proc. 13 Biennial Seminar Canadian Math. Congress. ed. J. R. Vanstone, Montreal, Canada, 1972, Vol. 1, pp. 69-92.

[15] Klingenberg, W., Lectures on Closed Geodesics, (second and revised edition), preprint, University Bonn, 1976.

Some Developments on Stability and Bifurcations of Dynamical Systems

by

J. Palis

We relate here some new results in the theory of structural stability and bifurcations of dynamical systems: vector fields and diffeomorphisms, including holomorphic vector fields. We also mention a number of known relevant results in order to present a better perspective on some aspects of this subject.

Two dynamical systems are equivalent if there is a homeomorphism of the ambient space sending orbits of one onto orbits of the other. This means that the topological behavior of the corresponding orbits is the same, in particular their α and ω-limit sets (assymptotic behavior) are topologically the same. Endowing the set of systems with some topology (usually the C^r topology), we say that a system is structurally stable if it is equivalent to all nearby systems.

After the definitions of structural stability and degree of instability introduced in 1937 in the works of Andronov, Pontrjagin and Leontovic, going through the results of Peixoto on two-manifolds [1] and Anosov in the globally hyperbolic case [2], it took some time before the existence of structurally stable systems on every (real) manifold was proved (Palis-Smale [3]). More recently, very general theorems on stability were established by Robbin [4] and Robinson [5], [6].

We point out that if one requires a diffeomorphism of the ambient space in the definition of equivalent systems, then very strong relations are imposed between the linear approximations of the systems at periodic orbits; in general no system can be stable in this sense. At this very early stage of the definition of stable systems one feels a good difference with the theory of stability of mappings. However, a basic goal in both theories (set forth mainly by Smale and Thom) consists in describing stable systems and mappings and see how "often" or how "generic" (open and dense or at least residual) are they in the space of all systems or all mappings. Going beyond that we discuss here how a system depending on a parameter can cease to be stable (a bifurcation point), what kind of system can appear afterwards and how stable is such a process. Then a conceptual correlation of both theories is again enlightening.

§1. Local Stability. Let $Diff^r(M)$ and $\chi^r(M)$ indicate the C^r diffeomorphisms and C^r vector fields of a C^∞ compact manifold M with the C^r topology, $r \geq 1$. One can ask the question when a diffeomorphism or a vector field is structurally stable near a periodic orbit. This question in the real case is completely understood through the concept of hyperbolicity.

Before we state a general theorem on local stability, let us remark the following interesting fact. We have said that $f \in Diff^r(M)$ is stable if for g C^r near f there is a homeomorphism sending orbits of f onto orbits of g. We could have required what seems to be a stronger property: a homeomorphism h which conjugates f and g; i.e., $hf = gh$. It turns out that the two concepts are identical [7].

A singularity $p \in M$ of a vector field X is hyperbolic if the derivative $DX(p)$ has no eigenvalue with zero real part. If all eigenvalues have negative (positive) real part we call it a sink or attractor (source or repellor); otherwise, it is a saddle. A periodic point $p \in M$ of a diffeomorphism f with period k is hyperbolic if $Df^k(p)$ has no eigenvalue with norm one. If all eigenvalues have norm less (bigger) than one we call the orbit a sink (source); otherwise it is a saddle. Finally, a periodic orbit γ of X is hyperbolic if $p \in \gamma$ is a hyperbolic fixed point for a Poincaré mapping associated to γ at p. This mapping is defined using a local cross section through p and considering the first return map via orbits of X. In all these cases we can associate immersed one to one submanifolds invariant by the system. the stable and unstable manifolds, consisting respectively of orbits whose ω and α-limit set is the periodic (or fixed) hyperbolic orbit [8].

A classical result by now is the following theorem of Grobman [9] and Hartman [10]; see also [11], [12].

Theorem - A necessary and sufficient conditions for a vector field or a diffeomorphism to be locally structurally stable near a periodic orbit is that the periodic orbit be hyperbolic.

This is a very satisfactory result not only because it characterizes local stability but also shows that it is a generic property in the following sense. If γ is a periodic orbit for $X \in \chi^r(M)$ or $f \in \text{Diff}^r(M)$, then near X or f there is an open set of vector fields or diffeomorphisms which are locally stable near γ.

For holomorphic vector fields near a singularity the result (perhaps surprisingly) is quite different. However, we shall make clear a good connection between the questions of local stability for holomorphic vector fields and global stability for real vector fields. Let us consider the set G of germs of holomorphic vector fields at the origin in C^n and having the origin as singularity.

Theorem ([13],[14]) - $F \in G$ is locally structurally stable at the origin if and only if (i) no two eigenvalues of $DF(0)$ are dependent over the reals and (ii) the convex hull of the eigenvalues of $DF(0)$ does not contain the origin in C.

This result implies the non generecity of locally stable holomorphic vector fields and similarly of locally stable $\mathbb{R}^2$-actions. That conditions (i) and (ii) were sufficient was proved by Guckenheimer [15]. Take a small sphere centered at the origin and transverse to the holomorphic field and consider the line field defined on the sphere by the intersections with the orbits of the holomorphic field. It turns out that the induced (real) vector field on the sphere is what we call Morse-Smale and, as we shall mention in the next section, such a vector field is structurally stable. From this the local stability of the holomorphic vector field follows immediately. On the other hand when the convex hull of the eigenvalues contains the origin in $\mathbb{C}$, the induced vector field on the sphere has a pair of periodic orbits with the stable manifold of one of them intersecting non transverselly the unstable manifold of the other. A key fact is that such a phenomenon persists under small perturbations. We shall mention in Section 3 that this is a (subtle) obstruction to topological equivalence between vector fields, unless the eigenvalues of the periodic orbits are specially related. These are some brick-ideas behind the block-results in [13], [14]; in fact much more is done there like a generic classification of linear holomorphic vector fields and the complete characterization of global stability for holomorphic vector fields on complex projective spaces.

§2. Global Stability - One main problem is to characterize global stability in terms of hyperbolicity and transversality. Naturally this question came up after several results on the stability of certain specially interesting kind of systems.

In what follows we usually consider a compact C^∞ manifold M without boundary. Let $f \in \mathrm{Diff}^r(M)$ and Λ a closed invariant set for f, i.e., $f\Lambda \subset \Lambda$. Also consider M with some riemannian metric. We say that Λ is hyperbolic if there is a continuous splitting of the tangent bundle of M restricted to Λ, $T_\Lambda M = E^s \oplus E^u$ and constants $0 < \lambda < 1$, $c > 0$ such that (i) $\|Df_x^n v\| \le c\lambda^n \|v\|$ for $v \in E_x^s$ and (ii) $\|Df_x^{-n} v\| \le c\lambda^n \|v\|$ for $v \in E_x^u$. In this case, $W_x^s = \{y;\ d(f^n x, f^n y) \to 0$ as $n \to \infty\}$ and $W_x^u = \{y;\ d(f^{-n}x, f^{-y}y) \to 0$ as $n \to \infty\}$ are immersed $\mathbb{R}^s$, $\mathbb{R}^u$ through $x \in \Lambda$, where $s = \dim E_x^s$ and $u = \dim E_x^u$ [8]. We call $W^s(x)$, $W^u(x)$ stable and unstable manifolds of x. An specially relevant closed invariant set for a diffeomorphism f is the non-wandering set $\Omega(f)$: $x \in \Omega(f)$ if given any neighborhood V of x and an integer $n_o > 0$ there is an integer n, $|n| > n_o$, such that $f^n V \cap V \neq \phi$.

We say that $f \in \mathrm{Diff}^r(M)$ satisfies Axiom A if $\Omega = \Omega(f)$ is hyperbolic and $\Omega = \overline{\mathrm{Per}(f)}$, i.e., the periodic points are dense in Ω. For such f, we say that it satisfies the strong transversality condition if $W^s(x)$ and $W^u(y)$ are transverse for any pair $x, y \in \Omega$. Important cases are Morse-Smale diffeomorphisms when Ω is finite and the Anosov diffeomorphisms when all of M is hyperbolic. The first exist on all manifolds. Similar concepts can be formulated for vector fields. In particular, the non-wandering set of a Morse-Smale vector field consists of a finite number of fixed or periodic orbits all of them hyperbolic and the transversality condition holds for their stable and unstable manifolds; for Anosov flows all of M is hyperbolic. These last systems were proved to be stable in [3] and in [2] (see also [16], [17]), respectivelly. The mot general result in this direction is

<u>Theorem</u> ([4],[6],[18]) - If $f \in \mathrm{Diff}^r(M)$ satisfies the Axiom A and the strong transversality condition then f is structurally stable.

A similar theorem is true for vector fields [5].

Good examples of Axiom A diffeomorphisms can be found in the classical survey of Smale [19]. Even before the above general result on stability, the following conjecture was posed in [3] where it was shown to be true when the nonwandering set is finite.

Conjecture - For $r \geq 1$, $f \in \mathrm{Diff}^r(M)$ is structurally stable if and only if f satisfies the Axiom A and the strong transversality condition.

It remains to show that Axiom A and the transversality condition are necessary for stability. However, some relevant facts are known about this converse. Robinson [20] proved that if f satisfies Axiom A then the transversality condition is necessary for stability. From Pugh's Closing Lemma [21], it follows easily that for $r = 1$ if f is stable then $\overline{\mathrm{Per}(f)} = \Omega(f)$. Thus for $r = 1$ it remains to prove that if f is stable then $\Omega(f)$ is hyperbolic.

I consider this question very interesting; it has been inspiring some relevant original ideas in a number of partial results. First, Franks [22] had the idea of making stronger the definition of structural stability in order to have the conjecture true both ways in $\mathrm{Diff}^r(M)$. He and Guckenheimer [23] for $r = 1$ and later Mañé [24] for any $r \geq 1$ succeeded in this project as follows. We say that $f \in \mathrm{Diff}^r(M)$ is absolutely stable if there exists $K > 0$ such that for g near f there is a homeomorphism h with $hf = gh$ and $\|h-1\| < K\|f-g\|_{C^0}$, where 1 means the identity on M. Then $f \in \mathrm{Diff}^r(M)$ is absolutely stable if and only if f satisfies the Axiom A and the strong transversality condition.

Going back to the original question, Pliss [25] showed that for $r = 1$ if f is stable then the number of periodic sinks must be finite. He also proved for $r = 1$ and $\dim M = 2$ that if f is stable and the Lebesgue measure of $\Omega(f)$ is zero then f satisfies the Axiom A [26]. Mañé [27] showed that for $r = 1$ and $\dim M = 2$ if f is stable and $\Omega(f) = M$ then f satisfies Axiom A and thus f is Anosov and $M = T^2$. More recently Lopes [28] proved in higher dimensions and for $r = 1$ that if f is stable then the number of hyperbolic attractors (not necessarily periodic) is finite.

Finally, we mention that structurally stable systems are not dense in general. Starting with Smale [29] there are several good counter examples to this fact; in this direction there are recent interesting papers by Guckenheimer [30] and Williams [31] on Lorentz attractors. Relevant cases of density of structural stability are the diffeomorphisms of the circle, vector fields on two-manifolds and in higher dimensions the gradient vector fields.

Bifurcations - Our point of view here will be the following. We consider arcs (one parameter families) of dynamical systems which are initially structurally stable, specially Morse-Smale diffeomorphisms, and then for some value of the parameter they go through a bifurcation, i.e., a topological change in the space of orbits.

We analyse how this transition occurs and describe which of these arcs are stable up to and beyond the first bifurcation point.

In this connection it is interesting to mention that Smale [32] has shown that any diffeomorphism is isotopic (connected by an arc) to a structurally stable one. Shub and Sullivan [33] also established conditions for a diffeomorphism to be isotopic to a Morse-Smale one, mainly in terms of the induced map on homology. Newhouse and Peixoto [34] constructed simple arcs (finite number of generic bifurcations) connecting any two Morse-Smale vector fields; a generalization appears in [35].

In order to break the stability of a system, we have to break the hyperbolicity condition on the non-wandering set or the transversality condition on the stable and unstable manifolds. We consider C^1 arcs $\xi: I = [0,1] \to \mathrm{Diff}^{\infty}(M)$ and denote $\xi_t = \xi(t)$. Starting with bifurcations of periodic orbits, we have three possibilities for a generic arc ξ_t.

a) ξ_b, $b \in I$, has $p \in M$ as fixed point and $D\xi_b(p)$ has one eigenvalue equal to 1, the other eigenvalues have norm different from 1.

b) ξ_b has p as fixed point and $D\xi_b(p)$ has one eigenvalue equal -1, the others have norm different from 1.

c) ξ_b has p as fixed point and $D\xi_b(p)$ has a pair of complex conjugate eigenvalues with norm 1, the others have norm different from one.

If p is periodic of period k we consider the eigenvalues of $D\xi_b^k(p)$. In all cases it is possible to give a rough description of the local phase portrait of ξ_t for $b-\delta < t < b+\delta$, $\delta > 0$ small. This goes as follows:

a) In the "central" direction, a pair of hyperbolic fixed points collapse and then disappear (or vice-versa)

b) In the "central" direction a hyperbolic attracting fixed point becomes non hyperbolic attracting and then becomes a hyperbolic repellor and a hyperbolic periodic attractor of twice the period is created.

c) In the "central" plane a hyperbolic attractor becomes a non hyperbolic attractor and then becomes a hyperbolic repellor and an invariant attracting (normally hyperbolic) circle appears.

Normal to these "central" directions we have hyperbolicity in all cases. Such periodic orbits are called quasi-hyperbolic. In case a), called saddle-node, the stable manifold of p is an immersed "half" euclidean space with the strong stable manifold (in the normal direction) as boundary. Similarly for the unstable manifold. In the other cases, called "flip" and "Hopf" bifurcations, the stable and unstable manifolds are immersed euclidean spaces.

Several authors contributed to the descriptions of these local unfoldings, like Sotomayor [36], [37], Brunovsky [38]; see Arnold [39] for more references and background and also [40], [41], [42].

To break transversality, we consider the following definition given by Sotomayor. Let N_1, N_2 be submanifolds of M and let $y \in N_1 \cap N_2$. Assume N_1 is locally flat at y and let $\mathbb{R}^\ell$ be its local "normal complement" in M, $\pi: M \to \mathbb{R}^\ell$ the natural local projection. We call N_1 and N_2 quasi-transversal at y if the derivative of π/N_2 has corank one at y and either (d.1) $\dim N_1 + \dim N_2 \geq \dim M$; then the intrinsic second derivative of π/N_2 is non-degenerate. This second derivative is defined from $K \times K$ into L, where K is the kernel of $D\pi/N_2$ at y and L is the cokernel or (d.2) $\dim N_1 + \dim N_2 = \dim M-1$.

Let us see some results from [43], [44] of a global nature for arcs ξ_t, $t \in I = [0,1]$, with ξ_o a Morse-Smale diffeomorphism and let $b \in I$ be the first bifurcation point. We assume that the limit set of ξ_b is a finite number of orbits. For generic arcs this is an open condition but it remains a relevant question if it is dense. Then the α-limit set or the ω-limit set of ξ_b is a finite number of periodic orbits. For ξ_b we have that either the periodic orbits are hyperbolic except one as in a), b) or c) and the transversality condition holds for their stable and unstable manifolds or all periodic orbits are hyperbolic and the transversality condition holds except along one orbit where a pair of stable and unstable manifold meet quasi-transverselly. Another important feature in this context is the existence or not of a cycle for the periodic orbits of ξ_b:

a sequence $p_1, p_2, \ldots, p_k, p_1$ of periodic orbits with $W^s(p_i) \cap W^u(p_{i+1}) \neq \neq \phi$, $1 \leq i \leq k-1$ and $W^s(p_k) \cap W^u(p_1) \neq \phi$.

For generic arcs ξ_t, $t \in I = [0,1]$, with ξ_o a Morse-Smale diffeomorphism and $b \in I$ the first bifurcation point, we have

Theorem - If ξ_b has no cycles then for an open and dense subset of t's in $[b, b+\delta]$, δ small, ξ_t is a Morse-Smale diffeomorphism. Moreover in cases a), b), d) the subset of t's in $[b, b+\delta]$ where ξ_t is not Morse-Smale is at most countable, and in case c) it is a Cantor set.

Theorem - If ξ_b has a cycle and an orbit of quasi transversal intersection between an stable and unstable manifolds as in (d.1), then there are infinitely many stable conjugacy classes for $t \in [b, b+\delta]$, $\delta > 0$ small, and they accumulate at ξ_b. Each of these classes contains infinitely many periodic orbits. The values of t in $[b, b+\delta]$ with ξ_t not stable is relatively small: given $\epsilon > 0$ there is δ so that the measure of this set is less than $\epsilon\delta$.

We also point out that if ξ_b has a cycle then cases b) and c) cannot occur.

In the last theorem, the set of t's in $[b, b+\delta]$ with ξ_t stable is not dense in general as recently shown by Newhouse; if $\dim M = 2$ there are values of t near b such that ξ_t has infinitely many periodic sinks (attractors). If ξ_b has a cycle and an orbit of quasi transversal intersection of stable and unstable manifolds as in (d.2), a general result as above is not yet available; partial results of a similar nature may be found in [44]. Similarly if the cycle involves a saddle node (case a), some partial and related results are known [45], [46], [47].

Finally, let us mention recent results from [47], characterizing structural stability for the arcs described above. An arc $\xi : I \to \mathrm{Diff}^\infty(M)$ is structurally stable if any nearby arc η is topologically equivalent to ξ in the following sense: for each $t \in I$ there is a homeomorphism h_t depending continuously on t such that $h_t \xi_t = \eta_{r(t)} h_t$ for some continuous reparametrization $r: I \to I$. In [47] we called such arcs continously stable to stress the continuous dependence of the conjugacies on the parameter.

The first basic question concerns the existence of universal models up to topological equivalence for the generic unfoldings of the saddle-node, flip and Hopf cases as in a), b), c). For singularities of vector fields one has to consider only the saddle-node and the Hopf cases, for each of these cases there is essentially only one model up to the "normal"hyperbolic part [48]. In general one cane reduce the question to the "central" part [49]. For periodic points of diffeomorphisms (our case) or equivalently for periodic orbits of vector fields the models are again unique in the saddle-node and flip cases up to the "normal" hyperbolic part [47]. However, the proof of this fact is much deeper for the saddle-node. On the other hand it follows from Herman [50] that no arc going through a Hopf bifucation is stable.

Another crucial fact in this context is that no arc is stable if it goes through a quasi-transversal orbit of intersection of stable and unstable manifolds of periodic orbits.

Thus we have a complete picture of the local unfoldings up to topological equivalence for generic arcs of diffeomorphisms.

For global results on the stability of arcs starting at a Morse-Smale diffeomorphism and going through a bifurcation point, we naturally consider two kinds of stability: stability up to and including the bifurcation point (left stability) and stability including a small interval of the parameter beyond the first bifurcation point.

Let ξ_t, $t \in I$, be a generic arc as above with ξ_0 a

Morse-Smale diffeomorphism and let $b \in I$ be its first bifurcation point.

<u>Theorem</u> - The arc ξ is left stable if and only if the stable and unstable manifolds of the periodic orbits of ξ_b meet transversally.

To characterize stability beyond the first bifurcation point we need the following concept. Let p be a saddle-node for $f \in Diff^{\infty}(M)$. In the stable manifold of p there is a unique locally invariant foliation $\mathfrak{F}^s$ so that the boundary of the stable manifold is a leaf of $\mathfrak{F}^s$ and f sends leaves to leaves. We say that the saddle-node is s-critical if there is a hyperbolic periodic orbit q of f such that $W^u(q)$ has a non transversal intersection with one of the leaves of $\mathfrak{F}^s$. Similarly for the unstable manifold of p. We say that p is not critical if none of these two possibilities occur.

<u>Theorem</u> - The arc ξ is (continuously) stable if and only if it is left stable, there is no cycle for ξ_b and the quasi-hyperbolic periodic orbit of ξ_b must be a non critical saddle-node or a flip.

A partial extension of this result for Axiom A diffeomorphisms appears in [51].

References

[1] M. Peixoto, Structural stability on two dimensional manifolds, Topology 1 (1962).

[2] D. Anosov, Geodesic flows on compact riemannian manifolds of negative curvature, Proc. Steklov Math. Institute 90 (1967).

[3] J. Palis and S. Smale; Structural stability theorems, Proc. Symp. in Pure Math. vol. XIV, Amer. Math. Soc. (1970).

[4] J. Robbin, A structural stability theorem, Annals of Math. 94 (1971).

[5] C. Robinson, Structural stability of vector fields, Annals of Math. 99 (1974).

[6] C. Robinson, Structural stability of C^1 diffeomorphisms, Journal of Diff. Equations (1976).

[7] I. Kupka, On two notions of structural stability, Journal of Diff. Geometry 9 (1974).

[8] M. Hirsch and C. Pugh, Stable manifolds and hyperbolic sets, Proc. Symp. in Pure Math. vol. XIV, Amer. Math. Soc. (1970).

[9] D. Grobman, Homeomorphisms of systems of differential equations, Dokl. Akad. Nauk. 128 (1959).

[10] P. Hartman, A lemma in the structural stability of differential equations, Proc. Amer. Math. Soc. 11 (1960).

[11] J. Palis, On the local structure of hyperbolic fixed points in Banach spaces, Anais da Acad. Bras. de Ciências (1968).

[12] C. Pugh, On a theorem of P. Hartman, Amer. Journal of Math. 91 (1969).

[13] C. Camacho, N. Kuiper and J. Palis, La topologie du feuilletage d'un champ de vecteurs holomorphe près d'une singularité, C.R. Sc. Paris t. 282 (1976).

[14] C. Camacho, N. Kuiper and J. Palis, The topology of holomorphic flows with singularities, to appear.

[15] J. Guckenheimer, Hartman's theorem for complex flows in the Poincaré domain, Compositio Math. 24 (1972).

[16] J. Moser, On a theorem of Anosov, Journal of Diff. Equations 5 (1969).

[17] J. Mather, Anosov diffeomorphisms, Appendix reference [19].

[18] W. Melo, Structural stability of diffeomorphisms on two manifolds, Inventiones Math. 21 (1973).

[19] S. Smale, Differentiable dynamical systems, Bull. Amer. Math. Soc. 73 (1967).

[20] C. Robinson, C^r structural stability implies Kupka-Smale, Dynamical Systems, Ed. M. Peixoto, Academic Press (1973).

[21] C. Pugh, An improved closing lemma and a general density theorem, Amer. Journal of Math. 89 (1967).

[22] J. Franks, Differentiable Ω-stable diffeomorphisms, Topology 11 (1972).

[23] J. Guckenheimer, Absolutely Ω-stable diffeomorphisms, Topology 11 (1972).

[24] R. Mañé, On infinitesimal and absolute stability of diffeomorphisms, Dynamical Systems - Warwick, Springer-Verlag Lecture Notes in Math. 468 (1975).

[25] V. Pliss, A hypothesis due to Smale, Differential Equations 8 (1972).

[26] V. Pliss, Properties of solutions of a periodic system of two differential equations having an integral set of zero measure, Differential Equations 8 (1972).

[27] R. Mañé, The stability conjecture on two dimensional manifolds, to appear.

[28] A. Lopes, Structural stability and hyperbolic attractors, IMPA thesis (1977), to appear.

[29] S. Smale, Structurally stable systems are not dense, Amer. Journal of Math. 88 (1966).

[30] J. Guckenheimer, A strange, strange attractor, to appear.

[31] R. Williams, The structure of Lorenz attractors, to appear.

[32] S. Smale, Stability and isotopy in discrete dynamical systems, Dynamical Systems, Ed. M. Peixoto, Academic Press (1973).

[33] M. Shub and D. Sullivan, Homology theory and dynamical systems, Topology 14 (1975).

[34] S. Newhouse and M. Peixoto, There is a simple arc joining any two Morse-Smale flows, Astérisque 31 (1976).

[35] S. Newhouse, On simple arcs between structurally stable flows, Dynamical Systems-Warwick, Springer-Verlag Lecture Notes in Math. 468 (1975).

[36] J. Sotomayor, Generic one-parameter families of vector fields on two dimensional manifolds, Publications Math. I.H.E.S. 43 (1973).

[37] J. Sotomayor, Generic bifurcations of dynamical systems, Dynamical Systems, Ed. M. Peixoto, Academic Press (1973).

[38] P. Brunovsky, On one parameter families of diffeomorphisms, Comment. Math. Univ. Carlinae 11 (1970).

[39] V. Arnold, Lectures on bifurcations in versal families, Russian Math. Surveys 27 (1972).

[40] J. Palis, Arcs of dynamical systems: bifurcations and stability, Dynamical Systems-Warwick, Springer-Verlag Lecture Notes in Math. 468 (1975).

[41] M. Peixoto, On bifurcations of dynamical systems, Intern. Congress of Math. vol. II, Vancouver (1974).

[42] F. Takens, Introduction to global analysis, Comm. Math. Inst. Univ. Utrecht (1973).

[43] S. Newhouse and J. Palis, Bifurcations of Morse-Smale dynamical systems, Dynamical Systems, Ed. M. Peixoto, Academic Press (1973).

[44] S. Newhouse and J. Palis, Cycles and bifurcation theory, Astérisque 31 (1976).

[45] V. Afraimovic and L. Silnikov, On attainable transitions from Morse-Smale systems to systems with many periodic motions, Izvestija 8 (1974).

[46] V. Afraimovic and L. Silnikov, On some global bifurcations connected with the disappearance of a fixed point of saddle-node type, Doklady 15 (1974).

[47] S. Newhouse, J. Palis and F. Takens, Stable arcs of diffeomorphisms, Bull. Amer. Math. Soc. 82 (1976) and to appear.

[48] A. Shoshitaishvili, Bifurcations of topological type at singular points of parametrized vectorfields, Funct. Analysis Appl. 6 (1972).

[49] J. Palis and F. Takens, Topological equivalence of normally hyperbolic dynamical systems, to appear.

[50] M. Herman, Les diffeomorphisms du cercle, to appear.

[51] C. Robinson, Global structural stability of a saddle-node bifurcation, to appear.

IMPA, Rio de Janeiro, Brazil.

Convexity and tightness of manifolds with boundary

by

Lucio L. Rodríguez

1. Introduction

In this paper we characterize convexity for compact manifolds with boundary in terms of tightness, and prove a result similar to those of Sacksteder in [9].

We say that an immersion $f\colon M \to \mathbb{R}^{n+1}$ of an n-manifold M is convex if (i) $f|\partial M$ consists of convex hypersurfaces of affine hyperplanes of $\mathbb{R}^{n+1}$, and (ii) f embeds M into the boundary of the convex hull of $f(M)$.

Given $f\colon M \to \mathbb{R}^{n+p}$ and a unit vector z in $\mathbb{R}^{n+p}$, we consider the height functions $z\cdot f$, where $(\cdot)$ denotes the usual inner product of $\mathbb{R}^{n+p}$. If c is any real number, let $i\colon M_{c,z} \to M$ denote the inclusion map of $M_{c,z} = \{x \in M\colon z\cdot f(x) \leq c\}$ into M. We say that f is k-tight if there exists a field F such that for all z in a dense subset E of the unit sphere S^{n+p-1} we have that $i_{*_j}\colon H_j(M_{c,z}) \to H_j(M)$ is injective for all real numbers c and all $j \leq k$, where $H_j(\ )$ denotes singular j-th homology with coefficients in F. When an immersion f of an n-manifold M is n-tight we say that it is tight. We say that f is substantial if $f(M)$ is not contained in any hyperplane of $\mathbb{R}^{n+p}$. A half-sphere is not tight, so that tightness is too strong a condition to characterize convexity. In fact, we will show: a substantial immersion $f\colon M \to \mathbb{R}^{n+1}$ of a manifold which is topologically an n-sphere with some n-discs removed is convex if and only if it is (n-2)-tight.

We will also prove the following result relating curvature to convexity: <u>if $f: M \to \mathbb{R}^{n+1}$ is a substantial immersion with convex boundary components and with non-negative sectional curvatures, then f is convex</u>.

2. Statements of results

For closed manifolds tightness is equivalent to minimal total absolute curvature, which Chern and Lashof [3] showed to be equivalent to convexity for immersions of manifolds homeomorphic to the sphere. However, there are many ways of extending this concept to manifolds with boundary. In [4] and [2], some notions of total absolute curvature for manifolds with boundary are defined, but they cannot be used to characterize convexity. On the other hand, the definition of k-tightness in terms of homology homomorphisms adapts itself naturally to manifolds with boundary. For our purposes we will give an equivalent definition of k-tightness in terms of the critical points of height functions on a manifold with boundary. These critical points and their indices are studied in Section 3, where the Morse inequalities and another formula are obtained. Section 4 deals with the equivalence of the two definitions of k-tightness.

In [1] Banchoff shows that for an n-manifold M embedded in R^n the homomorphisms i_{*_o} and $i_{*_{(n-2)}}: H_*(M_{c,z}) \to H_*(M)$ are injective if and only if i_{*_o} and $i_{*_{(n-2)}}: H_*(\partial M_{c,z}) \to H_*(\partial M)$ are injective for all c and z. We obtain the following generalization.

<u>Theorem 1</u> - <u>If M is a compact n-dimensional manifold with boundary embedded in $\mathbb{R}^n$, then M is tight if and only if its boundary ∂M is tight</u>.

Examples of tight (n+p)-manifolds embedded in $\mathbb{R}^{n+p}$ are closed tubular ϵ-neighborhoods of tight embeddings $f\colon M^n \to \mathbb{R}^{n+p}$ of closed manifolds. Knowledge of these (n+p)-manifolds could possibly give information about tight embeddings.

If the codimension p of an immersion $f\colon M \to R^{n+p}$ is greater than zero, then we could have ∂M tight without M being tight. However, if M is such that it can be differentiably embedded in $\mathbb{R}^n$, we do get that the tightness of f implies the tightness of $f|\partial M$. This is used in the proof of the main result.

Theorem 2 - *Let $f\colon M \to \mathbb{R}^{n+p}$ be a substantial immersion, with $p \geq 1$, of an n-manifold that can be embedded in $\mathbb{R}^n$, whose boundary components are topologically (n-1)-spheres. Then, f is convex if and only if f is (n-2)-tight.*

We remark that if we allow $p = 0$ then $f|\partial M$ is in fact convex and f is an embedding, so that in a sense it can also be considered convex. Before proving this theorem, in Section 6, we have some propositions reducing the codimension p of tight substantial immersions. The proof of Theorem 2 then rests on the following.

Theorem 3 - *Let $f\colon M \to \mathbb{R}^{n+1}$ be a substantial immersion of a compact connected n-manifold M whose boundary components are immersed by f as convex hypersurfaces in affine hyperplanes. If the Jacobian of the Gauss map, defined locally, is either singular or positive or negative definite at each point of M, then f is convex.*

The proof of this theorem is similar to that of Theorem 3 in Chern-Lashof [3]; a path has to be followed along which the tangent plane to M remains constant until a point in the boundary of the convex hull of $f(M)$ is reached. However, when the path hits a boundary point we must also take into account the Gauss map of the boundary. For that reason, the following lemma, which generalizes Theorem I of Hartman-Nirenberg [5], is essential. Suppose $M' \subset M$ is an (n-1)-dimensional submanifold of M, and $f: M \to \mathbb{R}^{n+1}$ is an immersion with $f(M')$ contained in some hyperplane H of $\mathbb{R}^{n+1}$; we assume for simplicity that M' is orientable. Let $g': M' \to S^{n-1} \subset H$ and $g: U \to S^n \subset \mathbb{R}^{n+1}$ be the Gauss maps of the two hypersurfaces, where U is a neighborhood of M' in M. Then, if we denote by TM_x and TM'_x the respective tangent spaces, we have

Main Lemma: *Let $D' \subset M'$ be the set of points on which both Gauss maps have singular Jacobians. Then, for every y in D', there exists a point z in $\partial D'$, the boundary of D' in M', such that*

(i) *$g'(z) = g'(y)$ and $TM'_z = TM'_y$ as affine hyperplanes of H, and*

(ii) *$g(z) = g(y)$ and $TM_z = TM_y$ as affine hyperplanes of $\mathbb{R}^{n+1}$.*

The statement of Theorem 3 is more general than what is needed in the proof of Theorem 2; however, this formulation can be useful in other situations. In particular, the condition that all sectional curvatures of M be non-negative implies the assumptions on the Jacobian of g in Theorem 3; hence, we have the following result similar to those of Sacksteder in [9].

Theorem 4 - *If $f: M \to \mathbb{R}^{n+1}$ is a substantial immersion of a compact connected n-manifold with non-negative sectional curvatures and with convex boundary components, then f is convex.*

The author would like to thank Manfredo do Carmo for some helpful suggestions.

3. Critical points for manifolds with boundary

In this section we will study the boundary critical points of a differentiable function on a manifold with boundary M. We will define their indices which, together with the indices of the interior critical points, will be related by some inequalities to the Betti numbers of M. Given a differentiable function $h:M \to \mathbb{R}$ on a manifold with boundary, we define the critical points in the interior of M in the usual way; that is, x is a critical point of h if $(\text{gradient } h)_x = 0$. We say that a boundary point x is a critical point of h if it is a critical point, in the usual way, of $h|\partial M$. We say that this boundary critical point is non-degenerate if i) it is non-degenerate as a critical point of $h|\partial M$, and ii) $Y\cdot(\text{grad } h)_x \neq 0$, where Y is the unit vector in the tangent space TM_x that is perpendicular to ∂M at x and points away from M.

We define the index of an interior non-degenerate critical point in the usual manner; see [7]. For a non-degenerate boundary critical point x, we say that its index is (i) k^- if the index of $h|\partial M$ is k an $(Y\cdot\text{grad } h)_x < 0$, and (ii) k^+ if the index of $h|\partial M$ at x is $k-1$ and $(Y\cdot\text{grad } h)_x > 0$. We let $\varphi_k^+(h)$ and $\varphi_k^-(h)$ equal the number of critical points at the boundary of index k^+ and k^-, respectively; and let $\mu_k(h)$ equal the number of interior non-degenerate critical points of index k.

Given a differentiable function $h: M \to R$ and a real number c, consider the set $M_c = \{x \in M: h(x) \leq c\}$. Let $H_k(X)$ denote the singular k-th homology groups of X with coefficients in a field F. If we assume that h has only non-degenerate critical points, then we have the following.

Lemma 3.1 - *If c is a critical value of h, then, for some $\varepsilon > 0$, $H_k(M_{c+\varepsilon}, M_{c-\varepsilon})$ is isomorphic to the direct sum of ℓ*

copies of the coefficient field F, where ℓ is equal to the number of interior critical points of index k at level c plus the number of boundary critical points of index k^- also at level c.

Proof: We can assume ϵ small enough so that c is the only critical value in $[c-\epsilon, c+\epsilon]$. We can then apply Lemma 3 in [6] to obtain that $H_k(M_{c+\epsilon}, M_{c-\epsilon}) \approx \bigoplus_{i=1}^{s} H_k(M_c, M_c - \{x_i\})$, where $\{x_1,\dots,x_l\}$ are the critical points al level c. If x_i is an interior critical point of index j, then, since by excision the problem is local, we can apply the critical point theory for manifolds without boundary to obtain that $H_k(M_c, M_c - \{x_i\}) = F$ if $j = k$, or zero otherwise; see [7].

If x_i is a boundary critical point, then we have two situations. First, we have the one in which x_i has index j^+, but, then, by excision, $H_k(M_c, M_c - \{x_i\}) \approx H_k(U, U - \{x_i\})$ for some small neighborhood U of x_i. However, since $-\mathrm{grad}\, h$ points inward at x, we have that both $U - \{x_i\}$ and U are contractible for suitable U. Hence, $H_k(M_c, M_c - \{x_i\}) = 0$.

If x_i is a boundary critical point of index j^-, then we have, by excision, that $H_k(M_c, M_c - \{x_i\} \approx H_k(U, U - \{x_i\})$. If U is small enough, we can deform U along the directions $-\mathrm{grad}\, h$ into $U \cap \partial M_c$, keeping $\partial M_c = \{x \in \partial M : h(x) \leq c\}$ fixed. Hence, $H_k(M_c, M_c - \{x_i\}) \approx H_k(U \cap \partial M_c, U \cap \partial M_c - \{x_i\})$. And, we are back in the case of an interior critical point; thus, $H_k(M_c, M_c - \{x_i\}) \approx H_k(\partial M_c, \partial M_c - \{x_i\}) = F$ if $j = k$, zero otherwise, since x_i is a critical point of index k of $f|\partial M$. Q.E.D.

Now order the critical points in ascending order $c_1 < c_2 < \dots < c_m$ where c_1 is the absolute minimum and c_m the absolute maximum; let $c_o = c_1 - 2\epsilon$. In the above lemma we can substitute $M_{c_{p+1}}$ for $M_{c_p+\epsilon}$, since the latter is a deformation retract of the former, by Lemma 2 in [6]. Observe that $M_{c_m+\varepsilon} = M$ and that $M_{c_o+\varepsilon} = \phi$.

Let $M_i = M_{c_i+\varepsilon}$, $i = 0,\ldots,m$. Looking at the exact sequence of the pair (M_p,M_{p-1}),

$$H_{k+1}(M_p,M_{p-1}) \xrightarrow{\partial^{p-1}_{k+1}} H_k(M_{p-1}) \longrightarrow H_k(M_p) \longrightarrow H_k(M_p,M_{p-1}) \xrightarrow{\partial^{p-1}_{k}} H_{k-1}(M_{p-1}) \rightarrow$$

$$\rightarrow \ldots \rightarrow H_o(M_p,M_{p-1}) \rightarrow 0$$

we see that, rank $\partial^{p-1}_{k+1} = \dim H_k(M_{p-1}) - \dim H_k(M_p) + \dim H_k(M_p,M_{p-1}) -$ $- \ldots \pm \dim H_o(M_p,M_{p-1})$. Therefore,

$$\sum_{i=0}^{k} (-1)^{k-i} \dim H_k(M_p,M_{p-1}) \geq \sum_{i=0}^{k} (-1)^{k-i} [\dim H_k(M_p) - \dim H_k(M_{p-1})]$$

with equality if and only if rank $\partial^{p-1}_{k+1} = 0$. Adding up over p, we obtain

$$\sum_{i=0} (-1)^{k-i} \sum_{p=1}^{m} \dim H_k(M_p,M_{p-1}) \geq$$

$$\geq \sum_{i=0}^{k} (-1)^{k-1} \sum_{p=1}^{m} [\dim H_k(M_p) - \dim H_k(M_{p-1})]$$

or

$$\sum_{i=0}^{k} (-1)^{k-1} (\mu_k + \varphi_k^-) \geq \sum_{i=0}^{k} (-1)^{k-i} \dim H_k(M) ,$$

by Lemma 3.1, since $\mu_k + \varphi_k^-$ equals the sum of the critical points of index k and k^- over all the levels. If we denote the dimension of $H_\ell(M)$, the ℓ-th Betti number of M, by $\overline{R}_\ell$, then we obtain the following Morse inequalities,

$$(3.1) \quad (\mu_k + \varphi_k^-) - (\mu_{k-1} + \varphi_{k-1}^-) + \ldots \pm (\mu_o + \varphi_o^-) \geq \overline{R}_k - \overline{R}_{K-1} + \ldots \pm \overline{R}_o$$

with equality if and only if $\sum_{i=0}^{m-1}$ rank $\partial^{i}_{k+1} = 0$.

Adding up the inequalities corresponding to k and k-1, we obtain,

$$(3.2) \qquad \mu_k + \varphi_k^- \geq \bar{R}_k, \quad \text{for all} \quad k,$$

with equality if and only if $\sum_{i=0}^{m-1} (\text{rank } \partial_{k+1}^i + \text{rank } \partial_k^i) = 0$.

Observe that $\mu_k(h|\partial M) = \varphi_k^-(h) + \varphi_{k+1}^+(h)$, so that formula (3.1) becomes for $k \doteq n-1$,

$$(3.3) \qquad \sum_{i=0}^{n-1} \varphi_i^-(h) + \varphi_{i+1}^+(h) = R_o - R_1 + \ldots \pm R_{n-1} = \chi(\partial M) \quad ,$$

where the R_i's are the Betti numbers of ∂M and $\chi(\partial M)$ is its Euler characteristic. Adding up equation (3.1), for $k = n$, and minus half equation (3.3) we obtain

$$(3.4) \qquad \sum_{i=0}^{n} (-1)^k \mu_k + \frac{1}{2} \sum_{i=0}^{n} (-1)^k \varphi_k = \chi(M) - \frac{1}{2} \chi(\partial M).$$

This formula will not be used hereafter. However, it can be applied in other situations; for example, to obtain a formula with the indices of gradient vector fields, and to prove the classical Gauss-Bonnet formula for surfaces with boundary.

Consider now an immersion $f\colon M \to \mathbb{R}^{n+p}$ and a unit vector z in $\mathbb{R}^{n+p}$. We have the height functions $h = z \cdot f\colon M \to R$, where $(\cdot)$ is the usual inner product of $\mathbb{R}^{n+p}$. It is well known that there exists an open dense subset E_1 of the unit sphere S^{n+p-1} with the property that $z \cdot f$ has only non-degenerate interior critical points if z is in E_1. Similarly, there exists an open dense $E_2 \subset S^{n+p-1}$ such that z in E_2 implies that $z \cdot (f|\partial M)$ has only non-degenerate critical points. Finally, the subset of the unit normal bundle of ∂M

that is perpendicular to the vector field Y, $B_1 = \{(x,v) \in \partial M \times \mathbb{R}^{n+p}: v\cdot v = 1$ and $v\cdot W = 0$ for all W in $f_*TM_*\}$ has dimension $n+p-2$. Therefore, its image by the map $g: B_1 \to S^{n+p-1}$ which sends (x,v) to the vector v has measure zero; thus, its complement E_3 is open and dense. We have $E = E_1 \cap E_2 \cap E_3$ open and dense; if z is in E then $z\cdot f$ has only non-degenerate interior critical points, and any boundary critical point x is also non-degenerate because z in E_3 implies that $(Y\cdot \text{grad}(z\cdot f))_x = Y_x\cdot z \neq 0$. In conclusion, there exist many functions on a manifold with boundary with only non-degenerate critical points of either kind.

4. Critical points and tightness

We will now give an equivalent definition of tightness in terms of the critical points of height functions.

Let $f: M \to \mathbb{R}^{n+p}$ be an immersion of a manifold with boundary in some Euclidean space. In the last section we saw that there exists an open dense subset E of S^{n+p-1} such that the height functions $z\cdot f$, for every z in E, have only non-degenerate boundary and interior critical points. To simplify notation we will write $\mu_k(z)$ for $\mu_k(z\cdot f)$, and so on. As in the last section, let $\overline{R}_k$ denote the k-th Betti number of M. The following applies for manifolds with or without boundary.

<u>Proposition 4.1</u> - <u>The immersion $f: M \to \mathbb{R}^{n+p}$ is k-tight if and only if $\mu_r(z) + \varphi_r^-(z) = \overline{R}_r$, for $r \leq k$ and all z in E.</u>

<u>Proof</u>: If we have equality between the number of critical points and the r-th Betti number, then by formula (3.2), we have that

that $\sum_{j=0}^{m-1} \ker i_r^j = \sum_{j=0}^{m-1} \operatorname{rank} \partial_{r+1}^j = 0$, where $i_r^j: H_r(M_j) \to H_r(M_{j+1})$ is induced by inclusion. Now given any real number c in $[c_o, c_{m+1}]$, we must have an integer ℓ such that $c_\ell \leq c < c_{\ell+1}$. By Lemma 1 in [6], we have that $\{x \in M: z \cdot f(x) \leq c\}$ is homotopically equivalent to M_ℓ; therefore, it is sufficient to show that the homomorphisms $j_\ell: H_r(M_\ell) \to H_r(M)$, induced by inclusions, are injective for $\ell = 0,1,\ldots,m-1$. However, $j_\ell = i_r^{m-1} \circ i_r^{m-2} \circ \ldots \circ i_r^\ell$ is a composition of injective maps.

Conversely, if j_ℓ is injective for $\ell = 0,1,\ldots,m-1$, we have that i_r^ℓ is injective for $\ell = 0,\ldots,m-1$, since $\ker i_r^\ell \subset \subset \ker j_\ell = 0$; consequently $\mu_r(z) + \varphi_r^-(z) = \overline{R}_r$. By the definition of k-tightness this result holds for all z in a dense subset E' of S^{n+p-1}. We must show that it holds for all z in E. The function $\mu_r(z) + \varphi_r^-(z)$ is locally constant on E; thus, since E' is dense, given any z in E there exists a z' in E' near z so that $\mu_r(z) + \varphi_r^-(z) = \mu_r(z') + \varphi_+^-(z') = \overline{R}_r$. Q.E.D.

As a consequence, we have that f is tight if and only if $\sum_{k=0}^{n} \mu_k(z) + \varphi_k^-(z) = \sum_{k=0}^{n} \overline{R}_k$, for all z in E.

5. Relation between tightness of M and that of ∂M

In this section we will prove Theorem 1. We will need the following duality formula.

Proposition 5.1 - *Let M be a compact n-manifold with boundary, and let $\overline{R}_k$ and R_k denote the Betti numbers of M and ∂M, respectively. If M can be embedded in $\mathbb{R}^n$, then*

(5.1) $$R_k = \overline{R}_k + \overline{R}_{n-k-1}, \qquad k = 0,1,\ldots,n-1$$

Proof: We can think of M as a subset of $\mathbb{R}^n$. By Alexander duality we have that $\bar{H}^k(\partial M) \approx \tilde{H}_{n-k-1}(\mathbb{R}^n - \partial M)$, where $\bar{H}^*$ denotes Alexander-Spanier cohomology and $\tilde{H}_*$ denotes reduced singular homology with coefficients in some field F. Now observe that $\mathbb{R}^n - \partial M$ is equal to the disjoint union of the interior of M and of $\mathbb{R}^n - M$. Thus, $\tilde{H}_{n-k-1}(\mathbb{R}^n - \partial M) \approx H_{n-k-1}(M) \oplus \tilde{H}_{n-k-1}(\mathbb{R}^n - M)$ because M and the interior of M have the same homotopy type. Using Alexander duality again, we obtain that $\bar{H}^k(\partial M) \approx H_{n-k-1}(M) \oplus \bar{H}^k(M)$.

Since both M and ∂M are deformation retracts of arbitrarily small neighborhoods of themselves in $\mathbb{R}^n$, we get that $\bar{H}^k(M) \approx H^k(M)$ and $\bar{H}^k(\partial M) \approx H^k(\partial M)$. Since F is a field, we have no torsion, so $H^k(M) \approx \mathrm{Hom}(H_k(M), F)$ and $H^k(\partial M) \approx \mathrm{Hom}(H_k(\partial M), F)$. Therefore, dimension $\bar{H}^k(M) = \bar{R}_k$ and dimension $\bar{H}^k(\partial M) = R_k$. Hence,

$$R_k = \bar{R}_k + \bar{R}_{n-k-1}, \qquad k = 0,1,\ldots,n-1$$

Q.E.D.

Proof of Theorem 1: Since M is embedded in $\mathbb{R}^n$, the height functions $z \cdot f$ have no interior critical points; that is, $\mu_k(z) = 0$ for all z in E and all k. Therefore, if M is tight, then $\varphi_k^-(z) = \bar{R}_k$ for all z in E and all k. We observe now that if z is in E so is -z, and $\varphi_{k+1}^+(z) = \varphi_{n-k-1}^-(-z)$. Thus, $\mu_k(z \cdot f|\partial M) = \varphi_k^-(z) + \varphi_{k+1}^+(z) = \varphi_k^-(z) + \varphi_{n-k-1}^-(-z) = \bar{R}_k + \bar{R}_{n-k-1} = R_k$, by the above proposition. Hence ∂M is k-tight for all k.

Conversely, if ∂M is k-tight for all k, we have that

$$\varphi_k^-(z) + \varphi_{n-k-1}^-(-z) = \varphi_k^-(z) + \varphi_{k+1}^+(z) = R_k = \bar{R}_k + \bar{R}_{n-k-1}$$

for all z in E; thus, since always $\varphi_\ell^-(z) \geq R_\ell$, we must have that $\varphi_k^-(z) = \bar{R}_k$ and $\varphi_{n-k-1}^-(-z) = \bar{R}_{n-k-1}$ for all z in E and $k \leq n-1$, and M is (n-1)-tight. Finally, we observe that $\bar{R}_n = \varphi_n^-(z) = 0$. Q.E.D.

<u>Proposition 5.2 - If</u> $f: M \to \mathbb{R}^{n+p}$ <u>is an (n-2)-tight immersion of an n-manifold with boundary that can be embedded in</u> $\mathbb{R}^n$, <u>then</u> $f|\partial M$ <u>is tight. Furthermore, if</u> $p = 1$, <u>the Jacobian of the Gauss map of</u> M, <u>defined locally, is either singular or positive or negative definite at each point of</u> M.

<u>Proof</u>: <u>Case 1</u>, $n > 2$.

Since f is (n-2)-tight, we have that $\mu_k(z) + \varphi_k^-(z) = \bar{R}_k$ and $\mu_{n-k-1}(-z) + \varphi_{n-k-1}^-(-z) = \bar{R}_{n-k-1}$, for $1 \le k \le n-2$. Combining these two formulas, we obtain that

$$\mu_k(z) + \mu_{n-k-1}(-z) + \varphi_k^-(z) + \varphi_{k+1}^+(z) = \bar{R}_k + \bar{R}_{n-k-1} = R_k \quad ,$$

by Proposition 5.1. Since we always have $\mu_k(z\cdot f|\partial M) = \varphi_k^-(z) + \varphi_{k+1}^+(z) \ge R_k$, we must have equalities, if $1 \le k \le n-2$. Furthermore, we obtain that $\mu_k(z) = 0$ for all z in E, $1 \le k \le n-2$. One of the Morse inequalities (3.1) says, considering ∂M as a manifold with boundary in its own right, that

$$\mu_1(z\cdot f|\partial M) - \mu_o(z\cdot f|\partial M) \ge R_1 - R_o$$

since $\mu_1(z\cdot f) = R_1$, $-\mu_o(z\cdot f|\varphi M) \ge -R_o$, but $\mu_o(z\cdot f|\partial M) \ge R_o$ always, so that $f|\partial M$ is 0-tight, hence (n-2)-tight. Since $\mu_{n-1}(z\cdot f|\partial M) = \mu_o(-z\cdot f|\partial M)$, $f|\partial M$ is also (n-1)tight. Finally we also observe that $\mu_{n-1}(z) = \mu_1(-z) = 0$ for all z in E.

<u>Case 2</u>: $n = 2$.

Formula (5.1) for $n = 2$ and $k=1$ says that $R_o = \bar{R}_o + \bar{R}_1 = 1 + \bar{R}_1$, because M is connected. This together with formula (3.1) gives us

$$\mu_2 - \mu_1 - \varphi_1^- + \mu_o + \varphi_o^- = \overline{R}_o - \overline{R}_1 = \overline{R}_o + \overline{R}_o - R_o = 2 - R_o \ .$$

On the other hand, $\mu_1(z\cdot f|\partial M) = \varphi_1^-(z) + \varphi_2^+(z) \geq R_1 = R_o$ so that $2 - R_o = \mu_2 - \mu_1 - \varphi_1^- + \mu_o + \varphi_o^- \leq \mu_2 - \mu_1 + \varphi_2^+ - R_o + \mu_o + \varphi_o^-$. Since M is 0-tight, $\mu_2(z) + \varphi_2^+(z) = \mu_o(-z) + \varphi_o^-(-z) = \overline{R}_o = 1$, so that $2 - R_o \leq 2 - \mu_1(z) - R_o$; hence, $\mu_1(z) = 0$, for all z in E, and we must have the equality $\mu_o(z\cdot f|\partial M) = \mu_1(-z\cdot f|\partial M) = R_1 = R_o$ for all z in E, so that $f|\partial M$ is tight.

In either case, we have $\mu_i(z) = 0$, for all z in E, and $i = 1,\ldots,n-1$. If $p = 1$, we can define, at least locally the Gauss map $g\colon U \subset M \to S^n$. It is well known that the Jacobian matrix of the Gauss map at x is equal to the Hessian of the height function $z\cdot f$, where $z = g(x)$. Therefore, if at a point x in M the Jacobian were to be non-singular but not definite, then it would continue to be so in a neighborhood V of x. But, then, $g(V)$ would intersect E, so that there exists $z = g(y) \in E$, such that y is a non-degenerate critical point of index i, $1 \leq i \leq n-1$, of $z\cdot f$, contradicting the fact that $\mu_i(z) = 0$, for $1 \leq i \leq n-1$. Q.E.D.

6. Immersions of the n-sphere with some n-discs removed

We aim to prove Theorem 2, but first we prove the following propositions about immersions of manifolds with non-empty boundary into $\mathbb{R}^{n+p}$. As before, we assume that they are compact and connected.

Lemma 6.1 - *If x in ∂M is a non-degenerate critical point of $z\cdot f$ of index n (zero), then $z\cdot f$ has a strict local maximum (minimum) at x.*

Proof: The proof is similar to that of Lemma 1 in [8] and we omit it.

We observe that if f is 0-tight, then $\varphi_0^-(z) + \mu_0(z) = \overline{R}_0 = 1$; hence, applying Lemma 6.1, we see that any local minimum of $z \cdot f$ is a global minimum. The same holds for a maximum, because a maximum of $z \cdot f$ is a minimum of $-z \cdot f$. Consider an immersion $f: M \to \mathbb{R}^{n+p}$, where M is a compact, connected manifold with boundary ∂M, whose connected components are $S_1, \ldots, S_k$. Let $\alpha, \alpha_1, \ldots \alpha_k$ denote the second fundamental forms of f, $f|S_1, \ldots, f|S_k$, respectively. For every x in M, let P_x be the smallest linear space that contains the image of the second fundamental form α if x is an interior point, or, of α_i if $x \in S_i$.

Proposition 6.2 - *Let f be 0-tight. If dimension $P_x \leq \ell$ for every x in M, then $f(M)$ is contained in some $(n+\ell)$-dimensional affine plane.*

Proof: Fix a vector z in E and let x be the non-degenerate global minimum of $z \cdot f$. If we assume, for simplicity, that $f(x) = 0$, we will show that $f(M) \subset P_x + f_* TM_x$, which has dimension at most $n+\ell$.

Suppose that $x \in S_i$. Let ω be any vector perpendicular to $P_x + f_* TM_x$. Since the image of α_i is contained in P_x, $(\omega \cdot \alpha_i) \equiv 0$. Let $\omega(\varepsilon) = (\omega + \varepsilon z))/\|\omega + \varepsilon z\|$, $\varepsilon > 0$. Then $\omega(\varepsilon) . \alpha_i = \varepsilon z \cdot \alpha_i$, and $\omega(\varepsilon) \cdot Y_x = \varepsilon z \cdot Y_x = \varepsilon (\text{grad } z \cdot f) \cdot Y_x < 0$, since x is a maximum. Therefore, since the indices of $\omega(\varepsilon) \cdot (f|S_i)$ and $z \cdot (f|S_i)$ at x can be determined in terms of the eigenvalues of $\omega(\varepsilon) \cdot \alpha_i$ and $z \cdot \alpha_i$, we obtain that $\omega(\varepsilon) \cdot f$ has index zero at x. Hence, by Lemma 6.1, x is a local minimum of $\omega(\varepsilon) \cdot f$. Now, there exists a unit vector $\omega(\varepsilon)'$ in E, a distance ε from $\omega(\varepsilon)$, such that $\omega(\varepsilon)' . (f|S_i)$ has a local minimum at a point x_ε in S_i a distance ε from x. Since we can assume that $\omega(\varepsilon)'.Y$ is still less than zero at x_ε, we have, by 0-tightness, that x_ε is a global minimum of $\omega(\varepsilon)'.f$; that is, $\omega(\varepsilon)'.f(x_\varepsilon) \leq \omega(\varepsilon)'.f(y)$, for every y in M. Letting ε tend to zero, we obtain that $\omega.f(x) \leq \omega.f(y)$, for every y in M. Since the same argument works for $-\omega$, we obtain that $\omega.f(y) = \omega.f(x) = 0$, for all y in M. But this holds for every ω perpendicular to $f_*TM_x + P_x$, so that we are finished if $x \in \partial M$.

If x is in the interior of M, then we proceed in a similar fashion; taking any ω perpendicular to $f_*TM_x + P_x$ we see that $\omega.\alpha \equiv 0$. We see that $\omega(\varepsilon).f = \varepsilon z.f$ so that $\omega(\varepsilon).f$ also has a critical point of index zero at x. We define $\omega(\varepsilon)'$ and proceed as before to conclude that $f(M) \subset f_*TM_x + P_x$. Q.E.D.

Corollary 6.3 - Let $f: M \to \mathbb{R}^{n+p}$ be a substantial immersion of an n-manifold M. If f is 0-tight, then $p \leq \frac{1}{2}n(n+1)$. If $\partial M \neq \phi$, and f is tight, then $p \leq \frac{1}{2}n(n-1)$.

Proof: The first assertion follows from Proposition 6.2 and the fact that, at an interior point, the space P_x is spanned by $\alpha(X_i,X_j)$, where $X_1,\ldots,X_n$ from a basis of TM_x. Since α is symmetric we have that dimension $P_x \leq \frac{1}{2}n(n+1)$. Similarly, at a boundary point x dimension $P_x \leq \frac{1}{2}n(n-1)$. If furthermore, f is a n-tight, we have that $\mu_n(z) = \mu_n(z) + \varphi_n^-(z) = \overline{R}_n = 0$, because $\partial M \neq \phi$; hence, $\mu_o(z) = \mu_n(-z) = 0$, for all z in E, and the global minimum x used in the proof of Proposition 6.2 must always be a boundary point. Since dimension $P_x \leq \frac{1}{2}n(n-1)$ at a boundary point, we arrive at the second assertion.

Q.E.D.

Corollary 6.4 - *Let $f: M \to \mathbb{R}^{n+p}$ be a substantial immersion of an n-manifold M which can be embedded in $\mathbb{R}^n$ and whose boundary components are homeomorphic to $(n-1)$-spheres. If f is $(n-2)$-tight, then $p \leq 1$.*

Proof: If $\mu_i(z) = 0$, for $i \neq 0,n$, then dimension $P_x \leq 1$, for an interior point x. Otherwise, if $\alpha(X,X) \neq \lambda\, \alpha(Y,Y)$, for some X and Y in TM_x and all λ in $\mathbb{R}$, then taking $z = (\alpha(X,X) - \alpha(Y,Y)/\|\alpha(X,X) - \alpha(Y,Y)\|$ we see that $\alpha \cdot z$ is not positive or negative definite; but this will imply that there exists a z' in E and $\ell \neq 0$, n such that $\mu_\ell(z') \neq 0$. In the proof of Proposition 5.2 we saw that $\mu_i(z) = 0$, for $i \neq 0,n$; thus, dimension $P_x \leq 1$ if x is an interior point. If $x \in \partial M$, then by hypothesis, x is in some $(n-1)$sphere $S^{n-1} \subset \partial M$. By Proposition 5.2, we have that $f|S^{n-1}$ is tight; since the i-th Betti numbers of S^{n-1} are zero if $i \neq 0$, $n-1$, we have that $\mu_i(f|S^{n-1}) = 0$ if $i \neq 0$, $n-1$. Hence, the above arguments show that $P_x \leq 1$.

Proof of Theorem 2 - By Corollary 6.4, $p \leq 1$. By Proposition 5.2, the Jacobian of the Gauss map is definite at each point where it is non-singular. Thus, we can apply Theorem 3 to obtain that f is convex. Conversely, if f is convex, then $\mu_i(z) = 0$, for $i \neq 0$, n, since $f(M)$ is always to one side of f_*TM_x. Also, since $f|\partial M$ are convex hypersurfaces, $\varphi_i^-(z) = 0$, $i = 1,\dots,n-2$. The type of arguments used in the first half of the proof of Proposition 5.2 then show that f is $(n-2)$-tight.

Q.E.D.

7. Singular points of the Gauss maps

In this section we prove the main lemma. We are considering

the following situation: $f: M \to R^{n+1}$ is an immersion of M as a hypersurface, such that for every boundary component S of M, $f(S)$ is contained in some n-plane. Let $g|N: N \to S^n$ and $g': S \to S^{n-1}$ be the Gauss maps of M, defined on a neighborhood N of S, and of S, respectively. For x in M, let $r(x)$ be the rank of the Jacobian map g_{*_x}; and, for x in S, let $r'(x)$ be the rank of g'_{*_x}. Let $r^*(x)$ equal the maximum integer k such that every neighborhood of x in M contains a point y with $r(y) = k$. Similarly, for x in S, define $r'^*(x)$ as the maximum integer k such that every neighborhood of x contains a point y in S with $r'(y) = k$.

Given a set L in M, we say that a connected k-submanifold P is a k-plane section of L if $P \subset L$ and $f(P)$ is part of a k-plane K in $\mathbb{R}^{n+1}$, such that if $P' \subset L$ is another connected k-submanifold with $P \subset P'$ and $f(P') \subset K$, then $P' = P$.

The idea of the proof of the main lemma is a follows. If $r'(x) = r'^*(x) = k$ for some x in S, then applying Lemma 2 of [3] to the map $f|S: S \to \mathbb{R}^n$, we have that through $f(x)$ passes a unique (n-k-1)-plane section P of $\{x \in S: r'(x) = r'^*(x) = k\}$ on which $g'(y)$ and the affine planes $f_*T\,\partial M_y$ are constant, and such that on the boundary of P, $\partial P = \overline{P}-P$, $r'(x) = k$. But we need further to be able to move to the boundary of P keeping also $g(y)$ and TM_y constant. For this purpose we will consider the restriction of g to P. A crucial step, at this point, is to show that the rank of $(g|P)_{*y}$ is strictly less than $n-k-1$; this is done in Lemma 7.1. Then, an adaptation to this situation of the proof of Lemma 2 in [5] shows that there exist plane-sections in P on which $g(y)$ and TM_y are constant. Patching up these results, we arrive at a point z with $r^*(z) = n$ or $r'^*(z) = n-1$, with $g(z) = g(x)$ and $TM_z = TM_y$.

Let ∇ denote the standard covariant derivative of $\mathbb{R}^{n+1}$. Since the tangent space of S^n at $g(x)$ can be identified with TM_x, we have that $g_*(X) = \nabla_X N$, for X in TM_x, where $N_x = g(x)$ is the unit normal to M at x. Since ∇ works also as the covariant derivative of $\mathbb{R}^n \subset \mathbb{R}^{n+1}$, we also have that $g'_*(X) = \nabla_X N'$, where $N'_x = g'(x)$ in $S^{n-1} \subset \mathbb{R}^n$ is the unit normal at x in S.

We have that $g_{*x}(X)\cdot Z = -(\nabla_X Z)\cdot N$, for every Z in TM. Therefore, X is in $\ker g_{*x}$ if and only if $(\nabla_X Z)_x$ is tangent for every Z in TM. Since $(\nabla_X Z)\cdot N = (\nabla_Z X + [X,Z])\cdot N$, we have that X is in $\ker g_{*x}$ if and only if, for any extension of X, $\nabla_{X_i} X$ is tangent to M, for some basis $X_1,\ldots,X_n$ of TM_x. We have a similar characterization of $\ker g'_{*x}$.

Let now $X_1,\ldots,X_{n-1}$ form a basis of TS_x and let W in TM_x be a vector not in TS_x. We claim that if W is in $\ker g_{*x}$ then $\ker g'_{*x} \subset \ker g_{*x}$. In fact, if we take X in $\ker g'_{*x}$, then, as we remarked above, $\nabla_{X_i} X$ is in $TS_x \subset TM_x$, for $i = 1,\ldots,n-1$, and $(\nabla_W X)\cdot N = -(\nabla_W N)\cdot X = g_{*x}(W)\cdot X = 0$, so that X is in $\ker g_{*x}$.

<u>Lemma 7.1</u> - <u>Let $P \subset S$ be a k-plane section of $\{y \in S: r'^*(y) = r'(y) = n-1-k\}$ such that $f(P) \subset K$, a k-plane in $\mathbb{R}^{n+1}$, $k \geq 1$. If x in P is not the limit of a sequence of points x_i in P with $r(x_i) = n$, then $\ker g_{*x} \cap \ker g'_{*x} = \ker g_{*x} \cap TP_x \neq 0$.</u>

<u>Proof</u>: We first observe that, by Lemma 2 in [3], this plane section P is unique, g' is constant on it and, consequently, $\ker g'_{*x} = TP_x \neq 0$.

<u>Case 1</u>: $N'_y = \pm Y_y$ for all y in a neighborhood $U \subset P$ of x, where Y_y in TM_y is perpendicular to S. In this case, $\nabla_X Y = \pm \nabla_X N' = \pm g'_*(X) = 0$ if $X \in \ker g'_{*x} = TP_x$. But we already know that if $X \in \ker g'_{*x}$ then $\nabla_X X_i$ are in $TS_x \subset TM_x$. Hence by the above remarks X is in $\ker g_{*x}$.

<u>Case 2</u>: Let $x = \lim_{i\to\infty} x_i$ with $x_i \in P$ and $N'_{x_i} \neq Y_{x_i}$, for all i.

We claim that $\ker g_{*x_i} \cap \ker g^!_{*x_i} \neq 0$, for every i.

Assuming this claim, we would have a sequence of unit vectors X_{x_i} in $\ker g_{*x_i} \cap \ker g^!_{*x_i}$. By taking a subsequence, we can assume that the X_{x_i}'s converge to a unit vector X_x in TM_x. By the continuity of the functions $(y,X_y) \to g_{*y}(X_y)$ and $(y,X_y) \to g^!_{*y}(X_y)$ on TS, we have that X_x is in $\ker g_{*x} \cap g^!_{*x}$.

Now we prove the claim. If $\ker g_{*x_i}$ were not contained in TS_{x_i}, then, by the remarks before the lemma, we would have that $\ker g^!_{*x_i} \subset \ker g_{*x_i}$. Since $\ker g^!_{*x_i} = TP_{x_i} \neq 0$, we would be done. Thus, assume that $\ker g_{*x_i} \subset TS_{x_i}$. If Z in TS_{x_i} is not in $\ker g^!_{*x_i}$, then $(\nabla_Z X_j)\cdot N' = c \neq 0$, for some j, $1 \leq j \leq n-1$. Let L in $\mathbb{R}^{n+1}$ be a unit vector perpendicular to $\mathbb{R}^n$; then, $N = aL + bN'$. We must have $b \neq 0$, because $N' \neq \pm Y$ at x_i. Thus, $(\nabla_Z X_j)\cdot N = a(\nabla_Z X_j)\cdot L + b(\nabla_Z X_j)\cdot N' = 0 + bc \neq 0$, since $\nabla_Z X_j \in \mathbb{R}^n$. Hence $Z \notin \ker g_{*x_i}$. Consequently, $\ker g_{*x_i} \subset \ker g^!_{*x_i}$. Now, by taking a subsequence, we can assume that $r(x_i) < n$, so that $\ker g_{*x_i} \neq 0$, and we are done. Q.E.D.

If P is a k-plane section of S, consider the set $V_\ell = \{x \in P: \text{dimension} \ker g_{*x} \cap TP_x = \ell\}$ and denote the interior of V_ℓ in P by $\operatorname{Int} V_\ell$.

<u>Lemma 7.2</u> - <u>Through each point p in $\operatorname{Int} V_\ell$ $1 \leq \ell \leq k$, there passes a unique ℓ-plane section Q of $\operatorname{Int} V_\ell$ on which $g(y)$ and TM_y are constant. Furthermore, we have that the closure in M of this section is contained in</u> V_ℓ.

<u>Proof</u>: The proof is an adaptation of that of Lemma 2 in [5]. Let $(x_1,..,x_n)$ and $(z_1,...,z_n)$ be the projection of neighborhoods V and $g(V)$ of M and S^n into TM_p and $TS^n_{g(p)}$

considered as subspaces of $\mathbb{R}^{n+1}$. Since $f(P)$ is contained in a k-plane K of TM_p, we can assume that $K = \{(y_1,\ldots,y_n): y_{k+1} = \ldots = y_n = 0\}$ and that $f|U = (x_1,\ldots,x_k,0,\ldots,0)$, where $U=P \cap V$. With respect to the above coordinate systems we can write g as $g_i(x_1,\ldots,x_n)$, $i = 1,\ldots,n$. Since the Jacobian of the Gauss map is symmetric we have that $\frac{\partial g_i}{\partial x_j} = \frac{\partial g_j}{\partial x_i}$.

The hypothesis of the lemma implies that $\tilde{g} = g|P$ has constant rank $s = k-\ell < k$. This implies that given any $p' \in V$ there exists a change of coordinates $\tilde{h}$ defined on a connected neighborhood U' of p' in P with $\overline{U}' \subset U$ such that $\tilde{g}' = \tilde{g}\circ\tilde{h}(y_1,\ldots,y_k) = (y_1,\ldots,y_s, \lambda(y_1,\ldots,y_s))$; we can take $\tilde{h}$ such that $\tilde{h}(y_1,\ldots,y_k) = (\phi(y_1,\ldots,y_k), y_{s+1},\ldots,y_k)$. We can extend $\tilde{h}$ to $V' \subset V$, where $U' = V' \cap P$, by defining $h(y_1,\ldots,y_n)$ as $(\tilde{h}(y_1,\ldots,y_k), y_{k+1},\ldots,y_n)$; hence $\frac{\partial h_i}{\partial y_j} = \delta^i_j$ if $i = s+1,\ldots,n$ and $j = 1,\ldots,n$. We define $g' = g\circ h$ and $\tilde{g}' = g'|U'$. Since the set $\{y \in V': y_{k+1} = \ldots = y_n = 0\}$ is equal to U', we have that $\partial/\partial y_i$ is in TP, $i = 1,\ldots,k$; hence $\frac{\partial g'_j}{\partial y_i} = \frac{\partial \tilde{g}'_j}{\partial y_i}$ throughout U', $i = 1,\ldots,k$, $j = 1,\ldots,n$. Since $\tilde{g}'$ is constant on the sets $\{y_i = C_i$, $i = 1,\ldots,s;$ $y_j = 0$, $j > k\}$ where $C_1,\ldots,C_s$ are constants, we have further that $\frac{\partial g'_j}{\partial y_i} = \frac{\partial \tilde{g}'_j}{\partial y_i} = 0$ in U' for $i = s+1,\ldots,k$, $j = 1,\ldots,n$.

Since the Gauss map is symmetric we have $\frac{\partial g_j}{\partial x_i} = \frac{\partial g_i}{\partial x_j}$ or, equivalently, $\sum_{i=1}^{n} dg_i \wedge dx_i = 0$. With the change of coordinates given by $x = h(y)$, the above equality becomes in the new coordinates,

$$0 = \sum_{i=1}^{n} \left[\left(\sum_{j=1}^{n} \frac{\partial g'_i}{\partial y_j}\, dy_j\right) \wedge \left(\sum_{j=1}^{n} \frac{\partial x_i}{\partial y_j}\, dy_j\right)\right].$$

Because, $\frac{\partial\tilde{g}'_\alpha}{\partial y_j} = \delta^\alpha_j$, $1 \le \alpha \le s$, $1 \le j \le k$, we have that the coefficient in U' of $dy_\alpha \wedge dy_j$, $\alpha \le s < j \le k$ is

$$0 = \sum_{i=1}^{n} \frac{\partial\tilde{g}'_i}{\partial y_\alpha} \cdot \frac{\partial x_i}{\partial y_j} - \sum_{i=1}^{n} \frac{\partial\tilde{g}_i}{\partial y_j} \cdot \frac{\partial x_i}{\partial y_\alpha} = \sum_{i=1}^{s} \frac{\partial\tilde{g}'_i}{\partial y_\alpha} \cdot \frac{\partial x_i}{\partial y_j} +$$

$$+ \sum_{i=s+1}^{n} \frac{\partial\tilde{g}'_i}{\partial y_\alpha} \cdot \delta^i_j = \sum_{i=1}^{s} \delta^i_\alpha \cdot \frac{\partial x_i}{\partial y_j} + \frac{\partial\tilde{g}'_j}{\partial y_\alpha} = \frac{\partial x_\alpha}{\partial y_j} + \frac{\partial\tilde{g}'_j}{\partial y_\alpha},$$

since $\frac{\partial x_i}{\partial y_j} = \frac{\partial h_i}{\partial y_j} = \delta^i_j$ if $s+1 \le i \le n$.

We saw above that $\tilde{g}'_j$ does not depend on the variables $y_{s+1},\ldots,y_k$; thus the same is true for $\frac{\partial x_\alpha}{\partial y_j} = -\frac{\partial\tilde{g}'_j}{\partial y_\alpha}$; also $\frac{\partial x_\alpha}{\partial y_\ell} = 0$ for $\ell > k$. Therefore, $x_\alpha(y_1,\ldots,y_k,0,\ldots,0) =$

$$= \sum_{i=s+1}^{k} y_i \frac{\partial x_\alpha}{\partial y_i}(y_1,\ldots,y_s,0,\ldots,0) + x_\alpha(y_1,\ldots,y_s,0,\ldots,0).$$

Therefore, the sets $h(\{y \in U' : y_i = c_i,\ c_i$ constants, $i=1,\ldots,s\})$ are the unique ℓ-plane sections of $U' \cap \text{Int } V_\ell$ on which g is constant; since U' is connected TM is constant also along them. Since these ℓ-plane sections are unique, we can patch them up to obtain a maximal one, that is, an ℓ-plane section of $\text{Int } V_\ell$.

Let Q be such an ℓ-plane section of $\text{Int } V_\ell$. We will show that if $q \in \overline{Q} - Q = \partial Q$ then $q \in V_\ell$, that is, the rank of $\tilde{g}$ at q is $k-\ell = s$. It is sufficient to show that the rank is greater or equal to s, since it is always less than or equal to s. Consider $q \in \partial Q$. Let W be a neighborhood of q in M small enough so that the projections of W and $g(W)$ into TM_q and $TS^n_{g(q)}$ are coordinate neighborhoods. Consider a component of $W \cap \text{Int } V_\ell$ that has q in its closure; it has the same properties as the neighborhood U in the beginning of the proof. For any point x in

U we had a neighborhood U' and a change of coordinates $\tilde{h}$ so that

$$\delta^{\alpha}_{\beta} = \frac{\partial \tilde{g}'_{\alpha}}{\partial y_{\beta}} = \sum_{i=1}^{n} \frac{\partial \tilde{g}_{\alpha}}{\partial x_i} \cdot \frac{\partial x_i}{\partial y_{\beta}} = \sum_{\gamma=1}^{s} \frac{\partial \tilde{g}_{\alpha}}{\partial x_{\gamma}} \cdot \frac{\partial x_{\gamma}}{\partial y_{\beta}} ,$$

since for $i > s$ we have $h_i = x_i = y_i$. Thus

$$\det\left(\frac{\partial \tilde{g}_{\alpha}}{\partial x_{\beta}}\right)_{1 \le \alpha \le \gamma \le s} = 1/\det\left(\frac{\partial x_{\alpha}}{\partial y_{\beta}}\right)_{1 \le \alpha, \beta \le s} ,$$

in the neighborhood U' of x. However, $\frac{\partial x_{\alpha}}{\partial y_{\beta}} = \sum_{i=s+1-}^{k} a_i y_i + b$,

where a_i and b are functions that do not depend on $y_{s+1}, \ldots, y_k$, and so are constant along Q. Since $y_{s+1} = x_{s+1}, \ldots, y_k = x_k$, we see that $\det(\frac{\partial x_{\alpha}}{\partial y_{\beta}})$ restricted to $U' \cap Q$ is a polynomial of degree $k-s$ in $x_{s+1}, \ldots, x_k$, and this expression does not depend on the choice of U' and h; thus, this polynomial is defined on all of $U \cap Q$. Therefore, $\det(\frac{\partial \tilde{g}_{\alpha}}{\partial x_{\beta}})$ is bounded away from zero in $\overline{U \cap Q}$. Finally we obtain that $\det(\frac{\partial \tilde{g}_{\alpha}}{\partial x_{\beta}})_q \neq 0$, and the rank of $\tilde{g}$ at $q \in \partial Q$ is equal to $s = k-\ell$. Q.E.D.

We proceed now to prove the main lemma. Recall that $D' = \{x \in \partial M: r'(x) < n-1 \text{ and } r(x) < n\}$. Let $\partial D'$ denote its boundary with respect to ∂M.

<u>Proof of the Main Lemma</u>: Consider the function $h: \overline{D}' \to \mathbb{R}$ defined on the closure of D' as $h(y) =$
$= \text{infimum}\{|g'(y) - g'(z)| + |g(y) - g(z)| + |g'(y) \cdot f(y) - g'(z) \cdot f(z)| +$
$+ |g(y) \cdot f(y) - g(z) \cdot f(z)| : z \in \partial D'\}$. This function is continuous; also the infimum is always attained since $\partial D'$ is compact. Hence $h(y)=0$ if and only there exist a z in ∂D such that $g(z) = g(y)$, $g'(z) = g'(y)$, and $TM_z = TM_y$ and $T\partial M_z = T\partial M_y$ as affine subspaces of

R^{n+1}. We have $h(\partial D') = 0$. The set $A = \{y \in D': h(y) > 0\}$ is open in D; since $A \cap \partial D' = \phi$, A is open in ∂M. We want to show that A is empty.

Suppose A is not empty. Let $a \in A$ such that $r'(a) =$ maximum $\{r'(x): x \in A\} = n-1-k$. We have that $r'(a) = r^*(a) = n-1-k$, since $r'(a)$ is the maximum and A is open; the point a is in D', so that $k \geq 1$. We can then apply Lemma 2 of [3] which says that there exists a k-plane section P of $A \cap \{x \in \partial M: r'(x) = r'^*(x) =$ $= n-1-k\}$ passing through a on which $g(y)$ and TM_y are constant. Furthermore, at a boundary point x of P, $r'(x) = n-1-k$. Hence, either $r'^*(x) > n-1-k$ or $x \in \partial A$; but $n-1-k$ was a maximum in A, so $\partial P \subset \partial A$.

Now we show that, given any point y in this plane section P, there exists a point $x \in \partial P$ such that $g(x) = g(y)$ and $TM_x=TM_y$, i.e., $h|\overline{P}$ is identically zero. Let $B = \{x \in P|$ there does not exist a z in ∂P such that $g(z) = g(x)$ and $TM_z = TM_x$ as affine hyperplanes of $R^{n+1}\}$. B is open in P. Suppose $B \neq \phi$ and let b be a point in $B \cap V_\ell$, where ℓ is equal to the smallest integer ℓ' with $B \cap V_{\ell'} \neq \phi$; then b is in $\text{Int}\, V_\ell$. Since $B \subset P \subset A \subset D'$, any y in B is not the limit of points x_i in P with $r(x_i) = n$. Hence, by Lemma 7.1, $\ell \geq 1$. We can then apply Lemma 7.2 to get an ℓ-plane section L_b of $\text{Int}\, V_\ell \cap B$ on which g and TM are constant. Now, if $y \in \partial L_b$ then $y \in \partial B$ because ℓ was a minimum in B. But, then y is not in B and there exists a $z_1 \in \partial P$ such that $TM_{z_1} = TM_y = TM_b$. Thus, $B = \phi$ and $h|\overline{P} \equiv 0$. In particular $h(a) = 0$, which contradicts the fact that a was in A. Hence, $A=\phi$ and we are done. Q.E.D.

8. Proof of Theorems 3 and 4

In this section we prove Theorem 3. We are considering an

n-manifold M whose boundary components, $S_1,\ldots,S_\ell$, are homeomorphic to (n-1)-spheres; we have a substantial immersion $f: \to \to \mathbb{R}^{n+1}$ such that $f|S_i$ is convex, $i = 1,\ldots,\ell$. We require further that the Jacobian of the Gauss map g, defined locally, by definite at the points where it is non-singular. Then, we show that f is an embedding into the boundary of the convex hull of $f(M)$, $\partial H(f(M))$. We denote the normals at points of M and ∂M as N and N'. We will only need N to be defined continuously locally or along ∂M. We recall that $D = \{x \in M: r(x) < n\}$ and $D' = \{x \in \partial M: r'(x) < n-1$ and $r(x) < n\}$, and that Y_x, is the unit vector in TM_x perpendicular to ∂M and pointing out of M.

Lemma 8.1 - *Under the assumptions of Theorem 3, let $r^*(x) = k < n$; we assume, furthermore, that whenever $x \in \partial M$ we have $(N'.Y)_x < 0$. Then, through x passes an (n-k)-plane section of a neighborhood of x in M on which g and TM are constant.*

Proof: Let x be an interior point of M such that $r^*(x) = k$; then $x = \lim_{i\to\infty} x_i$, where $r^*(x_i) = r(x_i) = k$. Hence, applying Lemma 2 in [3] to these points, we see that there exist (n-k)-sections L_{x_i} passing through these points. Taking a subsequence, if necessary, we can assume that the (n-k)-planes in $\mathbb{R}^{n+1}$ associated to the L_{x_i}'s converge to a plane through x, so that $L_x = \{z \in M: z = \lim_{i\to\infty} y_i$, $y_i \in L_{x_i}\}$ is mapped by f into part of a plane on which g is constant. We must have that $f(x)$ is in the interior, with respect to the limit plane, of $f(L_x)$. Otherwise, we would have $x = \lim_{i\to\infty} y_i$, with y_i being boundary points of L_{x_i}. But this would imply that $r^*(y_i) > k$, for every i, which in turn would imply that $r^*(x) > k$, a contradiction.

If $x \in \partial M$, we extend f and g locally across the boundary so that x is an interior point of some larger manifold. By the arguments in the first part of the proof we get an (n-k)-plane section passing through x. By the condition that $f|\partial M$ is convex and that $(N'\cdot Y)_x < 0$, we can guarantee that this plane section intersects M in an (n-k)-submanifold. Q.E.D.

Observe that this (n-k)-plane section P might be an (n-k)-manifold with boundary; this boundary is $P \cap \partial M$.

If x is a point of M, let $\{e_1,\ldots,e_n\}$ be an orthonormal base of the linear subspace of R^{n+1} parallel to f_*TM_x; if x is a boundary point, we always take e_1 to be $-Y_x$. Let $E_x =$ $= \{\Sigma a_i e_i : a_i \in R$ such that if $a_1 = a_2 = \ldots = a_{i-1} = 0$ then $a_i \geq 0\}$ and $B_x = \{z$ in f_*TM_x: $z = f(y)$ for some y in M with $g(y) =$ $= g(x)$ and $z = f(x) + a$, $a \in E_x\}$. If $x \in \partial M$ let S be the boundary component that contains x.

<u>Lemma 8.2</u> - <u>Under the assumptions of Theorem 3, if x is in the interior of M there exists a point y in $\overline{M-D} \cup \partial M$ with $f(y)$ in B_x; if x is in S with $f(S) \cap (f(x) + E_x) = \phi$, then there exists a point y in $\overline{M-D} \cup (\partial M - S)$ with $f(y)$ in B_x.</u>

<u>Proof</u>: Let z_1 be a maximum of the function $h_1(z) = z.e_1$ in B_x;

let z_2 be a maximum for the function $h_2(z) = z.e_2$ restricted to $\{z$ in B_x: $z\cdot e_1 = h_1(z_1)\}$. Obtain z_i inductively in this manner, so that z_n is a maximum for the function $h_n(z) = z\cdot e_n$ restricted to $\{z$ in B_x: $z\cdot e_i = h_i(z_i)$, $i = 1,\ldots,n-1\}$; we have that $z_n = f(y) \in B_x$. If y were in Int $D \cap$ Int M, then $r^*(y) < n$ and, by Lemma 8.1, there would exist an (n-k)-plane section of a neighborhood of y on which g is constant. Hence, there exists

$f(y') \in B_x$ such that $f(y') - f(y) \in E_x - \{0\}$; but this contradicts the maximality of $z_n = f(y)$, since $E_x + E_x = E_x$. If $y \in \partial M \cap \text{Int } D$, then, by the condition that $f(S) \cap (f(x)+E_x) = \phi$ we must have that $y \in \partial M - S$ or $y = x$. If $y = x$, then, again applying Lemma 8.1, we see that $z_n = f(y)$ would not be maximal. Q.E.D.

Proposition 8.3 - If $f: M \to \mathbb{R}^{n+1}$ is an immersion satisfying the assumptions of Theorem 3, then f is 0-tight.

Proof: The assumption that g_* is definite when it is non-singular implies that $\mu_k(z) = 0$ for $k = 1,\ldots,n-1$. Since the boundary components of ∂M are tight, $\varphi^-_{n-1}(z) + \varphi^+_n(z) = \mu_{n-1}(z\cdot f|\partial M) = \ell$, and $\varphi^-_i(z) = 0$, $i = 1,\ldots,n-2$.

Case 1: $n > 2$. The Morse inequalities gives us that $\mu_o + \varphi^-_o \geq \bar{R}_o = 1$, $0 = \mu_1 + \varphi^-_1 \geq \bar{R}_1$ and $\mu_1 + \varphi^-_1 - (\mu_o + \varphi^-_o) \geq \bar{R}_1 - \bar{R}_o$. Hence $\mu_o + \varphi^-_o = 1$.

Case 2: $n = 2$. Looking at the exact sequence $0 \to H_2(M,\partial M) \xrightarrow{\partial_2} H_1(\partial M) \xrightarrow{i_1} H_1(M) \xrightarrow{j_1} H_1(M,\partial M)$ with $\mathbb{Z}_2$ coefficients we have that dimension $H_2(M,\partial M) = \dim H_1(\partial M) + \text{rank } j_1 - \bar{R}_1$. Since $\dim H_1(\partial M) = \ell$, we have that $\bar{R}_1 \geq \ell - 1$. On the other hand, we have that $\varphi^-_1(z) + \varphi^+_2(z) = \mu_1(z\cdot f|\partial M) = \ell$. This together with formula 3.1 which becomes $\mu_o + \varphi^-_o - \varphi^-_1 + \mu_2 = \chi(M) = 1 - \bar{R}_1$ in this case, gets us that

$$\mu_o + \varphi^-_o + \mu_2 + \varphi^+_2 - \ell = 1 - \bar{R}_1 \leq 2 - \ell$$

Since every height function must always have a maximum and a minimum, we obtain that $\mu_o + \varphi^-_o = \mu_2 + \varphi^+_2 = 1$. Q.E.D.

What we have then, in the context of Theorem 3, is that if

x is a local maximum of a non-degenerate height function $z \cdot f$ then x is a global maximum. For a point $p \in \mathbb{R}^{n+1}$ and a unit vector v in S^n, we say that (p,v) is a top-vector of the immersion f if $p = f(y)$, for some y in M, and $f(x) \cdot v \leq v \cdot p$, for every x in M. It is clear that $f(x)$ is in the boundary of the convex hull of $f(M)$, $\partial H(f(M))$, if and only if there exists a unit vector v such $(f(x),v)$ is a top-vector of f. Observe also that if $p = \lim_{i \to \infty} p_i$ and $v = \lim_{i \to \infty} v_i$, and each (p_i, v_i) is a top-vector, then so is (p,v).

If x in the interior of M is a non-degenerate critical point of $z \cdot f$ of index n, then $z = \pm N_x$, where N is the normal to M, and the Jacobian of the Gauss map is non-singular at x. Hence, for every y in a neighborhood of x, $N_y \cdot f$ has a non-degenerate critical point of index n at y. Therefore, since arbitrarily near to x there exist points y such that $N_y \cdot f$ is non-degenerate, we have in the limit that $(f(x), N_x)$ is a top-vector.

If x in ∂M is a non-degenerate critical point of index n of $z \cdot f$, then $z \cdot (f|\partial M)$ has a non-degenerate critical point of index $n-1$ at x and $z \cdot Y_x > 0$. But, then, arbitrarily close to x and z, there exist x' and z' such that x' is a non-degenerate critical point of index n of $z' \cdot (f|\partial M)$ where $z' \cdot (f|\partial M)$ is non-degenerate. Since, for z' close to z, we also have that $z' \cdot Y_{x'} > 0$, we have that $(f(x),z)$ is also a top-vector.

<u>Proof of Theorem 3</u>: We have that $f|S_i$ is convex for every boundary component S_i of ∂M, $i = 1,\ldots,\ell$. Hence, we can consider the topological manifold $\tilde{M}$ obtained from M by attaching along S_i the convex set bounded by $f(S_i)$, $i = 1,\ldots,\ell$. Let $\tilde{f}\colon M \to \mathbb{R}^{n+1}$ be equal to f on M and equal to the identity on those convex sets. If we show that 1) $\tilde{f}$ is a topological immersion, and 2) $\tilde{f}(\tilde{M}) \subset \partial H(f(M))$, then $\tilde{f}$ is onto $\partial H(f(M))$ because

$\tilde{f}$ is open, and $\tilde{f}$ is a covering map. Since $\partial M(f(M))$ is homeomorphic to the n-sphere, $\tilde{f}$ is a homeomorphism, and, consequently, f is a diffeomorphism into the convex hypersurface $\partial H(f(M))$, concluding the proof of Theorem 3. We proceed now to prove assertions 1) and 2).

1) $\tilde{f}$ is an immersion.

The extended function $\tilde{f}$ could fail to be an immersion only at a boundary point y. And, in this case, only if $Y_y = N'_y$. We will show that this cannot occur. Otherwise, we would have, by the Main Lemma, that there exists a point x in $\overline{\partial M - D'}$ with $g'(x) = g'(y)$ and $g(x) = g(y)$, so that at x we also have $Y_x = N'_x$. We must have that $x = \lim_{i\to\infty} x_i$, $x_i \in \partial M{-}D'$; that is, with $r(x_i) = n$ or $r'(x_i) = n-1$. In the first case, we get that $r(y) = n$ in a neighborhood of x_i, so that we can assume that $r'^*(x_i) = r'(x_i)$. If $r'(x_i)$ were strictly less than $n-1$, then, by Lemma 8.1 applied to ∂M, we would have that through x_i passes a k-plane-section, $k\geq 1$. However, this contradicts the fact that g_* is non-singular and definite at x_i; hence, we can assume that $r'(x_i) = n-1$.

If S is the component of ∂M that contains x, then, since $f|S$ is convex, any vector Z normal to S at x_i with $Z.N'_{x_i} > 0$ is such that $Z\cdot f|S$ has a non-degenerate critical point of index $n-1$ at x_i. Thus, if $Z\cdot Y > 0$, then, by Lemma 6.1 $Z\cdot f$ has a local maximum at x_i. By continuity, we can assume $Y_{x_i} \neq - N'_{x_i}$, and the normal to M, N_{x_i}, can be chosen so that $N_{x_i}\cdot N'_{x_i} > 0$, if $N_{x_i} \neq Y$; thus, $N'_{x_i} = -\cos\theta_i\, Y_{x_i} + \sin\theta_i\, N_{x_i}$, $0 < \theta_i \leq \pi$. If we let $W_\alpha = \cos\alpha\, N_{x_i} + \sin\alpha\, Y_{x_i}$, then $W_\alpha\cdot Y_{x_i} = \sin\alpha > 0$ if $0 < \alpha < \pi$ and $N'_{x_i}\cdot W_\alpha = \cos\alpha \sin\theta_i - \cos\theta_i \sin\alpha = \sin(\theta_i - \alpha) > 0$ if $0 < \theta_i - \alpha < \pi$. Hence, by the arguments following Proposition 8.3, $(f(x_i), W_\alpha)$ is a top-vector for every α with $0 \leq \alpha \leq \theta_i$. Observe that $W_o = N_x$, and $W_{\theta_i} = L_i$ is the vector perpendicular to the hyperplane that contains $f(S)$.

We have that $\lim_{i\to\infty} \theta_i = \pi$. Observe that even though the choice of L_i and N_{x_i} depended on the point x_i, we can choose a subsequence such that $\lim_{i\to\infty} L_i = L$ and $\lim_{i\to\infty} N_{x_i} = N_x$. Hence, taking limits, $(f(x),W_o)$ and $f(x),W_\pi)$ are top-vectors. But $W_o = N_x$ and $W_\pi = -N_x$, which implies that $f(M)$ is contained in the hyperplane which contains $f(S)$, contradicting the fact that f is substantial. Hence, $\tilde{f}$ is an immersion.

2) $\tilde{f}(\tilde{M}) \subset \partial H(f(M))$.

We first consider two simple situations. The first is one in which $x \in M\text{-}D$ so that $x = \lim_{i\to\infty} x_i$ such that $r(x_i) = n$. Since g_* is definite, we have that $N_{x_i}\cdot f$ or $-N_{x_i}\cdot f$ has a local maximum at x_i. By the remarks following Proposition 8.3, we get that $(f(x),N_x)$ or $(f(x),-N_x)$ is a top-vector, and $f(x)$ is in $\partial H(f(M))$.

The other situation occurs when $x \in \partial M\text{-}D'$ and $Y_x \neq -N'_x$. In this case, we start by defining θ_x as in 1), and, by the same arguments, conclude that $(f(x),W_o)$ is a top-vector.

Now, we consider an arbitrary point x in M. If x is in D, then, by Lemma 8.2, there exists a y in $\partial M \cup \partial D$ such that $g(y) = g(x)$ and $TM_y = TM_x$ as affine hyperplanes of R^{n+1}. Hence, if we show that $(f(y), \pm g(y))$ is a top-vector, then so is $(f(x), \pm \pm g(x))$. Therefore, if y is in $\partial D \subset \overline{M\text{-}D}$, we are done as above. If y is in ∂M, we consider the following cases.

<u>Case 1</u>: $Y_y \neq -N'_y$. This is one of the above situations if $x \in \overline{\partial M\text{-}D'}$.

If $x \in D'$, then, by the Main Lemma, there exists a z in $\partial D' \subset \overline{\partial M\text{-}D'}$ with $g(y) = g(z)$, $g'(y) = g'(z)$, $TM_z = TM_y$ and $TS_z = TS_y$, so that $Y_z \neq N'_z$, and $(f(z),N_z)$ is a top-vector. Hence, $(f(y),N_y)$ is also a top-vector.

<u>Case 2</u>: $Y_y = -N'_y$, but $Y_z \neq N'_z$ for some z in the same boundary

component S which contains y. In this case, as in the proof of 1), we have that $(f(z), \pm L)$ is a top-vector. Since $f(S)$ is contained in the hyperplane perpendicular to L, we have that $f(y)\cdot L = f(z)\cdot L$, and $(f(y), \pm L)$ is also a top-vector.

Case 3: $Y_z = -N'_z$ for every z in S_1, the component of y. This implies that $TM_z = TM_x$ as affine hyperplanes, for every z in S, so that if $(f(z), \pm N_z)$ is a top-vector then $(f(x), \pm N_x)$ is also one. Recall that, before Lemma 8.2, we defined $(e_1)_x$ as $-Y_x$ if x is in ∂M. Let z_1 be a point in S, such that $(e_1)_x\cdot f(z_1) = \max(e_1)_x\cdot f(S_1)$. Define $z_i \in S$, inductively by the property that $(e_i)_x\cdot f(z_i)$ equals the maximum of $(e_i)_x\cdot f(z)$ when z varies over the set of z's such that $(e_j)_x\cdot f(z) = (e_j)_x\cdot f(z_j)$, $j \le i-1$. We have that $(e_1)_{z_n} = -Y_{z_n} = N'_{z_1} = -Y_{z_1} = (e_1)_x$; therefore, we can make E_{z_n} equal to E_x. By the definition of z_n, we get that $(E_{z_n} - \{0\}) \cap f(S_1) = \emptyset$. Hence, we can apply Lemma 8.2 to get a point $\omega \in \overline{M-D} \cup (\partial M - S_1)$ such that $f(\omega) \in B_{z_n}$. Since $E_{z_n} = E_x$ and $f(z_n) \in B_x$, we get that $f(\omega) \in B_x$, since $E_x + E_x = E_x$, so that it is enough to show that $(f(\omega), g(\omega))$ is a top-vector. If this point ω is in $\overline{M-D}$ then we are done.

If $\omega \in \partial M - S_1$, then we are in another component S_2 of ∂M; if case (1) or (2) holds we are finished; if case (3) holds then we proceed as above to get to a point ω' in $\overline{M-D} \cup (\partial M - S_2)$ such that $f(\omega') = f(\omega) + a'$, $a' \in E_x$. If $\omega' \in \overline{M-D}$, then $a' \neq 0$, and $\omega' \notin S_1 \cup S_2$. Proceeding in this fashion we must arrive either at a point $x' \in \overline{M-D}$ or a point $x' \in \partial M$ at which either case (1) or (2) holds, since there are only a finite number of boundary components. But then $(f(x'); \pm g(x'))$ is a top-vector; since $f(x') \in B_x$, $(f(x), \pm g(x))$ is a top-vector also.

Finally, if y is a point in $\tilde{M}-M$, then $N_y = L$, where

L is perpendicular to the component S of ∂M that bounds the component of y in $\tilde{M} - M$. If z is any point in S, we have, by one of the above cases, that $(f(z), \pm L)$ is a top-vector; thus $(f(y), \pm L)$ is also a top-vector. Q.E.D.

Proof of Theorem 4: If two of the eigenvalues of the second fundamental form of a hypersurface have different sign, then the sectional curvature of the plane spanned by the corresponding principal directions is negative. Hence, all the hypotheses of Theorem 3 are satisfied and f is convex. Q.E.D.

Note. Recently Wolfgang Kühnel informed me that he has obtained results (to appear in J. London Math.Soc.) similar to some of those contained here.

REFERENCES

[1] T.F. Banchoff, The two-piece-property and tight n-manifolds with boundary, Trans.Amer.Math.Soc. 161 (1971), 259-267.

[2] B.Y. Chen, On the total absolute curvature of manifolds immersed in Riemannian manifolds. III, Kodai Math.Sem.Rep. 22 (1970), 385-400.

[3] S.S. Chern & R. Lashof, On the total curvature of immersed manifolds. I, Amer.J.Math. 79 (1957), 306-318.

[4] N. Grossman, Relative Chern-Lashof theorems, J.Differential Geometry 7 (1972), 607-614.

[5] P. Hartman and L. Nirenberg, On spherical image maps whose Jacobian do not change sign, Amer.J.Math. 81 (1959), 901-920.

[6] N. Kuiper, Morse relations for curvature and tightness, in Proceedings of Liverpool Singularities Symposium II, Springer--Verlag, Berlin, 1971, 77-89.

[7] J. Milnor, Lectures on Morse theory, Ann. Math. Studies 51, Princeton University Press, Princeton, N.J., 1963.

[8] L. Rodriguez, The two-piece-property and convexity for surfaces with boundary, J. Differential Geometry 11(1976), 235-250.

[9] R. Sacksteder, On hypersurfaces with non-negative sectional curvatures, Amer.J.Math. 82 (1960), 609-630.

Instituto de Matemática Pura e Aplicada (IMPA)
Rio de Janeiro, Brasil.

EXISTENCE OF BRAKE ORBITS IN FINSLER MECHANICAL SYSTEMS

by

OTTO RAUL RUIZ M.

SUMMARY

The aim of this paper is to make a small contribution on the behavior of orbits in mechanical systems. It was motivated mainly by a paper written by Seifert (see reference), where it is proven that, in the analytic-Riemannian case, there exists periodic orbits of total energy E_0 that join points of a surface of potential energy E_0, when this surface bounds a region homeomorphic to the ball D^n. Professor S. Chern made an extension of Seifert's result to the case C^∞-Riemannian, in 1972.

The principal proposition of this paper shows that in the case of C^3 - Finsler mechanical systems, there exist brake-orbits of energy E_0. A brake-orbit is one which approaches with velocities decreasing to zero to fixed points, on both sides of any points of the orbit. As a corollary of this proposition we may extend the Seifert paper to the C^3 case, because for a Riemannian metric it is possible to prove that the E_0 - orbits cannot spend too much time in a neighborhood of the E_0 surface, without going out of it; therefore, we arrive at the E_0 surface in a finite time. Also, from the uniqueness theorems of solutions of differential equations, we can deduce that this orbit is periodic.

#0. Introduction.

The main results of this paper are presented in three propositions, the last one being a corollary of the second one, which in turn, besides motivating the name to the paper, shows how we can use it to extend Seifert's result (the analytic case) to the C^3-case.

Proposition A: Let (M_0, K, U) be a mechanical system such that M_0 is a domain in $\mathbb{R}^n$; K is the kinetic energy, $K = F^2$, where F is a Finsler metric, U the potencial energy. If E_0 is a regular value of U such that $B = \{x \mid U(x) \leq E_0\}$ is homeomorphic to the n-dimensional ball, then for $\delta > 0$ and small, we may join two points of the surface $U^{-1}(E_0 - \delta)$ with orbits of total energy E_0. We assume further that U is a C^3 function and K is C^3 for velocities different from zero.

Proposition B tells us about the existence of brake-orbits, in the same conditions of Proposition A. A brake-orbit is one which approaches with velocities decreasing to zero to fixed points, on both sides of any point of the orbit.

Proposition B: There exists at least one E_0-brake-orbit that approaches fixed points of the boundary of B.

Corollary. In the C^3-Riemannian case there exists a periodic orbit of energy E_0 which joins points of the boundary of B.

#1. Preliminaries.

We start by recalling some definitions and making some important remarks.

a) A metric F on a manifold M is a real-valued function from the tangent manifold. $F: TM \to \mathbb{R}$ satisfying

$$1)\ \ F(v_x) \neq 0 \ \text{ if } \ v_x \neq 0; \quad 2)\ \ F(\lambda v) = |\lambda| F(v) .$$

b) A metric is a <u>Finsler metric</u> if the functions $g_{ij}(x,\dot{x}) = \frac{1}{2} \frac{F^2(x,\dot{x})}{\partial \dot{x}^i \partial \dot{x}^j}$ define a positive quadratic form in every point $(x,\dot{x})$, $\dot{x} \neq 0$, i.e., $g_{ij}(x,\dot{x})\, \dot{z}^i \dot{z}^j > 0$ for all values $\dot{z}^i$, unless the $\dot{z}^i$ vanish simultaneously.

<u>We are using Einstein summation convention.</u>

We do not define the $g_{ij}(x,\dot{x})$ when $\dot{x} = 0$, because the g_{ij} are homogeneous of degree 0 in $\dot{x}$. Therefore, we cannot consider F^2 differentiable on the section 0 of TM, without reducing to the Riemannian case.

c) $K(x,\dot{x}) = F^2(x,\dot{x})$ is homogeneous of degree 2,

d) $g_{ij} = g_{ji}$, and since they define a positive quadratic form, there exists a symmetric tensor g^{ik} such that $g^{ij} g_{jk} = \delta^i_k$.

e) The definition of a Finsler metric is equivalent to the condition that the <u>indicatrix</u>, $F = \{x \mid F(x,\dot{x}) = 1\}$ is a convex set. This surface plays then the role of the unit sphere in the Euclidean geometry.

f) The convexity of the indicatrix is equivalent to the triangle inequality in $\dot{x}$: $F(x,\dot{y}+\dot{z}) \leq F(x,\dot{y}) + F(x,\dot{z})$. Therefore we may consider F as a norm in each $T_x M$.

<u>Notations</u>. (using Einstein convention)

a) $g_{ij,k} = \dfrac{\partial g_{ij}}{\partial x^k}$;

b) $g_{ij,k} = \dfrac{\partial g_{ij}}{\partial \dot{x}^k}$;

c) $[ij,k] = \frac{1}{2}\,(g_{ik,j} - g_{ij,k} + g_{jk,i})$;

d) $\Gamma^{\ell}_{ij} = g^{k\ell}[ij,k]$;

e) $U_k = \dfrac{\partial U}{\partial x^k}$;

f) $\dot{B}$, boundary of $\{x \mid U(x) < E_0 - \delta\} = B_\delta$.

#2. The Equations of Motion.

The Lagrange equations of motion are giving by

$$\frac{d}{dt}\frac{\partial(K-U)}{\partial \dot{x}^k} = \frac{\partial}{\partial x^k}(K-U). \qquad (*)$$

From Rund (6;p.23), we have the important relations

$$g_{ik;j}(x,\dot{x})\dot{x} = 0 , \qquad g_{ij;k}(x,\dot{x})\dot{x}^j = 0 .$$

These equations allow us to write $(*)$ on the form:

$$\ddot{x}^r + \Gamma^r_{ij}(x,\dot{x})\dot{x}^i\dot{x}^j + g^{rk}(x,\dot{x})U_k(x) = 0, \quad \dot{x} \neq 0. \qquad (**)$$

We also may note that this equation has the same form as in the Riemannian case, except that the g_{ij} depends on the velocity, $\dot{x} \neq 0$.

From Caratheodory (4), we remark that this equation always possesses a unique solution corresponding to any given initial point $(x,\dot{x},t)$, $\dot{x} \neq 0$.

#3. General Properties of Orbits of Total Energy E_0.

a) Every orbit of energy E_0 is all in $\overline{B}$. Indeed, $U(x) = E_0 - K(x,\dot{x})$, and since $K \geq 0$, then $U(x) \leq E_0$.

b) The components of the velocity are bounded during the whole motion, and are very small whenever the point is close to the boundary of B.

Let's take:

$$H = \{(x,z) \mid x \in \overline{B} \text{ and } |z| = \Sigma z_i^2 = 1\} .$$

Since H is compact and $K(x,z) > 0$ in H, there exists m such that $0 < m = \min K(x,z)$, $(z,x) \in H$. If $(x,\dot{x})$ is any point of energy E_0, there exists Z such that $\lambda Z = \dot{x}$ $|\dot{x}| = \lambda$, and $\lambda^2 \cdot m \leq \lambda^2 \cdot K(x,z) = K(x,\dot{x}) = E_0 - U(x) \leq E_0$. Therefore:

b_1) $|\dot{x}|^2 \leq E_{0/m}$,

b_2) $|\dot{x}|^2 \leq \dfrac{E_0 - U(x)}{m}$, and consequently, when $E_0 - U(x)$ is small, $\dot{x}$ is also small.

#4. Concept of Transversality.

In looking for a preliminary coordinate system in the neighborhood of B_δ, δ very small, we will use a special direction. The most natural is the trnasversal direction of Caratheodory. Suppose that (x,v_0) is any element with $v_0 \neq 0$. We construct the hypersurface in T_xM determined by $\{v; |v| = F(x,v) = |v_0|$. It is possible to see that the tangent hyperplane to this hypersurface in v_0 is given by

$$\sum_{i,j} g_{ij}(x,v_0)\, v_0^i\, v^j = |v_0|^2 \quad . \tag{1}$$

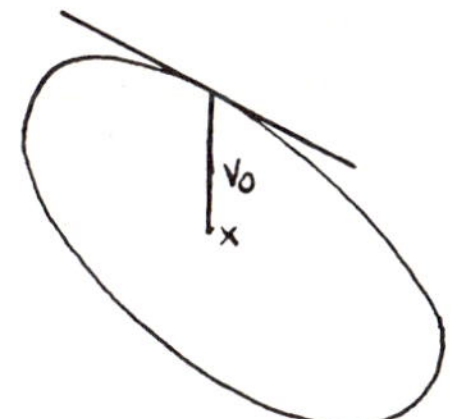

Any vector η contained in or parallel to this hyperplane is said to be transversal with respect to v_0.

Since such a vector may be represented by the difference of two vectors v_1 and v_2 satisfying (1), we see that the condition that <u>η be transversal to</u> v_0 in T_xM may be represented by the relation

$$\sum_{i,j} g_{ij}(x,v_0)\, v_0^i\, \eta^j = 0 \quad ,$$

which is not a symmetric relation.

It is also possible to prove, that, if we have one hyperplane there exists one, and only one transversal direction to this hyperplane.

#5. Introduction of a Coordinate System:

$$h: \dot{B}_\delta \times [-\epsilon,\epsilon] \to B, \quad B = \{x | U(x) < E_0\} \; .$$

We are going to define h in such a way that $h(q,t)$ is the orbit of E_0 -energy for q fixed and such that $h(q,0) = q \in \dot{B}_\delta$ and $h'(q,0)$ is transversal to $\dot{B}$ toward the interior.

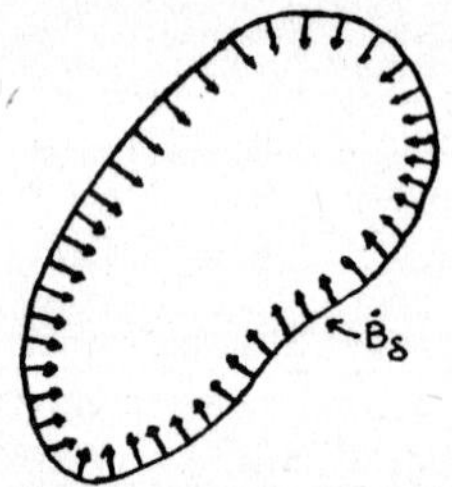

Since $h'(q,0) \neq 0$, we may consider this h as our embedding and, since $\dot{B}_\delta$ is compact, we may choose $\varepsilon < \delta$ such that $h: \dot{B}_\delta \times [-\varepsilon, \varepsilon] \to B$ is a coordinate system. $(q', q^2, \ldots, q^{n-1}, q^n = t)$, $-\varepsilon < t < \varepsilon$, may be considered as a local coordinate system.

Since the trajectories go to the interior, for small we may assume that

$$U_n(a) = \frac{\partial U}{\partial q^n}(a) = \frac{\partial U}{\partial t}(a) < 0 , \qquad a \in \dot{B} .$$

Furthermore,

$$U_i(a) = \frac{\partial U}{\partial q^i}(a) = 0, \qquad 0 < i \leq n-1 .$$

#6. The Jacobi Metric.

For our purposes we shall need the Jacobi Metric (J-metric):

$$dS^2 = (E_0 - U)\, g_{ij}(q,\dot{q})\, dq^i\, dq^j, \quad \text{on } B .$$

The main fact about this metric is that the geodesics parametrized by the arc lenght for the J-Metric, are those trajectories of the system with total energy E_0 (arc length is not necessarily positive). The differential equations of the geodesics are thus

$$\frac{d^2q^r}{dS^2} + \tilde{\Gamma}^r_{ij}(q,\dot{q})\,\dot{q}^i\dot{q}^j = 0\ , \qquad \text{where}$$

$$\tilde{\Gamma}^r_{ij} = \Gamma^v_{ij} + \frac{1}{E_0-U}\left(\delta^r_i(E_0-U)_j + \delta^r_j(E_0-U)_i - g^{\ell r} g_{ij}(E_0-U)\right) \ ,$$

and δ^r_i are the Kronecker symbols.

We may observe now that <u>to every geodesic fo finite length contained in</u> B_δ , $\delta > 0$ <u>corresponds a trajectory of finite time</u>. We may see this from the following equation:

$$\left(\frac{dS}{dt}\right)^2 = (E_0-U)\, g_{ij}(q,\dot{q})\,\frac{dq^i}{dt}\,\frac{dq^j}{dt} = (E_0-U)K \ ,$$

since

$$K = E_0-U \quad \text{and therefore} \ . \ \left(\frac{dS}{dt}\right)^2 = (E_0-U)^2 \ . \tag{2}$$

furthermore in B_δ, $E_0-U \geq \delta$ so that $\frac{dS}{dt} \geq \delta \Rightarrow \int dS \geq \delta \int dt$,

and therefore our assertion above is correct.

<u>Now we are going to change</u> t <u>by a function of</u> S, where S <u>is the arc length of the trajectories of total energy</u> E_0, <u>that start transversally to</u> $\dot{B}$ <u>toward</u> B_δ.

In particular, from (2) we obtain

$\left(\frac{dS}{dt}\right) = E_0 - U$, and therefore in $\dot{B}$ we have $\frac{dS}{dt} = \delta < 0$.

Therefore we can consider coordinate systems on a neighborhood of the boundary $\dot{B}$ with coordinates

$$y^1 = q^1 \ \dots \ y^{n-1} = q^{n-1} \qquad y^n = S$$

Next we will see that the transversality in $\dot{B}$ of our n-coordinate lines $(y^i = \text{constant}, \dots, y^{n-1} = \text{constant})$ to the surfaces $S = \text{constant}$, is preserved.

With this purpose it is natural to look for an extension of Gauss lemma for normal coordinates to Finsler case.

Theorem 1. Let $g_{ij}(y,\dot{y})$ be a Finsler metric and $y = y^1 \dots y^n$ a coordinate system. If $y = y(S,\lambda)$ represents a differentiable function of the arc length S, and a parameter λ, and if for λ = constant the equation represents a geodesic of arc length S, then the expression

$$\sum_{i,j} g_{ij}\left(y, \frac{\partial y}{\partial S}\right) \frac{\partial y^i}{\partial S} \frac{\partial y^j}{\partial \lambda}$$

is independent of S.

The proof is similar to the proof of Gauss lemma, using the identities

$$g_{ij;\ell}(y,y_S)\, y^i_S = 0 \quad .$$

In our case we may apply Theorem 1 if we consider $y^i = y^i(\lambda)$, $i \neq n$, $y^n = y^n(S) = S$. For constant λ the curves $y(\lambda,S)$ are geodesics. For constant S, we have curves on the surface y^n = constant.

From Theorem 1,

$$(E_0 - U) \sum_{i,j} g_{ij}\left(y, \frac{\partial U}{\partial S}\right) \frac{\partial y^i}{\partial S} \frac{\partial y^j}{\partial \lambda} \tag{3}$$

is independent of S.

Since for constant λ the geodesics $y(\lambda,S)$ were chosen transversal to the surface $y^n = 0$, $(\dot{B}_\delta)$ the expression (3) is 0 on $S = 0$, and by theorem 1, (3) is equal to 0, for all S in our neighborhood; therefore the surfaces $y^n =$ constant are transversal to the trajectories of energy E_0 which start transversally transversally to $\dot{B}_\delta$.

#7. Geodesic convexity.

Definition: We say that a surface $y^n = c$ is geodesic convex at the point p, in the direction of increasing of y^n if for every geodesic $y(t)$ that is tangent to the surface at p, we have that $y_n''(p) > 0$.

It is very easy to see that this condition is independent of the coordinate system, in the sense that if $z_1,\ldots,z_n$ is another coordinate system such that z_n increases in the same direction as y^n, in which the surface $y^n = c$ is described by $z_n = d$, we have also for the tangent geodesics at p that $z_n''(p) > 0$ if $y^n = c$ is geodesic convex for the coordinates y^n.

Now we make a study of the surfaces $y^n = c$ with respect to the property of geodesic convexity.

Because we will need to deform the Jacobi metric in the direction of y^n, we wish to have a way of recognizing this property for a metric $h_{ij}(y,\dot{y}) = \lambda(y^n)\cdot(E-U)\cdot g_{ij}(y,\dot{y})$ where λ is a differentiable function.

Since we are considering geodesics that are tangent to the surface $y^n = c$, we have $\dot{y}_n = 0$ on the contact point; hence

$$y^{n''} = - \sum_{ij=1}^{n-1} \tilde{\Gamma}^n_{ij}\,(y,\dot{y})\;\dot{y}^i\dot{y}^j \quad ,$$

where

$$\tilde{\Gamma}^n_{ij}(y,\dot{y}) = h^{nr}[ij,r]_h \ , \quad [ij,r]_h = h_{ir,j} - h_{ij,r} + h_{jr,i}$$

now

$$h_{ij,r} = \lambda(y^n)\ ((E_0-u)_j\ g_{ij} + (E_0-U)\ g_{ij,r}) + \frac{\partial\lambda(y^n)}{\partial y^r}\ (E_o-U)g_{ij}$$

If we consider surfaces $y^n = c$, for c very small, we may write, for $r \neq n$, $(E_0-U)_r = 0$. Furthermore we have for $r \neq n$,

$$\frac{\partial\lambda(y^n)}{\partial y^r} = 0; \quad \text{then, for} \quad r \neq n \ ,$$

$h_{ij,r} = \lambda(y^n)\ (E_0-U)\ g_{ij}$ and

$$h_{ij,n} = \lambda(y^n)(-U_r)g_{ij} + \lambda(y^n)(E_0-U)g_{ij,n} + \\ + \lambda'(y^n)(E_0-U)g_{ij}$$

Therefore, for $r \neq n$, $i \neq n$, $j \neq n$, we have $[ij,r]_h =$ $= \lambda(y^n)\ (E_0-U)\ [ij,r]_h$ and for $r = n$, $i \neq n$, $j \neq n$ we have

$$[ij,n]_h = \lambda(y^n)(E_0-U)\ [ij,n]_g + \lambda(y_n)U_n g_{ij} - \lambda'(y^n)(E_0-U)g_{ij} \ ;$$

now for $i \neq n \neq j$

$$\tilde{\Gamma}^n_{ij} = h^{nr}\ [ij,r]_h = \frac{g^{nr}}{\lambda(y^n)\ (E_0-U)}\ [ij,r]_h = [ij,r]_g +$$

$$+ \frac{g^{nn}}{\lambda(y^n)(E_0-U)}\ U_n g_{ij} - \frac{g^{nn}\ \lambda'(y^n)}{\lambda(y^n)(E_0-U)}\ (E_0-U)g_{ij} \quad ;$$

thus

$$\tilde{\Gamma}^n_{ij} = \Gamma^n_{ij} + \frac{g^{nn}}{\lambda(y^n)\ (E_0-U)} U_n g_{ij} - g^{nn} \frac{\lambda'(y^n)}{\lambda(y^n)} g_{ij} \ , \qquad (4)$$

for $i \neq n$, $j \neq n$.

In the particular case of our Jacobi metric

$$\lambda(y^n) = 1 \quad \text{and} \quad \tilde{\Gamma}^n_{ij} = \frac{g^{nn}}{E_0-U} U_n\, g_{ij} + \Gamma^n_{ij} \ ,$$

and in the contact point with $y_n = c$, we have

$$y''_n = - \sum_{i,j=1}^{n-1} \tilde{\Gamma}^n_{ij}\, y^i y^j = - \sum_{ij=1}^{n-1} \Gamma^n_{ij}\, \dot{y}^i \dot{y}^j \ + \qquad (5)$$

$$+ \frac{-U_n g^{nn}}{E_0-U} \sum_{i,j=1}^{n-1} g_{ij}\, \dot{y}^i \dot{y}^j$$

Now in Finsler spaces the norm $F^2\ (\gamma(t),\ \gamma'(t))$ for a geodesic $\gamma(t)$, is constant. If they are parametrized by the arc length the norm is equal to 1. Let $\overline{B}(1) = \{(y,\dot{y}) \mid y \in \overline{B}\ ,\ F^2(y,\dot{y}) = 1\}$ and $\overline{B} = \{y \mid U(y) \leq E_0\ ;$ clearly $\overline{B}(1)$ is a compact set.

The function $\sum_{i,j=1}^{n-1} \Gamma^n_{ij}\ (y,\dot{y})\ \dot{y}^i \dot{y}^j$ is continuous in this set $\overline{B}(1)$, therefore it is bounded in it.

The same is true for the function

$$\sum_{ij=1}^{n-1} -U_n(y) g^{nn}\ g_{ij}(y,\dot{y})\ \dot{y}^i \dot{y}^j \quad .$$

This last function is positive since

$$-U_n(y) > 0, \quad g^{nn}(y,\dot{y}) > 0 \quad \text{and} \quad \sum_{ij=1}^{n-1} g_{ij}(y,\dot{y})\ \dot{y}^i \dot{y}^j > 0 \ ,$$

because the g_{ij} defines a positive quadratic form. If we write again (5) as

$$y_n'' = - \sum_{i,j=1}^{n-1} \Gamma^n_{ij}(y,\dot{y})\, \dot{y}^i \dot{y}^j + \frac{-U_n g^{nn} \sum_{ij=1}^{n-1} g_{ij}(y,\dot{y})\, \dot{y}^i \dot{y}^j}{E_0 - U}$$

we can see that when E_0-U <u>is very small</u>, the sign of y_n'' is the same as that of the last term, and therefore positive, since we consider $E_0-U \to 0$.

#8. <u>Deformation of the Jacobi metric</u>.

We will need to construct a zone like the one in the picture where the geodesics are reflected to the exterior of this zone. We, therefore, need to invert the sense of convexity in one surface $y^n = \theta$.

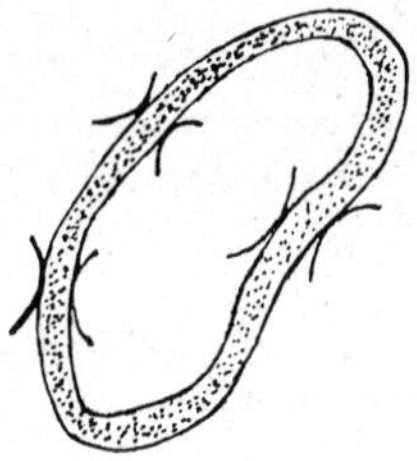

A good λ for this purpose is the following

$$\lambda(y^n) = \begin{cases} 1 & \text{if } \; y^n \geq j > 0 \\ \left(\dfrac{j - y^n}{j-\theta}\right)^k + 1 & \text{for } \; -\varepsilon \leq y^n \leq j \end{cases} ,$$

Where $\varepsilon > 0$ and k is going to be chosen sufficiently large in such a way that the surface $y^n = \theta$ be convex in the direction of decreasing of y^n .

We may see that $\lambda(r) = 1, \quad \lambda'(j) = \ldots = \lambda(k-1)_{(j)} =$

$\lambda(\theta) = 2$, and $\lambda'(\theta) = -k$. From (4) we may consider now the equation

$$y''_n = - \sum_{ij=1}^{n-1} \Gamma^n_{ij}(y,\dot{y})\, \dot{y}^i \dot{y}^j + \frac{g^{nn}(-U_n)}{(y_n)\ (E_0-U)} \sum_{i,j=1}^{n-1} g_{ij}\, \dot{y}^i \dot{y}^j +$$

$$+ \frac{\lambda'(y_n)}{\lambda(y_n)}\, g^{nn} \sum_{ij=1}^{n-1} g_{ij}\, \dot{y}^i \dot{y}^j \ .$$

In the particular case $y^n = \theta$, then we have $\lambda(\theta) = 2$ $\lambda'(\theta) = -k$. From the last arguments, E_0-U is now very small but fixed. The first and second terms are bounded, and

$$\frac{g^{nn}}{\lambda(\theta)} \sum_{ij=1}^{n-1} g_{ij} \dot{y}^i \dot{y}^j$$

is positive.

If we consider now k large enough the sign of $(y^n)''$ will be the same as the sign of the last therm, which is negative, since $\lambda'(\theta) = -k$. $k > 0$. Therefore, for this modified metric the surface $y^n = \theta$ is geodesic-convex in the sense of decreasing of y^n.

#9. Some Properties of the Geodesics of Finsler Spaces.

From Busemann [3] and Rund [6] we may see that Finsler spaces, share with Riemannian spaces the principal properties that are necessary for the rest of the proof.

1) The geodesic curves are uniquely determined as the shorter paths joining any two sufficiently neighboring points.

2) Given a compact set $K \subset M$ there exists a number $\delta > 0$ so that any two points of K with distance less than δ are joined by a unique geodesic of length less than δ; its length is $d(P,q)$. It

is called <u>the elementary arc</u> connecting P and Q, and d is the <u>elementary length</u>.

The proof of 2) is the same as in the Riemannian case. It depends on existence and uniqueness theorems of differential equations, and an application of the Inverse function theorem.

3) From Busemann [3]; p.24) we get the following: Let $P_v \to P_0$, $Q_v \to Q_0 \neq P_0$ and let g_v be a geodesic that carries the elementary arc P_vQ_v $v = 0, 1, \ldots$, then $g_v \to g_0$.

4) Let σ be a parameter that increases from 0 to 1 and is proportional to the arc length on the elementary arc PQ; then the point with the parameter σ is continuously dependent on σ, P and Q.

We call a polygon whose corners are in k and whose sides are elementary arcs, an <u>elementary polygon</u>.

<u>Remark</u>: In the next section we are going to use the fact that we can go a little behind the surface $y^n = S = 0$ or $\dot{B}$. In other words, it is possible to consider parallel surfaces $y^n = -\epsilon$, $\epsilon > 0$, with ϵ small.

#10. <u>Proof of the existence</u>.

We may consider that in the neighborhood of $\dot{B}_\delta$, $-\epsilon \le y^n < \alpha$, $\epsilon > 0$, $\alpha > 0$, all the surfaces $y^n = c$ are geodesic convex in the direction of increasing of y^n for the Jacobi metric. We name this set V_α.

Let $M = B_\delta \cup V_\alpha$. And $-\epsilon < \theta < \gamma < 0 < \beta < \alpha$.

Analogously we define V_θ, V_γ, V_β, V_α and we see that $V_\theta \subset V_\gamma \subset V_\beta \subset$ $\subset V_\alpha$. We also consider the following sets: $R = M - V_\alpha$; $\bar{L} = M - V_\beta$, $\bar{Y} = M - V_\gamma$ $\bar{D} = M - V_\theta$. Remark that $\bar{R} \subset \bar{L} \subset \bar{Y} \subset \bar{D}$ M. Also since $\gamma < 0$, $\bar{B}_\delta \subset Y$ and since $\beta > 0$, $\bar{L} \subset \mathring{B}_\delta$, where $\mathring{B}_\delta$ is the interior of B_δ.

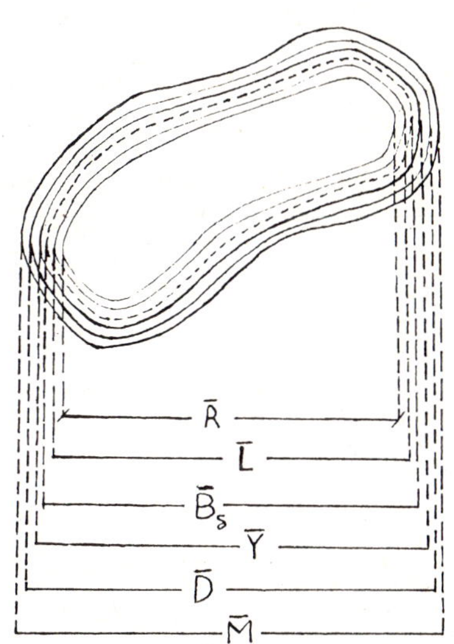

From the construction of the coordinates y^i, we may note that all these domains are homeomorphic to the n-dimensional ball. Then it makes sense to talk about diameters: The images of Euclidean diameters of an n-ball topologically mapped onto F.

We select now an elementary length on $\bar{R}$ so small that every elementary arc with end points on $\bar{R}$ is completely contained in $\bar{L}$.

We consider now the set of diameters of $\bar{R}$, and divide each of them in N parts, such that the distance between subsequent division points is not longer than the elementary length. We may also assume that the division points vary continuously with the diameter; this is possible according to the theorem about uniform continuity, since we only have to divide the diameters of the original euclidean ball in N euclidean equal parts.

Now, we substitute the diameters in $\overline{R}$, by elementary polygons that one gets connecting subsequent division points by elementary arcs.

Then we extend these elementary polygons to the boundary of $\overline{L}$, by segments $y^1 = \text{constant}, \dots y^{n-1} = \text{constant}$. In this way we get a family of curves $S_{\overline{L}}$ in $\overline{L}$.

In the same form, we can consider a family S_D of curves in $\overline{D}$.

Now, we are going to modify the metric h_{ij} in the zone M-Y, with the function $\lambda(y^n)$, in such a way that the boundary of $\overline{D}$ (i.e., the surface $y^n = \theta$), is geodesically convex in the direction of <u>decreasing</u> of y^n. According to this, we may see that the boundaries of the zone $\overline{D}$-L have the property of geodesic reflexion in the sense of the picture.

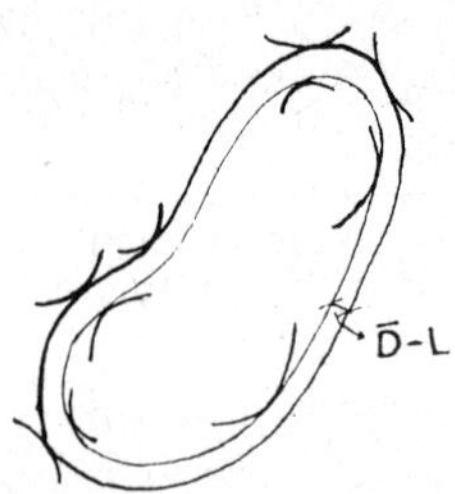

One of the most important points of the proof, is that we may choose in $\overline{D}$ an elementary length $\overline{d}$ for the modified metric, so small that an elementary arc with endpoints in $\overline{D}$-L, is included in $\overline{D}$-L.

This is possible from the property of reflection of geodesics on the boundary of $\overline{D}$-L. If there were no such $\tilde{d}$, there would be arbitrary short geodesic arcs with endpoints in $\overline{D}$-L, that stick out of $\overline{D}$-L. These geodesics would be tangent to surfaces

y^n = constant that are outside of $\bar{D}-L$, but arbitrarily close to $\bar{D}-L$, and this contradicts the geodesic convexity of concavity of the surfaces.

Now we apply to the family S_D, a well known shortening procedure that appeared in (8) and in (2) where it is used to construct a closed geodesic on a convex surface. This procedure consists of two steps.

1) First we divide v in S_D in q equal parts and take q so large, that the arcs are shorter than the elementary length $\tilde{d}$. We replace every arc, by the elementary arc that connects its endpoints.

2) Second, in the above elementary polygon, we connect the midpoints of each side, by a new elementary polygon, and extend it by half of the first and last arcs. We then have an elementary $q+1$ sided polygon $f(v)$ between the same points as v. $f(v)$ is not shorter than v only if v is a geodesic.

To $f(v)$ we can apply the same procedure again and get an elementary polygon $f^2(v)$. We can continue in this way, to obtain families $f(S_D)$, $f^2(S_D)$ ..., that have the property that each one covers $\bar{D}$ completely.

The proof of all this depends only on properties of the geodesics in section #9, and can be found in (8) and (2).

From this fact, we may find curves in the vth family whose intersection with L is non empty. Suppose that v_i is a curve that after application of the procedure f, i times, still strikes L. Then we have

$$f^i(v_i) \cap L \neq \phi \qquad (6)$$

Now, we proceed to prove that if $j < i$,

$$f^j(v_i) \cap L \neq \phi \text{ also .} \tag{7}$$

If not we have a situation like in the picture for $f^j(v_i)$ but by geodesic reflection we must have $f^i(v_i) \cap L \neq \phi$ which contradicts the definition of v_i.

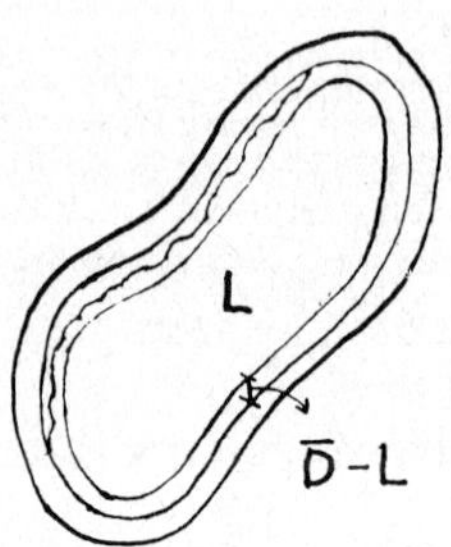

From the sequence $v_1, v_2 \ldots$ of elements of S_D, we choose now a subsequence that converges to a certain polygon v. To guarantee this, all we have to do is to choose our subsequence in such a way that the diametral points of the subsequence, i.e. the points of the boundary of D, converge.

Suppose that this subsequence is $v_{i_1}, \ldots, v_{i_j}, \ldots$. We are interested in proving that $f^i(v) \cap \bar{L} \neq \phi$ for all i. Suppose that this is false for a certain i ; then $f^i(v)$ lies in $\bar{D} - \bar{L}$. Then a sufficiently close neighboring approximation elementary polygon would lie also in $\bar{D} - \bar{L}$, then we have $f^i(v_{i_j}) \cap \bar{L} = \phi$ for some $i_j > i$, but this is in contradiction with (7) above.

Since $\bar{D}$ is compact, it is easy to observe that we may consider a subsequence of v, $f(v)$, $f^2(v)$, ... that converges to a limit polygon W. From the above argument, $f^i(v) \cap \bar{L} \neq \phi$, for all i, and therefore $W \cap \bar{L} \neq \phi$.

Now, W is a sided elementary polygon without corners, i.e. a geodesic. If W had a corner, then f(W) would be shorter than W, and this contradicts the fact that $f^i(v) \to W$ as $i \to \infty$, while the length of $f^i(v)$ monotonically decreases to the length of W, and converges to it.

This W is a finite geodesic for the modified metric that has its end points in the boundary of D.

We consider now a point in W and $\bar{L}$, ($\bar{L}$ is contained in the interior of B_δ), and follow it along both directions of the geodesic until the first point of intersection with the surface $\dot{B}_\delta$.

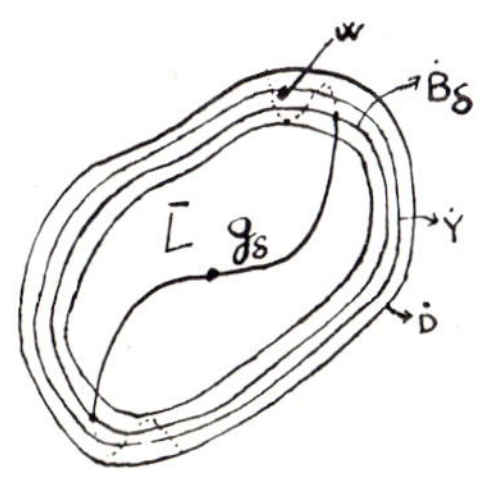

We call this segment g_δ. It is a geodesic for both metrics, the Jacobi and the modified metric, since the metrics agree in Y, and $Y \subset \bar{B}_\delta$. From part 8, since g_δ is a finite geodesic, there exists a finite interval $[t_1, t_2]$ and an E_0-trajectory ϕ_δ defined in it such that $\phi_\delta(t_1) \in \dot{B}_\delta$ and $\phi_\delta(t_2) \in \dot{B}$.

This completes the proof of the Proposition A.

<u>Proposition B</u>. <u>There exists at least one E_0-brake-orbit that approaches fixed points of $\dot{B}$.</u>

<u>Proof</u>. If we now introduce a different coordinate system on the boundary $\dot{B}$ by

$$\mu: \dot{B} \times [E_0 - h, E_0] \to B$$

such that, $\mu(q,U)$ is the gradient orbit of U, and $\mu(q,E_0) = q$. Since E_0 is a regular value of U, μ is a coordinate system.

If we call $z^1 = q^1, \ldots, z^{n-1} = q^{n-1}$ $z^n = U$, it is very easy to observe that for the geodesics that are tangent to the surfaces $z^n = c$ we have a similar expression to ( )

$$z^{n''} = - \sum_{i,j=1}^{n-1} \Gamma^n_{ij} \dot{z}^i \dot{z}^j + \frac{-U_n g^{nn}}{E_0 - U} \sum_{i,j=1}^{n-1} g_{ij} \dot{z}^i \dot{z}^j \tag{8}$$

and with the same ideas of #7, we can see that for E_0-U very small the sign of $z^{n''}$ is positive.

Therefore, our surfaces $U = c$, c close to E_0, are geodesic convex. We may suppose that this is true for $U \geq E_0 - \delta_0$.

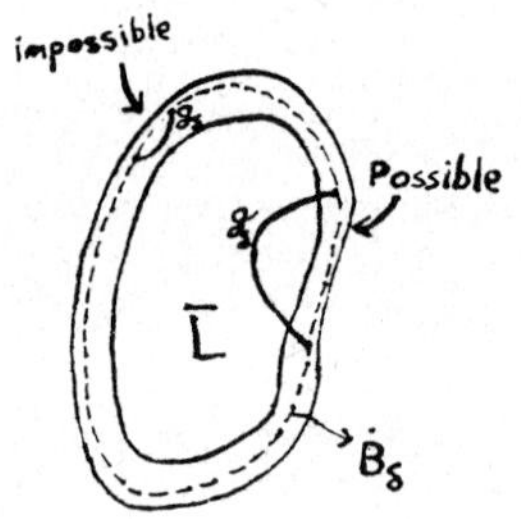

In other words, we have constructed a zone like the one in the picture, where the surfaces $z^n = c$ are geodesic convex. Therefore, our geodesics g_δ have to go outside of this zone, so they have to intercept the zone $\bar{L}\ \{x \mid V(x) \leq E_0 - \delta_0\}$.

Now we will consider in any g_δ a point P_δ in $\bar{L}$, and the tangential line element e_δ of g_δ that goes through P_δ.

We may now consider a convergent subsequence $\{(P_\delta, e_\delta)\}$. Let (P_0, e_0), the limit of which belongs to $\bar{L}$.

(P_0, e_0) defines a geodesic g_0 maximal on B for the Jacobi metric.

If g_0 has finite length, g_0 corresponds to a nice geodesic that joins two points of $\dot{B}$.

<u>Lemma</u>. g_0 <u>has finite length</u>.

We are going to consider a slightly different construction for the families S_{D_δ} such that we can see easily that we have a bound for the lengths of all the curves that are in any S_D.

First we consider the diameters of $\overline{B}$, and the elementary arc of $\overline{B}$ for the metric E_1-Jacobi, $E_1 > E$. The we divide the diameters in N parts such that the distance between subsequent points is not longer than the elementary arc of E_1-Jacobi metric. This is possible from the uniform continuity. We may also consider that the division points vary continuously with the diameter.

Now we substitute the diameters in $\overline{B}$ by elementary polygons for the E_1-Jacobi metric that one gets when one connects subsequent division points by elementary arcs.

The E_1-length of these curves is less than A, A equal to N times the E_1-elementary length of $\overline{B}$.
Now since

$$(E_0-U)\ \Sigma\, g_{ij}\, \dot{y}^i \dot{y}^j < (E_1-U)\ \Sigma\, g_{ij}\, \dot{y}^i \dot{y}^j \ .$$

The E_0-lengths of these curves are also bounded for the same number A.

For the construction of the families S_{D_δ}, it is possible to consider the parts of the curves that are in $\overline{R}_\delta$ and consider these curves as the new "diameters" in $\overline{R}_\delta$. Now we can consider the E_0-elementary arc of $\overline{R}_\delta$ and construct the elementary polygons over the "diameters" in $\overline{R}_\delta$. Then we extend the elementary polygons to the boundary of D_δ, by segments y^1 = constant ... y^{n-1} = constant. These curves form our new S_{D_δ}.

We may suppose also that for any δ, the segments y^1 = constant,...,y^{n-1} = constant, have length less than 1. Therefore our geodesics of S_{D_δ} have length less that A+1. The same is true for $f(S_{D_\delta})$, $f^2(S_{D_\delta})$...; therefore, our geodesics g_δ have lengths less than A+1 in the E_0-Jacobi metric.

For the modifying metric the length is less than 2(A+1). If we suppose that we can measure off, to any side of P_0, a length greater than 2(A+1), then the laid off closed geodesic of length 2(A+1), would lie in a certain Y_δ, and therefore also the geodesic acrcs of length 2(A+1) that are determined by the approximating pairs (e_δ, P_δ) with sufficiently small δ.

Then the endpoints of g_δ, δ small would lie in $Y_{\delta'}$, since g_δ is shorter than 2(A+1) but this is not the case for $\delta < \delta'$. Therefore, we may conclude that g_0 is a geodesic of finite length.

Therefore, there exists an interval $[S_1, S_2]$ such that $g(S_1)$ and $g(S_2)$ belong to $\dot{B}$. Hence, if ϕ is an orbit of energy E_0 and (a,b) is the maximal interval where ϕ is defined, then we have that

$$\lim_{t\to a} \phi(t) \in B \quad \text{and} \quad \lim_{t\to b} \phi(t) \in \dot{B} ,$$

and in fact

$$\lim_{t\to a} \phi(t) = \lim_{S\to S_1} g(S), \qquad \lim_{t\to a} \phi(t) = \lim_{S\to S_2} g(S)$$

this completes the proof of Proposition B.

#11. Extension of Seifert paper to the C^3-case.

Corollary: In the C^3-Riemannian case there exists a periodic orbit of energy E_0 that joins points of $\dot{B}$.

Proof: With the coordinates $z^1, \ldots, z^n$ we may write the n-equation of motion in the form

$$\ddot{z}^n + \Gamma^n_{ij}(z)\dot{z}^i\dot{z}^j + g^{n\ell}(z)U_\ell = 0 .$$

Since $U_\ell(z) = 0$ for $\ell \neq n$, then

$$\ddot{z}^n = -\Gamma^n_{ij}(z)\dot{z}^i\dot{z}^j - g^{nn}(z)\, U_n(z) .$$

Since $g^{nn}(z)$ and $-U_n(z)$ are positive in $\dot{B}$, by the compacity of $\dot{B}$, there exists a number h such that $0 < h < -g^{nn}(z)\, U_n(z)$, for z in $\dot{B}$.

Furthermore, we know that the E_0-orbits have small velocity close to D, and that $\Gamma^n_{ij}(z)$ is bounded in B; therefore there exists a neighborhood of $\dot{B}$ such that

$$\ddot{z}^n = -\Gamma^n_{ij}(z)\dot{z}^i\dot{z}^j - g^{nn}(z)\, U_n(z)$$

is greater than a number $\eta > 0$. Therefore, we may consider that there exists a neighborhood of $\dot{B}$ such that $|\dot{z}^n| < \varepsilon$ and $\ddot{z}^n > \eta > 0$ for the E_0-orbits.

If $z(t)$ is an E_0-orbit which is in this neighborhood for $t_1 \leq t \leq t_2$, we have:

$$2\varepsilon > \dot{z}^n(t_2) - \dot{z}^n(t_1) = \int_{t_1}^{t_2} \ddot{z}^n(t)dt > \eta(t_2 - t_1) \quad ;$$

then $t_2 - t_1 < \frac{2\varepsilon}{\eta}$.

In other words, we cannot stay too long in this neighborhood without going out. Then, from Proposition B, we may deduce that the orbit ϕ corresponding to the geodesic g, spends finite time arriving at the extremes of g. Therefore, there exists a finite interval $[t_1, t_2]$ such that $\phi(t_1)$ and $\phi(t_2)$ belong to $\dot{B}$.

We may suppose that $t_1 = 0$. Now, by uniqueness theorem of differential equations $\phi(t)$ and $\phi(-t)$ must be identical, and we may extend ϕ to a periodic solution on $(-\infty, +\infty)$.

Acknowledge

The author wants to express his gratitude to his wife, Lúcia de Ruiz, and to his thesis adviser, Professor Alan Weinstein.

REFERENCES

[1] Abraham and Marsden, "Foundations of Mechanics", W.A. Benjamin, inc., Reading, Massachusetts, 1967.

[2] Birhoff, G. D. Dynamical Systems, Colloq. Publ. IX Secon Ed., Amer.Math.Soc., Providence, R.I., 1966.

[3] Buseman, Metric methods in Finsler Spaces.

[4] Caratheodory, C., "Calculus of Variations and Partial Differential Equations, Holden-Day, San Francisco, 1965.

[5] Hermann, R., "Differential Geometry and the Calculus of Variations", Academic Press, New York and London, 1968.

[6] Rund, H., "The Differential Geometry of Finsler Spaces", Springer Verlag, 1959.

[7] Seifert, H., "Periodische Bewegungen Mecanischer System, Math. Z, 1948, ps. 197-216.

[8] Seifert und Threlfall, "Variationsrechnung im Grozen" Leipzig 1938, Hamburger Einzelschrift 24, Anm.20 S 97.

[9] Spivak, M., "Differential Geometry I and II", Publish or Perish, 1972.

Otto Raúl Ruiz Monroy
Departamento de Matemáticas
Universidad Nacional de Colombia
Bogotá, D.F. Colombia.

ON THE COMPLEX PROJECTIVE BUNDLE CONSTRUCTION

José Carlos de Souza Kiihl

Using the complex projective bundle construction one can define a homomorphism $P: \oplus \mathcal{U}_{*-2k}(BU_k) \longrightarrow \mathcal{U}_{*-2}(CP^\infty)$. In this paper we investigate this homomorphism and some of its applications. The main result presented here is an explicit formula to compute $H(P[1^k \oplus \eta_n])$, where H is the Hurewicz homomorphism, 1^k is the trivial complex k-plane bundle over CP^n, and η_n is the Hopf line bundle over CP^n.

1. INTRODUCTION

Given a smooth complex (k+1)-plane bundle $p: \zeta \to M^{2n}$ over a closed weakly complex manifold, we may consider the associated complex projective space bundle with fiber CP^k, $\pi: CP(\zeta) \to M$. Then the total space $CP(\zeta)$ itself is a closed $2(n+k)$-manifold. We define a weakly complex structure on $CP(\zeta)$ the following way: According to Borel and Hirzebruch [2], the tangent bundle to $CP(\zeta)$ is $\pi^*(\tau(M)) \oplus \tau_F$, where $\tau(M)$ is the tangent bundle to M and τ_F is the complex k-plane bundle of vectors tangent to the fibers. Thus we assign to $CP(\zeta)$ the weakly complex structure determined by that on $\tau(M)$ together with the complex structure on τ_F.

There is a complex line bundle $\eta \to CP(\zeta)$, a point in the total space is a pair consisting of a line in $p^{-1}(x)$, $x \in M$, together with a vector in this line. If we denote by $c_1(\eta) \in H^2(CP(\zeta),Z)$ the characteristic class of η, according to Dold [4], via the induced homomorphism π^*, $H^*(CP(\zeta),Z)$ is a free graded $H^*(M,Z)$-module with basis $1, c_1(\eta), c_1(\eta)^2, \ldots c_1(\eta)^k$.

By Borel and Hirzebruch [2], we have that the total Chern class of τ_F is given by

$$(1.1) \qquad c(\tau_F) = \sum_{i=0}^{k+1} (1-c_1(\eta))^{k+1-i} \pi^*(c_i)$$

where $c_i \in H^{2i}(M,Z)$ is the Chern class of ζ. Since τ_F is a complex k-plane bundle it follows that

$$(1.2) \qquad \sum_{j=0}^{k+1} (-1)^{k-j+1} \pi^*(c_j)c_1(\eta)^{k-j+1} = 0$$

Those formulas will be used later in this paper.

Using this construction we can define the following homomorphism

$$P: \bigoplus_{k=0}^{\infty} U_{*-2k}(BU_k) \to U_{*-2}(CP^{\infty})$$

by setting $P(\Sigma[\zeta]) = \Sigma[\eta \to CP(\zeta)]$ (where $U_*(X)$ denotes the complex bordism group of the space X).

In this paper we investigate $P[1^k \oplus \eta_n]$, where 1^k is the trivial complex k-plane bundle over CP^n, and η_n is the Hopf line bundle over CP^n ($\overline{\eta}_n = \xi_n$ the canonical line bundle).

2. THE HUREWICZ HOMOMORPHISM

We have the Hurewicz homomorphism $H: U_*(CP^{\infty}) \to H_*(MU)\hat{\otimes}H_*(CP^{\infty})$

(see Liulevicius [6]). The main result in this paper is a formula to compute $H(P[1^k \oplus \eta_n])$.

In [6] we have that the Hurewicz homomorphism

$$H\colon \pi_*(MU \wedge X^+) \longrightarrow H_*(MU) \hat{\otimes} H_*(X)$$

followed by the Thom isomorphism

$$\varphi_* \otimes 1\colon H_*(MU) \hat{\otimes} H_*(X) \to H_*(BU) \hat{\otimes} H_*(X)$$

is the normal characterisitic number map:

$$\nu_\#\colon \mathcal{U}_*(X) \to H_*(BU) \hat{\otimes} H_*(X)$$

where $\mathcal{U}_*(X)$ is identified to $\pi_*(MU \wedge X^+)$ via the Thom-Pontrjagin isomorphism, and X^+ is the space X with an extra base point. In other words, we have a commutative diagram

$$\begin{array}{ccc}
\pi_{2m}(MU \wedge X^+) & \xrightarrow{H} & [H_*(MU) \hat{\otimes} H_*(X)]_{2m} \\
\Vert & & \downarrow \varphi_* \otimes 1 \\
\mathcal{U}_{2m}(X) & \xrightarrow{\nu_\#} & [H_*(BU) \hat{\otimes} H_*(X)]_{2m}
\end{array} \tag{2.1}$$

where if $u \in \mathcal{U}_{2m}(X)$ is represented by $f\colon M^{2m} \to X$, M a connected manifold, then $\nu_\#(u)$ is the class in homology given by $(\nu_* \otimes f_*)\Delta_*(\mu)$, with μ being the fundamental class of M, and $\Delta\colon M \to M \times M$ the diagonal map, and $\nu\colon M \to BU$ the classifying map for the stable normal bundle.

In terms of cohomology classes: if $w \in H^{2i}(BU)$, $v \in H^{2j}(X)$, $i+j = m$, then

$$\langle w \otimes v, \nu_{\#}(u)\rangle = \langle \nu^{*}(w) \cup f^{*}(v), \mu\rangle .$$

This gives us a method to determine the Hurewicz homomorphism.

We also recall that

$$H_*(BU) \cong Z[b_1, b_2, \ldots, b_n, \ldots]$$

with $b_k \in H_{2k}(BU)$, and $\langle c_1^n, b_n\rangle = 1$, $\langle c^E, b_n\rangle = 0$ for $E \neq (n,0,\ldots,0)$, where $c_i \in H^{2i}(BU)$ are the generators of $H^*(BU)$.

We also know that

$$H_*(CP^\infty) \cong Z[y_1, y_2, \ldots, y_n, \ldots]$$

with $y_k \in H_{2k}(CP^\infty)$ and $\langle y^j, y_k\rangle = \delta_{jk}$, where $y = -c_1(\gamma_1)$ belonging to $H^2(CP^\infty)$ is the fundamental class in cohomology.

We will use the following notation: if S is an inhomogeneous sum we write S_i for the component of S of dimension $2i$.

We also know that $U_*(CP^\infty)$ has a U_*-basis given by $\{i_n\}_{n \geq 0}$, where i_n is the class of the inclusion map $CP^n \subset CP^\infty$.

In what follows we will identify $H_*(MU)$ and $H_*(BU)$ by the Thom isomorphism φ_*.

<u>PROPOSITION 2.2</u>: We have that

$$H(i_n) = \sum_{k=0}^{n} B_{n-k}^{-n-1} \otimes y_k$$

where $B = \sum_{i=0}^{\infty} b_i$ in $H_*(BU)$.

PROOF:

Suppose i_n is given by the inclusion map $f\colon CP^n \subset CP^\infty$, and $\nu\colon CP^n \to BU$ is the classifying map for the stable normal bundle. Then since

$$f_* \in \operatorname{Hom}(H_*(CP^n), H_*(CP^\infty)) \cong [H_*(CP^\infty) \otimes H^*(CP^n)]$$

we have

$$f_* = \sum_{j=0}^{n} y_j \otimes y^j$$

where $y = -c_1(\eta_n) \in H^2(CP^n)$ is the fundamental class, and η_n is the Hopf line bundle over CP^n, and $\langle y^k, y_j \rangle = \delta_{kj}$.

Now if $\xi\colon CP^n \to BU$ classifies the canonical line bundle, then

$$\xi_* = \sum_{i=0}^{\infty} b_i \otimes y^i$$

and if $\tau\colon CP^n \to BU$ classifies the stable tangent bundle, then $\tau_* = [(n+1)\xi]_* = (\xi_*)^{n+1}$. Therefore $\nu_* = (\xi_*)^{-n-1}$, that is

$$\nu_* = \sum_{i=0}^{n} B_i^{-n-1} \otimes y^i .$$

Thus

$$\nu_* \otimes f_* = \sum_{\substack{i,j \\ i+j \leq n}} B_i^{-n-1} \otimes y_j \otimes y^{i+j} \tag{2.3}$$

By the previous remark, $\nu_\#(i_n)$ is the class in homology given by $(\nu_* \otimes f_*)\,\Delta_*(\mu)$, where $\mu \in H_{2n}(CP^n)$ is the fundamental class. Thus $\nu_\#(i_n)$ is the coefficient of y^n in (2.3), that is

$$\nu_{\#}(i_n) = \sum_{i+j=n} B_i^{-n-1} \otimes y_j$$

or equivalently

$$\nu_{\#}(i_n) = \sum_{k=0}^{n} B_{n-k}^{-n-1} \otimes y_k .$$

By the commutativity of the diagram (2.1), and recalling that we are identifying $H_*(MU)$ and $H_*(BU)$ by the Thom isomorphism, we have the result.

From Adams [1], we have the following relations:

PROPOSITION 2.4: In $H_*(BU)$ we have the relations

$$m_n = \frac{1}{n+1} B_n^{-n-1}$$

$$b_n = \frac{1}{n+1} M_n^{-n-1}$$

where $B = \sum_{i\geq 0} b_i$ and $M = \sum_{i\geq 0} m_i$. Furthermore, the image of the bordism class $[CP^n] \in \mathcal{U}_{2n}$ in $H_{2n}(BU)$ by the Hurewicz homomorphism is $(n+1)m_n$. (Here we are again identifying $H_*(MU)$ and $H_*(BU)$ by the Thom isomorphism).

REMARK: A table for some of the B_s^r's, in terms of the m_n's, is given in [5].

We now present the main result of this paper.

THEOREM 2.5: For $k \geq 1$, we have

$$H(P[1^k \oplus \eta_n]) = \sum_{j=0}^{n+k} a_j \otimes y_j$$

where

$$a_j = \sum_{\substack{p+q=n+k-j \\ q \leq n}} \sum_{r=0}^{q} \sum_{s=0}^{p} (-1)^r \binom{r+s}{r} B_{q-r}^{-n-1} B_{r+s}^{-1} B_{p-s}^{-k}$$

with $B = \sum_{i=0}^{\infty} b_i$ in $H_*(BU)$ and $y_j \in H_{2j}(CP^\infty)$ such that $\langle y^k, y_j \rangle = \delta_{kj}$, $y = -c_1(\gamma_1) \in H^2(CP^n)$ being the fundamental class in cohomology.

PROOF:

Let $\eta_n \to CP^n$ denote the Hopf line bundle over CP^n. Then $P[1^k \oplus \eta_n]$ can be represented by $\eta \to CP(1^k \oplus \eta_n)$. We denote $M = CP(1^k \oplus \eta_n)$.

We have that the Chern class $c(1^k \oplus \eta_n) = 1-y$, and also have $c(\tau_F) = (1+t)^{k+1} + (1+t)^k(-y) = (1+t)^k(1+t-y)$ with the relation $t^{k+1} = t^k y$, and $t = -c_1(\eta) \in H^2(M)$, according to (1.1). We also have that $c(\tau(CP^n)) = (1+y)^{n+1}$. Therefore $c(\tau M) =$
$= (1+y)^{n+1}(1+t)^k(1+t-y)$.
Here we identify y with $\pi^*(y)$, since there is no danger of confusion.

Consider now the maps $f_i\colon M \to BU$, $i = 1,2,3$, such that $f_i^*\colon H^*(BU) \to H^*(M)$ give us

$$f_1^*(c_1) = y \quad \text{and} \quad f_1^*(c_n) = 0 \quad \text{for} \quad n > 1$$

$$f_2^*(c_1) = t \quad \text{and} \quad f_2^*(c_n) = 0 \quad \text{for} \quad n > 1$$

$$f_3^*(c_1) = t-y \quad \text{and} \quad f_3^*(c_n) = 0 \quad \text{for} \quad n > 1.$$

Then

$$f_{i_*} \in \mathrm{Hom}(H_*(M), H_*(BU)) \cong H_*(BU) \otimes H^*(M)$$

Hence if $\tau: M \to BU$ classifies the stable tangent bundle to M, we have $\tau_* = (f_{1*})^{n+1}(f_{2*})^k(f_{3*})$. And since

$$(t-y)^i = \sum_{\substack{r+s=i \\ r,s\geq 0}} (-1)^r \binom{r+s}{r} t^s y^r$$

we have

$$\tau_* = \sum_{p,q} \sum_{r=0}^{q} \sum_{s=0}^{p} (-1)^r \binom{r+s}{r} B_{q-r}^{n+1} B_{p-s}^{k} B_{r+s} \otimes t^p y^q$$

And since $\nu = (-1)\tau$, where $(-1): BU \to BU$ is such that $(-1)_*(B) = B^{-1}$, we have

$$\nu_* = \sum_{\substack{p+q=n+k \\ q\leq n}} \sum_{r=0}^{q} \sum_{s=0}^{p} (-1)^r \binom{r+s}{r} B_{q-r}^{-n-1} B_{p-s}^{-k} B_{r+s}^{-1} \otimes t^p y^q$$

Now suppose $f: M \to CP^\infty$ classifies the bundle $\eta \to M$, then $f^*(y) = t$ and we have that

$$f_* \in \operatorname{Hom}(H_*(M), H_*(CP^\infty)) \cong H_*(CP^\infty) \otimes H^*(M)$$

is given by

$$f_* = \sum_{j=0}^{\infty} y_j \otimes t^j .$$

And by the remark at the beginning of this section, $\nu_\#(P[1^k \oplus \eta_n])$ is the class in homology given by $(\nu_* \otimes f_*)\Delta_*(\mu)$. Since μ is dual to $t^k y^n$, we have that $\nu_\#(P[1^k \oplus \eta_n])$ is the coefficient of $t^k y^n$ in

$$\left(\sum_{\substack{p+q=n+k \\ q\leq n}} \sum_{r=0}^{q} \sum_{s=0}^{p} (-1)^r \binom{r+s}{r} B_{q-r}^{-n-1} B_{p-s}^{-k} B_{r+s}^{-1} \otimes t^p y^q \right) \left(\sum_{j=0}^{\infty} y_j \otimes t^j \right) =$$

$$= \sum_{j=0}^{\infty} \left(\sum_{\substack{p+q=n+k \\ q\leq n}} \sum_{r=0}^{q} \sum_{s=0}^{p} (-1)^r \binom{r+s}{r} B_{q-r}^{-n-1} B_{p-s}^{-k} B_{r+s}^{-1} \otimes y_j \otimes t^{p+j} y^q \right).$$

Since $t^{k+1} = t^k y$ we have $t^{k+j} y^{n-j} = t^k y^n$ for $j \geq 0$, therefore

$$\nu_{\#}(P[1^k \oplus \eta_n]) = \sum_{j=0}^{n+k} a_j \otimes y_j$$

with

$$a_j = \sum_{\substack{p+q=n+k-j \\ q \leq n}} \sum_{r=0}^{q} \sum_{s=0}^{p} (-1)^r \binom{r+s}{r} B_{q-r}^{-n-1} B_{p-s}^{-k} B_{r+s}^{-1}$$

Recalling that we are identifying $H_*(MU)$ and $H_*(BU)$ via the Thom isomorphism we have the result.

3. APPLICATIONS

An interesting consequence from the formula we got in the Theorem 2.5 is the Lemma 4.4 from Conner [3], for the case $\xi = 1^k \oplus \eta_n$.

First we recall some basic facts. Let $\xi \to M$ be a complex (k+1)-plane bundle over M. The total Chern class of ξ can be expressed in the factored form $\sum_{j=1}^{k+1} (1+t_j)$ and for each $m \geq 0$ we introduce the symmetric functions $S_m(\xi) = \sum_{j=1}^{k+1} (t_j)^m$. Each $S_m(\xi)$ can be uniquely expressed as a homogeneous polynomial in the Chern classes (the elementary symmetric functions), hence we have $S_m(\xi) \in H^{2m}(M,Z)$. For a Whitney sum we have

$$S_m(\xi_1 \oplus \xi_2) = S_m(\xi_1) + S_m(\xi_2) .$$

If $\tau \to M^{2m}$ is the tangent bundle to M, we will denote by $S_m[M] = \langle S_m(\tau), \mu \rangle$, where μ is the fundamental class of M.

We now have the result mentioned above.

COROLLARY 3.1: If $\eta_n \to CP^n$ denotes the Hopf line bundle over CP^n, and $1^k \to CP^n$ is the trivial k-plane bundle, then

$$S_{n+k}[CP(1^k \oplus \eta_n)] = -\left(\sum_{q=0}^{n} (-1)^q \binom{n+k}{q} + k \right) .$$

PROOF:

If we denote by $h: \mathcal{U}_* \to H_*(MU)$ the Hurewicz homomorphism (the case the space X is a point), then we have that

$$h[CP(1^k \oplus \eta_n)] = a_0 = \sum_{\substack{p+q=n+k \\ q \leq n}} \sum_{r=0}^{q} \sum_{s=0}^{p} (-1)^r \binom{r+s}{r} B_{q-r}^{-n-1} B_{p-s}^{-k} B_{r+s}^{-1}$$

We also have that $S_{n+k}[CP(1^k \oplus \eta_n)]$.

Since

$$B^{-n} = 1 - nb_1 - nb_2 - \dots \quad \text{(mod. decomposables)}$$

we have

$$h[CP(1^k \oplus \eta_n)] = \underbrace{\sum_{q=0}^{n} (-1)^q \binom{n+k}{q} (-b_{n+k})}_{\substack{r\,=\,q \\ s\,=\,p}} + \underbrace{(-kb_{n+k})}_{\substack{r=q=0 \\ s\,=\,0}} \quad \text{(mod.dec.)}$$

Therefore

$$S_{n+k}[CP(1^k \oplus \eta_n)] = -\left(\sum_{q=0}^{n} (-1)^q \binom{n+k}{q} + q \right)$$

as we want to prove.

In [7] Milnor has announced the following result.

THEOREM 3.2: $\mathcal{U}_*$ is the integral polynomial ring on classes x_n of dimension $2n$. A stably almost complex manifold M^{2n}

may be taken to be the 2n-dimensional generator if and only if

$$s_n[M^{2n}] = \begin{cases} \pm 1 & \text{if } i+1 \neq p^s \text{ for any prime } p \\ \pm p & \text{if } i+1 = p^s \text{ for some prime } p \text{ and integer } s. \end{cases}$$

PROOF:

See Stong [8].

Using this criterium and the formula in Corollary 3.1, one can decide when $[CP(1^k \oplus \eta_n)]$ is an indecomposable element and find this way explicit generators for the bordism ring $\mathcal{U}_*$. This investigation we will leave for another paper.

In the proof of Theorem 2.5 we saw that $t^{k+s}y^{n-s} = t^k y^n$. Thus we can rewrite the formula obtained there in a more convenient way. If we denote by S_i^r the sum of the coefficients of $t^{r-j}y^j$, $j \leq i$, then we have

$$(3.3) \qquad H(P[1^k \oplus \eta_n]) = \sum_{j=0}^{n+k} a_j \otimes y_j$$

where

$$a_j = \sum_{i=0}^{k+n-j} B_i^{-n-1} S_{n-i}^{n+k-j-i}$$

with the following conventions $S_i^r = 0$ if $i < 0$, and $S_i^r = S_r^r$ if $r \leq i$. Note that $S_0^0 = 1$.

REMARK: Tables for the S_i^r's, in terms of the m_n's, are given in [5].

An immediate consequence from the formula (3.3) is the following.

PROPOSITION 3.4: If $\eta_1 \to CP^1$ denotes the Hopf line bundle over CP^1, and $1^k \to CP^1$ is the trivial k-plane bundle, then

$$[CP(1^k \oplus \eta_1)] = [CP^1]\,[CP^k]$$

in $\mathcal{U}_*$.

PROOF:

In fact, we have that $h[CP(1^k \oplus \eta_1)] = a_0$. And by (3.3) we have that $a_0 = B_0^{-2} S_1^{k+1} + B_1^{-2} S_0^k = B_1^{-2} S_0^k = (2m_1)(k+1)m_k$ because $S_1^{k+1} = 0$, and $B_1^{-2} = 2m_1$, and $S_0^k = (k+1)m_k$. Since by Proposition 2.4 we have that $h^{-1}(n\, m_{n-1}) = [CP^{n-1}]$ we get the result.

So we see that $[CP(1^k \oplus \eta_1)]$ is not an indecomposable element in $\mathcal{U}_*$.

Also as a consequence from the formula given in Theorem 2.5, we can find the expressions for $P[1^k \oplus \eta_n]$ in terms of the generators $\{i_n\}$ of $\mathcal{U}_*(CP^\infty)$. In fact, since $P[1^k \oplus \eta_n] = \sum_r \alpha_r\, i_r$, we have that $H(P[1^k \oplus \eta_n]) = \sum_r h(\alpha_r)\, H(i_r)$. Since H is a monomorphism and we have the expressions for the $H(i_r)$'s the equality above determines $h(\alpha_r)$, hence α_r. In [5] we present some of those explicit expressions in terms of convenient $\mathcal{U}_*$-generators.

REFERENCES

1. Adams, J.F., Quillen's work on formal groups and complex cobordism, Dept. of Mathematics, University of Chicago, 1970 (mimeo.).

2. Borel, A., and Hirzebruch, F., Characteristic classes and homogeneous spaces I, Amer.J.Math. 80 (1958), 458-538.

3. Conner, P.E., The bordism class of a bundle space, Michigan Math. Journ. 14 (1967), 289-303.

4. Dold. A., Relations between ordinary and extraordinary homology, Colloquium on Algebraic Topology, Aarhus University, 1962, 2-9

5. Kiihl, J.C.S., On the bordism of weakly complex involutions, Thesis, University of Chicago, 1975.

6. Liulevicius, A.L., On characteristic classes, Lecture Notes, Aarhus University, 1968.

7. Milnor, J.W., On the cobordism ring Ω^*, Abstract 547-26, Notices Amer.Math.Soc. (1958), 457.

8. Stong, R.E., Notes on cobordism theory, Mathematical Notes, Princeton University Press, N.J. 1968.

Universidade Estadual de Campinas
Departamento de Matemática
Campinas - São Paulo - Brasil

SYMMETRIES, CONSERVATION LAWS AND VARIATIONAL PRINCIPLES

by Floris Takens

1. Introduction

By a well known theorem of Noether [7.], the solutions of a variational problem, which is invariant under a continuous symmetry group, satisfy certain "differential relations". These "differential relations" can in general be interpreted as conservation laws, like the laws of conservation of energy and momentum in classical mechanics, which are a consequence of the fact that the underlying variational principle is invariant under translations in the time and space directions. Also, under more restrictive assumptions, one can show that if the solutions of a variational problem respect some conservation law, the variational problem itself must have a corresponding symmetry.

The central problem of this paper is: if there are a number of symmetries present and if the corresponding conservation laws hold, is there then necessarily an underlying variational principle? Before trying to give a partial answer to this question, we have to give it a precise meaning. In particular we have to specify what type of equations we are going to consider, i.e., which equations are possibly going to have symmetries or whose "solutions" are possibly going to satisfy conservation laws. These equations, which we shall call source equations, can sometimes be obtained as Euler equation from a variational problem. We shall describe this now in detail.

The Euler equation as source equation. We consider smooth maps from $\mathbb{R}^n$ to $\mathbb{R}^m$ together with a smooth function L, a Lagrangian, which assigns to each k-jet of such a map $f\colon \mathbb{R}^n \to \mathbb{R}^m$ a real number. So, for $f\colon \mathbb{R}^n \to \mathbb{R}^m$, $L(j^k(f))$ is a real function on $\mathbb{R}^n$ and $L(j^k(f))(x)$ is determined by x, $f(x)$, $df(x),\dots,d^{(k)}f(x)$. In this paper we assume everything to be C^∞, hence also L as a function of k-jets. Consequently we assume $L(j^k(f))$ also to be smooth. The map $f\colon \mathbb{R}^n \to \mathbb{R}^m$ is a solution, or rather an extremal, of the variational problem defined by the Lagrangian L if for any bounded open subset $A \subset \mathbb{R}^n$ and any map $\tilde{f}\colon \mathbb{R}^n \to \mathbb{R}^m$ with support in A, $\int_A L(j^k(f + t\tilde{f})) = \int_A L(j^k(f)) + O(t^2)$.

The Euler equation [3] determines for each Lagrangian as above an unique m-tuple of functions $E_1,\dots,E_m$, assigning to each 2k-jet of a function $f\colon \mathbb{R}^n \to \mathbb{R}^m$ a real number, such that:

Whenever $A \subset \mathbb{R}^n$ is a bounded open set, $f, \tilde{f}\colon \mathbb{R}^n \to \mathbb{R}^m$ are maps and $\tilde{f}$ has support in A, then $\int_A L(j^k(f + t\tilde{f})) = \int_A L(j^k(f)) + t.\int_A \sum_{i=1}^m E_i(j^{2k}(f)).\tilde{f}^i + O(t^2)$ ($\tilde{f}^i$ is the ith component of $\tilde{f}$). Hence, f is an extremal if and only if $E_i(j^{2k}(f)) = 0$ for $i = 1,\dots,m$. In many cases however, if f is not an extremal, the map $E(f)$, which assigns to each $x \in \mathbb{R}^n$ the co-vector $(E_1(j^{2k}(f))(x),\dots,E_m(j^{2k}(f))(x))$ has an interesting interpretation:

I. In the case of "classical mechanics". Here we have $n = 1$, $k = 1$ and $L(y_1,\dots,y_m,\dot{y}_1,\dots,\dot{y}_m,t) = -\sum_{i=1}^m \tfrac{1}{2}M_i|\dot{y}_i|^2 + V(y_1,\dots,y_m)$ in the obvious coordinate system. The corresponding extremals are maps $\gamma\colon \mathbb{R} \to \mathbb{R}^m$ which describe the motion of a 1-dimensional m-body system with masses $M_1,\ldots,M_m$ and potential function V. These extrelals have the form $\gamma(t) = (\gamma_1(t),\ldots,\gamma_m(t))$ with

$$\frac{d}{dt}(M_i\cdot\gamma_i'(t)) + \frac{\partial V}{\partial y_i}(\gamma_1(t),\ldots,\gamma_m(t)) = 0;$$

the left-hand side of the above formula is the function E_i (from the Euler equation) applied to γ. If $\gamma\colon \mathbb{R} \to \mathbb{R}^n$ is arbitrary, so that $E(\gamma)$ need not be zero, then the ith component $E_i(j^2(\gamma))(t) = \frac{d}{dt}(M_i\cdot\gamma_i'(t)) + \frac{\partial V}{\partial y_i}(\gamma_1(t),\ldots,\gamma_m(t))$ is the extra force on the ith body, as a function of time, needed to have the n-body system moving according to the map γ. So $E(\gamma)$ may be considered as the cause or <u>source</u> of the movement γ.

II. In the case of "classical, i.e., Newtonian, gravitation". Now we have $n = 4$ ($\mathbb{R}^n = \mathbb{R}^4$ is spacetime and has coordinates $x_0,\ldots,x_3$) and $m = 1$ (y is the coordinate on $\mathbb{R}^m = \mathbb{R}^1$) and $k = 1$. $L(y,y_0,\ldots,y_3,x_0,\ldots,x_3) = -\frac{1}{2}\sum_{i=1}^{3} y_i^2$, where the 1-jet of a function $f\colon \mathbb{R}^4 \to \mathbb{R}$ in $x = (x_0,\ldots,x_3)$ is represented by $(y = f(x),\ y_0 = \frac{\partial f}{\partial x_0}(x),\ldots,y_3 = \frac{\partial f}{\partial x_3}(x), x_0,\ldots,x_3)$. The extremals are in this case function $f\colon \mathbb{R}^4 \to \mathbb{R}$ which represent gravitation potentials in the absence of matter. If $f\colon \mathbb{R}^4 \to \mathbb{R}$ is arbitrary, then $E(f) = \sum_{i=1}^{3} \frac{\partial^2 f}{\partial x_i^2}$ which is just the mass density if f is a gravitation potential. Hence also here $E(f)$, representing the mass density, may be regarded as the <u>source</u> of the gravitation potential f.

The above examples motivate the following

<u>(Provisional) definition.</u> A <u>source equation</u> of order 1 is a map E which assigns, for some integers n,m, to each map $f\colon \mathbb{R}^n \to \mathbb{R}^m$ a map $E(f)$ from $\mathbb{R}^n$ to the dual of $\mathbb{R}^m$ in such a way that $E(f)(x)$ is determined by the 1-jet of f in x.

Not all source equations can be obtained in this way as Euler equation from some Lagrangian. A source equation can be obtained from a variational problem, i.e., is the Euler equation of some Lagrangian if and only if it is self-adjoint, for a definition of this see [3] or § 4 of this paper. This last statement was conjectured and partly proven by A. Hirsch [4] in 1897; a complete proof was finally given by E. Tonti [9] in 1969. In this paper we shall give a different proof of this theorem based on P. Dedecker's ideas [2] who proved the above statement for the case n = 1. The extension of these ideas of Dedecker's turned out to be very useful for obtaining our other results.

<u>Symmetries and conservation laws.</u> Now we need to make precise what we mean by a "symmetry and its corresponding conservation law". We shall do this in a rather restricted setting: our treatment here is not even general enough to treat classical mechanics adequatly. However it gives the right idea and in the rest of this paper we have to generalize the notions, defined in this introduction, anyhow.

<u>(Provisional) definition.</u> A vectorfield X on $\mathbb{R}^n \times \mathbb{R}^m$ is called a <u>symmetry</u> (of the projection $\pi\colon \mathbb{R}^n \times \mathbb{R}^m \to \mathbb{R}^n$) if

1. there is a divergence free vectorfield $\pi(X)$ on $\mathbb{R}^n$ such that for each $(x,y) \in \mathbb{R}^n \times \mathbb{R}^m$, $d\pi(X(x,y)) = (\pi(X))(x)$;
2. there is, for each $x \in \mathbb{R}^n$ and $t \in \mathbb{R}$ such that the time t integral $D_{\pi(X),t}(x)$ is defined, a linear map $F_t(x)\colon \mathbb{R}^m \to \mathbb{R}^m$ such that the time t integral of $(x,y) \in \mathbb{R}^n \times \mathbb{R}^m$ equals $D_{X,t}(x,y) = (D_{\pi(X),t}(x), (F_t(x))(y))$.

If E is a source equation, defined on functions $f: \mathbb{R}^n \to \mathbb{R}^m$ and X is a symmetry on $\mathbb{R}^n \times \mathbb{R}^m$, then we say that X is a <u>symmetry of E</u> if

3. for each $f: \mathbb{R}^n \to \mathbb{R}^m$, $x \in \mathbb{R}^n$, $t \in \mathbb{R}$ $|t|$ small, $(E(f_t))(x) = (E(f)(D_{\pi(X),-t}(x))) \circ F_{-t}(x)$, where $f_t: \mathbb{R}^n \to \mathbb{R}^m$ is such that $D_{X,t}\{(x,f(x))\} = \{(x,f_t(x))\}$.

In this last condition we have of course to assume that $D_{\pi(X),-t}(x)$ exist. Since the condition is only considered for $|t|$ small this is justified.

The requirement in condition 1. that $\pi(X)$ is divergence free is related with the fact that certain objects are defined by using integrals over $\mathbb{R}^n$; hence symmetries should not change the canonical measure on $\mathbb{R}^n$. The linearity in the $\mathbb{R}^m$ direction in condition 2. is only a matter of technical convenience. Both restrictions will be removed later. Now we come to the much less obvious definition of "conservation law".

<u>(Provisional) definition.</u> If E is a source equation, defined for functions $f: \mathbb{R}^n \to \mathbb{R}^m$, and X a symmetry of $\pi: \mathbb{R}^n \times \mathbb{R}^m \to \mathbb{R}^n$, then we say that the <u>X-conservation law</u> holds for E, or that E has <u>X-conservation</u> if and only if for any $f: \mathbb{R}^n \to \mathbb{R}^m$ and any bounded open $A \subset \mathbb{R}^n$, $\int_A < E(f), -\frac{\partial f_t}{\partial t} >$ depends only on the germ of f along ∂A. $f_t: \mathbb{R}^n \to \mathbb{R}^m$, as in the previous defition, is such that $D_{X,t}\{(x,f(x))\} = \{(x,f_t(x))\}$; $\frac{\partial f_t}{\partial t}(x)$ is the tangent vector of $t \to f_t(x)$ at $t = 0$.

It is non-trivial to show that this definition is equivalent to the usual one; this fact follows from a result of Horndesci [5], see § 4 of this paper. I shall show in this introduction only that the above definition implies the usual one in the case of conservation of energy in a "classical mechanical system".

Let E be a source equation defined for maps $f: \mathbb{R} \to \mathbb{R}^n$ of order 2. We shall give E the following interpretation: the maps $f: \mathbb{R} \to \mathbb{R}^n$ with $E(f) \equiv 0$ represent the configuration f(s) as function of time s for "unperturbed" motions of some (mechanical) system. If, for some $f: \mathbb{R} \to \mathbb{R}^n$, $E(f)$ is non-zero, $E(f)(s)$ is the external force, as function of time, needed to obtain the motion f. We consider the symmetry $X = \frac{\partial}{\partial s}$ (infinitesimal translation in the time direction) and show that if E has X-conservation in the above sense then there is conservation of energy in the usual sense.

If $f: \mathbb{R} \to \mathbb{R}^n$ is any map, then $-\frac{\partial f_t}{\partial t}(s) = \frac{df}{ds}$ so $< E(f), -\frac{\partial f_t}{\partial t} > = < E(f), \frac{df}{ds} >$. This last expression, being the inner product of force and velocity, is the work done by the external forces per unit time. We assume now that E has X-conservation. This means that for each $f: \mathbb{R} \to \mathbb{R}^n$ and each bounded interval $(s_0, s_1) \subset \mathbb{R}$, $\int_{s_0}^{s_1} < E(f), \frac{df}{ds} >$ only depends on the germs of f in s_0 and in s_1. In other words, the work, done by the external forces, to bring the system from the state described by the germ of f at s_0 to the state described by the germ of f at s_1, is independent of the way in which these external forces menage to make this transition. This means that there is a function E, defined on germs of maps from $\mathbb{R}$ to $\mathbb{R}^n$ such that $\int_{s_0}^{s_1} < E(f), \frac{df}{ds} > = E(g(f,s_1)) - E(g(f,s_0))$ where $g(f,s_i)$ denotes the germ of f at s_i. E is unique up to to an aditive constant. From the fact that E is of order two, one can show that $E(g(f,s))$ is already determined by the 1-jet of f in s; so we may write $E(j^1(f)(s))$. This function E has the usual properties of energy and if $E(f) \equiv 0$ then $E(j^1(f)(s))$ is independent of s hence we have conservation of energy. It should be noted that for the construction of E such that $E(j^1(f)(s_1)) - E(j^1(f)(s_0)) = \int_{s_0}^{s_1} < E(f), - \frac{\partial f_t}{\partial t} >$, we only needed the fact that E has X-conservation in the above sense and made no use of the mechanical interpretation. It should be mentioned that, with the above definitions, Noethers theorem can be formulated as follows:

If E is a source equation, defined on functions $f: \mathbb{R}^n \to \mathbb{R}^m$, which is the Euler equation of some Lagrangian and if X is a symmetry of $\pi: \mathbb{R}^n \times \mathbb{R}^m \to \mathbb{R}^n$, then E is X-symmetric if and only if E has X-conservation.

A proof of this form of Noethers theorem is also included in § 4.

Statement of the main results. Having made our definitions precise, we can now formulate our main problem and some of our main results. Our main problem is the following: consider source equations of order 1, defined on maps $f: \mathbb{R}^n \to \mathbb{R}^m$. For which Lie-algebra X of symmetries of $\pi: \mathbb{R}^n \times \mathbb{R}^m \to \mathbb{R}^n$ is it true that each such source equation E, which is X-symmetric and has X-conservation for each $X \in X$, is the Euler equation of some variational problem?

Note that it is no restriction to require X to be a Lie algebra: if X_1 and X_2 are symmetries of π, which are also symmetries of E then the same holds for $\alpha.X_1 + \beta.X_2$, $\alpha,\beta \in \mathbb{R}$ and $[X_1,X_2]$. If moreover E has X_1- and X_2-conservation then, E also has $(\alpha.X_1 + \beta.X_2)$-conservation; in § 4 we shall show that E has in this case also $[X_1,X_2]$-conservation.

Our main results show that for some Lie-algebras X the answer to the above question is affirmative.

Theorem. Let X be a Lie-algebra of symmetries of $\pi: \mathbb{R}^n \times \mathbb{R}^1 \to \mathbb{R}^n$ such that for each constant vectorfield Z on $\mathbb{R}^n$ there is an $X \in X$ such that $\pi(X) = Z$. Each source equation E, defined on functions $f: \mathbb{R}^n \to \mathbb{R}^1$, of order ≤ 2, which is X-symmetric and has X-conservation for each $X \in X$, is the Euler equation of some variational problem.

Note that this result any be applied in the following way. Let $X_1 \subset X_2$ be two Lie-algebras of symmetries of $\pi: \mathbb{R}^n \times \mathbb{R}^1 \to \mathbb{R}^n$ such that the vectorfields $\pi(X)$, $X \in X_1$, are just all the constant vectorfields on $\mathbb{R}^n$. If we then know that some source equation E, defined on functions $f: \mathbb{R}^n \to \mathbb{R}^1$ and of order ≤ 2, is X-symmetric for each $X \in X_2$ and has X-conservation for each $X \in X_1$, we conclude that E has also X-conservation for all $X \in X_2$.

Theorem. Let X be a Lie algebra of symmetries on $\pi: \mathbb{R}^n \times \mathbb{R}^m \to \mathbb{R}^n$ such that for at least one $X \in X$, $\pi(X)$ is not identically zero. Let E be a linear source equation defined on functions $f: \mathbb{R}^n \to \mathbb{R}^m$, i.e., the co-vector $(E(f))(x)$ depends linearly on the "vector" $(x, f(x), df(x), d^2f(x), \ldots)$. If, for each $X \in X, E$ is X-symmetric and has X-conservation, then E is the Euler equation of a variational problem.

For n = 1 the above theorems were obtained in [8].

A more general setting. The setting we used up to now is not general enough to make our considerations applicable to some important examples. E.g., if we consider the N-body problem (say in 3-space) with Newtonian potential, then we have to deal with maps $\gamma: \mathbb{R} \to (\mathbb{R}^3 \times \ldots \times \mathbb{R}^3 - \Delta_N)$ where $\Delta_N = \{(x^1,\ldots,x^N) \mid x^i \in \mathbb{R}^3,\ x^i = x^j$ for some $i \neq j\}$. In other situations we have to consider sections of some differentiable bundle instead of mappings $f: \mathbb{R}^n \to \mathbb{R}^m$ (which are sections of the trivial bundle $\pi: \mathbb{R}^n \times \mathbb{R}^m \to \mathbb{R}^n$), for example:

Let W be a 4-manifold and $\pi: E \to W$ the differentiable fibration such that each fibre $\pi^{-1}(x)$, $x \in W$, consists of the non-degenerate quadratic functions on $T_x(W)$ with signature (+,-,-,-). A cross-section of π defines a non-definite "metric" on W. The source equation for relativistic gravitation theory can be obtained from a variational problem for such sections. In § 5 we shall also show for source equations of this type that, if they have enough symmetry and satisfy enough conservation laws, they are necessarily the Euler equation of some variational problem.

So in what follows, we consider smooth (local) sections $s: W \to E$ of some differentiable fibration $\pi: E \to W$ instead of maps $f: \mathbb{R}^n \to \mathbb{R}^m$. We used several times, even to define certain concepts, integration on $\mathbb{R}^n$. In the present setting this would correspond to integration on W. But on W we do not have (and do not want to have) a canonical measure. This means that we have to modify slightly our definitions of Lagrangian and of source equation so that, for instance, for a Lagrangian L and a section s, $L \circ s$ (or $L(j^k(s))$) can be integrated over W, i,e., is an n-form on W (n = dim (W)).

The generalization to sections of bundles gives some extra complications due to the possible non-triviality of the topology of the bundle (and especially of its fibre: if the fibre is contractible, not much extra complications arise). In the appendix we show an example how the topological non-triviality of the bundle can violate some theorems which hold in the more restricted case where we consider only maps from $\mathbb{R}^n \to \mathbb{R}^m$. Sometimes one can only conclude (due to the non-triviality of the bundles involved) that a source equation can locally be obtained as Euler equation of some variational problem.

The paper is organized as follows. In § 2 we consider infinite jet spaces with their functions vectorfields and differential forms. We show that the same formulas, relating the d-operator the Lie derivative and the substitution operator, as in the finite dimensional case, see for example [6], hold also on these infinite jet spaces. In § 3 we introduce special classes of differential forms, using the bundle structure, and prove among other things, some local exactness results related to the Poincaré lemma. In § 4 we use the results from the previous two sections to derive Noether's theorem, Tonti's theorem and also a more technical theorem which is the basis of the proofs of our main results in § 5. We also discuss in § 4 the relation between our definition of conservation law and the usual one.

Finally one might ask whether the functional analytic methods, used by Tonti to prove his theorem [9] and to analyse the mathematical structure of physical theories [10] and [11], could also be used here instead of the geometry of ∞-jet spaces. I think that this is the case but, since certain aspects, e.g., conservation laws, are more transparent if one uses jet spaces, I think that it is worthwile to have also the jet space formalism.

2. Jetbundles

We consider a differentiable bundle $\pi : E \to W$, i.e., E and W are smooth manifolds and π is a bundle projection as well as a smooth map which has everywhere maximal rank. The bundle of k-jets of (local) sections of π is denoted by $J^k(\pi)$. The elements of $J^k(\pi)$ are equivalence classes of $\{(s,w) \mid w \in W$ and s is a smooth local section defined on a neighbourhood of $w\}$. $(s,w) \sim (s',w')$ if and only if $w = w'$ and, for each smooth function $f: E \to \mathbb{R}$, the derivatives up to order k of $f \circ s$ and $f \circ s'$ in w are the same. It is clear that:

- $J^0(\pi) = E$;
- all the canonical projections $\pi^l_k: J^k(\pi) \to J^l(\pi)$, $k \geq l \geq 0$ and $\pi_k: J^k(\pi) \to W$, are differentiable fibre bundles;
- for each $\bar{s} \in J^k(\pi)$, $(\pi^k_{k+1})^{-1}(s)$ is an affine space.

We define the infinite jet space as inverse limit: $J^\infty(\pi) = \lim J^k(\pi)$; by E. Borel's theorem, each element of $J^\infty(\pi)$ is indeed the ∞-jet of some local section. On $J^\infty(\pi)$ we use the inverse limit topology: if $\bar{s} \in V \subset J^\infty(\pi)$, then V is a neighbourhood of $\bar{s}$ if there is for some k and a neighbourhood V_k of $\pi^k_\infty(\bar{s})$ in $J^k(\pi)$ such that $(\pi^k_\infty)^{-1}V_k \subset V$; $\pi^k_\infty: J^\infty(\pi) \to J^k(\pi)$ and $\pi_\infty: J^\infty(\pi) \to W$ are the canonical projections. We shall use a notion of C^∞-functions on $J^\infty(\pi)$ which was introduced by Boardman [1].

Definition (2,1). A function $f: J^\infty(\pi) \to \mathbb{R}$ is called C^∞ if there is for every $\bar{s} \in J^\infty(\pi)$ a neighbourhood V of $\bar{s}$ in $J^\infty(\pi)$ and a smooth $\tilde{f}: J^k(\pi) \to \mathbb{R}$ (for some k which may depend on $\bar{s}$) such that $f|V = \tilde{f} \circ \pi^k_\infty|V$. If, for some $\tilde{f}: J^k(\pi) \to \mathbb{R}$, $f = \tilde{f} \circ \pi^k_\infty$, we say that f is of type k; note that there are functions which are "of no type".

Definition (2,2). A (C^∞-)vectorfield on $J^\infty(\pi)$ is a map X which maps the set of C^∞-functions on $J^\infty(\pi)$ to itself and which satisfies:

- X is $\mathbb{R}$-linear;
- $X(f.g) = (Xf).g + f.(Xg)$.

The following theorem gives a more concrete description of such vectorfields. In this theorem we need the notion of "vectorfield along a map". A vectorfield along $f: M \to N$, N a finite dimensional smooth manifold, is a map $Z: M \to T(N)$ such that for each $m \in M$, $Z_{(m)} \in T_{f(m)}(N)$. We shall always deal with cases where also M is a smooth manifold and f, Z are smooth.

Theorem (2,3). Let X be a smooth vectorfield on $J^\infty(\pi)$. Then there are open subsets $U_{k,l} \subset J^k(\pi)$ for all $k \geq l \geq 0$ and vectorfields $X_{k,l}$ along $\pi^l_k|U_{k,l}$ such that:

1. $(\pi^k_{k+1})^{-1}U_{k,l} \subset U_{k+1,l}$;
2. $U_{k,l} \subset U_{k,l+1}$;
3. $X_{k,l} \circ \pi^k_{k+1} \mid (\pi^k_{k+1})^{-1}(U_{k,l}) = X_{k+1,l} \mid (\pi^k_{k+1})^{-1}(U_{k,l})$;
4. $d\pi^l_{l+1} \circ X_{k,l+1} \mid U_{k,l} = X_{k,l}$;
5. for each l, $\bigcup_{k \geq l} (\pi^k_\infty)^{-1}U_{k,l} = J^\infty(\pi)$;
6. for each smooth $f: J^l(\pi) \to \mathbb{R}$ and $\bar{s} \in U_{k,l}$, $X(f \circ \pi^l_\infty)(\bar{\bar{s}}) = (X_{k,l}(\bar{s}))(f)$ if $\pi^k_\infty(\bar{\bar{s}}) = \bar{s}$.

X is uniquely determined by $\{X_{k,l},U_{k,l}\}_{k\geq l\geq 0}$; each $\{X_{k,l},U_{k,l}\}_{k\geq l\geq 0}$ satisfying 1.,,,5. above determines a smooth vectorfield on $J^\infty(\pi)$. Two presentations $\{X_{k,l},U_{k,l}\}_{k\geq l\geq 0}$ and $\{X'_{k,l},U'_{k,l}\}_{k\geq l\geq 0}$ define the same vectorfield if

$X_{k,l} \mid U_{k,l} \cap U'_{k,l} = X'_{k,l} \mid U_{k,l} \cap U'_{k,l}$ for each $k \geq l \geq 0$.

The proof of this theorem, which is based on the fact that on finite dimensional manifolds $\mathbb{R}$-linear derivations on the C^∞ functions are just the usual vectorfields (see [6]), is left to the reader.

Definition (2,4). A smooth vectorfield X on $J^\infty(\pi)$ is said to be of type $(r_0,r_1,r_2,\ldots)$, $r_i \geq i$, if there is a presentation $\{X_{kl},U_{kl}\}_{k\geq l\geq 0}$ of X such that $U_{k,l} = J^k(\pi)$ whenever $k \geq r_l$.

Definition (2,5). Let $P \subset J^\infty(\pi)$ be some subset. Two vectorfields X_1 and X_2 are said to be equal along P, if for each smooth $f: J^\infty(\pi) \to \mathbb{R}$, $X_1(f) \mid P = X_2(f) \mid P$. For $\bar{s} \in J^\infty(\pi)$, we define $T_{\bar{s}}(J^\infty(\pi))$ to be the set of equivalence classes of vectorfields on $J^\infty(\pi)$, with the equivalence relation: $X_1 \sim X_2$ if $X_1(\bar{s}) = X_2(\bar{s})$. It is easy to see that $T_{\bar{s}}(J^\infty(\pi)) = \varprojlim T_{\pi^k_\infty(\bar{s})}(J^k(\pi))$; $d\pi^k_\infty$ denotes the corresponding projection.

Example (2,6) (symmetries, integrable vectorfields and deformations). Let $\bar{X}$ be a smooth vectorfield on E such that there is a vectorfield $\pi(\bar{X})$ on W with the property that for each $e \in E$, $d\pi(\bar{X}(e)) = \pi(\bar{X})(\pi(e))$. We shall call such a vectorfield a symmetry of π. From $\bar{X}$ we shall construct a vectorfield X on $J^\infty(\pi)$ of type $(0,1,2,\ldots)$. For this we construct a sequence of vectorfields $X^{(0)},X^{(1)},\ldots$ on $J^0(\pi),J^1(\pi),\ldots$ such that for each $k \geq l \geq 0$, and $\bar{s} \in J^k(\pi)$, $d\pi^l_k(X^{(k)}(\bar{s})) = X^{(l)}(\pi^l_k(\bar{s}))$. X is then described by the presentation $\{X_{k,l},U_{k,l}\}_{k\geq l\geq 0}$ with $U_{k,l} = J^k(\pi)$ and $X_{k,l} = X^{(l)} \circ \pi^l_k$.

For $\bar{s}_k \in J^k(\pi)$, $X^{(k)}(\bar{s}_k)$ is defined as follows. Let s be a local section, defined in a neighbourhood of $\pi_k(\bar{s}_k) = w$, of π representing $\bar{s}_k$. If $\mathcal{D}_{\bar{X},t}$ denotes the time t integral of $\bar{X}$, then, for a sufficiently small neighbourhood V of w and for $|t|$ sufficiently small, $\mathcal{D}_{\bar{X},t}\{(w',s(w')) \mid w' \in V\}$ is the image of a local section s_t of π defined on $\mathcal{D}_{\pi(\bar{X}),t}(V)$. Let $\bar{s}_k(t)$ be the k-jet of s_t at $\mathcal{D}_{\pi(X),t}(w)$ and define $X^{(k)}(\bar{s}_k)$ as the tangent vector of the curve $t \to \bar{s}_k(t)$ at $t = 0$. It is easy to see that $d\pi^1_k(X^{(k)}(\bar{s}_k)) = X^{(1)}(\pi^1_k(\bar{s}_k))$; hence X is well defined.

A vectorfield X on $J^\infty(\pi)$ which can be obtained in the above way from a symmetry is called integrable; if moreover it can be obtained from a symmetry $\bar{X}$ with $\pi(\bar{X}) \equiv 0$, it is called vertical integrable. Finally we say that a vectorfield X on $J^\infty(\pi)$ is a deformation if for each local section $s: V \to E$, $V \subset W$ an open subset, and for each closed subset $K \subset V$, there is a vertical integrable vectorfield $X_{s,K}$ such that $X \mid (j^\infty(s))(K) = X_{s,K} \mid (j^\infty(s))(K)$. $j^\infty(s): V \to J^\infty(\pi)$ is the map which assigns to each $w \in V$ the ∞-jet of s in w.

Example (2,7) (total vectorfields). Let Z be a vectorfield on W; we associate to Z a vectorfield Z_{tot}, the total vectorfield of Z, on $J^\infty(\pi)$ of type $(1,2,3,\ldots)$. Note that the notion of "total vectorfields" defined by Boardman in [1] is somewhat more general.

In order to define Z_{tot}, we shall construct a sequence $Z^{(k)}$ of vectorfields along π^k_{k+1}, $k = 0,1,\ldots$. For $\bar{s}_{k+1} \in J^{k+1}(\pi)$ we choose a local section s of π, defined in a neighbourhood of $w = \pi_{k+1}(\bar{s}_{k+1})$, representing $\bar{s}_{k+1}$. Let $\bar{s}_k(t)$ denote the k-jet of s in $D_{Z,t}(w)$. We define $Z^{(k)}(\bar{s}_{k+1}) \in T_{\pi^k_{k+1}}(J^k(\pi))$ to be the tangent vector of the curve $t \to \bar{s}_k(t)$ at $t = 0$. It is not hard to see that $Z^{(k)}(\bar{s}_{k+1})$ is well defined (i.e., independent of the particular choice of the section s). Using the sequence $Z^{(k)}$ we define a presentation of Z_{tot} as

$\{Z_{k,l}, U_{k,l}\}_{\underline{k \geq l \geq 0}}$ with

$U_{l,l} = \emptyset$,

$U_{k,l} = J^k(\pi)$ if $k > l$,

$Z_{k,l} = Z^{(l)} \circ \pi^{l+1}_k$ if $k > l$.

It is easy to see that $\{Z_{k,l}, U_{k,l}\}_{\underline{k \geq l \geq 0}}$ satisfies the conditions 1.,...,5. in theorem (2,3).

Example (2,8) (evolution vectorfields). Let $\bar{X}$ be a symmetry of π and X the corresponding integrable vectorfield on $J^\infty(\pi)$. X_{ev}, the evolution vectorfield of $\bar{X}$ (or X), is a vectorfield on $J^\infty(\pi)$ of type (1,2,3,...) defined by the following vectorfields $\tilde{X}^{(k)}$ along π^k_{k+1}, $k = 0,1,2,\ldots$.

For $\bar{s}_{k+1} \in J^{k+1}(\pi)$ we take a local section s representing it. As we mentioned before, $D_{\bar{X},t}(\mathrm{Im}(s))$ is the image of a local section which we call s_t and which is, for $|t|$ small, defined on a neighbourhood of $w = \pi_{k+1}(\bar{s}_{k+1})$. $\tilde{s}_k(t)$ denotes the k-jet of the section s_t in w. We define $X^{(k)}(s_{k+1}) \in T_{\pi^k_{k+1}(s_{k+1})}(J^k(\pi))$ to be the tangent vector of the curve $t \to s_k(-t)$ (note the - sign) at $t = 0$. The definition X_{ev} from the sequence $\{X^{(k)}\}$ is now done as in the previous example.

Remark (2,9). If $\bar{X}$ is a symmetry of π, X the corresponding integrable vectorfield and X_{ev} the corresponding evolution vectorfield, then $X = (\pi(\bar{X}))_{tot} - X_{ev}$. Note also that X_{ev} is a deformation.

Now we come to the definition of differential forms on $J^\infty(\pi)$. First we observe that the product of a smooth function and a vectorfield on $J^\infty(\pi)$ is again a vectorfield on $J^\infty(\pi)$: for f a function and X a vectorfield, f.X is defined by $(f.X)(h) = f.(X(h))$.

Definition (2,10). A p-form on $J^\infty(\pi)$ is a map ω which assigns to each p-tuple of vectorfields on $J^\infty(\pi)$ a function on $J^\infty(\pi)$ in such a way that:

1. $\omega(X_1,\ldots,X_p) = (-1)^{|\sigma|}\omega(X_{\sigma(1)},\ldots,X_{\sigma(p)})$ for any permutation $\sigma \in S_p$ and any vectorfields $X_1,\ldots,X_p$;
2. $\omega(f.X_1 + g.X_1', X_2,\ldots,X_p) = f.\ (X_1,\ldots,X_p) + g.\ (X_1', X_2,\ldots,X_p)$ for any functions f,g and vectorfields $X_1, X_1', X_2,\ldots,X_p$.

More important than this definition is the following characterization:

Theorem (2,11). Let ω be a p-form on $J^\infty(\pi)$. Then there is a sequence $\{\omega_i, U_i\}_{\underline{i \geq 0}}$ with $U_i \subset J^i(\pi)$ open subsets and ω_i a p-form on U_i such that

1. $(\pi^i_{i+1})^{-1}(U_i) \subset U_{i+1}$;

2. $\bigcup_{i\geq 0} (\pi^i_\infty)^{-1}(U_i) = J^\infty(\pi)$;

3. $\omega_{i+1} \mid (\pi^i_{i+1})^{-1} U_i = (\pi^i_{i+1})^*(U_i)$;

4. if $X_1,\ldots,X_p$ are vectorfields on $J^\infty(\pi)$, $\bar{s} \in J^\infty(\pi)$ and $\pi^k_\infty(\bar{s}) \in U_k$, then $\omega(X_1,\ldots,X_p)(\bar{s}) = (\omega_k(\pi^k_\infty(\bar{s})))(d\pi^k_\infty(X_1(\bar{s})),\ldots,d\pi^k(X_p(\bar{s})))$.

ω is completely determined by the presentation $\{\omega_i,U_i\}_{i\geq 0}$. If $\{\omega'_i,U'_i\}$ satisfies the above conditions 1., 2. and 3. then it defines a unique p-form on $J^\infty(\pi)$. $\{\omega_i,U_i\}$ and $\{\omega'_i,U'_i\}$ define the same p-form if and only if for each i, $\omega_i \mid U_i \cap U'_i = \omega'_i \mid U_i \cap U'_i$.

The proof of this theorem is straightforward and left to the reader.

Definition (2,12). A p-form is said to be of type (l,k), $0 \leq l \leq k$ if

1. a presentation $\{\omega_i,U_i\}_{i\geq 0}$ as in theorem (2,11) exists with $U_k = J^k(\pi)$;
2. for each p-tuple of vectorfields $X_1,\ldots,X_p$ on $J^\infty(\pi)$ with $d\pi^1_\infty(X_1,(\bar{s})) = 0$ for each $\bar{s} \in J^\infty(\pi)$, $\omega(X_1,\ldots,X_p) = 0$.

Some simple operations (2,13). Just as on finite dimensional manifolds we define for p and q forms ω, σ, the forms $d\omega$, $\omega \wedge \sigma$, $\iota_X\omega$ (X a vectorfield). We omit the formal definitions but point out that if ω is of type (l,k) then $d\omega$ is of type (k,k). If ω is of type (l,k), X of type $(r_0,r_1,r_2,\ldots)$ and if ω is a p-form with $p \geq 2$ then $\iota_X\omega$ is of type $(l,\max(r_1,k))$. If ω is a 1-form of type (l,k) and X a vectorfield of type $(r_0,r_1,\ldots)$ then the function $\iota_X\omega = \omega(X)$ is of type $\max(r_1,k)$.

General assumption (2,14). From now on we shall assume all functions vectorfields and differential forms on $J^\infty(\pi)$ to be of some type. As mentioned before in definition (2,1) this is a real restriction. It should be noted that with the above mentioned operations we stay in this class of objects.

Reduction principle (2,15). This reduction principle gives a method to extend theorems on finite dimensional manifolds to $J^\infty(\pi)$. Let $h \geq 0$ be some integer and let

$\mathcal{F}^h$ denote the set of functions on $J^\infty(\pi)$ of type h;

$\mathcal{X}^h$ denote the set of vectorfields on $J^\infty(\pi)$ each of which being of some type $(\alpha_0,\alpha_1,\ldots)$ satisfying $\alpha_h = h$;

Ω^h denote the set of differential forms of type (h,h).

Observe that $\mathcal{F}^h$ and Ω^h are in a canonical 1-1 corresponding with the functions respectively differential forms on $J^h(\pi)$. Let $p_{\mathcal{F}},p_\Omega$ denote the maps which assign to each element in $\mathcal{F}^h,\Omega^h$ the corresponding object on $J^h(\pi)$. We let p denote the map which assigns to each $X \in \mathcal{X}^h$ the vectorfield $p_{\mathcal{X}}(X)$ on $J^h(\pi)$ which is defined by $(p_{\mathcal{X}}(X))(\bar{s}_h) = d\pi^h_\infty(X(\bar{s}_\infty))$ whenever $\pi^h_\infty(\bar{s}_\infty) = \bar{s}_h$.

If we apply the operations d, $\wedge$ and ι_X to elements in $\mathcal{F}^h,\mathcal{X}^h,\Omega^h$ (of course we assume $X \in \mathcal{X}^h$) we do not leave this class of elements. Moreover we have

$\iota_{p_{\mathcal{X}}(X)}(p_\Omega\omega) = p_\Omega(\iota_X\omega)$,

$d(p_\Omega\omega) = p_\Omega(d\omega)$ etc.

In other words it makes no difference whether we apply the definitions given for $J^\infty(\pi)$ or the usual ones for the corresponding objects on $J^1(\pi)$. All the above statements follow directly from the definitions.

Lie derivatives (2.16). Usually one defines the Lie derivative L_X with respect to a vectorfield X by using the time t integral of X (locally and for small t). For vectorfields on $J^\infty(\pi)$ there is no existence and uniqueness theorem for integral curves of vectorfields (although for integrable vectorfields X on $J^\infty(\pi)$ the time t integral exists locally and for small t). Hence we have to be somewhat carefull.

I. We define the Lie derivative of a function by $L_X f = X(f)$; we note that if f is of type k and X of type $(r_0, r_1, \ldots)$, X(f) is of type r_k.

II. We define the Lie derivative of a vectorfield by $(L_X Y)(f) = X(Y(f)) - Y(X(f))$; we note that if X is of type $(r_0, r_1, \ldots)$ and Y of type $(s_0, s_1, \ldots)$ then $L_X Y$ is of type $(t_0, t_1, \ldots)$ with $t_k = \max\{r_{s_k}, s_{r_k}\}$. $L_X Y$ is also denoted by $[X,Y]$.

In case X is an integrable vectorfield on $J^\infty(\pi)$, the time t integral $\mathcal{D}_{X,t}$ exists locally and for small t; this integral is obtained by integrating the correponding symmetry $\overline{X}$, which gives $\mathcal{D}_{\overline{X},t}: E \to E$ and which includes $\mathcal{D}_{X,t}$. $\mathcal{D}_{X,t}$ has the following properties:

- if $f: J^\infty(\pi) \to \mathbb{R}$ is smooth then also $f \circ \mathcal{D}_{X,t}$ is smooth (where it is defined);
- if Y is a vectorfield on $J^\infty(\pi)$, $(\mathcal{D}_{X,t})_* Y$ is also a (locally defined) vectorfield: $((\mathcal{D}_{X,t})_* Y)(f) = Y(f \circ \mathcal{D}_{X,t})$

Since $\mathcal{D}_{X,t}$ is induced by $\mathcal{D}_{\overline{X},t}: E \to E$ and since $\mathcal{D}_{\overline{X},t}$ respects π, in the sense that $\pi \circ \mathcal{D}_{\overline{X},t} = \mathcal{D}_{\pi(\overline{X}),t} \circ \pi$ (everything only locally), $\mathcal{D}_{X,t}$ also has the following properties

- if Y is a total vectorfield then $(\mathcal{D}_{X,t})_* Y$ is also total; if $Y = (\pi(Y))_{tot}$ then $\pi((\mathcal{D}_{X,t})_* Y) = (\mathcal{D}_{\pi(\overline{X}),t})_* \pi(Y)$;
- if Y is integrable and if $\overline{Y}$ is the correponding symmetry, then $(\mathcal{D}_{X,t})_* Y$ is integrable and $(\mathcal{D}_{\overline{X},t})_* \overline{Y}$ is the correponding symmetry.

From the above definitions of $L_X Y$ and $L_X f$ it follows that if X is integrable, $L_X Y = \lim_{t \to 0} \frac{1}{t}[Y - (\mathcal{D}_{X,t})_* Y]$. This, combined with the properties of $(\mathcal{D}_{X,t})_*$, immediately gives:

If X is integrable and Z is total, then $[X,Z]$ is the total vectorfield corresponding to the vectorfield $[\pi(X), \pi(Z)]$ on W } (1)

and

If X and Z are integrable then $[X,Z]$ is integrable and $\pi([X,Z]) = [\pi(X), \pi(Z)]$. } (2)

also, from Boardman [1] § 1 it follows that

If X and Z are total vectorfields then also $[X,Z]$ is total and $\pi([X,Z]) = [\pi(X), \pi(Z)]$. } (3)

III. Finally we define the Lie derivative of a differential form by $(L_X \omega)(X_1, \ldots, X_p) = L_X(\omega(X_1, \ldots, X_p)) - \sum_{i=1}^{p} \omega(X_1, \ldots, [X, X_i], \ldots, X_p)$, for ω a p-form and vectorfields $X, X_1, \ldots, X_p$. The verification that $L_X \omega$, defined in this way, is again a p-form is straightforward. If X is of type $(r_0, r_1, \ldots)$ and ω of type $(1,k)$ then $L_X \omega$ is of type (r_k, r_k). The last statement can be proved as follows: from the definition of type for differential forms we know that a p-form σ has type (s,s) if and only if for each p-tuple of vectorfields $Y_1, \ldots, Y_p$, each of type $(0,1,2,\ldots)$, $\sigma(Y_1, \ldots, Y_p)$ is of type s. If we apply this to $L_X \omega$ we find, using the rules for the types of X(f) and $L_X(Y)$, that $L_X \omega$ is of type (r_k, r_k).

Like in the finite dimensional case we have $L_X\omega = \iota_X d\omega + d\iota_X\omega$ }(4)
The proof of this formula goes as follows: Let X be again of type $(r_0,r_1,r_2,\ldots)$ and let ω be of type (1,k). It is enough to prove that (4) holds after applying left and right hand side to any p-tuple of vectorfields $Y_1,\ldots,Y_p$ of type $(0,1,2,\ldots)$; choose a fixed such p-tuple. After this substitution, the validity of (4) does not change if we replace X by some X' if $(d\pi^k_\infty(X))(\bar{s}) = (d\pi^k_\infty(X'))(\bar{s})$ for each $\bar{s} \in J^\infty(\pi)$. We choose such X' of type $(t_0,t_1,\ldots)$ with $t_i = i$ for all $i \geq r_k$. Since now all vectorfields are in $\mathcal{X}^{r_k}$ and also ω and $L_X\omega$ are in Ω^{r_k}, the formula follows from the reduction principle (2,15) and the corresponding formula in the finite dimensional case, see [6].

Homotopies (2,17). Let Φ_t: $E \to E$ and ϕ_t: $W \to W$, $t \in I$, I an open interval in $\mathbb{R}$, be smooth 1-parameter families of diffeomorphisms such that for each t, $\pi \circ \Phi_t = \phi_t \circ \pi$. Since Φ_t, for each $t \in I$, maps images of sections to images of sections there are induced maps Φ^∞_t: $J^\infty(\pi) \to J^\infty(\pi)$ and Φ^k_t: $J^k(\pi) \to J^k(\pi)$. Hence, if ω is a differential form, say of type (k,k), on $J^\infty(\pi)$ then $\omega_t = (\Phi^\infty_t)^*\omega$ is well defined and also of type (k,k). To define ω_t one can even use the map Φ^k_t instead of Φ^∞_t. It follows now from the reduction principle and the corresponding formula in the finite dimensional case that $\frac{d}{dt}(\omega_t)\big|_{t=\bar{t}} = (\Phi^\infty_{\bar{t}})(L_{X_{\bar{t}}}\omega)$ where $X_{\bar{t}}$ is the integrable vectorfield on $J^\infty(\pi)$ defined as follows. For $e \in E$, $\bar{X}_{\bar{t}}$ is the tangent vector of the curve $t \to \Phi_{\bar{t}+t} \circ \Phi^{-1}_{\bar{t}}(e)$; X_t is a symmetry of π. $X_{\bar{t}}$ is the corresponding integrable vectorfield.

3. Special differential forms.

In this section we define and derive some properties of those classes of differential forms which are of special importance for variational problems. We use the same notation as in § 2, i.e., π: $E \to W$ is a differentiable fibration, $J^k(\pi)$ the bundle of k-jets of its local sections etc.. Also we shall assume that all functions vectorfields and differential forms are C^∞ and, as far as they are defined on $J^\infty(\pi)$, are of "some type", see (2,14). Let $n = \dim(W)$.

Definition (3,1). $\bar{H}^i_j(\pi)$ denotes the set of $(i + j)$-forms on $J^\infty(\pi)$ which have the property that they become zero when applied to $(i + 1)$ vertical vectorfields. In other words, $\omega \in \bar{H}^i_j(\pi)$ if and only if for each $\bar{s} \in J^\infty(\pi)$ and $X_1,\ldots,X_{i+j} \in T_{\bar{s}}(J^\infty(\pi))$ such that $d\pi_\infty(X_l) = 0$ for $l = 1,\ldots,i+1$, $\omega(\bar{s})(X_1,\ldots,X_{i+j}) = 0$. $\bar{H}^i_j(\pi) \supset \bar{H}^{i+1}_{j-1}(\pi)$; $H^i_j(\pi)$ is by definition equal $\bar{H}^i_j(\pi)\,/\,\bar{H}^{i+1}_{j-1}(\pi)$. We say that $\omega \in H^i_j(\pi)$ has type (l,k) if there is a representative of ω in $\bar{H}^i_j(\pi)$ with that type.

Theorem (3,2). Each $\omega \in H^i_j(\pi)$ of type (k,k) determines a map E_ω which assigns to each i-tuple of verticle integrable vectorfields $X_1,\ldots,X_i$ and each (local) section s of π a j-form on the domain of s. E_ω is defined by $E_\omega(s;X_1,\ldots,X_i) = (j^\infty(s))^*(\iota_{X_1,\ldots,X_i}\omega)$. E_ω completely determines ω. If E is some map which assigns to each i-tuple of vertical integrable vectorfields $X_1,\ldots,X_i$ and each local section s of π a j-form $E(s;X_1,\ldots,X_s)$ on the domain of s such that:

1. for each $w \in$ domain of s, $E(s;X_1,\ldots,X_s)(w)$ is completely determined by the k-jet of s in w and the k-jets of $\bar{X}_i \mid \mathrm{Im}(s)$ in $s(w)$, where $\bar{X}_i$ is the symmetry of π corresponding to X_i;

2. $E(s;X_1,\ldots,X_s)$ is $\mathbb{R}$-linear and antisymmetric in $X_1,\ldots,X_s$,
Then there is a unique $\omega \in H^i_j(\pi)$ of type (k,k) such that $E = E_\omega$.

The proof of this theorem is a consequence of the fact if $\bar{s} \in J^\infty(\pi)$, $\hat{X} \in T_{\bar{s}}(J^\infty(\pi))$ such that $d\pi_\infty(\hat{X}) = 0$, then there is a vertical integrable vectorfield X on $J^\infty(\pi)$ such that $X(\bar{s}) = \hat{X}$. The details are left to the reader.

Remark (3,3). 1. $d(\bar{H}^i_j(\pi)) \subset \bar{H}^{i+1}_j(\pi)$. Hence there are induced maps $H^i_j(\pi) \to H^{i+1}_j(\pi)$. These induced maps are also denoted by d.

2. For $\omega \in H^i_j(\pi)$, we define $D\omega \in H^i_{j+1}(\pi)$ by $d(E_\omega(s;X_1,\ldots,X_i)) = E_{D\omega}(s;X_1,\ldots,X_i)$ for any local section s and vertical integrable vectorfields $X_1,\ldots,X_i$. The meaning of E_ω and $E_{D\omega}$ is as in theorem (3,2). We shall call D the divergence operator. We shall use it in the next section in relation with the definition of conservation law.

3. If X is any deformation on $J^\infty(\pi)$ then ι_X maps $H^i_j(\pi)$ into $H^{i-1}_j(\pi)$; if X is "only vertical", this is also the case.

4. If $\omega \in H^0_j(\pi)$, then the j-form $(j^\infty(s))^*\omega$ shall also be denoted by $\omega \circ s$.

Definition (3,4). We have $n = \dim(W)$; a subset $A \subset W$ is called a bounded open set if A is open and $\bar{A}$ is compact.

We define the subsets $G^{00}_n(\pi)$, resp. $G^0_n(\pi)$, of $H^0_n(\pi)$ as follows. If $\omega \in H^0_n(\pi)$ then $\omega \in G^{00}_n(\pi)$, resp. $G^0_n(\pi)$, if and only if for each oriented bounded open set $A \subset W$ and each pair of sections s_0,s_1, defined on a neighbourhood of $\bar{A}$ satisfying:

1. s_0 and s_1 are equal in a neighbourhood of ∂A;
2. (only for $\omega \in G^0_n(\pi)$) there is a smooth homotopy s_t, $t \in [0,1]$, joining s_0 and s_1 in such a way that s_0 and s_t are equal on some neighbourhood of ∂A;

$\int_A \omega \circ s_0 = \int_A \omega \circ s_1$.

The subset $G^i_n(\pi) \subset H^i_n(\pi)$, $i > 0$, is defined as follows. If $\omega \in H^i_n(\pi)$ then $\omega \in G^i_n(\pi)$ if and only if for each oriented bounded open set $A \subset W$, each pair of sections s_0,s_1 defined on a neighbourhood of $\bar{A}$ and each pair of i-tuples of vertical integrable vectorfields $X^1_0,\ldots,X^i_0,X^1_1,\ldots,X^i_1$ such that

1. s_0 and s_1 are equal in a neighbourhood of ∂A;
2. for each $j = 1,\ldots,i$ X^j_0 and X^j_1 are equal in a neighbourhood of $j^\infty(s_0)(\partial A)$;

$\int_A (\iota_{X^1_0,\ldots,X^i_0}\omega) \circ s_0 = \int_A (\iota_{X^1_1,\ldots,X^i_1}\omega) \circ s_1$.

Remark (3,5). It should be observed that the above definition of $G^i_n(\pi)$, $i > 0$, is equivalent with the following. If $\omega \in H^i_n(\pi)$ then $\omega \in G^i_n(\pi)$ if and only if for each bounded oriented open set $A \subset W$, each section s, defined on a neighbourhood of $\bar{A}$, and each i-tuple of vertical integrable $X_1,\ldots,X_i$, with X_1 zero on a neighbourhood of $(j^\infty(s))\partial A$.

$$\int_A (\iota_{X_1,\ldots,X_i}\omega) \circ s = 0.$$

Since in this last definition only one section occures we may obtain a third equivalent definition of $G^i_n(\pi)$, $i > 0$, by allowing the $X_1,\ldots,X_i$ in the above definition to be deformations.

Finally also the definition of $G_n^i(\pi)$, $i > 0$, in (3,4) does not change if we allow $X_0^1,\ldots,X_0^i,X_1^1,\ldots,X_1^i$ to be deformations. Because in that case one may always choose vertical integrable vectorfields $\tilde{X}_0^1,\ldots,\tilde{X}_0^i,\tilde{X}_1^1,\ldots,\tilde{X}_1^i$ such that along $\mathrm{Im}(j^\infty(s_k))$, $X_k^l = \tilde{X}_k^l$ $k = 0,1$, $l = 1,\ldots,i$ and such that in in a small neighbourhood of $(j^\infty(s_k))(\partial A)$, $\tilde{X}_0^l = \tilde{X}_1^l$ for $l = 1,\ldots,i$.

Theorem (3,6). (a) If X is a deformation then ι_X maps $H_j^i(\pi)$ to $H_j^{i-1}(\pi)$ and $G_n^i(\pi)$ to $G_n^{i-1}(\pi)$ (where $H_j^{-1}(\pi)$ and $G_n^{-1}(\pi)$ are zero vectorspaces). Moreover ι_X maps $G_n^1(\pi)$ to $G_n^{00}(\pi)$.

(b) If $\Phi\colon E \to E$ and $\phi\colon W \to W$ are diffeomorphisms such that $\pi \circ \Phi = \phi \circ \pi$ and $\Phi^\infty\colon J^\infty(E) \to J^\infty(E)$ is the induced map, then $(\Phi^\infty)^*$ maps $H_j^i(\pi)$, $G_n^i(\pi)$, $G_n^{00}(\pi)$ all into themselves.

(c) If X is an integrable vectorfield then the Lie derivative L_X maps $H_j^i(\pi)$, $G_n^i(\pi)$, $G_n^{00}(\pi)$ all into themselves.

(d) The d-operator maps $G_n^i(\pi)$ to $G_n^{i+1}(\pi)$.

(e) If X is a deformation, then the Lie derivative L_X maps $H_j^i(\pi)$, $G_n^i(\pi)$, $G_n^{00}(\pi)$ all into themselves.

(f) If X is a total vectorfield, then the Lie derivative L_X maps $H_n^i(\pi)$ into $G_n^i(\pi)$ and $H_n^0(\pi)$ into $G_n^{00}(\pi)$.

Proof. (a). We observed already that for a deformation X, $\iota_X\overline{H}_j^i(\pi) \subset \overline{H}_j^{i-1}(\pi)$ hence also $\iota_X H_j^i(\pi) \subset H_j^{i-1}(\pi)$. $\iota_X G_n^i(\pi) \subset G_n^{i-1}(\pi)$ and $\iota_X G_n^1(\pi) \subset G_n^{00}(\pi)$ follow from definition (3,4) if one allows (see remark (3,5)) the vectorfields $X_0^1,\ldots,X_0^i,X_1^1,\ldots,X_1^i$ to be deformations.

(b) This follows from the fact that Φ, $(\Phi^\infty)_*$ preserve images of sections, resp., vertical integrable vectorfields.

(c) Let $\overline{X}$ be the symmetry corresponding to X. If ω is a differential form on $J^\infty(\pi)$ then $L_X\omega = \lim_{t\to 0} \frac{1}{t}(-(D_{\overline{X},t}^\infty)^*\omega + \omega)$ (where $D_{\overline{X},t}$ may only be locally defined). Observe that $D_{\overline{X},t}\colon E \to E$ is a diffeomorphism like Φ in (b). This implies (c).

(d) We shall prove this by induction with respect to i. Let $\omega \in G_n^0(\pi)$, X a vertical integrable vectorfield, A a bounded, oriented, open set in W and s a section, defined on a neighbourhood of $\overline{A}$. We have to show that $\int_A (\iota_X d\omega) \circ s$ depends only on the germs of s and X along ∂A and $(j^\infty(s))\partial A$ respectively. Observe that $\iota_X d\omega = L_X\omega$ and hence that $\int_A (\iota_X d\omega) \circ s = \frac{d}{dt}\int_A \omega \circ s_t \big|_{t=0}$ where s_t is the section such that $D_{\overline{X},t}\{\mathrm{Im}(s)\} = \{\mathrm{Im}\, s_t\}$; $\overline{X}$ is the symmetry corresponding to X. The fact that $d\omega \in G_n^1(\pi)$ now follows from the fact that $\frac{d}{dt}\int_A \omega \circ s_t$ depends only on the germ of s_t along ∂A and hence only on the germs of s and X along ∂A and $(j^\infty(s))(\partial A)$ respectively.

Now we come to the induction. Suppose the assertion holds for $G_n^{p-1}(\pi)$ and take $\omega \in G_n^p(\pi)$. Let $A \subset W$ be an open bounded oriented subset, s a section defined on a neighbourhood of $\overline{A}$ and $X_0,\ldots,X_p$ vertical integrable vectorfields with X_p zero on a neighbourhood of $(j^\infty(s))(\partial A)$. We have to show that $\int_A (\iota_{X_0,\ldots,X_p} d\omega) \circ s = 0$ (see remark

(3,5)). We know that $\iota_{X_0} d\omega = L_{X_0}\omega - d\iota_{X_0}\omega$. Hence, by (c) and the induction hypothesis, $L_{X_0}\omega - d\iota_{X_0}\omega \in G_n^p(\pi)$ and hence also $\iota_{X_0} d\omega$. By (3,5) we now have $\int_A (\iota_{X_1,\ldots,X_p}\iota_{X_0} d\omega) \circ s = 0$ and hence $\int_A (\iota_{X_0,\ldots,X_p}\omega) \circ s = 0$.

(e) Here we use that $L_X = \iota_X d + d\iota_X$. Then, if X is a deformation, the statement follows from (a) and (d).

(f) Let $\omega \in H_n^i(\pi)$, $A \subset W$ an oriented bounded open subset, s a section, defined on a neighbourhood of $\bar{A}$ and $X_1,\ldots,X_i$ vertical integrable vectorfields. Then, because $[X,X_i] \equiv 0$, see (2,16) formula (1),
$\int_A (\iota_{X_1,\ldots,X_i}(L_X\omega)) \circ s = \int_A (L_X(\iota_{X_1,\ldots,X_i}\omega)) \circ s = \int_A L_{\pi(X)}((\iota_{X_1,\ldots,X_i}\omega) \circ s)$
$= \int_A d\iota_{\pi(X)}[(\iota_{X_1,\ldots,X_i}\omega) \circ s]$. The value of this last integral clearly depends only on the germs of $\pi(X)$ and a along ∂A and the germs of $X_1,\ldots,X_i,\omega$ along $(j^\infty(s))(\partial A)$. This implies that $L_X\omega \in G_n^i(\pi)$ and, for $i = 0$, that $L_X\omega \in G_n^{00}(\pi)$.

This concludes the proofs.

Theorem (3,7). (Local exactness of G and H).

Let $U \subset E$ be an open subset such that there is a smooth 1-parameter family of maps $\Phi_t\colon E \to E$, $t \in [0,1]$, such that:

1. $\pi \circ \Phi_t = \pi$ for all $t \in [0,1]$;
2. Φ_1 = identity and Φ_t, for $t > 0$ is a diffeomorphism;
3. $\Phi_t(U) \subset U$ for all t;
4. $\Phi_0(U) = s(\pi(U))$ for some section s defined on a neighbourhood of $\overline{\pi(U)}$.

Then for each $\omega \in H_j^i(\pi)$, resp. $\in G_n^i(\pi)$, with $i \geq 1$, such that $d\omega \mid (\pi_\infty^0)^{-1}(U) = 0$, there is an $\eta \in H_j^{i-1}(\pi)$, resp. $\in G_n^{i-1}(\pi)$, such that on $(\pi_\infty^0)^{-1}(U)$, $\omega = d\eta$.

Proof. We define $\omega_t = (\Phi_t^\infty)^*\omega$ and have $\omega = \omega_1 = \int_0^1 \frac{\partial\omega_t}{\partial t}\,dt = \lim_{\tau\downarrow 0}\int_\tau^1 (\Phi_t^\infty)^*(\iota_{X_t} d\omega + d\iota_{X_t}\omega)dt$, where X_t is the integrable vectorfield as defined in (2,17). On U this last expression takes the form $\omega = \lim_{\tau\downarrow 0}\int_\tau^1 (\Phi_t^\infty)^* d\iota_{X_t}\omega\,dt = \lim_{\tau\downarrow 0} d(\int_\tau^1 (\Phi_t^\infty)^*\iota_{X_t}\omega\,dt$. All limits involved convergence nicely so we can take $\eta = \lim_{\tau\downarrow}\int_\tau (\Phi_t^\infty)^*\iota_{X_t}\omega\,dt$ and have on $(\pi_\infty^0)^{-1}(U)$ that $d\eta = \omega$. By (3,6) η is in $H_j^{i-1}(\pi)$, resp. $G_n^{i-1}(\pi)$.

Theorem (3,8). Let $F(\pi) \subset H_n^1(\pi)$ be the set of those forms in $H_n^1(\pi)$ which are of type $(0,k)$ for some k. Then $H_n^1(\pi) = G_n^1(\pi) \oplus F(\pi)$; the canonical projection $H_n^1(\pi) \to F(\pi)$ is denoted by Π.

Proof. It is easy to see that $G_n^1(\pi) \cap F(\pi) = 0$. We shall now construct the projection $\Pi\colon H_n^1(\pi) \to F(\pi)$ whose kernel is $G_n^1(\pi)$. Since this projection, if it exists, is unique, and even locally unique, it suffices to construct Π within some local coordinate neighbourhood.

Consider a neighbourhood $U \subset E$ which can be considered as the product of a coordinate neighbourhood in the base-space W, with coordinates $x_1,\ldots,x_n$, and a coordinate neighbourhood in the fibre, with coordinates $y_1,\dots,y_m$; $x_1,\dots,x_n,y_1,\dots,y_m$ are considered as coordinates in U. Let ω be any element of $H^1_n(\pi)$, $s: W \to E$ a (local) section of π (mapping the $x_1,\dots,x_n$ coordinate neighbourhood into U). Then for any vertical symmetry $\overline{X}$, which has on U the form $\sum_{j=1}^{m} \overline{X}_j(x_1,\dots,x_n,y_1,\dots,y_m)\frac{\partial}{\partial y_j}$, and associated vertical integrable X, $(\iota_X\ )\circ s$ has the form

$$\sum_{\substack{I\in\Lambda_{n,k}\\ j=1,\dots,m}} A^I_j(x)\ .\ \partial_I(\overline{X}_j(x,s(x)))dx_1\wedge\dots\wedge dx_n$$

where: $\Lambda_{n,k} = \{(i_1,\dots,i_l)\mid 1\le i_1\le i_2\le\dots\le i_l\le n,\ l\le k\}$,

$\partial_{(i_1,\dots,i_l)} = \dfrac{d^l}{dx_{i_1}\,dx_{i_2}\dots dx_{i_l}}$, for $(i_1,\dots,i_l)\in\Lambda_{n,k}$, $|(i_1,\dots,i_l)|$ is l.

The functions A^I_j depend on ω and on the section s but not on X, (or $\overline{X}$). Now we define $\Pi\omega$ by the requirement that, in the above notation, $(\iota_X(\Pi\omega))\circ s$ is

$$\sum_{\substack{I\in\Lambda_{n,k}\\ j=1,\dots,m}} (-1)^{|I|}(\partial_I(A^I_j(x)))\ .\ X_j(x,s(x))\ dx_1\wedge\dots\wedge dx_n.$$

By theorem (3,2) this determines, at least on U, the differential form $\Pi\omega$. It is clear that $\omega - \Pi\omega \in G^1_n(\pi)$ and that if $\omega\in F(\pi)$, $\Pi\omega = \omega$. So Π has the required properties. Since Π is, even locally, unique we have the global existence. This proves the theorem.

4. Some general theorems; applications of differential forms on jet bundles to variational calculus.

Also in this section $\pi: E\to W$ denotes a differentiable fibration. We observe that a variational problem for sections of π, in the sense of the introduction, is given by some $L\in H^0_n(\pi)$, $n = \dim(W)$; we call L the Lagrangian of the variational problem. A (local) section s is a solution (or extremal) of such a variational problem if for any bounded open oriented $A\subset W$, whose closure is contained in the domain of s, and for any smooth 1-parameter family of (local) sections s_t, $t\in\mathbb{R}$, with

1. domain (s_t) = domain (s) for all t;
2. s and s_t are equal on a neighbourhood of the complement of A;

$\int_A L\circ s_t = \int_A L\circ s + O(t^2)$.

Notice that this notion of Lagrangian is the right extention of the one given in the introduction: instead of assigning to each k-jet of a section a real number, we assign to each k-jet of a section in $w\in W$, an element of $\Lambda^n(T^*_w(W))$ so that we can still integrate.

This definition of extremal suggests the consideration of the form $E = \Pi dL\in F(\pi)$ which has a remarkable property:

Theorem (4,1) (Euler). For $L \in H^0_n(\pi)$ a Lagrangian, $E = \Pi dL$ is called its Euler form. If s: W → E is a local section, A an open bounded oriented subset of the domain of s and if s_t is a smooth 1-parameter family of sections as above, then

$$\int_A L \circ s_t = \int_A L \circ s + t \cdot \int_A (\iota_X E) \circ s + O(t^2).$$

X is a vertical integrable vectorfield with corresponding symmetry $\overline{X}$ such that for each w in the domain of s, $\overline{X}(s(w))$ is the tangent vector of $t \to s_t(w)$.

As a consequence, s is an extremal if and only if $(\iota_X E) \circ s$ is identically zero for each vertical integrable X (this is indeed necessary and sufficient because $E \in F(\pi)$); this criterion for extremals is called the Euler equation. For this reason E can also be called the "Euler equation" of L.

Proof. It is clear that with s, s_t, X as above, $\int_A L \circ s_t = \int_A L \circ s + t\int_A (L_X L) \circ s + O(t^2)$. So we have to prove that $\int_A (L_X L) \circ s = \int_A (\iota_X E) \circ s$. We know that $L_X L = \iota_X dL$, so $(L_X L - \iota_X E) = \iota_X \omega$ for some $\omega \in G^1_n(\pi)$. X is zero in a neighbourhood of $(j^\infty(s))(\partial A)$. Hence $\int_A (\iota_X \omega) \circ s = 0 = \int_A (L_X L - \iota_X E) \circ s$.

Definition (4,2). Let $\overline{X}$ be a symmetry of π and X the corresponding integrable vectorfield. $\overline{X}$ is said to be a symmetry of the Lagrangian $L \in H^0_n(\pi)$ if $L_X L \in G^{00}_n(\pi)$ and it is called a local symmetry if $L_X L \in G^0_n(\pi)$.

Remark (4,3). We observe that two Lagrangians L_1 and L_2 have the same Euler equation if and only if $L_1 - L_2 \in G^0_n(\pi)$. In one direction this is clear, namely $L_1 - L_2 \in G^0_n(\pi)$ implies $\Pi dL_1 - \Pi dL_2 = 0$. To prove the statement in the other direction, we choose some bounded open and oriented $A \subset W$ and a 1-parameter family of sections s_t, defined on a neighbourhood of $\overline{A}$ and constant on a neighbourhood of the complement of A. Then

$$\int_A (L_1 - L_2) \circ s_1 - \int_A (L_1 - L_2) \circ s_0 = \int_0^1 [\frac{d}{dt} \int (L_1 - L_2) \circ s_t]dt =$$

$= \int_0^1 \iota_{X_t} (\Pi dL_1 - \Pi dL_2) \circ s_t$, where the vertical integrable vectorfield X_t is as in theorem (4,1). If we assume that the Euler equations of L_1 and L_2 are equal, then this last integral is zero and hence $L_1 - L_2 \in G^0_n(\pi)$.

So we see that $\overline{X}$ is a local symmetry of L if and only if $D^*_{X,t}(E) = E$ for all t, $E = \Pi dL$. The adjective "local" only refers to the fact that the condition can be locally verified.

Definition (4,4). A source equation is a form in $F(\pi)$. Such a form assigns to a section s and a point x in its domain a covector in $(T_{s(x)}(\pi^{-1}(x)))^*$ and hence is a straightforward generaliztion of the corresponding notion in the introduction. Let E be such a source equation and let $\overline{X}$ be a symmetry of π, with corresponding integrable vectorfield X and evolution vectorfield X_{ev}. We say that $\overline{X}$ is a symmetry of E if $L_X E = 0$. E is said to have local $\overline{X}$-conservation if $L_{X_{ev}} E \in G^0_n(\pi)$ and $\overline{X}$-conservation if $\iota_{X_{ev}} E \in G^{00}_n(\pi)$.

Remark (4,5). Usually a conservation law of a source equation E is supposed to be an element $\Omega \in H^0_{n-1}(\pi)$ such that for each section s: W → E, with $(\iota_X E) \circ s$ zero for each vertical integrable X, $d(\Omega \circ s) = 0$.

Indeed, if W is 1-dimensional, Ω is a function on $J^\infty(\pi)$ (element of $H_0^0(\pi)$) which is then constant along sections of the above type. If W is 4-dimensional space time, and s is a section as above then $\Omega \circ s = \sum_{i=0}^{3}(-1)^i \Omega_i(x_0,\dots,x_3)dx_0 \wedge \dots \wedge \widehat{dx_i} \wedge \dots \wedge dx_3$ (the functions Ω_i depend of course also on s). $\Omega_0(x_0,\dots,x_3)$ is then the "density" of the conserved quantity at time x_0 in (x_1,x_2,x_3) and $\Omega_i(x_0,\dots,x_3)$ is the transport density of the conserved quantity in the x_i direction. $d(\Omega \circ s) = 0$ is equivalent with $\sum_{i=0}^{3} \frac{\partial \Omega_i}{\partial x_i} = 0$ which is the continuity equation.

In order to relate our conservation laws with the usual ones we wish to have, for each E in $H_n^1(\pi)$ which has $\overline{X}$ conservation, an element $\Omega \in H_{n-1}^0(\pi)$ such that for each (local) section s, $d(\Omega \circ s) = (\iota_{X_{ev}} E) \circ s$. This requirement is of course equivalent with $D(\Omega) = \iota_{X_{ev}} E$; see the definition in remarks (3,3) sub. 2. It is easy to see that for $\Omega \in H_{n-1}^0(\pi)$, $D(\Omega) \in G_n^{00}(\pi)$. By Horndesci [5], if $\omega \in G_n^0(\pi)$ there is <u>locally</u> an $\Omega \in H_{n-1}^0$ such that $D\Omega = \omega$. Note that in Horndesci's paper the condition $\omega \in G_n^0(\pi)$ is expressed as: ω is a Lagrangian with vanishing Euler equation; according to remark (4,3) this is really equivalent with our assumption. As to the global situation we have the following

<u>Conjecture (4,6).</u> For each $\omega \in G_n^{00}(\pi)$ there is some $\Omega \in H_{n-1}^0(\pi)$ with $D(\Omega) = \omega$. Also $G_n^0(\pi) / G_n^{00}(\pi)$ is canonically isomorphic with $H^n(E;\mathbb{R})$.

<u>Theorem (4,7).</u> Let $L \in H_n^0(\pi)$ be a Lagrangian, let $\overline{X}$ be a symmetry of π and let the source equation E be the Euler form ΠdL of L. Then the following three conditions are equivalent:

(a) $\overline{X}$ is a local symmetry of L;

(b) $\overline{X}$ is a symmetry of E;

(c) E has local $\overline{X}$-conservation.

Also the following two conditions are equivalent:

(α) $\overline{X}$ is a symmetry of L;

(β) E has $\overline{X}$-conservation

<u>Proof.</u> For (a) $\Leftrightarrow$ (b) see remark (4,3). We now prove (a) $\Leftrightarrow$ (c). By theorem (3,6,f) and remark (2,9), $L_X L \in G_n^0(\pi)$ if and only if $L_{X_{ev}} L \in G_n^0(\pi)$; X, X_{ev} is the integrable, resp. evolution, vectorfields corresponding to $\overline{X}$. $L_{X_{ev}} L = \iota_{X_{ev}} dL = \iota_{X_{ev}} E + \iota_{X_{ev}} (dL - E)$. $dL - E \in G_n^1$ so $L_X L \in G_n^0(\pi)$ if and only if $\iota_{X_{ev}} E \in G_n^0(\pi)$. This proves (a) $\Leftrightarrow$ (c).

Finally we have to prove (α) $\Leftrightarrow$ (β). This proof can be obtained from the above proof by replacing $G_n^0(\pi)$ everywhere by $G_n^{00}(\pi)$.

<u>Theorem (4,8) (Tonti).</u> Let $E \in F(\pi)$ be a source equation with $dE \in G_n^2(\pi)$ and let $U \subset E$ be an open subset as in theorem (3,7) (if the fibre of π is a vectorspace we may take $U = E$). Then there is $L \in H_n^0(\pi)$ such that on $(\pi_\infty^0)^{-1}(U)$, $\Pi dL = E$.

<u>Remark (4,9).</u> Note that if there exists such L with $\Pi dL = E$, then $dE = d(\Pi dL) = d(\Pi dL - dL)$. $\Pi dL - dL \in G_n^1(\pi)$ and hence $dE \in G_n^2(\pi)$. So $dE \in G_n^2(\pi)$ is a necessary and sufficient condition for E to be locally the Euler equation of a variational problem. If a source equation $E \in F(\pi)$ is such that $dE \in G_n^2(\pi)$, we call it <u>locally variational</u>; we call it <u>variational</u> if there is some $L \in H_n^0(\pi)$ with $\Pi dL = E$. Concerning the problem

whether locally variational implies variational, we have the following.

Conjecture (4,10). If $H^{n+1}(E;\mathbb{R}) = 0$ then every source equation which is locally variational is variational.

Proof of theorem (4,8). On U, or rahter on $(\pi^0_\infty)^{-1}(U)$, we may use the exactness theorem (3,7). So in $(\pi^0_\infty)^{-1}(U)$ we have $d(dE) = 0$ and $dE = G^1_n(\pi)$, so there is a $K \in G^1_n(\pi)$ with $dK = dE$. Hence $d(E - K) = 0$ and there is $L \in H^0_n(\pi)$ with $dL = E - K$ (everything on $(\pi^0_\infty)^{-1}(U)$). But then, since $K \in G^1_n(\pi)$, $E = \Pi dL$.

Theorem (4,11). Let $E \in F(\pi)$ be a source equation and let $\overline{X}$ be a symmetry of π with X, X_{ev} the corresponding integrable, resp. evolution vectorfield. If two of the following three statements hold, then the third holds:

(a) $L_X E = 0$;

(b) $\iota_{X_{ev}} E \in G^0_n(\pi)$;

(c) $\iota_{X_{ev}}(dE) \in G^1_n(\pi)$.

Proof. Because $L_X E \in F(\pi)$, $L_X E = 0$ is equivalent with $L_X E \in G^1_n(\pi)$. So, due to (3,6,f) $L_X E = 0$ is equivalent with $L_{X_{ev}} E = \iota_{X_{ev}} dE + d\iota_{X_{ev}} E \in G^1_n(\pi)$. This, together with the fact that for $\omega \in H^0_n(\pi)$, $d\omega \in G^1_n(\pi)$ if and only if $\omega \in G^0_n(\pi)$ (see (4,3)), proves the theorem.

Remark (4,12). One should observe the meaning of the conditions (a), (b) and (c) in (4,11). (a) is equivalent with: $\overline{X}$ is a symmetry of E, (b) is equivalent with: E has local $\overline{X}$-conservation, and condition (c) is part of the requirement that E is locally variational. Namely E is locally variational if for each deformation $\widetilde{X}$, $\iota_{\widetilde{X}} dE \in G^1_n(\pi)$, see (4,8), (3,4) and (3,5). In § 5 we shall show, in some cases, that if $\iota_{X_{ev}}(dE) \in G^1_n(\pi)$ for a few symmetries and if also some other qualitative requirements hold for E then $dE \in G^2_n(\pi)$.

We observe that if $\overline{X}^1$ and $\overline{X}^2$ are symmetries of π, then, if some source equation E is $\overline{X}^1$ and $\overline{X}^2$ symmetric, it is $[\overline{X}^1,\overline{X}^2]$-symmetric. A corresponding result holds for constervation:

Lemma (4,13). Let $\overline{X}^1$, $\overline{X}^2$ be symmetries of π. If some source equations $E \in F(\pi)$ is $\overline{X}^1$ and $\overline{X}^2$ symmetric and has local $\overline{X}^1$ and $\overline{X}^2$ conservation, then E is $[\overline{X}^1,\overline{X}^2]$-symmetric and has local $[\overline{X}^1,\overline{X}^2]$-conservation.

Proof. Let $X^1, X^2, X^1_{ev}, X^2_{ev}$, and X^1_{tot}, X^2_{tot} be the corresponding integrable, evolution and total vectorfields. Then (2,9) $X^i = X^i_{tot} - X^i_{ev}$. Corresponding to $[\overline{X}^1,\overline{X}^2]$ we have $[X^1,X^2] = [X^1_{tot} - X^1_{ev}, X^2_{tot} - X^2_{ev}] = [X^1_{tot},X^2_{tot}] + [X^1_{ev},X^2_{ev}]$ (since $[X^i_{ev},X^j_{tot}]$ is zero; the proof of this is analogous to the proof of (2,16) formula (1)). Hence the evolution vectorfield corresponding to $[\overline{X}^1,\overline{X}^2]$ is $-[X^1_{ev},X^2_{ev}]$; we have to show that $\iota_{[X^1_{ev},X^2_{ev}]} E \in G^0_n(\pi)$. This follows from $[X^1_{ev},X^2_{ev}] = -[X^1,X^2_{ev}] = L_{-X^1} X^2_{ev}$, so $\iota_{[X^1_{ev},X^2_{ev}]} E = L_{-X^1}(\iota_{X^2_{ev}} E) + \iota_{X^2_{ev}}(L_{X^1} E) = L_{-X^1}(\iota_{X^2_{ev}} E)$. $\iota_{X^2_{ev}} E \in G^1_n(\pi)$ and hence also $L_{-X^1}(\iota_{X^2_{ev}} E) \in G^1_n(\pi)$.

Finally it may be observed that many of the results in this section are also true in the following more general, context. Let X be a deformation. X is a symmetry of E if $L_X E = 0$, E has local X conservation if $\iota_X E \in G^0_n(\pi)$, $L \in H^0_n(\pi)$ has local X conservation if $L_X L \in G^0_n(\pi)$.

5. The main results.

In this section we want to show, in some cases, that symmetry and conservation may imply a source equation to be locally variational. Is $\overline{X}$ is any symmetry of π and if X_{ev} is the corresponding evolution field, then $\overline{X}$-conservation and $\overline{X}$-symmetry for a source equation E implies by theorem (4,11) that $\iota_{X_{ev}}(dE) \in G^1_n(\pi)$ and hence $\Pi(\iota_{X_{ev}}(dE)) = 0$. If we can prove that for any deformation Z, $\Pi(\iota_Z(dE)) = 0$, then we know that $dE \in G^2_n(\pi)$ and hence that E is locally variational.

In the following proofs we shall proceed as follows. We start with a source equation E and derive a formula, in local coordinates, for $(\iota_{Z_2}\Pi\iota_{Z_1} dE) \circ s$ where Z_1, Z_1 are vertical integrable and s is a section (since we work "along a fixed section" it is enough to consider vertical integrable vectorfields). The hypothesis (symmetry and conservation) imply that for certain Z_1 and all Z_2 (namely for $Z_1 \mid Im(s)$ = evolution of symmetry) the above expression is zero. We shall then prove that the expression is zero for any Z_1, Z_2, s which is equivalent with $\Pi\iota_Z dE = 0$ for all deformations Z.

Definition (5,1). Let E,W be finite dimensional vectorspaces and π: E → W a linear projection. (In this case we say that the bundle is linear). $E \in F(\pi)$ is called linear if the coefficient functions of E, as an (n + 1)-form on $J^k(\pi)$ (for some k) are linear. In other words, for $x_1, \ldots, x_n$ linear coordinates on W and $x_1\pi, \ldots, x_n\pi, y_1 \ldots, y_m$ linear coordinates on E, $E = \sum_{i=1}^{m} E_i dy_i \wedge dx_1 \wedge \ldots \wedge dx_n$ is linear if and only if $E_1, \ldots, E_m$ are linear functions on $J^k(\pi)$.

Theorem (5,2). Let π be a linear bundle and let $E \in F(\pi)$ be a linear source equation on π. Let $\overline{X}$ be a symmetry of π with $\pi(X)$ not identically zero. Then, if E has $\overline{X}$-symmetry and $\overline{X}$-conservation, E is variational.

Proof. We use x and y coordinates as in the above definition. Let $y_j = y_j(x_1, \ldots, x_n)$ j = 1,...,m be a section and let $\overline{Z}_1, \overline{Z}_2$ be two symmetries of π of the form $\overline{Z}_i = \sum_{j=1}^{m} \overline{Z}_{i,j}(x_1, \ldots, x_n)\frac{\partial}{\partial y_j}$; since we work along one section it is enough to consider the vertical integrable vectorfields corresponding to symmetries of the form Z_1, Z_2. Clearly $[\overline{Z}_1, \overline{Z}_2] = 0$ and hence $[Z_1, Z_2] = 0$; Z_i is the integrable vectorfield corresponding to $\overline{Z}_i$. Hence $(\iota_{Z_1 Z_2}(dE)) \circ s = \frac{d}{d\varepsilon}\{(\iota_{Z_2} E) \circ (s + \varepsilon\overline{Z}_1) - (\iota_{Z_1} E) \circ (s + \varepsilon\overline{Z}_2)\}\big|_{\varepsilon=0}$, where s denotes the section $(x_1, \ldots, x_n) \to (y_1(x_1, \ldots, x_n), \ldots, y_m(x_1, \ldots, x_n))$ introduced before and $s + \varepsilon Z_i$ the section $(x_1, \ldots, x_n) \to (y_1(x_1, \ldots, x_n) + \varepsilon . \overline{Z}_{i,1}(x_1, \ldots, x_n), \ldots, y_m(x_1, \ldots, x_n) + \varepsilon . \overline{Z}_{i,m}(x_1, \ldots, x_n))$. Because of the linearity of E, this last expression is independent of s and linear in $\overline{Z}_1, \overline{Z}_2$. If we assume that E is of type (k,k), we can write $(\iota_{Z_1, Z_2} dE) \circ s = \sum_{i=0}^{k} \{E_i^\alpha . d^i\overline{Z}_1 . \overline{Z}_2 - E_i^\beta d^i\overline{Z}_2 . \overline{Z}_1$, where $E_i^\alpha . d^i\overline{Z}_1 . \overline{Z}_2$ just stands for an expression which is linear in the ith derivatives of $\overline{Z}_1$ and the 0th derivatives of $\overline{Z}_2$. Hence, with the same notation we can write

$$(\iota_{Z_2}\Pi\iota_{Z_1} dE) \circ s = \sum_{i=0}^{k} E_i^\gamma . d^i\overline{Z}_1 . \overline{Z}_2 \quad \ldots\ldots\ldots \tag{1}$$

From the assumptions it follows that (1) is zero if $\overline{Z}_1$ is choosen so that, along Im s, $\overline{Z}_1 = X_{ev}$. Since (1) however is independent of s and since we can find for each $\overline{Z}_1$ a local section s such that along Im(s), $\overline{Z}_1 = X_{ev}$ (at least in $\pi^{-1}\{x \in W \mid \pi(\overline{X}) \neq 0\}$ = $V \subset E$), formula (1) gives zero for any $\overline{Z}_1, \overline{Z}_2$. Since the expression is invariant under translation, it gives zero everywhere. Hence E is locally variational. Since the fibre of π is a vectorspace, E is variational.

Theorem (5,3). Let π: $E \to W$ be a linear fibration with $\dim(E) = \dim(W) + 1$. Let X be a Lie algebra of symmetries of π such that for each constant vectorfield Z on W, there is an $\overline{X}$ in X with $\pi(\overline{X}) = Z$. Let E be a source equation on π of order ≤ 2, i.e. as differential form has type (0,2). If E is $\overline{X}$-invariant and has $\overline{X}$-symmetry for each $X \in X$, then E is variational.

Proof. We introduce the following coordinates.

$x_1,\ldots,x_n$ are linear coordinates on W;

$x_1,\ldots,x_n,y$ are linear coordinates on E, we identify x_i with $x_i\pi$;

$x_1,\ldots,x_n,y,y_1,\ldots,y_n$ are linear coordinates on $J^1(\pi)$, the 1-jet $j^1(s)(x)$ has coordinates $x = (x_1,\ldots,x_n)$, $s(x)$, $\frac{\partial s}{\partial x_1}(x),\ldots,\frac{\partial s}{\partial x_n}(x)$;

$x_1,\ldots,x_n,y,y_1,\ldots,y_n,y_{11},\ldots,y_{ij},\ldots,y_{nm}$ $i \leq j$ are linear coordinates on $J^2(\pi)$, y_{ij} has in $j^2(s)(x)$ the value $\frac{\partial^2 s}{\partial x_i \partial x_j}(x)$.

We may write E as $E(x_1,\ldots,x_n,y,y_1,\ldots,y_n,y_{11},\ldots,y_{nn})\ dy \wedge dx_1 \wedge \ldots \wedge dx_n$. Without loss of generality we may assume that X containes the vectorfields $\frac{\partial}{\partial x_1},\ldots,\frac{\partial}{\partial x_n}$ (if not one has to apply a nonlinear transformation $(x_1,\ldots,x_n,y) \to (x_1,\ldots,x_n,Y(x_1,\ldots,x_n,y))$. Hence we may assume that E does not depend on $x_1,\ldots,x_n$. Next we choose vertical symmetries $\overline{Z}_1,\overline{Z}_2$: $\overline{Z}_i = \overline{Z}_i(x_1,\ldots,x_n)\frac{\partial}{\partial y}$ and a section s. By a straightforward calculation we find:

$$(\iota_{Z_2}\Pi\iota_{Z_1} dE) \circ s(x) = \overline{Z}_2(x) \cdot J(s,\overline{Z}_1,x) \cdot dx_1 \wedge \ldots \wedge dx_n \text{ with}$$

$$J(s,\overline{Z}_1,x) = 2 \sum_{i=1}^{n} \{ \frac{\partial E}{\partial y_i}(j^2(s)(x)) - \tfrac{1}{2} \frac{d}{dx_i}(\frac{\partial E}{\partial y_{ii}}(j^2(s)(x)) -\tfrac{1}{2} \sum_{j=1}^{n} \frac{d}{dx_j}(\frac{\partial E}{\partial y_{ij}}(j^2(s)(x)))\} \frac{\partial \overline{Z}_1}{\partial x_i}(x) +$$
$$\sum_{i=1}^{n} \frac{d}{dx_i} \{\frac{\partial E}{\partial y_i}(j^2(s)(x)) - \tfrac{1}{2} \frac{d}{dx_i}(\frac{\partial E}{\partial y_{ii}}(j^2(s)(x)) - \tfrac{1}{2} \sum_{j=1}^{n} \frac{d}{dx_j} \frac{\partial E}{\partial y_{ij}}(j^2(s)(x)))\} \cdot \overline{Z}_1(x).$$

Where, for $i \leq j$, $\frac{\partial}{\partial y_{ji}}$ is by defintion the same as $\frac{\partial}{\partial y_{ij}}$ and where $\frac{d}{dx_i}$ denotes the total derivative with respect to x_i or, what is the same, it denotes the differential operator defined by the total vectorfield (see (2,7)) with projection $\frac{\partial}{\partial x_i}$. $J(s,\overline{Z}_1,x)$ gets a simpeler expression if we define $f_i(s,x) = \frac{\partial E}{\partial y_i}(j^2(s)(x)) - \tfrac{1}{2}\frac{d}{dx_i}\frac{\partial E}{\partial y_{ii}}(j^2(s)(x)) - \tfrac{1}{2}\sum_{j=1}^{n}\frac{d}{dx_j}(\frac{\partial E}{\partial y_{ij}}(j^2(s)(x)))$; so that $f_i(s,x)$ depends on x and the 3-jet of s in x (and hence can be considered as function on $J^3(\pi)$). Now $J(s,\overline{Z}_1,x) = 2\sum_{i=1}^{n} f_i(s,x) \cdot \frac{\partial \overline{Z}_1}{\partial x_i}(x) + \sum_{i=1}^{n}(\frac{d}{dx_i} f_i(s,x)) \cdot \overline{Z}_1(x)$.

To obtain this nice form for $J(s,\overline{Z}_1,x)$ it was essential to have the fibre of π one dimensional and the order of E to be ≤ 2.

From the assumptions we know that for $\overline{Z}_1(x) = \frac{\partial s}{\partial x_j}(x)$, $J(s,\overline{Z}_1,x) = 0$ and hence

$$2\sum_{i=1}^{n} f_i(s,x) \cdot \frac{\partial^2 s}{\partial x_i \partial x_j}(x) + \sum_{i=1}^{n}\frac{d}{dx_i}(f_i(s,x)) + \frac{\partial s}{\partial x_j}(x) = 0 \tag{2}$$

for all s and all $j = 1,\ldots,n$. If we can conclude from this that $f_i \equiv 0$ we are done.

To prove this last fact we proceed as follows:

case 1. Let s be a section such that on $U \subset W$, $\left(\frac{\partial^2 s}{\partial x_i \partial x_j}\right)_{i,j}(x)$ is everywhere invertible as a matrix; let $A^s_{ij}(x)$ be the inverse, and define $F^s(x)$ as $\sum_{i=1}^{n} \frac{d}{dx_i}(f_i(s,x))$. Then we conclude from (2) that $2f_i(s,x) + F^s(x) \cdot \frac{\partial s}{\partial x_j}(x) \cdot A^s_{ji} = 0$. Hence:
$2F^s(x) + n \cdot F^s(x) + \sum_{i,j}\{\frac{\partial s}{\partial x_j}(x)A^s_{ji}(x) \cdot \frac{\partial F^s}{\partial x_i}(x) + \frac{\partial s}{\partial x_j}(x) \cdot \frac{\partial A^s_{ji}}{\partial x_i}(x) \cdot F^s(x)\} = 0.$
This means that of F^s is zero on some open subset and if γ is an integral curve of the vectorfield $\sum_{i,j=1}^{n} \frac{\partial s}{\partial x_j}(x) \cdot A^s_{ji}(x) \cdot \frac{\partial}{\partial x_i}$ which passes through this open subset, then F^s is zero along γ. Also we see from the above equation that if $F^s(x)$ is zero then also $f_i(s,x)$

case 2. Choose $s(x_1,\ldots,x_n) = c_0 + \sum_{i=1}^{n} \pm \frac{1}{2}(x_i - \overline{x}_i)^2$. One has $\frac{\partial^2 s}{\partial x_i \partial x_j} = \pm \delta_{ij}$ hence $A^s_{ij} = \pm \delta_{ij}$ so $2F^s + \sum_{j=1} \pm (x_j - \overline{x}_j) \cdot (\pm) \cdot \frac{\partial F^s}{\partial x_j} + n \cdot F^s = 0$ or $-(n + 2)F^s(x) = \Sigma(x_j - \overline{x}_j)\frac{\partial F^s}{\partial x_j}(x)$. This implies that $F^s \equiv 0$. in a neighbourhood of $(\overline{x}_1,\ldots,\overline{x}_n)$, and hence everywhere.

case 3. Now we take any x and any s but so that $\frac{\partial^2 s}{\partial x_i \partial x_j}(x)$ is invertible and $ds(x) \neq 0$. There is a section $\overline{s}$ as in case two with the same 2-jet in x. With a C^2 small pertubation, we obtain from $\overline{s}$ a section $\tilde{s}$ which agrees with s on a neighbourhood of x. The vectorfields, as defined in case 1, are C^0 close for the sections $\overline{s}$ and $\tilde{s}$. For $\overline{s}$ we have $F^{\overline{s}} \equiv 0$ and the above vectorfield for $\overline{s}$ is linear and has a source (away from x). Hence also for $\tilde{s}$ we have $F^{\tilde{s}} \equiv 0$. Hence near x $F^s \equiv 0$.

By continuity we conclude now that $F^s(x)$ and hence $f_i(s,x)$ is identically zero for all s and x. Hence E is locally variational; because the fibre of π is a vectorspace E is variational.

Theorem (5,4). Let W be an n-dimensional manifold and p,q nonnegative integers such that $p + q = n = \dim(W)$. $\pi: E \to W$ denotes the bundle whose fibre $\pi^{-1}(x)$ consists of the non-degenerate quadratic functions on $T_x(W)$ with signature (p,q) (i.e., positive definite on a p-dimensional subspace and negative definite on a q-dimensional subspace). Let X_W be the Lie algebra of all vectorfields on W. Each $X \in X_W$ induces a vectorfield $\overline{X}$ on E which is characterised by:

1. $\overline{X}$ is a symmetry of π with $\pi(\overline{X}) = X$;
2. If $s: W \to E$ is a section, i.e., s is a, possibly non-definite, metric, then the section corresponding to the metric $(D_{X,t})^*(s)$ has image $(D_{\overline{X},-t})(\mathrm{Im}(s))$.

We denote the Lie algebra of vectorfields on E, which can be obtained in this way from X_W, by X_E.

If $E \in F(\pi)$ is a source equation of order 2, which is $\overline{X}$-symmetric and has $\overline{X}$-conservation for each $\overline{X} \in X_E$, then E is locally variational.

Proof. Let $x_1,\ldots,x_n$ be a local coordinate system on W and let $x_1,\ldots,x_n,y_1,\ldots,y_m$ $(m = \frac{1}{2}n \cdot (n + 1))$ be a local coordinate system on $\pi^{-1}(U)$ such that on each fibre $\pi^{-1}(x)$, $y_1,\ldots,y_m$ is the restriction of a linear coordinate system on $T^*_x(W) \odot T^*_x(W) \supset \pi^{-1}(x)$; $\odot$ denotes the symmetric tensor product. Since sections of π are essentially pseudo-Riemannian metrics on W, we denote these sections by g. If g is such a section, $X \in X_W$, then the evolution field corresponding to X along the image of $j^\infty(g)$ "is just

$L_X g$", the Lie derivative of g with respect to X.

Let $\overline{Z}_1, \overline{Z}_2$ be vertical symmetries of π of the form $\overline{Z}_i(x_1,\ldots,x_n,y_1,\ldots,y_m) = \sum_{j=1}^{m} \overline{Z}_{i,j}(x_1,\ldots,x_n)\frac{\partial}{\partial y_j}$; such a $\overline{Z}_i$ may be considered as section of $T^*(W) \otimes T^*(W)$. As in the proof of (5,2), (5,3) we find that $\iota_{Z_2} \Pi \iota_{Z_1} dE \circ g = \sum_{j=1}^{m} \overline{Z}_{2,j}(x) \cdot J^j(x,g,Z_1) \cdot dx_1 \wedge \ldots \wedge dx_n$. A further calculation shows that each $J^j(x,g,\overline{Z}_1)$ can be written as the sum of two terms $J_1^j(x,g,\overline{Z}_1)$ and $J_1^j(x,g,\overline{Z}_1)$ with the following properties.

$J_1^j(x,g,\overline{Z}_1)$ depends on the 4-jet of g in x and on the 1-jet of $\overline{Z}_1$ in x; for g fixed $J_1^j(x,g,\overline{Z}_1)$ depends linearly on the 1-jet of $\overline{Z}_1$. $J_2^j(x,g,\overline{Z}_1)$ depends on the 2-jet of g and on the 2nd derivatives of $\overline{Z}_1$. For g fixed $J_2^j(x,g,\overline{Z}_1)$ depends linearly on the 2nd derivatives of $\overline{Z}_1$ (in each point).

We have to show that J_1^j and J_2^j are identically zero. From the assumptions it follows that if $\overline{Z} = L_X g$ (along im g) for some vectorfield X on W, then $J^j(x,g,\overline{Z}) = 0$. In order to prove that J_1^j and J_2^j are identically zero, we shall prove the following two facts:

<u>fact 1.</u> Let $x \in W$, $g: W \to E$ a section of π and $\overline{Z} = \sum_{j=1}^{n} \overline{Z}_j(x_1,\ldots,x_n)\frac{\partial}{\partial y_j}$ a symmetry of π with $j^1(\overline{Z})(x) = 0$. Then there is a section $\tilde{g}$, with the same 2-jet in x as g, and a vectorfield X on W such that $j^2(\overline{Z})(x) = j^2(L_X g)(x)$ (along image of g).

<u>fact 2.</u> Let $x \in W$, $g: W \to E$ a section and $Z = \sum_{j=1}^{n} \overline{Z}_j(x_1,\ldots,x_n)\frac{\partial}{\partial y_j}$ a symmetry of π. Then there is a vectorfield X on W such that $j^1(\overline{Z})(x) = j^1(L_X g)(x)$ (along the image of g).

<u>Proof of fact 1.</u> First we construct a vectorfield X on W with $X(x) \neq 0$ and such that the 1-jet of $L_X g$ in x is zero. For this we take a 1-parameter family of maps $\Theta_s: \mathbb{R}^n \to W$ $s \in \mathbb{R}$, defined in a neighbourhood of $0 \in \mathbb{R}^n$, such that $\Theta_0(0) = x$, $\frac{d}{ds}\Theta_s(0)|_{s=0} \neq 0$ and such that for each s, $\Theta_s^* g$ has the same 1-jet in 0. Such a family Θ_s can be obtained by choosing Θ_s so that it defines normal coordinates [6], i.e., such that each line $t \to \Theta_s(tx_1,\ldots,+x_n)$ is a geodesic, and such that $\Theta_s^*(g)$, for each s, has diagonal form in the origin with only $\pm$ 1 on the diagonal. We then define X (locally) by: X(x) is the tangent vector of the curve $x \to \Theta_s \circ \Theta_0^{-1}(x)$ at s = 0.

Next we take local coordinates $\tilde{x}_1,\ldots,\tilde{x}_n$ on W near x with respect to which X has the form $\frac{\partial}{\partial x_1}$. In these coordinates, partial differentiation with respect to $\tilde{x}_1$ and the Lie derivative L_X are the same. So now it is clear that by keéping the 2-jet of g fixed, but by changing the 3-jet of g, we can get for $L_X g$ any 2-jet whose 1-jet is zero. This proves fact 1.

<u>Proof of fact 2.</u> Here we need that for any section g of π, any $x \in W$ and any section $\overline{Z}$ of $T^*(W) \otimes T^*(W)$, there is a vectorfield X such that $L_X(g)$ and $\overline{Z}$ have the same 1-jet in x. This however is an easy consequence of the existence of normal coordinates.

Appendix.

We consider the linear bundle $\pi: \mathbb{R}^2 \to \mathbb{R}$, $\pi(x,y) = x$. On $J^2(\pi)$ we have coordinates $x,y,\dot{y},\ddot{y}$, defined on the obvious way. We consider the source equation $E = (\ddot{y} + 1)dy \wedge dx$. If x is interpreted as time and y as (1-dimensional) space coordinate then E corresponds to the notion of a mass-point with mass 1 on which a constant force is acting. E is the Euler equation of the Lagrangian $L = (-\frac{1}{2}\dot{y}^2 + y)dx$. E is invariant both under the x- and under the y-translation; hence we can also consider E on the bundles obtained by identifying x with x + 1 and/or y with y + 1. In this way we get simple examples of source equations E, E_1, E_2 and E_3 on the bundles $\pi: \mathbb{R}^2 \to \mathbb{R}$, $\pi_1: S^1 \times \mathbb{R} \to S^1$, $\pi_2: \mathbb{R} \times S^1 \to \mathbb{R}$ and $\pi_3: S^1 \times S^1 \to S^1$. They all look locally the same but, as we shall see, their global properties are quite different.

Since E is the Euler equation of L, $dE \in G_1^2(\pi)$; hence also $dE_i \in G_1^2(\pi_i)$. Since L is invariant under translation in the x-direction, also E_1 is the Euler equation of some Lagrangian. Next we observe that E is also the Euler equation of $\tilde{L} = (-\frac{1}{2}\dot{y}^2 - x\dot{y})dx$. $\tilde{L}$ is invariant under translation in the y-direction so also E_2 can be derived from a variational principle.

Next, because E, E_1, E_2 and E_3 are invariant under translation in the x-direction we check whether the corresponding conservation law holds. The evolution vectorfield corresponding to this symmetry is $\dot{y}\frac{\partial}{\partial y} + \ddot{y}\frac{\partial}{\partial \dot{y}} + \dots = X_{ev}$. So $\iota_{X_{ev}} E = (\ddot{y} + 1)\dot{y}dx$ (and the analogue for E_1, E_2 and E_3). In order to decide whether $\iota_{X_{ev}} E$ is in $G_1^0(\pi)$ or in $G_1^{00}(\pi)$ we calculate

$$\int_{[x_1,x_2]} (\iota_{X_{ev}} E) \circ s = \int_{x_1}^{x_2} (s''(x) + 1)s'(x)dx = \tfrac{1}{2}(s'(x))^2 + s(x) \Big|_{x_1}^{x_2} .$$

This means that $\iota_{X_{ev}} E$ and also $\iota_{X_{ev}} E_1$ are in $G_1^{00}(\pi)$, resp. $G_1^{00}(\pi_1)$. Since, in the bundles π_2 and π_3 the y-coordinate is only defined modulo 1, " $s(x)\Big|_{x_1}^{x_2}$ " is only determined up to an integer by the behaviour of s near x_1 and x_2. This implies that $\iota_{X_{ev}} E_2$ and $\iota_{X_{ev}} E_3$ are in $G_1^0(\pi_2)$, resp. $G_1^0(\pi_3)$, but not in $G_1^{00}(\pi_2)$, resp. $G_1^{00}(\pi_3)$.

Finally I want to show that E_3 is not the Euler equation of any Lagrangian. Suppose the converse, i.e., E_3 is the Euler equation of some $\overline{L}$. Let T_a be the automorphism induced on $J^2(\pi)$ by the translation $x \to x + a$. Then, since $T_a^*(E) = E$, E is also the Euler equation of $\int_0^1 T_a^*(\overline{L})da = \overline{L}_{inv}$. $\overline{L}_{inv}$ is now invariant under translation in the x-direction and hence E should have (global) $\frac{\partial}{\partial x}$-conservation. But this we just disproved. So we have in this example:

If the fibre is S^1, the conservation of energy ($\frac{\partial}{\partial x}$-conservation) is violated;

If the fibre <u>and</u> the base manifold are S^1 then there is also no underlying variational principle;

In one case (fibre S^1 and base $\mathbb{R}$) there is a variational principle, there is symmetry of the source equation (with respect to translation in the x-direction) but no corresponding conservation of energy.

References.

1. J.M. Boardman, Singularities of differentiable maps, Publ. I.H.E.S., 33 (1967), 21-57.

2. P. Dedeker, Sur un problème inverse du calcul des variations, Bull. Sci. Acad. roy. Belg. V, 36 (1950), 63-70.

3. P. Funk, Variationsrechnung und ihre Anwendung in Physik und Technik, Springer, N.Y., 1970.

4. A. Hirsch, Uber eine charakteristische Eigenschaft der Differentialgleichungen der Variationsrechnung, Math. Ann. 49 (1897), 49-72.

5. G.W. Horndesci, Sufficiency conditions under which a Lagrangian is an ordinary divergence, Aequ. Math. 12 (1975), 232-241.

6. S. Kobayashi and K. Nomizu, Foundations of differential geometry, Intersc., N.Y., 1963.

7. E. Noether, Invariante Variationsprobleme, Nachr. kgl. Ges. Wiss. Göttingen, math-phys. Kl. 1918, 235-257.

8. F. Takens, Variational and conservative systems, preprint Univ. of Groningen, Holland (1976).

9. E. Tonti, Variational formulation of nonlinear differential equations (I), Bull. Sci. Acad. roy. Belg. V, 55 (1969), 137-165.

10. E. Tonti, On the mathematical structure of a large class of physical theories, Atti accad. naz. d. Lincei (8) 52 (1972), 48-56.

11. E. Tonti, A mathematical model for physical theories I, II, Atti accad. naz. d. Lincei (8) 52 (1972), 175-181, 350-356.

University of Groningen

Classification of generic quadratic vector fields

with no limit cycles

by

Geovan Tavares dos Santos

1. Introduction

In this paper we study polynomial vector fields in two variables. The results we prove are on stability, genericity and classification of such vector fields, mainly on quadratic vector fields. First we want to present an overall view of the subject.

In 1881, Poincaré described an adequate way to study polynomial vector fields. For this, he took the plane R^2 tangent to the sphere S^2 at the north pole and projected the plane over it from its center. This projection, called central projection, provides the extension of the vector field to infinite which is now the equator S^1 of the sphere.

Let $\mathfrak{X}$ be the space of all vector fields $X(x,y) = (P(x,y), Q(x,y))$, where degree $P \leq n$ and degree $Q \leq n$, with the topology of the coefficients. If $X \in \mathfrak{X}$, we will denote by $\pi(X)$ the extension to S^2 of the central projection of X. Let Σ_∞ be the set of all $X \in \mathfrak{X}$ such that $\pi(X)$ is structurally stable in a neighborhood of infinite. In 1969, Gonzalez [11] proved the first results about stability and genericity of polynomial vector fields: a) $X \in \Sigma_\infty$ if and only if the singularities and closed orbits of $\pi(X)$ in a neighborhood of infinite are hyperbolic and b) Σ_∞ is open and dense in $\mathfrak{X}$. In 1974, de La Rocque Palis [16] did study linear vector fields $X(x) = Ax$, $x \in R^n$ and A a $n \times n$ real matrix, proving the following results: a) $\pi(X)$ is structurally stable if and only if $\pi(X)$ is Morse-Smale; b) $\Sigma = \{\pi(X)\colon X$ is linear and $\pi(X)$ is Morse-Smale$\}$ is open and dense in $\pi(\mathfrak{X}) = \{\pi(X)\colon X$ is linear$\}$.

As far as the classification of polynomial vector fields is concerned, two aspects are present: the algebraic and the topological one. The classification is algebraic if starting from algebraic relations on the coefficients of P and Q it is possible to determine the phase space of $X = (P,Q)$; the topological one intends to determine the phase space of polynomial vector fields which are invariant by topological equivalence.

In the case where P and Q are second degree homogeneous polynomials, the classification problem has been completely solved. In the algebraic aspect, Liagina [14] solved entirely the question, while the topological classification has been made by Marcus [17] using non-associative algebras and by Argemi [3] using the directional blowing-up method.

The first attempt to classify algebraically linear vector fields was made by Gonzalez [11] who obtained some results in R^3. The algebraic classification as well the topological one for generic linear vector fields in R^n were made by de La Rocque Palis [16] who proved the following results: a) $\pi(X(x) = Ax)$ is Morse-Smale if and only if the eigenvalues of A have distinct (except for pairs of conjugate complex eigenvalues) non zero real parts; b) two Morse-Smale vector fields, $\pi(X(x) = Ax)$ and $\pi(Y(x) = Bx)$ are topological equivalent if and only if its phase diagrams are isomorphic.

When P and Q are quadratic polynomials the first attempt of algebraic classification was made, as far as we know, by Sharpe [22] in 1910, who starting from the isoclines (curves of R^2 where $\frac{dy}{dx} = \frac{\dot{y}}{\dot{x}} =$ constant) of the vector field $X(x,y) = (\dot{x},\dot{y}) = (P,Q)$ tried to determine the phase space of X. There are several papers aiming towards an algebraic classification of quadratic polynomial vector fields and in the case where X has a center the problem has been solved through the work of several authors, see Coppel [6].

Our work is mainly concerned with the topological classification of generic quadratic polynomial vector fields. Before we go into the description of our work, we want to indicate some applications of polynomial vector fields.

Several phenomena have their evolution in time described through equation of the type

$$(*)\qquad \begin{aligned} \dot{x} &= x(a_o + a_1x + a_2y) \\ \dot{y} &= y(b_o + b_1x + b_2y). \end{aligned}$$

For example, the equation of Endem-Fowler in Astrophysics and the equation of Blasius in hydrodynamics [6]; the equations of Volterra-Lotka in Ecology [25] and [15]; the equations of Richardson in Psychology [21]. From the mathematical viewpoint the qualitative behaviour of (*) was completely described by Bautin [6].

The equations

$$(**)\qquad \begin{aligned} \dot{x} &= y \\ \dot{y} &= -A_oy - A_1x + B \end{aligned}$$

are very common in applications. For example, if A_o, A_1 and B are constants, we have the equations of a harmonic oscillator or a capacity-indutance-resistance system in electricity. In the equations of a charged electrical conductor, A_o, A_1 and B are affine linear functions. In the Van der Pohl equation A_1 is a quadratic polynomial, A_o and B are constant. Any equation of the type $\ddot{x} + A_o\dot{x} + A_1x = B$ can, by the change of variable $\dot{x} = y$, be transformed in (**). Other applications can be found in [1] and [2].

After this brief survey of what has been proved by many researchers on the subject, we will go into the description of the theorems proved in this paper.

Let PMS be the set of all vector fields $\pi(X)$ in S^2, where $X = (P,Q)$, P and Q polynomials of degree less than or equal to n, such that: a) the only recurrences of $\pi(X)$ are hyperbolic singularities and closed orbits; b) $\pi(X)$ has no saddle connections, except those along the infinite S^1.

Theorem A proves that PMS is open and dense in the set of all polynomial vector fields of degree less than or equal to n. In Theorem B we prove that $\pi(X)$ belongs to PMS if and only if $\pi(X)$ is structurally stable. On proving both theorems we use, as much as possible, techniques of analogue theorems for C^r vector fields on S^2. Independently C. Pugh has announced these results in [16].

In Theorem C, we list the kinds of singularities permitted for a vector field $\pi(X = (P,Q))$, where P and Q are second degree polynomials. The proof of this theorem is based mainly on two lemmas. The first lemma says that the way we extend X to infinite establishes some symmetry between the singularities of $\pi(X)$ along infinite. The second one shows that it is possible to determine the indices of the singularities of $X = (P,Q)$ from the relative position between $P^{-1}(0)$ and $Q^{-1}(0)$.

Theorem D gives a topological classification of the structurally stable vector fields with no limit cycles. In proving this theorem we show, either the impossibility of existence of a given phase space in the set of all quadratic vector fields or we start from a quadratic homogeneous vector field (which gives the topological behaviour of the vector field along infinite) to construct all the structurally stable ones which have this homogeneous part.

The following question was proposed by Coppel [6]: when given a phase space $E(X)$ of a C^r vector field X in S^2 is it possible to find a quadratic vector field $\tilde{X}$ such that the phase space $E(\pi(\tilde{X})) = E(X)$? Theorem D gives a partial answer to this question.

A complete answer to this question depends upon the (still open) Hilbert 16th problem: there exists a positive integer number N, which depends only on the degrees of P and Q such that given a polynomial vector field $X = (P,Q)$, the number of limit cycles of X is less or equal to N.

At the end of this paper we propose some problems we find interesting.

This work corresponds to my thesis at IMPA. I would like to thank my adviser J. Palis who suggested me this problem; F. Takens and C. Pugh with whom I had so many helpful conversations. I also would like to thank all my colleagues and teachers for their suppprt and advice.

I acknowledge the financial support of IMPA and CNPq during my Doctoral program.

2. Statement of the results.

Let $\mathfrak{X}$ be the set of all vector fields in R^2 of the form $X = (P,Q)$, where P and Q are polynomials of degree less than n in two variables. If we set in $\mathfrak{X}$ the topology of the coefficients it will become a complete finite dimensional vector space.

The central projection is the map which associates to each $x = (x_1,x_2)$ of R^2, two points in S^2,

$$f(x) = (x_1,x_2,1)/\Delta(x) \quad \text{and} \quad g(x) = -(x_1,x_2,1)/\Delta(x)$$

where $\Delta(x) = \sqrt{x_1^2+x_2^2+1}$. The vector fields $df(x)X(x)$ and $dg(x)X(x)$ extend in a natural way to S^2 and one obtains an analytic vector field $\pi(X)$ (see Proposition 3.1).

Let $\pi(\mathfrak{X}) = \{\pi(X): X \in \mathfrak{X}\}$ with the topology induced by $\mathfrak{X}$. The vector fields which belongs to $\pi(\mathfrak{X})$ are called Poincaré vector fields.

Definition 2.1 - A vector field $X \in \mathfrak{X}$ is called structurally stable if there exists a neighborhood $\mathcal{U}$ of X such that for every Y in $\mathcal{U}$ there is a homeomorphism $h: S^2 \to S^2$ mapping trajectories of $\pi(X)$ onto trajectories of $\pi(Y)$.

Two vector fields X and Y are equivalent if there exists a homeomorphism, like in definition above, between $\pi(X)$ and $\pi(Y)$.

Let $W^s(x)$ and $W^u(y)$ be the stable and unstable manifolds of a vector field X through the point $x,y \in R^2$ (for precise definition of $W^s(x)$ and $W^u(y)$ see [13]).

Now, let Ω_o and Ω_∞ be the non-wandering sets of $\pi_o(X) = \pi(X)/S^2-S^1$ and $\pi(X)$ in S^1.

Definition 2.2 - A vector field $X \in \mathfrak{X}$ is called Poincaré-Morse-Smale if:

a) $\Omega (= \Omega_o \cup \Omega_\infty)$ is finite and hyperbolic

b) 1. $x,y \in \Omega_o$, then $W^s(x) \pitchfork W^u(y)$

2. $x \in \Omega_o$, $y \in \Omega_\infty$, then $W^s(x) \pitchfork W^u(y)$

3. $x,y \in \Omega_\infty$, then $W^s(x) \pitchfork W^u(y)$ in S^2-S^1.

Let PMS be the set of all Poincaré-Morse-Smale vector fields and Σ_n be the set of all Poincaré vector fields which are structurally stable.

Theorem A

PMS is open and dense in $\mathfrak{X}$.

Theorem B

$X \in \Sigma_n$ if and only if $X \in$ PMS.

We will now be restricted to quadratic polynomial vector fields, i.e. those one in the form $X = (P,Q)$, where P and Q are second degree polynomials.

Let H^+ and H^- the upper and lower hemisphere. The way we built up $\pi(X)$ has as consequence that the number of singularities of $\pi(X)$ on H^+, H^- and that of X on R^2 are the same.

We denote by p and s, the singularities of index one and the saddles of $\pi_o(X)$. P, S, F are the notation for sinks, saddles and sources of $\pi(X)$ along the infinite S^1. Since each singularity of $\pi(X)$ at infinite determines the opposite to it, we will give always one or three successive singularities (see Lemma 4.1).

Theorem C

If $X \in \Sigma_2$ then the singularities of $\pi(X)$ on $\bar{H}^+$ (= topological closure of H^+) fit together in one of following ways:

1.F 2.psF 3.$p_1p_2s_1s_2F$ 4.p_1p_2S 5.$p_1p_2p_3sS$

6.$s_1s_2F_1PF_2$ 7.$ps_1s_2s_3F_1PF_2$ 8.SPF 9.$psSPF$

10.$p_1p_2s_1s_2SPF$ 11.$p_1p_2S_1S_2F$ 12.$p_1p_2p_3sS_1S_2F$.

Theorem D

There are 25 equivalence classes of structurally stable quadratic vector fields with no limit cycles.

3. Proofs of Theorems A and B.

We will split the proof of the Theorems A and B in some lemmas and propositions.

If f and g are like in Section 2, we have the following

Proposition 3.1 [11] - The induced vector fields, $df(x)X(x)$ and $dg(x)X(x)$, on S^2-S^1 can be extended to a analytical vector field $\pi(X)$ defined on the whole sphere S^2, after multiplying by a factor y_3^{n-1}, and in such a way that the equator S^1 is invariant.

Definition 3.1 - $(PMS)_\infty = \{X \in \mathfrak{X}: \Omega_\infty(X)$ is hyperbolic$\}$.

Theorem 3.1 [11] - $(PMS)_\infty$ is open and dense in $\mathfrak{X}$.

Let $\mathcal{P}_o = \{X \in \mathfrak{X}:$ the singularities of $\pi_o(X)$ are hyperbolic$\}$.

Lemma 3.1 - $\mathcal{P}_o$ is open and dense in $\mathfrak{X}$.

Proof: It is enough to show that X can be approximated by a vector field which has hyperbolic singularities. Let $X = (P,Q)$ and $f: R^2 \to R^2$ the mapping defined by $(x,y) \mapsto (P(x,y), Q(x,y))$. By Sard's theorem, we can find (ϵ_1,ϵ_2) which is a regular value of f, with ϵ_1 and ϵ_2 small. The mapping $\tilde{f}: R^2 \to R^2$, $(x,y) \mapsto (P(x,y) - \epsilon_1, Q(x,y) - \epsilon_2)$ has $(0,0)$ as a regular value. The vector field $\tilde{X} = (\tilde{P},\tilde{Q})$, with $\tilde{P} = P - \epsilon_1$, $\tilde{Q} = Q - \epsilon_2$ is close to X and has a finite number of singularities. Let $p_1,\dots,p_k$ be the non-hyperbolic singularities. Now, we consider a 2×2 matrix ξ, $|\xi|$ small enough, such that, $\tilde{X}\circ T + \xi$ has $(0,0)$ as a hyperbolic singularity. Then p_1 is a hyperbolic singularity of $X_1 = (\tilde{X}\circ T + \xi)\circ T^{-1}$. The vector field X_1 has r, $r < k$, non-hyperbolic singularities, and it is close to X. The same argument

applied finitely many times produces a vector field $\bar{X} = (\bar{P},\bar{Q})$, which is close to X and has hyperbolic singularities.
This completes the proof.

Corollary - If $X \in \mathfrak{X}$ then the boundary of the strips of closed orbits of $\pi(X)$ are saddle connections and singularities.

Let $\Gamma_o = \{X \in \mathfrak{X}:$ a) and b1) of Definition 2.2 holds for $\pi(X)\}$.

Lemma 3.3 - Γ_o is dense in $\mathcal{P}_o \cap (PMS)_\infty$.

Proof: For r sufficiently large, let $D(r)$ be an open disc such that the vector field X has all of its singularities and saddle connections in $D(r)$, [19]. It follows from the corollary of Lemma 3.2 that the closed orbits of Y are limit cycles and so finite number, [8]. $Z = Y_\beta = (\tilde{P} - \beta\tilde{Q},\ \tilde{Q} + \beta\tilde{P})$ with β small, $\tilde{P} = P - \alpha Q$ and $\tilde{Q} = Q + \alpha P$, has a finite number of hyperbolic limit cycles, [1]. If Z still has saddle connections, we do the same argument again to obtain a vector field $W = Z_\delta = (\bar{P} - \delta\bar{Q},\ \bar{Q} + \delta\bar{P})$ with $\bar{P} = \tilde{P} - \beta\tilde{Q}$ and $\bar{Q} = \tilde{Q} + \delta\tilde{P}$. The vector field $\pi(W)$ is close to $\pi(X)$ and $\pi(W) \in \Gamma_o$; so the proof is finished.

Proposition 3.1 [24] - Let $X = (P,Q) \in \mathfrak{X}$ and γ be a connection between its saddles p_1 and p_2. If $X_\alpha = (P_\alpha,Q_\alpha)$, $\alpha \in R$, is a one-parameter family such that $X_o = X$, then, u beeing a parameter for γ, $\sigma(X) = \frac{\partial P}{\partial x} + \frac{\partial Q}{\partial y}$ and $\det(X, \frac{\partial X_\alpha}{\partial \alpha}) = \det\begin{pmatrix} P & Q \\ \frac{\partial P_\alpha}{\partial \alpha} & \frac{\partial Q_\alpha}{\partial \alpha}\end{pmatrix}$ we have

$$\int_{-\infty}^{\infty} \exp\left(-\int_0^u \sigma(X)du\right)\cdot\det\left(X, \frac{\partial X_\alpha}{\partial\alpha}\right)du \neq 0$$

if and only if X_α does not have saddle connections between p_1 and p_2, for small α.

Lemma 3.4 - PMS is dense in Γ_o.

Proof: Let $p_1 \in S^1$ and p_2, two saddles of $\pi(X)$ having no connection on S^2-S^1. Since $W^s(p_1)$ does not accumulate on $W^u(p_1)$, there exists a plane $\beta \neq xOy$ plane, which contains p_1 and $-p_1$, such that the saddle connection is contained in one of the hemisphere of $S^2 - \beta \cap S^2$. Let $\tilde{\beta}$ be the parallel plane to β tangent to S^2. The central projection of S^2 over $\tilde{\beta}$ is given by $\varphi(u) = \frac{u}{\langle u,v_o\rangle}$, $u \in S^2$, $v_o = (x_o,y_o,z_o) \in S^2$ is a normal vector to $\tilde{\beta}$ and $\langle\ ,\ \rangle$ is the usual inner product of R^3. If we take the vectors $(y_o,-x_o,0)$ and $(0,-z_o,y_o)$ as a basis of β, the coordinate functions of the central projection $f(x) = (x_1,x_2,1)/\Delta(x)$ are given by

$$f_1(x,y) = \frac{x}{y_o\langle(x,y,1),v_o\rangle}, \qquad f_2(x,y) = \frac{1}{y_o\langle(x,y,1),v_o\rangle}.$$

$\pi(X)$ has the following expression

$$(A(x,y)P(x,y) + B(x,y)Q(x,y),\ \ C(x,y)P(x,y) + D(x,y)Q(x,y)),$$

except for a positive factor which permits the extension of the vector field to infinite, where

$$A(x,y) = \frac{\langle(x,y,1),v_o\rangle - xx_o}{y_o\langle(x,y,1),v_o\rangle^2}, \qquad B(x,y) = -\frac{x}{\langle(x,y,1),v_o\rangle^2}$$

$$C(x,y) = -\frac{x_o}{y_o\langle(x,y,1),v_o\rangle^2}, \qquad D(x,y) = -\frac{1}{\langle(x,y,1),v_o\rangle^2}$$

Setting $X_\alpha = (P - \alpha Q,\ Q + \alpha P)$, we have $\det(\pi(X), \frac{\partial(\pi(X_\alpha))}{\partial\alpha}) =$ $= (AD-BC)(P^2+Q^2) = \frac{1}{y_o\langle(x,y,1),v_o\rangle^3}(P^2(x,y)+Q^2(x,y))$ which is $\neq 0$ if $y_o \neq 0$. Now the proof of the lemma follows from the last proposition.

Proof of Theorem A: PMS is open in $\mathfrak{X}$, since for every $X \in$ PMS we have no cycles for $\pi(X)$ and $\Omega(X)$ is finite, [18]. The density is a consequence of the lemmas before and from the fact that $\mathcal{P}_o \cap (\mathrm{PMS})_\infty$ is dense in $\mathfrak{X}$.

Proof of Theorem B: From [19], [11] and [23] we have that $X \in \Sigma_n$ implies $X \in$ PMS. The other way we can apply the canonical region method of [18], noticing that the only new regions are the given below where we can use the very same method.

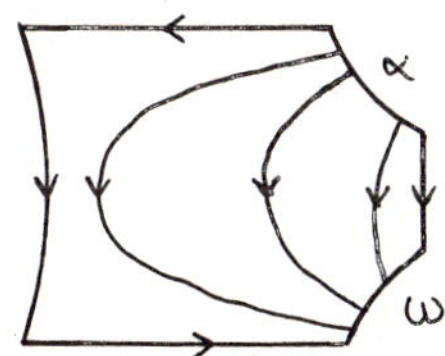

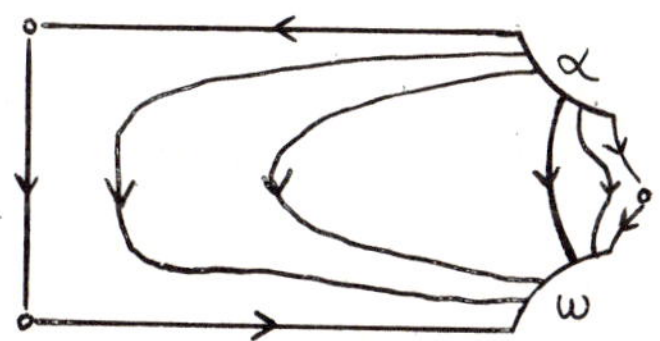

4. Proof of Theorem C

In the first place we will study what kind of singularities $\pi(X)$ has at infinite.

For this, if $X = (P,Q)$, let us denote by P_2 and Q_2 the degree two homogeneous parts of P and Q.

Lemma 4.1 - If $X \in \Sigma$ then, at infinite, $\pi(X)$ has two or six diammetrically opposed singularities having eigenvalues of contrary signs.

Proof: Since Σ is open and X is structurally stable, we can suppose that $(0,1,0)$ and $(0,-1,0)$ are not singularities of $\pi(X)$. These singularities are locally of the form $(z_1,0)$, where z_1 is a root of the equation $Q_2(1,z_1) - z_1 P_2(1,z_1) = 0$. From the structural stability of X, it follows that the roots of this third degree equation are simple, then, we would have one or three roots. Now, the way we built $\pi(X)$ shows that the eigenvalues of the singularities as stated in the lemma are of contrary signs, and the proof is finished.

From this lemma it follows that we can given one or three successive singularities of $\pi(X)$ at infinite.

Corollary - If $X \in \Sigma$ then the singularities of $\pi(X)$ in S^1 are of the following types:

a) F; b) S; c) F_1PF_2; d) SPF; e) S_1S_2F; f) $S_1S_2S_3$.

The following lemma is a consequence of the affine classification of quadrics [23].

Lemma 4.2 - $\mathcal{C} = \{X = (P,Q);\ P^{-1}(0)$ and $Q^{-1}(0)$ are ellipses or hyperbolas$\}$ is open and dense in $\mathfrak{X}$.

If $X = (P,Q) \in \Sigma_2$ we know that $P^{-1}(0) \cap Q^{-1}(0)$ has at most four points.

Let $\Sigma_2(i) = \{X \in \Sigma_2;\ \pi(X)$ has i singularities in $H^+\}$.

Lemma 4.3 - $\Sigma_2 = \Sigma_2(0) \cup \Sigma_2(2) \cup \Sigma_2(4)$.

Proof: Let us suppose that $P^{-1}(0) \cap Q^{-1}(0)$ has one or three points.

By Lemma 4.2 we can take $X \in \mathcal{C}$. If one of the points in $P^{-1}(0) \cap Q^{-1}(0)$ is a tangency point we can make a translation and so modifying the number of singularities of X, which is not possible since $X \in \Sigma_2$. Neither $P^{-1}(0)$ nor $Q^{-1}(0)$ are ellipses, since if it was we would have necessarily none, two or four points in the intersection $P^{-1}(0) \cap Q^{-1}(0)$. Now, if $P^{-1}(0)$ and $Q^{-1}(0)$ are hyperbolas, the number of points in $P^{-1}(0) \cap Q^{-1}(0)$ is the same as in the intersection of their assymptotes. Since this number is odd, it would exist two parallel assymptotes, then with a small perturbation we could increase the number of singularities of X, which is impossible and then proving the lemma.

We will describe now the way the singularities of $\pi(X)$ on S^2-S^1 fit together. For this it is enough to consider the vector field $X = (P,Q)$ in R^2.

If $P^{-1}(0)$ intersects a simple closed curve of R^2 in a finite number of points we will denote these points by α_i and by β_i the points at the intersection of $Q^{-1}(0)$ and the same curve. $[\alpha_i\beta_i]$ will denote the oriented polygon determined by the intersection $(P^{-1}(0) \cup Q^{-1}(0)) \cap C$, where C is a simple closed curve. $I(X,C)$ is the index of the vector field X relative to the curve C.

Lemma 4.4 - Let p_1 and p_2 be the singularities of X and C be a simple closed curve which contains only these singularities in its interior:

a) If $\{P^{-1}(0) \cup Q^{-1}(0)\} \cap C = [\alpha_1\beta_1\alpha_2\beta_2\alpha_3\beta_3\alpha_4\beta_4]$ then either $I(X,C) = 2$ or $I(X,C) = -2$.

b) If $\{P^{-1}(0) \cup Q^{-1}(0)\} \cap C = [\alpha_1\alpha_2\beta_1\alpha_3\beta_2\beta_3\alpha_4\beta_4]$ then $I(X,C) = 0$.

Proof: a) If $Q(\alpha_1) > 0$ and $P(\beta_4) > 0$ then $P(\beta_1) < 0$, $Q(\alpha_2) < 0$, $P(\beta_2) > 0$, $Q(\alpha_3) > 0$, $P(\beta_3) < 0$ and $Q(\alpha_4) < 0$ (fig.4.1). For each curve segment $[\alpha_i,\beta_i]$ the vector field $\frac{X}{\|X\|}$ runs through one quadrant of S^1, since neither $P(x,y)$ nor $Q(x,y)$ is zero in the open curve segment (α_i,β_i). Running through the curve C, we have $I(X,C) = 2$. If we had chosen $Q(\alpha_1) > 0$ and $P(\beta_4) < 0$ then $I(X,C) = -2$. Case b) is analogue to a).

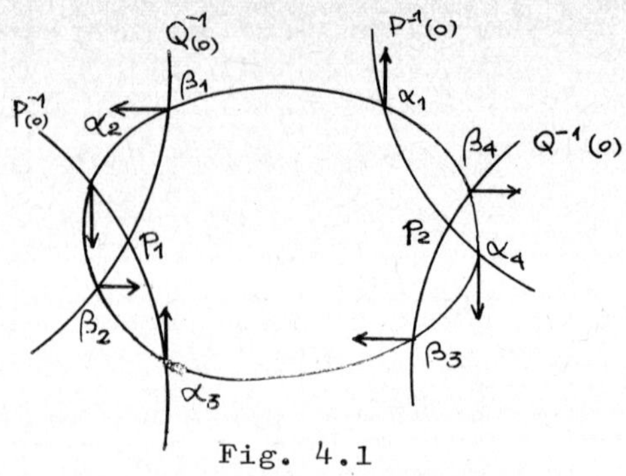

Fig. 4.1

Corollary - If $p_1, p_2 \in P^{-1}(0) \cap Q^{-1}(0)$ are on the same branch of $P^{-1}(0)$ then the sum of the indices of p_1 and p_2 is zero.

Take $X \in \Sigma(4)$ and let R be the oriented polygon whose vertices are the singularities of X.

Lemma 4.5 - The singularities of $X \in \Sigma(4)$ are distributed on R in one of the following ways:

a) saddles alternated with singularities of index one;

b) one singularity of index one and three saddles;

c) one saddle and three singularities of index one.

Proof: Since X is structurally stable we can take $X \in \mathcal{C}$. If $P^{-1}(0)$ or $Q^{-1}(0)$ is an ellipse, we apply the last corollary to obtain case a). If $P^{-1}(0)$ and $Q^{-1}(0)$ are hyperbolas we have three choices: 1) all the singularities of X are on the same branch of $P^{-1}(0)$; 2) only two singularities are in the same branch of $P^{-1}(0)$; 3) one branch of $P^{-1}(0)$ has just one singularity of X. By Lemma 4.3 we have that 1, 2 and 3 imply a, b and c, and the proof is finished.

Let I_o = index of $\pi_o(X)$ and I_∞ = index of $\pi(X)$ at infinite.

Proof of Theorem C: From the Euler characteristic of the sphere S^2 it follows that $2I_o + I_\infty = 2$ and then $S_1S_2S_3$ is not possible. Now the proof of the theorem follows combining this fact with Lemmas 4.1 and 4.3.

Remark: A different proof of Lemma 4.5 can be found in [5].

5. Proof of Theorem D.

The proof of Theorem D will be divided in several cases according the kind of singularities that $X \in \mathfrak{X}$ has at infinite given by Theorem C. Before we start the proof we give lemmas and theorems necessary to it.

Definition 5.1 - A singularity p of $X \in \mathfrak{X}$ is a center if there exists a neighborhood V of p such that $V-\{p\}$ is the union of closed orbits of X.

Lemma 5.1 - Let $X = (P,Q) \in \mathfrak{X}$ and $X^\theta = (P - \theta Q,\ Q + \theta P)$, $0 < |\theta| < \epsilon$, with ϵ small.

a) If p is a center for X then p is a hyperbolic singularity for X^θ.

b) Inside the regions where the centers of X are defined, X^θ has no limit cycles.

Proof: a) If we make a translation of the origin of R^2 to p, then we can write $X = (\alpha y + h_1(x,y),\ -\alpha_2 x + h_2(x,y))$, where h_1 and h_2 are second degree homogeneous polynomials, thence $h_i(0,0) = \frac{\partial h_i}{\partial x_j}(0,0) = 0$, $i,j = 1,2$.

$$X^\theta = (\alpha\theta x + \alpha y + h_1(x,y) - \theta h_2(x,y),\ -\alpha x + \alpha\theta y + h_2(x,y) + \theta h_1(x,y))$$

so the jacobian matrix for X^θ at $(0,0)$ is

$$\begin{pmatrix} \alpha\theta & \alpha \\ -\alpha & \alpha\theta \end{pmatrix}$$

which implies that $(0,0)$ is a hyperbolic singularity for X^θ.

b) The orbits of X and X^θ are transversal.

Definition 5.2 - A point $p \in R^2$ is a contact between $X \in \mathfrak{X}$ and the straight line L if $X(p)$ is tangent to L.

Lemma 5.2 [5] - If a straight line L is not composed of orbits of $X \in \mathfrak{X}$, then X and L has at most two contacts.

If two vector fields $X,Y \in \mathfrak{X}$ are topologically equivalent we will use the notation $\pi(X) \cong \pi(Y)$. The phase space of $\pi(X)$ will be given always in $\bar{H}^+$.

We now begin the proof of Theorem D.

Case a: The vector field is type F at infinite.

$$\begin{aligned} \dot{x} &= -2xy \\ \dot{y} &= x^2+y^2 \end{aligned}$$ is type F at infinite and hamiltonian, whose first integral is $\varphi(x,y) = x(\frac{x^2}{3} + y^2)$.

a1) Let $\varphi_1(x,y) = \varphi(x,y) + \epsilon x$, ϵ positive and small. The vector field $X_1 = X_{\varphi_1}$, whose first integral is φ_1, is given by

$$\begin{aligned} \dot{x} &= -2xy \\ \dot{y} &= x^2+y^2+\epsilon . \end{aligned}$$

Since $x^2+y^2+\epsilon > 0$, for all $(x,y) \in R^2$, the vector field X_1 has no singularities in R^2. $\pi(X_1)$ is structurally stable; its phase space is given in Figure 5.a1.

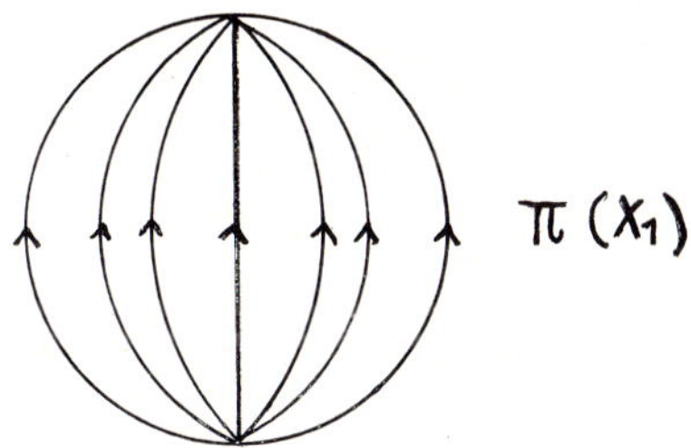

Fig. 5.a1

a2) Set $\varphi_2(x,y) = x(\frac{x^2}{3}+y^2) + \frac{\alpha}{2}x^2 + \frac{\beta}{2}y^2$, $\alpha < 0$, $\beta > 0$, whose associated vector field X_{φ_2} is

$$\dot{x} = -y(\beta + 2x)$$

$$\dot{y} = (x + \frac{\alpha}{2})^2 + y^2 - \frac{\alpha^2}{4}.$$

The straight line $x = -\frac{\beta}{2}$ does not meet the circle $\dot{y} = 0$, which implies that X_{φ_2} has only two singularities, a saddle and a center, since it is hamiltonian. The rotated vector field $X_2 = X^{\theta}_{\varphi_2}$, θ positive and small, has no saddle connections and, by Lemma 5.1, its singularities are hyperbolic. X_2 has no limit cycles and, by Theorem B, it is structurally stable. Figure 5.a2 gives the phase space of $\pi(X_2)$.

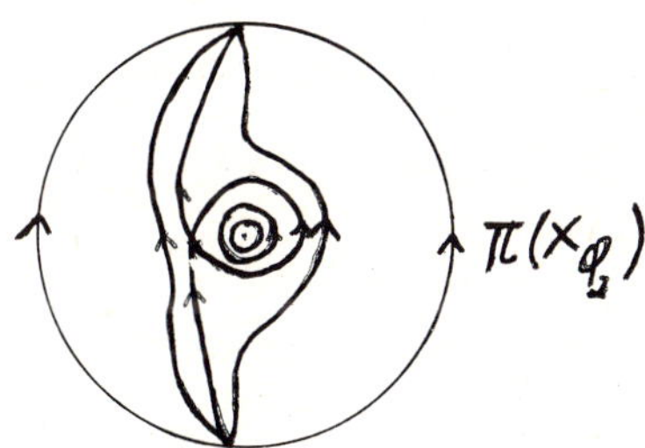

Fig. 5.a2

a3) $\tilde{\varphi}(x,y) = x(\frac{x^2}{3} + y^2 + \lambda y)$, $\lambda < 0$, is the integral of the vector field $X_{\tilde{\varphi}}$, which is given by

$$\dot{x} = -x(\lambda + 2y)$$
$$\dot{y} = x^2 + (y + \frac{\lambda}{2})^2 - \frac{\lambda^2}{4} .$$

This vector field has four singularities, two saddles and two centers. The saddles are on the same level $\tilde{\varphi}^{-1}(0)$ and its connections have the equations $x = 0$, $0 \le y \le -\lambda$ and $\frac{x^2}{3} + y^2 + \lambda y = 0$. Let $X_\epsilon =$ $= X_{\tilde{\varphi}} + (\epsilon x, 0)$, $\epsilon > 0$. With this perturbation, we brake the saddle connection $\frac{x^2}{3} + y^2 + \lambda y = 0$ but, since $X_\epsilon\big|_{x=0} = X_{\tilde{\varphi}}\big|_{x=0}$, the sadddle connection $x = 0$, $0 \le y \le -\lambda$ is the same one for X_ϵ and $X_{\tilde{\varphi}}$. At the centers of $X_{\tilde{\varphi}}$, the jacobian matrix is

$$\begin{pmatrix} 0 & \alpha_i \\ -\alpha_i & 0 \end{pmatrix}$$

while at the singularities of index one X_ϵ has jacobian

$$\begin{pmatrix} \epsilon_1^i & \alpha_i + \epsilon_2^i \\ -\alpha_i + \epsilon_3^i & \epsilon_4^i \end{pmatrix}$$

with ϵ_i^j small, $j = 1,2$; $i = 1,2,3,4$. Thence the singularities of X_ϵ are all hyperbolic. The vector field X_ϵ has no limit cycles since $\operatorname{div} X_\epsilon = \epsilon > 0$. Now, the rotated vector field $X_3 = X_\epsilon^\theta$, $\theta > 0$, is structurally stable and has no limit cycles. The phase space of $\pi(X_3)$ is given in Figure 5.a3

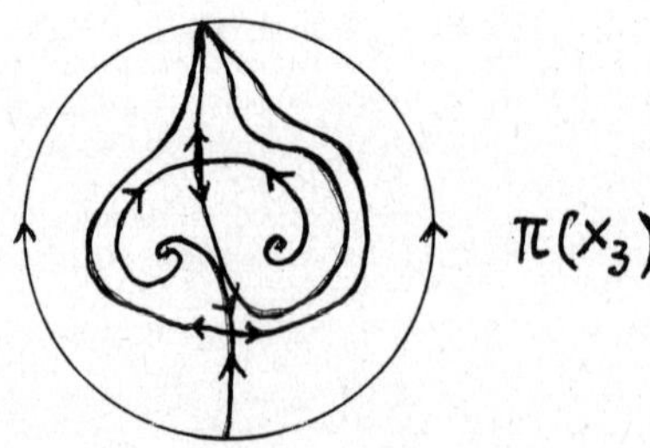

Fig. 5.a3

a4) We now consider $\bar{\varphi}(x,y) = \tilde{\varphi}(x,y) - \frac{\lambda\epsilon}{2} y^2 - \frac{2\epsilon}{3} y^3$. The hamiltonian vector field $X_{\bar{\varphi}}$, which has $\bar{\varphi}$ as a first integral, is

$$\dot{x} = (\epsilon y - x)(\lambda + 2y)$$
$$\dot{y} = x^2 + (y + \frac{\lambda}{2})^2 - \frac{\lambda^2}{4}.$$

$X_{\bar{\varphi}}$ has four singularities, two saddles and two centers. As a consequence of Lemma 3.2, the existence of connection between the two saddles implies $\pi(X_{\bar{\varphi}}) \cong \pi(X_{\tilde{\varphi}})$, then, by the corollary of Lemma 5.2 this saddle connection is the straight line segment them. This straight line segment is contained in the straight line $x = \epsilon y$, where $\dot{x} = 0$ and $\dot{y} < 0$. The saddles of $X_{\bar{\varphi}}$ are then in different levels of $\bar{\varphi}$, which gives a contradiction. Using part b) of Lemma 5.1 on the rotated vector field $X_4 = X^{\theta}_{\bar{\varphi}}$, $\theta > 0$, we obtain a vector field $\pi(X_4)$ which has no limit cycles and, by Theorem B, it is structurally stable. Figure 5.a4 gives the phase space of $\pi(X_4)$.

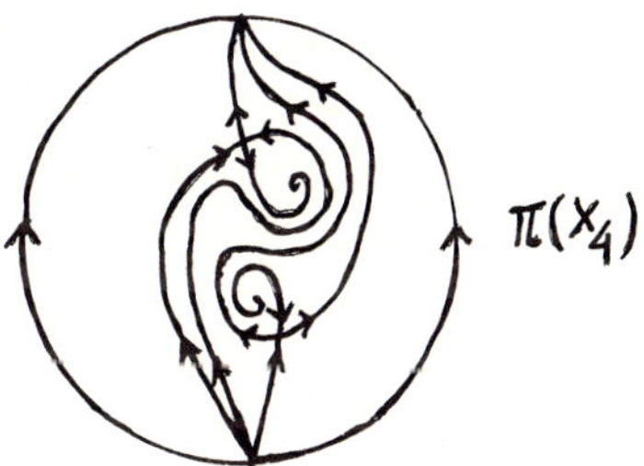

Fig. 5.a4

a5) We take the perturbation of $X_{\bar{\varphi}}$ given by $X_{\bar{\varphi},\gamma} = X_{\bar{\varphi}} + (\gamma x, 0)$.

$X_{\bar{\varphi},\gamma}$ has no limit cycles since $\operatorname{div} X_{\bar{\varphi},\gamma} = \gamma \neq 0$ and, by an argument analogue to the one used in a3), this vector field has hyperbolic singularities. $\pi(X_5)$, with $X_5 = X_{\bar{\varphi},\gamma}$, $\gamma > 0$, is structurally stable with no limit cycles. The phase space of $\pi(X_5)$ is given in Figure 5.a5.

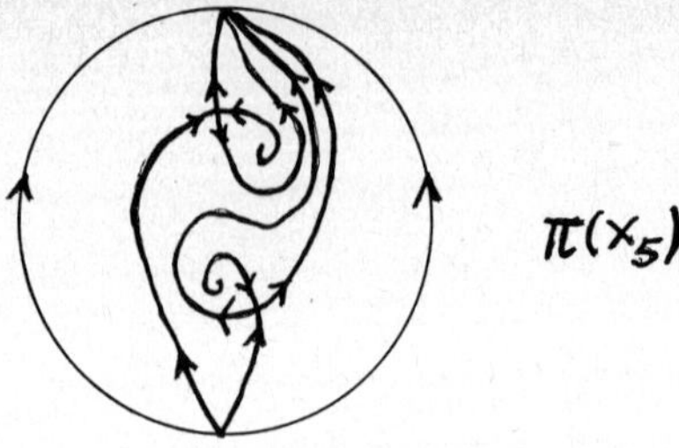

Fig. 5.a5

a6) Beeing δ positive, we consider the perturbation of $X_{\tilde{\varphi}}$ given by $X_\delta = X_{\tilde{\varphi}} + (\delta(\frac{x^2}{3} + y^2 + \lambda y),0)$. Since $X_\delta = X_{\tilde{\varphi}}$ on $\frac{x^2}{3} + y^2 + \lambda y = 0$, the saddle connection of $X_{\tilde{\varphi}}$, which is not a straight line, remains; but the straight line segment $x = 0$, $0 \leq y \leq -\lambda$ is not an orbit of X_δ, since over it we have $y^2 + \lambda y < 0$. $\operatorname{div} X_\delta = \frac{2\delta x}{3}$ implies that every closed orbit intersects the y-axis, then we have the existence of contacts which are not singularities, between this axis and X_δ; but this is not possible since $(0,0)$ and $(0,-\lambda)$ are singularities of this vector field. Taking the rotated vector field $X_6 = X_\delta^\theta$, $\theta > 0$, we obtain $\pi(X_6)$ structurally stable with no limit cycles. The next figure gives the phase space of $\pi(X_6)$.

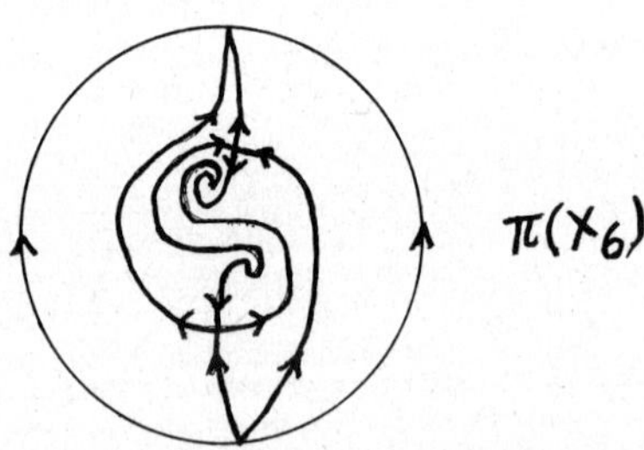

Fig. 5.a6

Case b: The vector field is type S at infinite.

Let $\bar{H}$ be the homogeneous vector field

$$\dot{x} = 2xy$$
$$\dot{y} = y^2 - x^2$$

which is type S at infinite. The integral curves of this vector field satisfy the equation $(x-a)^2 + y^2 = a^2$.

b1) Let Y_ε be the vector field

$$\dot{x} = 2xy$$
$$\dot{y} = y^2 - x^2 + \varepsilon, \quad \varepsilon > 0.$$

Since Y_ε leaves invariant the y-axis and its integral curves satisfy the equation $(x-a)^2 + y^2 = a^2 - \varepsilon$, then the singularities of index one are centers. The rotated vector field $Y_1 = Y_1^\theta$, $\theta > 0$, has no limit cycles and $\pi(Y_1)$ is structurally stable. Figure 5.b1 gives the phase space of $\pi(Y_1)$.

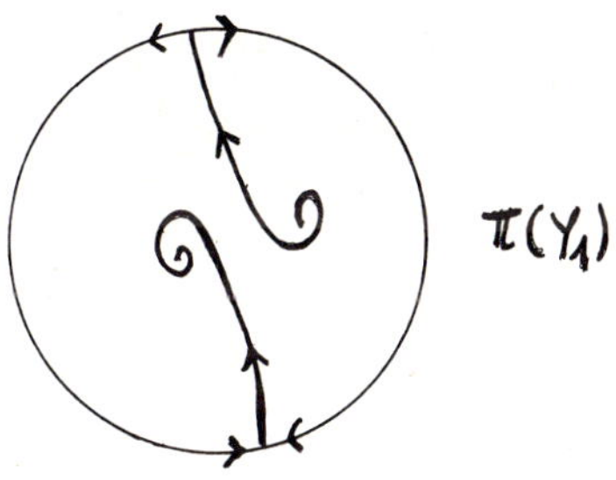

Fig. 5.b1

b2) The vector field $\bar{Y}_2 = A + \bar{H}$, where $A(x,y) = (-\lambda x, \lambda y)$, $\lambda > 0$, leaves invariant the y-axis and it is tranversal to the x-axis, except at $(0,0)$. $\bar{Y}_2$ has two singularities at the y-axis, one saddle and one sink, and two singularities of index one on each of the half-planes, $x > 0$ and $x < 0$. By direct calculation we can show that $\bar{Y}_2$ has hyperbolic singularities. Since div $\bar{Y}_2 = 4y$, every limit cycle cross the x-axis, which implies in the existence of contacts between $\bar{Y}_2$ and $y = 0$, and this is not possible by Lemma 5.2. Let $Y_2 = \bar{Y}_2^{\theta}$ be the rotated vector field of $\bar{Y}_2$. $\pi(Y_2)$ is structurally stable and has no limit cycles. See figure below for the phase space of $\pi(Y_2)$.

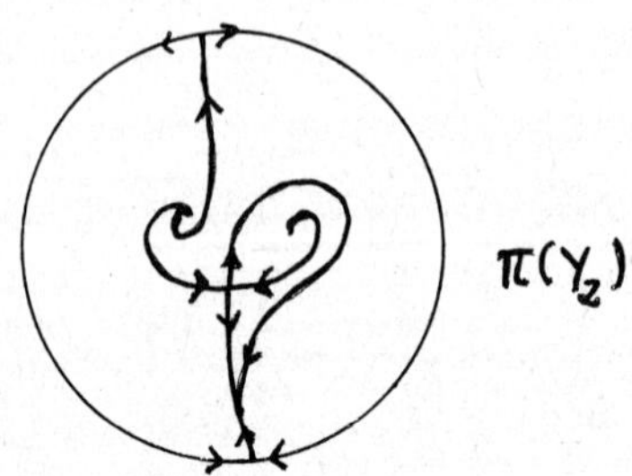

Fig. 5.b2

Case c: The vector field is type FPF at infinite.

The homogeneous vector field

$$\dot{x} = x^2 + 2xy$$
$$\dot{y} = -2xy - y^2$$

has type FPF singularities at infinite. This vector field is hamiltonian and its first integral is $\psi(x,y) = xy(x+y)$.

c1) Let Z_1 be the vector field

$$\dot{x} = x^2 + 2xy$$
$$\dot{y} = -2xy - y^2 - \epsilon, \quad \epsilon > 0.$$

Z_1 has the y-axis as an invariant line and two hyperbolic saddles in R^2, since $I_\infty = 6$ implies $I_o = -2$. The phase space for $\pi(Z_1)$ is given in Figure 5.c1.

Fig. 5.61

c2) Now, we take the perturbation of ψ given by $\tilde{\psi}(x,y) = xy(x+y-\varepsilon)$ whose associated vector field $Z_{\tilde{\psi}}$ is

$$\dot{x} = -\varepsilon x + x^2 + 2xy$$

$$\dot{y} = \varepsilon y - 2xy - y^2.$$

The vector field $\bar{Z}_2 = Z_{\tilde{\psi}} + (\delta xy, 0)$, $\delta > 0$, has saddle connections only along $xy = 0$. The same argument, as the one used in subcase a3, proves that the singularity of $\bar{Z}_2$ obtained as a perturbation of the center of $Z_{\tilde{\psi}}$ is a hyperbolic sink. Since $\operatorname{div} \bar{Z}_2 = \delta y$ we have no limit cycles for $\bar{Z}_2$. Let $\bar{Z}_2^\theta$ be the rotated vector field of $\bar{Z}_2$; $\pi(Z_2)$ and $\pi(Z_3)$, with $Z_2 = \bar{Z}_2^\theta$, $\theta > 0$, and $Z_3 = \bar{Z}_2^\theta$, $\theta < 0$, are structurally stable vector fields with no limit cycles. Figures 5.c2.1 and 5.c2.2 give the phase spaces of $\pi(Z_2)$ and $\pi(Z_3)$.

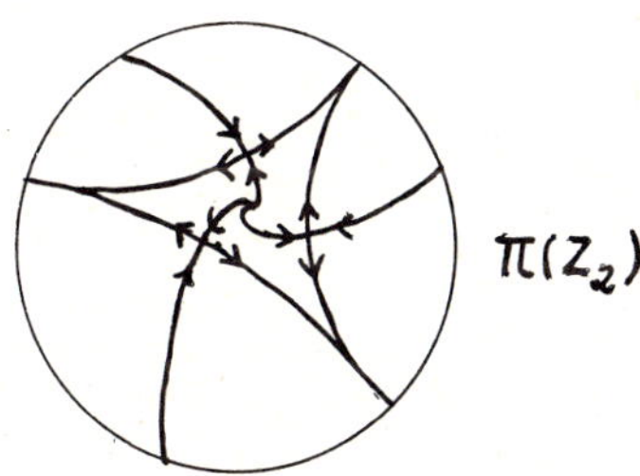

Fig. 5.c2.1

Fig. 5.c2.2

c3) Let $\bar{Z}_4 = Z_{\bar{\psi}} + (\eta(x+y-\epsilon), \eta(x+y-\epsilon))$ with $\eta > 0$. $x+y-\epsilon = 0$ is the only saddle connection of $\bar{Z}_4$, which has hyperbolic singularities and it has no limit cycles since $\operatorname{div} \bar{Z}_4 = 2\eta \neq 0$. The rotated vector field $Z_4 = \bar{Z}_4^{\theta}$, $\theta > 0$, has no limit cycles and $\pi(Z_4)$ is structurally stable; its phase space is given in Figure 5.c3.

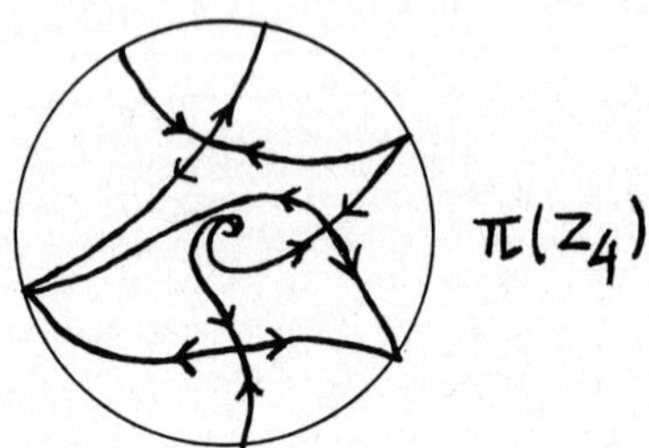

Fig. 5.c3

Let $Z_{\bar{\psi}}$ be the hamiltonian vector field associated with the function $\bar{\psi}(x,y) = \psi(x,y) + \frac{\lambda\epsilon}{2} x^2 - \frac{2\lambda}{3} x^3$, whose explicit form is

$$\dot{x} = x(x+2y-\epsilon)$$
$$\dot{y} = -y(2x+y-\epsilon) + \lambda x(2x+\epsilon), \quad \lambda < 0.$$

This vector field has three hyperbolic saddles and the straight line $x = 0$ contains two of them: $(0,0)$ and $(0,a)$. $\bar{\psi}^{-1}(0)$ is composed by the straight line $x = 0$ and the hyperbola $\frac{2\lambda}{3} x^2 - xy - y^2 - \frac{\lambda\epsilon}{2} x + \epsilon y = 0$. If it exists a saddle connection between one of the saddles in $x = 0$ and the third one, then, by Lemma 5.2 and the fact that $Z_{\bar{\psi}}$ is hamiltonian, $\bar{\psi}^{-1}(0)$ is composed of three straight lines, which is not possible. The two branches of the hyperbola are $W^s(0,a)$ and $W^u(0,0)$. The phase space of $\pi(Z_{\bar{\psi}})$ is given below.

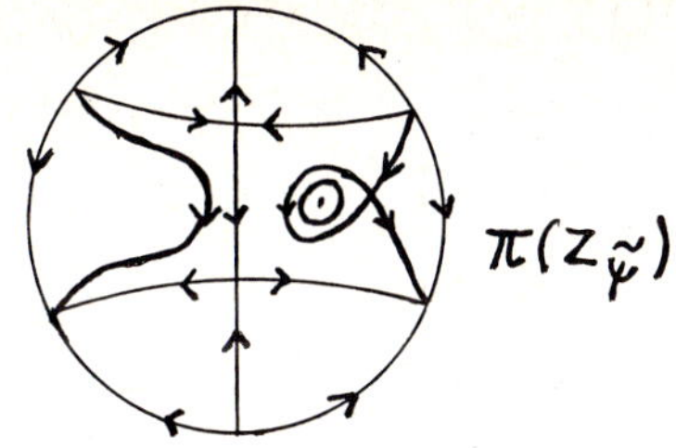

c4) Set $\bar{Z}_5 = Z_{\bar{\psi}} + (\delta x, 0)$, $\delta > 0$. This vector field has a unique saddle connection along the straight line $x = 0$, a hyperbolic sink and has no limit cycles since $\operatorname{div} \bar{Z}_5 = \delta \neq 0$. If $Z_5 = \bar{Z}_5^{-\theta}$, then $\pi(Z_5)$ is structurally stable and has no limit cycles. The phase space of $\pi(Z_5)$ is given in Figure 5.c4.

Fig. 5.c4

Let G be the phase space given in the next figure.

Lemma 5.c - There is no $X \in \mathfrak{X}$ such that the phase space of $\pi(X)$ is G.

Proof: Let $L_{p,-p}$ be the straight line between p and $-p$, where $p \in (F_1, -F_2) \subset S^1$ (then $-p \in (-F_1, F_2) \subset S^1$) and $\mathcal{L}_{p,-p}$ be the union of straight lines parallel to $L_{p,-p}$. Every straight line $L \subset \mathcal{L}_{p,-p}$ intersects $W^u(s_1)$ and $W^s(s_2)$, as well $W^u(s_1)$ and $W^s(s_2)$, from which it follows that it has two contacts with any vector field $X \in \mathfrak{X}$ whose phase space is G. But $\mathcal{L}_{p,-p} = R^2$, then there is no singularity inside the region determined by $(-P)s_1Ps_2$, so we get a contradiction proving the lemma.

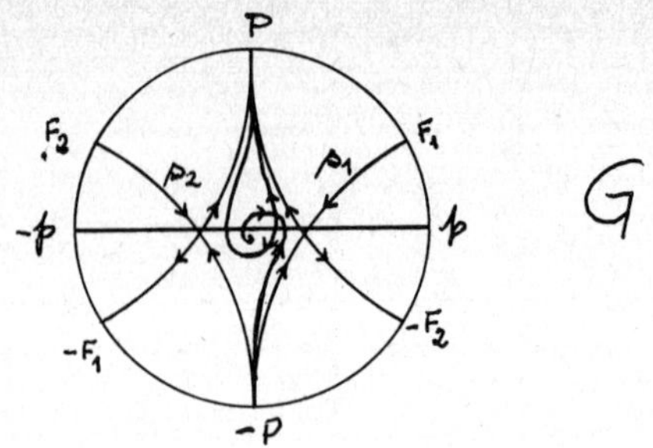

<u>Case d:</u> <u>The vector field is type</u> SPF <u>at infinite</u>.

Consider the following regions:

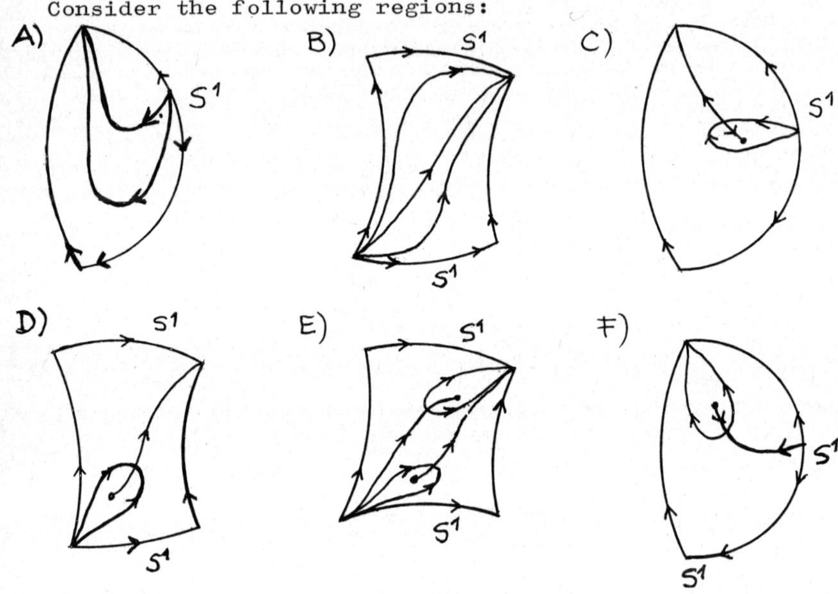

where the indication S^1 means an arc of the infinite S^1.

<u>Definition 5.d</u> - Two regions $\mathcal{U}$ and $\mathcal{V}$ in the phase space of a vector field X are symmetrical if $X|\mathcal{U}$ and $X|\mathcal{V}$ are topologically equivalent by an orientation reversing homeomorphism.

<u>Lemma 5.d</u> - Let $X \in \Sigma_2$ be a vector field with type SPF singularities at infinite.

a) If A and its symmetric occur in the phase space of X, then it has no singularities.

b) If either the region C or F occur in the phase space of X, then the regions D, E, the symmetric of C and the symmetric of F cannot occur in the same phase space.

<u>Proof</u>: a) Let $p \in (P,S) \subset S^1$ and define $\mathcal{L}_{p,-p}$ as in Lemma 5.c.

By a similar argument to the one used in that lemma we can prove part a) (see fig. 5.d).

b) We now suppose that the phase space of $\pi(X)$ is the union of the regions C, B and the symmetric of C. Let s_1 and s_2 be the saddles of X, $\mathcal{L}_{s_2}$ the union of all straight lines through s_2 and $L_{s_1s_2} \subset \mathcal{L}_{s_2}$ the straight line through s_1 and s_2. We consider two cases: 1) $L_{s_1s_2}$ is tangent at s_1 to some invariant manifold of s_1; 2) $L_{s_1s_2}$ is transversal at s_1 to all invariant manifolds. In the first case we take the straight line $L_1 \subset \mathcal{L}_{s_2}$ close to $L_{s_1s_2}$ such that near to s_2 we have two contacts between X and L_1. This is impossible by Lemma 5.2. The second case can be proved by a similar argument. If we change C by the region F, analogue arguments can be used to proof the non existence of the phase space realizing $X \in \Sigma_2$.

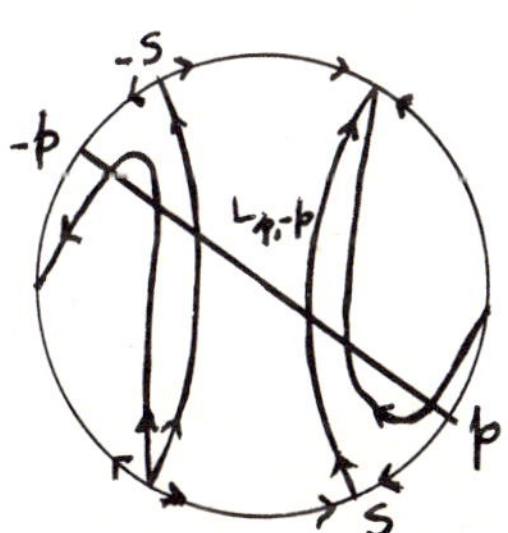

Fig. 5d

We will now describe how to obtain all the phase spaces which are possible in this case.

Take the homogeneous vector field

$$\dot{x} = x^2 + y^2$$
$$\dot{y} = 2xy$$

which has type SPF singularities at infinite.

d1) Let $\tilde{W}_1$ be the vector field

$$\dot{x} = x^2 + y^2 + \varepsilon$$
$$\dot{y} = 2xy, \quad \varepsilon > 0.$$

$y = 0$ is the connection between the saddles at infinite and we have no singularities since $x^2 + y^2 + \varepsilon > 0$. Taking the rotated vector field $W_1 = \tilde{W}_1^{\theta}$, $\theta > 0$, we obtain $\pi(W_1)$ with no limit cycles and structurally stable. Figure 5.d1 gives the phase space of $\pi(W_1)$.

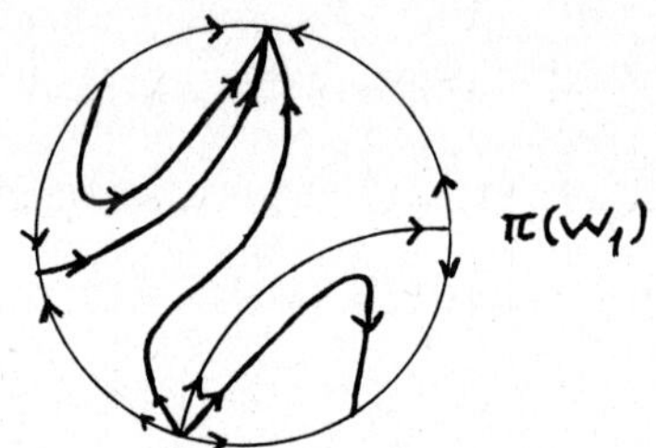

Fig. 5.d1

d2) Take $\tilde{W}_2$ to be the vector field

$$\dot{x} = x^2 + y^2 - \varepsilon^2$$
$$\dot{y} = 2(x+\alpha)y, \quad \text{with} \quad \varepsilon > 0 \quad \text{and} \quad \alpha+\varepsilon < 0.$$

$\tilde{W}_2$ leaves invariant the x-axis and on it we have two singularities, one saddle and one sink, both hyperbolic ones. It follows that $\pi(\tilde{W}_2)$ has no limit cycles and has a connection between one saddle on the plane and one saddle at infinite which is the extreme of the half line $y = 0$, $x > 0$. If we set $W_2 = \tilde{W}_2^{\theta}$, the vector field $\pi(W_2)$ (see fig. 5.d2) is structurally stable and has no limit cycles.

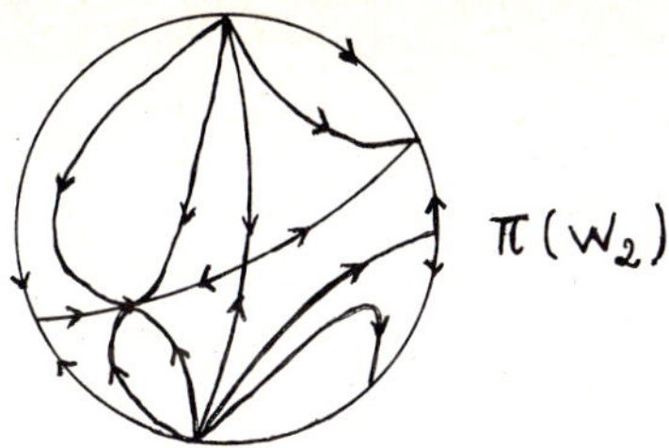

Fig. 5.d1

d3) We now consider the vector field $\tilde{W}_3$ given by

$$\dot{x} = \epsilon y + x^2 + y^2$$

$$\dot{y} = -\epsilon x + 2xy, \quad \epsilon > 0.$$

This vector field leaves invariant the straight line $y = \frac{\epsilon}{2}$ and its singularities are $(0,0)$ and $(0,-\epsilon)$. The singularity $(0,0)$ is a center since the vector field is symmetric under the transformation $(x,y) \mapsto (-x,y)$. $\pi(W_3)$, with $W_3 = \tilde{W}_3^{\theta}$, $\theta > 0$, is structurally stable having no limit cycles. Now, let $\tilde{W}_4 = \tilde{W}_3 + (\delta x, 0)$, $\delta > 0$; this vector field has $y = \frac{\epsilon}{2}$ as a connection between the saddles at infinite and since $\tilde{W}_3$ and $\tilde{W}_4$ are transversal, except along the horizontal lines, it has no limit cycles. Take $W_4 = \tilde{W}_4^{\theta}$, $\theta > 0$. $\pi(W_4)$ has no limit cycles and it is stable. Figure 5.d3.1 and 5.d3.2 give the phase spaces of $\pi(W_3)$ and $\pi(W_4)$.

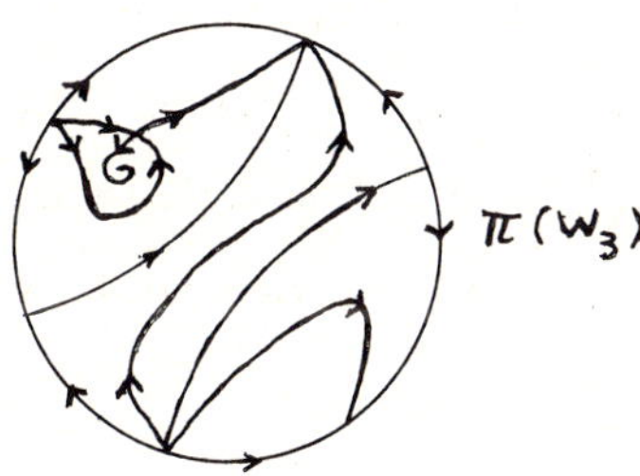

Fig. 5.d3.1

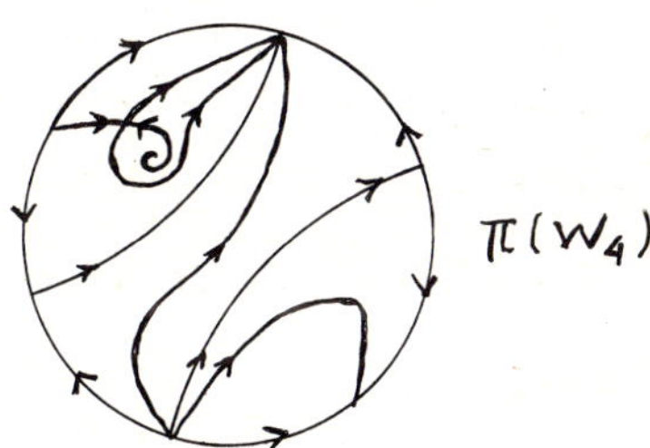

Fig. 5.d3.2

d4) The linear vector field

$$A = \begin{pmatrix} \lambda & -2\lambda \\ 0 & -\lambda \end{pmatrix}$$

leaves invariant the x-axis and the straight line $y = x$. Let $\tilde{W}_5$ be the vector field $(\dot{x},\dot{y}) = A(x,y) + (x^2+y^2,\ 2xy)$. Since $\tilde{W}_5$ keep leaving invariant the x-axis and $y = x$, it will appear, besides $(0,0)$ which is a hyperbolic saddle, two singularities one source and one sink, each one in a different invariant straight line and both hyperbolic. In the regions $y > 0$, $y > x$, $\tilde{W}_5$ has a hyperbolic saddle and, since the source and the sink are situated on invariant straight lines, it has no limit cycles. By rotation we obtain two vector fields, $W_5 = \tilde{W}_5^{\theta}$, $\theta > 0$ and $W_6 = \tilde{W}_5^{\theta}$, $\theta < 0$, such that $\pi(W_5)$ and $\pi(W_6)$ are structurally stable and have no limit cycles. Figures 5.d4.1 and 5.d4.2 give the phase spaces of $\pi(W_5)$ and $\pi(W_6)$.

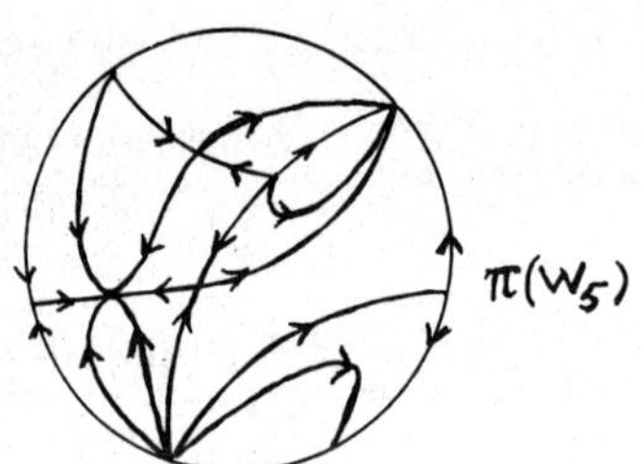

Fig. 5.d4.1

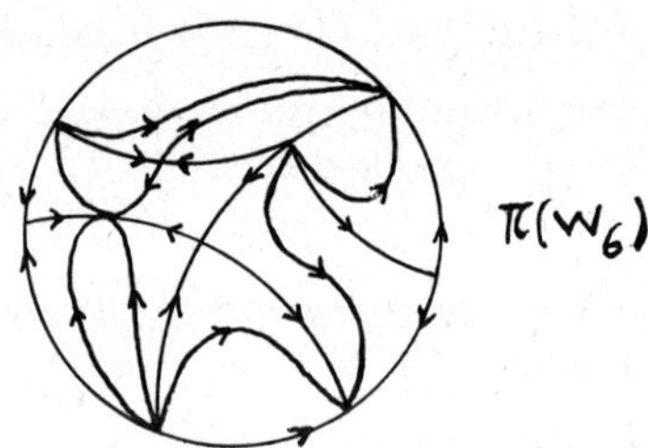

Fig. 5.d4.2

Case e: The vector field is type S_1S_2F at infinite.

Consider the homogeneous vector field R given by

$$\dot{x} = x(2y+x)$$
$$\dot{y} = y(y+2x)$$

whose singularity is type S_1S_2F at infinite.

e1) Take the vector field $\bar{R}_1 = R + (\varepsilon^2,\varepsilon^2)$. This vector field leaves invariant the straight line $y = x$ and has the singularities $(\varepsilon,-\varepsilon)$ and $(-\varepsilon,\varepsilon)$. Beeing symmetrical under the transformation $(x,y)\longmapsto(-y,-x)$ these singularities are centers. Let $R_1 = \bar{R}_1^\theta$, $\theta > 0$, a rotation of R_1. $\pi(R_1)$ is structurally stable and has no limit cycles. Its phase space is given in Figure 5.e1.1. Now we fix ε and take the vector field R_2 given by

$$\dot{x} = x(2y+x) + \varepsilon^2 + \delta(x+y)$$
$$\dot{y} = y(y+2x) + \varepsilon^2 + \delta(x+y).$$

If $\delta > 0$ and small, R_2 has two singularities. The orbits of R_2 and $\bar{R}_1$ are transversal except along the line $y + x = 0$ where they coincide. In the regions $y + x > 0$ and $y + x < 0$, the vector field points outside the closed orbits of $\bar{R}_1$ which implies that R_2 has no limit cycles. By direct calculations we can verify the $(-\varepsilon,\varepsilon)$ and $(\varepsilon,-\varepsilon)$ are hyperbolic sources. $\pi(R_2)$ is structurally stable and has no limit cycles. Its phase space is given in Figure 5.e1.2.

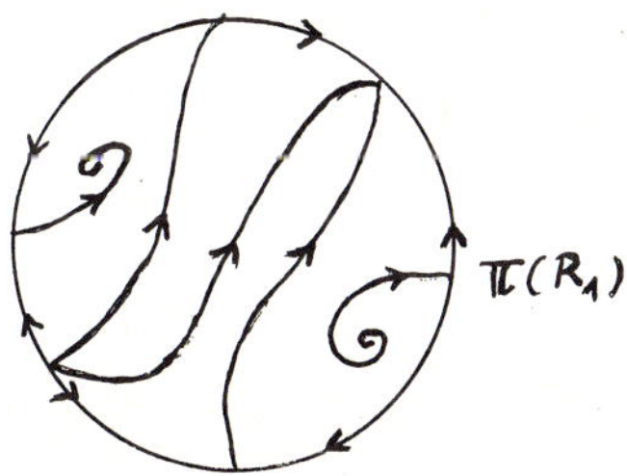

Fig. 5.e1.1

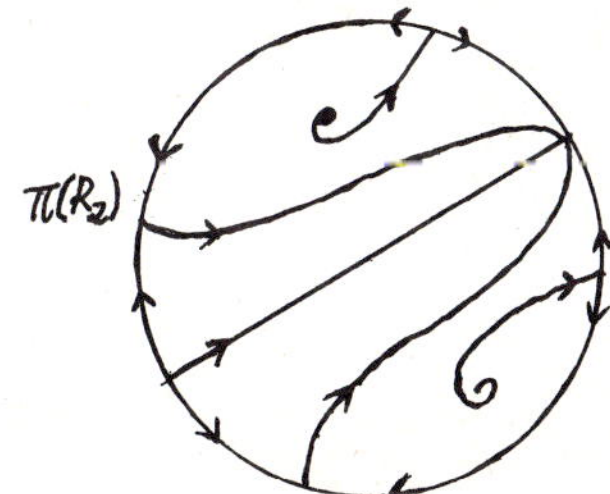

Fig. 5.e1.2

e2) Let $R_3 = R + (-\varepsilon^2,0)$. R_3 has two singularities, $(\varepsilon,0)$ and $(-\varepsilon,0)$, leaving invariant the x-axis. These singularities are one sink and one source, both hyperbolic. The invariance of the x-axis implies that we have neither limit cycles nor saddle connections for R_3. $\pi(R_3)$ is structurally stable and has no limit cylces. Its phase space is given in Figure 5.e2.

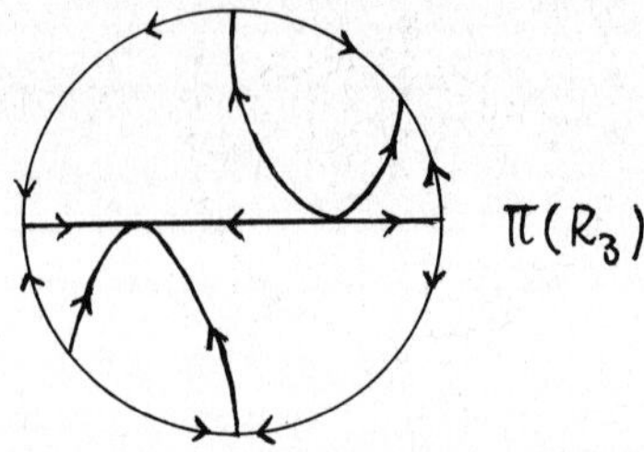

Fig. 5.e2

e3) We take now the vector field $\bar{R}_4$:

$$\dot{x} = x(\lambda+x+2y)$$
$$\dot{y} = y(-\lambda+2x+y), \quad \lambda > 0.$$

The singularities of $\bar{R}_4$ are $(0,0)$, $(-\lambda,0)$, $(0,\lambda)$ and $(\lambda,-\lambda)$. Since the vector field is symmetric under the transformation $(x,y) \longmapsto (-y,-x)$ and the eigenvalues of the last singularity are complex, than this singularity is a center. The remaining singularities are a saddle, a source and a sink. The vector field $\bar{R}_{4,\delta} = \bar{R}_4 + (\delta x(x+y), \delta x(x+y))$ keep leaving invariant the y-axis and using the same argument as in subcase e1), we have no limit cycles. Fix $\delta > 0$ and consider $R_4 = \bar{R}_4^{\theta}$, $\theta > 0$, and $R_5 = \bar{R}_{4,\delta}^{\theta}$, $\theta < 0$. $\pi(R_4)$ and $\pi(R_5)$ are structurally stable, have no limit cycles and their phase spaces are given in 5.e3.1 and 5.e3.2. Now fix $\delta < 0$ and take $R_6 = \bar{R}_{4,\delta}^{\theta}$, $\theta > 0$ then obtaining $\pi(R_6)$ with no limit cycles and structurally stable. Its phase space is given in Figure 5.e3.3.

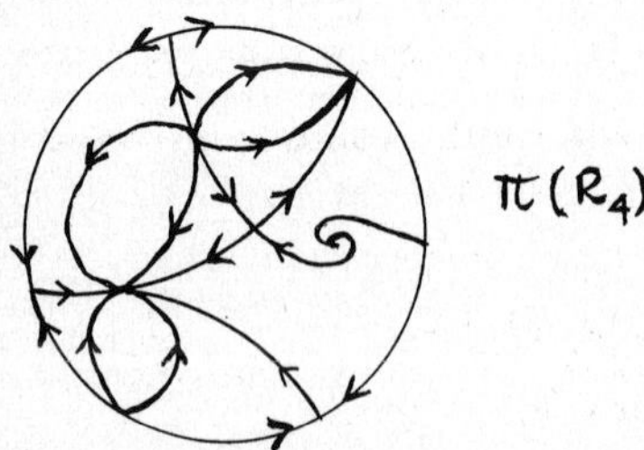

Fig. 5.e3.1

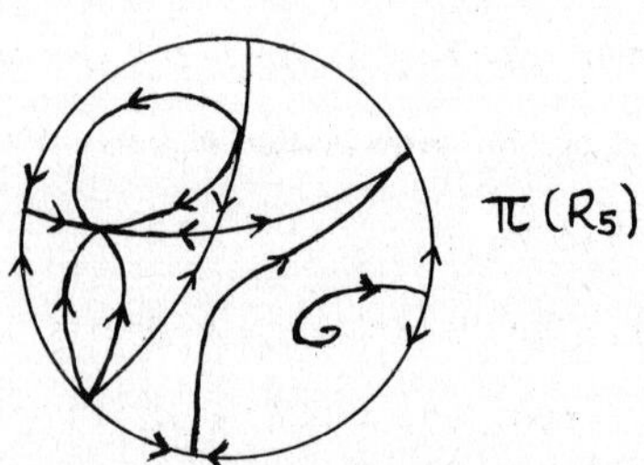

Fig. 5.e3.2

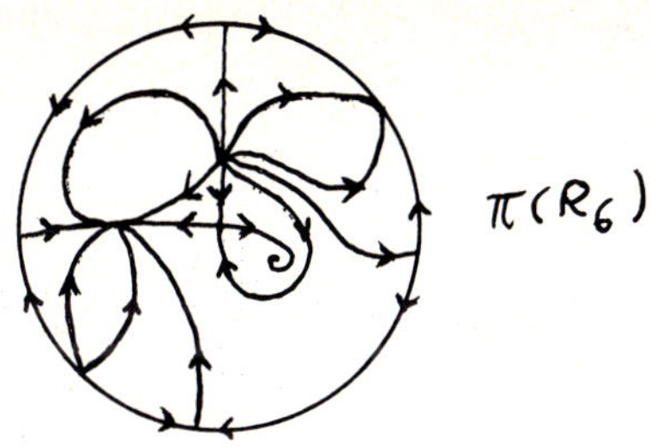

Fig. 5.e3.3

With next lemma we finish the proof of case e.

Lemma 5.e - There is no vector field $X \in \Sigma$ satisfying the following conditions:

a) X has type S_1S_2F singularities at infinite and one saddle s in the finite plane.

b) $\bar{W}^s(s)$ (closure of $W^s(s)$) is a unique singularity and $\bar{W}^u(s) = \{p,-F\}$, where p is a sink of X in the finite plane.

Proof: Let $q \in (S_1,S_2) \subset S^1$ and $\mathcal{L}_{q,-q}$ be the union of all straight lines parallel to $L_{q,-q}$. Using similar arguments to the one used in Lemma 5.2 we can prove the lemma.

Cases a trough e prove Theorem D.

Now we state some problems suggested by the theorems we just prove. Let $X(x,y) = (P(x,y),Q(x,y))$ such that $\pi(X) \in$ PMS. On S^2 take the open neighborhoods $A_1(X),\ldots,A_p(X)$ in such way that given a limit cycle γ of $\pi(X)$ then $\gamma \subset A_i(X)$ for some i. Define $\pi(X)$ to be equivalent to $\pi(Y)$ if the following conditions are satisfied:

a) There is a homeomorphism $h: S^1 - \bigcup_{i=1}^{p} A_i(X) \to S^2 - \bigcup_{i=1}^{p} A_i(Y)$ mapping orbits of $\pi(X)$ into orbits of $\pi(Y)$.

b) The number of limit cycles of $\pi(X)$ in $A_i(X)$ is congruent module 2 to the number of limit cycles of $\pi(Y)$ in $A_i(Y)$.

<u>Problem 1</u> - Classify the PMS vector fields module the equivalence defined above.

<u>Problem 2</u> - Given a vector field $X = (P,Q,R)$ where P, Q and R are polynomials in three variables, find necessary and sufficient conditions for $\pi(X)$ be structurally stable in a neighborhood of infinite.

REFERENCES

[1] Andronov, A.A.; Leontovich, E.A.; Gordin I.I.; Maier, A.G. Theory of dynamical systems on the plane. Israel Program for Scientif. Translations. Jerusalem, 1971.

[2] Andronov, A.A.; Vitt, A.A.; Khaikin, S.E. - Theory of oscillators, Addison Wesley, 1966.

[3] Argemi, J. - Sur les points singuliers multiple des systèmes dynamiques dans R^2. Ann. Mat. Pura ed Appl. Ser. IV, 79, 1968.

[4] Bendixson, I. - Sur les courbes definies par des equations differentielles, Acta Math. Scandin. 24, 1901.

[5] Coddington, E.A.; Levinson, N. - Theory of ordinary differential equations, McGraw Hill, 1955.

[6] Coppel, W.A. - A survey of quadratic systems, Jour. of diff. equations, 2, 1966.

[7] Dickson, R.J.; Perko, L.M. - Bounded quadratic systems in the plane, Jur. of diff. equations, 7, 1970.

[8] Dulac, H. - Sur les cycles limites, Bull. Soc. Math. de France, 51, 1923.

[9] Duff, G.F.D. - Limit cycles and rotated vector fields, Annals of Math. 57, 1953.

[10] Dumortier, F. - Singularities of vector fields on the plane, Thesis, Vrie Universiteit Brussel, 1973.

[11] Gonzalez Velasco, E.A. - Generic properties of polynomials vector fields at infinity, Trans. of the AMS, 143, 1969.

[12] Hilbert, D. - Mathematical Problems, Bull. A.M.S. 8, 1902.

[13] Hirsch, W.M.; Pugh, C.C. - Stable manifolds and hyperbolic sets, Global Analysis, Proc. of Symp. in Pure Math. of the A.M.S. vol. XIV, 1970.

[14] Liagina, L.S. - Integral curves of the equation $y' = \frac{ax^2+bxy+cy^2}{dx^2+exy+fy^2}$. Uspekhi Mat. Nauk 6, nº 2 (42), 1951 (in Russian).

[15] Lotka, A.J. - Elements of Mathematical Biology, Dover, 1956.

[16] Manning, A. (Ed.) - Dynamical Systems - Warwick 1974, Springer Verlag Notes 468, 1975.

[17] Marcus, L. - Quadratic differential equations and non-associative algebras, Contribution to the theory of non-linear oscillations, vol. 5 (Edited by S. Lefeschetz).

[18] Palis, J. - On Morse-Smale dynamical systems, Topology 8, 1969.

[19] Peixoto M.C.; Peixoto, M.M. - Structural stability in the plane with enlarged boundary conditions, An. Acad. Bras. de Ciências, 31, 1959.

[20] Poincare, H. - Memoire sur les courbes definies par une equation differentielle, J. Mathematiques (3), 7, 1881.

[21] Richardson, L.F. - Generalized Foreigh Politics (a study in group psychology), Cambridge University Press, 1939.

[22] Sharpe, F.R. - The Topography of certain curves defined by a differential equation, Annals of Math. 11, 1910.

[23] Struik, D.J. - Analytic and Projective Geometry, Addison Wesley, 1953.

[24] Sotomayor, J. - Estabilidade Estrutural de 1ª ordem e variedades de Banach, Tese, IMPA, 1964.

[25] Volterra, V. - Théorie Mathématique de la lutte pour la vie, Gauthier-Villars, 1931.

Departamento de Matemática

Universidade Federal da Paraiba

João Pessoa - Paraiba - Brasil

THE CLASSIFYING RING OF SL(2,**C**)

by

Juan Tirao

Introduction. If G is a connected semi-simple Lie group, say with finite center, it is known that many of the fundamental questions concerning the infinite dimensional representation theory of G, reduce to questions about the structure and finite dimensional representation theory of the ring $\mathcal{G}^K$. Here $\mathcal{G}$ is the universal enveloping algebra, over **C**, of the Lie algebra of G, and $\mathcal{G}^K$ is the centralizer in $\mathcal{G}$ of a maximal compact subgroup K of G.

To study $\mathcal{G}^K$ one has an injective anti-homomorphism

$$P: \mathcal{G}^K \to \mathcal{K}^M \otimes A,$$

where M is the centralizer in K of a Cartan subalgebra $\mathfrak{a}$ of the symmetric pair (G,K), $\mathcal{K}$ and A are the universal enveloping algebras, over **C**, corresponding to K and $\mathfrak{a}$, respectively, $\mathcal{K}^M$ is the centralizer of M in $\mathcal{K}$ and $\mathcal{K}^M \otimes A$ is given the tensor product algebra structure. The map P may be viewed as generalizing the famous Harish-Chandra's homomorphism

$$\gamma: \mathcal{G}^K \to A.$$

In [3] a subalgebra $\mathcal{B}$ of $\mathcal{K}^M \otimes A$ containing $P(\mathcal{G}^K)$ was defined. However P is not an anti-isomorphism of $\mathcal{G}^K$ onto $\mathcal{B}$. But it was proved that $\mathcal{G}^K$ suitably localized and completed is indeed isomorphic to $\mathcal{B}$ suitably localized and completed.

The subalgebra $\mathcal{B}$ may be taken as the ring of invariants in $\mathcal{K}^M \otimes A$ of the Weyl group W. The operators associated to the elements of W being related to the Kunze-Stein intertwining operators.

In this paper we solve the problem of computing the image of $\mathcal{G}^K$ under P when G = SL(2,**C**). The image of P is characterized in the following way: The algebra $\mathcal{K}^M \otimes A$ is replaced by a new subalgebra $\mathcal{B}$ containing $P(\mathcal{G}^K)$ and stable under W; then we prove that $P(\mathcal{G}^K) = \mathcal{B}^W$, the ring of Weyl group invariants in $\mathcal{B}$.

The definition of $\mathcal{B}$ should be tracked down to some equivalences among irreducible subquotients of certain representations of the principal series of G (see Theorem 1.5).

It turns out that $P(G^K)$ is a polynomial ring in three variables. This is a particular case of a more general result of Cooper for groups with G^K commutative (see [1]).

This study of the image of P: $G^K \to K^M \otimes A$, when G = SL(2,**C**), should prove useful for the further development of this problem in the general case of a semi-simple Lie group. Indeed, it was the simplest example of the proper Lorentz group, or its universal covering SL(2,**C**), which elucidated many of the fundamental properties and methods of the infinite dimensional representation theory of a semi-simple Lie group.

Finally I want to thank my students Oscar Brega and Jorge Boggino for their helpful companionship in various stages of preparation of the manuscript.

1. Let $G = SL(2,\mathbf{C})$ be the group of all 2×2 complex matrices of determinant 1. Then G is the unique connected simply connected semi-simple Lie group of complex dimension 3. The group G contains $K = SU(2)$ as a maximal compact subgroup. This group consists of all matrices

$$\begin{pmatrix} a & b \\ -\overline{b} & \overline{a} \end{pmatrix}$$

of determinant 1. The diagonal matrices

$$\begin{pmatrix} a & 0 \\ 0 & a^{-1} \end{pmatrix} \quad a \in \mathbf{R},\ a > 0,$$

form an abelian subgroup A of G of dimension 1. The subgroup of all triangular matrices

$$\begin{pmatrix} 1 & z \\ 0 & 1 \end{pmatrix} \quad z \in \mathbf{C},$$

will be denoted by N.

Every matrix $g \in G$ can be represented in a unique way in the form

$$g = \kappa(g)a(g)n(g)$$

with $\kappa(g) \in K$, $a(g) \in A$, $n(g) \in \mathbf{N}$. Moreover, the map $(k,a,n) \longmapsto kan$ is a diffeomorphism of $K \times A \times N$ onto G. The decomposition $G = KAN$ is an Iwasawa decomposition for G.

A computation shows that if

$$g = \begin{pmatrix} \alpha & \gamma \\ \beta & \delta \end{pmatrix}$$

then

$$\kappa(g) = c^{-1}\begin{pmatrix} \alpha & -\overline{\beta} \\ \beta & \overline{\alpha} \end{pmatrix}, \qquad a(g) = \begin{pmatrix} c & 0 \\ 0 & c^{-1} \end{pmatrix}$$

with $c = (\alpha\overline{\alpha} + \beta\overline{\beta})^{1/2}$.

The centralizer M of A in K consists of all matrices

$$m_\theta = \begin{pmatrix} e^{i\theta} & 0 \\ 0 & e^{-i\theta} \end{pmatrix} \qquad \theta \in \mathbf{R}.$$

The set $\hat{M}$ of all equivalence classes of finite dimensional irreducible representations of M is parametrized by the integers, with

$$\sigma_\ell(m_\theta) = e^{i\ell\theta}, \ \ell \in \mathbf{Z}.$$

The group of characters of A is parametrized by the complex numbers, in the following way: if

$$a = \begin{pmatrix} t & 0 \\ 0 & t^{-1} \end{pmatrix}, \ t > 0$$

then $a^\lambda = t^\lambda$, $\lambda \in \mathbf{C}$.

We shall consider a family U^λ of continuous representations of G parametrized by **C** and realized on $L^2(K)$. Given $\lambda \in \mathbf{C}$ and $g \in G$, define $U^\lambda(g)$ by the prescription

$$(1.1) \qquad (U^\lambda(g)f)(k) = a(g^{-1}k)^{-(\lambda+2)}f(\kappa(g^{-1}k)), \ f \in L^2(K)$$

(see Warner [6], p. 445).

Let R denote the right regular representation of K on $L^2(K)$; then $R(m)U^\lambda(g) = U^\lambda(g)R(m)$ $(g \in G,\ m \in M)$. This follows upon observing that

$$\begin{aligned} gkm &= \kappa(gk)a(gk)n(gk)m \\ &= \kappa(gk)ma(gk)(m^{-1}n(gk)m), \ k \in K \end{aligned}$$

where $m^{-1}n(gk)m \in N$. Put

$$Q(\ell) = \frac{1}{2\pi}\int_0^{2\pi} e^{-i\ell\theta}R(m_\theta)d\theta, \ \ell \in \mathbf{Z}.$$

The operators $Q(\ell)$ are projections in $L^2(K)$ and all commute with $U^\lambda(g)$, $g \in G$. We have

$$L^2(K) = \sum_{\ell \in \mathbf{Z}} Q(\ell) L^2(K),$$

the sum being a unitary direct sum. The closed subspaces $Q(\ell)L^2(K)$ are U^λ-stable; the representation $U^\lambda | Q(-\ell)L^2(K)$ will be denoted by $U^{\ell,\lambda}$. Note that the representation space of $U^{\ell,\lambda}$ is the subspace of all $f \in L^2(K)$ such that

$$f(km_\theta) = e^{-i\ell\theta} f(k), \quad k \in K.$$

The series of representations $U^{\ell,\lambda}$ ($\ell \in \mathbf{Z}$, $\lambda \in \mathbf{C}$) is the so called nonunitary principal series of G.

Every finite dimensional irreducible representation of K is determined, up to equivalence, by its degree (i.e. by the dimension of the representation space). Conversely, for any non-negative integer n, there exists an irreducible representation of K of dimension n + 1. Thus, the set $\hat{K}$ of all equivalence classes of finite dimensional irreducible representations of K can be parametrized by the non-negative integers. A representation of degree n + 1 can be realized in the space V_n of all homogeneous polinomials $p(z,w)$ of degree n. The operator representing the element

$$k = \begin{pmatrix} a & b \\ -\bar{b} & \bar{a} \end{pmatrix}$$

in this space is given by the formula

$$(k.p)(z,w) = p(\bar{a}z - bw, \bar{b}z + aw). \tag{1.2}$$

For each n = 0, 1, 2, ..., let V_n^* be the dual space of V_n, and let A_n be the map from $V_n^* \otimes V_n$ to $C(K)$ defined by $A_n(\nu \otimes p)(k) = \nu(k.p)$ for $k \in K$, $p \in V_n$, $\nu \in V_n^*$. We note that

$$A_n(k_1\nu \otimes k_2 p)(k) = (k_1\nu)(kk_2 p) = \nu(k_1^{-1}kk_2 p)$$

$$= A_n(\nu \otimes p)(k_1^{-1}kk_2); \; k, k_1, k_2 \in K.$$

Now, identifying the elements of $V_n^* \otimes V_n$ with their A_n-images in $C(K)$ we have (Peter-Weyl theorem)

$$L^2(K) = \sum V_n^* \otimes V_n,$$

a unitary direct sum.

Let $p^n_\ell(z,w) = z^\ell w^{n-\ell}$, $\ell = 0, 1, \ldots, n$. Then $p^n_0, \ldots, p^n_n$ is a basis of V_n. Let $\nu^n_0, \ldots, \nu^n_n$ be the corresponding dual basis and set $f^n_{j,\ell} = A_n(\nu^n_j \otimes p^n_\ell)$. As a function on K, $f^n_{j,\ell}$ is given by

$$f^n_{j,\ell}(k) = \sum_{p+q=j} \binom{\ell}{p}\binom{n-\ell}{q} \bar{a}^p \bar{b}^q (-b)^{\ell-p} a^{n-\ell-q},$$

with

$$k = \begin{pmatrix} a & b \\ -\bar{b} & \bar{a} \end{pmatrix}.$$

If

$$m_\theta = \begin{pmatrix} e^{i\theta} & 0 \\ 0 & e^{-i\theta} \end{pmatrix}$$

then

$$m_\theta p^n_\ell = e^{i(n-2\ell)\theta} p^n_\ell.$$

Therefore $R(m_\theta)f^n_{j,\ell} = A_n(\nu^n_j \otimes m_\theta p^n_\ell) = e^{i(n-2\ell)\theta} f^n_{j,\ell}$ and consequently

$$Q(-m)L^2(K) = \sum_{\ell \geq 2^{-1}(m+|m|)} \sum_j \mathbf{C} f^{2\ell-m}_{j,\ell}, \tag{1.3}$$

a unitary direct sum. The orthogonality follows from

$$L(m_\theta)f^n_{j,\ell} = A_n(m_\theta \nu^n_j \otimes p^n_\ell) = e^{i(2j-n)\theta} f^n_{j,\ell}$$

where L denotes the left regular representation of K on $L^2(K)$.

It is well known that G = KAK; a proof goes like this: given $g \in G$, let $b = g^*g$ (* denotes conjugate transpose); b is a Hermitian positive definite matrix. Therefore there exist $v \in K$ and $a \in A$ such that $vbv^* = a^2$. Letting $u = gv^*a^{-1}v$, we have $u \in K$ and $g = (uv^*)av$ as desired.

By virtue of this, the representations of G are determined by their restrictions to K and A. Clearly $U^\lambda|K = L$ and

$$L^2(K) = \sum V^*_n \otimes V_n,$$

K acting on the left factors V_n^*. Thus the representation U^λ will be determined once the effect of $U^\lambda(a)$ for each $a \in A$, is known. This is equivalent to computing the action of the infinitesimal operator $U^\lambda(H)$ on the algebraic direct sum $\sum V_n^* \otimes V_n$, H being the matrix

$$\begin{pmatrix} 1 & 0 \\ 0 & -1 \end{pmatrix}.$$

Lemma 1.1. *If* $k = \begin{pmatrix} a & b \\ -\bar{b} & \bar{a} \end{pmatrix}$ *is an element of* K, *then*

$$(U^\lambda(H)f^n_{j,\ell})(k) = (n + \lambda + 2)(a\bar{a} - b\bar{b})f^n_{j,\ell}(k)$$

$$+ \sum_{p+q=j} \binom{\ell}{p}\binom{n-\ell}{q}(2q - 2p + 2\ell - n)\bar{a}^p\bar{b}^q(-b)^{\ell-p}a^{n-\ell-q}.$$

Proof. Let

$$a_t = \exp tH = \begin{pmatrix} e^t & 0 \\ 0 & e^{-t} \end{pmatrix}.$$

We have

$$a_t^{-1}k = \begin{pmatrix} e^{-t}a & e^{-t}b \\ -e^t\bar{b} & e^t\bar{a} \end{pmatrix}$$

$$\kappa(a_t^{-1}k) = c^{-1}\begin{pmatrix} e^{-t}a & e^t b \\ -e^t\bar{b} & e^{-t}\bar{a} \end{pmatrix}$$

and

$$a(a_t^{-1}k) = \begin{pmatrix} c & 0 \\ 0 & c^{-1} \end{pmatrix}$$

where $c = (e^{-2t}a\bar{a} + e^{2t}b\bar{b})^{1/2}$. Then

$$(U^\lambda(a_t)f^n_{j,\ell})(k) = c^{-(\lambda+2)}f^n_{j,\ell}(\kappa(a_t^{-1}k))$$

$$= c^{-(\lambda+2)-n}\sum_{p+q=j}\binom{\ell}{p}\binom{n-\ell}{q}e^{t(2q-2p+2\ell-n)}\bar{a}^p\bar{b}^q(-b)^{\ell-p}a^{n-\ell-q}.$$

Now differentiating with respect to t and letting $t = 0$ we get the desired result upon observing that at $t = 0$ we have $c = 1$ and $dc/dt = b\bar{b} - a\bar{a}$. Q.E.D.

Lemma 1.2. *If* $n \geq 2$ *and* $j(n - j)\ell(n - \ell) \neq 0$ *there exist numbers* A, B *and* C *(which depend on* j, ℓ, n, λ*) such that*

$$U^{\lambda}(H)f^{n}_{j,\ell} = Af^{n+2}_{j+1,\ell+1} + Bf^{n}_{j,\ell} + Cf^{n-2}_{j-1,\ell-1}.$$

Moreover

$$C = \frac{2\ell(n - \ell)(\lambda - n)}{n(n + 1)}.$$

Proof. Since $a\bar{a} + b\bar{b} = 1$ we can write

$$(U^{\lambda}(H)f^{n}_{j,\ell})(k) = (n + \lambda + 2) \sum_{p+q=j} \binom{\ell}{p}\binom{n-\ell}{q}\bar{a}^{p+1}\bar{b}^{q}(-b)^{\ell-p}a^{n-\ell-q+1}$$

$$+ (n + \lambda + 2) \sum_{p+q=j} \binom{\ell}{p}\binom{n-\ell}{q}\bar{a}^{p}\bar{b}^{q+1}(-b)^{\ell-p+1}a^{n-\ell-q}$$

$$+ \sum_{p+q=j} \binom{\ell}{p}\binom{n-\ell}{q}(2q - 2p + 2\ell - n)\bar{a}^{p+1}\bar{b}^{q}(-b)^{\ell-p}a^{n-\ell-q+1}$$

$$- \sum_{p+q=j} \binom{\ell}{p}\binom{n-\ell}{q}(2q - 2p + 2\ell - n)\bar{a}^{p}\bar{b}^{q+1}(-b)^{\ell-p+1}a^{n-\ell-q}$$

as a homogeneous function of degree $n + 2$.

Similarly we can write

$$(Af^{n+2}_{j+1,\ell+1} + Bf^{n}_{j,\ell} + Cf^{n-1}_{j-1,\ell-1})(k) = \sum_{\substack{p+q=j \\ -1\leq p\leq j}} \Big\{ A\binom{\ell+1}{p+1}\binom{n-\ell+1}{q}$$

$$+ B\binom{\ell}{p}\binom{n-\ell}{q} - B\binom{\ell}{p+1}\binom{n-\ell}{q-1} + C\binom{\ell-1}{p-1}\binom{n-\ell-1}{q} + C\binom{\ell-1}{p+1}\binom{n-\ell-1}{q-2}$$

$$- 2C\binom{\ell-1}{p}\binom{n-\ell-1}{q-1}\Big\}\bar{a}^{p+1}\bar{b}^{q}(-b)^{\ell-p}a^{n-\ell-q+1},$$

where as usual the combinatorial number $\binom{r}{s}$ is zero whenever $s < 0$ or $s > r$.

Now, these two homogeneous functions of degree $n+2$ will agree on K if and only if they agree for all $(a,b) \in \mathbf{C} \times \mathbf{C}$. This happens if and only if they have the same coefficients. (Recall that if $f\colon \mathbf{C}^n \times \mathbf{C}^n \to \mathbf{C}$ is holomorphic in the first n variables and antiholomorphic in the last n variables, then $f(z,z) = 0$ for all $z \in \mathbf{C}^n$ implies $f = 0$). Thus we consider the linear equations

$$A\binom{\ell+1}{p+1}\binom{n-\ell+1}{q} + B\binom{\ell}{p}\binom{n-\ell}{q} - B\binom{\ell}{p+1}\binom{n-\ell}{q-1} + C\binom{\ell-1}{p-1}\binom{n-\ell-1}{q}$$

$$+ C\binom{\ell-1}{p+1}\binom{n-\ell-1}{q-2} - 2C\binom{\ell-1}{p}\binom{n-\ell-1}{q-1} = (n+\lambda+2)\binom{\ell}{p}\binom{n-\ell}{q}$$

$$+ (n+\lambda+2)\binom{\ell}{p+1}\binom{n-\ell}{q-1} + (2q-2p+2\ell-n)\binom{\ell}{p}\binom{n-\ell}{q}$$

$$- (2q-2p+2\ell-n-4)\binom{\ell}{p+1}\binom{n-\ell}{q-1},$$

where $p+q = j$ and $-1 \leq p \leq j$. Expanding and eliminating denominators we obtain the stronger conditions:

$$A\ell(\ell+1)(n-\ell)(n-\ell+1) + B\ell(n-\ell)(p+1)(n-\ell+1-j+p) - B\ell(n-\ell)(\ell-p)(j-p)$$

$$+ Cp(p+1)(n-\ell-j+p)(n-\ell-j+p+1) + C(\ell-p)(\ell-p-1)(j-p)(j-p-1)$$

$$- 2C(p+1)(\ell-p)(j-p)(n-\ell+1-j+p) - (n+\lambda+2)\ell(n-\ell)(p+1)(n-\ell+1-j+p)$$

$$- (n+\lambda+2)\ell(n-\ell)(\ell-p)(j-p) - (2\ell-n+2j-4p)\ell(n-\ell)(p+1)(n-\ell+1-j+p)$$

$$+ (2\ell-n+2j-4-4p)\ell(n-\ell)(\ell-p)(j-p) = 0,$$

for $-1 \leq p \leq j$. The left hand side can be viewed as a polynomial in p of degree ≤ 4. The coefficient of p^4 is

$$2C - 2C = 0.$$

The coefficient of p^3 is also equal to zero. The coefficients of p^2, p, p^0 are respectively of the form

$$(n^2+n)C - \ell(n-\ell)(2\lambda-2n)$$

$$\ell(n-\ell)(n+2)B + \quad * \; C + \quad *$$

$$\ell(\ell+1)(n-\ell)(n-\ell+1)A + \quad * \; B + \quad * \; C + \quad *.$$

Since, by assumption $n\ell(n-\ell) \neq 0$ there exists a unique solution A, B, C annhilating the coefficients of p^2, p and p^0. Moreover, $C = 2\ell(n-\ell)(\lambda-n)n^{-1}(n+1)^{-1}$. This proves the lemma. Q.E.D.

Theorem 1.3. *We have*

$$U^\lambda(H)f^n_{j,\ell} = Af^{n+2}_{j+1,\ell+1} + Bf^n_{j,\ell} + Cf^{n-2}_{j-1,\ell-1} \tag{1.4}$$

where

$$A = \frac{2(j+1)(n-j+1)(n+\lambda+2)}{(n+1)(n+2)},$$

$$B = \frac{(n-2\ell)(n-2j)\lambda}{n(n+2)} \quad \text{if} \quad n \neq 0 \quad \text{and} \quad B = 0 \quad \text{if} \quad n = 0,$$

$$C = \frac{2\ell(n-\ell)(\lambda-n)}{n(n+1)} \qquad (n \geq 2,\ j(n-j)\ell(n-\ell) \neq 0).$$

Proof. We know that $U^\lambda(H)f^n_{j,\ell}$ is always a finite linear combination of the $f^r_{s,t}$ with $r - 2t = n - 2\ell$ (see (1.3)). Moreover, since M centralizes A we can also assume that $r - 2s = n - 2j$. Thus

$$U^\lambda(H)f^n_{j,\ell} = \sum_t a_t f^{n+2t}_{j+t,\ell+t} . \tag{1.5}$$

Set $F = 2^{-1}(-u - iv)$, where

$$u = \begin{pmatrix} 0 & 1 \\ -1 & 0 \end{pmatrix} \quad \text{and} \quad v = \begin{pmatrix} 0 & i \\ i & 0 \end{pmatrix} . \tag{1.6}$$

Considering the corresponding one-parameter subgroups of K

$$\exp tu = \begin{pmatrix} \cos t & \sin t \\ -\sin t & \cos t \end{pmatrix}, \quad \exp tv = \begin{pmatrix} \cos t & i\sin t \\ i\sin t & \cos t \end{pmatrix},$$

an easy calculation yields

$$U^\lambda(F)\bar{a}^p\bar{b}^q(-b)^{\ell-p}a^{n-\ell-q} = -\frac{1}{2}(U^\lambda(u) + iU^\lambda(v))\bar{a}^p\bar{b}^q(-b)^{\ell-p}a^{n-\ell-q}$$

$$= p\bar{a}^{p-1}\bar{b}^q(-b)^{\ell-p+1}a^{n-\ell-q} + q\bar{a}^p\bar{b}^{q-1}(-b)^{\ell-p}a^{n-\ell-q+1}.$$

Of course, we identify the matrix

$$\begin{pmatrix} a & b \\ -\overline{b} & \overline{a} \end{pmatrix} \in K$$

with the pair $(a,b) \in \mathbf{C} \times \mathbf{C}$. Hence $U^\lambda(F)$ operates on the algebraic direct sum $\sum V_n^* \otimes V_n$ as the differential operator

$$-b \frac{\partial}{\partial \overline{a}} + a \frac{\partial}{\partial \overline{b}}.$$

In particular we have

$$U^\lambda(F) f^n_{j,\ell} = (n - j + 1) f^n_{j-1,\ell} \quad (1 \leq j \leq n)$$

$$= 0 \qquad \text{if} \quad j = 0.$$

Iterating we also have

$$(U^\lambda(F))^k f^n_{j,\ell} = \frac{(n - j + k)!}{(n - j)!} f^n_{j-k,\ell} \quad (k \leq j).$$

Now by Lemma 1.1

$$(U^\lambda(F))^{j+2}(U^\lambda(H) f^n_{j,\ell}) = 0,$$

because $U^\lambda(H) f^n_{j,\ell}$ is of degree $j + 1$ in $\overline{a}$ and in $\overline{b}$. Therefore in (1.5) we have $a_t = 0$ if $t \geq 2$.

If $n \geq 2$ and $j(n - j)\ell(n - \ell) \neq 0$ we already know (Lemma 1.2) that $U^\lambda(H) f^n_{j,\ell}$ is a linear combination of $f^{n+2}_{j+1,\ell+1}$, $f^n_{j,\ell}$ and $f^{n-2}_{j-1,\ell-1}$. It is now clear that if $n < 2$ or $j(n - j)\ell(n - \ell) = 0$, then $U^\lambda(H) f^n_{j,\ell}$ is a linear combination of $f^{n+2}_{j+1,\ell+1}$ and $f^n_{j,\ell}$.

It only remains to compute the coefficients A and B in (1.4). For this we apply $U^\lambda(F)^{j+1}$ to both sides of (1.4). We have

$$(U^\lambda(F))^{j+1}(U^\lambda(H) f^n_{j,\ell})(k) = (n + \lambda + 2)(U^\lambda(F))^{j+1}((a\overline{a} - b\overline{b}) f^n_{j,\ell}(k))$$

$$= \frac{(n + \lambda + 2)2(j + 1)n!}{(n - j)!} f^{n+2}_{0,\ell+1}.$$

On the other hand

$$(U^\lambda(F))^{j+1}(Af^{n+2}_{j+1,\ell+1} + Bf^n_{j,\ell} + Cf^{n-2}_{j-1,\ell-1}) = A\,\frac{(n+2)!}{(n-j+1)!}\,f^{n+2}_{0,\ell+1}.$$

Hence

$$A = \frac{2(j+1)(n-j+1)(n+\lambda+2)}{(n+1)(n+2)}.$$

To obtain B we compute

$$(1.7)\quad (U^\lambda(F))^j(U^\lambda(H)f^n_{j,\ell} - Af^{n+2}_{j+1,\ell+1})(k) = (n+\lambda+2)\Big(\frac{-2j(n-1)!}{(n-j)!}ab f^n_{1,\ell}(k)$$

$$+ \frac{n!}{(n-j)!}(a\bar{a} - b\bar{b})f^n_{0,\ell}(k)\Big) + \Big(-b\frac{\partial}{\partial\bar{a}} + a\frac{\partial}{\partial\bar{b}}\Big)^j \sum_{p+q=j} \binom{\ell}{p}\binom{n-\ell}{q}$$

$$(2q - 2p + 2\ell - n)\bar{a}^p\bar{b}^q(-b)^{\ell-p}a^{n-\ell-q} - A\frac{(n+1)!}{(n-j+1)!}\,f^{n+2}_{1,\ell+1}(k).$$

To calculate the second term on the right we first observe that

$$\Big(-b\frac{\partial}{\partial\bar{a}} + a\frac{\partial}{\partial\bar{b}}\Big)^j = \sum_s \binom{j}{s}(-b)^s a^{j-s}\Big(\frac{\partial}{\partial\bar{a}}\Big)^s\Big(\frac{\partial}{\partial\bar{b}}\Big)^{j-s}.$$

Then

$$\Big(-b\frac{\partial}{\partial\bar{a}} + a\frac{\partial}{\partial\bar{b}}\Big)^j \sum_{p+q=j} \binom{\ell}{p}\binom{n-\ell}{q}(2q - 2p + 2\ell - n)\bar{a}^p\bar{b}^q(-b)^{\ell-p}a^{n-\ell-q}$$

$$= j!\sum_s \binom{\ell}{s}\binom{n-\ell}{j-s}(2j - 4s + 2\ell - n)(-b)^\ell a^{n-\ell}$$

$$= \frac{(2j-n)(n-2\ell)(n-1)!}{(n-j)!}\,(-b)^\ell a^{n-\ell},$$

where we have used the identities

$$\sum_s \binom{\ell}{s}\binom{n-\ell}{j-s} = \binom{n}{j},\quad \sum_s s\binom{\ell}{s}\binom{n-\ell}{j-s} = \ell\binom{n-1}{j-1}.$$

Going back to (1.7) we obtain

$$(U^\lambda(F))^j(U^\lambda(H)f^n_{j,\ell} - Af^{n+2}_{j+1,\ell+1})(k) = \frac{(n+\lambda+2)(n-1)!(n-2\ell)(n-2j)}{(n+2)(n-j)!}$$

$$\times\ (-\bar{b}(-b)^{\ell+1}a^{n-\ell} + \bar{a}(-b)^\ell a^{n+1-\ell}) + \frac{(2j-n)(n-2\ell)(n-1)!}{(n-j)!}(-b)^\ell a^{n-\ell}$$

$$= \frac{(n-1)!(n-2j)(n-2\ell)\lambda}{(n+2)(n-j)}\,(-b)^\ell a^{n-\ell},$$

where we have used the equation $a\bar{a} + b\bar{b} = 1$. Hence

$$U^{\lambda}(F)^{j}(U^{\lambda}(H)f^{n}_{j,\ell} - Af^{n+2}_{j+1,\ell+1}) = \frac{(n-1)!(n-2j)(n-2\ell)\lambda}{(n+2)(n-j)!} f^{n}_{0,\ell}.$$

On the other hand,

$$U^{\lambda}(F)^{j}(Bf^{n}_{j,\ell} + Cf^{n-2}_{j-1,\ell-1}) = B\frac{n!}{(n-j)!} f^{n}_{0,\ell}.$$

Therefore, if $n > 0$

$$B = \frac{(n-2j)(n-2\ell)\lambda}{n(n+2)},$$

and if $n = 0$, $B = 0$. This completes the proof of the theorem. Q.E.D.

Corollary 1.4. *If* $\lambda \neq \pm(|m| + 2t)$, $t \in \mathbf{Z}_+$, *then* $U^{m,\lambda}$ *is a (topologically) irreducible representation of* G. *If* $\lambda = |m| + 2t$ *then*

$$(1.8) \qquad \sum_{s \geq t} (\sum_{j} \mathbf{C}\, f^{|m|+2s}_{j,2^{-1}(m+|m|)+s})$$

is the unique proper closed $U^{m,\lambda}(G)$*-invariant subspace of* $Q(-m)L^2(K)$. *If* $\lambda = -|m| - 2t$ *then* $Q(-m)L^2(K)$ *contains also a unique proper closed* $U^{m,\lambda}(G)$*-invariant subspace, namely*

$$(1.9) \qquad \sum_{0 \leq s < t} (\sum_{j} \mathbf{C}\, f^{|m|+2s}_{j,2^{-1}(m+|m|)+s}).$$

Proof. Assume that $\lambda \neq \pm(|m| + 2t)$, $t \in \mathbf{Z}_+$. To prove that $U^{m,\lambda}$ is irreducible it suffices to show that the corresponding coefficients A and C are never zero (cf. Theorem 1.3). This follows by inspection.

Suppose now that $\lambda = |m| + 2t$ for some $t \in \mathbf{Z}_+$. Then, the coefficients A are always different from zero and the coefficients C are also different from zero except those corresponding to $U^{\lambda}(H)f^{|m|+2t}_{j,2^{-1}(m+|m|)+t}$. This implies that (1.8) is a proper closed invariant subspace, and that it is unique.

Similarly when $\lambda = -|m| - 2t$ for some positive integer t, the C's never vanish and the A's are zero if and only if they correspond to $U^{\lambda}(H)f^{|m|+2t-2}_{j,2^{-1}(m+|m|)+t-1}$. Therefore (1.9) is the unique proper $U^{m,\lambda}(G)$-invariant subspace of $Q(-m)L^2(K)$. The corollary is proved. Q.E.D.

We shall denote by $W_{m,\lambda}$ and $V_{m,\lambda}$ the closed invariant subspaces defined respectively in (1.8) and (1.9). The corresponding irreducible subrepresentations of G will be denoted by $\overline{U}^{m,\lambda}$ and $S^{m,\lambda}$. In the case $\lambda = -|m| - 2t$, $t \in \mathbf{Z}_+$, the subquotient of $U^{m,\lambda}$ on the space $Q(-m)L^2(K)/V_{m,\lambda}$ will also be denoted by $\overline{U}^{m,\lambda}$. Note that this subquotient is irreducible.

Recall that λ denotes a complex number, m an integer and t a positive integer. Equivalence of representations will mean infinitesimal equivalence (see Warner [6], p. 326).

Theorem 1.5.

(i) *If* $\lambda \neq \pm(|m| + 2t)$ *then* $U^{m,\lambda} \simeq U^{-m,-\lambda}$

(ii) *If* $\lambda = \pm(|m| + 2t)$ *then* $\overline{U}^{m,\lambda} \simeq \overline{U}^{-m,-\lambda}$

(iii) *If* $m = \pm(|\lambda| + 2t)$, $\lambda \in \mathbf{Z}$ *then* $U^{m,\lambda} \simeq \overline{U}^{\lambda,m}$.

(i) is a classical result (see Naimark [5], p. 246) and (iii) is of special interest for us. Set $\overline{U}^{m,\lambda} = U^{m,\lambda}$ whenever $\lambda \neq \pm(|m| + 2t)$ and let L denote the integral lattice generated by (1,1) and (-1,1). Then we can rephrase Theorem 1.5 in the following way:

Theorem 1.5'.

(i) $\overline{U}^{m,\lambda} \simeq \overline{U}^{-m,-\lambda}$ *for all* $(m,\lambda) \in \mathbf{Z} \times \mathbf{C}$

(ii) $\overline{U}^{m,\lambda} \simeq \overline{U}^{\lambda,m}$ *for all* $(m,\lambda) \in L$.

In this respect is is helpful to look at the diagram

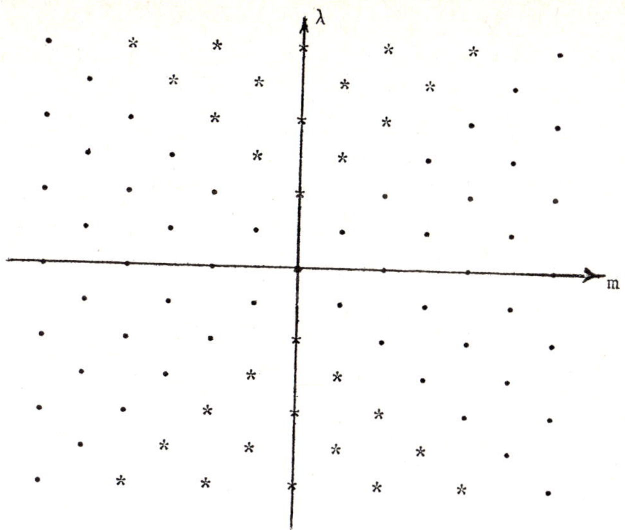

Proof of Theorem 1.5. (i) To show that $U^{m,\lambda}$ is equivalent to $U^{-m,-\lambda}$ is necessary and sufficient to exhibit an algebraic isomorphism

$$(1.10) \qquad Q: \sum_{\ell \geq 2^{-1}(m+|m|)} \sum_j \mathbf{C} f^{2\ell-m}_{j,\ell} \to \sum_{\ell \geq 2^{-1}(-m+|m|)} \sum_j \mathbf{C} f^{2\ell+m}_{j,\ell}$$

(algebraic direct sums) such that $QU^{m,\lambda}(k) = U^{-m,-\lambda}(k)Q$ for all $k \in K$ and $QU^{m,\lambda}(H) = U^{-m,-\lambda}(H)Q$. The first condition implies that Q is of the form

$$Qf^{2\ell-m}_{j,\ell} = q_\ell f^{2\ell-m}_{j,\ell-m}, \quad \ell \geq 2^{-1}(m + |m|), \; 0 \leq j \leq 2\ell - m,$$

where q_ℓ is a number. The second condition gives

$$(1.11) \qquad (2\ell - m + \lambda + 2)q_{\ell+1} = (2\ell - m - \lambda + 2)q_\ell, \quad \ell \geq 2^{-1}(m+|m|).$$

Now it is clear that the existence of the isomorphism Q is precisely equivalent to the hypothesis $\lambda \neq \pm(|m| + 2t)$.

(ii) Obviously to prove the second assertion we only have to consider $\lambda = -|m| - 2t$. In this case, a linear map Q as in (1.10) commuting with the actions of K and H will be given by constants q_ℓ,

$\ell \geq 2^{-1}(m + |m|)$, such that

$$q_\ell = 0 \quad \text{if} \quad 2^{-1}(m + |m|) \leq \ell < 2^{-1}(m + |m|) + t$$

and

$$(1.12) \qquad (2\ell - m + \lambda + 2)q_{\ell+1} = (2\ell - m - \lambda + 2)q_\ell \quad \text{if} \quad \ell \geq 2^{-1}(m + |m|) + t.$$

Thus if $Q \neq 0$ we have

$$\ker(Q) = \sum_{0 \leq s < t} \sum_j \mathbf{C} f^{|m|+2s}_{j,2^{-1}(m+|m|)+s}$$

and

$$\operatorname{im}(Q) = \sum_{s \geq t} \sum_j \mathbf{C} f^{|m|+2s}_{j,2^{-1}(-m+|m|)+s}.$$

From this (ii) follows directly.

(iii) In this case we need a linear map

$$Q: \sum_{\ell \geq 2^{-1}(m+|m|)} \sum_j \mathbf{C} f^{2\ell-m}_{j,\ell} \to \sum_{\ell \geq 2^{-1}(\lambda+|\lambda|)} \sum_j \mathbf{C} f^{2\ell-\lambda}_{j,\ell}$$

commuting with K and H. Now Q is of the form

$$Q f^{2\ell-m}_{j,\ell} = q_\ell f^{2\ell-m}_{j,2^{-1}(2\ell-m+\lambda)}, \quad \ell \geq 2^{-1}(m + |m|),\ 0 \leq j \leq 2\ell - m,$$

with

$$(1.13) \qquad (2\ell - m + \lambda + 2)q_{\ell+1} = (2\ell + 2)q_\ell, \quad \ell \geq 2^{-1}(m + |m|).$$

If $m = |\lambda| + 2t$, $\lambda \in \mathbf{Z}$, a nonzero solution of (1.13) corresponds to an isomorphism Q with

$$\operatorname{im}(Q) = \sum_{s \geq t} \sum_j \mathbf{C} f^{|\lambda|+2s}_{j,2^{-1}(\lambda+|\lambda|)+s}.$$

This gives $U^{m,\lambda} \cong \overline{U}^{\lambda,m}$, as desired.

If $m = -|\lambda| - 2t$, $\lambda \in \mathbf{Z}$, we can define a linear map

$$Q: \sum_{\ell \geq 2^{-1}(\lambda+|\lambda|)} \sum_j \mathbf{C} f^{2\ell-\lambda}_{j,\ell} \to \sum_{\ell \geq 2^{-1}(m+|m|)} \sum_j \mathbf{C} f^{2\ell-m}_{j,\ell}$$

which commutes with K and H by taking

$$Qf^{2\ell-\lambda}_{j,\ell} = q_\ell f^{2\ell-\lambda}_{j,2^{-1}(2\ell-\lambda+m)}, \quad \ell \geq 2^{-1}(\lambda + |\lambda|), \quad 0 \leq j \leq 2\ell - \lambda,$$

where

$$(1.14) \qquad (2\ell - \lambda + m + 2)q_{\ell+1} = (2\ell + 2)q_\ell, \; \ell \geq 2^{-1}(\lambda + |\lambda|).$$

This condition is equivalent to

$$q_\ell = 0 \quad \text{if} \quad 2^{-1}(\lambda + |\lambda|) \leq \ell < 2^{-1}(\lambda + |\lambda|) + t$$

and

$$(2\ell - \lambda + m + 2)q_{\ell+1} = (2\ell + 2)q_\ell \quad \text{if} \quad \ell \geq 2^{-1}(\lambda + |\lambda|) + t.$$

Therefore if $Q \neq 0$ we have

$$\ker(Q) = \sum_{0 \leq s < t} \sum_j \mathbf{C} f^{|\lambda|+2s}_{j,2^{-1}(\lambda+|\lambda|)+s}$$

and

$$\operatorname{im}(Q) = \sum_{s \geq 0} \sum_j \mathbf{C} f^{2s-m}_{j,s}.$$

This gives $\overline{U}^{\lambda,m} \simeq U^{m,\lambda}$. This completes the proof of the theorem. Q.E.D.

The continuity properties of the various linear isomorphisms Q which establish the infinitesimal equivalences stated in Theorem 1.5 can be easily discussed with the help of

Lemma 1.6. *Given complex numbers* a, b, $b \neq -1, -2, \ldots,$ *let*

$$u_N = \prod_{n=1}^{N} \left(\frac{a+n}{b+n}\right).$$

The sequence $\{u_N\}$ *is bounded if and only if* $\operatorname{Re} a \leq \operatorname{Re} b$.

Proof. It is clearly enough to consider the case when $a_1 = \text{Re } a$ and $b_1 = \text{Re } b$ are positive.

If $a_1 \leq b_1$, there exists $c > 0$ such that

$$\frac{a\bar{a} + 2a_1 n + n^2}{b\bar{b} + 2b_1 n + n^2} \leq 1 + \frac{c}{b\bar{b} + 2b_1 n + n^2} \leq 1 + \frac{c}{n^2}.$$

Therefore, $\{|u_N|^2\}$ is convergent and hence $\{u_N\}$ is bounded.

If $a_1 > b_1$, there exists $\varepsilon > 0$ such that

$$a\bar{a} + 2a_1 n + n^2 \geq b\bar{b} + 2b_1 n + n^2 + \varepsilon(b\bar{b} + 2b_1 n)$$

for n sufficiently large. Hence

$$\frac{a\bar{a} + 2a_1 n + n^2}{b\bar{b} + 2b_1 n + n^2} \geq 1 + \frac{\varepsilon(b\bar{b} + 2b_1 n)}{b\bar{b} + 2b_1 n + n^2} \geq 1 + \frac{\varepsilon b_1}{n}$$

for n large. This implies the existence of N_o such that

$$|u_N|^2 \geq |u_{N_o}|^2 \prod_{N_o < n \leq N} \left(1 + \frac{\varepsilon b_1}{n}\right) \quad \text{if} \quad N > N_o.$$

Thus $\{u_N\}$ is unbounded in this case. The lemma is proved. Q.E.D.

It is well known that Q is continuous if and only if the corresponding defining sequence $\{q_\ell\}$ is bounded. Hence (1.11), (1.12), (1.13) and (1.14) together with Lemma 1.6 imply that the equivalences of Theorem 1.5 extended to equivalences in the strong sense, only when $\text{Re } \lambda = 0$. This corresponds precisely to the case when both representations of G are unitary.

2. We shall denote by $\mathfrak{g}$ the Lie algebra of G, and by $\mathfrak{k}$, $\mathfrak{a}$ and $\mathfrak{n}$ the subalgebras corresponding to the subgroups K, A and N respectively. Also, G will denote the universal enveloping algebra, over **C**, of $\mathfrak{g}$ and K, A, N will be the universal enveloping algebras over **C** of $\mathfrak{k}$, $\mathfrak{a}$ and $\mathfrak{n}$, respectively, regarded as canonically embedded in G. We have

$$G = KAN = KA(\mathbf{C}1 \oplus Nn) = KA \oplus Gn.$$

Let P: $G \to KA$ denote the corresponding projection map. We give KA an algebra estructure by identifying it with the algebra $K \otimes A$, and regard P as a map P: $G \to K \otimes A$. Let G^K and K^M denote the centralizers of K in G and of M in K, respectively. Then we know that P defines and injective homomorphism of G^K into $K^M \otimes A$ (see [3], Proposition 3.1).

It is our purpose, to introduce in this section a subalgebra of $K^M \otimes A$ which contains $P(G^K)$, and to prove equality later.

To this end we make the following definitions:

Definition 2.1. *Given* $t \in \mathbf{Z}$, *let*

$$U^t = \bigoplus_{m \in \mathbf{Z}} U^{m,m-2t}.$$

Definition 2.2. *Given* $t \in \mathbf{Z}_+$, *let* A_t: $\sum V_n^* \otimes V_n \to \sum V_n^* \otimes V_n$ *be the linear map defined by*

$$A_t f_{j,\ell}^n = \frac{\ell!}{(\ell - t)!} f_{j,\ell-t}^n$$

(of course we interpret $A_t f_{j,\ell}^n = 0$ *when* $\ell < t$*).*

It may help to look at the diagram which displays the direct sum nature of A_t. In fact A_t is defined as the direct sum of the corresponding intertwining operators introduced in the proof of Theorem 1.5.

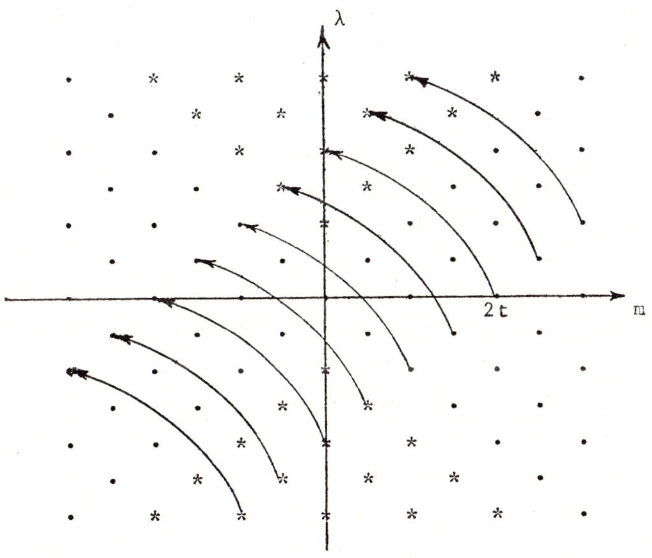

Set $E = 2^{-1}(u - iv)$ (see (1.6) for the definition of u and v). Let $C^{\infty}(K)$ denote the set of C^{∞} complex-valued functions on K equipped with the usual Fréchet structure (Schwartz topology). One notes that U^t defines a continuous representation of G on $C^{\infty}(K) \cap Q(-m)L^2(K)$ for each $m \in \mathbf{Z}$.

Proposition 2.3. *For all* $t \in \mathbf{Z}_+$ *we have:*

(i) $A_t = A_1^t$.

(ii) $A_1 = -E$, E *acting on* $\sum V_n^* \otimes V_n$ *as a left invariant vector field.*

(iii) A_t *extends to a continuous endomorphism of* $C^{\infty}(K)$.

(iv) $A_t U^t(g) = U^{-t}(g) A_t$ *on* $C^{\infty}(K) \cap Q(-m)L^2(K)$, *for all* $g \in G$.

Proof. By induction on t one proves that

$$A_1^t f_{j,\ell}^n = \frac{\ell!}{(\ell - t)!} f_{j,\ell-t}^n \quad \text{for all} \quad f_{j,\ell}^n,$$

and obtains (i).

A computation similar to that started in (1.6) to obtain $U^{\lambda}(F)$, shows that the action of E as a left invariant vector field is given by the differential operator

$$a \frac{\partial}{\partial b} - \bar{b} \frac{\partial}{\partial \bar{a}}.$$

From this we obtain that $Ef_{j,\ell}^n = -\ell f_{j,\ell-1}^n$, which proves (ii).

(iii) follows from (i) and (ii) upon observing that E, being a differential operator on K, defines a continuous endomorphism of $C^{\infty}(K)$.

To prove (iv) we first note that on $\sum V_n^* \otimes V_n$ we have $A_t U^t(k) = U^{-t}(k) A_t$ $(k \in K)$ and $A_t U^t(H) = U^{-t}(H) A_t$, by the very definition of A_t. Let P(n) denote the orthogonal projection of $Q(-m)L^2(K)$ onto $\sum_j \mathbf{C} f_{j,\ell}^n$, $n = 2\ell - m$, $2\ell \geq m + |m|$ (see (1.3)). Then for all f in the algebraic direct sum $\sum_{j,\ell} \mathbf{C} f_{j,\ell}^n$ $(n = 2\ell - m,\ 2\ell \geq m + |m|)$ we have

$$P(n) A_t U^t(\exp sH) f = A_t P(n) \sum_{i \geq 0} \frac{s^i}{i!} U^t(H)^i f =$$

$$= \sum_{i\geq 0} \frac{s^i}{i!} A_t P(n) U^t(H)^i f$$

$$= P(n) \sum_{i\geq 0} \frac{s^i}{i!} U^{-t}(H)^i A_t f$$

$$= P(n) U^{-t}(\exp sH) A_t f,$$

since f and $A_t f$ are analytic vectors. This being true for all n says that $A_t U^t(g) f = U^{-t}(g) A_t f$ for all $g \in G = KAK$ and all $f \in \sum_{j,\ell} \mathbf{C} f^n_{j,\ell}$. Now (iv) follows by continuity. Q.E.D.

Corollary 2.4. *For all* $g \in G$ *and all* $t \in \mathbf{Z}_+$ *we have*

$$E^t(U^t(g) f^n_{j,\ell}) = U^{-t}(g)(E^t f^n_{j,\ell}). \tag{2.1}$$

We shall regard the elements of $\mathcal{G}$ as left invariant differential operators on G; then the elements of $\mathcal{G}^K$ are also right invariant under K. The algebra $\mathcal{A}$ is just the symmetric algebra, over $\mathbf{C}$, of $\mathfrak{a}$; hence each linear mapping $\mu: \mathfrak{a} \to \mathcal{K}^M$ extends uniquely (because $\mathcal{K}^M$ is abelian, see Lemma 2.7 below) to a homomorphism $D \longmapsto D(\mu)$ of $\mathcal{A}$ into $\mathcal{K}^M$ satisfying $1(\mu) = 1$. Now given μ we can also consider the homomorphism $\mathcal{K}^M \otimes \mathcal{A} \to \mathcal{K}^M$ defined by

$$Y \otimes D \longmapsto (Y \otimes D)(\mu) = YD(\mu), \quad (Y \in \mathcal{K}^M, \quad D \in \mathcal{A}).$$

Recall that

$$H = \begin{pmatrix} 1 & 0 \\ 0 & -1 \end{pmatrix} \tag{2.2}$$

is a basis of $\mathfrak{a}$, and let

$$X = -i \begin{pmatrix} i & 0 \\ 0 & -i \end{pmatrix} \in i\mathfrak{k}.$$

Accordingly a linear map $\mu: \mathfrak{a} \to \mathcal{K}^M$ will be identified with $\mu(H)$. Observe that $X \in \mathcal{K}^M$. Now we are in a position to state the main result of this section:

Theorem 2.5. *For all* $D \in G^K$ *and all* $t \in \mathbf{Z}_+$ *we have*

$$E^t P(D)(X + 2t - 2) = P(D)(X - 2t - 2)E^t$$

(this is an equation in K*).*

Proof. From Definition 2.1 and (1.1) it follows that

$$(U^t(g)f^n_{j,\ell})(k) = (U^{2\ell-n,2\ell-n-2t}(g)f^n_{j,\ell})(k)$$

$$= a(g^{-1}k)^{-(2\ell-n-2t+2)} f^n_{j,\ell}(\kappa(g^{-1}k))$$

for all $g \in G$, $k \in K$.

Given $\lambda \in \mathbf{C}$ and $f \in C^\infty(K)$, we let

$$F^\lambda_f(g) = a(g)^{-(\lambda+2)} f(\kappa(g)), \quad g \in G.$$

With this notation (2.1) becomes

$$E^t F^{2\ell-n-2t}_{f^n_{j,\ell}} = F^{2\ell-n}_{E^t f^n_{j,\ell}}, \tag{2.3}$$

whenever $t \in \mathbf{Z}_+$.

For any $D \in G$ it easily follows (see the proof of Theorem 3.2 in [3]) that

$$DF^\lambda_f = P(D)(-\lambda - 2)F^\lambda_f.$$

Therefore, it we apply $D \in G^K$ to both sides of (2.3) we obtain

$$DE^t F^{2\ell-n-2t}_{f^n_{j,\ell}} = E^t DF^{2\ell-n-2t}_{f^n_{j,\ell}}$$

$$= E^t P(D)(-2\ell + n + 2t - 2)F^{2\ell-n-2t}_{f^n_{j,\ell}}$$

from the left hand side, and

$$DF^{2\ell-n}_{E^t f^n_{j,\ell}} = P(D)(-2\ell + n - 2)F^{2\ell-n}_{E^t f^n_{j,\ell}}$$

from the right hand side of (2.3). Hence, upon restriction to K we obtain

$$E^t P(D)(-2\ell + n + 2t - 2)f^n_{j,\ell} = P(D)(-2\ell + n - 2)E^t f^n_{j,\ell}, \tag{2.4}$$

since $F^\lambda_f | K = f$.

Since $Xf^n_{j,\ell} = (n - 2\ell)f^n_{j,\ell}$ (2.4) can be written in the following way

$$E^t P(D)(X + 2t - 2)f^n_{j,\ell} = P(D)(X - 2t - 2)E^t f^n_{j,\ell}$$

($E^t f^n_{j,\ell}$ is a scalar multiple of $f^n_{j,\ell-t}$). This completes the proof of the theorem. Q.E.D.

Let $\mathcal{B}$ be the set of all elements $B \in K^M \otimes A$ such that

$$E^t B(X + 2t - 2) = B(X - 2t - 2)E^t$$

for all $t \in \mathbf{Z}_+$. Clearly $\mathcal{B}$ is a subalgebra of $K^M \otimes A$, and according to Theorem 2.5 it contains the image of $P: G^K \to K^M \otimes A$.

Lemma 2.6. *Let $\xi = X^2 + 4EF - 2X$. Then ξ lies in the center of K, K^M is generated by ξ and X, and they are algebraically independent.*

Proof. The first assertion is a simple verification. To prove the second take

$$b = \sum a_{rsu} E^r F^s X^u \in K^M.$$

Since iX is in the Lie algebra of M we have $Xb = bX$, which implies that $a_{rsu} = 0$ if $r \neq s$. Therefore $b \in K^M$ is of the form

$$b = \sum b_{ru} E^r F^r X^u.$$

The converse is also true. All this follows from the bracket relations

$$[X,E] = 2E, \quad [X,F] = -2F, \quad [E,F] = X.$$

Hence, to prove that K^M is equal to the subalgebra $\mathbf{C}[\xi,X]$ generated by ξ and X, it suffices to show that $E^rF^r \in \mathbf{C}[\xi,X]$, $r \in \mathbf{Z}_+$. This is clear for $r = 1$ since $4EF = \xi - X^2 + 2X$. The general result follows by induction, because

$$4E^rF^r = E^{r-1}(\xi - X^2 + 2X)F^{r-1}$$

$$= E^{r-1}F^{r-1}(\xi - (X - 2r + 2)^2 + 2(X - 2r + 2)).$$

To prove that ξ and X are algebraically independent assume that $\sum a_{ru}\xi^rX^u = 0$. Then

$$\sum a_{ru}\xi^rX^uf^n_{j,\ell} = \sum a_{ru}(n^2 + 2n)^r(n - 2\ell)^uf^n_{j,\ell} = 0.$$

Therefore $\sum a_{ru}(n^2 + 2n)^r(n - 2\ell)^u = 0$ for all $0 \leq \ell \leq n$ and $n = 0, 1, \ldots$. This implies $a_{ru} = 0$, and the lemma is proved. Q.E.D.

They Weyl group W of the pair $(\mathfrak{g},\mathfrak{a})$ acts on K^M, on $\mathfrak{a}$ and on the complex dual $\mathfrak{a}'_{\mathbf{C}}$ of $\mathfrak{a}$, via the adjoint representation. Considering A as the algebra of polynomial functions on $\mathfrak{a}'_{\mathbf{C}}$ we can define a translated action of W on $K^M \otimes A$ letting

$$(wB)(\lambda - \rho) = B(w^{-1}(\lambda) - \rho)^w, \quad \lambda \in \mathfrak{a}'_{\mathbf{C}}$$

($\rho: \mathfrak{a} \to \mathbf{C}$ denotes the linear map defined by $\rho(H) = 2$). In this way W acts by automorphism of the algebra $K^M \otimes A$.

Proposition 2.7. (i) *The subalgebra B is stable under the translated action of the Weyl group.*

(ii) *B is a polynomial algebra in three algebraically independent generators, namely,*

$$\xi \otimes 1, \quad X \otimes 1 + 1 \otimes (H + 2), \quad X \otimes (H + 2).$$

Proof. An element $B' \in K^M \otimes A$ can be written uniquely in the form (see Lemma 2.6)

$$B' = \sum b_{rsu}\, \xi^r X^s \otimes H^u. \tag{2.5}$$

Let $B \longmapsto B'$ denote the automorphism of $K^M \otimes A$ generated by $b \otimes H \longmapsto b \otimes (H - 2)$. Then (cf. (2.5))

$$B \in \mathcal{B} \quad \text{if and only if} \quad b_{rsu} = b_{rus}. \tag{2.6}$$

In fact

$$E^t B(X + 2t - 2) = E^t \sum b_{rsu}\xi^r X^s (X + 2t)^u = \sum b_{rsu}\xi^r (X - 2t)^s X^u E^t$$

and

$$B(X - 2t - 2)E^t = \sum b_{rsu}\xi^r X^s (X - 2t)^u E^t.$$

Hence, $B \in \mathcal{B}$ if and only if

$$\sum b_{rsu}\xi^r (X - 2t)^s X^u = \sum b_{rsu}\xi^r X^s (X - 2t)^u, \quad t \in \mathbf{Z}_+$$

which is easily seen to be equivalent to $b_{rsu} = b_{rus}$.

Now, in our case $W \simeq \mathbf{Z}_2$; let $w \in W$ be the generator of W. Then if B' is as in (2.5) we have

$$(wB)'(\lambda) = (wB)(\lambda - \rho) = B(-\lambda - \rho)^w$$

$$= \left(\sum b_{rsu}\xi^r (-X)^s \otimes (-H)^u\right)(\lambda), \quad \lambda \in \mathfrak{a}'_{\mathbf{C}},$$

because $wX = -X$ and $wH = -H$. Therefore

$$(wB)' = \sum b_{rsu}(-1)^{s+u}\xi^r X^s \otimes H^u. \tag{2.7}$$

At this point (i) follows at once (see (2.6)).

From (2.6) we also obtain that the algebra $\mathcal{B}' = \{B' : B \in \mathcal{B}\}$ is generated by

$$\xi \otimes 1, \quad X \otimes 1 + 1 \otimes H, \quad X \otimes H,$$

which are algebraically independent. This proves (ii). Q.E.D.

Theorem 2.8. *Denote by B^W the set of invariants in B for the Weyl group W. Then*

(i) *The image of G^K under P is contained in B^W.*

(ii) *B^W is generated by the following algebraically independent elements:*

$$\xi \otimes 1, \quad (X \otimes 1 + 1 \otimes (H + 2))^2, \quad X \otimes (H + 2).$$

Proof. (i) follows from the definition of B and Corollary 3.3 in [3]. The second assertion is a direct consequence of Proposition 2.7 (ii) and (2.7). Q.E.D.

3. Let $g = k + p$ be the Cartan decomposition corresponding to (G,K). If subscript $\mathbf{C}$ denotes complexification one also has $g_{\mathbf{C}} = k_{\mathbf{C}} + p_{\mathbf{C}}$. The subalgebra a is a maximal abelian subspace of p; let $a_{\mathbf{C}} \subset p_{\mathbf{C}}$ be its complexification.

For any vector space V, let S(V) denote the symmetric algebra over V. For every non-negative integer i, let $S^i(V)$ denote the homogeneous subspace of S(V) of degree i.

The inclusion map $k_{\mathbf{C}} + a_{\mathbf{C}} \to k_{\mathbf{C}} + p_{\mathbf{C}} = g_{\mathbf{C}}$ induces contravariantly the restriction homomorphism

$$S(g'_{\mathbf{C}}) \to S((k_{\mathbf{C}} + a_{\mathbf{C}})'), \tag{3.1}$$

where $g'_{\mathbf{C}}$ denotes the dual of $g_{\mathbf{C}}$ and $(k_{\mathbf{C}} + a_{\mathbf{C}})'$ the dual of $k_{\mathbf{C}} + a_{\mathbf{C}}$.

Let $G_{\mathbf{C}}$ be the adjoint group of $g_{\mathbf{C}}$. The Lie algebra of $G_{\mathbf{C}}$ is $\mathrm{ad}g_{\mathbf{C}}$. Now let $K_{\mathbf{C}}$ be the subgroup of $G_{\mathbf{C}}$ corresponding to $\mathrm{ad}k_{\mathbf{C}}$. Then $S(g'_{\mathbf{C}})$ is, in a natural way, a $K_{\mathbf{C}}$-module; $S(g'_{\mathbf{C}})^{K_{\mathbf{C}}}$ will denote the corresponding ring of $K_{\mathbf{C}}$-invariants. Also if $M'_{\mathbf{C}}$ is the normalizer of $a_{\mathbf{C}}$ in $K_{\mathbf{C}}$, then $M'_{\mathbf{C}}$ operates on $k_{\mathbf{C}} + a_{\mathbf{C}}$ and hence on $S((k_{\mathbf{C}} + a_{\mathbf{C}})')$; $S((k_{\mathbf{C}} + a_{\mathbf{C}})')^{M'_{\mathbf{C}}}$ will denote the ring of $M_{\mathbf{C}}'$-invariants in $S((k_{\mathbf{C}} + a_{\mathbf{C}})')$. Since $M_{\mathbf{C}}' \subset K_{\mathbf{C}}$ the homomorphism (3.1) restricted to $S(g_{\mathbf{C}}')^{K_{\mathbf{C}}}$ induces a homomorphism

$$\pi: S(g_{\mathbf{C}}')^{K_{\mathbf{C}}} \to S((k_{\mathbf{C}} + a_{\mathbf{C}})')^{M_{\mathbf{C}}'}. \tag{3.2}$$

It is known that π is injective (see [3], Theorem 6.1). Now we shall determine its image, which will be used to prove that $P(G^K) = B^W$.

The $K_{\mathbf{C}}$-module $S(k_{\mathbf{C}}')$ ($k_{\mathbf{C}}'$ being the dual of $k_{\mathbf{C}}$) is completely reducible, therefore

$$S(k_{\mathbf{C}}') = \bigoplus_{m \geq 0} W_m, \tag{3.3}$$

where W_m is the set of all elements $f \in S(k_{\mathbf{C}}')$ which transform under $K_{\mathbf{C}}$ according to the irreducible representation of dimension $m + 1$. We also have a natural isomorphism between $S((k_{\mathbf{C}} + a_{\mathbf{C}})')$ and $S(k_{\mathbf{C}}') \otimes S(a_{\mathbf{C}}')$.

With this identification and (3.3) in mind, we can state the following result.

Theorem 3.1. *The homomorphism* (3.2) *establishes an isomorphism of* $S(g_{\mathbf{C}}')^{K_{\mathbf{C}}}$ *onto*

$$\mathrm{im}(\pi) = \Big(\bigoplus_{j \geq 0} \bigoplus_{0 \leq i \leq j} W_{2i} \otimes S^j(a_{\mathbf{C}}')\Big)^{M_{\mathbf{C}}'}.$$

To prove the theorem we need some preparation. First of all we shall quote Theorem 6.1 of [3]. For any $y \in p_{\mathbf{C}}$ let

$$\det(t - \mathrm{ad}\,y) = \sum a_i(y) t^i$$

be the characteristic polynomial of ad y. One has $a_i = 0$ for all $i < \ell = \dim A + \dim M$. Let

$$b(x + y) = a_\ell(y) \quad \text{for } x \in k_{\mathbf{C}}, \quad y \in p_{\mathbf{C}}.$$

Then $b \in S^{n-\ell}(g_{\mathbf{C}}')^{K_{\mathbf{C}}}$, $n = \dim g$. Let $S(g_{\mathbf{C}}')_b^{K_{\mathbf{C}}}$ and $S((k_{\mathbf{C}} + a_{\mathbf{C}})')_{b_o}^{M_{\mathbf{C}}'}$ ($b_o = \pi(b)$) be the localizations of $S(g_{\mathbf{C}}')^{K_{\mathbf{C}}}$ and $S((k_{\mathbf{C}} + a_{\mathbf{C}})')^{M_{\mathbf{C}}'}$ by b and b_o, respectively. Then the injection map $k_{\mathbf{C}} + a_{\mathbf{C}} \to k_{\mathbf{C}} + p_{\mathbf{C}}$ induces an isomorphism π of $S(g_{\mathbf{C}}')_b^{K_{\mathbf{C}}}$ onto $S((k_{\mathbf{C}} + a_{\mathbf{C}})')_{b_o}^{M_{\mathbf{C}}'}$.

The space $p_{\mathbf{C}}$ is an irreducible $k_{\mathbf{C}}$-module of complex dimension 3, therefore isomorphic to $k_{\mathbf{C}}$. Set

$$b = \begin{pmatrix} 0 & i \\ -i & 0 \end{pmatrix} \quad \text{and} \quad c = \begin{pmatrix} 0 & 1 \\ 1 & 0 \end{pmatrix},$$

then, b and c are in p. Also set $X' = iH$ (see (2.2)), $E' = 2^{-1}(b + ic)$ and $F' = 2^{-1}(-b + ic)$. Then $\{X', E', F'\}$ is a basis of $p_{\mathbf{C}}$ over $\mathbf{C}$ and the linear map defined by

$$X \longmapsto X', \quad E \longmapsto E', \quad F \longmapsto F'$$

is a $k_{\mathbf{C}}$-isomorphism of $k_{\mathbf{C}}$ onto $p_{\mathbf{C}}$. Moreover

$$[X',E'] = -2E, \quad [X',F'] = 2F, \quad [E',F'] = -X.$$

Lemma 3.2. *Let* (y_1, y_2, y_3) *be the linear coordinates on* $p_{\mathbf{C}}$ *defined by the basis* $\{X', E', F'\}$. *Then*

$$b(y_1, y_2, y_3) = 16(y_2y_3 + y_1^2)^2.$$

Note that $y_2y_3 + y_1^2$ *is irreducible.*

Proof. Given $y = y_1X' + y_2E' + y_3F'$, the matrix of ady: $g_{\mathbf{C}} \to g_{\mathbf{C}}$ with respect to the basis $\{X,E,F,X',E',F'\}$ is:

$$A = \begin{pmatrix} 0 & C \\ -C & 0 \end{pmatrix} \quad \text{where} \quad C = \begin{pmatrix} 0 & y_3 & -y_2 \\ 2y_2 & -2y_1 & 0 \\ -2y_3 & 0 & 2y_1 \end{pmatrix}.$$

Now

$$A^2 = \begin{pmatrix} B & 0 \\ 0 & B \end{pmatrix} \quad \text{with} \quad B = -C^2.$$

The characteristic polynomial of B is

$$\det(t - B) = t^3 + 8t^2(y_2y_3 + y_1^2) + 16t(y_2y_3 + y_1^2)^2.$$

Multiplying three rows and three columms of $t - A$ by -1 we see that

$$\det(t - A) = \det(t + A).$$

Therefore

$$(\det(t - A))^2 = \det(t^2 - A^2) = (\det(t^2 - B))^2.$$

Since $\det(t - A)$ and $\det(t^2 - B)$ are both monic polynomials in t we have

$$\det(t - A) = \det(t^2 - B),$$

from where the lemma follows. Q.E.D.

The natural isomorphism between $S(\mathfrak{g}'_{\mathbf{C}})$ and $S(\mathfrak{k}'_{\mathbf{C}}) \otimes S(\mathfrak{p}'_{\mathbf{C}})$, allows us to view the elements in $S(\mathfrak{g}'_{\mathbf{C}})$ as polynomial functions on $\mathfrak{p}_{\mathbf{C}}$ with values in $S(\mathfrak{k}'_{\mathbf{C}})$.

Let $Z = E + F \in \mathfrak{k}_{\mathbf{C}}$. Then Z is semi-simple, and hence diagonalizable in $S(\mathfrak{k}'_{\mathbf{C}})$. Let $S(\mathfrak{k}'_{\mathbf{C}})_r$ denote the eigenspace of Z corresponding to the eigenvalue r.

Proposition 3.3. *Let* $u \in (S(\mathfrak{k}'_{\mathbf{C}}) \otimes S^{4q+s}(\mathfrak{p}'_{\mathbf{C}}))^{K_{\mathbf{C}}}$. *Then the rational function* u/b^q *on* $\mathfrak{g}_{\mathbf{C}}$ *is a polynomial, if and only if*

$$\frac{u}{b^q}(y) \in \bigoplus_{n=-s}^{s} S(\mathfrak{k}'_{\mathbf{C}})_{2n} \qquad (3.4)$$

for all $y \in \mathfrak{p}_{\mathbf{C}}$ *such that* $b^q(y) \neq 0$.

Proof. The eigenvalues of $\operatorname{ad}Z$ in $\mathfrak{k}_{\mathbf{C}}$ are 2, 0, -2. Therefore

$$S(\mathfrak{k}'_{\mathbf{C}}) = \bigoplus_{j=-\infty}^{\infty} S(\mathfrak{k}'_{\mathbf{C}})_{2j}.$$

Accordingly, we can write uniquely $u = \sum u_j$ with

$$u_j \in (S(\mathfrak{k}'_{\mathbf{C}})_{2j} \otimes S^{4q+s}(\mathfrak{p}'_{\mathbf{C}}))^{K_{\mathbf{C}}}.$$

Clearly u/b^q is a polynomial on $\mathfrak{g}_{\mathbf{C}}$, if and only if u_j/b^q is a polynomial for all j.

For each $y \in \mathfrak{p}_{\mathbf{C}}$ there exist

$$\lim_{t\to+\infty} e^{-2t} \operatorname{Ad}(\exp tZ)y = y_+$$

and

$$\lim_{t\to-\infty} e^{2t} \operatorname{Ad}(\exp tZ)y = y_-,$$

since the eigenvalues of adZ in $\mathfrak{p}_{\mathbf{C}}$ are 2, 0, -2.

If u/b^q is a polynomial on $\mathfrak{g}_{\mathbf{C}}$, and $b^q(y) \neq 0$, $y \in \mathfrak{p}_{\mathbf{C}}$, we have

$$(3.5)\qquad \left(\frac{u_j}{b^q}\right)(y_+) = \lim_{t\to+\infty} \frac{u_j(e^{-2t} \operatorname{Ad}(\exp tZ)y)}{b^q(e^{-2t} \operatorname{Ad}(\exp tZ)y)} = \lim_{t\to+\infty} \frac{e^{-2ts}e^{2tj}u_j(y)}{b^q(y)}.$$

If $u_j \neq 0$ we can certainly choose $y \in \mathfrak{p}_{\mathbf{C}}$ such that $b^q(y)u_j(y) \neq 0$. If this is the case (3.5) implies that $j \leq s$. Similarly letting $t \to -\infty$ we obtain that $j \geq -s$.

Conversely, if (3.4) holds and $b^q(y) \neq 0$, $y \in \mathfrak{p}_{\mathbf{C}}$, then

$$\lim_{t\to+\infty} \frac{u(e^{-2t} \operatorname{Ad}(\exp tZ)y)}{b^q(e^{-2t}\operatorname{Ad}(\exp tZ)y)} = \frac{u_s(y)}{b^q(y)}.$$

But

$$(3.6)\qquad b(y_+) = \lim_{t\to+\infty} b(e^{-2t} \operatorname{Ad}(\exp tZ)y) = \lim_{t\to+\infty} e^{-8t} b(y) = 0,$$

therefore $u(y_+) = 0$ whenever $q > 0$.

It is easy to see that any $y \in \mathfrak{p}_{\mathbf{C}}$ different from zero is regular, and that any $y \in \mathfrak{p}_{\mathbf{C}}$ such that $b(y) = 0$ is nilpotent. The converse is also true (see Kostant-Rallis [2], Proposition 11). Hence the principal nilpotents are precisely those $y \in \mathfrak{p}_{\mathbf{C}}$, $y \neq 0$ such that $b(y) = 0$. (see [2] for definitions). Moreover, these elements make a single $K_{\mathbf{C}}$-orbit (see [2], Theorem 3) since $\{y \in \mathfrak{p}_{\mathbf{C}} : b(y) = 0\}$ is an irreducible algebraic set (Lemma 3.2).

Clearly there are y's in $\mathfrak{p}_{\mathbf{C}}$ such that $b(y) \neq 0$ and $y_+ \neq 0$. Now, when $q > 0$, (3.6) implies that if $b(y) = 0$, $y \in \mathfrak{p}_{\mathbf{C}}$, then $u(y) = 0$. From this it follows easily (by Hilbert Nullstellensatz) that u/b^q is a polynomial. The Proposition is proved. Q.E.D.

Lemma 3.4. *The following two conditions are equivalent:*

(i) $A \in \bigoplus_{n=-s}^{s} S(k'_{\mathbf{C}})_{2n}, \quad X \cdot A = 0.$

(ii) $A \in \bigoplus_{n=0}^{s} W_{2n}, \quad X \cdot A = 0.$

Proof. In the $k_{\mathbf{C}}$-irreducible module V_{2n} of dimension $2n + 1$, Z is diagonalizable with eigenvalues $2n, 2n - 2, \ldots, -2n$. Thus (i) follows from (ii).

Given $A \in V_{2n}$, let A_{2n} and A_{-2n} denote the components of A along the eigenspaces of Z corresponding to the eigenvalues $2n$ and $-2n$, respectively. We shall prove that if $X \cdot A = 0$ and $A \neq 0$ then A_{2n} and A_{-2n} are also different from zero. This will clearly show that (ii) is a consequence of (i).

We can identify $k_{\mathbf{C}}$ with $\underline{s\ell}(2,\mathbf{C})$ by letting

$$X = \begin{pmatrix} 1 & 0 \\ 0 & -1 \end{pmatrix}, \quad E = \begin{pmatrix} 0 & 1 \\ 0 & 0 \end{pmatrix}, \quad F = \begin{pmatrix} 0 & 0 \\ 1 & 0 \end{pmatrix}.$$

Then

$$\mathbf{Z} = \begin{pmatrix} 0 & 1 \\ 1 & 0 \end{pmatrix}.$$

Setting

$$g = \frac{\sqrt{2}}{2} \begin{pmatrix} 1 & 1 \\ -1 & 1 \end{pmatrix} \in SU(2),$$

we have $\mathrm{Ad}(g)Z = X$.

We can realize V_{2n} as the space of all homogeneous polynomials in z and w of degree $2n$. Then (see (1.2))

$$g(z^{\ell}w^{2n-\ell}) = 2^{-n}(z - w)^{\ell}(z + w)^{2n-\ell} \quad (\ell = 0, \ldots, 2n). \tag{3.7}$$

If (g_{ij}) denotes the matrix associated to g with respect to the basis $\{z^{\ell}w^{2n-\ell}\}$, then from (3.7) it follows that

$$(3.8) \qquad g_{0,\ell} = (-1)^{\ell} 2^{-n}, \quad g_{2n,\ell} = 2^{-n} \quad (\ell = 0, \ldots, 2n).$$

Since $X \cdot A = 0$ and $A \neq 0$ we can assume that A corresponds to $z^n w^n$. The j-th column of (g_{ij}) gives the coefficients of $z^j w^{2n-j}$ in the basis $\{g^{-1} . z^{\ell} w^{2n-\ell}\}$ which diagonalizes Z. Hence (3.8) proves our assertion. Q.E.D.

Proof of Theorem 3.1. Given $u \in (S(k_{\mathbf{C}}') \otimes S^s(p_{\mathbf{C}}'))^{M_{\mathbf{C}}'}$, Proposition 3.2 establishes that

$$u(y) \in \bigoplus_{n=-s}^{s} S(k_{\mathbf{C}}')_{2n}$$

for all $y \in p_{\mathbf{C}}$. When $y \in a_{\mathbf{C}}$ we have $X.u(y) = 0$ because $M_{\mathbf{C}}$ centralizes $a_{\mathbf{C}}$. Therefore (Lemma 3.4)

$$u(y) \in \bigoplus_{n=0}^{s} W_{2n}$$

for all $y \in a_{\mathbf{C}}$. This proves that

$$\operatorname{im}(\pi) \subset \Big(\bigoplus_{j \geq 0} \bigoplus_{0 \leq i \leq j} W_{2i} \otimes S^j(a_{\mathbf{C}}')\Big)^{M_{\mathbf{C}}'}.$$

Conversely, if $v \in (W_{2i} \otimes S^s(a_{\mathbf{C}}'))^{M_{\mathbf{C}}'}$, $0 \leq i \leq s$, then (Theorem 6.1 in [3]) there exist $q \geq 0$ and

$$u \in (S(k_{\mathbf{C}}') \otimes S^{4q+s}(p_{\mathbf{C}}'))^{K_{\mathbf{C}}}$$

such that $\pi(u/b^q) = v$.

For $y \in a_{\mathbf{C}}$ and $k \in K_{\mathbf{C}}$ we have

$$\Big(\frac{u}{b^q}\Big)(k.y) = k.\Big(\frac{u}{b^q}\Big)(y) = k.v(y) \in W_{2i}.$$

Since $K_{\mathbf{C}}.a_{\mathbf{C}}$ is dense in $p_{\mathbf{C}}$ (see [2], § I.1) it follows that

$$\Big(\frac{u}{b^q}\Big)(y) \in \bigoplus_{n=-s}^{s} S(k_{\mathbf{C}}')_{2n}$$

($i \leq s$) for all $y \in p_{\mathbf{C}}$ such that $b^q(y) \neq 0$. Now, Proposition 3.3 shows that u/b^q is in fact a polynomial on $g_{\mathbf{C}}$. The theorem is proved. Q.E.D.

4. Let h be a Lie algebra and $j \subset h$ any subalgebra in h. Let $J \subset H$ be the corresponding universal enveloping algebras. Let $H_{(n)}$ be the usual filtration of H. Thus $H_{(n)}$ is spanned by $x_1, \ldots, x_j$, $x_i \in h$, $j \leq n$, and the identity. Then one gets a new filtration of H by putting

$$H_n = JH_{(n)}$$

(see [3], §7). Moreover, the filtration H_n defines a valuation ν on H (in the sense of ring theory): if $0 \neq a \in H$ let $\nu(a) = \min n$ such that $a \in H_n$ and let $\nu(0) = -\infty$. See Theorem 1.7 of [3].

We can identify the universal enveloping algebra of the direct sum $k_{\mathbf{C}} \oplus a_{\mathbf{C}}$ with $K \otimes A$. Since $k_{\mathbf{C}}$ is a subalgebra in $k_{\mathbf{C}} \oplus a_{\mathbf{C}}$ it defines a valuation ν_o on $K \otimes A$. Now A is graded $A = \oplus A_i$. If $\nu_o(u) = d$, $u \in K \otimes A$ then there exists a unique $\tilde{u} \in K \otimes A_d$ such that

$$\nu_o(u - \tilde{u}) < d.$$

We also let ν be the valuation on G defined by $k_{\mathbf{C}}$.

Let λ: $S(g_{\mathbf{C}}) \to G$ denote the symmetrization mapping. We note that λ is defined on $S(p_{\mathbf{C}})$ by regarding $S(p_{\mathbf{C}}) \subset S(g_{\mathbf{C}})$. Let q be the orthogonal complement of a in p with respect to the Killing form of g, and let $q_{\mathbf{C}} \subset p_{\mathbf{C}}$ be the complexification of q. Then

$$G = (K \otimes A) \oplus (K \otimes \lambda(q_{\mathbf{C}} S(p_{\mathbf{C}}))).$$

Let F: $G \to K \otimes A$ denote the projection onto the first summand in this decomposition. Then

$$\nu_o(P(v)) = \nu_o(F(v)) \quad \text{and} \quad \widetilde{P(v)} = \widetilde{F(v)} \tag{4.1}$$

for all $v \in G^K$ (see Proposition 7.2 in [3]).

Let δ: $g_{\mathbf{C}} \to g'_{\mathbf{C}}$ be the isomorphism defined by the Killing form of $g_{\mathbf{C}}$. We may extend δ to an algebra, $G_{\mathbf{C}}$-isomorphism of symmetric algebras

$$\delta: S(g_{\mathbf{C}}) \to S(g'_{\mathbf{C}}).$$

Now let

$$\sigma^n: G_{(n)} \to S^n(g'_{\mathbf{C}})$$

be the linear map defined by composing λ^{-1}: $G_{(n)} \to S_{(n)}(g_{\mathbf{C}}) = \sum_{j=0}^{n} S^j(g_{\mathbf{C}})$ with δ: $S_{(n)}(g_{\mathbf{C}}) \to S_{(n)}(g'_{\mathbf{C}})$ and then with the projection $S_{(n)}(g'_{\mathbf{C}}) \to S^n(g'_{\mathbf{C}})$. It is clear that σ^n is a $G_{\mathbf{C}}$-linear map. Let σ_o^n: $(K \otimes A)_{(n)} \to S^n((k_{\mathbf{C}} \oplus a_{\mathbf{C}})')$ be defined similarly, so that σ_o^n is a $G_{\mathbf{C}}$-linear map. It then follows that

$$
\begin{array}{ccc}
G_{(n)} & \xrightarrow{\sigma^n} & S^n(g'_{\mathbf{C}}) \\
F\downarrow & & \downarrow\pi \\
(K \otimes A)_{(n)} & \xrightarrow{\sigma_o^n} & S^n((k_{\mathbf{C}} \oplus a_{\mathbf{C}})')
\end{array} \tag{4.2}
$$

is a commutative diagram (see (7.4) in [3]).

Let ρ be the usual valuation on G. Thus if $u \in G$, $\rho(u) = -\infty$ if $u = 0$, otherwise $\rho(u)$ is the least $n \in \mathbf{Z}$ such that $u \in G_{(n)}$. Note that if $u \in G_{(n)}$ then $\sigma^n(u) \neq 0$ if and only if $\rho(u) = n$. One defines the valuation ρ_o on $K \otimes A$ similarly. Now since π is injective on $S(g'_{\mathbf{C}})^{K_{\mathbf{C}}}$ it then follows from (4.2) that for any $u \in G^K$

$$\rho(u) = \rho_o(Fu).$$

The following lemma should be compared with Lemma 7.5 of [3].

Lemma 4.1. *For any* $0 \neq f_o \in B^W$ *there exists* $w \in G^K$ *such that*

$$\nu_o(f_o - P(w)) < \nu_o(f_o).$$

Proof. Clearly $\tilde{u} \neq 0$ for any $0 \neq u \in K \otimes A$ and $\rho_o(\tilde{u}) \geq \nu_o(\tilde{u}) = \nu_o(u)$. Put $\beta(u) = \rho_o(\tilde{u}) - \nu_o(u)$.

We will prove the lemma by induction on $\beta(f_o)$. Let

$$\tau\colon S(g'_{\mathbf{C}}) \to G$$

be the $G_{\mathbf{C}}$-linear map defined by putting $\tau = \lambda \circ \delta^{-1}$. Thus $\tau(S_{(n)}(g'_{\mathbf{C}})) = G_{(n)}$ and $\sigma^n \circ \tau$ is the identity on $S^n(g'_{\mathbf{C}})$.

Now assume $\beta(f_o) = 0$. Thus $\tilde{f}_o \in A_d$ where $d = \nu_o(f_o)$. But then $\tilde{f}_o$ and hence $\sigma_o^d(\tilde{f}_o) \in S^d(a'_{\mathbf{C}})$ are Weyl group invariant. Thus there exists (see Theorem 3.1) $\xi \in S^d(g'_{\mathbf{C}})^{K_{\mathbf{C}}}$ such that $\pi(\xi) = \sigma_o^d(\tilde{f}_o)$. If $w = \tau(\xi)$ one has $w \in G^K$. But by (4.2) one has $\sigma_o^d(F(w)) = \sigma_o^d(\tilde{f}_o)$. Thus $\rho_o(\tilde{f}_o - F(w)) < d$. But $\nu_o(\tilde{f}_o - F(w)) \leq \rho_o(\tilde{f}_o - F(w)) < d$. Thus $\tilde{f}_o = \widetilde{F(w)} = \widetilde{P(w)}$, see (4.1). Then $\nu_o(f_o - P(w)) < d$ proving the lemma for $\beta(f_o) = 0$.

Now let $\beta(f_o) > 0$ and assume the theorem is true for smaller values. Again set $d = \nu_o(f_o)$.

Now put $m = \rho_o(\tilde{f}_o)$ so that $m - d = \beta(f_o)$. Since $\tilde{f}_o$ is Weyl group invariant it follows that $\sigma_o^m(\tilde{f}_o) \in S^m((k_{\mathbf{C}} \oplus a_{\mathbf{C}})')$ is $M_{\mathbf{C}}'$-invariant. Moreover $\sigma_o^m(\tilde{f}_o) \in \bigoplus_{0\leq i\leq d} W_{2i} \otimes S^d(a_{\mathbf{C}}')$. In fact, if K_r denote the eigenspace in K of ad(Z) corresponding to the eigenvalue r, then $X \in K_2 \oplus K_0 \oplus K_{-2}$ and $\xi \in K_0$. Hence $\xi^p X^q \in \bigoplus_{i=-q}^{q} K_{2i}$. Then $\tilde{f}_o \in \bigoplus_{i=-d}^{d} K_{2i} \otimes H^d$ (see Proposition 2.7 (ii)), and therefore

$$\sigma_o^m(\tilde{f}_o) \in \bigoplus_{0\leq i\leq d} W_{2i} \otimes S^d(a_{\mathbf{C}}')$$

(see Lemma 3.4) as stated.

Now by Theorem 3.1 there exists $\psi \in S(g_{\mathbf{C}}')^{K_{\mathbf{C}}}$ such that $\pi(\psi) = \sigma_o^m(\tilde{f}_o)$. Furthermore since $\sigma_o^m(\tilde{f}_o) \in S(k_{\mathbf{C}}') \otimes S^d(a_{\mathbf{C}}')$ it follows from the injectivity of $\pi|S(g_{\mathbf{C}}')^{K_{\mathbf{C}}}$ that $\psi \in S(k_{\mathbf{C}}') \otimes S^d(p_{\mathbf{C}}')$. It follows therefore, it we put $u = \tau(\psi) \in G^K$, that $\nu(u) \leq d$. On the other hand by (4.2) one has

$$\sigma_o^m(F(u)) = \sigma_o^m(\tilde{f}_o) \neq 0. \tag{4.3}$$

But since $0 \neq \sigma_o^m(F(u)) \in S(k_{\mathbf{C}}') \otimes S^d(a_{\mathbf{C}}')$ it follows that $\nu_o(F(u)) \geq d$. Thus

$$\nu_o(F(u)) = \nu(u) = d \tag{4.4}$$

(see [3] Proposition 7.2). Also $\nu_o(P(u)) = \nu_o(F(u)) = d$ by (4.1).

If $\nu_o(f_o - P(u)) < d$ we are done. Assume therefore, that $\nu_o(f_o - P(u)) = d$. Setting $x = f_o - P(u)$ one has $0 \neq x \in B^W$ and $\tilde{x} = \tilde{f}_o - \widetilde{P(u)} \in K \otimes A_d$. Moreover $\rho_o(\tilde{f}_o) = m$ and by (4.3) $m = \rho(u) = \rho_o(F(u)) \geq \rho_o(\widetilde{P(u)})$. Thus $m \geq \rho_o(\tilde{x})$.

On the other hand we assert that $\sigma_o^m(F(u)) = \sigma_o^m(\widetilde{F(u)})$. Indeed $\sigma_o^m(F(u)) \in S(k_{\mathbf{C}}') \otimes S^d(a_{\mathbf{C}}')$ and $\sigma_o^m(\widetilde{F(u)}) \in S(k_{\mathbf{C}}') \otimes S^d(a_{\mathbf{C}}')$ since $\nu_o(F(u)) = d$ by (4.4). However one necessarily has $\sigma_o^m(F(u) - \widetilde{F(u)}) \in S(k_{\mathbf{C}}') \otimes S_{(d-1)}(a_{\mathbf{C}}')$ since $\nu_o(F(u) - \widetilde{F(u)}) < d$. Thus $\sigma_o^m(F(u)) = \sigma_o^m(\widetilde{F(u)}) = \sigma_o^m(\tilde{f}_o)$. But $\sigma_o^m(\widetilde{P(u)}) = \sigma_o^m(\widetilde{F(u)})$ by (4.1). Therefore $\sigma_o^m(\tilde{x}) = 0$ so that $\rho_o(\tilde{x}) < m$. But then $\beta(x) = \rho_o(\tilde{x}) - d < m - d = \beta(f_o)$. The induction assumption then applies to x so that there exists $v \in G^K$ such that $\nu_o(x - P(v)) < \nu_o(x)$. Now put $w = u + v \in G^K$. Then $\nu_o(f_o - P(w)) < \nu_o(x) = \nu_o(f_o)$. The lemma is proved. Q.E.D.

Theorem 4.2. *The map* $P: G^K \to B^W$ *is a surjective isomorphism.*

Proof. To prove the theorem we just need to show that given $0 \neq u_o \in B^W$ there exists $u \in G^K$ such that $P(u) = u_o$. This will be proved by induction on $\nu_o(u_o)$.

If $\nu_o(u_o) = 0$ then by Lemma 4.1 there exists $u \in G^K$ such that $\nu_o(u_o - P(u)) < 0$; i.e. $u_o = P(u)$.

Now assume $\nu_o(u_o) > 0$ and assume surjectivity for smaller values. Again by Lemma 4.1 there exists $w \in G^K$ such that $\nu_o(u_o - P(w)) < \nu_o(u_o)$. Set $x = u_o - P(w) \in B^W$. By the inductive assumption there exists $v \in G^K$ such that $P(v) = x$. But then $u = v + w$ lies in G^K and satisfies $P(u) = u_o$. Q.E.D.

To close this work we shall indicate how the characterization of $P(G^K)$ leads to the determination of all irreducible representations of G.

Let U be a continuous topologically irreducible representation of G on a Banach space E such that the multiplicity of any element in $\hat{K}$ is finite. This is equivalent to assume that U is a topologically completely irreducible (TCI) representation of G on E (see Warner [6], §4).

In our case, every $\delta \in \hat{K}$ occurs at most once in every TCI Banach representation of G. This happens because K^M, and hence G^K, is abelian. Thus the representation of G^K on E_δ (the δ-primary component of E as K-module) is given by a homomorphism $\sigma: G^K \to \mathbf{C}$. Moreover if $E_\delta \neq 0$ the homomorphism σ determines the representation U up to infinitesimal equivalence.

Given $\lambda \in \mathfrak{a}'_{\mathbf{C}}$, let $\lambda: A \to \mathbf{C}$ denote also the evaluation homomorphism at λ, and let $\rho \in \mathfrak{a}'_{\mathbf{C}}$ be defined by $\rho(H) = 2$. Then the image of $K^M \otimes A$ under the homomorphism $\delta \otimes (\lambda - \rho): K \otimes A \to \mathrm{End}(E_\delta)$ leaves invariant every M-primary component $E_{\delta,\gamma}$ of E_δ, $\gamma \in \hat{M}$. Since $\dim E_{\delta,\gamma} = 1$ one has accordingly a homomorphism $\sigma_{\delta,\gamma,\lambda}: K^M_\delta \otimes A \to \mathbf{C}$. Now, a theorem of Lepowsky (see [4], Theorem 8.7) says that given $\sigma: G^K \to \mathbf{C}$ as above, there exist $\gamma \in \hat{M}$ and $\lambda \in a'_{\mathbf{C}}$ such that $\sigma = \sigma_{\delta,\gamma,\lambda} \circ P$.

Thus σ corresponds to a homomorphism of $P(B^W)$ into $\mathbf{C}$ defined by:

$$\xi \otimes 1 \longmapsto n(n+2)$$

$$(X \otimes 1 \oplus 1 \otimes (H+2))^2 \longmapsto (m+\lambda)^2$$

$$X \otimes (H+2) \longmapsto m\lambda$$

(see Theorem 2.7) where $\lambda = \lambda(H)$, $\dim E_\delta = n+1$ and $\sigma(X) = m$. But this is precisely the homomorphism corresponding to the representation $U^{m,\lambda}$, as an easy computation shows.

Thus one has

Theorem 4.3. (cf. Naimark [5], p. 246). *Every TCI Banach representation of G is defined by a pair of numbers* (m,λ) *with* $m \in \mathbf{Z}$ *and* $\lambda \in \mathbf{C}$; *the pairs* (m,λ) *and* $(-m,-\lambda)$ *define one and the same representation.*

When $\lambda \neq \pm(|m| + 2t)$, $t \in \mathbf{Z}_+$, *the representation is infinitesimally equivalent to the representation* $U^{m,\lambda}$ *of the principal series; when* $\lambda = |m| + 2t$, $t \in \mathbf{Z}_+$, *the representation is infinitesimally equivalent to* $U^{\lambda,m}$; *when* $\lambda = -(|m| + 2t)$, $t \in \mathbf{Z}_+$, *the representation is equivalent to the finite-dimensional, spinor representation* $S^{m,\lambda}$ *of the group* G. (See Theorem 1.5).

REFERENCES

1. A. Cooper, *The classifying ring of groups whose classifying ring is commutative*, Ph.D. Thesis, Massachusetts Institute of Technology, 1975.

2. B. Kostant and S. Rallis, *Orbits and representations associated with symmetric spaces*, Amer. J. Math. 93 (1971), 753-809.

3. B. Kostant and J. Tirao, *On the structure of certain subalgebras of a universal enveloping algebra*, Trans. Amer. Math. Soc. 218 (1976), 133-154.

4. J. Lepowsky, *Algebraic results on representations of semi-simple Lie groups*, Trans. Amer. Math. Soc. 176 (1973), 1-44.

5. M. A. Naimark, *Les représentations linéaires du groupe de Lorentz*, Dunod, Paris, 1962.

6. G. Warner, *Harmonic analysis on semi-simple Lie groups*. I, Springer-Verlag, New York, 1972.

IMAF, UNIVERSIDAD N. DE CORDOBA, VALPARAISO Y R. MARTINEZ, 5000 CORDOBA, ARGENTINA.

Current address: Departamento de Matemática, Universidade de Brasília, 70.000 - Brasília, Brasil.

On the Finite Solvability of Plateau's Problem

Friedrich Tomi

The class of minimal surfaces which arises in the study of the classical Plateau problem is the class of conformal, branched, minimal immersions. By definition, such a surface is a map f from some two-dimensional domain D into $\mathbb{R}^3$, $f = f(x,y)$, satisfying the following conditions:

(1) $f_x \cdot f_y = 0, \ |f_x| = |f_y|$

(2) $\Delta f = 0.$

The problem of Plateau consists in proving the existence of such maps f whose boundary values $f|\partial D$ parametrize a given system of contours. We restrict our attention to the simplest case when D is the unit disc and the prescribed contour is a simple closed curve.

Using the complex derivative F of f, $F := f_w = f_x - if_y$, (1) and (2) can be expressed in the equivalent form

(1') $F \cdot F \equiv F_1^2 + F_2^2 + F_3^2 = 0$

and

(2') $F_{\overline{w}} = 0.$

A branch point of order k of f is a point $w \in \overline{D}$ where F vanishes of order k. Since F is holomorphic, clearly, interior branch points $w \in D$ are isolated. If the bounding curve of f is sufficiently differentiable the same is true for boundary branch points.

If $\gamma : S^1 \to \mathbb{R}^3$ is any Jordan curve then we denote by $M(\gamma)$ the set of minimal surfaces spanned by γ, i.e. the set of mappings f satisfying (1), (2), the boundary condition

(3) $\exists$ homeomorphism $\tau: S^1 \to S^1$ such that $f|S^1 = \gamma \circ \tau$,

and the normalization condition

(4) $f(\exp(\frac{2}{3}k\pi i)) = \gamma(\exp(\frac{2}{3}k\pi i))$, $k=1,2,3$.

We know from the classical existence theorems of Douglas and Radò [14] that $M(\gamma)$ is not empty. The regularity theory of Plateau's problem, accomplished only rather recently, tells us that $M(\gamma) \subset C^{k+\alpha}$ provided that $\gamma \in C^{k+\alpha}$ where k is any positive integer and $0<\alpha<1$ [12]. In the following we shall work in the class of $C^{4+\alpha}$-curves. It is convenient to use the following

Notation. Γ is the set of $C^{4+\alpha}$-embeddings of S^1 into $\mathbb{R}^3$. By Γ_o we denote the set of all $\gamma \in \Gamma$ such that all $f \in M(\gamma)$ do not have any boundary branch points. The set of proper curves, Γ_p, is the set of all $\gamma \in \Gamma_o$ such that all $f \in M(\gamma)$ have at most simple (i.e first order) branch points. Finally, Γ_f, the set of fine curves, is the set of all $\gamma \in \Gamma_o$ such that all $f \in M(\gamma)$ are immersed.

For special classes of fine curves the unique solvability of Plateau's problem is known, namely for curves which admit a one-to-one convex projection onto some plane and for curves of total curvature not bigger than 4π. The first result is due to Radò [14], the second one to Nitsche [11]. Up to know, the just mentioned cases are the only ones where an explicit estimate of the number of solutions is known.

Let us now briefly describe the other known finiteness results. The author proved that the number of surfaces of least area spanned in any analytic or any Γ_o-curve is always finite [16]. Using Sobolev spaces instead of $C^{4+\alpha}$, Tromba proved the existence of an open and dense subset of Γ_f for which $M(\gamma)$ is finite [18]. On a conference on variational calculus at Oberwolfach, Böhme recently presented a more general result which assures the finiteness of $M(\gamma)$ for an open and dense subset of Γ_o.

It is the purpose of this paper to prove, for all proper curves, the finiteness of a certain subset of $M(\gamma)$ which includes all local minima of surface area with respect to the C^o-topology. Our result, unfortunately, does not give any information on the number of instable minimal surfaces. On the other hand it gives a simpler and more geometric description of the class of curves to which it applies as this is the case in Böhme's and Tromba's results, where the curves with finite $M(\gamma)$ are defined as the regular values of some map. In the final section of the paper we shall discuss a nice, purely geometric, sufficient criterion for a curve to be proper. Moreover, we shall not only prove the finiteness of the number of stable minimal surfaces spanned into each specific $\gamma\in\Gamma_p$ but the local boundedness of this number on Γ_p. The local finiteness is trivial in the situation considered by Böhme and Tromba since they restrict to "non-degenerate" minimal surfaces.

The plan of the paper is as follows. In the first section we prove the non-existence of certain "blocks" in $M(\gamma)$, i.e. non-trivial connected components of $M(\gamma)$. This is used in section 2 in order to establish our finiteness result. In the final third section we apply the preceding results to a special geometric situation.

1. The non-existence of certain blocks.

In this section we shall give an improved version of the method used in [16]. As in that paper we consider (non-conformal) minimal surfaces of the form

(5) $\quad f + u\,N(f),$

where f is an immersion, N(f) its normal field and u some real function with zero coundary values.

It is well known [4] that every surface h in a sufficiently small neighborhood of f and having the same boundary as f can uniquely be reparametrized **in** such a way that it takes the form (5). We shall call this special representation of h <u>the normal representation of h with respect to f</u>.

Let us denote by J the open subset of $C^{4+\alpha}(\bar{D}, \mathbb{R}^3)$ consisting of immersions, by X the closed subspace of $C^{2+\alpha}(\bar{D}, \mathbb{R})$ of functions vanishing on ∂D, and by Ω the open set of all $(f,u)\in J\times X$ such that $f+uN(f)$ is immersed. It is easily seen that the surface area of $f+uN(f)$, denoted by $A(f,u)$, is an analytic functional on Ω. (As for the concept of analytic mappings between Banach spaces the reader is referred to [6]). The partial derivative of A with respect to u can be written in the form

$$D_u A(f,u)v = \int \phi(f,u)v\, d\omega(f),$$

where $d\omega(f)$ is the surface measure of f and

$$\phi:\Omega \to Y := C^{\alpha}(\bar{D}, \mathbb{R})$$

a certain analytic map [17].

Let us recall the basic formulas

(6) $\quad D_u\phi(f,o)v = -\Delta_f v + 2K_f v$

and

$$\beta(f)\langle v,w\rangle = \int (-\Delta_f v + 2K_f v)w\, d\omega(f),$$

where Δ_f and K_f are the Laplace-Beltrami operator and the Gaussian curvature of the surface f, respectively, and where

(7) $\beta(f) := D_u^2 A(f,o)$.

For any $\gamma \in \Gamma$, let us denote by $M^*(\gamma)$ the set of all immersed surfaces in $M(\gamma)$. We wish to study the local behavior of $M^*(\gamma)$ if γ changes. Let $f_o \in M^*(\gamma_o)$ be fixed. We can construct an analytic (linear) extension operator

$$E:\Gamma \to C^{4+\alpha}(\bar{D}, \mathbb{R}^3)$$

such that $E\gamma_o = f_o$. For example,

$$E\gamma = H(\gamma \circ \gamma_o^{-1} \circ (f_o|S^1))$$

is appropiate, where H denotes harmonic extension. For abbreviation, let us put

$$A := D_u\phi(f_o, 0).$$

Then the following Lemma holds which is basic for all what follows.

Lemma 1. (I) Assume that the operator $A:X \to Y$ is injective. Then there exist neighborhoods U of $0 \in X$ and V of $\gamma_o \in \Gamma$ and an analytic map

$$G:V \to C^{2+\alpha}(\bar{D}, \mathbb{R}^3)$$

such that, for each $\gamma \in V$, $G(\gamma)$ is the only minimal surface of the form $f = E\gamma + u\, N(E\gamma)$ with $u \in U$.

(II) In case $n := \dim \operatorname{Ker} A > 0$ there exist neighborhoods U and V as above, a neighborhood W_o of $0 \in \mathbb{R}^n$ and analytic mappings

$$\varphi: V \times W_o \to \mathbb{R},$$

$$G: V \times W_o \to C^{2+\alpha}(\bar{D}, \mathbb{R}^3)$$

with the properties that, for each $\gamma \in V$,

i) $G(\gamma,\cdot)$ is an embedding of W_o

ii) the image of the set of critical points of $\varphi(\gamma,\cdot)$ under the map $G(\gamma,\cdot)$ is exactly the set of all minimal surfaces of the form

$f = E\gamma + uN(E\gamma)$ with $u \in U$.

<u>Proof.</u> The surface $f = E\gamma + uN(E\gamma)$ is minimal if and only if $\phi(E\gamma, u) = 0$. Therefore, in particular, we have $\phi(E\gamma_o, 0) = 0$. Since the operator $A = D_u\phi(E\gamma_o, 0)$, as a second order elliptic differential operator with Dirichlet boundary conditions is Fredholm of index 0, it must be an isomorphism if it is injective [10]. Therefore, the first part of the lemma is an immediate consequence of the implicit function theorem. In order to prove the second part let us put

$$X_o := \ker A, \; Y_1 := \text{range } A.$$

Let $\{u_1, \ldots, u_n\}$ be an orthonormal base of X_o with respect to the scalar product

$$\langle u, v\rangle = \int uv\, d\omega(f_o)$$

and let

$$P: Y \to \text{Ker } A \subset Y$$

be the corresponding orthogonal projection, i.e.

$$Pu = \sum_{k=1}^{n} \langle u, u_k\rangle u_k.$$

Since A is of index 0 and symmetric with respect to the above scalar product it follows that

$$Y_1 = \text{range } (I-P).$$

The closed linear subspace

$$X_1 := (I-P)X$$

of X is complementary to X_o and, therefore, $X \cong X_o \times X_1$ algebraically and topologically. For sufficiently small neighborhoods W_o, W_1 and V of $0 \in X_o$, $0 \in X_1$, and $\gamma_o \in \Gamma$, respectively, we can define a map

$$\psi: V \times W_o \times W_1 \to Y_1$$

by

$$\psi(\gamma, u_o, u_1) := (I-P)\phi(E\gamma, u_o + u_1).$$

Since, obviously, $\psi(\gamma_o,0,0) = 0$ and the partial derivative

$$D_{u_1}\psi(\gamma_o,0,0) = (I-P)A|X_1 = A|X_1$$

is an isomorphism from X_1 onto Y_1 we can conclude from the implicit function theorem that, for sufficiently small neighborhoods V, W_o, and W_1, there exists an analytic map

$$g: V \times W_o \to W_1$$

with the property that

$$\psi^{-1}(0) = \{(\gamma,u_o,g(\gamma,u_o)) | \gamma \in V,\ u_o \in W_o\} \tag{8}$$

and

$$D_{u_o} g(\gamma_o,0) = 0. \tag{9}$$

Now we identify X_o with $\mathbb{R}^n$ and define the maps φ and G by

$$\varphi(\gamma,u_o) := A(E\gamma,u_o+g(\gamma,u_o)),$$

$$G(\gamma,u_o) := E\gamma+(u_o+g(\gamma,u_o))N(E\gamma),$$

and the neighborhood U of $0 \in X$ by

$$U := W_o + W_1.$$

We have to show that all asserted properties of φ, G, and U are satisfied. Let $f = E\gamma+uN(E\gamma)$ be any minimal surface with $\gamma \in V$ and $u \in U$. Writing $u = u_o+u_1$ with $u_o \in W_o$ and $u_1 \in W_1$ we see immediately that $\psi(\gamma,u_o,u_1) = 0$. It follows, therefore, from (8) that $u_1 = g(\gamma,u_o)$ and, consequently, $f = G(\gamma,u_o)$. The fact that u_o is a critical point of φ is the well known variational property of minimal surfaces. On the other hand, let us assume that, for a certain $\overline{\gamma} \in V$, $\overline{u}_o$ is a critical point of $\varphi(\overline{\gamma},\cdot)$. We must show that the surface $G(\overline{\gamma},\overline{u}_o) = E\overline{\gamma}+\overline{u}N(E\overline{\gamma})$ with $\overline{u} = \overline{u}_o+g(\overline{\gamma},\overline{u}_o)$ is minimal i.e. that the equation $\phi(E\overline{\gamma},\overline{u}) = 0$ holds. As a consequence from the definition of G we have at once

$$(I-P)\phi(E\overline{\gamma},\overline{u}) = 0.$$

Since $\bar{u}_o$ is a critical point of $\varphi(\bar{\gamma},\cdot)$ we obtain

$$0 = D_u A(E\bar{\gamma},\bar{u})(u_o+D_{u_o} g(\bar{\gamma},\bar{u}_o)u_o)$$

$$= \int\phi(E\bar{\gamma},\bar{u})(P+D_{u_o} g(\bar{\gamma},\bar{u}_o))u_o d\omega(E\bar{\gamma})$$

for all $u_o \in X_o$ which can be written as

$$Q\phi(E\bar{\gamma},\bar{u}) = 0,$$

where the linear operator $Q:Y \to Y$ is defined as

$$Qv = \sum_{k=1}^{n} \int v(P+D_{u_o} g(\bar{\gamma},\bar{u}_o))u_k d\omega(E\bar{\gamma})u_k$$

In view of the splitting $Y = PY \oplus (I-P)Y$ we have the inequality

$$\|v\| \leq C(\|Pv\| + \|(I-P)v\|),\ v\in Y,$$

with a certain constant C. It follows that

$$(1-C\|P-Q\|)\|v\| \leq C(\|Qv\| + \|(I-P)v\|)$$

which means that the equations $Qv = 0$ and $(I-P)v = 0$ imply $v = 0$, provided only that $\|P-Q\| < 1/C$. It can beseen from (9), however, that $\|P-Q\|$ can be made arbitrarily small by making the neighborhoods V and W_o sufficiently small. This proves the lemma.

<u>Corollary 1.</u> Assume that $f_o \in M^*(\gamma_o)$ is non-isolated in $M^*(\gamma_o)$ and that, furthermore, the form $\beta(f_o)$ is non-negative. Then there exists a $C^{2+\alpha}$-neighborhood U of f_o such that $U \cap M^*(\gamma_o)$, in normal representation with respect to f_o, is a one-dimensional analytic manifold.

<u>Proof.</u> Since f_o is not isolated it follows from Lemma 1, part (I), that the operator A must have a kernel of positive dimension. The bilinear form corresponding to A is, however, non-negative by assumption. It is then a classical result [5] that dim ker A = 1. Now we can apply part (II) of Lemma 1. From the non-isolatedness of f_o it

follows that the critical points of the real analytic function of one variable $\varphi(\gamma_o,\cdot)$ must accumulate at O. Hence this function is constant, i.e. the set of its critical points is an interval. It follows then from Lemma 1 that, in normal representation with respect to f_o, $M^*(\gamma_o)$, in a neighborhood of f_o, is an embedded interval. This proves the Corollary.

Lemma 2. Let C be a connected subset of $M^*(\gamma_o)$ and assume that, for some $f_o \in M^*(\gamma_o)$, the bilinear form $\beta(f_o)$ is non-negative. Then it follows that $\beta(f) \geq 0$ for all $f \in C$.

Proof. Since, by continuity, the set of all $f \in C$ such that $\beta(f) \geq 0$ is clearly closed we need only to show that this set is also open as a subset of C. To this purpose let us pick any $f \in C$ such that $\beta(f) \geq 0$ and let us assume that there exists a sequence (f_n) in C such that $f_n \to f(n \to \infty)$ and $\beta(f_n)$ is not non-negative. Since none of the f_n can be isolated we conclude from Lemma 1 that the operators $A := D_u\phi(f,0)$ and $A_n := D_u\phi(f_n,0)$ have zero as an eigenvalue. Using the variational principles for eigenvalues [5] and the assumption on $\beta(f_n)$ we see that there must be negative eigenvalues of A_n. Therefore, denoting by λ_2 the second (positive) eigenvalue of A, the sum of the multiplicities of all eigenvalues of A_n less than $\lambda_2/2$ must be greater or equal to 2. It follows then from Kato's perturbation theory [9] that the corresponding sum of multiplicities for A must also be ≥ 2. This is, however, a contradiction since the only eigenvalue of A less than $\lambda_2/2$ is $\lambda_1 = 0$ and has multiplicity 1.

We are now ready to prove the main result of this section,

Theorem 1. Let C be a compact connected component of $M^*(\gamma)$ and suppose that for some $f \in C$ the form $\beta(f)$ is non-negative. Then C is an isolated point.

Proof. Let us assume that C is not an isolated point. Then, clearly, no point of C is isolated. As in [16] let us consider the oriented volume

$$V(f) := \int f \cdot (f_x \times f_y) dxdy$$

as a function on C. Since C is compact, there must be a point $f_o \in C$ where V attains its maximal value on C. According to Lemma 2, we have $\beta(f_o) \geqq 0$. Therefore, we may conclude from Corollary 1 that there exists an analytic curve of minimal surfaces

$$f_o + u(t)N(f_o), \ |t| < \varepsilon, \ u(0) = 0, \ u'(0) \neq 0,$$

which, after conformal reparametrization, must belong to C. Since the volume is invariant under change of parameters we finde that the function $t \to V(f_o + u(t)N(f_o))$ has a local maximum at $t = 0$. Computing the derivative we obtain

$$(10) \qquad \int u'(0) d\omega(f_o) = 0.$$

The elements $u(t) \in X$ being solutions of the equation $\phi(f_o, u(t)) = 0$, it follows that $u'(0)$ belongs to the kernel of $A := D_u\phi(f_o, 0)$. Since $\beta(f_o) \geqq 0$, clearly, 0 is the lowest eigenvalue of A. It is, however, classicaly known that the eigenfunctions of an operator of the form (6) corresponding to the lowest eigenvalue do not change sign. This stands in contradiction to (10) and proves the theorem.

2. The Finiteness Result

In this section we shall consider a certain subset of the set $M^*(\gamma)$ of all immersed minimal surfaces spanned in γ, namely

$$M_+^*(\gamma) := \{f \in M^*(\gamma) \mid \beta(f) \geqq 0\},$$

the form $\beta(f)$ being defined in (7).

According to results of Osserman, Alt, Gulliver, and Gulliver-Lesley, [1, 2, 7, 8, 13], for analytic γ or $\gamma \in \Gamma_o$, all minimal surfaces in $M(\gamma)$ representing C^o-local minimal of Dirichlet's integral (or of surface area) are immersed up to the boundary and, therefore, contained in $M_+^*(\gamma)$. Hence $M_+^*(\gamma)$ contains all minimal surfaces which can actually be produced by soap film experiments.

In order to deduce the finiteness of $M_+^*(\gamma)$ from Theorem 1 we need to know that $M^*(\gamma)$ is compact. In view of the estimate

$$(11) \qquad \|f\|_{C^{4+\alpha}} \leqq \text{const}(\gamma) \qquad \forall\ f \in M(\gamma)$$

which follows from the regularity theory of minimal surfaces the compactness of the whole set $M(\gamma)$ with respect to some $C^{4+\delta}$-topology, $0 < \delta < \alpha$, is secured. It suffices, therefore, to show that $M^*(\gamma)$ is closed as a subset of $M(\gamma)$. Unfortenately, we cannot prove this in general, but only for a proper curve γ. The closedness of $M^*(\gamma)$ for such curves, in fact, follows from Böhme's work on the stratification of minimal surfaces [3]. Since Böhme's paper, in the moment, is not easily accessible we prove the statement we need in the following

Lemma 3. Let f be a conformal minimal surface with a simple branchpoint at $w_o \in D$ and let $(f^{(n)})$ be a sequence of conformal minimal surfaces converging to f in C^o. Then, for all sufficiently large n, $f^{(n)}$ has a simple branch point w_n and $w_n \to w_o$ $(n \to \infty)$.

Proof. Let us put $F := f_w$, $F^{(n)} := f_w^{(n)}$. Then, clearly,

$F^{(n)} \to F (n \to \infty)$ uniformly in some neighborhood of w_o. From the relation

$$F^2 = F_1^2 + F_2^2 + F_3^2 = 0$$

and the fact that F has a simple zero at w_o it can be seen that at least two of the three coordinate functions of F can only have a simple zero at w_o. Let us assume that, for example, F_1 and F_2 have this property. Then we define the holomorphic functions

$$\Phi := F_1 - iF_2, \quad \Psi := F_1 + iF_2,$$

$$\Phi^{(n)} := F_1^{(n)} - iF_2^{(n)}, \quad \Psi^{(n)} := F_1^{(n)} + iF_2^{(n)}$$

which satisfy the relation

$$\Phi^{(n)}\Psi^{(n)} = -F_3^{(n)2}. \tag{12}$$

Clearly, one of the two functions Φ and Ψ must have a simple zero at w_o. We can assume that Φ has this property. Then, for sufficiently large n, the functions $\Phi^{(n)}$ have simple zeros w_n, $w_n \to w_o (n \to \infty)$. Since, according to (12), the product $\phi^{(n)}\psi^{(n)}$ is a square and, consequently, must have a zero of at least second order at w_n, it follows that $\psi^{(n)}(w_n) = 0$. Hence we obtain $F^{(n)}(w_n) = 0$ and the lemma is proved.

Corollary 2. For proper curves γ, $M^*(\gamma)$ is closed in $M(\gamma)$.

Proof. According to Lemma 3, any accumulation point $f \in M(\gamma) \setminus M^*(\gamma)$ of $M^*(\gamma)$ can only have boundary branch points or interior branch points of at least second order. Such f, however, are excluded by the definition of proper curves.

Proposition 1. Let γ be a proper curve. Then, all surfaces in $M_+^*(\gamma)$ are isolated points of $M(\gamma)$. Therefore, $M_+^*(\gamma)$ is a finite set.

Proof. Let us assume that there exists some $f \in M_+^*(\gamma)$ which is not isolated. From Corollary 1 it follows then that the connected component C of f in $M^*(\gamma)$ contains more than one point. From Corollary 2 and the remarks preceding Lemma 3 we know that $M^*(\gamma)$ and, consequently, C are compact in some $C^{4+\delta}$-topology, $0 < \delta < \alpha$. Using this topology, we can apply Theorem 1 and conclude that C must be a point, contradicting our assumption. The finiteness of $M_+^*(\gamma)$ follows then at once from the compactness of this set.

Lemma 4. Let γ_o be a proper curve. Then, for every $C^{4+\delta}$-neighborhood U of $M_+^*(\gamma_o)$, $0 < \delta < \alpha$, there exists a $C^{4+\alpha}$-neighborhood V of γ_o such that $M_+^*(\gamma) \subset U$ for all $\gamma \in V$.

Proof. Let us assume that for a given U the corresponding V does not exist. Then there exist a sequence of curves $\gamma_n \in \Gamma$, $\gamma_n \to \gamma_o$ $(n \to \infty)$ and a sequence of minimal surfaces $f_n \in M_+^*(\gamma_n)$, $f_n \notin U$. Since in the estimate (11) the right hand side is locally bounded on Γ we know that the sequence (f_n) is bounded in the $C^{4+\alpha}$-norm. Hence, replacing (f_n) by a suitable subsequence, we may assume that (f_n) converges in $C^{4+\delta}$ to some $f_o \in M(\gamma_o) \cap \complement U$. The curve γ_o being proper, f_o can only have simple, interior branch points. Since, however, f_o is the limit of immersed minimal surfaces, it follows from Lemma 3 that f_o must be immersed. Then, by continuity we also have $\beta(f_o) \geq 0$, proving that $f_o \in M_+^*(\gamma_o)$, an obvious contradiction.

Now we are ready to prove our main result,

Theorem 2. The number of elements of $M_+^*(\gamma)$ is locally bounded on Γ_p, the set of proper curves.

Proof. Let $\gamma_o \in \Gamma_p$. In view of Proposition 1 and Lemma 4 we need only to show that, for any $f_o \in M_+^*(\gamma_o)$, there exist $C^{4+\delta}$-neighborhoods U of f_o and V of γ_o such that the number of elements of $M(\gamma) \cap U$ is uniformly bounded for all $\gamma \in V$. Let us first consider the trivial case when $\beta(f_o)$ is positive definite. Then, according to Theorem 1 (I), there are neighborhoods U and V such that, for any $\gamma \in V$, $M(\gamma) \cap U$ contains exactly one element. In case $\beta(f_o)$ is degenerate we can apply the second part of Theorem 1 and choose neighborhoods U and V and a connected neighborhood W_o of $0 \in \mathbb{R}$ in such a way that, for each $\gamma \in V$, the elements of $M(\gamma) \cap U$ are in a one-to-one correspondance with the critical points of the analytic function $\varphi(\gamma,\cdot)$ on W_o, i.e. with the zeros of the derivative $\varphi'(\gamma,\cdot)$ on W_o. Since f_o is isolated in $M(\gamma_o)$ the function $\varphi'(\gamma_o,\cdot)$ cannot vanish identically and, therefore, has a development of the form

$$\varphi'(\gamma_o,t) = c_q t^q + \text{higher order terms}$$

with some $q \in \mathbb{N}$, $q \geq 3$, and $c_q \neq 0$. In this situation, however, we may apply the infinite dimensional version of the Weierstrass preparation theorem [15] and conclude that, for sufficiently small V and W_o, the zeros of φ' in $V \times W_o$ are identical with the zeros of a Weierstrass polynomial of the form

$$t^q + a_1(\gamma)\, t^{q-1} + \ldots + a_{q-1}(\gamma)t + a_q(\gamma)$$

where $a_1(\gamma_o) = \ldots = a_q(\gamma_o) = 0$. Since, for each $\gamma \in V$, the number of zeros of that polynomical is less or equal to q, the number q is a bound for the number of elements in $M(\gamma) \cap U$. The theorem is proved.

Corollary 3. For every compact subset C of Γ_p there exists a bound for the number of elements of $M_+^*(\gamma)$, $\gamma\in C$.

3. A special class of proper curves.

In this section we want to illustrate our Theorem 2 and Corollary 3 in a special geometric situation. We need some simple definitions.

If P is a plane in $\mathbb{R}^3$ and $\gamma\in\Gamma$ then we call any connected component C of $\gamma(S^1) \cap P$ an essential intersection of γ and P if, in each neighborhood of C, there are points of γ on both sides of P. Let us remark that the number of essential intersections of a curve γ with a fixed plane P is either zero, an even number, or infinite. We say that γ satisfies an interior k-point Radò-condition if through each point of $\mathbb{R}^3\setminus\gamma(S^1)$ there exists a plane with at most k essential intersections with γ. The curve γ satisfies a boundary k-point Radò-condition if through each point of $\gamma(S^1)$ there exists a plane which has at most k essential intersections with γ. It was shown by Radò that the conditions just introduced imply a priori bounds on the orders of branch points of minimal surfaces spanned in γ. In particular, we have

Lemma 5 [12,§366,373]. If γ satisfies an interior 4-point and a boundary 0-point Radò condition then it follows that γ is proper.

Now we want to construct an exhaustive family of compact subsets of the subclass of proper curves defined by the Radò conditions of Lemma 5. To this purpose we define the quantity

$$L(\gamma) := \sup\{\frac{|p-q|}{|\gamma(p)-\gamma(q)|} \mid p,q\in S^1\}$$

which is finite for every immersion γ of S^1. It can then easily be seen that, for any $R > 0$,

$$A_R := \{\gamma\in\Gamma \mid \|\gamma\|_{C^{4+\alpha}} + L(\gamma) \leq R,\ \gamma \text{ satisfies an interior 4-point and a boundary 0-point Radò condition}\}$$

is a compact subset of Γ_p, endowed with some $C^{4+\delta}$ topology, $0 < \delta < \alpha$. Hence, from Corollary 3 we immediately obtain

Corollary 4. For any $R>0$ there exists a number $N(R)$ such that for any $\gamma\in A_R$, the set $M_+^*(\gamma)$ has at most $N(R)$ elements.

References

1. Alt,H.W.: Verzweigungspunkte von H-Flächen. I.Math.Z. 127, 333-362 (1972)

2. Alt,H.W.: Verzweigungspunkte von H-Flächen. II.Math.Ann. 201, 33-55 (1973)

3. Böhme,R.: Stabilität von Minimalflächen gegen Störung der Randkurve. Habilitationsschrift. Göttingen 1974

4. Böhme,R., Tomi,F.: Zur Struktur der Lösungsmenge des Plateauproblems. Math.Z. 133, 1-29 (1973)

5. Courant,R., Hilbert,D.: Methods of Mathematical Physics I. New York-London: Interscience 1953

6. Fucik,S., Necas,J., Soucek,J., Soucek,V.: Spectral Analyis of Nonlinear Operators. Lecture Notes in Mathematics 346. Berlin-Heidelberg-New York: Springer (1973)

7. Gulliver,R.: Regularity of minimizing surfaces of prescribed mean curvature. Ann. of Math. 97, 275-305 (1973)

8. Gulliver,R., Lesley,F.D.: On boundary branch points of surfaces. Arch.Rat.Mech.Analysis 52, 20-25 (1973)

9. Kato,T.: Perturbation Theory for Linear Operators. Berlin-Heidelberg-New York: Springer (1976)

10. Mizohata,S.: The Theory of Partial Differential Equations. Cambridge: University Press 1973

11. Nitsche,J.C.C.: A new uniqueness theorem for minimal surfaces. Arch.Rat.Mech.Anal. 52, 319-329 (1973)

12. Nitsche,J.C.C.: Vorlesungen über Minimalflächen. Berlin-Heidelberg-New York: Springer 1975

13. Ossermann,R.: A proof of the regularity everywhere of the classical solution to Plateau's problem. Ann. of Math. 91, 550-569 (1970)

14. Radò,T.: On the problem of Plateau. Berlin: Julius Springer 1933.

15. Ramis,J.P.: Sous-ensembles analytiques d'une variété banachique complexe. Berlin-Heidelberg-New York: Springer 1970

16. Tomi,F.: On the local uniqueness of the problem of least area. Arch.Rat.Mech.Analysis 52, 312-318 (1973)

17. Tomi,F.: A perturbation theorem for surfaces of constant mean curvature. Math.Z. 141, 253-264 (1975)

18. Tromba,A.J.: On the number of simply connected minimal surfaces spanning a curve in $\mathbb{R}^3$. Preprint

Friedrich Tomi
Fachbereich Mathematik
der Universität des Saarlandes
6600 Saarbrücken

The Set of Curves of Uniqueness for Plateau's Problem Has a Dense Interior

by

A.J. Tromba*

§I. Introduction

A well known argument of Paul Levy asserts that there exists rectifiable Jordan curves in $\mathbb{R}^3$ for which the number of geometrically distinct solutions to the Plateau problem is non-denumerably infinite. A natural question is whether boundaries of uniqueness, that is boundaries where the absolute infimum over simply connected surfaces is unique, are in some sense generic. A partial answer is given by the following theorem of F. Almgren.

Theorem 1 - Let β denote the set of Jordan curves in $\mathbb{R}^n$. For $\Gamma,\Gamma' \in \beta$, $d(\Gamma,\Gamma') = \inf\{\|\gamma-\gamma'\|_\infty\}$ where $\|\ \|_\infty$ is the usual sup norm and where γ and γ' run over all possible homomorphisms of S^1 with Γ and Γ' respectively.

Then the set of Jordan curves for which the absolute infimum solution to the Plateau problem (the so-called Douglas solution) is unique is dense in β.

In his lectures on minimal submanifolds [10] Blaine Lawson proves

Theorem 2 - The set of Jordan curves for which the Douglas solution to the Plateau problem is not unique is dense in β.

Lawson states that it is an open question whether the curves of uniqueness are generic; that is whether they form a set of second

---

* Research partially supported by the National Science Foundation.

category in β and whether or not in the set β_k of regular Jordan curves of class C^k, $k \geq 1$, with an appropriate "C^k-topology" the set of curves of uniqueness is open and dense.

We do have an answer to this question in the case $k \geq 7$. Although we have no formal counterexample it is our belief that curves of uniqueness are not open in any reasonable topology. We shall show that the set of curves of uniqueness has a dense interior.

§II. Backgrond theory and the global analytical formulation of the Plateau problem

We shall outline the approach to the Plateau problem given in [19].

Let M be a smooth Banach manifold and $K: T^2M \to TM$ a connection map [5]. In [18] the author defines a smooth vector field $X: M \to TM$ to be Fredholm with respect to K if for each $p \in M$ the covariant derivative of X with respect to K, $\nabla X(p)$, which is a linear map of T_pM to itself is linear Fredholm. By the index of X we mean the dim ker $\nabla X(p)$ - dim coker $\nabla X(p)$. If M is connected this index does not depend on p. If M is not connected, the index is constant on components and we shall require it to be the same for all components. A Fredholm vector field is Palais-Smale if $\nabla X(p)$ is of the form $I+C$, where C is a compact linear map. Palais-Smale vector fields have index zero.

Definition - Suppose M is a smooth Banach manifold with $X: M \to TM$ a smooth vector field. A zero $m \in M$ is non-degenerate if the Frechet derivative $X_*(m): T_mM \circlearrowleft$ is an isomorphism. (At a zero the Frechet derivative agrees with the covariant derivative of X with respect to any connection on M.)

As a trivial consequence of this definition it follows that non-degenerate zeros are isolated.

Let $H^{r+2}(S^1,R^n)$ by the Sobolev Hilbert space of H^{r+2} maps of the unit circle S^1 into R^n, with $r \geq 5$. Let $\mathcal{G} = \mathrm{Emb}(S^1,R^3)$ be the open submanifold of $H^{r+2}(S^1,R^3)$ which consists of embeddings of S^1 into R^3. Let Γ be the image of such an embedding $\alpha \in \mathcal{G}$. Set η^α to be the component of $H^2(S^1,\Gamma)$ {the C^r Hilbert manifold of H^2 maps from S^1 to Γ} determined by the embedding α. Let $\mathfrak{m}^\alpha$ be the open submanifold of η^α consisting of the diffeomorphisms. For every $u \in H^2(S^1,\Gamma) \subset H^2(S^1,R^3)$ we can extend $u = (u_1,u_2,u_3)$ harmonically to the disc $\mathcal{D}$. Define the smooth Dirichlet energy functional $E: \eta^\alpha \to R$ by

$$E(u) = \frac{1}{2} \sum_{i=1}^{3} \int_{\mathcal{D}} \left[\left(\frac{\partial u_i}{\partial x}\right)^2 + \left(\frac{\partial u_i}{\partial y}\right)^2 \right] dxdy. \tag{1}$$

We shall think of η^α as the manifold of harmonic surfaces spanning $\alpha(S^1)$. Moreover if $u \in H^2(S^1,\Gamma)$ the harmonic extension of u to $\mathcal{D}$ is in $H^{5/2}(\mathcal{D},\mathbb{R}^3)$ and we may think of η^α as the submanifold of harmonic surfaces of $H^{5/2}(\mathcal{D},\mathbb{R}^3)$ spanning α. We can write the Dirichlet functional as an integral over $S^1 = \partial\mathcal{D}$ as follows. Let $\theta \to \frac{\partial u}{\partial r}(\theta)$ represent the partial derivative with respect to the polar coordinate r of the harmonic extension of u to $\mathcal{D}$ evaluated at $r = 1$. This agrees with the normal derivative on S^1 of the harmonic extension. Let $\langle\ ,\ \rangle$ denote the $\mathbb{R}^2$ inner product. Then applying Green's theorem to (1) we find that

$$E(u) = \frac{1}{2} \int_S \langle \frac{\partial u}{\partial r}, u \rangle \, d\theta .$$

Denote by $\bar{\mathfrak{m}}^\alpha$ the closure of $\mathfrak{m}^\alpha$ in η^α.

J. Douglas showed in this pionnering work [3] that the harmonic extensions of the critical points of E in $\bar{\mathfrak{m}}^\alpha$ are the simply connected minimal surfaces spanning Γ. For a somewhat simpler proof see [19].

<u>Theorem 3</u> - There exists a smooth connection K_α on the second

tangent bundle $T^2\eta^\alpha$ and a smooth vector field $X^\alpha\colon \eta^\alpha \to T\eta^\alpha$ which is Palais-Smale with respect to the connection K_α and whose zeros are precisely all the critical points of E. Moreover,

$$X^\alpha(E) = dE(u)(x^\alpha(u)) \geq 0.$$

<u>Proof</u> (A Sketch): Let $h,k \in T_u\eta^\alpha$. Define the weak Riemannian structure $\langle\langle\ ,\ \rangle\rangle\colon T\eta^\alpha \times T\eta^\alpha \to R$ by setting

$$\langle\langle h,k\rangle\rangle = \int_{S^1} \langle \frac{\partial h}{\partial r}, k\rangle \, d\theta .$$

By definition $\eta^\alpha \subset H^2(S^1,\mathbb{R}^3) = H$ is a submanifold. For each $u \in \eta^\alpha$

$$H = T_u\eta^\alpha \oplus (T_u\eta^\alpha)^\perp \tag{2}$$

where $(T_u\eta^\alpha)^\perp$ is the weak orthogonal complement of $T_u\eta^\alpha$ in H. In a standard way one can use the splitting (2) to define the connection K on η^α.

The vector field X^α is the "gradient" of E with respect to $\langle\langle\ ,\ \rangle\rangle$. It is necessary to check that this weak gradient gives a smooth C^{r-1} vector field on η^α and we use regularity theorems to accomplish this. The connection K_α and the vector field X^α are both defined naturally in terms of the weak Riemannian structure. By a direct computation we show that X^α is Palais-Smale with respect to K_α.

Let G be the three dimensional non-compact Lie group of bijective homomorphic maps of the disc onto itself. The functional E_α and the vector field X^α of Theorem 1 will be equivariant with respect to the action of G. Therefore, there is no hope that the critical points of E_α in $\bar{\mathfrak{m}}^\alpha$ will be isolated or non-degenerate in the sense described earlier since the orbits of any critical point will consist or critical points. Let $u \in \bar{\mathfrak{m}}^\alpha$ be a zero of X^α, and let $\mathcal{O}_G(u)$ denote the orbit of u under G. Then $\mathcal{O}_G(u)$ is an immersed three-dimensional submanifold of $\mathcal{N}^\alpha$ consisting of critical points of E_α. We call $\mathcal{O}_G(u)$ a <u>critical submanifold</u> on $\mathcal{N}^\alpha$. The covariant derivative of X^α at u induces a homomorphism of the normal bundle of $\mathcal{O}_G(u)$ into itself. A critical point u, or more precisely $\mathcal{O}_G(u)$, is <u>non-degenerate</u> if this induced homomorphism is an isomorphism.

Let Q_1, Q_2, Q_3 be three fixed points on S^1. For $\rho \in \mathcal{G}$ let $\mathcal{S}^\rho$ be the codimension three submanifold of $\mathcal{N}^\rho$ consisting of those maps in $\mathcal{N}^\rho$ which send Q_i to $\rho(Q_i)$. Each orbit of G intersects $\mathcal{S}^\rho$ in one and only one point and $\mathcal{S}^\rho$ is transverse to the orbits of points in $\mathfrak{M}^\rho$. Consequently if we restrict the Dirichlet functional E to $\mathcal{S}^\rho$ the action of G will have been "factored out".

Let $\mathcal{J} = \mathcal{J}m^1(\mathcal{D}, \mathbb{R}^3)$ be the set of C^1 immersions of the closed unit disc into $\mathbb{R}^3$ and let $\mathcal{C}$ be any fixed closed set in $\mathcal{J} \cap H^{5/2}(\mathcal{D}, \mathbb{R}^3)$. It is essentially shown in [19] that

<u>Theorem 4</u> - For an open dense set of embedding $\mathcal{V} \subset \mathcal{G}$ the zeros of X^α, $\alpha \in \mathcal{V}$ in $\mathcal{C} \cap \mathfrak{M}^\alpha$ are non-degenerate and therefore isolated points in $\mathcal{C} \cap \mathcal{S}^\alpha$. Moreover for $\alpha \in \mathcal{V}$ there are only finitely many such points in $\mathcal{C} \cap \mathcal{S}^\alpha$,

and

<u>Theorem 5</u> - If $\alpha \in \mathcal{V}$ is fixed then given any closed set $\mathcal{C} \subset \mathcal{J} \cap H^{5/2}$ there exists a closed set $\mathcal{J} \cap H^{5/2} \supset \mathcal{C}_1 \supset \mathcal{C}$ with $\mathcal{C}$ contained in the interior $\mathcal{C}_1^o$ of $\mathcal{C}_1$ and such that the minimal surfaces spanning α in $\mathcal{C}_1 \cap \mathcal{S}^\alpha$ are in $\mathcal{C} \cap \mathcal{S}^\alpha$. Moreover if γ is sufficiently close to α then the minimal surfaces in $\mathcal{C}_1^o$ are C^{r-1} smooth functions of the parameter γ, and therefore the number of minimal surfaces spanning γ in $\mathcal{C}_1^o$ is equal to the number spanning α in $\mathcal{C}$.

§III. Statement and Proof of the Main Theorem

Let β_k denote the set of H^k, $k \geq 7$ embedded Jordan curves in $\mathbb{R}^3$. Therefore β_k would be the set of images of embeddings in $H^k(S^1,\mathbb{R}^3)$. We topologize β_k with a metric ρ as follows. For $\Gamma,\Gamma' \in \beta_k$

$$\rho(\Gamma,\Gamma') = \inf \{\|\gamma-\gamma'\|_{H^k}\}$$

where $\| \|_{H^k}$ represents the H^k norm and where γ,γ' run over all possible diffeomorphisms of S^1 with Γ and Γ' respectively.

Theorem 6 (Main) - Let $\mathcal{J}_k \subset \beta_k$, $k \geq 7$, be the set of curves Γ for which E_Γ has a unique minimum. Then $\mathcal{J}_k$ has a dense interior.

Before proving this we shall need some preliminary results.

Theorem 7 - As before let $\alpha \in \mathcal{C} = \mathrm{Emb}(S^1,\mathbb{R}^3)$ be the open submanifold of $H^{r+2}(S^1,\mathbb{R}^3)$ for $r \geq 1$ which consists of embeddings of S^1 into $\mathbb{R}^3$. Let

$$\mathcal{K}_\alpha = \{w \in \bar{\mathfrak{m}}^\alpha \mid E(w) = \inf_{u \in \bar{\mathfrak{m}}^\alpha} E(u)\} = \mathcal{K}_\Gamma, \Gamma = \alpha(S^1)\}.$$

Then for an open dense set $\mathfrak{W} \subset \mathcal{C}$, if $\alpha \in \mathfrak{W}$ the $\mathcal{K}_\alpha$ consists of immersions and $w|S^1$ is an embedding.

Proof: For α analytic the fact that $\mathcal{K}_\alpha$ consists of immersions which are diffeomorphisms when restricted to S is essentially due to Osserman, Royden, Gulliver, Lesley and Alt [1], [6], [7].

Tomi [17] also proves that if α is analytic $\mathcal{K}_\alpha$ consists of a finite set.

We shall prove the theorem by contradiction. It suffices to show that each analytic α has a neighborhood W_α with the property that for $\gamma \in W_\alpha$, $\mathcal{K}_\alpha$ consists of immersions. Suppose this is false. Let $\gamma_n \to \gamma$ in $\mathcal{C}$ and let u_n be a sequence of minimal surfaces in K_{γ_n} which are not immersions. By regularity results for minimal surfaces first proved by Hildebrandt [8], and then improved by Nitsche [12], and Heinz and Tomi [9], it follows that u_n has a convergent subsequence u_{n_j} which converges to some $u \in \mathcal{K}_\alpha$. Since each u_{n_j} is not an immersion it follows that u is not an immersion, but $\mathcal{K}_\alpha$ consists only of immersions, a contradiction. The following result is well known.

Theorem 8 - Let a Jordan curve Γ bound a minimal surface u and let $\tilde{\Gamma}$ be any real analytic curve lying on u but not entirely on Γ. Then $\tilde{\Gamma}$ bounds exactly one minimal surface, of absolute minimum energy E.

Let $\alpha \in \mathfrak{W}$ be analytic and let $u \in \mathcal{K}_\alpha$ be one of the finite number of minimal surfaces in this set. Then u is an analytic map from $\mathcal{D}$ to $\mathbb{R}^3$. Let $\mathcal{D}_s$ be the disc of radius $s < 1$ and let $\gamma_s = u|\partial\mathcal{D}_s$.

Theorem 9 - For s close to 1 γ_s is an analytic embedding of $\partial\mathcal{D}_s$ into $\mathbb{R}^3$ which spans a unique minimal surface.

Proof: By the Osserman, Royden, Gulliver, Lesley, Alt result [1], [7], [6] and the theorem of H. Lewy [11] γ_1 is an analytic embedding of S^1 into $\mathbb{R}^3$ with image $\alpha(S^1)$. For s sufficiently close to 1, γ_s is an embedding of $\partial\mathcal{D}_s$ into $\mathbb{R}^3$ which lies on u and therefore by Theorem 8 spans a unique minimal surface, of absolute minimum energy E.

Theorem 10 - Let $\mathcal{P} \subset \beta_k$ by the subset of analytic curves (images of real analytic embeddings) and $\mathcal{P}_o$ those analytic curves of uniqueness. Then $\mathcal{P}_o$ is dense in β_k.

Proof: Since $\mathcal{P}$ is clearly dense in β_k for any k it suffices to show that $\mathcal{P}_o$ is dense in $\mathcal{P}$ in the β_k topology. Let $\Gamma \in \mathcal{P}$ and suppose $u\colon \mathcal{D} \to \mathbb{R}^3$ is a Douglas solution to the Plateau problem. Therefore $E(u) = \inf_v E(v)$ where the inf is taken over all surfaces spanning Γ. Now $\gamma_1 = u|S^1$ is an analytic embedding, and for s

sufficiently close to 1, γ_s is also an analytic embedding which is C^∞ (in fact analytically) close to γ_1. Using the previous Theorem 9 we see that for $s < 1$, γ_s spans a unique minimal surface of minimum energy. This concludes the proof of this theorem.

We now come to the proof of the main theorem. By the last result the curves of uniqueness are dense in β_k for any k. Again let $\mathfrak{J}_k$ denote this dense set. We shall show that for $k \geq 7$, $\mathfrak{J}_k$ has a dense interior. Let $\Gamma \in \beta_k$ and $\alpha \in H^k(S^1,R^3)$ be an embedding whose image is Γ.

We know from the regularity results that $\mathcal{K}_\Gamma$ is a closed subset of $\mathcal{J} \cap H^{5/2}(\mathcal{D},\mathbb{R}^3)$, and $\mathcal{K}_\Gamma \cap \mathcal{S}^\alpha$ is a compact subset of $H^{5/2}(\mathcal{D},\mathbb{R}^3) \cap \mathcal{S}^\alpha$. By Theorem 8 if Γ is the image of an element of the open dense set $\mathfrak{W}$ we have that $\mathcal{K}_\Gamma \cap \mathcal{S}^\alpha$ is a compact subset of $\mathcal{J} \cap H^{5/2}(\mathcal{D},R^3) \cap \mathcal{S}^\alpha$.

Let $\tilde{\mathfrak{W}}$ be the sets of images of elements of $\mathfrak{W} \subset \mathfrak{a}$. Then $\tilde{\mathfrak{W}}$ is open and dense in β_k.

<u>Lemma 1</u> - Let Γ be a fixed element of $\tilde{\mathfrak{W}}$. There is a closed subset $\mathcal{C} \subset \mathcal{J} \cap H^{5/2}(\mathcal{D},\mathbb{R}^3)$ with the property that $\mathcal{K}_\Gamma \subset \mathcal{C}^o$ and for all Γ' sufficiently close to Γ in β_k, $k \geq 3$ $\mathcal{K}_{\Gamma'} \cap \mathcal{S}^{\alpha'} \subset \mathcal{C}^o \cap \mathcal{S}^{\alpha'}$ where α and α' denote some parametrization of Γ and Γ' in $H^k(S^1,R^3)$.

<u>Proof</u>: Let α be some fixed parametrization of Γ. It clearly suffices to show there exists a set $\mathcal{C}$ so that for all α' close to α in H^k $\mathcal{K}_{\Gamma'} \cap \mathcal{S}^{\alpha'} \subset \mathcal{C} \cap \mathcal{S}^{\alpha'}$. Since $\mathcal{K}_\Gamma$ is a closed subset of $\mathcal{J} \cap H^{5/2}$ there exists an open set $U \subset \mathcal{J} \cap H^{5/2}$ containing $\mathcal{K}_\Gamma$ with $\bar{U} \subset \mathcal{J} \cap H^{5/2}$. Set $\mathcal{C} = \bar{U}$.

Using the regularity theorems for minimal surfaces it follows as in Theorem 7 that $\mathcal{K}_{\Gamma'} \cap \mathcal{S}^{\alpha'} \subset \mathcal{C} \cap \mathcal{S}^\alpha$ for all α' close enough to α.

Lemma 2 - Let $\mathcal{C}$ be as in Lemma 1 and let α be a parametrization of Γ. Then for an open dense set $\mathcal{V}_k \subset \beta_k$, $k \geq 7$, the zeros of $X^{\alpha'}$ in $\mathcal{C} \cap \mathfrak{m}^{\alpha'}$ are non-degenerate where α' is a parametrization of any $\Gamma' \in \mathcal{V}_k$. In particular if $\Gamma' \in \mathcal{V}_k$, $K_{\Gamma'}$ consists only of non-degenerate critical points which are minima for E_Γ.

Proof: This is essentially a paraphrasing of Theorem 4 after setting $\mathcal{V}_k$ = images of the elements of $\mathcal{V}$ in this theorem and noting that the degeneracy or non-degeneracy of a critical point of X^α in $\mathfrak{m}^\alpha$ depends only on the image curve Γ and not on its particular representation α.

Lemma 3 - Let Γ be a fixed curve in $\tilde{\mathfrak{W}}$, and let $\mathcal{C} \subset \mathcal{J} \cap H^{5/2}$ be a closed set given by Lemma 1 and let U_Γ be a neighborhood of Γ in β_k $(k \geq 7)$ so that Lemmas 1 and 2 hold for $\Gamma' \in U_\Gamma$. Then if $\Gamma' \in \mathcal{P}_o \cap \mathcal{V}_k \cap U_\Gamma$, Γ' is a curve of uniqueness. Moreover if Γ'' is sufficiently close to Γ' in β_k, Γ'' is also a curve of uniqueness.

Proof: That Γ' is a curve of uniqueness follows at once since $\mathcal{P}_o$ consists only of curves of uniqueness. Since $\Gamma' \in \mathcal{V}_k$, $K_{\Gamma'} \cap \mathcal{S}^{\alpha'}$, α' any parametrization of Γ', consists of a unique non-degenerate zero of $X^{\alpha'}$. Therefore if $\Gamma'' \in U_\Gamma$ is close enough to Γ' it follows that $K_{\Gamma''} \cap \mathcal{S}^{\alpha''} \subset \mathcal{C}^o \cap \mathcal{S}^{\alpha'}$ and by Theorem 5 that $K_{\Gamma''} \cap \mathcal{S}^{\alpha''}$ consists of a single point. This proves Lemma 3.

Therefore $\mathcal{P}_o \cap \mathcal{V}_k \cap U_\Gamma$ is a dense set in U_Γ and every $\Gamma' \in \mathcal{P}_o \cap \mathcal{V}_k \cap U_\Gamma$ has a neighborhood consisting of curves of uniqueness. Therfore there exists an open set $\mathcal{V}_\Gamma \subset \mathcal{P}_o \cap \mathcal{V}_k \cap U_\Gamma$ consisting of curves of uniqueness which is clearly dense in U_Γ. Taking the union $\mathcal{Y}$ of all such $\mathcal{Y}_\Gamma$ we obtain an open dense set of curves of $\tilde{\mathfrak{W}}$ of uniqueness. Since $\tilde{\mathfrak{W}}$ is open and dense it follows that $\mathcal{Y}$ is open and dense. Thus $\mathcal{J}_k$ has a dense interior which is what was to be proved.

References

[1] H.W. Alt, Verzweigungspunkte von H-Flächen #1, Math. Zeitschrift vol. 127, 33-362.

[2] Beeson, M. "Non continuous dependence of surface of least area on the boundary curve" (to appear).

[3] Douglas, J., "Solution to the problem of Plateau", Trans. Amer. Math. Soc., 33 (1931), 263-321.

[4] Courant, R., Dirichlet's Principle, Interscience, New York, 1950.

[5] Eliasson, H., "Geometry of manifolds of maps", J. Diff. Geometry, 1 (1967), 169-194.

[6] Gulliver, R., and Lesley, F., "On boundary branch points of minimizing surfaces", Arch. Rat. Mich. Anal., 52 (1973), 20-25.

[7] Gulliver, R., Osserman, R., and Royden, H., "A theory of branched immersions", American Journal of Math., Vol. XCV, nº 4 (1973), 750-812.

[8] Hildebrandt, S., "Boundary behavior of minimal surfaces", Arch. Rat. Mech. Anal. 35 (1969), 47-82.

[9] Heinz, E., and Tomi, F., Zu einen Satz von Hildebrandt über das Randverhalten von Minimal Flächen. Math. Z. 111, (1969), 372-386.

[10] Lawson, B., "Lectures on Minimal Surfaces", vol. 1, Instituto de Matemática Pura e Aplicada, Conselho Nacional de Pesquisas, Brazil.

[11] Lewy, H., "On the boundary behavior of minimal surfaces", Proc. Natl. Acad. Sci., 37 (1951), 103-110.

[12] Nitsche, J.C.C., "The boundary behavior of minimal surfaces, Kellog's theorem and branch points on the boundary", Inventiones Math., 8 (1969), 313-333.

[13] Osserman, R., A Survey of Minimal Surfaces, Van Nostrand, Princeton, 1969.

[14] Osserman, R., "A proof of the regularity everywhere of the classical solution to Plateau's problem", Ann. Math. 91 (1970), 550-569.

[15] Rado, T., "On the problem of Plateau", Ergebnisse der Mathematik und Ihrer Grenzgebeite, Springer Verlag, Berlin, 1933.

[16] Rado, T., "The problem of least area and the problem of Plateau", Math. Z., 32 (1930), 763-796.

[17] Tomi, F., "On the local uniqueness of the problem of least area", Arch. Rat. Mech. Anal., 52 ( ), 312-318.

[18] Tromba, A., "Fredholm vector fields and a transversality theorem" (to appear in J. Functional Anal.).

[19] Tromba, A., "On the number of solutions to Plateau's problem" (to appear as a Memoir of the AMS).

University of California
at
Santa Cruz

Geometric properties of generic differentiable manifolds

By C.T.C. Wall

This is an amplified version of a course of three lectures given at IMPA in Rio in July 1976, on the happy occasion of ELAM III. The objective was to survey applications of singularity theory to differential geometry. However, in preparing the survey I noticed that the majority of such applications made crucial use of transversality and the notion of generic property, so I decided to restrict myself to these. This lends a certain unity to the otherwise rather different constructions studied below, which itself makes the topic easier to assimilate.

The talks were rather hurriedly prepared (some of the questions discussed having been raised at the conference itself), and some errors in them are corrected below. This account too is written under pressure, as I kept discovering new results and new sources while engaged on the exposition, so the reader will find a number of loose ends to tidy up. Indeed, this survey raises a large number of interesting unsolved problems.

This paper is divided into three chapters (following the three lectures), each with further subdivisions. In the first, I describe the applications to differential geometry and give examples. I have been at pains to give a large number of examples, as I feel these are more instructive (and sometimes more useful) than the somewhat abstract general theorems. Working these out involves a lot of routine algebra, much of which I have suppressed. I have tried to show that this philosophy yields interesting generalisations to higher dimensions of results familiar for curves and surfaces. This chapter is divided into sections A - F, and Theorem 3 of section C (for example) is elsewhere referred to as Theorem C3.

Most of the proofs are given in chapter 2, on transversality, and it is here that the unity of treatment is an advantage since they have a great deal in common, only needing a little special argument for each. The deeper results rely on the main theorems of stratification theory, which are recalled in chapter 3.

The reader is assumed to have at least a nodding acquaintance with both subjects. For all the results I need on differential geometry I refer to the little book by Milnor (1963) on Morse theory, denoted [M] below. Results from singularity theory will be introduced as necessary; some basic notation is recalled at the start of chapter 2.

The detailed table of contents is as follows:

§1 Description of Results

A Projections 1

B Distance functions 4

C Lagrangean singularities: the exponential map and the Gauss map 7

D Osculating bundles 13

E Embeddings, metrics and curvature 21

F Geodesics, conjugate locus and cut locus 26

§2 Transversality

General discussion 30

Proof of Theorems A and B 34

Equivalence of lagrangean maps 36

Proof of Theorems D and E 40

Generic conditions on metrics 43

§3 Stratifications

Definitions and transversality 48

Stratifying maps 51

Multitransversality 52

The topological stability theorem 53

Families of functions 56

Bibliography 60

§1 Description of Results

(A) Projections

When a bent and twisted piece of wire is looked at from a general viewpoint, it presents the appearance of a smooth curve with (perhaps) some double points, of transverse self-intersection. But from certain transitional positions of the eye, one may see a triple point, tacnode or cusp. These appear when the eye crosses certain surfaces, and are resolved as follows when the eye moves to one side or the other :

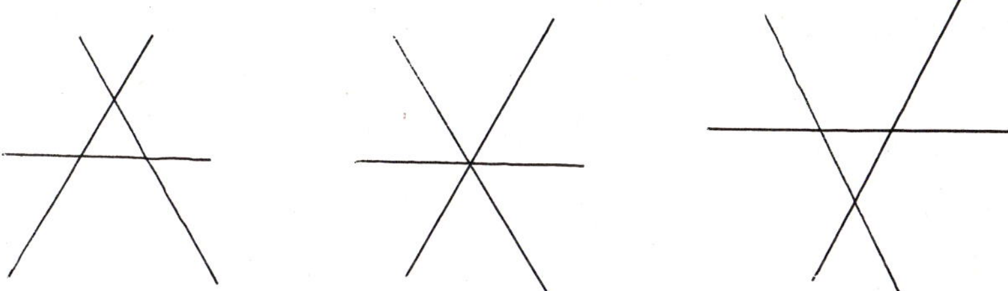

Triple point (the surface generated by trisecants of the curve)

Tacnode (the surface of chords AB such that the tangents at A and at B are coplanar - say, T-secants)

Cusp (the surface generated by tangents).

For positions of the eye on certain curves in space, one may see either two of the above or one of five further types of singularity, as follows :

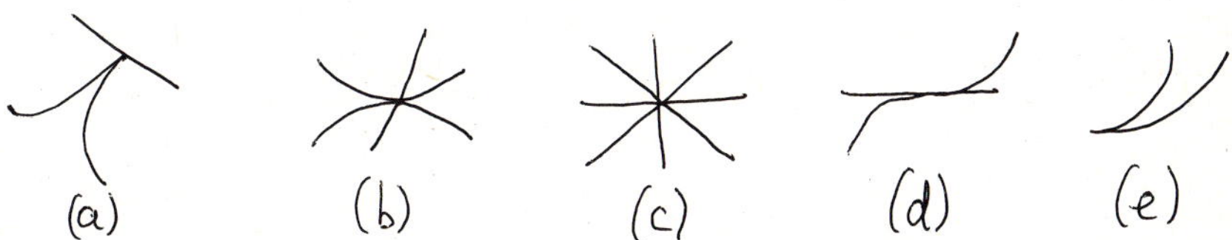

The first three consist of the three above, but with an extra curve passing through the singular point; (d) is two curves with 3-point (or inflexional) contact, and (e) is a rhamphoidal cusp. We may see (a), (b) by looking along a tangent or T-secant which meets the curve again; (c) by looking along a tetrasecant and (e) by looking along the tangent at a point of the curve where the torsion vanishes. And a general T-secant has just one point from which the two branches of the curve will appear to have inflexional contact.

As the eye moves in a small circle round one of these exceptional curves, it will cross some of the transitional surfaces, and the apparent shape of the curve varies as indicated in the diagrams below.

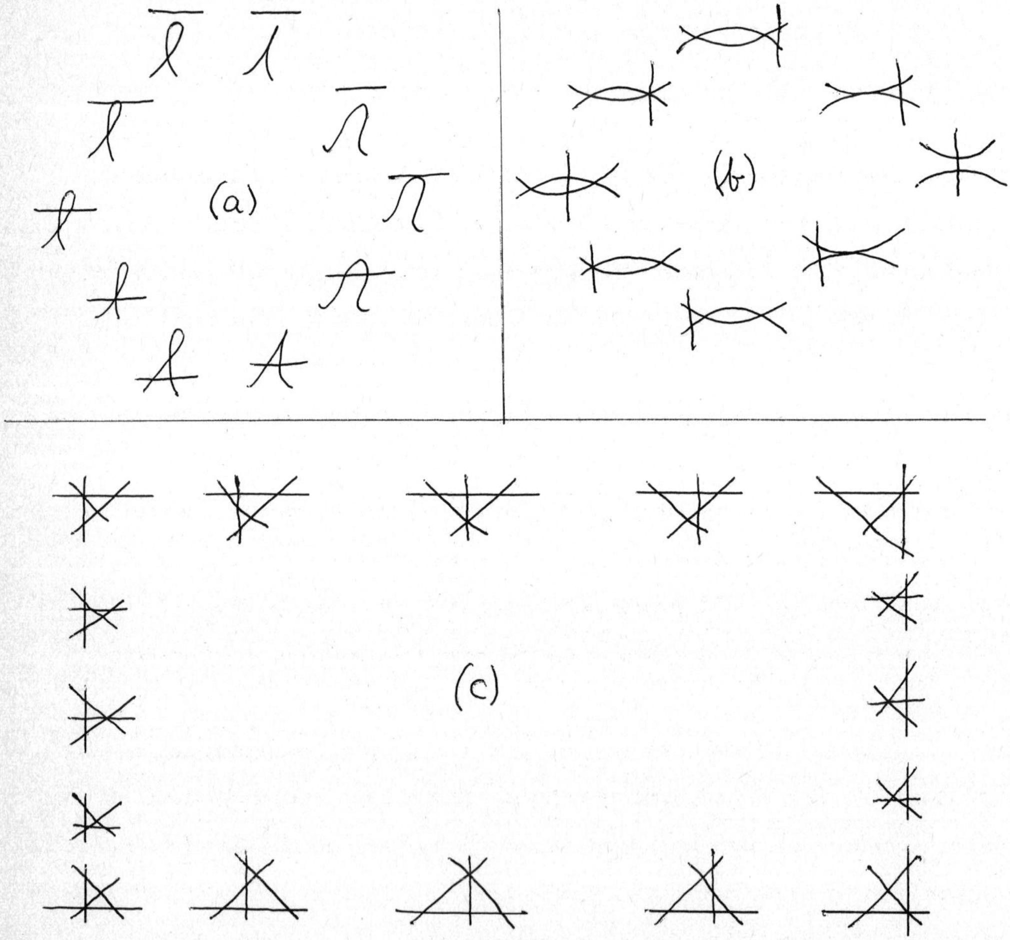

(a)

(b)

(c)

From certain isolated positions of the eye, one may see either three singularities from our first list, or one from each list, or one of two further types:

For (f), two curves have third order contact, and the other is transverse; the curves in (g) have fourth order contact. Just as (d) is seen from special points on general T-secants, (f) is seen from special points on T-secants which meet the curve again. And for certain exceptional T-secants, we obtain (g) instead of (d).

However, for a <u>general</u> curve, this is the complete list of possibilities. This is one of the simplest theorems of the kind I will discuss. Its characteristic feature is that instead of studying an arbitrary space-curve, we restrict to those in general position - in a sense to be made precise by the proof. The idea of studying such generic embeddings (or maps), and the ideas of many of the proofs, are due to Rene Thom, following some pioneering ideas of Hassler Whitney.

We next formulate a somewhat more general and precise version of the above statement. Let E be a euclidean space, E^0 a hyperplane separating E into open half-spaces $E^{\pm}$, M a manifold and $i : M \to E$ an embedding (or, indeed, any map) whose image we constrain to lie in E^-. Then for each point P (position of the eye) in E^+, we can project $i(M)$ from P into E^0, obtaining a map $H_i(-, P)$ of M into E^0. This yields a family H_i of maps $M \to E^0$, parametrised by $P \in E^+$. We call i projection-generic if H_i is a generic family of maps. This notion will be discussed below, in terms of stratifications: for the time being, we merely observe that the definition will require a finite number of transversality conditions.

Theorem Projection-generic maps form a residual (hence dense) set in the space of all maps $M \to E^-$.

This result was proved by my student, José Soares David. The restrictions $i(M) \subset E^-$, $P \in E^+$ are not strictly necessary, but some device is needed to ensure $P \notin i(M)$.

The entire construction is clearly projectively invariant. Write PE for the compactification of E by the projective space of unoriented directions in E: then we only use the projective structure of PE. Assigning to the pair (x, y) of distinct points in PE the line xy, or its image in the projective tangent space at y, gives a map from $(PE \times PE - \Delta)$ to the projective tangent bundle. Thus H_i becomes a section over $M \times PE$ (or rather the subset $i(m) \neq y$) of the bundle induced from the projective tangent bundle of PE. Here, if we restrict y to a compact subset K, and $i(M)$ to an open set U disjoint from K, the projection-generic maps are open and dense in $C^\infty(M, U)$. Note also that this formulation encompasses affine projections in E, as well as central projections.

Observe the difference in character between this result and that obtained by Mather (1973a). Mather showed that for any embedding $i : M \to E$, most projections are generic. Here we insist on considering the family of all projections, and show this is generic for most embeddings (or maps) i.

(B) Distance functions

A similar, but better known result comes from considering distances instead of projections, so the projection from P, $M \to E^0$, is replaced by distance from P, $M \to \mathbb{R}$ or better (to allow $P \in i(M)$) by the square of the distance. Thus we have a smooth family of real valued functions on M given by $G_i(m, P) = \|i(m) - P\|^2$.

A further refinement allows us again to compactify E. Take a base point $0 \in E$ (so we regard E as euclidean vector space), and write (following Looijenga, (1974) - but note the unfortunate misprint in his formula relating G_i to $\tilde{G}_i$) $S(E)$ for the unit sphere in $E \times \mathbb{R}$, which is one of the standard compactifications of E. Now define

$$\tilde{G}_i : M \times S(E) \to \mathbb{R} \qquad \text{by}$$

$$\begin{aligned} \tilde{G}_i(m, e, s) &= s\,\|i(m)\|^2 - 2\langle e, i(m)\rangle \\ &= s\; G_i(m, e/s) - s^{-1}\|e\|^2 \end{aligned}$$

for $\|e\|^2 + s^2 = 1$. For $s \neq 0$, this is essentially the same as G_i; for $s = 0$ we obtain the affine functions on $i(M)$. We call i <u>distance - generic</u> if both $\tilde{G}_i$ and its restriction to the sphere $S_0(E)$ defined by $s = 0$ are generic families of functions on M.

<u>Theorem</u> <u>Distance - generic embeddings form a dense open set in</u> Emb (M, E).

This was proved by Looijenga (1974). The idea was suggested by early work of Thom, most conveniently to be found in Chapter 5 of Thom (1972a), especially the remarks on umbilics.

We now discuss applications of this theorem: these are mostly drawn from the papers of Porteous (1971 a), (1971 b). I prefer to use the notation introduced by Arnold in his classification (1972a) of simple singularities, generalising Thom's 'elementary catastrophes',

$$A_n \qquad x_1^{n+1} + \Sigma_2^k \pm x_i^2 \qquad (n \geq 1)$$

$$D_n \qquad x_1^2 x_2 \pm x_2^{n-1} + \Sigma_3^k \pm x_i^2 \qquad (n \geq 4)$$

$$E_6 \;\; x_1^3 \pm x_2^4 \qquad\qquad E_7 \;\; x_1^3 + x_1 x_2^3 \qquad\qquad E_8 \;\; x_1^3 + x_2^5$$

<u>Remark</u> In the case of functions of two variables, some of these are pictured above (though they were obtained in (A) from functions $\mathbb{R}^1 \to \mathbb{R}^2$, so the interpretation is different): A_1 is an ordinary crossing, A_2 a cusp, A_3 a tacnode, A_4 a rhamphoid cusp (e), A_5 is illustrated in (d) and

A_7 in (g). D_4 is an ordinary triple point, and we see D_5 in (a), D_6 in (b) and D_8 in (f).

It is well known that for families of functions depending on at most 5 parameters, all critical points are in general simple: for at most 4 parameters we have the elementary catastrophes A_1, A_2, A_3, A_4, A_5, D_4 and D_5 ; then we allow also A_6, D_6 and E_6.

Now let $i : M \to E$ be generic : consider a point $(m, P) \in M \times E$. Then $G_i(-, P)$ has a critical point at m if and only if P lies on the normal plane at m, i.e. (m, P) belongs to the normal bundle. The point is not ordinary (of type A_1) iff we have a critical point of the exponential map of the normal bundle. The critical values of this map are the foci of $i(M)$. Most foci determine critical points of type A_2 : the surface of foci (or evolute of $i(M)$) has however a 'cuspidal edge', a generic point of which determines points of type A_3 ; this edge in turn is a manifold with cuspidal edge ....

Where the rank of the exponential map drops by 2 , we say we have an umbilic, and these are generally of type D_4 . There are two types: elliptic and hyperbolic, separated by the (parabolic) locus of D_5 - points. It follows from transversality theorems that these strata in $M \times E$ (or in $M \times S(E)$) fit together locally in the same way as in the universal example (the jet space), though it should be noted that it is customary to describe the figures in the euclidean space E , and this projection will complicate the descriptions. Thus for example at elliptic D_4 points, three sheets of the A_3 locus have a common boundary, while at hyperbolic D_4 points there is only one such sheet.

A comparison with traditional differential geometry, especially of surfaces, will be found in Porteous (1971b). An application of a very different nature is given by Thom (1972b). He considers for each P just the closest point (or points) of $i(M)$, and obtains interesting global theorems. It should be noted that the theorem of Looijenga implies conjecture CL of that paper. Other cognate ideas have been pursued by J. Bolton, S. Carter and S. Robertson - see for example Carter and Robertson (1967) and Carter (1973).

It would be interesting also to see some geometrical discussion of the higher order singularities. Next after the simple come the simple elliptic

$$T_{333} \text{ or } \tilde{E}_6 \qquad x_1^3 + x_2^3 + x_3^3 + \lambda x_1 x_2 x_3$$
$$T_{244} \text{ or } \tilde{E}_7 \qquad x_1^4 + x_2^4 + \lambda x_1^2 x_2^2$$
$$T_{236} \text{ or } \tilde{E}_8 \qquad x_1^3 + x_2^6 + \lambda x_1^2 x_2^2 ,$$

each depending on a single modulus: $\tilde{E}_r$ appears generically for the first time in r parameter families. See Siersma (1973) and (1974). Arnol'd went on to classify (1973b) all unimodular singularities: his (1975) survey article contains a beautiful account of these and related matters.

(C) Lagrangean singularities; the exponential map and the Gauss map.

Note particularly from the above that it is not true that the exponential map $N \to E$ from the normal bundle of a generically embedded $i(M)$ has the same types of singularity as a general map. The types of its singularities can be inferred from the theorem. A similar conclusion holds for the exponential map of the tangent bundle. I now present a fuller discussion, modelled for the most part on Arnol'd (several papers, particularly Arnol'd (1967) and Arnol'd (1972b) summarised in his (1975) survey) but see also Weinstein (1971) and Guckenheimer (1973) and particularly Hörmander (1971). I first recall some standard differential geometry.

A symplectic structure on an even dimensional manifold N^{2m} is defined by an (exterior) 2-form ω which is closed - $d\omega = 0$ - and nondegenerate - ω^m is nowhere zero. A classical result of Darboux (1883) asserts that near any point of N one can find local coordinates with respect to which $\omega = \Sigma_1^m dp_i \wedge dq_i$. The main example is the cotangent bundle of a manifold M^m. Each $u \in T^*M$ is a cotangent vector at $P = \pi(u)$, so $\pi^* u$ is a cotangent vector at u. The collection of these yields a 1-form θ on T^*M. If $q_1, \ldots, q_m$ are local coordinates on M, and $p_1, \ldots, p_m$ are the local coordinates on T^*M defined by the $\partial/\partial q_i$, then we have $\theta = \Sigma_1^m p_i \, dq_i$. Thus $\omega = d\theta = \Sigma_1^m dp_i \wedge dq_i$ (in local coordinates) defines a canonical symplectic structure.

Given a symplectic structure on N^{2m}, an m- dimensional submanifold L is called lagrangean if ω restricts to 0 on L. This notion, though quite old in applied mathematics, seems to have been first formalised in Arnold (1967). For an example, in the cotangent bundle T^*M, any (phase) function $f : M \to \mathbb{R}$ has a derivative df at each point of M, which defines a section of $\pi : T^*M \to M$. In local coordinates, this is given by $p_i = \partial f/\partial q_i$. Thus on the graph Γ_f of the section, θ restricts to $\Sigma\, \partial f/\partial q_i\, dq_i = df$, and so $\omega = d\theta$ to $ddf = 0$. Thus Γ is lagrangean. Note also the theorem of Weinstein (1971) (which we do not need) that any lagrangean inclusion $L^m \subset N^{2m}$ extends locally to a symplectic diffeomorphism $T^*L \to N$.

More important to us is the following generalisation of the notion of phase function. Let $F : M^m \times \mathbb{R}^k \to \mathbb{R}$ be a map such that 0 is a regular value of d_2F. Here, $d_2F : M^m \times \mathbb{R}^k \to \mathbb{R}^k$ is given by $d_2F(q, x) = (\partial F/\partial x_1, \ldots, \partial F/\partial x_k)$. Then F is called a generalised phase function. The critical set $C = C(F) = (d_2F)^{-1}(0)$ is a smooth m-manifold, called the catastrophe set and at any point of it, the matrix $(\partial^2F/\partial x^2, \partial^2F/\partial x\partial q)$ has rank k.

Lemma 1 *For any generalised phase function F, the map $\Phi : C \to T^*M$ given by $\Phi(q, x) = (q, \partial F/\partial q)$ is an immersion, and $\Phi(C)$ is lagrangean.*

Proof. To prove Φ an immersion is equivalent to proving that $\Phi \times d_2F : M \times \mathbb{R}^k \to T^*M \times \mathbb{R}^k$ is an immersion, at least along C. But this map has Jacobian $\begin{pmatrix} 0 & \partial^2F/\partial x\partial q & \partial^2F/\partial x^2 \\ I & \partial^2F/\partial q^2 & \partial^2F/\partial q\partial x \end{pmatrix}$ which has maximal rank on C, by the remark above.

Next, $\Phi^*\theta = \Phi^*(\Sigma\, p_i\, dq_i) = \Sigma\, \partial F/\partial q_i\, dq_i$, so along C (where $\partial F/\partial x_i = 0$) this equals dF, and again $\Phi^*\omega = \Phi^* d\theta = d\Phi^*\theta = d^2F = 0$.

Not only does this construction produce examples of lagrangean manifolds: it produces them all. And this construction is in some sense canonical. This follows from the next result, a slight modification of ones due to Arnold (1967), Hörmander (1971) and Weinstein (1971).

<u>Proposition 2</u> (i) <u>Any lagrangean manifold is given (at least locally) by the above construction.</u>

(ii) <u>Given a smooth family (parametrised by an open set in $\mathbb{R}^d$) of lagrangean manifolds, we can find a corresponding smooth d-parameter family of generalised phase functions.</u>

Observe that we are realising lagrange submanifolds, and not all the immersions with them as images.

<u>Proof.</u> (i) Let $L \subset T^*M$ be lagrangean, and $P \in L$. Take local coordinates p_i, q_i as above. Choose $f^1 : M \to \mathbb{R}$ with a nondegenerate critical point at $\pi(P)$, and consider the group of local diffeomorphisms of T^*M given by $h_t(p, q) = (p + tdf^1_q, q)$. Then for all except at most m values of t, we can take the p_i as local coordinates on $h_t(L)$ at $h_t(P)$: for the Jacobian of the corresponding projection has the form $A + tH(f^1)$, and the Hessian $H(f^1)$ is non-singular. We may suppose $t = 1$ a suitable value.

Thus on $h_1(L)$ we can write locally $q_i = g_i(p)$. The lagrangean condition $0 = \Sigma\, dp_i \wedge dq_i$ becomes $0 = \Sigma_{i,j}\, \partial g_i/\partial p_j \;dp_i \wedge dp_j$, so is equivalent to $\partial g_i/\partial p_j = \partial g_j/\partial p_i$. There is thus (locally) a function $f^2 = f^2(p)$ with $g_i = \partial f^2/\partial p_i$.

Now define $F : \mathbb{R}^m \times \mathbb{R}^m \to \mathbb{R}$ by

$$F(q, x) = \Sigma\, q_i x_i - f^1(q) - f^2(x).$$

Then C is given by $0 = \partial F/\partial x_i$, i.e. $q_i = \partial f^2/\partial x_i$, and our nondegeneracy condition is satisfied. Finally, Φ yields

$$p_i = \partial F/\partial q_i = x_i - \partial f^1/\partial q_i ,$$

so we recover the original manifold L.

(ii) Write $z = (z_1, \ldots, z_d)$ for coordinates in the parameter space, so we have lagrangean manifolds L_z which we study for z near 0 in a neighbourhood of $P \in L_0$.

Now apply the argument above. First choose the function f^1. Since the p_i can be taken as local coordinates on L_0 at P, they can also be used (in a suitable neighbourhood) on each L_z with z small. If $q_i = g_i(p, z)$ on

L_z, the lagrangean condition implies $\partial g_i/\partial p_j = \partial g_j/\partial p_i$, so we can find a function $f^2(p, z)$ (determined up to an arbitrary function of z) with $g_i = \partial f^2/\partial p_i$. But then

$$F(q, x, z) = \Sigma\, q_i x_i - f^1(q) - f^2(x, z)$$

gives the desired d-parameter family of generalised phase functions.

It follows from the proposition that we will get, in some sense, a generic lagrangean map from a generic phase function. However, to obtain an explicit general position result for lagrangean singularities, it is essential to be more precise.

Following Arnol'd (1972b) we consider a lagrangean map $\phi : L \to M$ as obtained from a lagrangean immersion $\gamma : L \hookrightarrow T^*M$ by projecting on the base. Correspondingly, define the lagrangeans (L_i, M_i, γ_i) $(i = 1, 2)$ to be equivalent if there exist diffeomorphisms $\alpha : L_1 \to L_2$, $\beta : M_1 \to M_2$ and a symplectic diffeomorphism $\beta' : T^*M_1 \to T^*M_2$ covering β such that $\gamma_2 \circ \alpha = \beta' \circ \gamma_1$. An obvious modification yields a definition of equivalence for lagrangean map germs which, as Arnol'd points out, is more useful than the global definition in this case. See Arnol'd (1972b) also for examples showing that this is stronger than C^∞-equivalence of map-germs, and that not every smooth map-germ can be so obtained.

Generalised phase functions $F : M^m \times E^k \to \mathbb{R}$ will be considered as m-parameter families of maps $f : E^k \to \mathbb{R}$; and we will study f under the relation of right equivalence. We can set up an abstract category, but it is enough to define two such functions to be equivalent if there is a commutative diagram

$$\begin{array}{ccc} M_1 \times E_1 & \xrightarrow{pr_1 \times F_1} & M_1 \times \mathbb{R} \\ \downarrow H & & \downarrow \tilde{G} \\ M_2 \times E_2 & \xrightarrow{pr_1 \times F_2} & M_2 \times \mathbb{R}, \end{array}$$

where $\tilde{G}$ is of the form $\tilde{G}(q, t) = (G(q), t + g(q))$ for some diffeomorphism $G : M_1 \to M_2$, and H is a diffeomorphism. Again, we shall only be interested in the corresponding equivalence relation for germs. Observe in passing that Guckenheimer (1973) omits the above restriction on $\tilde{G}$, and hence his assertion that "III - equivalence implies II- equivalence" is incorrect.

It is necessary for our purposes to extend slightly the above notion of equivalence to allow k to vary. Although there are doubtless far more abstract ways to express it, it will suffice to prescribe that F is equivalent to the maps $F': M \times E \times \mathbb{R} \to \mathbb{R}$ defined by $F'(q, x, t) = F(q, x) \pm t^2$. The following result seems to fill an awkward gap in the literature.

Theorem 3 Germs of generalised phase functions define equivalent germs of lagrangean maps if and only if they are equivalent.

A sketch of the proof, and further discussion, will be given in §2.

It follows from this result that any singularity - type of generalised phase functions, invariant under equivalence, defines a singularity type of lagrangean maps. In particular, the right equivalence class of the function $f(x) = F(0, x)$ is an invariant. Examples were given in the preceding section: we note in particular that generic singularities will be C^∞ - stable if $m \geq 5$ and C^0 - stable in general.

It is perhaps worth illustrating the point that the above differ from the generic singularities of an unrestricted map $\mathbb{R}^m \to \mathbb{R}^m$. By the proof of Lemma 2, we can take coordinates x_i in L such that the map is given by $q_i = \partial f^2/\partial x_i$ for some smooth function $f^2(x)$. Thus the Jacobian matrix of the map is everywhere symmetric. Hence $\Sigma^i f$, the set of points where df has kernel rank i, has codimension $\frac{1}{2}i(i + 1)$ for a generic lagrangean map: for a generic map its codimension is i^2. This argument is due to Weinstein (1970). Indeed, Arnol'd (1972b) gives an example due to Palamodova of a map germ which cannot be considered as lagrangean.

The example which introduced this discussion was the exponential map of the normal bundle of a manifold embedded in euclidean space: here we were able to write down an explicit generalised phase function (G_i or $\tilde{G}_i$). In fact exponential maps are always projections of lagrangean submanifolds. If M is a

Riemannian manifold, the geodesic flow ϕ_t in $TM \cong T^*M$ preserves the symplectic structure, and hence takes the fibre T_pM - clearly a lagrangean submanifold - onto another. The exponential map $\exp : T_pM \to M$ is the composite

$$T_pM \subset TM \overset{\phi_1}{\to} TM \overset{\pi}{\to} M ,$$

so is identified with the projection of the lagrangean manifold $\phi_1(T_pM)$. It is not so simple here to write down a generalised phase function since the distance depends on the path along which it is calculated, so we have to use the path space of M. We return to this example in (F) below.

We now turn to the Gauss map $\gamma : M^m \to S^m$ of a submanifold $M^m \subset \mathbb{R}^{m+1}$ of a submanifold $M^m \subset \mathbb{R}^{m+1}$ of codimension one. I say that this, too, can be regarded as the projection of a lagrangean manifold, and will have corresponding singularities. Indeed, suppose M given (locally) by the equation $0 = f(q_0, \ldots, q_m)$, where $df \neq 0$. Then consider the generalised phase function

$$F : \mathbb{R}^{m+1} \times \mathbb{R} \to \mathbb{R}$$

defined by $F(q, t) = tf(q)$. Since $df \neq 0$ along M, this satisfies our nondegeneracy condition. We have

$$C = \{(q, t) : f(q) = 0\} = M \times \mathbb{R} ,$$

and $\Phi : C \to T^*\mathbb{R}^{m+1} \cong \mathbb{R}^{m+1} \times \mathbb{R}^{m+1}$ is given by

$$\Phi(q, t) = (q, t\,\partial f/\partial q) .$$

This gives a 'conic lagrangean manifold' in the sense of Guckenheimer (1973). In that paper, the first projection of these manifolds is studied: this has everywhere rank $\leqslant m$, so special remarks are needed. Here we note that the Gauss map is the projection on the second set of coordinates, made projective by factoring out t.

<u>Proposition 4</u> <u>The Gauss map has generically the same types of singularity as any lagrangean map.</u>

To see this, it suffices to observe that in suitable local coordinates we can take $f(q) = q_0 + g(q_1, \ldots, q_m)$. Then $q_1, \ldots, q_m$ are local coordinates on M; Φ reduces to

$$\Phi(q_1, \ldots, q_m, t) = (-g(q_1, \ldots, q_m), q_1, \ldots, q_m ; t; t\,\partial g/\partial q_1, \ldots, t\,\partial g/\partial q_m).$$

If we take inhomogeneous coordinates in the projective space, setting the zero'th coordinate = 1, the Gauss map reduces to

$$G(q_1, \ldots, q_m) = (\partial g/\partial q_1, \ldots, \partial g/\partial q_m).$$

But this is a generic lagrangean map, as we saw above.

Noting the description in (B) of the simplest types of singularity that occur generically, we observe that G has corank 1 along a submanifold N^{m-1}; $G(N^i)$ has a cuspidal edge $G(N^{i-1})$ for $m - 1 \geqslant i \geqslant 1$. G has corank 2 along a submanifold L^{m-3} of "umbilic points", and so on.

The study of lagrangean maps lies at the heart of catastrophe theory. Essentially no study has been made of generic singularities of maps with other types of structure (though for equivariant maps, at least the foundations have been laid - Poenaru (1975))- e.g. maps preserving a G-structure (symplectic, measure, ...) or maps with non-negative Jacobian, c.f. Titus (1971). It is to be expected that interesting theorems remain to be discovered in this area.

(D) Osculating bundles

We can also study structures of higher order. For example, the kth order tangent bundle T_k^+M is defined (Pohl (1962)) as the bundle of differential operators (on functions) of order $\leqslant k$, and so has basis (in local coordinates)

$$1, \partial/\partial x_i, \partial^2/\partial x_{i_1}\partial x_{i_2}, \ldots, \partial^k/\partial x_{i_1} \ldots \partial x_{i_k} \quad (i_1 \leqslant \ldots \leqslant i_k)$$

and fibre dimension $\binom{m+k}{k}$. This is dual to the vector bundle $J^k(M, \mathbb{R})$ of k-jets of functions on M. Write T_kM for the subbundle (codimension 1) of operators annihilating constant functions.

If E is an affine space, any (smooth) map $i : M \to E$ induces a bundle map

$$\tau_k^+(i) : T_k^+ M \to TE \times \mathbb{R}$$

by applying the differential operators to functions which are the restrictions of affine functions on E; by restriction we obtain $\tau_k(i) : T_kM \to TE$. Now i is an immersion, or nondegenerate, if $T(i) = \tau_1(i)$ is fibrewise injective. Call i nondegenerate of order k if $\tau_k(i)$ is fibrewise injective. In this case, the image of $\tau_k(i)$ is a subbundle of $(TE)|M$ called the kth order osculating bundle of M.

We now present the elementary proof of Pohl (1962) that

Lemma 1 *For any (paracompact) smooth manifold M^m, and positive integer k, there exist an affine space E, and kth order nondegenerate embedding $i: M \to E$.*

Proof. First consider the case $M = \mathbb{R}^m$. Here it suffices to define $\phi : \mathbb{R}^m \to \mathbb{R}^N$, where $N = \binom{m+k}{k}$, to have as coordinates all the monomials of degree $\leqslant k$ in the coordinate functions on $\mathbb{R}^m$ (we could even exclude the trivial monomial). It is clear that ϕ is nondegenerate of order k.

But any manifold M^m can be embedded in a number space $\mathbb{R}^p$ for some p, say by $\psi : M \to \mathbb{R}^p$. If $\phi : \mathbb{R}^p \to \mathbb{R}^N$ is nondegenerate of order k, so is the composite $\phi \circ \psi : M \to \mathbb{R}^N$, since both the maps

$$T_k M \xrightarrow{T_k(\psi)} T_k \mathbb{R}^p \xrightarrow{T_k(\phi)} T_1 \mathbb{R}^N$$

are fibrewise injective, and their composite is $\tau_k(\phi \circ \psi)$. Note also that we may suppose ψ bounded; then $\phi \circ \psi$ will be bounded too.

The concept of kth order nondegeneracy can be generalised as follows (Pohl (1962) and Feldman (1965)). For any manifold V^v, there are natural exact sequences

$$0 \to T_{k-1} V \to T_k V \to 0^k \, TV \to 0$$

of vector bundles, where 0^k denotes the symmetric kth power. A symmetric linear connection γ on V induces splittings of all these sequences, and hence a composite map $\gamma_k : T_k V \to T_1 V$. Thus if V has such a connection - in particular if V is Riemannian - a map $\phi : M \to V$ induces

$$T_k M \xrightarrow{T_k\phi} T_k V \xrightarrow{\gamma_k} T_1 V ,$$

and we call f kth order nondegenerate if the composite is fibrewise injective.

There are several variants of the basic results concerning generic properties of these maps: I will attempt to indicate the main variations without being pedantic. First, call a submanifold A of $\mathrm{Hom}(T_k \mathbb{R}^m, \mathbb{R}^v)$ a model singularity if it is invariant by the product group $L_k(m) \times O_v$, where $L_k(m)$ is the group of invertible k-jets from $(\mathbb{R}^m, 0)$ to itself. Given such an A, we define $A(M, V)$ (for a manifold M^m and Riemannian V^v) to be the subbundle of $\mathrm{Hom}(T_k M, T_1 V)$ consisting of composites of maps of fibres

$$T_k M \xrightarrow{T_k(\xi)} T_k(\mathbb{R}^m) \xrightarrow{\alpha} \mathbb{R}^v \xrightarrow{\eta} T_1 V,$$

where ξ is a local chart in M, $\alpha \in A$ and η is an isometry.

Theorem 2 (i) Feldman (1965); Little (1969) For any model singularity S, the maps f with $\tau_k(f) \pitchfork A(M, V)$ are dense in $C^\infty(M, V)$.

(ii) The maps $\tau_k f$ have the same generic singularity types as unrestricted bundle maps $T_k M \to TV$.

The second assertion here is not a formal statement. We will indicate at the end of the paper what needs doing to make it precise. But the general philosophy is simple: that the 'integrability condition' stating that the map $T_k M \to T_1 V$ is induced by a function f is irrelevant for the study of singularities (so the situation resembles those in (A) and (B) above, but not (C)). Observe also that we do not need the full strength of the Riemannian hypothesis. If V has a vector space structure instead, the obvious formulation differs from the above in replacing O_v by GL_v. But in fact this gives a weaker result. On the other hand, suppose all points of A restrict to injective maps $T_1 \mathbb{R}^m \to \mathbb{R}^v$. Then f is an immersion near each point $P \in M$ such that $\tau_k(f)(P) \in A(M, V)$, so the Riemannian structure on V induces one near such a point of M. Hence we can sharpen the result by insisting only that A is invariant under the subgroup of $L_k(m)$ of maps whose tangent at 0 is an isometry, and then modifying the definition of $A(M, V)$ so that $d\xi$ has to be an isometry at the chosen point. I will refer to this modification if (i) above as Theorem 2 (iii).

Indeed, the most interesting and accessible model singularities arise when we assume that $\alpha_{k-1} : T_{k-1} \mathbb{R}^m \to \mathbb{R}^v$ is injective, and look at the induced map of quotients, $O^k \mathbb{R}^m \to \mathrm{Coker}\, \alpha_{k-1}$. We can identify this cokernel with $\mathbb{R}^q$, where $q = v - \binom{m+k-1}{k-1} + 1$, and are then studying $O_m \times O_q$ - invariant submanifolds of $\mathrm{Hom}(O^k \mathbb{R}^m, \mathbb{R}^q) = \mathrm{Sym}^k(\mathbb{R}^m, \mathbb{R}^q)$. In particular, let A_0 be the set of non-injective maps $T_k \mathbb{R}^m \to \mathbb{R}^v$ (this is not a manifold, but is a finite union of such).

Since non-injective homomorphisms $\mathbb{R}^a \to \mathbb{R}^b$ have codimension $b - a + 1$ when $b \geq a$, we deduce the density assertion of

<u>Corollary 3</u> <u>kth order nondegenerate immersions</u> $M \to V$ <u>are open and dense in</u> $C^\infty(M, V)$, <u>provided</u> $v \geq \binom{m+k}{k} + m - 1$.

This sharpens Lemma 1, but we will use this lemma in our proof of Theorem 2.

We turn to discuss examples, and will proceed from the general to the particular. We confine ourselves to the latter case above.

In particular, if we restrict to the case of $(k-1)^{st}$ order nondegenerate maps $\phi : M^m \to V^v$, then the orthogonal complement to the $(k-1)^{st}$ order osculating bundle is called the $(k-1)^{st}$ order normal bundle, and $\tau_k(\phi)$ induces a bundle homomorphism $\odot^k TM \to N_{k-1}(V/M)$ called the <u>kth fundamental form</u>. We are, in effect, studying singularities of this.

One might expect to construct invariant manifolds by writing equations in invariant functions. Now if $e_1, \ldots, e_m$ is an orthonormal base of $\mathbb{R}^m$, $\{e_{i_1} \circ \ldots \circ e_{i_k} : i_1 \leq \ldots \leq i_k\}$ is an orthonormal base of $\odot^k \mathbb{R}^m$, so our homomorphism $\beta_k : \odot^k \mathbb{R}^m \to \mathbb{R}^q$ is determined by the vectors $a_{i_1 \ldots i_k} = \beta_k(e_{i_1} \circ \ldots \circ e_{i_k})$. Applying invariant theory for O_q yields the matrix of scalar products $\langle a_{i_1 \ldots i_k}, a_{j_1 \ldots j_k} \rangle$, with the proviso that its rank is $\leq q$. (For SO_q we also have scalar q^{tuple} products of these vectors, and formulae for the product of any two such in terms of elements of our matrix.) If we ignore the rank restriction, the matrix entries yield the representation $\odot^2(\odot^k \mathbb{R}^m)$ of O_m. Invariant theory will not be easy. In case $k = 2$, $\odot^2(\odot^2 \mathbb{R}^m)$ is a sum of two irreducible GL_m-modules: $\odot^4 \mathbb{R}^m$ (corresponding to the partition (4)), and the kernel of the obvious epimorphism, corresponding to the partition (22). Each of these splits further into 2, 3 or 4 irreducible O_m-modules (see under section (E) below), and now the 'symbolic method' will, <u>in principle</u>, yield a determination of the invariants. In practice, this looks hopeless.

The easy case is $m = 1$, when all symmetric powers are 1-dimensional. kth order nondegenerate immersions are open and dense if $v \geq k + 1$; if $v = k$, a generic map is $(k-1)^{st}$ order nondegenerate, but the kth order condition fails at isolated points.

If $m = 2$, it is also feasible to discuss some general results. $0^k \mathbb{R}^2$ has dimension $k+1$, and is dual to the space of homogeneous polynomials of degree k in x and y. This space has a natural partition, corresponding to the multiplicities of the irreducible linear factors of the polynomial. Dualising β, we have a linear system of polynomials. Results useful for the classification of such systems have been obtained by Iarrobino (1975). One can give a partial classification in the case $(q = 2)$ of pencils. If the two polynomials f_0 and f_1 defining the pencil have no common factor, then in general there are $(2k-2)$ values of λ for which $\lambda f_0 + f_1$ has a repeated factor. But if (for example) $f_1(x, y) = c\Pi(x - a_i y)^{r_i}$, with the a_i distinct, then $\lambda = 0$ counts with multiplicity $\Sigma(r_i - 1)$. One thus has a classification by a type of Segre symbol, modified to allow for the possibility of common factors of f_0 and f_1. However, not all such symbols can occur, and it seems unlikely that, even working over $\mathbb{C}$, the pencils with given symbol form a connected (irreducible) manifold in all cases. Let us conclude this discussion with one example, where $k = 3$ so we are studying second order nondegenerate immersions $M^2 \to V^7$. There are two types of pencil of codimension 1, represented by $< t^3 \pm tu^2, t^2u >$ and $< t^3 \pm tu^2, u^3 >$ and three types of codimension 2, represented by

$< t^2u, tu^2 >$ (or $< t(t^2 tu^2), u(t^2 + u^2) >$), $< t^3, u^3 >$ (or $< t^3 - 3tu^2, u^3 - 3ut^2 >$)

and $< t^3, tu^2 >$.

We now concentrate on the most basic case $k = 2$, which amounts to a study of the second fundamental form. Here the geometry is better understood, and there are several different approaches. The most geometric starts from the natural quadratic map $\mathbb{R}^m \to 0^2(\mathbb{R}^m)$: the image of the unit sphere S^{m-1} is the Veronese manifold in $0^2\,\mathbb{R}^m$, and we have a linear image (or non-orthogonal projection) of this.

The study of these leads to an eigenvalue problem, which also arises when we look at the focal set (see (B) above). For any unit normal vector $f \in \mathbb{R}^q$, we have a quadratic form $x \to < \beta(x \circ x), f >$ on $\mathbb{R}^m$: this determines (in general)

m orthogonal principal directions, given by unit vectors $e_i \in \mathbb{R}^m$, and (real) eigenvalues λ_i such that

$$<\beta(x \circ x), f> = \Sigma \lambda_i <x, e_i>^2 .$$

The corresponding foci are seen at once to be the $\lambda_i^{-1} f$. There are m (in general distinct) on each radius line in $\mathbb{R}^q$, and they form a primal of order m. The vector v belongs to this primal if and only if the quadratic form

$$x \mapsto <\beta(x \circ x), v> - \Sigma <x, e_i>^2$$

is degenerate, so we have an equation (for v) given by the vanishing of the determinant of a symmetric $m \times m$ matrix whose entries are linear functions of v. For example, if $m = q = 3$ the cubic surface we obtain must be 4-nodal (or a specialisation). In general, the reduction from studying ϕ to studying this equation amounts to "doing the O_m-invariant theory first". Again, the final invariant problem is too difficult in general, but in the $m = q = 3$ case, for example, we can define model singularities by listing types of cubic surface as classified by Schläfli (1863) and Cayley (1870).

One can work out the full details in the case $m = 2$. These are given in Little (1969); also several earlier papers (e.g. Kommerell (1905), Moore & Wilson (1916) and Perepelkine (1935)).
Set $\beta(e_1 \circ e_1) = v_0 + v_1$, $\beta(e_1 \circ e_2) = v_2$, $\beta(e_2 \circ e_2) = v_0 - v_1$: then O_q-invariant theory yields the symmetric matrix $a_{ij} = <v_i, v_j>$. If we rotate coordinates in the source, v_0 is fixed but v_1, v_2 are rotated by twice the amount : say

$$v_1' = v_1 \cos\alpha + v_2 \sin\alpha$$

$$v_2' = -v_1 \sin\alpha + v_2 \cos\alpha .$$

Then a_{00} and $a_{11} + a_{22}$ are invariant, and

$a_{01}' = a_{01} \cos\alpha + a_{02} \sin\alpha$, $(a_{11}' - a_{22}') = (a_{11} - a_{22}) \cos 2\alpha + 2a_{12} \sin 2\alpha$

$a_{02}' = -a_{01} \sin\alpha + a_{02} \cos\alpha$, $2a_{12}' = -(a_{11} - a_{22}) \sin 2\alpha + 2a_{12} \cos 2\alpha$.

So $a_{01}^2 + a_{02}^2$ is invariant, and

$$(a'^2_{01} - a'^2_{02}) = (a^2_{01} - a^2_{02}) \cos 2\alpha + 2a_{01}\, a_{02} \sin 2\alpha ,$$

$$2a'_{01}\, a'_{02} = -(a^2_{01} - a^2_{02}) \sin 2\alpha + 2a_{01}\, a_{02} \cos 2\alpha .$$

It follows that the ring of invariants is generated by

$$I_1 = a_{00},\quad I_2 = a_{11} + a_{22},\quad I_3 = a^2_{01} + a^2_{02},$$

$$I_4 = (a_{11} - a_{22})^2 + 4\,a^2_{12},\quad I_5 = (a_{11} - a_{22})(a^2_{01} - a^2_{02}) + 4\,a_{12}\, a_{01}\, a_{02}$$

and $$I_6 = (a_{11} - a_{22})\, a_{01}\, a_{02} - a_{12}(a^2_{01} - a^2_{02})$$

with the unique syzygy $I_5^2 + 4\,I_6^2 = I_4\, I_3^2$.

In particular, $\det(a_{ij}) = \Delta = \frac{1}{4} I_1(I_2^2 - I_4) + \frac{1}{2}(I_5 - I_2 I_3)$.

Thus for $q \geqslant 4$, we are complete ; for $q = 3$ there is a further SO_3 - invariant $[v_0, v_1, v_2]$ with square Δ ; for $q = 2$ we have $\Delta = 0$, and may expect further SO_2 - invariants.

These can be obtained as follows. Using the canonical identification of $\mathbb{R}^2$ with $\mathbb{C}$, regard the v_r as complex numbers. Rotation in $\mathbb{C}$ multiplies them all by the same $e^{i\theta}$; the effect of rotation in the source (given above) is to fix v_0, and multiply $v_1 \pm i\, v_2$ by $e^{\mp i\alpha}$. The ring of invariants is thus generated by $J_1 = |v_0|^2$, $J_2^{\pm} = |v_1 \pm i\, v_2|^2$, J_3 and J_4 where $J_3 + 2i\, J_4 = \bar{v}_0^2\, (v_1^2 + v_2^2)$. Thus $J_3^2 + 4\, J_4^2 = J_1^2\, J_2^+\, J_2^-$. One may verify that $I_1 = J_1$, $2I_2 = J_2^+ + J_2^-$, $4\, I_3 = J_1(J_2^+ + J_2^-) + 2J_3$, $I_4 = J_2^+\, J_2^-$, $4\, I_5 = 2\, J_1\, J_2^+\, J_2^- + J_3(J_2^+ + J_2^-)$ and $4\, I_6 = J_4(J_2^+ - J_2^-)$.

The simplest way to relate these invariants to the geometry of the ellipse is to note that v_1 and v_2 represent conjugate directions for the ellipse. If we choose coordinates so that these are principal, they are orthogonal; so we may suppose $v_1 = (a, 0)$, $v_2 = (0, b)$ the semiaxes and $v_0 = (x, y)$ the position vector of the centre. Then

$$J_1 = x^2 + y^2 , \qquad J_2^{\pm} = (a \mp b)^2 ,$$

$$J_3 = (x^2 - y^2)(a^2 - b^2) \quad \text{and} \quad J_4 = -xy(a^2 - b^2) .$$

We can thus recover the semiaxes from $J_2^{\pm}$ - e.g. the area of the ellipse is $\pi/4\, (J_2^- - J_2^+)$; the squared distance of the centre from the origin is J_1, and the inclination θ of this vector to the semiaxes satisfies $\tan 2\theta = -2J_4/J_3$.

We also obtain the invariant equations for various conditions :

The ellipse degenerates to a repeated line segment : $J_2^+ = J_2^-$

The ellipse is a circle : $J_2^+ = 0$ (hence $J_3 = J_4 = 0$)

The ellipse is a point : $J_2^+ = J_2^- = 0$ (hence $J_3 = J_4 = 0$)

0 lies on the ellipse : $J_3 + \frac{1}{2}J_1(J_2^+ + J_2^-) - \frac{1}{8}(J_2^+ - J_2^-)^2 = 0$

0 lies on the director circle : $2J_1 = J_2^+ + J_2^-$

0 is at the centre : $J_1 = 0$ (hence $J_3 = J_4 = 0$)

0 is at a focus : $J_1^2 = J_2^+ J_2^- = J_3$ (hence $J_4 = 0$)

0 is on a principal axis : $J_4 = 0$

The ellipse degenerates to a line through 0 : $J_2^+ = J_2^-$, $J_3 + J_1J_2^- = 0$ (hence $J_4 = 0$)

The ellipse is a repeated segment with end point 0 : $J_1 = J_2^+ = J_2^-$, $J_3 = J_1J_2^-$ ($J_4 = 0$).

For the higher dimensional case we may take, correspondingly

$v_1 = (a, 0, 0, \ldots)$, $v_2 = (0, b, 0, \ldots)$, $v_0 = (x, y, z, 0, \ldots)$.

Then $I_1 = x^2 + y^2 + z^2$, $I_2 = a^2 + b^2$, $I_3 = a^2x^2 + b^2y^2$, $I_4 = (a^2 - b^2)^2$, $I_5 = (a^2 - b^2)(a^2x^2 - b^2y^2)$ and $I_6 = abxy(a^2 - b^2)$. Here one may make the corresponding calculations and some new ones: the most substantial is that the ellipse lies on a circular cone with vertex 0 if and only if the discriminant of

$$\mu^3 + 2\mu^2(I_1 - I_2) + \mu(I_4 - I_2^2 + 4(I_1 I_2 - I_3)) + I_5 - I_2 I_3 + 2I_1(I_2^2 - I_4)$$

vanishes.

After this digression, we can return to our surface. For a surface in $\mathbb{R}^4$, the mean curvature vector is v_0 (given in terms of the invariant theory of the second fundamental form) with squared length J_1. The Gauss curvature is $J_1 - \frac{1}{2}(J_2^+ + J_2^-)$. The normal curvature is $\frac{1}{2}(J_2^+ - J_2^-)$. The invariants J_3 and J_4 do not have standard names.

In any case, invariants are not altogether convenient for our purpose since we cannot always compute the codimension of a submanifold by counting the invariant conditions that define it. This can be clearly seen in terms of the theory of Mumford (1965). Let W be a submanifold whose general point is semistable, and has k-dimensional isotopy group. If the principal isotropy group has dimension

p , and d independent conditions on the invariants are needed to define W , then W has codimension $d + (k - p)$. Away from semistable points, we deduce nothing. Fortunately the set of non semi-stable points frequently has sufficiently large codimension that we may generically avoid it.

For the present problem, we conclude that we have in general nonsingular embedded curves C_1, C_2 in the surface corresponding to where the curvature ellipse degenerates (C_1) or passes through 0 (C_2) ; these meet transversely. There are further isolated points corresponding to when the ellipse is a circle, has 0 as centre or as focus. More degenerate cases do not appear.

For a generic surface in $\mathbb{R}^5$, the ellipse will only degenerate at isolated points. The condition $(\Delta = 0)$ that the plane of the ellipse should contain the origin defines a smooth curve which contains these points. It contains also exceptional points corresponding to the ellipse passing through 0. For a surface in $\mathbb{R}^6$, $\Delta = 0$ generally only yields isolated points, and beyond that the locus is empty. Note that the points $\Delta = 0$ are those of second order degeneracy. We can easily see that the curve is smooth, for the locus where v_0, v_1 and v_2 define a matrix of rank 2 is singular only where this rank drops to 1 , which we may avoid in the current problem.

(E) Embeddings, metrics and curvature

As further instances of the principle of "irrelevance of integrability" we now cite the interrelations between embeddings (or immersions), metrics and curvature. The celebrated isometric embedding theorem of Nash (1956) (see also Jacobowitz, (1972)) is beyone our scope, but the local theorem which is the first step in the main proof fits neatly into our context.

The space $\mathrm{Emb}(M, \mathbb{R}^n)$ of embeddings can be considered as global cross-sections of a sheaf; the k-jets of sections from a bundle which can be denoted $J^k_{emb}(M, \mathbb{R}^n)$. Similarly, Riemannian structures on M are global sections of the bundle $\mathrm{Riem}(M)$ of symmetric second-order tensors; and we have the associated bundle $J^k(\mathrm{Riem}\ M)$ of k-jets of sections. Given an embedding (or immersion) $M \to \mathbb{R}^n$, we have an induced Riemannian metric on M . Induction is local : the function space map $\mathrm{Emb}(M, \mathbb{R}^n) \to \Gamma \text{ Riem } (M)$ comes from maps $J^k_{emb}(M, \mathbb{R}^n) \to J^{k-1}(\text{Riem } M)$ of total spaces of bundles.

Theorem 1 *Each of the maps*

$$\mathrm{Emb}(M, \mathbb{R}^n) \to \Gamma \text{ Riem } (M), \qquad J^k_{emb}(M, \mathbb{R}^n) \to J^{k-1}(\text{Riem } M)$$

has surjective differential at f *if* f *is second order nondegenerate.*

Thus second order nondegeneracy plays an essential role in the theory. Since (Corollary E3) a generic map is second order nondegenerate for $n \geq \frac{(m+1)(m+2)}{2} + m - 1 = \frac{1}{2}m(m+5)$, one would expect isometric embeddings in the same range of dimensions. For the purely local theory, $\frac{1}{2}m(m+1)$ suffices. However, even in the compact case (where the technicalities of the analysis are less extreme), Nash's best result is $\frac{1}{2}m(3m+11)$. It was announted by Gromov (1970) that this could be improved to $\frac{1}{2}m(m+7)+5$.

According to the theorem, 'generic embeddings yield generic metrics'. There is no obvious notion of singularity of a metric, but structure certainly appears with the curvature tensor. Let Curv M denote the bundle of covariant tensors of degree 4 with the usual symmetry properties (see below). Associating to each Riemannian structure its curvature tensor gives maps

$$J^{k+2}(\text{Riem } M) \to J^k(\text{Curv } M).$$

These are not (as I mistakenly said in Rio) all submersions.

Proposition 2 *There is an induced submersion*

$$J^2(\text{Riem } M) \to J^1(\text{Riem } M) \times_M J^0(\text{Curv } M).$$

Observe, in fact, that it is natural to think of the curvature tensor as a metric tensor. But we clearly cannot vary the curvature and metric independently. At present, I cannot see a universal method for handling this problem. However, one method which works is to consider the covariant derivatives of the curvature tensor. Our account is inspired by that of Epstein (1975).

From the metric tensor, with components $g_{ij} = g_{ij}(x)$ we construct first the inverse g^{ij}, the coefficients

$$\Gamma^i_{jk} = \tfrac{1}{2} g^{hi}(\partial_k g_{hj} + \partial_j g_{hk} - \partial_h g_{jk})$$

of affine connexion, then those of the curvature tensor

$$R_{ijk\ell} = g_{ih}\{\partial_k \Gamma^h_{j\ell} - \partial_\ell \Gamma^h_{jk} + \Gamma^r_{j\ell}\Gamma^h_{rk} - \Gamma^r_{jk}\Gamma^h_{r\ell}\}$$

where the summation convention is used throughout, and ∂_k denotes $\partial/\partial x_k$. One easily verifies the symmetry properties

$$R_{ijk\ell} = - R_{jik\ell}, \quad R_{k\ell ij} = R_{ijk\ell}, \quad R_{ijk\ell} + R_{ik\ell j} + R_{i\ell jk} = 0 .$$

See for example [M, p.53]

The tensors (on a vector space V) with these symmetry properties form an irreducible $GL(V)$ - module $K(V)$ with Young diagram the partition (2, 2): this is the fibre of our bundle $Curv(M)$. We will shortly investigate it more fully.

The covariant derivatives of R satisfy further symmetry properties known as the Bianchi identities, which I shall not write down. The symmetry properties for $\Delta^k R$ define a space $K_k(V)$ of tensors on V of degree $(k + 4)$ which is an irreducible $GL(V)$-module corresponding to the partition $(k + 2, 2)$. We have

<u>Theorem 3</u> <u>The natural map</u>

$$J^{k+2}(\text{Riem } M) \to J^1(\text{Riem } M) \times \prod_{0 \leq i \leq k} K_i(TM)$$

<u>is a submersion for all</u> $k \geq 0$.

The proof is essentially identical to that of Epstein's Theorem 2.6 (loc.cit.). I shall not repeat it here, nor shall I give illustrative examples, but will content myself with the special case $k = 0$ (Proposition 2).

Let us define a <u>model singularity</u> S to be any O_v-invariant submanifold of $K(\mathbb{R}^v)$. For a Riemannian manifold V^v, let $S(V)$ be the associated subbundle of $Curv(V)$.

<u>Theorem 4</u> <u>For any model singularity</u> S, <u>a generic metric on</u> V^v <u>has the property that the associated curvature tensor</u> $R : V \to Curv\ V$ <u>is transverse to</u> $S(V)$.

To illustrate this, we must construct some model singularities. Now although $K(\mathbb{R}^v)$ is GL_v-irreducible, it is not O_v-irreducible. Indeed, the contraction which defines the Ricci tensor is an O_v-linear surjection $\mathrm{Ric} : K(\mathbb{R}^v) \to \mathbb{R}^v \circ \mathbb{R}^v$, provided $v \geq 3$. (Here $\circ$ denotes, as usual, symmetric tensor product.) Using the symmetries, we see that all contractions vanish on the kernel of Ric. Hence (Weyl (1946), p.158) this kernel is irreducible over the orthogonal group; over the proper orthogonal group it is irreducible $(v \geq 5)$ or a sum of two irreducible representations of the same rank $(v = 4)$. We can also see this directly by computing with weights.

Start by listing weights for the natural action of O_v on $\mathbb{R}^v$

$$v = 2n \quad \{u_i, u_i^{-1} : 1 \leq i \leq n\}$$

$$v = 2n+1 \quad \{u_i, u_i^{-1} : 1 \leq i \leq n\} \cup \{1\} .$$

Now set $\delta = \Pi\{(u_r^{\frac{1}{2}} u_s^{\frac{1}{2}} - u_r^{-\frac{1}{2}} u_s^{-\frac{1}{2}})(u_r^{\frac{1}{2}} u_s^{-\frac{1}{2}} - u_r^{-\frac{1}{2}} u_s^{\frac{1}{2}}) : 1 \leq r < s \leq n\}$ if $v = 2n$, and this multiplied by $\Pi_1^n (u_r^{\frac{1}{2}} - u_r^{-\frac{1}{2}})$ if $v = 2n+1$. Then δ is the alternating sum formed from

$$u_1^{n-1} u_2^{n-2} \ldots u_{n-1} \quad (v = 2n) \quad u_1^{n-1/2} u_2^{n-3/2} \ldots u_{n-1}^{3/2} u_n^{1/2} \quad (v = 2n+1) .$$

We then verify that δ multiplied by the character of $\ker(\mathrm{Ric})$ is the alternating sum formed from

$$u_1^{n+1} u_2^{n} u_3^{n-3} \ldots u_{n-1} \quad (v = 2n) \quad u_1^{n+3/2} u_2^{n+1/2} u_3^{n-5/2} \ldots u_n^{1/2} \quad (v = 2n+1).$$

This is irreducible save that for $v = 4$ $(n = 2)$, the sum formed from $u_1^3 u_2^2$ has 8 terms, so is formed from two sums over the Weyl group (of order 4 in this case).

Now for a 2-manifold, Curv M has 1-dimensional fibre. Thus, in general, the curvature vanishes along a smooth curve, and changes sign on crossing it. For a 3-manifold, Ric is an isomorphism. Now in any dimension, since the metric tensor is positive definite and the Ricci tensor symmetric, we can diagonalise this orthogonally (if one wished to generalise to the pseudo-Riemannian case, it would be necessary to invoke the classification (see e.g. Hodge and Pedoe (1952)) of pairs of quadratic forms). Thus the structure is determined by the eigenvalues. Interesting conditions arise from equalities between the eigenvalues, and vanishing of some eigenvalues and (perhaps) vanishing of their sum - the trace. In order to compute correctly the codimensions of these manifolds in the space of symmetric matrices, it is necessary to recall that the isotropy group in O_v of a matrix depends on equalities between its eigenvalues.

For a general Riemannian 3-manifold M^3 we have a smooth surface C where one eigenvalue is 0 and a curve D where two are equal. C meets D transversely at points with eigenvalues $(0, \lambda, \lambda)$; but at points $(0, 0, \lambda)$ C is locally diffeomorphic to a quadric cone with D a smooth 'time-like' curve through the vertex. There is another smooth surface C' where the trace vanishes: it meets both C and D transversely.

For a 4-manifold we obtain similar generic properties of the Ricci curvature, though of course all dimensions are raised by one, so C' may meet $C \cap D$ (transversely), and we have isolated points E with eigenvalue pattern $(\lambda, \lambda, \mu, \mu)$, where two sheets of D meet transversely. However, this is not the complete story here.

The first two symmetry relations defining $K(V)$ exhibit it as a quotient of $O^2(\lambda^2 V)$ - c.f. Ruse (1946). Now in the case $v = 4$, $\lambda^2 V$ is already reducible over SO_4, as we see using duality in the exterior algebra

$$e_1 \wedge e_2 \leftrightarrow e_3 \wedge e_4, \quad e_1 \wedge e_3 \leftrightarrow - e_2 \wedge e_4, \quad e_1 \wedge e_4 \leftrightarrow e_2 \wedge e_3$$

into ± 1 eigenspaces: $\lambda^2 V = W^+ \oplus W^-$ say. This yields the well known double covering $SO(V) \rightarrow SO(W^+) \times SO(W^-)$. Now

$$O^2(\lambda^2 V) = O^2 W^+ \oplus O^2 W^- \oplus (W^+ \otimes W^-) .$$

Here we can split off a single invariant from each of $O^2 W^{\pm}$ corresponding to the scalar product; the rest are irreducible. Since $\dim O^2(\lambda^2 V) = 21$ and $\dim K(V) = 20$, we must have

$$K(V) \cong \mathbb{R} \oplus (O^2 W^+/\mathbb{R}) \oplus (O^2 W^-/\mathbb{R}) \oplus (W^+ \otimes W^-),$$

and the surjection Ric onto $O^2 V \cong \mathbb{R} \oplus (O^2 V/\mathbb{R})$ must be an isomorphism on first and last summands, zero on the others. These isomorphisms of SO_4-modules can also be discovered by computing weights.

Since the composite $\mathbb{R} \to O^2 W^{\pm} \to O^2(\lambda^2 V) \to K(V)$ is nonzero, we can canonically identify an element of $K(V)$ with a triple of quadratic forms - on V, W^+ and W^- - all having the same trace. One can now extend the earlier discussion to consider equalities or zeroes of any of the corresponding eigenvalues of these three forms.

In general $\dim K(V^v) = \frac{1}{12} v^2(v^2 - 1)$, so that

$$\dim \operatorname{Ker}(\operatorname{Ric}) = \frac{1}{12}(v + 2)(v + 1)v(v - 3) .$$

For $v = 5$, these numbers are 50 and 35 . Thus more insight will be needed to analyse the structure here.

(F) Geodesics, conjugate locus and cut locus

The final group of problems I wish to discuss lie deeper, and the known results concerning them remain incomplete. Consider a closed (i.e. compact unbounded) Riemannian manifold M^m with base point P, and the exponential map $\exp_p : T_p M \to M$. We saw above that in some sense this had to be treated as a Lagrangean map. The points of $T_p M$, which correspond to geodesics in M, must be considered as critical points of a function on some 'unfolding space'.

We are thus led to consider spaces of paths on M. Write $L^\alpha(M)$ for the space of (continuous) maps $f : I \to M$, where the affix α may be varied to indicate the type of path considered and the topology on the space. Two natural choices are $\alpha = $ P.D., the space of piecewise C^∞-paths ([M, p.67], topology defined on p.88) and $\alpha = $ A.C., the space of absolutely continuous paths (see e.g. Palais (1963)). Evaluating f at the end-points 0, 1 of I yields projection maps, and hence

$$Ev : L^\alpha(M) \to M \times M .$$

We will restrict later to paths starting at $P : f(0) = P :$ and write $L^\alpha_P(M) = Ev^{-1}(P \times M)$; to $Ev^{-1}(P \times Q) = L^\alpha_{PQ}(M)$; or to 'closed paths' $f(0) = f(1)$, and write $L^\alpha_\Delta(M) = Ev^{-1}\, \Delta(M)$. Then Ev induces a (terminal) evaluation $ev_P : L^\alpha_P(M) \to M$.

The definitions of these path spaces depend on the differential structure, but not on the metric. The <u>energy function</u> with respect to a given metric g is defined by

$$E_g(f) = \int_0^1 ||df(t)||^2 \, dt ,$$

where $||df(t)||$ is the length of the tangent vector $df(dt)$ at $f(t)$ with respect to the metric g. The arc-length

$$L_g(f) = \int_0^1 ||df(t)|| \, dt$$

satisfies $L_g(f)^2 \leq E_g(f)$, with equality only when f is parametrised proportional to arc-length [M, p.70]. The critical points of $E_g(f)$ are the geodesics, parametrised proportional to arc-length (we will omit this phrase from now on) [M, p. 72]. By mapping each geodesic to its initial vector $(f(0), df0(dt))$, we obtain an isomorphism of the space of geodesics onto the tangent bundle TM ([M, p.56]; since M is compact, it is complete).

Two points $P = f(0)$ and $Q = f(1)$ are <u>conjugate</u> along the geodesic f if and only if f is a degenerate critical point of the restriction of E_g to $L^{\alpha}_{PQ}(M)$. The null space of the Hessian of E_g at f can be identified with the space of Jacobi fields along f which vanish at both end points, so has dimension $< m$ [M, pp.78-79]. Thus the locus of points Q conjugate to P, or <u>conjugate locus</u> of P, is the set of critical values of $\exp_P$. The first natural question concerns the generic structure of this locus. There is a difficulty here: since $\exp_P$ is <u>not</u> a proper map its critical values will tend to accumulate. One would expect, however, that their local structure is generically that for any lagrangean map. A partial result in this direction is given by Weinstein (1970); see Warner (1965) and Dos Santos (1967) for earlier discussions. Weinstein's approach is surprisingly indirect.

<u>Theorem 1</u> <u>For a generic metric on</u> M, <u>the map</u> $\exp_P$ <u>has only singularities which are generic for lagrangean maps.</u>

Indeed, one would expect more; though this raises further technical difficulties.

<u>Conjecture 2</u> <u>For a generic metric on</u> M, $\mathrm{Exp} = \{\exp_P : P \in M\}$ <u>is a generic</u> m-<u>parameter family of lagrangean maps.</u>

Next, consider the problem of closed geodesics: critical points of the restriction of E_g to $L^\alpha_\Delta(M)$, which correspond to the fixed points of the "time one map" of the geodesic flow on TM. The correct result here was proved by Abraham (1970). To formulate it, note that a closed geodesic determines a map $S^1 \to M$, and any point of S^1 can be taken as base point.

Theorem 3 Abraham (1970) *For a residual set of metrics g on M, the critical points of E_g on L^α_Δ consist of*

(i) *the constant maps, giving a nondegenerate critical manifold isomorphic to M, and*

(ii) *a set of nondegenerate critical manifolds, each isomorphic to S^1, and corresponding to a single closed geodesic.*

Note also that each geodesic $\phi : S^1 \to M$ gives rise to $\phi_n : S^1 \to M$ with $\phi_n(z) = \phi(z^n)$ for each $n \in \mathbb{Z} - \{0\}$. This possible recurrence causes considerable complications in the proof, which we shall not give.

Generically, the set of closed geodesics is thus countable, and none will pass through a given point $P \in M$. Generic properties of these closed geodesics have been studied by Klingenberg and Takens (1972); see also Klingenberg (1976). Since TM is symplectic, the Poincare map given by continuation of Jacobi fields round a closed geodesic γ is an automorphism of the symplectic space $\mathbb{R}^{2m}$. Then its eigenvalues are generically multiplicatively independent. Moreover, for any two points conjugate on γ one may suppose the multiplicity is 1. I conjecture that these are generally of type A_2, but for a discrete set of pairs, of type A_3.

A geodesic $\gamma : (I, 0, 1) \to (M, P, Q)$ gives an extreme value of the energy function; and also of the length function on $L^\alpha_{P,Q}(M)$; it is called *minimal* if we obtain the absolute minimum. Since M is compact, this minimum must be attained. If we extend γ to $(\mathbb{R}_+, 0) \to (M, P)$, then for each $t > 0$ the restriction to $[0, t]$ defines (after reparametrising, to agree with our conventions) a geodesic γ_t. There is a real number $t(\gamma) > 0$ such that for $t < t(\gamma)$, γ_t is minimising (see e.g. [M, p.62]) and for $t > t(\gamma)$ it is not (by an obvious argument; $t(\gamma) < \infty$ by compactness).

The cut locus is the locus of points $\gamma(t(\gamma))$. Such points may occur in two ways. If $c(\gamma)$ is the parameter of the first conjugate point to P along γ, then for $t < c(\gamma)$, γ_t gives a local minimum and for $t > c(\gamma)$ it does not [M, p.83]. Thus Q belongs to the cut locus of P if and only if either

(a) there exist at least two distinct minimal geodesics from P to Q, or

(b) there is only one, but P and Q are conjugate along it.

The cut locus X is thus defined as a point set in M. Examples have been constructed by Gluck and Singer (1976) (see also talk at this conference) to show that almost any compact ANR can occur as a cut locus, and that pathologies occur for any M of dimension ≥ 2. On the other hand, the cut locus of an analytic manifold is subanalytic, **and** hence by Hironaka (1973) triangulable. In the case of surfaces, this result is due to Myers (1935 - 6).

This is more typical

Theorem 4 (i) Buchner (1974) *If M^m is a closed manifold of dimension $m \leq 5$, then for a generic C^∞-Riemannian metric on M, the cut locus is triangulable, and is C^∞-stable under small perturbations of the metric.*

For $m > 5$, we lose the C^∞-stability. However, we shall indicate how to combine results from the theses of Buchner (1974) and Looijenga (1974) to obtain

Theorem 4 (ii) *For any closed manifold M^m, there is a dense open set U of C^∞-Riemannian metrics such that for any $\alpha \in U$, the corresponding cut locus is triangulable, and C^0-stable under small perturbations.*

In fact the ideas should allow one to go further and show that metrics with nontriangulable cut loci form a subset of infinite codimension, and even

Conjecture 5 *For M closed, a generic metric on M has $\Gamma = \{(P, Q) \in M \times M : Q$ in the cut locus of $P\}$ triangulable, also the cut locus of each $P \in M$, with C^0-stability in an appropriate sense.*

§2 Transversality

General discussion

The key to each of the results described above is a form of Thom's (1956) transversality theorem. I need to be more formal now, so recall a few standard definitions; however, we shall assume familiarity with these notions and with their elementary consequences.

Let N^n, P^p be smooth manifolds. If $f : N \to P$ is a smooth map, and W a smooth submanifold of P, we say f is <u>transversal</u> to W and write $f \pitchfork W$ if for each $x \in N$ with $f(x) = y \in W$ we have $df(T_xN) + T_yW = T_yP$ on tangent spaces. If f is only defined in a neighbourhood of $x \in N$, its <u>germ</u> is the equivalence class defined by restrictions to small neighbourhoods of x. This germ has <u>source</u> x and <u>target</u> $f(x)$. Two germs with the same source and target have the same <u>k-jet</u> if, in local coordinates, all partial derivatives of order $\leqslant k$ agree at x. The set $J^k(N, P)$ of all such k-jets has a natural structure of smooth bundle over $N \times P$ with fibre $J_0^k(n, p)$ the set of all k-jets $\mathbb{R}^n \to \mathbb{R}^p$ with source and target 0. This fibre can be identified with the vector space of polynomial maps of degree $\leqslant k$, with zero constant term, which has dimension $p(\binom{n+k}{k} - 1)$; but the bundle is not a vector bundle if $k > 1$. For any subset X of N - e.g. a single point - we will write $J_X^k(N, P)$ for the subbundle of jets with source in X. Similarly $J_{X,Y}^k$ will restrict the target to belong to $Y \subset P$. For a bundle ξ with projection $\pi : P \to N$, $J^k(\xi)$ denotes the bundle of k-jets of maps $f : N \to P$ which are (local) sections of π.

Following Mather (1970b) we write $N^{(r)} \subset N^r$ for the subset of r tuples of distinct points, and denote its preimage under the product projection $(J^k(N, P))^r \to N^r$ by ${}_rJ^k(N, P)$: the space of <u>multijets</u> of multiplicity r and order k. An (everywhere defined) smooth map $f : N \to P$ induces, by taking its jet at each point, a smooth section $j^kf : N \to J^k(N, P)$, hence also ${}_rj^kf : N^{(r)} \to {}_rJ^k(N, P)$.

Denote by $C^\infty(N, P)$ the space of smooth maps $N \to P$ with the <u>Whitney topology</u> (essentially due to Whitney (1936)). A base of open sets consists of the

$$\mathcal{F}(U) = \{f : N \to P,\ j^r f(N) \subset U\}$$

with U open in $J^r(N, P)$ (some r). This is a Baire space (Whitney, (1936) and Mather, (1970b)), so a countable intersection of dense open sets is dense. A set containing such a subset is called <u>residual</u>. Now the standard transversality theorem runs as follows.

<u>Theorem 1</u> (a) (Thom, (1956)) <u>Let</u> W <u>be any smooth submanifold of</u> $J^k(N, P)$. <u>Then</u> $\{f : j^k f \pitchfork W\}$ <u>is residual in</u> $C^\infty(N, P)$.

(b) (Mather, (1970b)) <u>If</u> $W \subset {}_rJ^k(N, P)$ <u>is a smooth submanifold,</u> $\{f : {}_rj^k f \pitchfork W\}$ <u>is residual.</u>

There are other mild extensions, for example

(c) <u>If</u> $\pi : P \to N$ <u>is the projection of a smooth fibre bundle</u> ξ, <u>and</u> $W \subset J^k(\xi)$ <u>a smooth submanifold, then the set of</u> f <u>with</u> $j^k f : N \to J^k(\xi)$ <u>transverse to</u> W <u>is residual in the space of smooth sections of</u> ξ.

The following is the basic idea underlying all proofs of transversality theorems. Let $F : N \times U \to P$ be a submersion, and let $Q \subset P$ be a submanifold. Then $F^{-1}(Q)$ is a submanifold of $N \times U$, and for each point $u \in U$, $f_u = F|(N \times \{u\})$ is transverse to Q if and only if $N \times \{u\}$ meets $F^{-1}(Q)$ transversely. Interchanging the rôles of F and the projection $\pi: N \times U \to U$ in this argument, we see that the above are equivalent to having $\pi|F^{-1}(Q)$ transverse to $\{u\}$, considered as submanifold of U - i.e. to u being a regular value of $\pi|F^{-1}(Q)$. But by the theorem of Sard (1958) this holds for all u except those in a subset of zero measure. Thus our method for deforming $f = f_0$ to a map transverse to Q will be to embed f in a family F which gives a submersion - in fact $F \pitchfork Q$ is all we need for the argument.

To extend this simple argument even to the above theorem is not easy: c.f. the involved arguments in Thom (1954). Thom (1956) used Baire category arguments to reduce (not very straightforwardly) to the local case, where it is easy to find submersions. Abraham (1963) gave an abstract approach to this reduction through function spaces, which clarified matters somewhat. We shall return to this, but begin by giving a (new) direct proof of (a), more in keeping with the spirit of these lectures.

By Lemma D1, there exists a kth order nondegenerate embedding $i : N^n \to \mathbb{R}^{n_0}$, whose image may be taken to lie in the unit disc. Then the space U of affine functions on $\mathbb{R}^{n_0}$ defines, by restriction, a mapping $U \to C^\infty(N, \mathbb{R})$ such that for each $x \in N$, U maps onto $J^k_x(N, \mathbb{R})$.

Now choose a closed embedding $j : P^p \to \mathbb{R}^{p_0}$ with a submersive retraction $\rho : U \to P$ of the ϵ-neighbourhood U of P. Then for any map $f : N \to P$, and affine $\phi : \mathbb{R}^{n_0} \to \mathbb{R}^{p_0}$, with norm (defined as the sum of the norms of the linear and constant parts) $||\phi|| \leq \epsilon$, $j \circ f + \phi \circ i$ has image in U, so can be composed with ρ. Thus if B_ϵ denotes the ball of radius ϵ in $\mathrm{Aff}(\mathbb{R}^{n_0}, \mathbb{R}^{p_0})$ we can define $\Phi : N \times B_\epsilon \to P$ by

$$\Phi(x, \phi) = \rho\{j(f(x)) + \phi(i(x))\}.$$

Considering this as a family of maps $N \to P$, it induces $j^k\Phi : N \times B_\epsilon \to J^k(N, P)$. I claim that this is a submersion. Once this is established, the proof that transversal maps are dense concludes as in the easy case.

But since $\mathrm{Aff}(\mathbb{R}^{n_0}, \mathbb{R})$ submerses onto $J^k_x(N, \mathbb{R})$, it follows that $\mathrm{Aff}(\mathbb{R}^{n_0}, \mathbb{R}^{p_0})$ submerses onto $J^k_x(N, \mathbb{R}^{p_0})$ and hence, since ρ is a submersion, onto $J^k_x(N, P)$. It is now immediate that $j^k\Phi$ is a submersion.

This proves density. As to openness, observe that transversality fails if and only if for some $x \in N$, $j^kf(x) \in W$ and $dj^kf\,x$ is not transverse to W. Now dj^kfx is determined by $j^{k+1}fx$, and non-transversality gives an algebraic subvariety, hence (proper) closed subset, of the $(k+1)$-jets lying over a given k-jet. Thus if W is compact, we have a closed family of $(k+1)$-jets to avoid, which defines an open subset of $C^\infty(N, P)$. A general W is a countable union of compact manifolds, so we have a countable intersection of dense open subsets of $C^\infty(N, P)$, which is residual.

This last argument can be avoided if W is closed: it is, however, essential when we come to multijets. To prove (b), note that we can cover $N^{(r)}$ with countably many compact sets $K = K_1 \times \ldots \times K_r$, where the K_i are disjoint, and W with compact submanifolds W_α. Then ${}_rj^kf \pitchfork W$ if and only if, for each K and α, ${}_r j^k f | K \pitchfork W_\alpha$. As before, each of these conditions defines an open set in $C^\infty(N, P)$ and we have to show density.

Choose a smooth partition of unity $1 = \Sigma_1^r \lambda_i$ with $\lambda_i \geq 0$, $\lambda_i(K_i) = 1$. Now define $\Psi : N \times B_\epsilon^r \to P$ by

$$\Psi(x;\ \phi_1, \ldots, \phi_r) = \rho\{j(f(x)) + \Sigma_1^r \lambda_i(x)\phi_i(i(x))\}.$$

For $x \in K_i$, this reduces to $\Phi(x, \phi_i)$. It follows (using our earlier result) that Ψ induces a map

$$N^{(r)} \times B_\epsilon^r \to {}_r J^k(N, P)$$

with surjective differential along $K \times B_\epsilon^r$, and the argument concludes as before.

It seems appropriate here to say more about Abraham's result. A smooth manifold A, with a map $\alpha : A \to C^\infty(X, Y)$, is called a <u>manifold of mappings</u> if evaluation $ev\,\alpha : A \times X \to Y$ is smooth. Here, X and Y are assumed finite dimensional; A can have infinite dimension.

<u>Proposition</u> (Abraham,(1963)) <u>If</u> $ev\,\alpha : A \times X \to Y$ <u>is transverse to the submanifold</u> W <u>of</u> Y, <u>then</u> $\{a \in A :\ \alpha(a) \pitchfork W\}$ <u>is residual in</u> A.
The proof involves mainly point - set considerations, together with the ideas above.

In this result, the topology on function spaces was the compact - open C^∞ - topology. Later developments, especially Quinn (1970) and Abraham (1970) have replaced this by the Whitney topology, and made some progress towards removing the finite - dimensionality restriction on X and Y.

We illustrate the use of this result by reconsidering Theorem 1 above. Take $X = N^{(r)}$, $Y = {}_r J^k(N, P)$ and W as given; and let $A = C^\infty(N, P)$. The map α associates to $f : N \to P$ its multijet extension $\alpha(f) = {}_r j^k f$. The theorem follows from the proposition, together with the assertion that $ev\,\alpha$ is a submersion. Although the proof of this assertion involves essentially the same calculations as above, much less care is now needed since the assertion is purely local. Thus the general case may be deduced from the case $r = 1$ by the simple observation that we can use independent variations of f at any r distinct points (the formal proof uses partitions of unity, as above). Moreover, we can also deduce Theorem 1 (c), since the local problem is the same as that for (a).

We observed above that in Theorem 1 (a), if W is closed then the set of transversal maps is open. Even this result is by no means adequate for all our applications, and we postpone the discussion of conditions yielding open sets of maps until the next chapter.

<u>Proof of Theorems A and B</u>

We are now ready to give the proofs of the transversality theorems which form the first part of the results described in Sections A and B. Recall that for Theorem A, from a map $i : M \to E^-$ we constructed a family H_i of maps $M \to E^0$ depending on a parameter $p \in E^+$. Taking jets gives $j^k H_i(p) : M \to J^k(M, E^0)$; similarly we have the multijet extension ${}_r j^k H_i(p) : M^{(r)} \to {}_r J^k(M, E^0)$ which depends smoothly on $p \in E^+$, so defines a map which we denote by

$${}_r j^k H_i : M^{(r)} \times E^+ \to {}_r J^k(M, E^0).$$

<u>Theorem A (i)</u> (Soares David) <u>For any submanifold</u> W <u>of</u> ${}_r J^k(M, E^0)$, <u>the set of maps</u> i <u>with</u> ${}_r j^k H_i \pitchfork W$ <u>is residual in</u> $C^\infty(M, E^-)$.

<u>Proof</u> It follows from the above considerations that it will suffice to embed i in a family of maps $\Psi : M \times U \to E^-$ such that the induced map

$${}_r j^k H_\Psi : M^{(r)} \times E^+ \times U \to {}_r J^k(M, E^0)$$

is a submersion, at least when restricted to a chosen $K = K_1 \times \ldots \times K_r$ in $M^{(r)}$. Since we can piece together independent deformations of i on $K_1, \ldots, K_r$ using a partition of unity, and the space of multijets is just a product, we seek a family $\Phi : M \times A \to E^-$ such that

$$j^k H_\Phi \times pr_2 : M \times A \times E^+ \to J^k(M, E^0) \times E^+$$

is a submersion. For then so is the product

$$M^{(r)} \times A^r \times (E^+)^r \to {}_r J^k \times (E^+)^r ,$$

and we can restrict to $K \subset M^{(r)}$, identify A^r with the desired U, and restrict to the diagonal in $(E^+)^r$.

We first need a formula for H. Choose an affine function h on E such that E^-, E^0 and E^+ are defined by $h < 0$, $h = 0$ and $h > 0$ respectively. Then H_i is given by

$$H_i(x, P) = \frac{h(P)\, i(x) - h(i(x))P}{h(P) - h(i(x))} \quad .$$

Now in our proof of Thom's transversality theorem, if V^0 is the vector space of translations of E^0, we found a family of maps (deforming zero)

$$\Phi_0 : M \times A \to V^0$$

such that $j^k\Phi_0 : M \times A \to J^k(M, V^0)$ is a submersion. (In fact $A = \mathrm{Aff}(W, V)$ where $M \subset W$ is kth order nondegenerate.) We define Φ by

$$\Phi(m, v) = i(m) + \Phi_0(m, v) \; .$$

Then

$$H_\Phi(m, P, v) = \frac{h(P)\,\Phi(m, v) - h(\Phi(m, v))P}{h(P) - h(\Phi(m, v))}$$

$$= \frac{h(P)(i(m) + \Phi_0(m, v)) - h(i(m))P}{h(P) - h(i(m))}$$

since h takes the same value on points related by translation by V^0. If we fix m and P and vary v, this has the form $\alpha\Phi_0(m, v) + \beta$, where $\alpha > 0$. By the choice of Φ_0, this gives a submersion of A onto $J^k_m(M, E^0)$. This concludes the proof.

The next result is similar, but a little more complicated. Recall that E is now a euclidean vector space, $S(E)$ the unit sphere in $E \times \mathbb{R}$, $S_0(E) = S(E) \cap E$. Given an embedding $i : M \to E$ we define $\tilde{G}_i : M \times S(E) \to \mathbb{R}$ by

$$\tilde{G}_i(m, e, s) = s\,\|i(m)\|^2 - 2\langle e, i(m)\rangle$$

where $e \in E$, $s \in \mathbb{R}$ and $\|e\|^2 + s^2 = 1$. Again, regarding this as a family of functions on M we define

$${}_rj^k\tilde{G}_i : M^{(r)} \times S(E) \to {}_rJ^k(M, \mathbb{R}) \, ,$$

and denote by ${}_rj^k\tilde{G}_i^0$ its restriction to $M^{(r)} \times S_0(E)$. Then the transversality theorem in question is

<u>Theorem B (1)</u> Looijenga(1974) <u>For any submanifold</u> W <u>of</u> ${}_rJ^k(M, \mathbb{R})$ <u>invariant under addition of constants, the set of embeddings</u> $i : M \to E$ <u>with</u> ${}_rj^k\tilde{G}_i \pitchfork W$ <u>and</u> ${}_rj^k\tilde{G}_i^0 \pitchfork W$ <u>is residual.</u>

Invariance here means that for any $(q_1, \ldots, q_r) \in W$ and $c \in \mathbb{R}$, then $(q_1 + c, \ldots, q_r + c) \in W$.

Proof We again start with $\Phi : M \times U \to E$ such that $j^k\Phi : M \times U \to J^k(M,E)$ is a submersion, compute the corresponding $j^k\tilde{G}_\Phi$, and hope to obtain a submersion $M \times S(E) \to J^k(M, \mathbb{R})$. To differentiate, it is easier to consider a 1-parameter family $i_t : M \to E$ (coming from a curve in U) with $i_0 = i$, so that

$$\tilde{G}_{i_t}(m, e, s) = s \| i_t(m) \|^2 - 2 < e, i_t(m) > .$$

Differentiating with respect to t at $t = 0$ gives

$$\begin{aligned} \partial/\partial t\, \tilde{G}_{i_t}(m, e, s) &= 2s < i(m), \partial/\partial t\, i_t(m) > \\ &\quad - 2 < e, \partial/\partial t\, i_t(m) > \\ &= 2 < s\, i(m) - e, \partial/\partial t\, i_t(m) > . \end{aligned}$$

Now if $e \neq s\, i(m)$, inner product is a nonzero linear map $E \to \mathbb{R}$ inducing a submersion $J^k(M, E) \to J^k(M, \mathbb{R})$. Thus at such points, $j^k\tilde{G}_\Phi$ (and, if $s = 0$, $j^k\tilde{G}^0_\Phi$) is indeed a submersion.

However, if $e = s\, i(m)$ (which cannot occur if $s = 0$, for then $\|e\| = 1$) we see that the image of $j^k\tilde{G}_\Phi$ is contained in the subset of jets with zero target. Nevertheless, since i is an immersion, we can take a subset of the coordinates of $s\, i(m) - e$ as local coordinates on M. Since these generate the maximal ideal in the ring of germs of C^∞ functions at $m \in M$, the above formula defines a submersion $J^k_m(M, E) \to J^k_{(m,0)}(M, \mathbb{R})$. But this is still sufficient to obtain transversality to a W invariant under addition of constants.

For the case $r > 1$ of multijets we observe - as usual - that deformations on disjoint compact sets can be chosen independently. Here we need the extra remark that since i is injective, for a fixed $(e, s) \in S(E)$ the equation $e = s\, i(m)$ can hold for at most one value of $m \in M$. Thus the image of the corresponding $d({}_rj^k\tilde{G}_\Psi)$ is at worst the subset of the tangent space of ${}_rJ^k(M, \mathbb{R})$ where one coordinate has target zero. But this subspace meets W transversely by our hypothesis.

Equivalence of Lagrangean maps

We begin this section with the proof of Theorem C3. Recall that this asserts (roughly speaking) that a pair of generalised phase functions are right equivalent as unfoldings if and only if the corresponding lagrangean maps are equivalent. Although there is no formal proof in the literature, the result is virtually identical with Theorem 3.1.6 of Hörmander (1971), as is pointed out by Pham (1976).

First suppose given a commutative diagram

$$\begin{array}{ccccc} M_1 \times E_1 & \xrightarrow{pr_1 \times F_1} & M_1 \times \mathbb{R} & \xrightarrow{pr_1} & M_1 \\ \downarrow H & & \downarrow \tilde{G} & & \downarrow G \\ M_2 \times E_2 & \xrightarrow{pr_1 \times F_2} & M_2 \times \mathbb{R} & \xrightarrow{pr_1} & M_2 \end{array}$$

where $pr_2\tilde{G}(q, t) = t + g(q)$ and G, H are diffeomorphism germs. Identify M_1 and M_2 via G : we may thus suppose $M_1 = M_2 = M$, G the identity and $H(q, x) = (q, h(q, x))$. Differentiating the identity $F_2(H(q, x)) = F_1(q, x) + g(q)$ with respect to x, we obtain

$$\partial F_1/\partial x_i(q, x) = \Sigma_j\ \partial F_2/\partial x_j(H(q, x))\ \partial h_j/\partial x_i(q, x)\ .$$

But since H is a diffeomorphism, $\partial h_j/\partial x_i$ is a nonsingular matrix, so we deduce that (q, x) belongs to the catastrophe set C_1 of F_1 if and only if $H(q, x) \in C_2$; thus H induces a diffeomorphism of C_1 on C_2.

Now differentiate with respect to q, yielding

$$\partial F_1/\partial q_i(q, x) + \partial g/\partial q_i = \Sigma_j\ \partial F_2/\partial x_j(H(q, x))\ \partial h_j/\partial q_i(q, x) + \partial F_2/\partial q_i(H(q, x))\ .$$

On the catastrophe set, the term involving $\partial F_2/\partial x_j$ disappears. Hence the symplectic diffeomorphism of T^*M given (in the usual coordinates) by

$$(q_i, p_i) \mapsto (q_i, p_i + \partial g/\partial q_i)$$

defines the required lagrangean equivalence.

Next suppose $F' : M \times E \times \mathbb{R} \to \mathbb{R}$ defined by $F'(q, x, t) = F(q, x) + \lambda t^2$ (λ a nonzero constant). It is clear that F' has catastrophe set $C \times 0$, and yields the same lagrangean map as F.

The nontrivial assertion of the theorem is the converse implication. We first reduce to the case where the partial derivatives $\partial^2 F/\partial x_i\, \partial x_j$ vanish at the base point. By a linear coordinate change, we may suppose this matrix diagonal: suppose $\partial^2 F/\partial x_i^2 = 2\lambda_i \neq 0$ for $k_0 < i \leq k$, and all other $\partial^2 F/\partial x_i \partial x_j$ vanish. Then by an extension of the Morse lemma due to Gromoll and Meyer (1969) there is a change of coordinates

$$x_i \to g_i(q, x) \qquad (k_0 < i \leq k),$$

with the same 1 - jet as the identity, reducing F to the form $F_0(q, x_1, \ldots, x_{k_0}) + \Sigma_{k_0+1}^{k} \lambda_i x_i^2$. Thus F is equivalent to F_0, which has the desired property.

It is thus sufficient to consider functions F with $\partial^2 F/\partial x_i \partial x_j = 0$ at the origin. Moreover, since $F(q, x)$ is equivalent to $F(q, x) - F(q, 0)$, we may suppose $F(q, 0) = 0$ also. The terms of second degree in the Taylor expansion of F are then of the form $\Sigma\, b_{ij}\, q_i\, x_j$, and the nondegeneracy condition $d_2 F \pitchfork 0$ states that the matrix $B = (b_{ij})$ has rank k. Hence $k \leq m$, and by a linear change of the x_j we reduce to the case $b_{ij} = \delta_{ij}$. The tangent space to C at 0 is then given by $q_1 = \ldots = q_k = 0$, so the rank of the lagrangean map at the base point is $(m - k)$. Note in passing that we have now determined the minimum number of parameters x_i for a germ of generalised phase function at the given point. Further we have local coordinates $x_1, \ldots, x_k, q_{k+1}, \ldots, q_m$ in C and modulo second order terms, Φ takes this point to

$$((0, \ldots, 0, q_{k+1}, \ldots, q_m), (x_1, \ldots, x_k, 0, \ldots, 0))$$

in T^*M.

With this preparation, we are ready for the main part of the proof. Let F_0, F_1 be generalised phase functions defining equivalent lagrangean maps and satisfying the auxiliary conditions

$$0 = \partial^2 F_t/\partial x_i \partial x_j = F_t = \partial F_t/\partial q_i = \partial^2 F_t/\partial q_i \partial q_j$$

at the origin for $t = 0, 1$. The lagrangean equivalence gives diffeomorphism germs

$$\alpha : C_0 \to C_1 \qquad \text{of catastrophe sets}$$
$$\beta : M_0 \to M_1 \qquad \text{of base manifolds}$$
$$\beta' : T^*M_0 \to T^*M_1 \qquad \text{covering } \beta$$

with $\Phi_1 \circ \alpha = \beta' \circ \Phi_0$. Again we identify M_0 and M_1 via β. Making linear changes in the x_i as above, we can now normalise $\partial^2 F_t/\partial q_i \partial x_j = \delta_{ij}$. Moreover, for each coordinate x_i on $M_1 \times E^k$, we can extend $x_i \circ \alpha$ from C_0 to a function x_i' on $M_0 \times E^k$. From the above description of C_i, we deduce that $dx_i' - dx_i$ is a linear combination of $dq_1, \ldots, dq_k$ at 0. Thus we have a diffeomorphism, respecting projection on M, which takes C_0 to C_1 by α. Identify the copies of $M \times E^k$ by this diffeomorphism; then $C_0 = C_1$. As a final normalisation, since β' is now given by

$$\beta'(q, p) = (q, p + \partial g/\partial q)$$

for some function $g(q)$, we replace F_1 by the equivalent function $F_1(q, x) - g(q)$; this reduces β' to the identity.

We may thus suppose that F_0 and F_1 have the same catastrophe set and the same lagrangian. Now consider $F_t = tF_1 + (1 - t)F_0$. Since $\partial^2 F_t/\partial q_i \partial x_j = \delta_{ij}$, F_t is a generalised phase function; since $\partial F_t/\partial x$ vanishes along C, its catastrophe set coincides (locally) with C. Finally, since $\partial F_0/\partial q = \partial F_1/\partial q$ on C, each equals $\partial F_t/\partial q$, so F_t has the same lagrangean.

For a point $(q, x) \in C$, both $\partial F_t/\partial x$ (which vanishes) and $\partial F_t/\partial q$ are independent of t. Hence $\partial \dot{F}_t/\partial x = \partial \dot{F}_t/\partial q = 0$ along C, so $\dot{F}_t$ is constant along C, hence zero. As the ideal of C is defined by the $\partial F_t/\partial x_i$, we have $\dot{F}_t \in \langle \partial F_t/\partial x_1, \ldots, \partial F_t/\partial x_k \rangle$. The rest of the proof is now a standard "infinitesimal stability implies stability" argument (see Mather (1969) for the original such), or can be more simply deduced (following Hörmander) from the implicit function theorem.

The theorem allows us the following approach to classification. For large enough k (and $k = m$ is always large enough), germs of lagrangean maps into M^m correspond to germs of maps from M^m into $C^\infty(\mathbb{R}^k, \mathbb{R})$, modulo right-equivalence. We obtain generic conditions by imposing transversality with respect to invariant submanifolds of the jet spaces $J^{\ell}(\mathbb{R}^k, \mathbb{R})$. A fuller discussion of this is given in §3 below.

One may contrast this with other approaches. The basic distinction between the above and the treatment by Porteous (1971) is that he considers F as a map, not as a family of maps. For simple cases this leads to the same list of possibilities, but it appears to be inadequate for the general theory proposed here.

If the map $M^m \to J^{\ell}(\mathbb{R}^k, \mathbb{R})$ is transverse to the orbit (under right equivalence) of a map f, we have a versal unfolding of f, and may look for C^{∞}-stability. One may enquire about the relation between versality and the basic nondegeneracy condition $(d_2 F \pitchfork 0)$ defining generalised phase functions. Regard F as an unfolding of $f(x) = F(0, x)$, and reduce (as usual) to the case $\partial^2 F/\partial x_i \, \partial x_j = 0$, i.e. $j^2 f = 0$. Then $d_2 F \pitchfork 0$ if and only if $\partial^2 F/\partial x_i \, \partial q_j$ has rank k, or equivalently the $\partial F/\partial q_j \, (0, x)$ generate the vector space $\mathfrak{m}_x / \mathfrak{m}_x^2$. The condition for versality is that they generate $\mathfrak{m}_x / J_f$, where J_f is the ideal $\langle \partial f/\partial x_1, \ldots, \partial f/\partial x_k \rangle$. This is a much stronger condition.

Proof of Theorems D and E

The only result in section D that remains to be proved is Feldman's transversality theorem D2. But this is an easy corollary of Thom's transversality theorem as follows.

The natural map $J^k(M, V) \to \mathrm{Hom}(T_k M, T_1 V)$ induced by a Riemannian structure on V is an isomorphism, as is seen at once from the duality between $T_k M$ and $J^k(M, \mathbb{R})$. Thus for any submanifold W of $\mathrm{Hom}(T_k M, T_1 V)$ we know that the maps f with $\tau_k(f) \pitchfork W$ are residual; and this holds in particular if $W = A(M, V)$ is constructed from a model singularity A. This proves Theorem D2; note that in (i) $A(M, V)$ is a subbundle and in (ii) the construction was more complicated, but this does not matter.

We actually used a minor modification of the result in our examples. An $O_m \times O_v$ - invariant submanifold B of $\mathrm{Sym}^k(\mathbb{R}^m, \mathbb{R}^v) = \mathrm{Hom}(O^k \mathbb{R}^m, \mathbb{R}^v)$ defines, as in (iii), a submanifold $B(M, V)$ of $\mathrm{Sym}^k(TM, TV)$, and I claim that for a generic immersion $f : M \to V$, its kth fundamental form $\beta_k(f)$ meets $B(M, V)$

transversely. For the metric on M induces an immersion $0^k TM \to T_k M$, hence a submersion $\mathrm{Hom}(T_k M, TV) \to \mathrm{Sym}^k(TM, TV)$, and $\beta_k(f) \pitchfork B(M, V)$ if and only if $\tau_k(f) \pitchfork$ the preimage of $B(M, V)$ in $\mathrm{Hom}(T_k M, TV)$.

For Theorem E1, consider an immersion $f : M^m \to \mathbb{R}^n$. Take local coordinates $x_1, \dots, x_m$ in M; let $y_1, \dots, y_n$ denote the standard coordinates on $\mathbb{R}^n$. Then the metric induced by f is

$$\Sigma_1^n \, dy_\nu \, dy_\nu = \Sigma_{\nu=1}^n \Sigma_{i,j=1}^m \partial y_\nu/\partial x_i \;\; \partial y_\nu/\partial x_j \;\; dx_i \, dx_j ,$$

so we can write $g_{ij} = \Sigma_{\nu=1}^n \; \partial y_\nu/\partial x_i \;\; \partial y_\nu/\partial x_j$. For a 1-parameter family f_t $(t \in \mathbb{R})$ of immersions through $f = f_0$, with $u = \partial f/\partial t \big|_{t=0}$ we have (on differentiating and setting $t = 0$)

$$\dot{g}_{ij} = \partial/\partial t \;\; g_{ij}(f_t)\big|_{t=0} = \Sigma_{\nu=1}^n \left(\partial y_\nu/\partial x_i \;\; \partial u_\nu/\partial x_j + \partial u_\nu/\partial x_i \;\; \partial y_\nu/\partial x_j \right). \tag{1}$$

Clearly for suitable deformations f_t the u_ν may be arbitrary smooth functions. But any deformation of the image manifold can be achieved by a motion in the normal direction (we can absorb the tangential part in a diffeomorphism of M). It is thus reasonable to consider deformations satisfying

$$\Sigma_{\nu=1}^n \;\; u_\nu \;\; \partial y_\nu/\partial x_i = 0 \quad \text{for each } i . \tag{2}$$

Differentiating this, we obtain

$$\Sigma_{\nu=1}^n \left(\partial u_\nu/\partial x_j \;\; \partial y_\nu/\partial x_i + u_\nu \;\; \partial^2 y_\nu/\partial x_i \, \partial x_j \right) = 0$$

for all i, j. Substituting this in (1), we have

$$\Sigma_{\nu=1}^n \;\; u_\nu \;\; \partial^2 y_\nu/\partial x_i \, \partial x_j = -\tfrac{1}{2}\dot{g}_{ij} . \tag{3}$$

Now for prescribed $\dot{g}_{ij}$, (2) and (3) give a system of linear equations for u, and they will have a solution for arbitrary $\dot{g}_{ij}$ if and only if the matrix of the system is nonsingular. Considering this as an equation for germs at a point, this means that $\tau_2(f) \;(\partial/\partial x_i)$ and $\tau_2(f)\;(\partial^2/\partial x_i \, \partial x_j)$ are linearly independent - i.e. f is second order nondegenerate.

The above argument is due to Nash. Clearly it covers the case of jets as well as that of germs. It seems natural to expect the implication to be reversible: that if $J^k_{emb}(M, \mathbb{R}^n) \to J^{k-1}(\text{Riem } M)$ is a submersion at f for all k, then f is second order nondegenerate. On account of the restriction imposed on u, the above proof cannot be reversed, however, and the best I have been able to prove with an asymptotic argument is that the vectors $\tau_2(f)(\partial/\partial x_i)$ and $\tau_2(f)(\partial^2/\partial x_i \partial x_j)$ must span a subspace of dimension at least $\frac{1}{2}m(m+1)$ everywhere. I now expect this to be a necessary and sufficient condition.

We next prove Proposition E2, that the natural map

$$J^2(\text{Riem } M) \to J^1(\text{Riem } M) \times_M J^0(\text{Curv } M)$$

is a submersion. It will suffice to consider a local perturbation $g'_{ij} = g_{ij} + \Sigma_{k,\ell}\, a_{ijk\ell}\, x_k x_\ell$ leaving the 1-jet fixed, and calculate the effect on the curvature tensor. We may assume $a_{ijk\ell}$ symmetric in i and j, and in k and ℓ. Calculating (using the summation convention) we have

$$\Gamma'^{\,i}_{jk} - \Gamma^i_{jk} = \tfrac{1}{2}g^{hi}\,(2\,a_{hjk\ell}\,x_\ell + 2\,a_{hkj\ell}\,x_\ell - 2\,a_{jkh\ell}\,x_\ell)$$

(modulo higher terms), so that at the origin

$$R'_{ijk\ell} - R_{ijk\ell} = a_{i\ell jk} - a_{j\ell ik} - a_{ikj\ell} + a_{jki\ell}.$$

Now if $\rho_{ijk\ell}$ has the symmetry properties of the curvature tensor, we take

$$a_{ijk\ell} = \rho_{ikj\ell} + \rho_{jki\ell}.$$

This has the above symmetry, and

$$\begin{aligned} R'_{ijk\ell} - R_{ijk\ell} &= \rho_{ij\ell k} - \rho_{ji\ell k} - \rho_{ijk\ell} + \rho_{jik\ell} \\ &\quad + \rho_{\ell jik} - \rho_{\ell ijk} - \rho_{kji\ell} + \rho_{kij\ell} \\ &= -4\rho_{ijk\ell} + 2\rho_{ik\ell j} + 2\rho_{i\ell jk} \\ &= -6\rho_{ijk\ell} \end{aligned}$$

using the symmetry properties of ρ. This yields a local cross-section, so we do indeed have a submersion.

Theorem E4 follows from Proposition E2 in the same way as Theorem D above was proved. Namely, the pullback A of $J^0(\text{Curv } M)$ to the total space of $J^0(\text{Riem } M)$ can be considered as an O_m-bundle. Our model singularity S

defines an associated subbundle. Since $J^2(\text{Riem } M)$ submerses onto $J^1 \text{ Riem } M \times_M J^0 \text{ Curv } M$, it submerses onto A, so we have an induced submanifold of $J^2(\text{Riem } M)$. A generic metric defines a section transverse to this submanifold (by the basic transversality theory). But this metric yields a section $\alpha(M)$ of $J^0(\text{Riem } M)$, and clearly the curvature tensor $R(\alpha(M))$ is transverse to $S(M)$.

Generic conditions on metrics

In section F above we introduced the path spaces $L^\alpha(M)$, $L^\alpha_P(M)$; the evaluation maps $Ev : L^\alpha(M) \to M \times M$ and $ev_P : L^\alpha_P(M) \to M$ and the energy integral with respect to a metric g, $E_g : L^\alpha(M) \to \mathbb{R}$. The underlying philosophy is to consider these as families of functions E_g on the fibre, and expect that for generic g we have a generic family.

Since, as mentioned above, transversality theory is not yet developed in strong enough form for infinite dimensional manifolds to be applicable to this problem, it is necessary to use finite dimensional approximations. A good exposition is given in [M, §16] : it depends on the metric g. As we only need to consider small variations of g, it suffices to proceed as follows. Since M is compact, we can find a number $\delta > 0$ such that any two points within a distance δ are joined by a unique minimal geodesic. Let $U_\Delta = \{(x, y) \in M \times M : \rho_g(x, y) < \delta/2\}$. Then there is a neighbourhood U_g of the metric g such that for any $g' \in U_g$ and any $(x, y) \in U_\Delta$, there is a unique minimal geodesic for g' joining x to y. Now set

$$X_r = \{(m_1, \dots, m_r) \in M^r : (m_i, m_{i+1}) \in U_\Delta \text{ for } 1 \leq i < r\},$$

$$Y_r = \{(t_1, \dots, t_r) \in I^r : 0 = t_1 < t_2 < \dots t_r = 1\}.$$

Then we can define

$$j : X_r \times Y_r \times U_g \to L^{PD}(M)$$

by letting $j\{(m_1, \dots, m_r), (t_1, \dots, t_r), g'\}$ be the path taking the interval $[t_i, t_{i+1}]$ to the minimal geodesic (for g') joining m_i to m_{i+1}.

We consider $X_r \times Y_r$ (or, more crudely, X_r) as approximating $L^{PD}(M)$. The fact that the embedding j depends on g' does not matter for our purposes, since the dependence is C^∞. For full details, one may expect to restrict to a suitable open submanifold of $X_r \times Y_r$, defined e.g. by $L_g < c$. Let us write $L^F(M)$ to denote $X_r \times Y_r$, where r is supposed "sufficiently large".

The key to interesting results is again a transversality theorem. We continue the discussion in general terms, so that one can see where difficulties occur and why Buchner's theorem works. Now $\dim L^F(M) = mr + r - 2$, so $L^F_{PQ}(M)$ has dimension $N = (m + 1)(r - 2)$. As we wish to vary r, we must allow N to vary too.

Write $I^k(n)$ for the subspace of singular jets $z \in J^k_0(n, 1)$ - i.e. jets with $j^2 z = 0$. Define a model singularity to be a submanifold W of $I^k(n)$ for some n and k, invariant under right equivalence - i.e. composition with invertible k-jets from $(\mathbb{R}^n, 0)$ to itself. By the splitting theorem of catastrophe theory (the version needed here is that of Gromoll and Meyer (1969)), any (singular) jet f can be written, after change of coordinates, as

$$f_0(x_1, \ldots, x_r) - \Sigma^s_{r+1} x_i^2 + \Sigma^n_{s+1} x_j^2$$

where $j^3 f_0 = 0$: r is called the <u>corank</u> and $(s - r)$ the (Morse) index. Further, the right equivalence class of f determines that of f_0 and the integers r, s. In general, we define an embedding of $I^k(n)$ in $I^k(n + 1)$ by

$$i(f)(x_1, \ldots, x_{n+1}) = f(x_1, \ldots, x_n) + x^2_{n+1} .$$

Then it follows from the results just cited that for any model singularity W, the saturation W_{n+1} of $i(W)$ under right equivalence is a model singularity in $I^k(n + 1)$, meeting $I^k(n)$ transversely in W. Iterating this, we obtain model singularities W_N in $I^k(N)$ for all $N \geq n$, and the result of Gromoll and Meyer allows us to include the case $N = \infty$ here.

The type of transversality theorem we seek is as follows. Collecting the k-jets $j^k(E_g | L^\alpha_{PQ}(M)) : L^\alpha_{PQ}(M) \to J^k(L^\alpha_{PQ}(M), \mathbb{R})$ as Q varied defines a map we will denote by $j^k E_g : L^\alpha_P(M) \to B$. Here B is fibred over $L^\alpha_P(M)$, and we write $W(B)$ for the associated bundle with fibre W. Then we want to show that for generic g, $j^k E_g$ is transverse to $W(B)$. We first seek to reduce to the finite dimensional approximation. Now the critical points of E_g are the geodesics, each of which belongs to (the image of) $L^F(M)$ for r large enough. The essential point is now [M, 16.2] that the critical points of the restriction are the same; the corank of the restriction at each one is the same, and so is its nullity (in particular, these are both finite). Since the normal space to W_N in $I^k(N)$ does not depend on N, we see that the desired transversality holds in L^α if and only if it holds for L^F.

We thus consider a geodesic γ (for g_0) from P to Q which are conjugate along γ (so that we have a degenerate critical point), and seek a family U of metrics, deforming g_0, such that the map from $g \in U$ to the germ at γ of $E_g | L^\alpha_{PQ}$ has k-jet transverse to W_∞. We shall define U by varying g_0 in a carefully chosen small open set: it is now time to consider the geometry.

Choose a point R on PQ, not conjugate to P along γ, and possessing a neighbourhood U_R through which γ only passes once - i.e. $\{t \in I : \gamma(t) \in U_R\}$ is connected, and γ embeds it in U_R. Since $\gamma(I)$ is compact, this choice is always possible except if γ runs at least twice round a closed geodesic in M. This case is thus excluded from the following argument.

The space of Jacobi fields along γ vanishing at P has dimension n [M, p.77]; choose a basis $e_1, \ldots, e_n$ with e_n as in [M, 14.2]. Let $e_1, \ldots, e_r$ be a basis of those vanishing at Q. Since P, R are not conjugate along γ, these are all linearly independent at, hence near R. Choose local coordinates on a neighbourhood $U_R^2 \subset U_R$ of R such that

i) γ meets U_R^2 along the x_n-axis

ii) along $\gamma \cap U_R^2$ we have $e_i = \partial/\partial x_i$

(it is easily seen that this is possible, e.g. by constructing an n-parameter variation of γ through geodesics). Choose bump functions $\rho = \rho(x_n)$, $\sigma = \sigma(x_1, \ldots, x_{n-1})$, vanishing outside a neighbourhood of 0 and satisfying

iii) $\rho\sigma$ has support U_R^3 contained in U_R^2

iv) $\int_{-\infty}^{\infty} \rho(x)\,dx = 1$

v) $\sigma \equiv 1$ in a (small) neighbourhood of 0 .

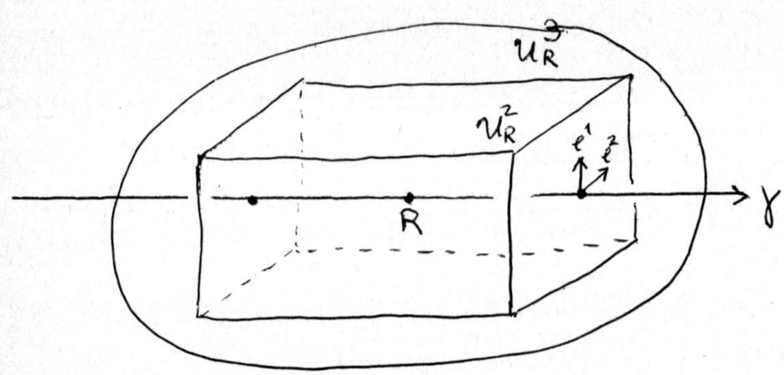

We are now ready to construct (following Buchner (1974)) the desired variation of the metric. First observe that since the null space of the Hessian of $E_g | L_P^\alpha$ is spanned by the Jacobi fields e_i $(1 \leqslant i \leqslant r)$, it follows by the splitting theorem that it suffices to study the restriction to a subspace of L_{PQ}^α of paths which cross U_R^3 following lines parallel to the x_n - axis, and with the remaining coordinates as parameters; and transversality will certainly be achieved if the parameter space submerses onto $J_{0,\,0}^k(n - 1, 1)$.

Take U to be a neighbourhood of 0 in $J_{0,0}^k(n - 1, 1)$, and define a variation $\{g_f\}$ of the metric g_0, depending on $f \in U$, by

$$g_f \equiv g_0 \quad \text{outside } U_R^3 .$$

In U_R^3, all components of $g_f - g_0$ vanish, except

$$(g_f - g_0)_{n,n} = \sigma(x_1, \ldots, x_{n-1})\rho(x_n)\, f(x_1, \ldots, x_{n-1}).$$

Then $E_g(\gamma) = \int_0^1 \|\dot\gamma(t)\|^2 \, dt$ and the integral is changed only in U_R^3, where $\dot\gamma(t) = e_n = \partial/\partial x_n$ in our coordinates. Thus

$$\begin{aligned} E_{g_f}(\gamma) - E_{g_0}(\gamma) &= \int (g_f - g_0)_{n,n}\, dx_n \\ &= \sigma(x_1, \ldots, x_{n-1}) f(x_1, \ldots, x_{n-1}) \int \rho(x_n)\, dx_n \\ &= \sigma(x_1, \ldots, x_{n-1})\, f(x_1, \ldots, x_{n-1}) \end{aligned}$$

using (iii) and (iv); and by (v) this has the same germ as f at 0 .

Since the space of metrics is a Baire space, we have proved the first part of

Buchner's transversality theorem Let W be a model singularity, and K_P^α be a closed subset of $L_P^\alpha(M)$ containing no twice traversed closed geodesics (for g_0). Then g_0 has a neighbourhood U_g such that the set of $g \in U$ with $j^k E_g$ transverse on K_P^α to $W(B)$ is dense in U. The corresponding result holds also for multijets.

The extension to multijets follows, as usual, by constructing independent variations in disjoint regions of M.

This result is neither the weakest nor the strongest that could be contemplated. It is certainly much stronger than the assertion that for given P and Q and generic g, $E_g | L_{PQ}^\alpha(M)$ is nondegenerate with distinct critical values. On the other hand, it excludes the possibility of varying P, as well as the study of closed geodesics. These seem to be natural questions for further study.

If we apply this result to the problems raised in F above, we find that theorem F1 follows, since in general none of the closed geodesics (in general, countable in number) will pass through P. These should not be difficult to fill in using Abraham's theorem F3. Conjectures F2 and F5 require us to vary P also, and theorem F 3 appears to lie deeper. The result is just adequate for the study of the cut locus with respect to a fixed point P, as a geodesic running round a closed loop more than once is certainly not minimal. So for each $Q = \gamma(t(\gamma))$ on the cut locus of P, the theorem is applicable to say that any desired transversality conditions on the $\gamma[0, t(\gamma)]$ will hold for a generic metric.

§3 Stratifications

The original thrust of the theory of singularities as e.g. in Whitney (1955) was to find local normal forms for smooth mappings, up to differentiable change of coordinates. However, it was early realised that this was not always possible, even for 'generic' mappings, and the paper of Thom and Levine (1959, reprinted 1971) gives an example, and raises the question of topological equivalence. The fundamental theorem here is the density of topologically stable mappings. The article by Thom (1962) already contains the broad outlines of an attack on this, by a concept of 'stratification', and further work by Whitney (1965) and Thom (1969) went some way to making this explicit. However it was Mather, who had already in a brilliant series of papers (1968a, 1969, 1968b, 1970a, 1970b, 1971) elucidated the problem of C^∞-stability, who went on to clarify Thom's indications on stratifications and their application to C^0-stability. His work has not yet appeared in full in print, but the stratification theory is given in the notes (1970c) and there are further brief accounts by Chenciner (1973) and Mather himself (1973b), (1976). Subsequently some developments and simplifications - due mainly to Looijenga - have appeared (Looijenga (1974) and Gibson, Wirthmüller, du Plessis and Looijenga (1976)). This last reference is currently the best available introduction to the subject: we shall refer to it as [GWPL].

In this lecture I will give a very brief summary of the main properties of stratifications and of their applications to these problems, and hence to those in the first two chapters.

Definitions and transversality

A *stratification* $\mathcal{S}$ of a subset X of a manifold M is a locally finite partition of X into (smooth) submanifolds of M. The parts are called strata. The stratification is *regular* if it satisfies the conditions, due to Whitney (1965).

(A) *If* $S, S' \in \mathcal{S}$, $x \in S$, *and* $\{y_n\}$ *is a sequence of points of* S' *tending to* x *with* $T_{y_n}(S') \to \tau$, *then* $T_x S \subset \tau$.

(B) _If_ $S, S' \in \mathcal{S}$, $x \in S$, _and_ $\{x_n\}, \{y_n\}$ _are sequences in_ S, S' _respectively tending to_ x _and such that_ $T_{y_n}(S') \to \tau$ _and the direction_ $x_n y_n$ _(in local coordinates) tends to a limit_ $\ell \in T_x M$, _then_ $\ell \subset \tau$.

If $\mathcal{S}$ satisfies (A) alone, it is called _cohesive_ (Feldman, 1965). Moreover, condition (B) implies condition (A) (Mather, 1970c) and also (as we will see) other natural properties such as the 'frontier condition' viz. that the frontier of a stratum is a union of lower dimensional strata. An alternative regularity condition (W) has been proposed by Verdier (1976).

A map $f : V \to M$ is called transverse to $\mathcal{S}$ ($f \pitchfork \mathcal{S}$) if it is transverse to each stratum. Now part of the reason for the above definitions is explained by

Lemma 1 _The set_ $Y \subset J^1(V, M)$ _of jets transverse to_ $\mathcal{S}$ _is open (for all_ V) _if and only if_ X _is closed and_ $\mathcal{S}$ _is cohesive._

Proof. _Necessity_ If Y is open, its intersection with the set of jets with $df = 0$ is open in the set $V \times M$ of all such jets. But this intersection is $V \times W$, where $W = (M - X) \cup \{\text{Int } S : S \in \mathcal{S}\}$ and Int denotes interior in M. Then W is open, $(M - X) \cup \text{Int } X$ is open, $X - \text{Int } X$ is closed and so X is closed. Now if (A) fails for any pair of strata choose $x \in S$ and a sequence $\{y_n\}$ of points of S' with $y_n \to x$, $T_{y_n}(S') \to \tau$, and $T_x S \not\subset \tau$. Now let A be a subspace of $T_x M$ with

$$A + T_s S = T_x M \neq A + \tau .$$

Then a 1-jet with $df(T_y V) = A$ (and such exist if $\dim V \geq \dim A$) is transverse to S, but nearby jets with target y_n will not be transverse to S'.

Sufficiency Suppose the conditions satisfied, and let $y \in Y$. If the target of y is not in X, it has a neighbourhood avoiding X, whence $y \in \text{Int } Y$. Otherwise, y has target $x \in S$ for some $S \in \mathcal{S}$. If $y \notin \text{Int } Y$, there is a sequence $y_n \to y$ with $y_n \notin Y$, so y_n has target $x_n \in X$. Since the strata are locally finite we may suppose (extracting a subsequence) that all x_n belong to the same stratum S', and (by compactness of the Grassmannian) that $T_{x_n}(S')$ tends to a limit $\tau \subset T_x M$. By property (A), $T_x S \subset \tau$.

Now take local coordinates so that we can identify tangent spaces at nearby points. If f represents y and df has image $A \subset T_x M$ then $A + T_x S = T_x M$ since y is a transversal jet. Since $y_n \to y$, any f_n representing y_n has $\mathrm{Im}(df_n) = A_n$ containing a subspace 'close' to A. But $A + \tau = T_x M$, and as $T_{x_n}(S') \to \tau$ it follows that $A_n + T_{x_n}(S') = T_{x_n}(M)$ for sufficiently large n. Thus y_n is a transversal jet, contrary to hypothesis.

<u>Corollary 1</u> <u>If</u> $\mathcal{S}$ <u>is a cohesive stratification of a closed subset</u> X <u>of</u> M, <u>then for any</u> V, $\{f \in C^\infty(V, M) : f \pitchfork \mathcal{S}\}$ <u>is open in</u> $C^\infty(V, M)$.

<u>Corollary 2</u> <u>If</u> $\mathcal{S}$ <u>is a cohesive stratification of a closed subset of</u> $J^k(V, M)$, <u>then</u> $\{f \pitchfork \mathcal{S}\}$ <u>is open in</u> $C^\infty(V, M)$.

These are immediate consequences of the above, with the definition of the topology on $C^\infty(V, M)$.

To give an idea of the strength of condition (B), we cite

<u>Thom's first isotopy lemma (1964)</u> <u>Let</u> $\mathcal{S}$ <u>be a regular stratification of the closed subset</u> X <u>of</u> M^m. <u>Then for any</u> $x \in S^s \in \mathcal{S}$ <u>there exist neighbourhoods</u> U_1 <u>of</u> x <u>in</u> S, U_2 <u>of</u> 0 <u>in</u> $\mathbb{R}^{m-s}$ <u>and an open embedding</u> $h : U_1 \times U_2 \to M$ <u>such that for any</u> $S' \in \mathcal{S}$, $h^{-1}(S') = U_1 \times T'$ <u>for some</u> $T' \subset U_2$.

One may suppose $h(x, 0) = x$. It is not in general possible also to choose h differentiable.

Further basic properties of stratifications (for amplification, see [GWPL, Part 1]) are:

The product of regular stratifications is regular

The transverse intersection of regular stratifications is regular

The pullback of a regular stratification by a map transverse to it is regular.

Any semialgebraic (or, eventually, subanalytic) set has a regular stratification.

Clearly stratifications are not unique. However, a stratification $\mathcal{S}$ of X defines a filtration $T_i(X)$ = union of strata of dimension $\leq i$. Then $\mathcal{S}$ is called <u>canonical</u> if, for each i, $T_i(X) - T_{i-1}(X)$ is the largest subset of $T_i(X)$ which is a smooth i-manifold and over which strata of higher dimension are regular. This property is invariant under diffeomorphisms, and it localises (it is not preserved by transverse intersection or by pullback). A semialgebraic set has a unique canonical stratification.

Stratifying maps

It is now necessary to relativise the above. A stratification of a smooth map $f : N \to P$ consists of stratifications $\mathcal{S}$ of N, $\mathcal{S}'$ of P (or, again, of certain subsets) such that for any stratum $S \in \mathcal{S}$ there exists $S' \in \mathcal{S}'$ with F inducing a submersion $S \to S'$. And $(\mathcal{S}, \mathcal{S}')$ is called regular if for any $S, T \in \mathcal{S}$, point $x \in S$ and sequence $\{y_n\}$ in T tending to x and with $\operatorname{Ker} d(f|T)_{y_n} \to \tau$, we have $\operatorname{Ker} d(f|S)x \subset \tau$. This condition ("sans éclatement") is due to Thom (1962).

Now not every smooth map, even if defined by polynomials, admits a stratification. Indeed, these stratifications are the key tool in the study of topological stability. But Thom (1962) gives an example of a 1 - parameter family of maps, all with distinct topological types. Thus some condition is necessary, and the following is most convenient.

(G) The restriction $f|\Sigma$ of f to its critical set Σ is proper and finite - to - one.

Now a polynomial map satisfying (G) admits regular stratifications, and there is even a theory of canonical stratification. This can be built up from a stratification $\mathcal{C}$ of $f(\Sigma)$ such that for any $U, V \in \mathcal{C}$,

(i) $f^{-1}U \cap \Sigma$ is smooth, and $f : (f^{-1}U \cap \Sigma) \to U$ is a local diffeomorphism,

(ii) $f^{-1}V \cap \Sigma$ and $f^{-1}V - \Sigma$ are regular over $f^{-1}U \cap \Sigma$.

Again the theory is invariant under diffeomorphisms and (appropriate) localisation, so the polynomial condition can be relaxed.

The raison d'être of the above is

Thom's second isotopy lemma (1964) Let $f : N \to P$ be a proper stratified map, $\pi : P \to \mathbb{R}$ a proper submersion such that each stratum (of N or P) submerses onto $\mathbb{R}$. Then f is topologically trivial, i.e. there is a commutative diagram

$$\begin{array}{ccc} (\pi f)^{-1}0 \times \mathbb{R} & \xrightarrow{f \times 1} & \pi^{-1}0 \times \mathbb{R} \\ \downarrow h & & \downarrow h' \\ N & \xrightarrow{f} & P \end{array}$$

<u>with</u> h, h' <u>homeomorphisms.</u>

Full details of the proof may be found in [GWPL, Part 2].

<u>Multitransversality</u>

Recall that subspaces V_i of a vector space W are in (surjective) <u>general position</u> if the diagonal map $W \to \oplus (W/V_i)$ is surjective, or equivalently if $\oplus V_i \to \oplus W$ is transverse to the diagonal. Observe that there can be at most $\dim W$ subspaces $V_i \neq W$. Correspondingly, we say that submanifolds M_r of a manifold N are in general position with respect to a map $f : N \to P$ if whenever points $x_i \in M_{r_i}$ have a common image $f(x_i) = y$, the $df(T_{x_i} M_{r_i})$ are in general position in $T_y P$. In particular, a stratification $\mathcal{S}$ of a subset X of N has <u>regular intersections</u> with respect to f if the set of strata is in general position with respect to f. If we exclude points where $(f|M_r)$ is submersive, this implies that preimages have at most $p = \dim P$ points; so it suffices to check collections of at most $(p + 1)$ points of N.

The powers P^r have a natural <u>diagonal stratification</u> $\mathcal{D} = \mathcal{D}(P^r)$ as follows: the stratum S_{pi} consists of r-tuples with precisely i distinct coordinates in P $(1 \leq i \leq r)$. Now $(\mathcal{S}, X)$ has regular intersections with respect to $F : N \to P$ if and only if for each stratum U of $\mathcal{S}^r$ (stratifying $X^r \subset N^r$), $F^r|U \pitchfork \mathcal{D}$. It suffices to take $r = p + 1$.

<u>Lemma</u> <u>If</u> $\mathcal{S}$ <u>has regular intersections with respect to</u> f, <u>there is an induced stratification</u> $\mathcal{S}_f$ <u>of</u> X <u>such that</u> $S, S' \in \mathcal{S}$ <u>imply</u> $S \cap f^{-1}f(S')$ <u>a union of strata of</u> $\mathcal{S}_f$.

In fact, we can take the nonempty sets of the form

$$A_{I,J} = \{x \in X : f(x) \in f(S_i) \iff i \in I,\ x \in S_j \iff j \in J\}$$

as $J \subset I$ run through subsets of $\mathcal{S}$, as the strata of $\mathcal{S}_f$.

Warning : $\mathcal{S}$ regular does not imply $\mathcal{S}_f$ regular.

Now suppose $\mathcal{S}$ stratifies a subset X of $J^k(N, P)$. We say that $f : N \to P$ is *multitransverse* to $\mathcal{S}$ if $j^k f \pitchfork \mathcal{S}$, and the stratification $(j^k f)^{-1}\mathcal{S}$ has regular intersections with respect to f . To show that this is a generic property, we reformulate it in terms of multijet transversality. Now we saw above that the intersections are regular if and only if, for some $r > p$ and for each stratum U of $((j^k f)^{-1}\mathcal{S})^r$, $f^r|U \pitchfork \mathcal{D}(P^r)$. Now $\mathcal{D}(P^r)$ pulls back to a stratification $\pi^*\mathcal{D}(P^r)$ of $(J^k(N, P))^r$. Further, $\mathcal{S}^r$ meets $\pi^*\mathcal{D}(P^r)$ transversely provided $\mathcal{S}$ satisfies

(SB) *For each* $S \in \mathcal{S}$, *the projection* $S \to P$ *is a submersion.*

Then assuming (SB), $\mathcal{S}^r \cap \pi^*\mathcal{D}(P^r)$ is a stratification and f is multitransverse to $\mathcal{S}$ if and only if, for some $r > p$, $r j^k f \pitchfork (\mathcal{S}^r \cap \pi^*\mathcal{D}(P^r))|N^{(r)}$. Note that we may restrict to r^{tuples} of distinct points at the source as this was done in the definition of multitransversality.

Although it was fairly straightforward to give conditions under which transversality defined an open set of maps, as soon as we impose transversality conditions on multijets the problem becomes much more subtle. The classical examples where openness holds are the transversality conditions defining C^∞-stable maps in the 'nice dimensions' (Mather (1970b)), and these are extended below. I know of no other results in this area. The problem is essentially one of 'filling in the diagonal' to get from $N^{(r)}$ to N^r. Thus one expects that if the conditions on jets of multiplicity $< r$ imply that the conditions on r-jets hold near the diagonal, then the conditions taken together define an open set of maps. But even this simple attempt to get started is not correct in detail - see Trotman (1977) for discussion of this mistake in an analogous situation.

The Topological Stability Theorem.

The heart of the theory comes in the use of canonical stratifications of stable maps to define a stratification of (a large open set of) a suitable jet space. This depends on the notions of contact equivalence of jets (introduced and studied in Mather (1968b)) and unfoldings (originally introduced by Thom in a slightly different context - see below). See [GWPL, part 3] for a suitable introduction; another excellent account is given by Martinet (1976). As we are only concerned here to give a rough outline of how the stratification is defined, we shall not attempt even to outline this subsidiary theory here.

A stable (i.e. C^∞-stable) map-germ F is locally equivalent to a polynomial map which satisfies (G), so has a canonical stratification. It is not hard to see that the germ of this stratification can be defined to depend only on the map-germ. We write cod F for the codimension (in the source manifold N) of the stratum containing the source, x of F.

Now suppose $f : N \to P$ has a stable unfolding $F : N \times U \to P \times U$ - i.e. $F(x, u_0) = (f(x), u_0)$ and F is stable as a map (not as a family). Then cod F does not depend on the choice of F, so may be written as cod f.

Write $W^\ell(n, p)$ for the set of ℓ-jets $(\mathbb{R}^n, 0) \to (\mathbb{R}^p, 0)$ admitting two representatives which are not contact equivalent - i.e. (moreorless) have non-isomorphic stable unfoldings. Then $W^\ell(n, p)$ is an algebraic subvariety (hence closed) of $J^\ell(n, p)$ whose codimension tends to ∞ with ℓ. We can now define $\mathrm{cod} : (J^\ell(n, p) - W^\ell(n, p)) \to \mathbb{N}$ in the obvious way; it yields a partition $A^\ell(n, p)$.

The key technical results are now the following (reference: [GWPL, part 4])

<u>The partition</u> $A^\ell(n, p)$ <u>of</u> $J^\ell(n, p) - W^\ell(n, p)$ <u>is a regular stratification.</u>

<u>If</u> $f : (N, x_0) \to (P, y_0)$ <u>is a stable map-germ, with</u> $j^\ell f_{x_0} \notin W^\ell$, <u>then</u> f <u>is multitransverse with respect to</u> $A^\ell(n, p)$, <u>and</u> $[(j^\ell f)^{-1} A^\ell(n, p)]_f$ <u>has the same germ at</u> x <u>as the canonical stratification of</u> f.

Thus the canonical stratification can be recovered. One may now go on to show

<u>Given a stable unfolding</u> F <u>of the map-germ</u> f

$$\begin{array}{ccc} (N, x_0) & \xrightarrow{\quad f \quad} & (P, y_0) \\ \downarrow i & & \downarrow j \\ (N', x_0') & \xrightarrow{\quad F \quad} & (P', y_0') \end{array} ,$$

<u>with canonical stratification</u> $(\mathcal{S}, \mathcal{J})$, <u>then</u> $i \pitchfork \mathcal{S} \iff j^\ell f \pitchfork A^\ell$; <u>and in this case,</u> $j^\ell f$ <u>is multitransverse to</u> A^ℓ <u>and</u> $[(j^\ell f)^{-1} A^\ell]_f = i^{-1}\mathcal{S}$ <u>stratifies</u> f.

In fact possession of a stable unfolding is essentially equivalent to (G). The usual term is 'finite singularity type'. Note, incidentally, that to make the above precise a little care is necessary in choosing representatives of germs. The proper smooth mappings $f : N \to P$ such that $j^\ell f(N) \cap W^\ell(N, P) = \phi$ and f is multitransverse to $A^\ell(N, P)$ form an open subset of $C^\infty(N, P)$. If $n < \operatorname{cod} W^\ell(N, P)$, it is dense.

This is the openness theorem referred to above. For such f, we can join any nearby g to f by a family f_t of maps satisfying the condition. There is then a diagram

$$\begin{array}{ccc} (N \times \mathbb{R},\ x_0) & \xrightarrow{\ f_t\ } & (P \times \mathbb{R},\ y_0) \\ \downarrow i & & \downarrow j \\ (N',\ x_0') & \xrightarrow{\ F\ } & (P',\ y_0') . \end{array}$$

As each $N \times t$ is transverse to $\mathcal{S}$, the induced stratification of $N \times \mathbb{R}$ (and $P \times \mathbb{R}$) has each stratum submersing onto $\mathbb{R}$. Topological triviality of the family $\{f_t\}$ can then be deduced from the second isotopy lemma. Hence f is topologically stable.

We say that $f : N \to P$ has generic singularities if, for some ℓ (which may be chosen with $n < \operatorname{cod} W^\ell(N, P)$) f satisfies the above. This defines a dense open subset of $C^\infty(N, P)$, consisting of C^0-stable maps. A generic family $F : N \times U \to P \times U$ may now be defined (for example) as one which (as a map) has generic singularities. Unfortunately, there does not yet exist in general any C^0-stability theory for families, so any definition is provisional.

Since our family is an unfolding of each of its maps, and $A^\ell(N \times U, P \times U)$ is related to $A^\ell(N, P)$ by letting an unfolding F of f have the same 'codimension', the stratification induced on $N \times U$ from $A^\ell(N \times U, P \times U)$ is the same as that obtained by restricting to fibres and pulling back from $A^\ell(N, P)$. Thus the above is equivalent to the apparently stronger condition that $N \times U \to J^\ell(N, P)$ is multitransverse to $A^\ell(N, P)$. It ensures that F is C^0-stable as a map, but not necessarily as a family. We take as a better provisional definition, that $N \times U \to P \times U \to U$ is multitransverse to $A^\ell(N, P)$.

See below for the formulation of this condition.

Families of functions

In the special case $P = \mathbb{R}$, however, there is a theory due to Looijenga (1974) of stability of families, which will thus yield the notion of generic singularities of a family needed for sections B, C and F. I will conclude by outlining how this may be derived by modifying the theory above. For the problems discussed in sections D and E, where we had sections of a bundle rather than general maps, multijets do not appear and there is no obvious stratification of the corresponding jet space, or even criterion (corresponding to C^0-stability) to say which stratifications are interesting. Perhaps a more thorough understanding of examples would suggest the right direction to proceed here.

We think of a family of functions $N \to P$ in terms of a pair of maps $N \times U \xrightarrow{F} P \times U \xrightarrow{\pi} U$, where the maps to U are the projection maps. A (regular) stratification of (F, π) consists of (regular) stratifications $\mathcal{S}$ of $N \times U$, $\mathcal{S}'$ of $P \times U$ and $\mathcal{S}''$ of U such that, if we write $\Sigma = \Sigma(F)$ for the critical set (of points of $N \times U$ where dF is not surjective), $C = F(\Sigma)$ and $V = F^{-1}(C) - \Sigma$, then

(i) If $X \in \mathcal{S}''$, then (components of) $\pi^{-1}X \cap C$ and $\pi^{-1}X - C$ are strata of $\mathcal{S}'$; $(\pi F)^{-1}X \cap \Sigma$, $(\pi F)^{-1}X \cap V$ and $(\pi F)^{-1}X - (\Sigma \cup V)$ are strata of $\mathcal{S}$,

(ii) For each $S \in \mathcal{S}$ with $S \subset \Sigma$, $F|S$ is a covering of some stratum of $\mathcal{S}'|C$; and for $S' \in \mathcal{S}'|C$, $\pi|S'$ is a covering of some stratum of $\mathcal{S}''$.

As to existence of stratifications, consider the case when N, P and U are euclidean spaces and F a polynomial map. Then a stratification of (F, π) exists provided F satisfies

(G') $\pi \circ F \mid \Sigma(F)$ is proper and finite-to-one.

Again, this modification of (G) is moreorless necessary, as well as sufficient. For $p = 1$, it is also generic and the theory can proceed. For $p > 1$ it is not generally satisfied, and this is one reason why an alternative theory is needed for this case.

When stratifications exist, we have appropriate modifications of the notion of canonical stratification, and of the second isotopy lemma.

Next, if $\mathcal{S}$ stratifies $X \subset M$ and we have maps $M \xrightarrow{F} P \times U \xrightarrow{\pi} U$ (with π as usual the projection), we say $\mathcal{S}$ has <u>regular intersections</u> with respect to (F, π) if, for all r and each stratum T of $\mathcal{S}^r$, $F^r|T$ is transverse to $\mathcal{D}((P \times U)^r) \cap (\pi^r)^{-1} \mathcal{D}(U^r)$. Note that although $\mathcal{D}(P \times U)^r$ is <u>not</u> transverse to $(\pi^r)^{-1} \mathcal{D}(U^r)$, the intersections of strata are clearly manifolds. Now $\mathcal{S}$ has regular intersections with respect to (F, π) if and only if $\mathcal{S}$ has regular intersections with respect to F, and for any union T of members of $\mathcal{S}_F$ with disjoint F-images, $\mathcal{S}_F|T$ has regular intersections with respect to $\pi \circ F$. There is then a refinement $\mathcal{S}_{F,\pi}$ of $\mathcal{S}_F$ minimal with respect to the condition that if $X, Y \in \mathcal{S}_{F,\pi}$ then FX and FY are equal or disjoint, and so are πFX and πFY.

We now come to multitransversality. With the above notation, if we write $F(x, u) = (f_u(x), u)$ then the assignment $(x, u) \mapsto j^\ell f_u(x)$ defines a map $N \times U \to J^\ell(N, P)$ which we denote $j^\ell F_\pi$. Now if $\mathcal{S}$ stratifies $X \subset J^\ell(N, P)$ we say (F, π) is <u>multitransverse</u> to $\mathcal{S}$ if $j^\ell F_\pi \pitchfork \mathcal{S}$ and $(j^\ell F_\pi)^{-1} \mathcal{S}$ has regular intersections with respect to (F, π). As before one can see that provided $\mathcal{S}$ satisfies (SB), (F, π) is multitransverse to $\mathcal{S}$ if and only if F is multitransverse to $\mathcal{S}$ and, for some $r > \dim U\ (\dim P + 1)$, $rj^\ell F_\pi \pitchfork (\mathcal{S}^r \cap \pi^{-1}\mathcal{D}(P^r))|N^{(r)}$. Thus here too, multitransversality is generic. (The fact that the condition is expressed in terms of $j^\ell F_\pi$ does not matter, for if $f : N \times U \to P$ is the first projection of F, the condition translates to a multitransversality condition on f since restriction to fibres defines a submersion $J^\ell(n + k, p) \to J^\ell(n, p)$.)

We come to the construction of another canonical stratification, and again the use of unfoldings is crucial. The notion of universal unfolding used above was that $N \times U \to P \times U$, as a map, is C^∞-stable : this amounts (see especially Martinet (1976)) to versality for contact equivalence. Here we shall use right equivalence (and this is another reason why we must restrict to $p = 1$). This was formally introduced by Mather (1968b), and developed in some unpublished notes which were in circulation about 1970. Perhaps the most convenient reference is Wassermann (1974); but see also the papers of Arnol'd.

The results we need from the theory are as follows:

There is an algebraic subset $X^\ell(n, 1)$ of $J^\ell(n, 1)$, whose codimension tends to infinity with ℓ for fixed n, such that jets z not in $X^\ell(n, 1)$ are sufficient - i.e. any two realisations f, f' are right equivalent. Any such germ f has a universal unfolding (F, π), and there is a smooth change of coordinates making (F, π) polynomial. Moreover, condition (G') is satisfied so (F, π) has a canonical stratification.

We may now, as before, define $\operatorname{cod} z = \operatorname{cod} f = \operatorname{cod} F$ to be the codimension of the stratum containing 0. Then :

The partition $\mathcal{B}^\ell(n, 1)$ *of* $J^\ell(n, 1) - X^\ell(n, 1)$ *defined by the function* cod *is a regular stratification.*

If (F, π) *is the universal unfolding of a realisation* f *of* $z \in J^\ell(n, 1) - X^\ell(n, 1)$ *then (locally)* (F, π) *is multitransverse to* $\mathcal{B}^\ell(n, 1)$ *and* (F, π) *has a canonical stratifications* $(\mathcal{S}, \mathcal{S}', \mathcal{S}'')$ *with*

$$\mathcal{S} = [(j^\ell F_\pi)^{-1} \mathcal{B}^\ell(n, 1)]_{F,\pi} .$$

This can now be globalised. Let N be a compact manifold; let $X(N)$ be the set of those C^∞ functions on N possessing a critical point $x \in N$ such that for all ℓ, $j^\ell f(x) \in X^\ell(N, \mathbb{R})$. Then - with an appropriate extension of the terminology to infinite dimensional manifolds -

There is a regular stratification $\mathcal{B}(N, \mathbb{R})$ *of* $C^\infty(N) - X(N)$ *into submanifolds of finite codimension.*

Now $X(N)$ has 'infinite codimension', so for any finite dimensional manifold U, a generic family $G : U \to C^\infty(N)$ will avoid $X(N)$, and then be transverse to $\mathcal{B}(N, \mathbb{R})$: indeed, if U is compact, the set of such maps is open (as well as dense). Suppose U compact and $N \times U \overset{F}{\to} \mathbb{R} \times U \overset{\pi}{\to} U$ corresponding to a map G transverse to $\mathcal{B}(N, \mathbb{R})$. Then :

(F, π) *is a topologically stable family of functions.* It has a stratification $(\mathcal{S}, \mathcal{S}', \mathcal{S}'')$ *with* $\mathcal{S}'' = G^{-1}\mathcal{B}(N, \mathbb{R})$.

Perhaps I should emphasise again that all the results in this section are copied from Looijenga (1974). As in the previous section, there is no explicit description of the canonical stratifications $\mathcal{B}$ or A, and finding them seems a tough problem. One would expect (following the early work of Cerf (1970)) that if $f \in C^\infty(N) - X(N)$ has critical points P_i $(1 \leq i \leq r)$, and the germ of f at P_i belongs to a stratum with codimension a_i, then the codimension of the stratum of $\mathcal{B}(N, \mathbb{R})$ containing f should be $\sum_1^r (a_i - n) + (r - s)$, where s is the number of distinct critical values.

Finally, observe that Theorem E4 (ii) is a consequence of these results. For with a suitable (compact) $L_P^F(M)$, we saw in §2 that the critical points of E_g on the fibres are generically the same as for $L_P^{PD}(M)$. Now the restrictions of E_g to the fibres define a family of functions

$$L_P^F(M) \xrightarrow{ev \times E_g} M \times \mathbb{R} \xrightarrow{\pi} M ,$$

and by Buchner's transversality theorem we may suppose this multitransverse to the appropriate $\mathcal{B}^\ell(N, 1)$. Then we have a stratification of the family, which is topologically stable. Since the cut locus is a union of certain strata in M, hence triangulable, the theorem follows.

Bibliography

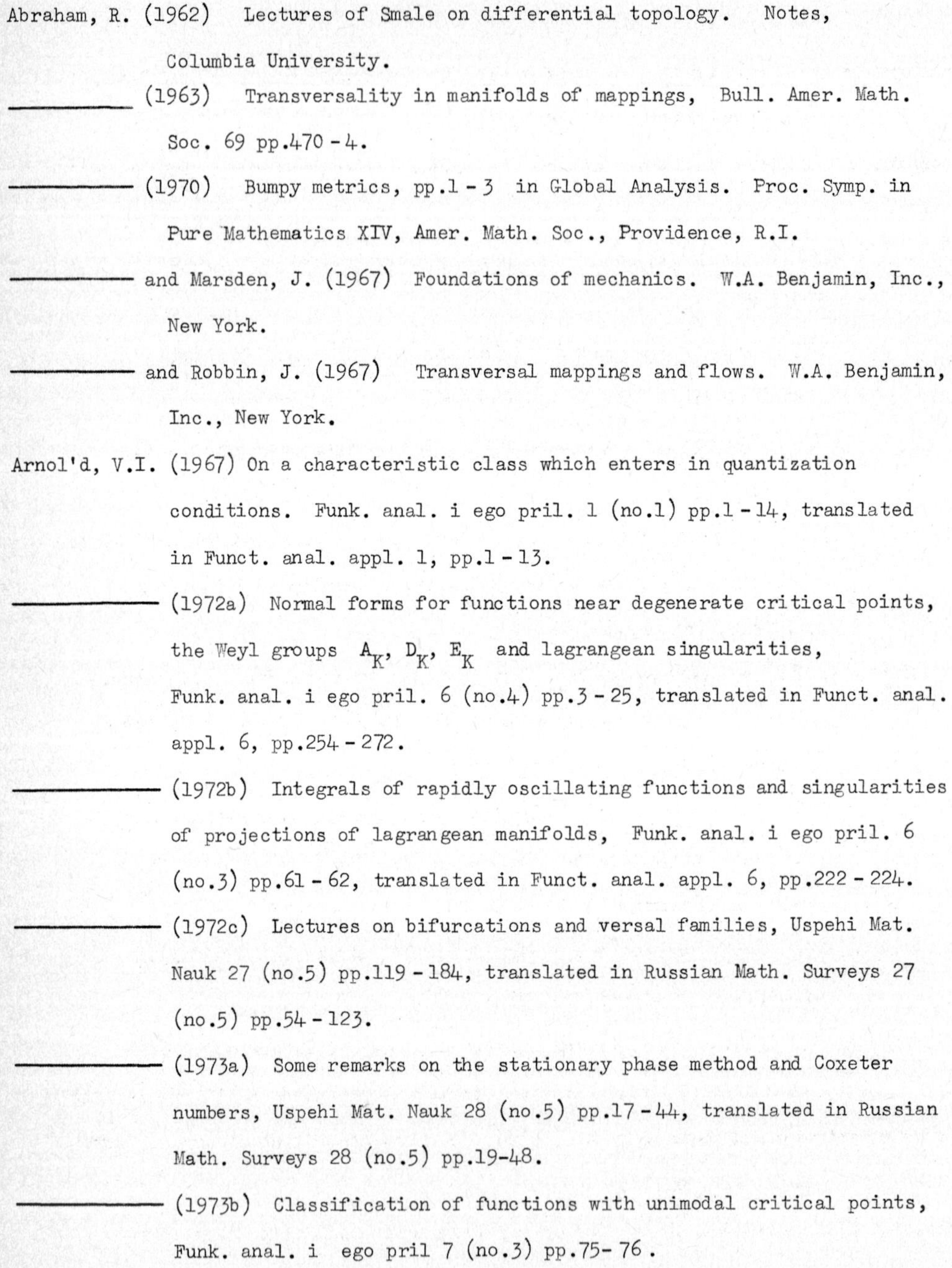

Abraham, R. (1962) Lectures of Smale on differential topology. Notes, Columbia University.

——— (1963) Transversality in manifolds of mappings, Bull. Amer. Math. Soc. 69 pp.470 - 4.

——— (1970) Bumpy metrics, pp.1 - 3 in Global Analysis. Proc. Symp. in Pure Mathematics XIV, Amer. Math. Soc., Providence, R.I.

——— and Marsden, J. (1967) Foundations of mechanics. W.A. Benjamin, Inc., New York.

——— and Robbin, J. (1967) Transversal mappings and flows. W.A. Benjamin, Inc., New York.

Arnol'd, V.I. (1967) On a characteristic class which enters in quantization conditions. Funk. anal. i ego pril. 1 (no.1) pp.1 - 14, translated in Funct. anal. appl. 1, pp.1 - 13.

——— (1972a) Normal forms for functions near degenerate critical points, the Weyl groups A_K, D_K, E_K and lagrangean singularities, Funk. anal. i ego pril. 6 (no.4) pp.3 - 25, translated in Funct. anal. appl. 6, pp.254 - 272.

——— (1972b) Integrals of rapidly oscillating functions and singularities of projections of lagrangean manifolds, Funk. anal. i ego pril. 6 (no.3) pp.61 - 62, translated in Funct. anal. appl. 6, pp.222 - 224.

——— (1972c) Lectures on bifurcations and versal families, Uspehi Mat. Nauk 27 (no.5) pp.119 - 184, translated in Russian Math. Surveys 27 (no.5) pp.54 - 123.

——— (1973a) Some remarks on the stationary phase method and Coxeter numbers, Uspehi Mat. Nauk 28 (no.5) pp.17 - 44, translated in Russian Math. Surveys 28 (no.5) pp.19-48.

——— (1973b) Classification of functions with unimodal critical points, Funk. anal. i ego pril 7 (no.3) pp.75- 76 .

Arnol'd, V.I. (1974) Normal forms of functions near degenerate critical points, Uspehi Mat. Nauk 29 (no.2) pp.11 - 49, translated in Russian Math. Surveys 29 (no.2) pp.10 - 50.

——————— (1975) Critical points of smooth functions, pp.19 - 39 in Proc. ICM (Vancouver, 1974) vol. 1.

Buchner, M.A. (1974) Stability of the cut locus. Doctoral thesis, Harvard University.

Carter, S. (1973) The focal set of a hypersurface with abelian fundamental group, Topology 12, pp.1 - 4.

——————— and Robertson, S.A. (1967) Relations between a manifold and its focal set, Invent. Math. 3, pp.300 - 307.

Cayley, A. (1870) A memoir on cubic surfaces, Phil. Trans. Roy. Soc. 159, pp.231 - 326.

Cerf, J. (1970) La stratification naturelle des espaces de fonctions différentiables réelles et le théorème de pseudo - isotopie, Publ. Math. I.H.E.S. 39 , pp.5 - 173.

Chenciner, A. (1974) Travaux de Thom et Mather sur la stabilité topologique, pp. in Séminaire Bourbaki (1972/73) Springer lecture notes no. 383.

Darboux, G. (1883) Mémoire sur les solutions singulières des équations aux dérivées partielles de premier ordre, Mém de l'Institut des Savants Etrangers.

Dos Santos, N.M. (1967) On the conjugate locus of a Riemannian manifold, An. Acad. Brasil Ci. 39, pp.19 - 26.

Duistermaat, J.J. (1974) Oscillatory integrals, Lagrange immersions, and unfoldings of singularities, Comm. Pure Appl. Math. 27, pp.207 - 281.

——————— and Hörmander, L. (1972) Fourier integral operators II, Acta Math. 128, pp.183 - 269.

Epstein, D.B.A. (1975) Natural tensors on Riemannian manifolds, Jour. Diff. Geom. 10, pp.631 - 645.

Feldman, E.A. (1965) The geometry of immersions I. Trans. Amer. Math. Soc. 120, pp.185 - 224.

Gibson, C.G., Wirthmüller, K., du Plessis, A.A. and Looijenga, E.J.N. (1976) (referred to as [GWPL]) Topological stability of smooth mappings, Springer lecture notes no.552.

Gluck, H. and Singer, D. (1976) Deformations of geodesic fields, Bull. Amer. Math. Soc. 82, pp.571 - 574.

Gromoll, D. and Meyer, W. (1969) On differentiable functions with isolated critical points, Topology 8, pp.361 - 369.

Gromov, M.L. (1970) Isometric embeddings and immersions, Doklady Akad. Nauk 192, pp.1206 - 1209, translated in Soviet Math. Doklady 11, pp.794 - 797.

Guckenheimer, J. (1973) Catastrophes and partial differential equations, Ann. Inst. Fourier 23, pp.31 - 59.

Hironaka, H. (1973) Subanalytic sets, pp.1 - 41 in Number theory, algebraic geometry and commutative algebra, Kinokuniya, Tokyo.

Hodge, W.V.D. and Pedoe, D. (1952) Methods of algebraic geometry, vol.II. Cambridge University Press.

Hörmander, L. (1971) Fourier integral operators I, Acta Math. 127, pp.79 - 183.

Iarrobino, A. (1975) Vector spaces of forms I: ancestor ideals. Preprint, University of Texas (Austin), 60pp.

Jacobowitz, H. (1972) Implicit function theorems and isometric embeddings, Ann. of Math. 95, pp.191 - 225.

Klingenberg, W. (1976) Existence of infinitely many closed geodesics, Jour. Diff. Geom. 11, pp.299 - 308.

——— and Takens, F. (1972) Generic properties of geodesic flows, Math. Ann. 197, pp.323 - 334.

Kobayashi, S. (1967) On conjugate and cut loci, pp.96 - 122 in Studies in global geometry and analysis, Math. Ass. Amer. Studies in Math., Prentice-Hall, Englewood Cliffs, N.J.

Kommerell, K. (1905) Riemannschen Flächen im ebenen Raum von vier Dimensionen, Math. Ann. 60, pp.548 - 596.

Libermann, P. (1961) Calcul tensoriel et connexions d'ordre supérieur, Tercer Coloquio Brasileiro de Matematica, Fortaleza, Ceara.

Little, J.A. (1969) On singularities of submanifolds of higher dimensional euclidean spaces, Ann. Mat. Pura Appl. 83, pp.261 - 336.

Łojasiewicz, S. (1969) Semianalytic sets. Notes, I.H.E.S., Bures-sur-Yvette, 153 pp.

Looijenga, E.J.N. (1974) Structural stability of smooth families of C^∞ - functions. Doctoral thesis, Universiteit van Amsterdam.

Malgrange, B. (1974) Intégrales asymptotiques et monodromie, Ann. Sci. Ec. Norm. Sup. 7, pp.405 - 430.

Martinet, J. (1969) Sur les singularités des formes différentielles. Doctoral thesis, Université de Grenoble.

——— (1976) Déploiements versels des applications différentiables et classification des applications stables, pp.1 - 44 in Singularités d'applications différentiables (Plans-sur-Bex, 1975) Springer lecture notes no.535.

Mather, J. (1968a) Stability of C^∞ mappings I : the division theorem. Ann.of Math. 87, pp.89 - 104.

——— (1968b) Stability of C^∞ mappings III : finitely determined map-germs. Publ. Math. I.H.E.S. 35, pp.127 - 156.

——— (1969) Stability of C^∞ mappings II : infinitesimal stability implies stability. Ann. of Math. 89, pp.254 - 291.

——— (1970a) Stability of C^∞ mappings IV : classification of stable germs by $\mathbb{R}$ algebras. Publ. Math. I.H.E.S. 37, pp.223 - 248.

——— (1970b) Stability of C^∞ mappings V : transversality. Advances in Math. 4, pp.301 - 336.

——— (1970c) Notes on C^0 - stability. Notes, Harvard University, 74pp.

——— (1971) Stability of C^∞ mappings VI : the nice dimensions, pp.207 - 253 in Proceedings of Liverpool Singularities Symposium I, Springer lecture notes no.192.

Mather, J. (1973a) Generic projections, Ann. of Math. 98, pp.226 - 245.

———— (1973b) Stratifications and mappings, pp.195 - 223 in Dynamical Systems (ed. M.M. Peixoto). Academic Press, New York.

———— (1976) How to stratify mappings and jet spaces, pp.128 - 176 in Singularites d'applications differentiables (Plans-sur-Bex, 1975) Springer lecture notes, no.535.

Milnor, J. (1963) (referred to as [M]) Morse theory, Princeton University Press.

Moore, C.L.E. and Wilson, E.B. (1916) Differential geometry of two-dimensional surfaces in hyperspaces, Proc. Amer. Acad. Arts.Sci. 52, pp.267 - 368.

Mumford, D. (1965) Geometric invariant theory, Springer Verlag, Berlin.

Myers, S.B. (1935-6) Connections between differential geometry and topology I, II, Duke Math. Jour. 1, pp.376 - 391; 2, pp.95 - 102.

Nash, J. (1956) The imbedding problem for Riemannian manifolds, Ann. of Math. 63, pp.20 - 63.

Palais, R. (1963) Morse theory on Hilbert manifolds, Topology 2, pp.299 - 340.

Perepelkine, D. (1935) Sur la courbure et les espaces normaux d'une V_m dans R_n, Rec. Math. (Mat. Sbornik) 42, pp.81 - 100.

Poenaru, V. (1970) Un théorème des fonctions implicites pour les espaces d'applications C^∞, Publ. Math. I.H.E.S. 38, pp.93 - 124.

———— (1974) Analyse différentielle, Springer lecture notes no.371.

———— (1975) Stabilité structurelle équivariante, notes, Université Paris Sud, 80pp.

———— (1976) Singularités C^∞ en présence de symétrie, Springer lecture notes, no. 510.

Pohl, W.F. (1962) Differential geometry of higher order, Topology 1, pp.169 - 211.

———— (1971) Singularities in the differential geometry of submanifolds, pp.104 - 113 in Proc. Liverpool Singularities Symposium II, Springer lecture notes no. 209.

Porteous, I. (1971a) Geometric differentiation - a Thomist view of differential geometry, pp.122 - 127 in Proc. Liverpool singularities symposium II, Springer lecture notes no.209.

Porteous, I. (1971b) The normal singularities of a submanifold, Jour. Diff. Geom. 5, pp.543 - 564.

Quinn, F. (1970) Transversal approximation on Banach manifolds, pp.213 - 222 in Global analysis, Proc. Symp. in Pure Math. XV, Amer. Math. Soc., Providence, R.I.

Ruse, H.S. (1946) The five dimensional geometry of the curvature tensor in a Riemannian V_4, Oxford Quart. Jour. Math. 17, pp.1 - 15.

Sard, A. (1958) Images of critical sets. Ann. of Math. 68, pp.247 - 259.

Schläfli, L. (1863) On the distribution of surfaces of the third order into species, in reference to the presence or absence of singular points, and the reality of their lines, Phil. Trans. Roy. Soc. 153, pp.193 - 241.

Siersma, D. (1973) The singularities of C^∞ functions of right codimension less than or equal to eight, Indag. Math. 35, pp.31 - 37.

——— (1974) Classification and deformation of singularities, Doctoral thesis, Universiteit van Amsterdam.

Singer, D. and Gluck, H. (1976) The existence of non-triangulable cut loci, Bull. Amer. Math. Soc. 82, pp.599 - 602.

Thom, R. (1954) Sur quelques propriétés globales des variétés différentiables, Comm. Math. Helv. 28, pp.17 - 86.

——— (1956) Un lemme sur les applications différentiables, Bol. Soc. Mat. Mex. 1, pp.59 - 71.

——— (1962a) La stabilité topologique des applications polynomiales, l'Ens. Math. 8, pp.24 - 33.

——— (1962b) Sur la théorie des enveloppes, Jour. Math. Pures Appl. 41, pp. 177 - 192.

——— (1964) Local topological properties of differentiable mappings, pp. 191 - 202 in 'Differential analysis', Oxford University Press.

——— (1969) Ensembles et morphismes stratifiés, Bull. Amer. Math. Soc. 75, pp.240 - 284.

Thom, R. (1972a) Stabilité structurelle et morphogénèse, W.A. Benjamin, Inc., Reading, Mass. (Translation into English published 1975, same publisher).

——— (1972b) Sur le cut-locus d'une variété plongée, Jour. Diff. Geom 6, pp.577-586.

——— and Levine, H. (1959) Singularities of differentiable mappings, notes, University of Bonn; republished as pp.1-89 in Proceedings of Liverpool singularities symposium I, Springer lecture notes no.192.

Titus, C.J. (1971) Transformation semigroups and extensions to sense-preserving mappings, preprint no.35, Aarhus Universitet.

Trotman, D.J.A. (1977) Counterexamples in stratification theory: two discordant horns. To appear in Singularities of real and complex mappings, Proc. Nordic summer school in math. (Oslo, August 1976), 11pp.

Verdier, J.-L. (1976) Stratifications de Whitney et théorème de Bertini-Sard, Invent. Math.36, pp.295-312.

Wall, C.T.C. (1975) Regular stratifications, pp.332-344 in Dynamical Systems - Warwick 1974, Springer lecture notes no.468.

Warner, F. (1965) The conjugate locus of a Riemannian manifold, Amer. Jour. Math. 87, pp.575-604.

Wassermann, G. (1974) Stability of unfoldings. Springer lecture notes no.393.

Weinstein, A. (1970) The generic conjugate locus, pp.299-302 in Global analysis, Proc. Symp. in Pure Math. XV, Amer. Math. Soc. Providence, R.I.

——— (1971) Symplectic manifolds and their lagrangean submanifolds, Advances in Math. 6, pp.329-346.

Weyl, H. (1946) The classical groups, Princeton University Press.

Whitney, H. (1936) Differentiable manifolds, Ann. of Math. 37, pp.645-680.

——— (1965) Tangents to an analytic variety, Ann. of Math. 81, pp.496-549.

A CATASTROPHE MODEL FOR THE STABILITY OF SHIPS

E.C. Zeeman

1. INTRODUCTION.

Catastrophe theory provides a new way of looking at the statics of a ship, and this in turn lends a new simplicity to the global non-linear dynamics. The weight of the ship and the position of the centre of gravity are taken as parameters. Then the set of equilibrium positions form a smooth manifold that maps onto the parameter space. It is the singularities of this map that are recognisable as elementary catastrophes.

For example heeling and capsizing are fold catastrophes. At the metacentre there is a cusp catastrophe. The point of inflexion of the lever arm curve is caused by another cusp catastrophe. The increased likelihood of capsizing when overloaded, or when the crest of a wave is amidships, is due to a swallowtail catastrophe. The evolution of hull shape from canoe to modern ship is characterised by a butterfly catastrophe. On the metacentric locus there are also hyperbolic umbilic catastrophes. The sudden onset of heavy rolling due to non-linear resonance with the wave is a dynamic fold catastrophe.

For the naval architect this approach in terms of canonical forms offers a qualitative geometry that is complementary to the classical approach [2,3,9,11]. Whether or not such formulation will be of any use remains to be seen, because to make quantitative predictions it is still necessary to choose coordinates and approximations as in the classical theory. However as a general principle it is always advantageous to retain the dynamics in a conceptually simple form for as long as possible, so that the important qualitative features can be kept in the forefront of the mind unobscured by detail, allowing the eventual approximation to be tailored to the job on hand. Catastrophe theory should be used like the zoom lens on a microscope, for gaining a global view of the problem, and thus enhancing the discrimination with which one selects the necessary tool from classical mathematics to zoom in and solve it.

For the mathematician the interest of this example lies in the rich variety of the mathematics that it brings together; it is a prototype revealing catastrophe theory as a natural generalisation of Hamiltonian dynamics. Besides having a parameter space, an elementary catastrophe model has a state space, a potential, and a dynamic minimising the potential. Here the state space is the usual phase space of Hamiltonian dynamics, namely the cotangent bundle of the configuration space of the ship, while the potential is none other than the Hamiltonian, which is a Lyapunov function for, and therefore locally minimised by, the Hamiltonian flow with any type of damping. The subtlety comes from the bifurcations of the dynamic over the parameter space, which are governed by the elementary catastrophes in the evolute of the buoyancy locus.

We begin the paper by giving in Sections 2 - 6 a brief elementary sketch of the classical linear theory of ship motions, partly for the benefit of readers unfamiliar with the topic, and partly to see later how the model generalises it. The reader familiar with the linear theory is recommended to proceed at once to Section 7, where we describe the significance of the geometry. The model itself is introduced in Section 10, at first for rolling only, and then successively enlarged to include pitching, heaving and loading. At the end in Section 14 we place the treatment in its proper group-theoretical setting.

An unsolved problem is how best to incorporate into the model the forcing terms of wind and wave, so as to prove global existence theorems concerning induced oscillations, resonance and capsizing. The model already sheds some light on capsizing. In the conclusion we tentatively suggest two areas for further exploration.

CONTENTS

1. Introduction.
2. Linear theory of rolling.
3. Quantitative estimates.
4. Rolling in a seaway.
5. Resonance.
6. Pitching and heaving.
7. Cusp catastrophe at the metacentre.
8. Global metacentric locus.
9. Lever arm curve.
10. Catastrophe model for rolling.
11. Global dynamics.
12. Model including pitching.
13. Model including heaving.
14. Model including loading.
15. Conclusion.

2. LINEAR THEORY OF ROLLING.

The classical linear theory of how a ship behaves like a pendulum hanging from the metacentrè dates back to Euler [7] in 1737 and Bouguer [4] in 1746. For simplicity we begin by confining ourselves to the artificial 2-dimensional problem of rolling only. When we come to the catastrophe model, the notation will allow us to generalise with ease to the full 3-dimensional situation.

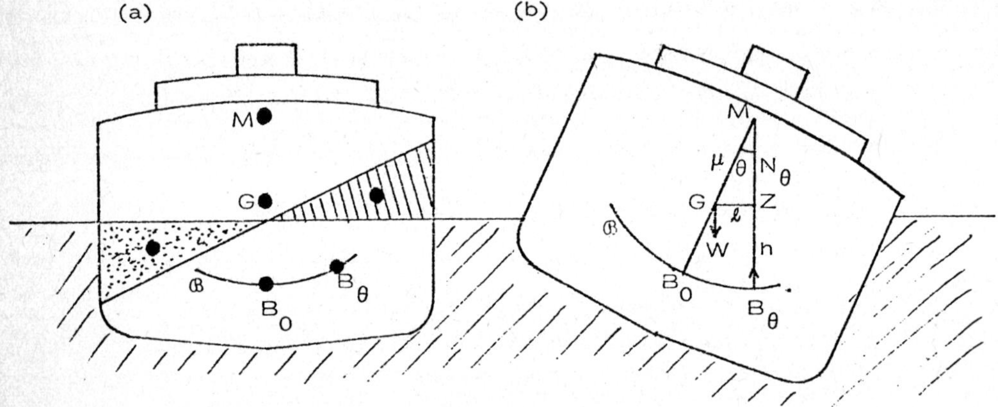

Figure 1. The buoyancy locus and metacentre.

Let G = centre of gravity of ship.

B_0 = centre of buoyancy when ship floats vertically

= centre of gravity of water displaced.

B_θ = centre of buoyancy when ship is at angle θ.

$\mathcal{B}$ = buoyancy locus = $\{B_\theta; -\pi < \theta \leq \pi\}$

N_θ = normal to $\mathcal{B}$ at B_θ.

M = metacentre = centre of curvature of $\mathcal{B}$ at B_0.

μ = GM = metacentric height.

Lemma 1. $\mathcal{B}$ is a convex closed curve. When the ship is heeled at angle θ, the normal N_θ is vertical.

Proof. Let A_θ denote the water displaced at angle θ. The area of A_θ is independent of θ, since by Archimedes principle the weight of water displaced equals that of the ship. A_θ is obtained from A_0 by adding the immersed wedge, shown shaded in Figure 1, and subtracting the emerged wedge, shown dotted. Since the wedges have equal area, B_0B_θ is parallel to the line joining the centres of gravity of the two wedges. As $\theta \to 0$ this line tends to the horizontal in Figure 1(a), and hence the tangent to $\mathcal{B}$ at B_0 is horizontal in Figure 1(a). The above argument is independent of the symmetry at $\theta = 0$, and hence applies to any θ. Therefore the tangent to $\mathcal{B}$ at B_θ is horizontal in Figure 1(b), and hence the normal N_θ is vertical.

For small θ, the inclination of B_0B_θ to the horizontal has the same sign as θ, and hence the curvature of $\mathcal{B}$ at B_0 is upwards. The same argument applies to any θ, and hence $\mathcal{B}$ is convex.

Corollary. For small θ, the buoyancy force passes approximately through the metacentre M.

Proof. The buoyancy force acts vertically upwards at B_θ, and therefore along the normal N_θ. For small θ the normals pass approximately through the centre of curvature, M. In fact since B_0 is a point of symmetry of $\mathcal{B}$, the distance from M to N_θ is of order θ^3.

Remark. For the 3-dimensional problem the same result holds, with the proviso that $\mathcal{B}$ is now a convex closed surface; the proof is the same. Note that the result is independent of the shape of hull, and holds not only for ordinary ships but also for catamarans and icebergs, for example.

The righting couple in Figure 1(b) consists of the weight W of the ship acting downwards at G, and the buoyancy force W acting upwards at B_θ. Let ℓ denote that <u>lever arm</u> of this couple :

$$\ell = \text{distance from G to } N_\theta$$
$$= GZ, \text{ in Figure 1(b)} ,$$

where Z is the foot of the perpendicular from G to N_θ. Newton's law of motion gives :

$$I\ddot{\theta} = -W\ell \quad \ldots\ldots\ldots\ldots\ldots\ldots\ldots \text{(1)}$$

where I is the moment of inertia of the ship (and the entrained water) about G. From the Corollary to Lemma 1 and Figure 1(b),

$$\ell = \mu \sin\theta, \text{ to second order in } \theta,$$
$$= \mu\theta, \text{ again to second order.}$$

Hence the approximate linear equation is

$$I\ddot{\theta} = -W\mu\theta. \quad \ldots\ldots\ldots\ldots\ldots\ldots \text{(2)}$$

This has rolling solution

$$\theta = \theta_o \cos\frac{2\pi t}{T} \quad \ldots\ldots\ldots\ldots\ldots \text{(3)}$$

where θ_o is the amplitude and T the period of the roll. The amplitude is arbitrary, and the period is given by

$$T = 2\pi\sqrt{\frac{I}{W\mu}} \;. \quad \ldots\ldots\ldots\ldots\ldots \text{(4)}$$

The viscosity of the water has a damping effect, and the simplest way to put this into the dynamics is to add a small damping term $2\varepsilon\dot{\theta}$ to equation (2), where ε is a small positive constant :

$$\ddot{\theta} + 2\varepsilon\dot{\theta} + \frac{W\mu}{I}\theta = 0 \;. \quad \ldots\ldots\ldots \text{(5)}$$

This has the effect of multiplying the solution (3) by a decay factor $e^{-\varepsilon t}$, and lengthening the period by a factor of order ε^2, which we can ignore.

3. QUANTITATIVE ESTIMATES.

By (2) the larger the metacentric height the more stable is the ship. However by (4) the larger the metacentric height the shorter is the period of roll, and the more uncomfortable is the ship; therefore choice of μ is an important feature of ship design.

What is at first surprising is how small μ can be compared with the size of the ship : for example a liner 300 metres long may have a metacentric height of only half a metre. Therefore we shall digress briefly to make some very rough estimates of μ and T, in order to give a quantitative feel for the problem complementary to the qualitative feel given by the subsequent catastrophe theory. We shall see in Section 7 that if G is too near M small alterations in the position of G may seriously increase the danger of capsizing.

Let $2a$ = beam of ship = width at water line,

A = area below water-line,

(x,y) = coordinates of B_θ relative to B_0,

$\rho = B_0M$ = radius curvature of $\mathcal{B}$ at B_0.

<u>Lemma 2.</u> $\rho = \dfrac{2a^3}{3A}$.

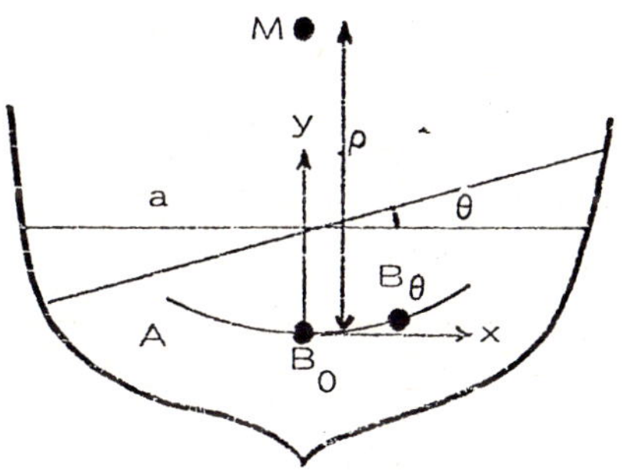

Figure 2. Computing ρ.

<u>Proof.</u> Let $t = \tan\theta$. Let 0^n denote order t^n. Each wedge has area $\frac{1}{2}a^2t + 0^2$, and the coordinates of the centres of gravity of the wedges relative to the mid point of the water line are :

$$\left(\pm \frac{2a}{3} + 0^1,\ \pm \frac{at}{3} + 0^2\right).$$

The coordinates (x,y) of B_θ are given by taking moments of the wedges about B_0 :

$$Ax = \left(\frac{1}{2}a^2t\right)\frac{4a}{3} + 0^2 = \frac{2a^3t}{3} + 0^2$$

$$Ay = \left(\frac{1}{2}a^2t\right)\frac{2at}{3} + 0^3 = \frac{a^3t^2}{3} + 0^3$$

Therefore putting $\rho = \dfrac{2a^3}{3A}$, $\quad x = \rho t + 0^2$

$$y = \frac{1}{2}\rho t^2 + 0^3 .$$

Therefore the equation of $\mathcal{B}$ is

$$y = \frac{x^2}{2\rho} + O(x^3),$$

which has radius of curve ρ at the origin, as required.

Define a ship to be <u>wall-sided</u> if the sides are parallel at the water-line, as in Figure 1. Let β denote the maximum angle of heel for which the water-line still meets the parallel part of the sides (in Figures 1 and 3, θ is about 25°).

<u>Lemma 3.</u> <u>In a wall-sided ship the buoyancy locus for $|\theta| < \beta$ is precisely the parabola $x^2 = 2\rho y$.</u>

<u>Proof.</u> In the proof of Lemma 2, leave out all the O^n's.

In order to estimate the metacentric height and period of roll we now make a couple of very rough quantitative assumptions. Assume (i) that A is approximately a rectangle with the draught equal to a third of the beam, as in Figure 3. Therefore

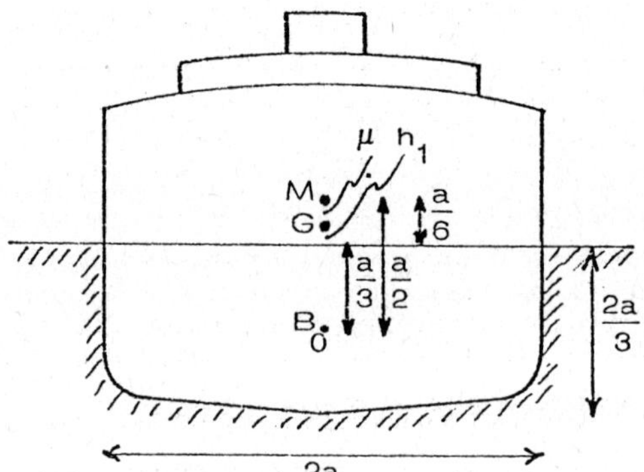

Figure 3. Estimate of μ.

$$A = 2a \times \frac{2a}{3} = \frac{4a^2}{3}$$

$$B_0M = \rho = \frac{2a^3}{3A} \text{ (by Lemma 2)}$$

$$= \frac{a}{2}.$$

Meanwhile B_0 is $\frac{a}{3}$ below the water-line, and so M is $\frac{a}{6}$ above. Therefore if h_1 = height of G above the water-line,

$$\mu = \frac{a}{6} - h_1 \quad \ldots\ldots\ldots\ldots\ldots\ldots \quad (6)$$

Assume (ii) that the moment of inertia I is the same as that of a solid disk of radius a :

$$I = \frac{W}{g}\frac{a^2}{2} \quad \ldots\ldots\ldots\ldots\ldots\ldots \quad (7)$$

where g = gravitational acceleration = 9.81 m/sec^2. We can now compute the period of roll:

$$T = 2\pi\sqrt{\frac{I}{W\mu}} \text{ by (4)}$$

$$= 2\pi\sqrt{\frac{a^2}{2g\mu}} \text{ by (7)}$$

$$= 1.42 \frac{a}{\sqrt{\mu}} \quad \ldots\ldots\ldots\ldots\ldots\ldots \quad (8)$$

Although the above assumptions are crude the resulting orders of magnitude are not unreasonable for both naval and merchant ships [2 pages 107,335]. Where ships tend to differ is in the height of G above the water-line, and so let us work out a couple of examples of a destroyer and a liner in order to illustrate the contrast. In each case we assume typical values for the beam and position of G, and deduce the resulting metacentric height and period of roll.

Table 1		Destroyer	Liner
Assume	Beam, 2a	10m	30m
	Height of G above water-line, h_1	0	2m
Deduce	Metacentric height, μ, by (6)	0.8m	0.5m
	Period of roll, T, by (8)	8 secs	30 secs

Notice that the greater metacentric height of the destroyer gives a greater righting couple, and hence makes her more stable, so that she can perform tighter manoeuvres, as well as causing a faster roll. By contrast the lesser metacentric height of the liner makes her less stable, although this does not matter so much since she does not have to indulge in manoeuvres; meanwhile she benefits from the increased comfort of the slower roll and smaller accelerations, which ensure that the passengers are less likely to be seasick.

4. ROLLING IN A SEAWAY.

The forced oscillations induced by periodic waves, and the resonance that occurs when the period of the waves coincides with the natural period T of rolling, were first studied by Bernoulli [3] in 1759.

Typical Atlantic ocean waves usually have a wave-length between 50 and 100 metres [2, page 177]. In deep water the wave-length determines both the speed and the period of the wave. For example a 60 metre wave has a speed of 35 Km/hour (= 19 knots) and a period of 6 seconds. The height of the wave from trough to crest is independent of the wave-length, and in the Atlantic, for instance, wave heights of over 6m occur on an average of 55 days in the year. In a 60m. wave of height 6m the maximum inclination of water surface to the horizontal is approximately 18°.

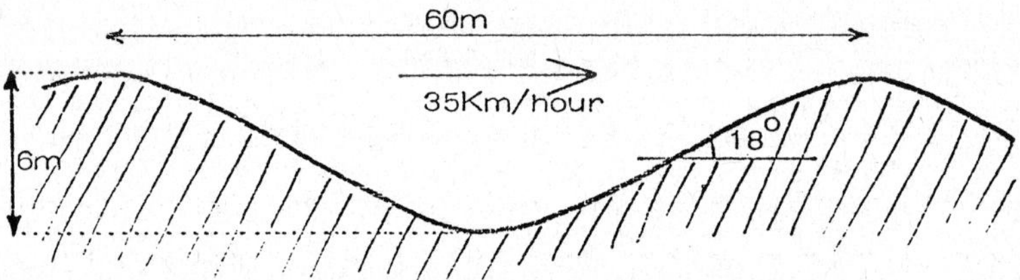

Figure 4. Typical Atlantic wave of period 6 seconds.

What is the effect of the waves on the ship? Suppose that in our 2-dimensional example the water surface is inclined at an angle α to the horizontal, as in Figure 5. The buoyancy force is the sum of all the pressures on the hull that would have sufficed to keep the displaced water in equilibrium with its surface inclined at angle α; hence the buoyancy force acts at $B_{\theta+\alpha}$ perpendicular to the water surface, in other words through M (approximately). Therefore the equation (2) for rolling must be modified to

$$I\ddot{\theta} = -W\mu(\theta+\alpha) \quad \ldots\ldots\ldots\ldots\ldots (9)$$

Suppose now that we have a beam sea, in other words waves are coming from the side with period τ and maximum inclination α_o to the horizontal. The inclination α will be a periodic function of time, t, approximately equal to

$$\alpha = \alpha_o \cos\frac{2\pi t}{\tau} \quad \ldots\ldots\ldots\ldots (10)$$

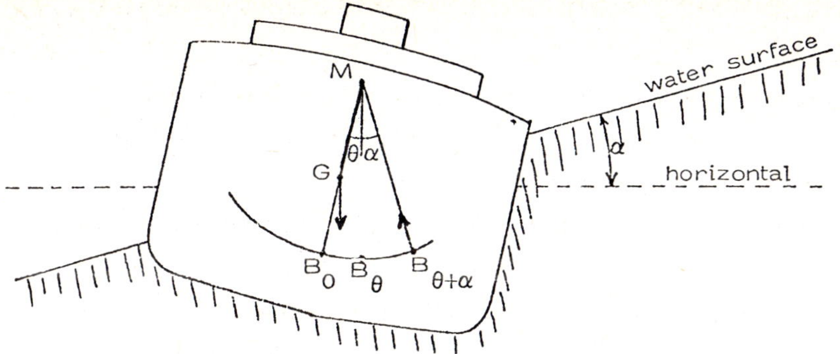

Figure 5. Inclined water surface.

Substituting (4) and (10) in (9) we obtain

$$\left(\frac{T}{2\pi}\right)^2 \ddot{\theta} + \theta = -\alpha = -\alpha_o \cos\frac{2\pi t}{\tau} . \quad \ldots\ldots (11)$$

The particular integral of (11) of period τ will give the rolling induced by the waves. The general solution of (11) will in fact be the sum of this particular integral together with a natural roll (3) of period T; however if we were to add a damping factor as in (5), then the latter would decay exponentially leaving only the particular integral, which is therefore an attractor of the dynamic. The particular integral will in fact be an amplification of the wave

$$\theta = \lambda\alpha = \lambda\alpha_o \cos\frac{2\pi t}{\tau} . \quad \ldots\ldots\ldots\ldots (12)$$

where λ is a constant amplifying factor. We can compute λ by substituting (12) into (11) :

$$\left[-\left(\frac{T}{\tau}\right)^2 + 1\right]\lambda\alpha = -\alpha .$$

Therefore $$\lambda = \left[\left(\frac{T}{\tau}\right)^2 - 1\right]^{-1} \quad \ldots\ldots\ldots\ldots\ldots (13)$$

We can now compute the effect of a typical Atlantic beam sea upon our destroyer and liner. Assume $\tau = 6$ secs and $\alpha = 18^o$ as in Figure 4.

Table 2	Destroyer	Liner
Period of roll T, by Table 1	8 secs	30 secs
Amplifying factor λ, by (13)	$\frac{9}{7}$	$\frac{1}{24}$
Amplitude of induced roll, by (12)	23°	1°

Thus while the destroyer will be wallowing in the seaway the liner will hardly notice it. As Barnaby observes [2 p.337].

> "This is the great paradox of naval architecture – that the more stable the vessel really is, the more unstable she appears in a seaway."

He reinforces his observation with a revealing anecdote [2 p.355] that gives life to our computations :

> "This can be illustrated by the case of two yachts that were virtual sister ships, differing only in metacentric height. The captain of the first yacht reported her to be a magnificent sea-boat, extremely dry, and "stiff as a church" in a sudden squall [like our destroyer]. Her owner thought she was much too quick and lively in a seaway. As yachts have to be built for owners rather than for captains, the second vessel was given less metacentric height [like our liner]. The same captain transferred to the new yacht, and, in accordance with the best seafaring tradition, greatly preferred his old ship. His new command was wetter and more sluggish in a seaway, and, for these reasons, he considered her a worse sea-boat. The new owner was delighted, and said his yacht was "stiff as a church" in a seaway."

5. RESONANCE.

Before leaving the seaway we make some remarks concerning resonance. Suppose as before that the ship encounters waves of fixed period.

(i) If the wave-period, τ, happens to coincide with the natural roll-period, T, then the amplifying factor reaches a maximum. Because of damping this maximum is finite. Formula (13) gives $\lambda = \infty$, but this is wrong because damping has been ignored. If, as in (5), a damping term of $2\varepsilon\dot{\theta}$ is added to (9), then for $\tau = T$ the resonant solution has a $\frac{\pi}{2}$-phase-shift,

$$\theta = -\lambda\alpha_o \sin\frac{2\pi t}{T}$$

with amplifying factor $\lambda = \frac{\pi}{\varepsilon T}$. However the induced roll in this case may be so large that the linear theory is no longer valid. Indeed in heavy seas resonance may cause capsizing.

(ii) Let V, v denote the speeds of ship and wave. If the ship's course is such as to cause resonance we call it a <u>sensitive course</u>.

<u>Lemma 4.</u> <u>If $V > v(1+\frac{\tau}{T})$ there are 4 sensitive courses.</u>

<u>Proof.</u> Figure 6 shows how the period with which the ship encounters the waves depends upon the course. The blob represents the velocity of the waves, and the circle represents the different courses of the ship at speed V. The dashed line indicates the two courses perpendicular to the velocity of the waves (beam sea with period τ). The dotted lines indicate the two courses on which the ship will ride with the waves (period ∞). The other four lines indicate the sensitive courses. In order to encounter the waves with period T, the relative velocity of the ship must have a component $\pm\frac{v\tau}{T}$ in the direction of the waves. Therefore the ship's velocity is $v(1\pm\frac{\tau}{T})$ in this direction. The condition for the four

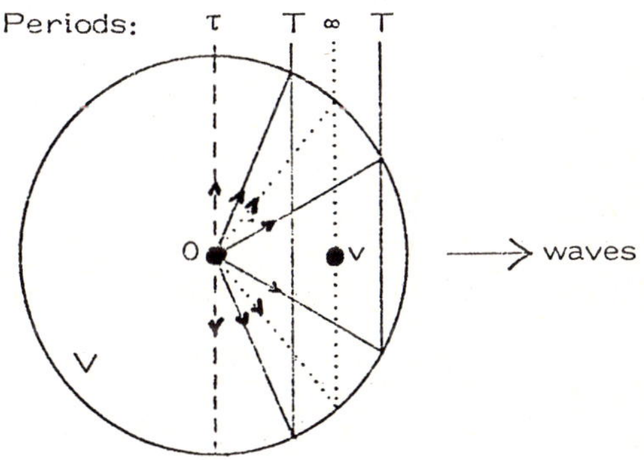

Figure 6. Sensitive courses.

directions to exist is that this component be less than V. This completes the proof. On the two sensitive courses closest to the wave direction the ship overtakes the waves, while on the other two ship is overtaken by the waves.

(iii) The non-linearity of the accurate rolling equation makes a wall-sided ship behave like a hard spring for small angles of roll, and a soft spring for large angles of roll (see Section 9 and Figure 19 below). The latter is liable to produce a Duffing effect [8,15] near resonance, as shown in Figure 7. If there is a Duffing effect, then when the ship reduces speed on a sensitive course overtaking the waves the amplitude of induced roll may at first increase slightly, and then drop suddenly when a critical speed V_1 is reached. Conversely, if the ship increases speed again a hysteresis delay will occur, during which the roll will remain deceptively small, until a higher critical speed V_2 is reached,

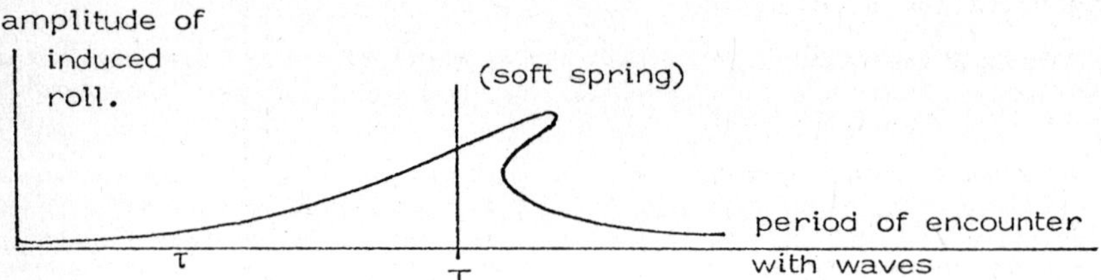

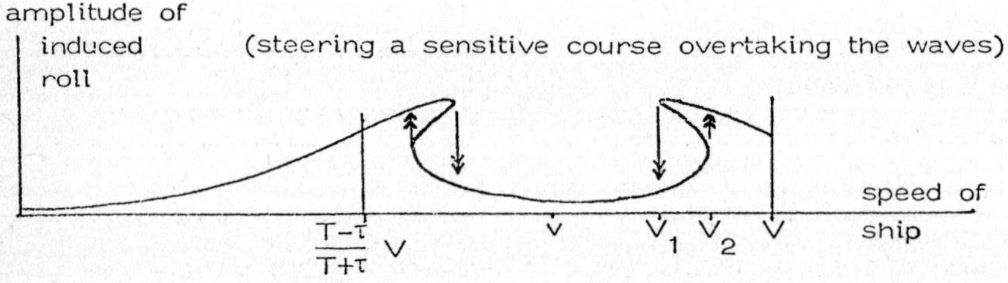

Figure 7. Duffing effects.

when the amplitude will suddenly increase again. Such a catastrophic jump could be dangerous because the dynamic stability of the resulting resonance might lead to capsizing before remedial action had time to take effect. Similar catastrophes occur near $(\frac{T-\tau}{T+\tau})V$, which in the case of the liner above equals $\frac{2}{3}V$. Note these catastrophes are different from those in the main model in Sections 7 - 14 below.

6. PITCHING AND HEAVING.

Passing from the 2-dimensional problem to the 3-dimensional problem, the buoyancy locus $\mathcal{B}$ becomes a convex closed surface rather than a convex closed curve. Therefore $\mathcal{B}$ has two principle directions of curvature at B_0, and two centres of curvature. One is the metacentre M for rolling about the longitudinal axis that we have already discussed, and the other is the metacentre M^* for pitching about the transverse axis. We now estimate the period of pitching, using the same notation as before, only with asterisks.

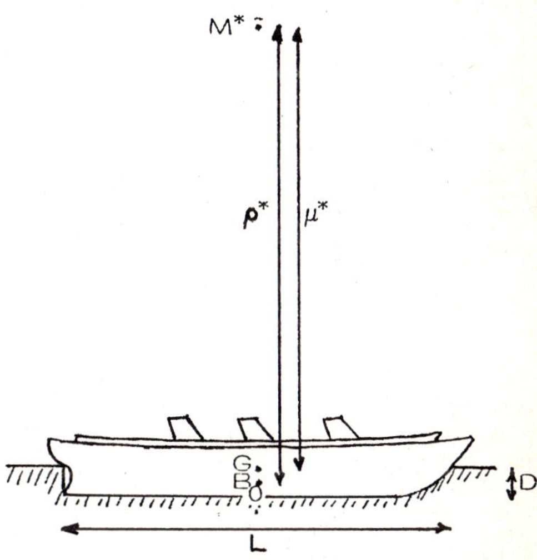

Figure 8. Pitching metacentre.

For simplicity assume (i) that the area below water is a rectangle with length L and draught D. Then by Lemma 2 the radius of curvature is

$$\rho^* = \frac{2(L/2)^3}{3LD} = \frac{L^2}{12D}.$$

Assuming (ii) that $L > 24D$, then $\rho^* > 2L \gg B_0G$, and therefore B_0G can be ignored in the estimate of μ^* :

$$\mu^* = \rho^* = \frac{L^2}{12D} \quad \ldots\ldots\ldots\ldots\ldots \text{(14)}$$

Assume (iii) that the moment of inertia about the transverse axis is the same as that of a rod of length L. Therefore

$$I^* = \frac{W}{g}\frac{(L/2)^2}{3} = \frac{WL^2}{12g} \quad \ldots\ldots\ldots (15)$$

By (4) the period of pitching is

$$T^* = 2\pi\sqrt{\frac{I^*}{W\mu^*}}$$

$$= 2\pi\sqrt{\frac{D}{g}} \quad \text{by(14) and (15)} \quad \ldots . (16)$$

$$= 2\sqrt{D}, \text{ approximately.}$$

Applying this to our two ships :

		Destroyer	Liner
Assume	Length, L	100m	300m
	Draught, D	3.3m	10m
Deduce	Pitching metcentric height μ^*, by (14)	250m	750m
	Period of pitching T^*, by (16)	4 secs	6 secs

Since μ^* is several hundred times larger than μ, the ship is far more stable with respect to pitching than to rolling, and hence the period is short, and amplitude kept small. The accelerations involved may be greater, and hence pitching is sometimes more uncomfortable than rolling.

Heaving refers to oscillations up and down. Let q denote the height of the ship above the equilibrium position. For a wall-sided ship of draught D, the volume of displaced water is reduced by a factor of approximately $\frac{q}{D}$.

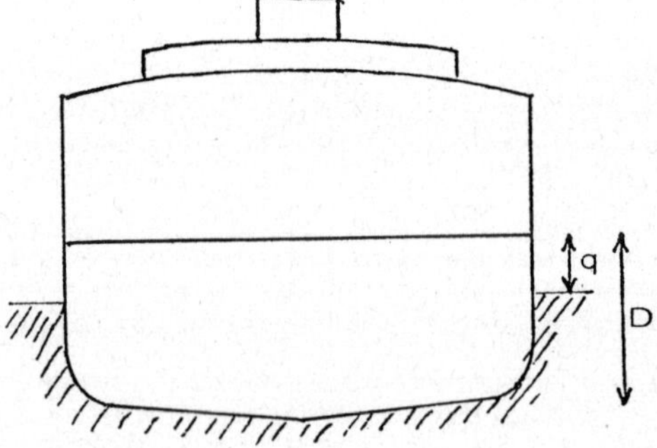

Figure 9. Heaving.

Therefore the buoyancy force is reduced by the same factor, and so there is a net downward restoring force of $\frac{Wq}{D}$. Therefore by Newton's law

$$\frac{W}{g}\ddot{q} = -\frac{Wq}{D}.$$

Therefore $$\ddot{q} = -\frac{g}{D}q \quad \ldots\ldots\ldots\ldots\ldots\ldots\ldots (17)$$

Hence the period for heaving is the same as that for pitching (16). In practice of course the periods differ slightly, because our assumptions are too imprecise. However the proximity of the periods implies that pitching and heaving will be coupled, and the classical theory of the coupling originated with Krilov in 1893 (see [3]).

The other 3 normal modes of oscillation, yawing (about the vertical axis), swaying (from side to side) and surging (fore and aft) differ from rolling, pitching and heaving in that buoyancy does not provide a natural restoring force; therefore these modes tend to occur only as induced, or secondary effects. We express the difference in group-theoretic terms in Lemma 12 below. This completes our elementary sketch of the classical linear theory, and we now begin the catastrophe theory, which is the main business of the paper.

7. CUSP CATASTROPHE AT THE METACENTRE.

For simplicity return to the 2-dimensional problem of rolling. For large angles the linear theory is no longer valid because the buoyancy force no longer goes through M. We need to look at not just one centre of curvature, but all of them. Therefore define the <u>metacentric locus</u> $\mathfrak{M}$ of the ship to be the locus of centres of curvature of the buoyancy locus; in other words $\mathfrak{M}$ is the evolute of $\mathfrak{B}$.

Now B_0 is a point of symmetry of $\mathfrak{B}$ where the radius of curvature is stationary, and hence M is a cusp point of $\mathfrak{M}$. The geometric question arises : Which way does the cusp branch, upwards or downwards? In Figure 10 we show two cases arising from different shapes of hull, on the left a modern wall sided ship, and on the right an old-fashioned canoe, shaped like an ellipse with major axis horizontal.

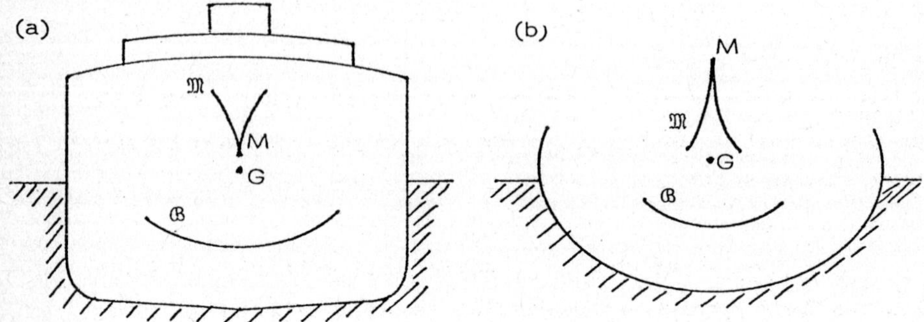

Figure 10. Cusps in ship (a) and canoe (b).

Lemma 5. The cusp branches upwards in the ship (a) and downwards in the canoe (b).

Proof. In a wall-sided ship 𝔅 is locally a parabola $x^2 = 2\rho y$, by Lemma 3. This has evolute

$$27\rho x^2 = 8(y-\rho)^3 ,$$

which is a cusp branching upwards. The result for an elliptical-shaped hull follows from the Corollary to Lemma 7 below.

We will now explain the physical significance of which way the cusp branches. Suppose that the position of the centre of gravity of the ship changes for some reason : for example the canoeist might lean over the side, or stand up, or put up a mast. In a liner the passengers might crowd to one side to see something interesting. A cargo boat might load or unload, or restow its cargo (see Figures 27 and 28 below). In heavy weather the cargo might shift* by itself, or slosh about if liquid, or the ship might

---

* Capsizing and the shifting of cargo are still hazards. During 1975 according to Lloyds Casualty Return [10] 125 merchant ships foundered mostly in heavy weather (not to mention another 211 lost, missing, burnt, wrecked, or in collison). Of the 125 foundered 13 are known to have capsized and sunk, 14 others developed a list before sinking, and in 9 cases the list was known to be due to the cargo shifting.

accumulate ice to windward, or sea-water on deck. Fishing vessels may be tempted by a good catch to take on more than is advisable. For simplicity assume for the moment that the position of G changes without altering the total weight, so that the buoyancy locus remains the same (in Section 14 we allow for change of weight).

Question : given the position of G, at what angles can the ship float in equilibrium?

Answer : by Lemma 1 it will be those angles θ, such that G lies on the normal N_θ to $\mathcal{B}$ at B_θ. But the normals to $\mathcal{B}$ are tangents to $\mathfrak{M}$; therefore the angles are obtained by drawing tangents from G to the cusp.

In Figure 11 we plot the graph of θ as a function of G, for the two boats pictured in Figure 10. In each case the position of G is represented by a point in the horizontal plane, C. The value of θ is represented by a point on the vertical axis $\mathbb{R}$, and for simplicity we assume $|\theta| < \beta$, where β is some suitable bound (in the full model in Section 10 below, we allow θ to be arbitrary). On the vertical line above each position of G we plot the corresponding equilibrium values of θ, and as G varies these points trace out a smooth surface, which we call the equilibrium surface, E. We shall prove in Theorem 2 below that in each case E is a cusp-catastrophe. By definition

$$E - \{(C,\theta);\ C \in N_\theta,\ |0| < \beta\}$$
$$= \{N_\theta \times \theta;\ |\theta| < \beta\}, \subset C \times \mathbb{R}\ .$$

Therefore E is a smooth ruled surface, consisting of horizontal lines parallel to the normals, one for each θ. In other words E is the normal bundle of $\mathcal{B}$. Over the outside of the cusp E is single-sheeted because if G lies outside (as in Figure 11(a)) there is only one tangent from G to the cusp. On the other hand over the inside of the cusp E is triple-sheeted, because if G lies inside (as in Figure 11(b)), there are three tangents from G to the cusp.

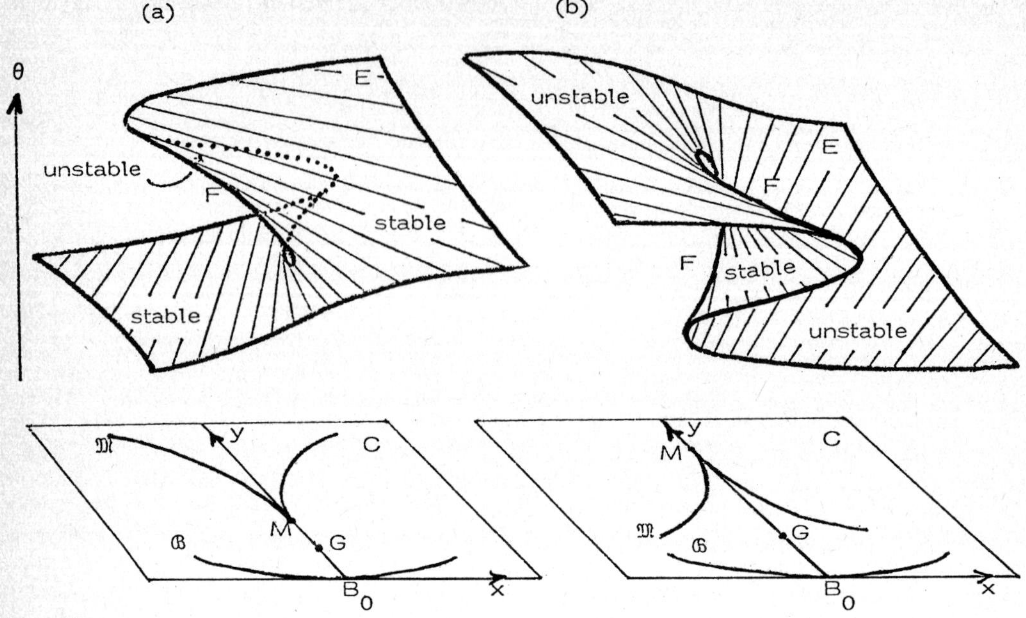

Figure 11. The cusp-catastrophe in the ship (a) and the canoe (b).

If E is projected down onto C it becomes folded along a curve, called the fold curve F, which projects onto the cusp. Hence the cusp is a bifurcation set. F separates E into two components, one representing stable equilibria and the other unstable equilibria. For example if G lies on the axis of symmetry (the y-axis) then by equation (2) in Section 1 the equilibrium position $\theta = 0$ is stable or unstable according as to whether the metacentric height μ is positive or negative, in other words whether G is below or above M.

If, further, G lies inside the cusp then the $\theta = 0$ equilibrium is represented by a point on the middle sheet of E, whilst the other two equilibria are represented by points on the upper and lower sheets of E. We call these heeling angles if the cusp branches upwards, as in case (a), and capsizing angles if the cusp branches downwards, as in case (b).

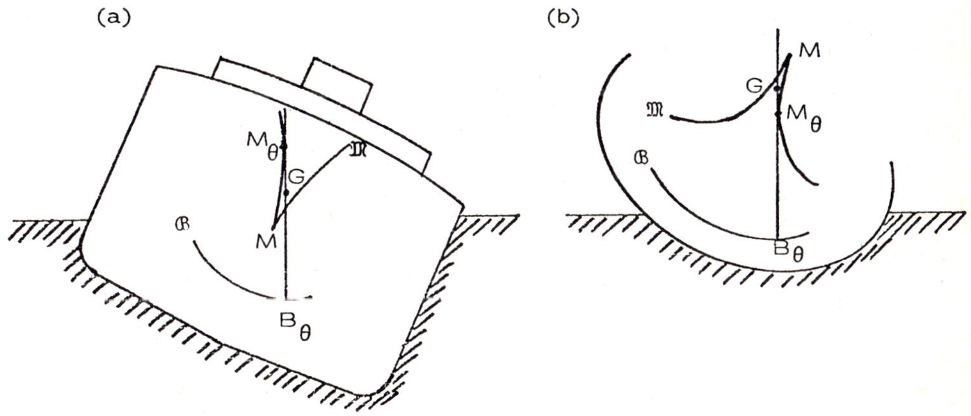

Figure 12 (a) Heeling angle. (b) Capsizing angle.

Here the difference between the two boats becomes apparent because :

Lemma 6. Heeling angles are stable, whereas capsizing angles are unstable.

Proof. Let θ be a heeling or capsizing angle. If M_θ denotes the corresponding metacentre, the centre of curvature of $\mathcal{B}$ at B_θ, then the tangent from G touches the cusp at M_θ. Figure 12 shows that in case (a) G lies below M_θ, and so the ship behaves stably like a pendulum hanging from M_θ. By contrast in case (b) G lies above M_θ, and so the canoe behaves unstably, balanced precariously over M_θ; any perturbation reducing θ will produce a righting couple that returns the canoe upright, whereas any perturbation the opposite way will produce an opposite couple, that will cause the canoe to turn turtle.

Therefore in Figure 11(a) the upper and lower sheets of E are stable, while the middle sheet over the inside of the cusp is unstable. In Figure 11(b) it is the other way round, and so in this case E is called a dual cusp-catastrophe. The difference is emphasised in Figure 13 which shows in each case the section of E over the y-axis, rotated through 90° so as to make the y-axis vertical. The section in each case consists of the line $\theta = 0$ bisecting a curve (which is a parabola modulo θ^4). The stable equilibria are shown firm, and the unstable dotted. In case (a) the curve is stable and rising and represents heeling angles, while in case (b) it is unstable and falling and represents capsizing angles. The specific angles are given by the intersection of the curve with the horizontal line through G.

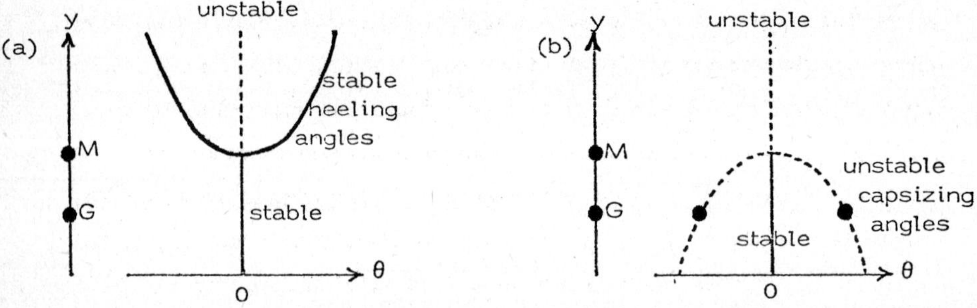

Figure 13. Sections of E over the y-axis for (a) ship and (b) canoe.

Now imagine an experiment in which the centre of gravity is raised up the y-axis past M. In case (a) when G passes M the ship will heel over to one side or the other, and will settle into stable equilibrium at the heeling angle. This happens for instance in certain old cargo boats when they unload, because they used to be designed with negative metacentric height when empty (see

[2, p.71] and Figures 27, 28 below). It also tends to happen in toys like model gondolas for the same reason, and sometimes the heel can be corrected by loading the model with a little ballast.

Suppose that G is on the y-axis above M with the ship heeled to the right, represented by a point on the upper sheet of Figure 11(a). If G is now moved to the left the ship will stay heeled to the right until G crosses the left side of the cusp, when it will suddenly heel over to the left; this is represented in Figure 11(a) by the point crossing the fold curve and jumping catastrophically onto the lower sheet. Conversely if G is now moved to the right then the ship will delay until G crosses the right side of the cusp when it will suddenly heel back again. Therefore in case (a) the cusp is a <u>heeling bifurcation set.</u>

By contrast case (b) is more dangerous because if G is raised up the y-axis past M the canoe will suddenly capsize. Worse still, if G happens to be off-centre when it is raised then it will cross the cusp sooner because the cusp branches downwards, and hence capsize sooner. This imperfection-sensitivity explains why when standing up in a canoe it is advisable to keep perfectly in the centre. Similarly if G is moved sideways, then as soon as G crosses either side of the cusp the canoe will capsize. A graphic description of such an event is given by Gerald Durrell [6, p.163].

> "Peter nodded, braced himself, clasped the mast firmly in both hands, and plunged it into the socket. Then he stood back, dusted his hands, and the Bootle-Bumtrinket, with a speed remarkable for a craft of her circumference, turned turtle."

Evidently raising the mast raised G close to the metacentre where the cusp was very narrow, and stepping back was sufficient to cause G to cross the cusp - or maybe only just to reach the cusp, and it was actually the dusting of the hands that gave the final perturbation across it. Durrell goes on to explain how the problem was solved :

> "For the rest of the morning he kept sawing

bits off the mast until she eventually floated upright, but by then the mast was only about three feet high."

Summarising :

Theorem 1. An upward branching cusp is a heeling bifurcation set, and a downward branching cusp is a capsizing bifurcation set.

Notice that this theorem only refers to the statics, because although we have used local righting couples to determine the local nature of the stability, the global dynamics has been ignored. We shall return to the dynamics again in Section 10.

8. GLOBAL METACENTRIC LOCUS.

We have yet to prove that the cusp in a canoe branches downwards. The easiest way to tackle this is to investigate the global metacentric locus of a completely elliptical ship, (like a submarine before it submerges).

Lemma 7. The buoyancy locus of an ellipse is a similar ellipse.

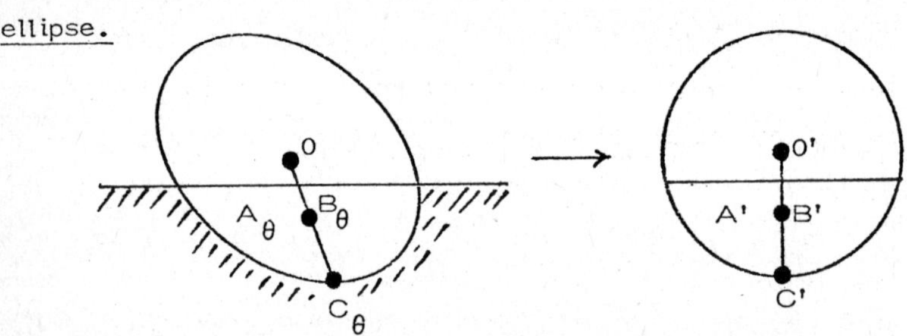

Figure 14.

Proof. Let O denote the centre of the ellipse, and C_θ the lowest point when heeled at angle θ. The line OC_θ bisects all horizontal chords, therefore bisects the region A_θ below the water line, and hence contains the centre of buoyancy B_θ. Map the ellipse onto a circle by an affine area-preserving map, and let A',B',C',O' denote the images of $A_\theta, B_\theta, C_\theta$, O. Then B' is the centre of gravity of A', and since the area of A' is independent of θ

the ratio $O'B'/O'C'$ = constant, k say, independent of θ. Since the map is affine $OB_\theta/OC_\theta = k$. Therefore, as C_θ traces out the ellipse, B_θ traces out a similar ellipse k times the size. This completes the proof. Notice that the result is independent of the weight (or density) of the ship.

Corollary. The metacentric locus of an ellipse has 4 cusps as shown in Figure 15.

For it is merely the evolute of the buoyancy locus, which by the lemma is a similar ellipse. In particular this completes the proof of Lemma 5, for in an elliptical shaped canoe with major axis horizontal, the metacentre M is the topmost cusp, and hence the cusp branches downwards.

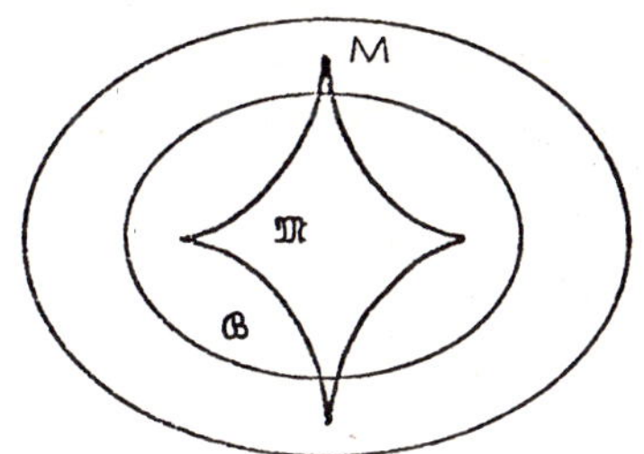

Figure 15.

Remark. In the 3-dimensional situation exactly the same proof shows that the buoyancy locus of an ellipsoid is a similar ellipsoid. The evolute of an ellipsoid, however, is more difficult to visualise because it consists of two sheets corresponding to the two metacentres, one for rolling and one for pitching. It can be regarded as 2 spheres, pinched along 3 elliptical cusped edges, one of which contains 4 hyperbolic umbilics [5].

Theorem 2. The equilibrium surface E has a cusp catastrophe at the metacentre M.

Proof. If the buoyancy locus $\mathfrak{B}$ is generic, then from general theory its evolute $\mathfrak{M}$ will have a generic cusp at M, and its normal bundle E will have a cusp catastrophe. However we cannot be sure that the curves in question are generic without checking the explicit formulae for wall-sided and elliptical ships.

The equation of E in (x,y,θ)-space is formally the same as that of the normal N_θ in (x,y)-space, with the proviso that θ is reinterpreted as a coordinate rather than a parameter. In case (a) of the wall-sided ship $\mathcal{B}$ is locally a parabola

$$x^2 = 2\rho y,$$

by Lemma 3. The normal N_θ is given by

$$x + (y-\rho)\tan\theta - \frac{1}{2}\rho\tan^3\theta = 0.$$

As a surface this is differentially equivalent (in the sense of [13]) to

$$x + (y-\rho)\theta - \theta^3 = 0,$$

which is a canonical cusp catastrophe at $(0,\rho)$ with x,y as normal and splitting factors.

In case (b) of the canoe, $\mathcal{B}$ is an ellipse by Lemma 7, and the equation of the ellipse with radius of curvature ρ at the origin and eccentricity e (where e is the ratio of the vertical axis to horizontal axis) is :

$$x^2 + \left(\frac{y}{e}\right)^2 = 2\rho y.$$

The normal N_θ is given by

$$x + (y-\rho)\tan\theta + \rho(1-e^2)\tan\theta\,[1-e(e^2+\tan^2\theta)^{-\frac{1}{2}}] = 0$$

Since $e < 1$, this is differentially equivalent, as a surface, to

$$x + (y-\rho)\theta + \theta^3 = 0,$$

which is a canonical cusp-catastrophe at $(0,\rho)$ with $-x,-y$ as normal and splitting factors. This completes the proof.

Remark. If in case (b) the eccentricity e is increased until $e > 1$, this converts the horizontal axis of the ellipse into the minor axis, and changes the sign of θ^3, converting the cusp from downwards branching into upwards branching, as in case (a).

The question now arises : what is the complete metacentric locus for a modern wall-sided ship? Where does $\mathfrak{M}$ go to after the initial upward branching of the cusp? How does $\mathfrak{M}$ compare globally with the 4-cusp evolute of an elliptical hull shown in Figure 15? We start by looking at a rectangle :

Theorem 3. In a rectangular hull of density ½ the buoyancy locus is the union of 4 pieces of parabolas, and the metacentric locus has 8 cusps.

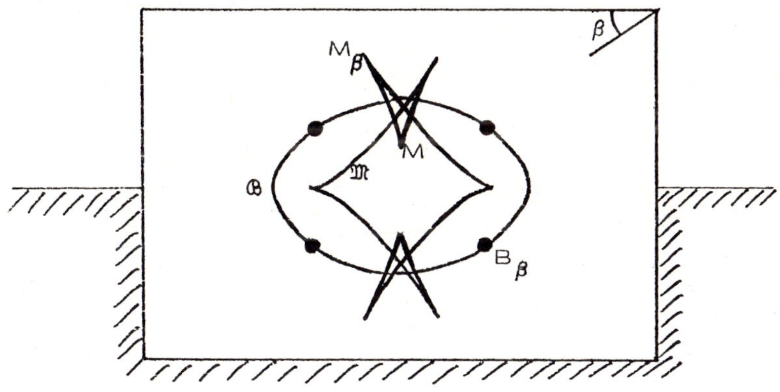

Figure 16. The metacentric locus of a rectangle.

Proof. The rectangle is a wall-sided ship for $|\theta| < \beta$, where β is the inclination of the diagonal to the horizontal. Therefore for $|\theta| < \beta$, 𝔅 is a piece of a parabola by Lemma 3, and contributes an upwards branching cusp to 𝔐 by Lemma 5. There are 4 pieces, corresponding to the 4 sides of the rectangle. Two pieces of 𝔅 join at B_β, and here the two parabolas have the same tangent by Lemma 1, the same radius of curvature by Lemma 2, and hence the same centre of curvature M_β. Therefore two pieces of 𝔐 touch at M_β, producing a parabolic cusp. Therefore 𝔐 is continuous, containing 4 generic cusps (of index $\frac{3}{2}$) separated by 4 non-generic parabolic cusps (of index 2), as shown in Figure 16.

Remark 1. If the density is reduced (or increased) the 4 parabolas in 𝔅 are separated by 4 pieces of rectangular hyperbolas. When the density reaches ½tan β then 4 swallowtails appear giving raise to another 8 cusps in 𝔐, making 16 in all (see Figures 25 and 28 below).

Remark 2. The non-genericity of the 4 parabolic cusps is due to the non-smoothness of the corners of the rectangle. If the corners are rounded-off in a C^∞-fashion, then the 4 parabolic cusps become generic, and the qualitative shape of $\mathfrak{M}$ is stable under small perturbations. Now the cross-section of a large modern ship can be regarded as a perturbation of a rounded-off rectangle. Therefore :

Conjecture. Large modern ships have metacentric locus $\mathfrak{M}$ similar to that in Figure 16. Detailed computations for individual ships show the top three cusps [11, p.135].

Remark 3. The evolution of hull shape from ellipse to rounded rectangle will cause a bifurcation of $\mathfrak{M}$ from the 4 cusps in Figure 15 to the 8 cusps in Figure 16. The reader familar with the elementary catastrophes will recognise immediately canonical sections of the butterfly catastrophe [12].

Problem. Prove, for an explicit isotopy of hull shape, that the bifurcation is an unbiased butterfly, in other words is equivalent to the symmetry section of the butterfly catastrophe given by putting the bias factor (the coefficient of θ^3) equal to zero. The governing potential at the bifurcation point should be

$$-k\theta^6 + i\theta^4 + \frac{1}{2}(\rho-y)\theta^2 - x\theta,$$

where x,y are coordinates of G, ρ the radius of curvature, $k > 0$, and i the isotopy parameter running from $i < 0$ for the canoe to $i > 0$ for the ship. The unbias is due to the symmetry of the ship, and a full butterfly should be obtained by introducing a bias factor measuring lop-sidedness of hull.

Remark 4. Globally the bifuraction set of the modern ship is not so very different from that of a canoe, and therefore the ship is not so safe as Theorem 1 would at first sight suggest. Therefore it is necessary to take another qualitative look at the heeling and capsizing. We assume the ship has metacentric locus similar to Figure 16, as conjectured above.

Theorem 4. The only heeling part of the metacentric locus $\mathfrak{M}$ is the cusp at M, shown dotted in Figure 17(a); the rest is capsizing. The equilibrium surface E is a section of a dual* butterfly catastrophe, as shown in Figure 17(b). The stable equilibria are shown shaded. Therefore for stability G must lie below $\mathfrak{M}$.

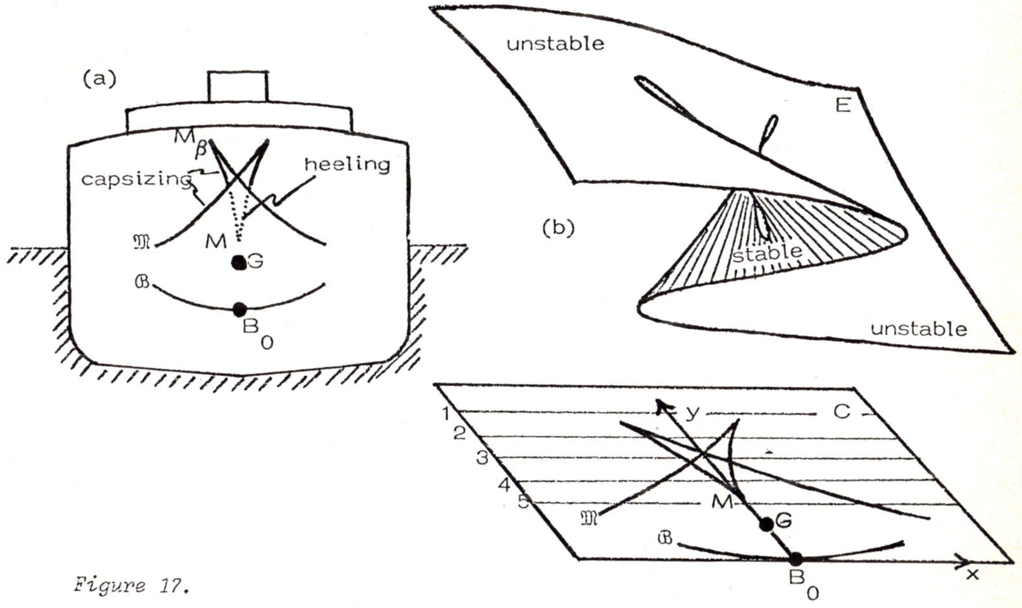

Figure 17.

(a) Metacentric locus $\mathfrak{M}$ *is part heeling (dotted) and part capsizing (firm).*

(b) Equilibrium surface E *is a section of a butterfly catastrophe.*

Proof. The bifurcation set in Figure 16 determines that E is a butterfly section, as shown in Figure 17(b) (see [12,13]). The identification of stable and unstable components of E is deduced from Figure 11(a), which is a subset of Figure 17(b). Hence E is a dual butterfly. The heeling and capsizing parts of $\mathfrak{M}$ are determined by consideration of the 5 sections of E over the 5 lines

---

* The dual butterfly [13] has germ $-\theta^6$, as opposed to the butterfly which has germ $+\theta^6$. This is the only application I know of the dual butterfly.

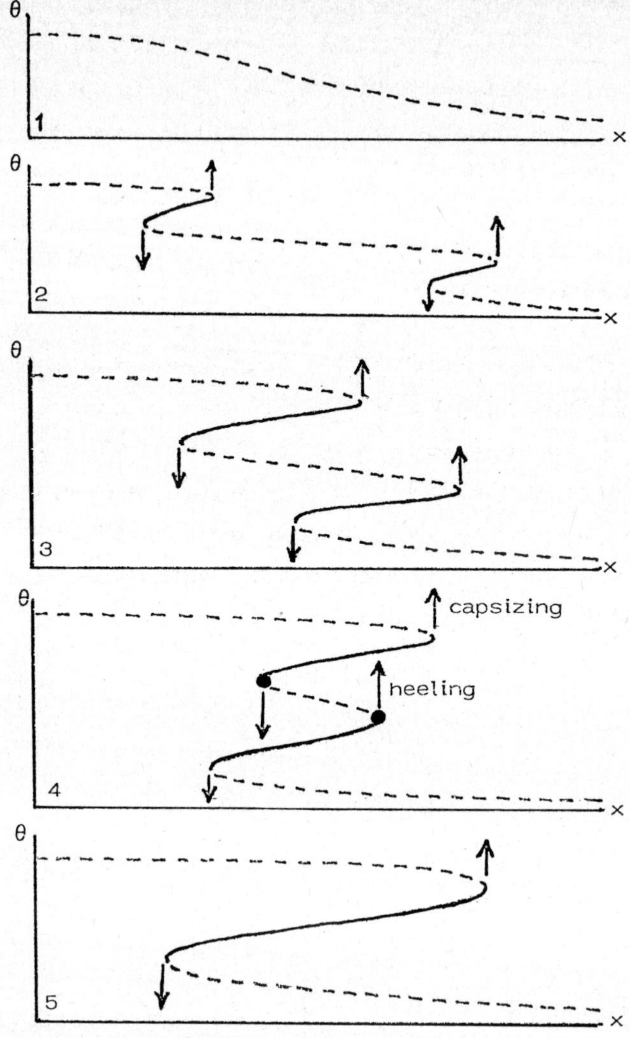

Figure 18. Sections of E.

parallel to the x-axis in Figure 17(b), as follows. The 5 sections are shown in Figure 18, with firm lines indicating stable equilibria, and dashed lines unstable equilibria. The nature of the stability determines which way the couple acts upon θ, and hence determines the direction of the catastrophic jump at each fold point, as indicated by the arrows. The catastrophe is heeling if the arrow tip stands on another stable sheet, and this only occurs for the middle two arrows of the fourth section. Hence the roots of those arrows (indicated by blobs) are the only two heeling parts of $\mathfrak{M}$. All the other parts of $\mathfrak{M}$ induce capsizing catastrophes. This completes the proof of Theorem 4.

9. LEVER ARM CURVE.

Assume G fixed. Recall that ℓ denotes the lever arm of the righting couple (see Figure 1(b)). The graph of ℓ for $0 \leq \theta \leq \frac{\pi}{2}$ is called the lever arm curve, and is illustrated in Figure 19 for the two boats shown in Figure 10. The slope of the lever arm curve at the origin is equal to the metacentric height, μ, because for small θ the linear approximation of the lever arm is $\ell = \mu\theta$.

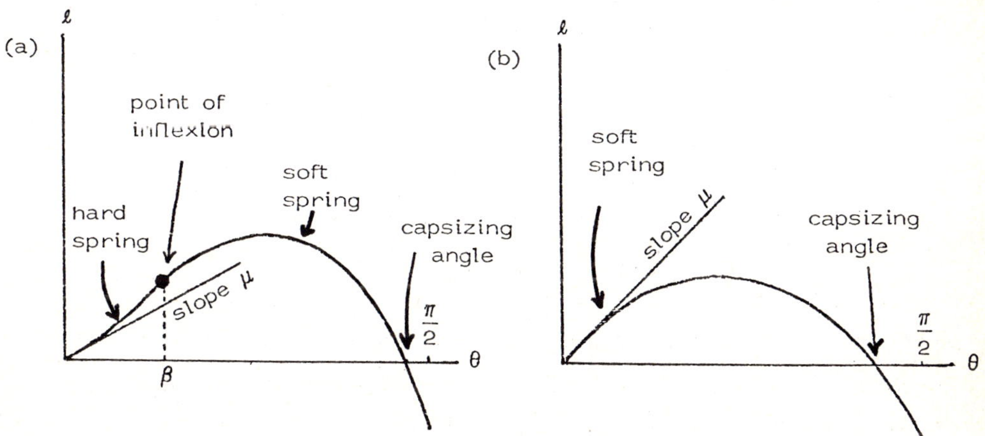

Figure 19. Lever arm curves for (a) ship and (b) canoe.

Lemma 8. In a wall-sided ship (a) the curvature of the lever arm curve is initially positive (like a hard spring), whereas in a canoe (b) it is negative (like a soft spring).

Proof. In the case of a wall-sided ship the normal N_θ is given by

$$x \cos\theta + (y-\rho)\sin\theta - \frac{1}{2}\rho \tan^2\theta \sin\theta = 0$$

by the proof of Theorem 2. Hence the distance ℓ from $G = (0,\rho-\mu)$ to N_θ, choosing the sign to be positive, is

$$\ell = \mu \sin\theta + \frac{1}{2}\rho \tan^2\theta \sin\theta$$

$$= \mu\theta + (\frac{\rho}{2} - \frac{\mu}{6})\theta^3 + O(\theta^5) .$$

Since $\mu < \rho$ the coefficient of θ^3 is positive, and hence for θ positive and small, the curvature is positive. In the canoe the normal N_θ is given by

$$x \cos\theta + (y-\rho)\sin\theta + \rho(1-e^2)\sin\theta[1-e(e^2+\tan^2\theta)^{-\frac{1}{2}}] = 0 .$$

Hence

$$\ell = \mu \sin\theta - \rho(1-e^2)\sin\theta[1-e(e^2+\tan^2\theta)^{-\frac{1}{2}}]$$

$$= \mu\theta - \theta^3[\frac{\rho}{2} e^{-2}(1-e^2) + \frac{\mu}{6}] + O(\theta^5) .$$

Since $e < 1$ the coefficient of θ^3 is negative, and hence, for θ small and positive, the curvature is negative.

Remark 1. In Figure 19 the difference in sign between the initial curvatures can be intuitively explained by which way the cusps branch in Figure 10. For in case (a) the upward branching causes the hard spring, while in case (b) the downward branching causes the soft spring. To be precise the conditions are slightly different : the cusp branches up or down as $e \gtrless 1$, whereas the curvature is hard or soft as $e \gtrless (1-\mu/3\rho)^{-\frac{1}{2}}$, and since $\mu < \rho$ this constant lies between 1 and $\sqrt{3/2}$.

Remark 2. In Figure 19(a) the change of curvature from hard to soft at the point of inflexion can be intuitively explained by the butterfly section in Figure 16. In the case of a rectangular hull, the initial hard spring is caused by the upward branch MM_β of $\mathfrak{M}$, the point of inflexion occurs at the angle β of the cusp M_β, and the subsequent soft spring is caused by the subsequent downward branch of $\mathfrak{M}$, to the capsizing angle, as indicated in Figure 12(b). In the wall-sided ship Figure 17 the same holds, except that the smoothness of hull causes the angle at which the inflexion occurs to be displaced slightly below that at which the cusp occurs.

Remark 3. The dynamical importance of the lever arm curve was first recognised by Atwood in 1796. Its use for judging stability in the design of ships was first proposed by Reed in 1868, and today various key features of the curve are widely used as stability criteria by naval architects and marine authorities [9]. What is new in this paper is the relationship between seemingly ad hoc features of the curve and generic properties of the metacentric locus arising from canonical sections of the butterfly catastrophe.

10. CATASTROPHE MODEL FOR ROLLING.

In Sections 2 – 6 we have discussed the local linear dynamics, and in Sections 7 – 9 the non-linear statics. We now weld the two together in order to study the global non-linear dynamics.

Definition : An elementary catastrophe model* is a parametrised system of gradient-like differential equations, specified by four things :

(i) a parameter space C,

(ii) a state space X,

(iii) an energy function $H: C \times X \to \mathbb{R}$, and

(iv) a dynamic D on X, parametrised by C, that locally minimises H.

---

* In the language of [14] this is at structure level 2.

The function H determines the equilibrium manifold, $E \subset C \times X$, by the equation $\nabla_X H = 0$. The catastrophe map $\chi: E \to C$ is induced by projection. The bifurcation set is the image in C of the singularities of χ. If H is generic then E has the same dimension as C, and the only singularities of χ are elementary catastrophes, by the classification theorem [12,13].

We first construct the model for the 2-dimensional rolling problem only, where it is easy to understand and visualise, and then in subsequent sections extend it to 3-dimensions to include pitching, heaving and loading.

(i) Define the parameter space to be the plane C containing our 2-dimensional ship. The parameter $G \in C$ is the position of the centre of gravity of the ship (relative to the hull).

(ii) Define the configuration space to be the unit circle, S. The configuration of the ship is uniquely determined by the angle $\theta \in S$. Define the state space, $X = T^*S$, to be the cotangent bundle* of S. The state of the ship is given by $(\theta, \omega) \in T^*S$, where ω is the angular momentum. As before let

W = weight of ship

I = moment of inertia of ship and entrained water.

$h = h(G,\theta)$ = height of G above B_θ, at angle θ

$= ZB_\theta$, in Figure 1(b)

Lemma 9.

The potential energy of the system is: $P = P(G,\theta) = Wh$.

The kinetic energy of the system is: $K = K(\omega) = \dfrac{\omega^2}{2I}$.

Proof.

Let h_1 = height of G above the water line,

h_2 = depth of B_θ below the water line.

---

* For a general treatment of Hamiltonians on cotangent bundles see [1].

Then taking the water line as zero potential,

Wh_1 = potential energy of the ship,

Wh_2 = potential energy of the displaced water.

Therefore P = total potential energy $= Wh_1 + Wh_2 = Wh$.

The angular momentum $\omega = I\dot{\theta}$, and therefore as required

$$K = \text{kinetic energy} = \frac{1}{2}I\dot{\theta}^2 = \frac{\omega^2}{2I} .$$

(iii) Define the energy function of the model to be the Hamiltonian $H: C \times T^*S \to \mathbb{R}$ given by

$$H = P + K = Wh + \frac{\omega^2}{2I}$$

We can now deduce the equilibrium surface and bifurcation set from H, as follows.

Lemma 10. $\frac{\partial h}{\partial \theta} = \ell$.

Proof. Let M_θ, ρ_θ be the centre and radius of curvature of $\mathcal{B}$ at B_θ. Let $\mu_\theta = GM_\theta$, $\alpha = G\hat{M}_\theta B_\theta$. Then

$$h(G,\theta) = \rho_\theta - \mu_\theta \cos\alpha$$

$$h(G,\theta+\varphi) = \rho_\theta - \mu_\theta \cos(\alpha+\varphi) + O(\varphi^2)$$

$$\frac{\partial h}{\partial \theta} = \left[\frac{\partial}{\partial\varphi} h(G,\theta+\varphi)\right]_{\varphi=0}$$

$$= [\mu_\theta \sin(\alpha+\varphi) + O(\varphi)]_{\varphi=0}$$

$$= \mu_\theta \sin\alpha = \ell .$$

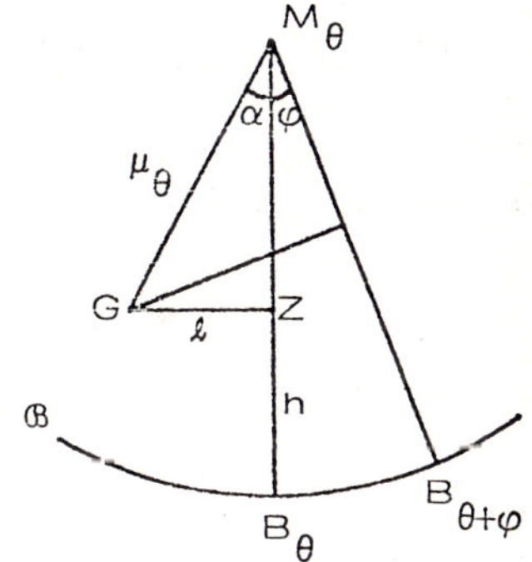

Figure 20.

The buoyancy locus $\mathcal{B}$ has coordinate θ, and is contained in the ambient space C. Therefore the normal bundle $N\mathcal{B}$ of $\mathcal{B}$ is defined by

$$N\mathcal{B} = \{N_\theta \times \theta\} \subset C \times S .$$

The geodesic spray is the natural map $N\mathcal{B} \to C$ of the normal bundle into the ambient space induced by projection $C \times S \to C$. The image of the singularities of the geodesic spray is the evolute of $\mathcal{B}$, which we have called the metacentric locus $\mathfrak{M}$.

Theorem 5. The equilibrium surface E is the normal bundle $N\mathcal{B}$ the buoyancy locus. The catastrophe map $\chi: E \to C$ is the geodesic spray. The bifurcation set is the metacentric locus, $\mathfrak{M}$.

Proof. E is given by $\nabla_X H = 0$, in other words by the equations

$$\frac{\partial H}{\partial \theta} = \frac{\partial H}{\partial \omega} = 0 .$$

Now $\frac{\partial H}{\partial \omega} = \frac{\omega}{I}$, and hence $\omega = 0$. Therefore $E \subset C \times S \subset C \times T^*S$, where S is identified with the zero-section of T^*S. Also $\frac{\partial H}{\partial \theta} = W\frac{\partial h}{\partial \theta} = W\ell$, by Lemma 10, and hence $\ell = 0$. Therefore $G \in N_\theta$, the normal to $\mathcal{B}$ at B_θ. Therefore

$$E = \{(G,\theta,0); G \in N_\theta\} = \{N_\theta \times \theta\} = N\mathcal{B} .$$

The catastrophe map χ merely says "forget θ", mapping each normal to itself, and giving the geodesic spray. Hence the bifurcation set, which is defined to be the image of the singularities of χ, equals $\mathfrak{M}$. This completes the proof of Theorem 5. To complete the model there remains to define the dynamic.

Assuming G fixed, the Hamiltonian dynamic on T^*S is uniquely determined from H by the intrinsic symplectic structure of the cotangent bundle (Newton's law of motion is built-in [1]). Explicitly the dynamic is given by the Hamiltonian equations

$$\dot{\theta} = \frac{\partial H}{\partial \omega} = \frac{\omega}{I}$$

$$\dot{\omega} = -\frac{\partial H}{\partial \theta} = -W\frac{\partial H}{\partial \theta} = -W\ell .$$

Therefore

$$I\ddot{\theta} = \dot{\omega} = -W\ell .$$

which is the same as equation (1) in Section 2. The resulting Hamiltonian flow is the accurate global non-linear generalisation of the approximate local simple harmonic rolling solution (3). However as yet we have not included any friction, because the Hamiltonian flow is conservative, conserving the energy H.

(iv) Define the dynamic D of the catastrophe model to be the Hamiltonian dynamic with non-zero damping. There is no need to be any more specific about the nature of the damping other than saying that energy is dissipated, because this ensures that H decreases along the orbits of D. Therefore H is a Lyapunov function for D. Therefore D locally minimises H, and depends upon the parameter G, as required. The model is complete.

11. GLOBAL DYNAMICS.

In order to understand the catastrophe dynamic D we fix G and draw the phase portrait of the resulting flow in Figure 21(c). The phase portrait is the family of orbits on T^*S. Since T^*S is a cylinder, we cut the cylinder along the generator $\theta = \pi = -\pi$, and lay it out flat, with the understanding that the two sides should be identified. In Figure 21 the dotted parts to the right of $\theta = \pi$ are merely the periodic repeats of the left hand sides of the portraits.

Before drawing the portrait of the damped flow, we first draw two portraits of the undamped Hamiltonian flow for two different positions of G in Figure 21(a) and (b). The latter is easier to understand because the Hamiltonian orbits are contained in the energy levels of H, which are themselves 1-dimensional since T^*S is 2-dimensional.

Figure 21(a) shows the Hamiltonian flow for G on the axis of symmetry below M, as in the case of the ship or the canoe in Figure 10. The 4 equilibria are given by $\omega = 0$ and

$\theta_2 = 0$, stable vertical

θ_1, θ_3, unstable capsizing angles (see Figure 12(b))

$\theta_4 = \pi$, stable turned turtle.

We are assuming that the ship does not sink if it capsizes, but is capable of floating upside down in stable equilibrium. The other three equilibria, $\theta_1, \theta_2, \theta_3$ correspond to the three points above G on the three sheets of E in Figure 11(b) for the canoe and

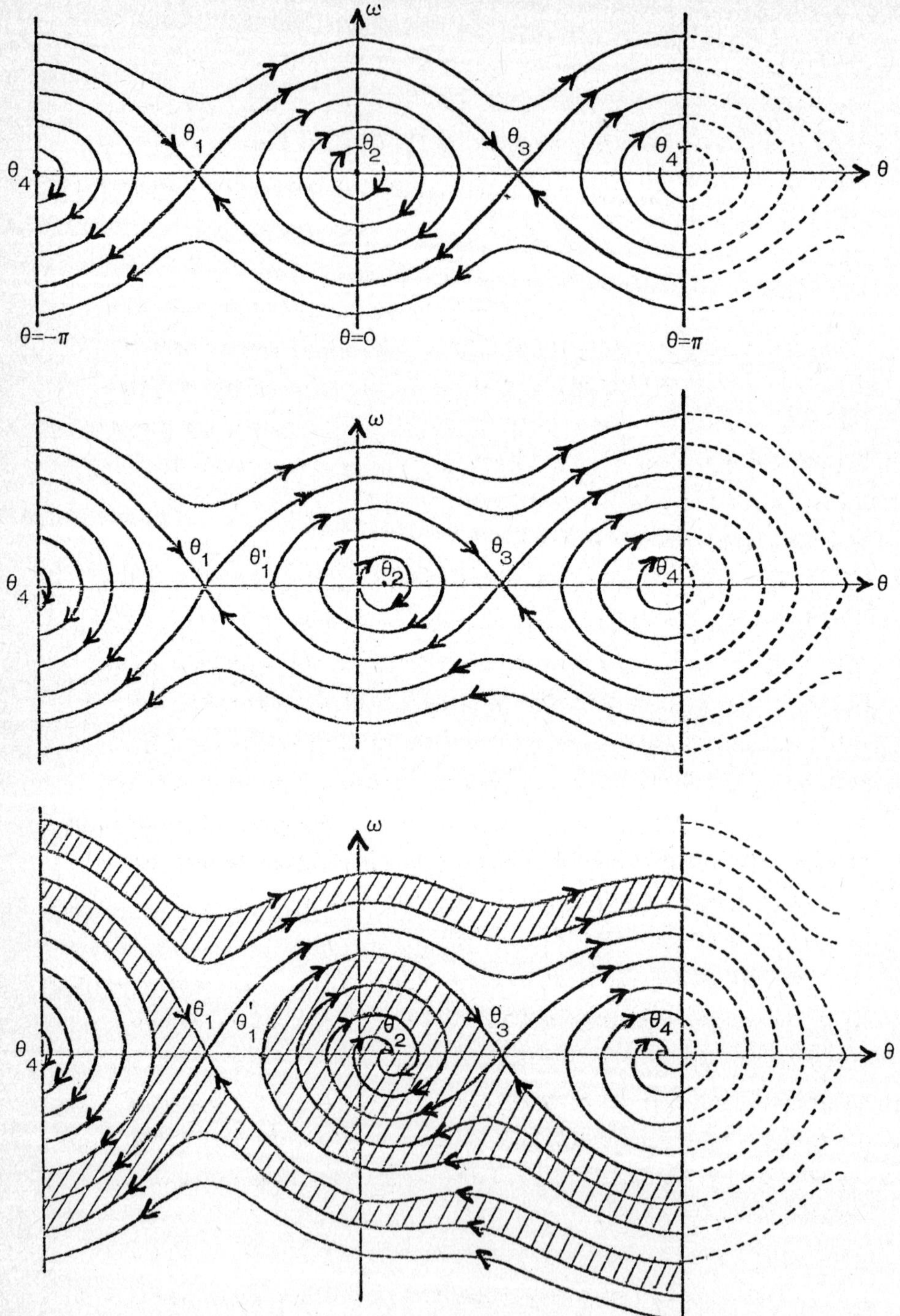

Figure 21. Phase portraits. (a) *Hamiltonian flow, G central.*
(b) *Hamiltonian flow, G offset to right.*
(c) *Damped flow, G offset to right.*

Figure 17(b) for the ship. The closed orbits round θ_2 represent rolling, as in the solution (3) of Section 2. The closed orbits round θ_4 represent rolling while turned turtle. The uppermost orbit going from left to right (which is closed since $\pi = -\pi$) represent the ship rolling over and over clockwise; the lowermost orbit represents the same but anticlockwise.

Figure 21(b) shows the Hamiltonian flow when G has been displaced slightly to the right. As a result, in stable equilibrium the ship will heel slightly to the right : $\theta_2 > 0$. Figures 11(b) and 17(b) show that the capsizing angles will also change slightly, θ_3 decreasing and $|\theta_1|$ increasing. Therefore it takes less energy to capsize to the right than to the left, because the θ_3-energy-level lies inside that of θ_1.

There is also a dynamic capsizing phenomenon, as follows. Let θ'_1 be the intersection of the θ-axis with the θ_3-energy-level, such that $\theta_1 < \theta'_1 < 0$. We call θ'_1 the dynamic capsizing angle. If the ship is displaced to angle θ such that $\theta_1 < \theta < \theta'_1$, then it will not capsize to the left, but on the return roll to the right it will roll right over and then capsize, because the orbit will lie outside the θ_3-energy-level. Consequently, from the point of view of resonance, the dynamic capsizing angle is more dangerous than the static capsizing angle, because it is smaller

$$|\theta'_1| < \theta_3 < |\theta_1| .$$

On the other hand, if the ship rolls to the right almost as far as θ_3, then the recovery will take much longer, because, for small ε, the time to roll back from $\theta_3-\varepsilon$ upright again will vary as $|\log \varepsilon|$, as can be seen by considering the linear approximation at θ_3. Therefore the ship will tend to hang perilously in the brink of capsizing for a long time, vulnerable to the chance wave or squall that might tilt her over the brink.

Figure 21(c) shows the damped flow for the same position of G. Since the dynamic is now dissipative the two stable equilibria

θ_2 and θ_4 are now attractors. If this figure was superimposed upon the figure above, all the damped orbits would cross the energy levels towards the attractors. The basin of attraction of the upright attractor θ_2 is shown shaded, and the complement is the basin of attraction of the turned turtle attractor θ_4. The capsizing angles are qualitatively unchanged since, being saddle-points of the Hamiltonian flow, they were already structurally stable. Similarly all the remarks about capsizing hold good.

Problem. What is the best way to introduce the effect of wind and wave into the model, so as to generalise the induced linear rolling of Section 4 to the non-linear situation? The effect of periodic waves can be simulated by introducing a new cyclic parameter that translates the phase-portrait to and fro parallel to the θ-axis, corresponding to adding a forcing term representing the varying water surface, as in Figure 5 and equation (9). The induced roll is then an attracting closed orbit lying over the parameter-cycle. Perhaps the effect of the wind could be similarly modelled by translating the phase-portrait parallel to the ω-axis, to simulate the impulse transmitted by the wind into angular momentum.

12. MODEL INCLUDING PITCHING.

We first enlarge the model to 3-dimensions to incorporate pitching as well as rolling, as follows.

(i) The parameter space for G is now 3-dimentional, $C = \mathbb{R}^3$.

(ii) The configuration space is now the unit sphere, $S = S^2$. The configuration of the ship is uniquely determined by the spherical coordinate $\theta \in S$. The state space is the cotangent bundle, $X = T^*S$, which is now a non-trivial 4-dimensional bundle. The state is given by $(\theta,\omega) \in T^*S$, where $\omega \in T^*_\theta S$ now represents the horizontal component of angular momentum at θ (the vertical yawing component of angular momentum is automatically excluded – see Lemma 12 below).

(iii) The energy is the Hamiltonian $H = P+K$, where the potential energy $P = Wh$, exactly as before. Meanwhile the kinetic energy $K = K_\theta(\omega)$ is the usual quadratic form on $T^*_\theta S$, defined as follows. If $(\omega_1, \omega_2, \omega_3)$ are coordinates of ω relative to the principal axes of inertia of the ship (which are tilted by θ), and (I_1, I_2, I_3) are the corresponding moments of inertia, then

$$K_\theta(\omega) = \Sigma\, \omega_i^2/2I_i \ .$$

(iv) As before, the dynamic is the damped Hamiltonian dynamic, and this completes the 3-dimension model.

Theorem 6. Theorem 5 holds for the 3-dimensional model.

Proof. E is defined by $\nabla_\theta H = \nabla_\omega H = 0$. Since the kinetic energy $K_\theta(\omega)$ is positive definite in ω, $\nabla_\omega H = \nabla_\omega K = 0$ implies $\omega = 0$. Therefore $E \subset C \times S$ as before.

Let Figure 20 represent the vertical plane through G and B_θ. Although θ is now a spherical coordinate, we can still define the angle α, and give meaning to $\theta+\varphi$, where φ is an angle. As in the proof of Lemma 10,

$$\left[\frac{\partial}{\partial\varphi} h(G, \theta+\varphi)\right]_{\varphi=0} = \ell \ .$$

$$\begin{aligned} \nabla_\theta H = 0 \ &\Rightarrow \nabla_\theta h = 0 \\ &\Rightarrow \text{left-hand side vanishes} \\ &\Rightarrow \ell = 0 \\ &\Rightarrow G \in N_\theta \\ &\Rightarrow E = N\mathcal{B}, \text{ as required.} \end{aligned}$$

The rest of the Theorem follows naturally, as before.

The geometry of the 3-dimensional model is more complicated, for this time E is a 3-manifold folded over $\mathfrak{M}$, and $\mathfrak{M}$ is a surface with cusped edges and singular points. In particular, as the ship pitches, the rolling metacentre M traces out a curve on $\mathfrak{M}$, which we call the pitching curve, $\mathfrak{P}$, and which, due to the bilateral symmetry of the ship, is a cusped edge of $\mathfrak{M}$ containing various singular points. In an ellipsoidal ship $\mathfrak{B}$ is a similar ellipsoid, and $\mathfrak{P}$ is an ellipse containing 4 hyperbolic umbilics, and these are the only singular points of $\mathfrak{M}$ (see [5]). For any floating shape, $\mathfrak{M}$ always has at least 4 hyperbolic umbilics, because, by counting indices, the number of hyperbolic umbilics minus the number of elliptic umbilics is twice the Euler characteristic of $\mathfrak{B}$, which is 2.

In a wall-sided ship there are 4 other singular points on $\mathfrak{P}$, namely 4 unbiased-butterflies, where the butterfly sections (shown in Figures 16 and 17) bifurcate back into single cusps fore and aft. These butterfly points explain why large pitching enhances the danger of capsizing, because if the pitching angle passes beyond them then the rolling-capsizing-angle decreases. In other words pitching has the same qualitative effect as isotoping a ship-shaped section into a canoe-shaped section (see the Problem after Theorem 3 Remark 3), and the resulting decrease in capsizing angle can be seen by comparing Figures 12(b) and 17(a). Moreover large pitching can occur at the same time as resonant rolling, if the ship is less than half the wave-length and steers a sensitive course (see Section 5 and Figure 7). Therefore the geometry of the butterfly may be relevant to the study of capsizing of small vessels in heavy seas. Of the 13 merchant ships that were known to have capsized and sunk during 1975, 11 were small cargo or fishing vessels of 350 tons or less [10]. In Section 14 below we discuss the effect on the capsizing angles of heaving, or having the crest of a wave amidships.

13. MODEL INCLUDING HEAVING.

Next we enlarge the model to incorporate heaving. The enlarged model automatically contains all the coupling between the three modes of oscillation, rolling, pitching and heaving.

(i) As before, the parameter space is $C = \mathbb{R}^3$.

(ii) The configuration of the ship is given by $(\theta,q) \in S\times\mathbb{R}$ where θ is the spherical coordinate, and q the vertical coordinate of heaving. Let $B_{\theta,q}$ denote the resulting centre of buoyancy. The state is given by

$$(\theta,q,\omega,p) \in T^*(S\times\mathbb{R}),$$

where p denotes the vertical linear momentum and ω is as before.

(iii) The potential energy is given by

$$P = p(G,\theta,q) = Wh_1 + Vh_2,$$

where
h_1 = height of G above water line
h_2 = depth of $B_{\theta,q}$ below water line
V = weight of water displaced
} all functions of θ,q.

The kinetic energy is given by

$$K = K_\theta(\omega) + K(p) = \Sigma\, \omega_i^2/2I_i + gp^2/2W \quad .$$

As before the energy

$$H : C\times T^*(S\times\mathbb{R}) \longrightarrow \mathbb{R}$$

is the Hamiltonian, $H = P+K$, and the dynamic is the damped Hamiltonian dynamic. We define the buoyancy locus to be the same as before, consisting only of those centres of buoyancy for which there is no heaving,

$$B = \{B_\theta\} = \{B_{\theta,q};\ q=0\} \ .$$

<u>Theorem 7.</u> <u>Theorem 5 holds.</u>

<u>Proof.</u> E is given by $\nabla_\theta H = \nabla_\omega H = \frac{\partial H}{\partial q} = \frac{\partial H}{\partial p} = 0$. Now $\frac{\partial H}{\partial p} = \frac{gp}{W}$, and so $p = 0$. We shall show $\frac{\partial H}{\partial q} = 0$ implies $q = 0$. Hence the problem reduces to the previous case, $E \subset C\times S$, and the result follows from Theorem 6. Let

$k_i = h_i(\theta,0)$

$U = W - V$.

u = height of centre of emerged slice above water line.

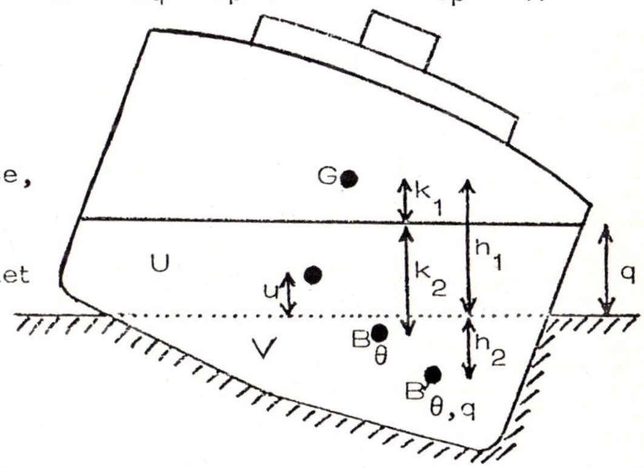

Figure 22.
Heaving, while rolling and pitching.

Then $$h_1 = k_1 + q \quad .$$

By taking moments of the displaced water about the water-line,

$$W(k_2 - q) = Vh_2 - Uu \quad .$$

Therefore

$$P = Wh_1 + Vh_2 = W(k_1 + q) + W(k_2 - q) + Uu$$
$$= W(k_1 + k_2) + Uu \quad .$$

Therefore

$$\frac{\partial H}{\partial q} = \frac{\partial P}{\partial q} = \frac{\partial U}{\partial q} u + U \frac{\partial u}{\partial q}, \text{ since } k_1 + k_2 \text{ independent of } q.$$

From Figure 22,

$$U, u \gtreqless 0 \text{ as } q \gtreqless 0, \text{ and } \frac{\partial U}{\partial q} \geq 0, \frac{\partial u}{\partial q} > 0.$$

Therefore

$$\frac{\partial H}{\partial q} \gtreqless 0 \text{ as } q \gtreqless 0.$$

Therefore

$$\frac{\partial H}{\partial q} = 0 \text{ implies } q = 0, \text{ as required.}$$

Lemma 11. For small heaving while upright the Hamiltonian dynamic reduces to the linear theory.

Proof. In the case of a wall-sided ship of draught D we have approximately

$$U = \frac{Wq}{D}, \quad u = \frac{q}{2}, \text{ and therefore } Uu = \frac{Wq^2}{2D} \quad .$$

The Hamiltonian equations give

$$\dot{q} = \frac{\partial H}{\partial p} = \frac{gp}{W}, \quad \dot{p} = -\frac{\partial H}{\partial q} = -\frac{\partial}{\partial q}(Uu) = \frac{Wq}{D}$$

Therefore $\ddot{q} = \frac{g}{D}q$, as in equation (17) of Section 6.

14. MODEL INCLUDING LOADING.

Finally we enlarge the model to incorporate loading by allowing the weight of the ship to act as another parameter. At the same time we place the treatment in a more elegant group-theoretical setting.

(i) The parameter is (W,G), where W is the weight and G the centre of gravity. The condition for the ship to float is $W \in \overline{W}$, where $\overline{W}$ denotes the open internal $0 < W < \mathcal{W}$, and $\mathcal{W}$ is the weight that would just sink it. Therefore the parameter space is 4-dimensional, $C = \overline{W} \times \mathbb{R}^3$.

(ii) Let $\mathfrak{G}$ be the 6-dimensional group of Euclidean motions in $\mathbb{R}^3$. Choose an arbitrary reference position of the ship, and let $\mathfrak{G}$ act on this reference position by right action. The potential energy of the water displaced is determined by $g \in \mathfrak{G}$, and the potential energy of the ship determined by g and the parameter (W,G), as in the last section. The kinetic energy is the classical positive definite quadratic on $T^*\mathfrak{G}$ for a rigid body. The sum gives the Hamiltonian

$$H : C \times T^*\mathfrak{G} \longrightarrow \mathbb{R}.$$

Remark 1. The kinetic energy depends not only upon the weight but also upon the inertia tensor I of the ship. Therefore both H and the dynamic depend on I, which we could take as another 6-dimensional parameter. However in equilibrium the kinetic energy vanishes, and so I does not affect the equilibrium manifold. Therefore to identify the latter it is not necessary to specify I, which can be arbitrary.

Remark 2. The above H is no good for a catastrophe potential because it is not generic, and would give a 7-dimensional equilibrium manifold, which is 3 dimensions too big. We therefore need to factor out by the symmetries of H, as follows. $\mathfrak{G}$ acts on the left of $T^*\mathfrak{G}$ by :

$$h(g,t) = (hg,(T_g^*h)^{-1}t), \quad h \in \mathfrak{G},\ g \in \mathfrak{G},\ t \in T_g^*\mathfrak{G} ,$$

and hence on the right of H by :

$$(Hh)(c,g,t) = (c,h(g,t)) .$$

Define the symmetry group

$$\mathrm{Sym}(H) = \{h \in \mathfrak{G};\ Hh = H\} .$$

Let $\mathfrak{H}$ be the 3-dimensional subgroup of $\mathfrak{G}$ preserving the horizontal (consisting of horizontal translations and rotations about vertical axes).

Lemma 12. Sym (H) = $\mathfrak{H}$.

Proof. The action of $\mathfrak{G}$ does not affect kinetic energy, and so we are only concerned with potential energy. $\mathfrak{H}$ does not alter potential energy and so $\mathfrak{H} \subset \mathrm{Sym}(H)$. Conversely $\mathrm{Sym}(H) \subset \mathfrak{H}$, because $\mathfrak{G}/\mathfrak{H}$ is spanned by the 1-parameter subgroups of rolling, pitching and heaving, which do alter H by the proof of Theorems 6 and 7.

We can now construct the model. Define the state space to be $(T^*\mathfrak{G})/\mathfrak{H}$, which is a 6-vector bundle over the 3-manifold $\mathfrak{G}/\mathfrak{H}$. In other words the state of the ship is determined by its 6 coordinates of momentum and its 3 coordinates of roll, pitch and heave (while its other 3 coordinates of latitude, longitude and course are automatically ignored). Define the potential to be the induced Hamiltonian

$$H : C \times (T^*\mathfrak{G})/\mathfrak{H} \longrightarrow \mathbb{R} .$$

Define the dynamic to be the damped Hamiltonian dynamic. This completes the model, and we now proceed to identify the equilibrium manifold E.

As before let S denote the unit sphere. Let

$$\theta : \mathfrak{G}/\mathfrak{H} \longrightarrow S$$

denote the projection given by defining $\theta(\mathfrak{H}g)$ to be the direction of the image of the vertical under g. (This is well-defined since $\mathfrak{H}$ preserves the horizontal, and hence also the vertical).

Each $W \in \overline{W}$ determines a buoyancy locus $\mathcal{B}^W \subset \mathbb{R}^3$. As W varies from 0 to $\mathcal{W}$, $\{\mathcal{B}^W\}$ is a decreasing nested family of convex closed surfaces filling the convex hull of the ship's hull, running from the boundary B^0 to the single point $\mathcal{B}^{\mathcal{W}}$ at the submerged centre of buoyancy.

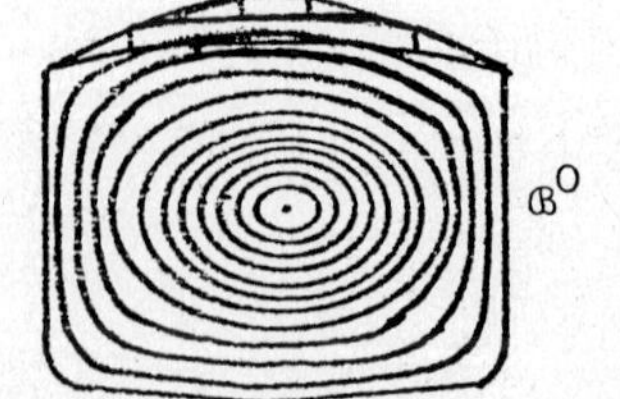

Figure 23. Buoyancy loci $\{\mathcal{B}^W\}$.

For each W, $\mathcal{B}^W$ has evolute $\mathfrak{M}^W$, normal bundle $N\mathcal{B}^W$, and geodesic spray

$$N\mathcal{B}^W = \{N_\theta^W \times \theta\} \xrightarrow{\subset} \mathbb{R}^3 \times S \xrightarrow{\text{proj}} \mathbb{R}^3 .$$

Define the <u>complete</u> buoyancy locus to be the 3-manifold

$$\mathcal{B} = \{W \times \mathcal{B}^W\} \subset \overline{W} \times \mathbb{R}^3 = C ,$$

with complete metacentric locus,

$$\mathfrak{M} = \{W \times \mathfrak{M}^W\} \subset C$$

and complete normal bundle and geodesic spray

$$N\mathcal{B} = \{W \times N\mathcal{B}^W\} \xrightarrow{\subset} C \times S \xrightarrow{\text{proj}} C .$$

<u>Theorem 8.</u> <u>θ induces a diffeomorphism from the equilibrium manifold E onto the complete normal bundle $N\mathcal{B}$ such that the diagram is commutative.</u>

$$\begin{array}{ccc} E & \xrightarrow{\cong} & N\mathcal{B} \\ \downarrow \subset & & \downarrow \subset \\ C \times \mathfrak{G}/\mathfrak{H} & \xrightarrow{1 \times \theta} & C \times S \end{array}$$

<u>Therefore the catastrophe map $\chi : E \to C$ is equivalent to the geodesic spray $N\mathcal{B} \to C$, and the bifurcation set is the complete metacentric locus $\mathfrak{M}$.</u>

<u>Proof.</u> Since kinetic energy is positive definite, equilibria lie in the zero section $\mathfrak{G}/\mathfrak{H}$ of the state space $(T^*\mathfrak{G})/\mathfrak{H}$. Fix W, and let E^W denote the corresponding section of the equilibrium manifold. Let φ^W denote the diffeomorphism

$$\varphi^W : \mathbb{R}^3 \times \mathfrak{G}/\mathfrak{H} \longrightarrow \mathbb{R}^3 \times S \times \mathbb{R}$$
$$(G, \mathfrak{H}g) \longmapsto (G, \theta, q) ,$$

where $\theta = \theta(\mathfrak{H}g)$ and q is the height of Gg above the level that G would be at angle θ, were a weight W of water displaced. (This is well defined since $\mathfrak{H}$ preserves that level). Then $\varphi^W E^W = N\mathcal{B}^W \times 0$ by Theorem 6. Therefore there are diffeomorphisms :

$$\begin{array}{ccccc} E^W & \xrightarrow{\cong} & N\mathcal{B}^W \times 0 & \xrightarrow{\cong} & N\mathcal{B}^W \\ \downarrow \subset & & \downarrow \subset & & \downarrow \subset \\ \mathbb{R}^3 \times \mathfrak{G}/\mathfrak{H} & \underset{\cong}{\xrightarrow{\varphi^W}} & \mathbb{R}^3 \times S \times \mathbb{R} & \xrightarrow{\text{project}} & \mathbb{R}^3 \times S \end{array}$$

Premultiplying by W, and taking the union for all $W \in \overline{W}$, give the result. Projecting onto C yields the required equivalence.

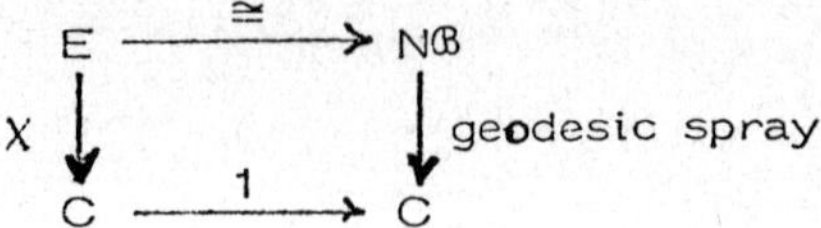

Both sides have the same image of singularities, and hence the bifurcation set is $\mathfrak{M}$. This completes the proof of Theorem 8.

<u>Theorem 9. The metacentric loci $\mathfrak{M}^W$ and $\mathfrak{M}^{\mathcal{W}-W}$ are similar. $\mathfrak{M}^W$ is obtained by reflecting $\mathfrak{M}^{\mathcal{W}-W}$ in the submerged buoyancy centre $\mathfrak{B}^{\mathcal{W}}$ and scaling by $\frac{\mathcal{W}-W}{W}$.</u>

<u>Proof.</u> Let V_θ^W denote the volume displaced by weight W at angle θ. Then the volume of the ship can be written as a disjoint union

$$V^{\mathcal{W}} = V_\theta^W \cup V_{T\theta}^{\mathcal{W}-W}$$

where T is the antipodal map of S. Therefore the buoyancy centre B_θ^W is obtained by reflecting $B_{T\theta}^{\mathcal{W}-W}$ in $B^{\mathcal{W}}$ and scaling by $\frac{\mathcal{W}-W}{W}$. Therefore this similarity maps B^W to $B^{\mathcal{W}-W}$, and hence $\mathfrak{M}^W$ to $\mathfrak{M}^{\mathcal{W}-W}$.

Figure 24.

<u>Corollary 1. $\mathfrak{M}^{\mathcal{W}/2}$ is symmetrical about $\mathfrak{B}^{\mathcal{W}}$.</u>

<u>Corollary 2. $\mathfrak{M}^{3\mathcal{W}/4}$ is one third the size of $\mathfrak{M}^{\mathcal{W}/4}$.</u>

The Corollaries are illustrated in Figure 25 for a typical 2-dimensional section of a ship having beam:draught ratio = 3 when $W = \mathcal{W}/2$. The position of G is fixed, and the properties are the same as those of the liner in Section 2. The metacentric locus in Figure 25(b) has 8 cusps, and when the weight increases or decreases 4 swallowtails appear creating 8 more cusps, 16 in all.

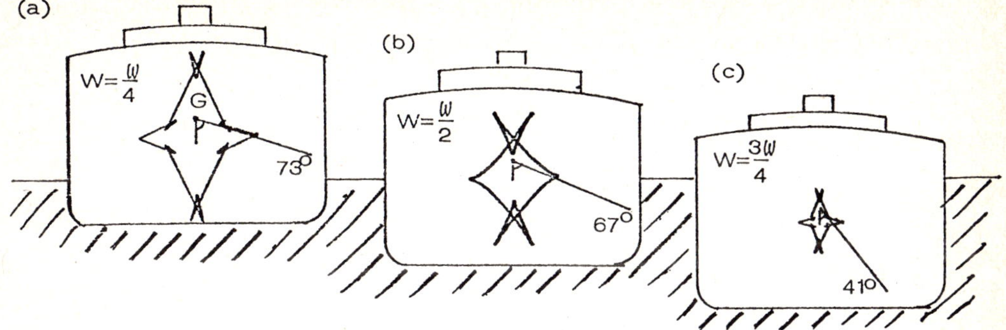

Figure 25. Decrease in size of $\mathfrak{M}^W$*, and capsizing angle, with weight.*

As the weight increases the ship becomes more susceptible to capsizing, not only because the bifurcation set shrinks closer to G, but also because the capsizing angle drops dramatically. The latter is caused by the appearance of the swallowtails. Figure 26 shows graphs of metacentric height and capsizing angle as functions of W, assuming G fixed. The dynamic capsizing angle can drop even further if there is a shift of cargo (see Figure 21).

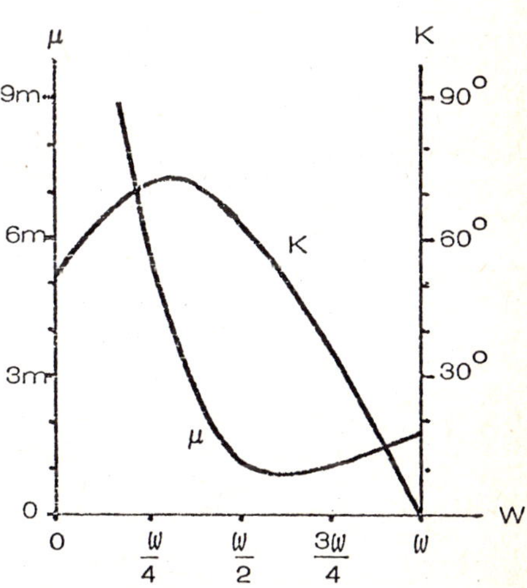

Figure 26. Metacentric height μ *and capsizing angle* K *as functions of weight* W.

This geometry is also relevant to the capsizing of small ships in following seas (small means the ship length is less than the wave-length), for when the crest of the wave is amidships, the buoyancy amidships carries more of the weight.

Therefore although the ship may be designed as in Figure 25(b), she may unwittingly find herself riding on the crest of a wave as in Figure 25(c). The usual approach of linear theory is to emphasise the loss of stability due to the loss of metacentric height [9], but Figure 26 suggests that the non-linear drop in capsizing angle may in fact be more dangerous. A similar phenomenon may occur in a head sea, if the high frequency of encountered waves happens to resonate with heaving.

We conclude by considering the loading of cargo onto the ship. As W increases G may move, and so the parameter (W,G) follows a path p in the parameter space C. If p crosses $\mathfrak{M}$ the ship may heel, or right itself, or capsize. To illustrate this consider two such loading paths shown in Figure 27. The weight of the ship is $\mathcal{W}/6$ when empty and $5\mathcal{W}/6$ when full. During path p_1 the cargo is stowed so as to keep G fixed at the centre $\mathcal{B}^{\mathcal{W}}$ of the ship, whereas during p_2 the cargo is loaded from the bottom upwards, as in a tanker. Meanwhile the dotted line shows how the height of the metacentre M^W depends upon W. The latter is drawn for a narrow ship, (with beam:draught ratio = 2 when W = $\mathcal{W}/2$) because the narrowness lowers the path of M^W so as to cut p_1. Therefore p_1 suffers the two catastrophes of heeling and righting, whereas p_2 suffers none because it skirts below the dotted line.

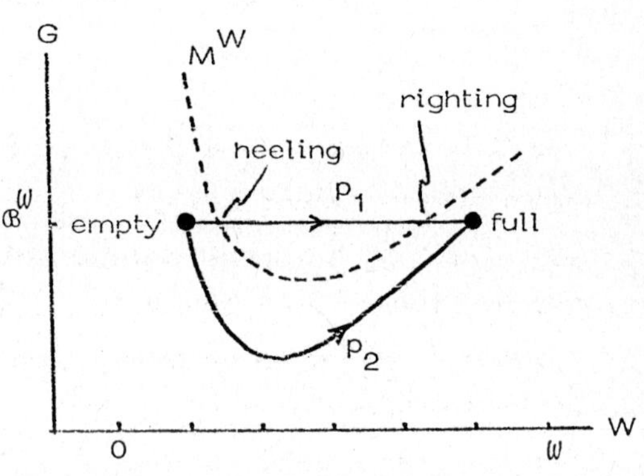

Figure 27. Two loading paths.

The corresponding behaviour of the ship is illustrated in Figure 28. In each case the metacentric locus $\mathfrak{M}^W$ is sketched,

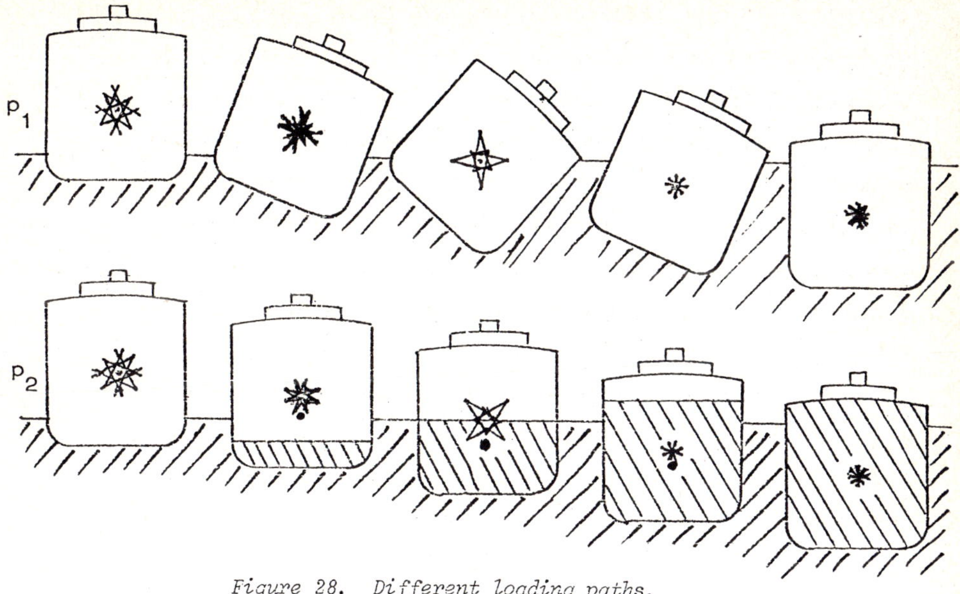

Figure 28. Different loading paths.

and G is indicated by a blob. Certain heavy narrow old cargo boats have a loading path like p_2, but the heaviness of the empty boat displaces the empty position in Figure 27 to the right across the dotted line; therefore such a boat heels with negative metacentric height when empty, and then suddenly rights itself during loading. If there is not enough cargo then the boat must take on ballast before sailing.

15. CONCLUSION

Could this geometrical approach to be of any practical use? It is difficult to say at this stage, but we tentatively suggest two areas for further exploration. Firstly the singularities involved are stable and qualitatively simple phenomena, sitting halfway between the geometric complexity of the shape of the ship and the dynamic complexity of its behaviour in the seaway. If details of a number of individual ships were analysed in the light of these singularities, it might give some further insight into how design can influence performance, and eventually might even suggest more sophisticated stability criteria.

Secondly there is at present a lack in the theory of non-linear coupling. Consequently there is a lack of mathematical language in which to express and communicate the intuition, which experienced pilots possess, of how to handle a ship in heavy weather. It is not enough to say that capsizing is probably due to that one-in-a-million freak wave, because capsizings do occur more frequently than we would wish [10], and it might be that the intuitive knowledge of one experienced pilot could have saved another. A case in point is Lindemann's discovery of how to get out of a flat spin in an aircraft, by pushing the stick fully forward and kicking hard on the opposite rudder. What was at one time a situation dreaded by all fliers, is now a routine recovery procedure taught to all beginners. Analogously in ship stability, the qualitative simplicity of the singularities that we have been discussing might eventually lead to a better understanding of the routine procedures for handling that freak wave.

REFERENCES.

1. R. Abraham & J.E. Marsden, Foundations of mechanics, Benjamin, New York, 1967.

2. K.C. Barnaby, Basic naval architecture, Hutchinson, London, 1963.

3. S.N. Blagoveshchensky, Theory of ship motions (English trans.), Dover, New York, 1962.

4. P. Bouguer, Traite du Navire, de sa construction et de ses mouvements, Paris, 1746.

5. A. Cayley, On the centro-surface of an ellipsoid, Trans. Camb. Phil. Soc. 12 (1873) 319-365.

6. G. Durrell, My family and other animals, Penguin, England, 1956.

7. L. Euler, Scientia Navalis, St. Petersburg, 1749.

8. P.J. Holmes & D.A. Rand, The bifurcations of Duffing's equation : and application of catastrophe theory, J. Sound & Vib, 44 (1976) 237-253.

9. C. Kuo, International Conference on Stability of Ships and Ocean vehicles, University of Strathclyde, Glasgow 1975.

10. Lloyd's Register of Shipping, Casualty Return, Lloyd's, London 1975.

11. A.M. Robb, Theory of Naval Architecture, Griffin, London, 1952.

12. R. Thom, Structural stability and morphogenesis, (trans : D D.H. Fowler), Benjamin, New York, 1975, (French edition 197

13. D.J.A. Trotman & E.C. Zeeman, The classification of elementary catastrophes of codimension ≤ 5, Structural stability, the Theory of catastrophes, and Applications in the Sciences, Springer Lecture Notes in Math. 525 (1976) 263-327.

14. E.C. Zeeman, Levels of structure in catastrophe theory, Proc. Int. Cong. Math. Vancouver (1974) 2, 533-546.

15. E.C. Zeeman, Duffing's equation in brain modelling, Bull. Inst. Math. & Appl., 12 (1976), 207-214.

NOTE ON SMOOTHING SYMPLECTIC AND VOLUME PRESERVING DIFFEOMORPHISMS

Eduard Zehnder

NOTATIONS, DEFINITIONS, RESULTS

a) Notations:

In the following, "manifold" means a C^∞-manifold, locally compact and Hausdorff, with countable base. Local charts will always be C^∞ and simply connected. With $\Omega_k^n(M)$ we denote the n-forms on a manifold M of class C^k, $1 \leq n \leq \dim(M)$, $1 \leq k \leq \infty$. A symplectic manifold is a pair (M,σ) consisting of a manifold M and a 2-form $\sigma \in \Omega_\infty^2(M)$, which is closed, $d\sigma = 0$, and non degenerate. The latter condition requires that the linear map $v \to C_v\sigma(m) := \sigma(m)(v,.)$ from T_mM into $(T_mM)^*$ is an isomorphism for all $m \in M$. A symplectic manifold is necessarily orientable and of even dimension, $\dim(M) = 2n$, $n \geq 1$. Locally this dimension is the only symplectic invariant of a symplectic manifold, for a well-known theorem by Darboux says that in every point $m \in M$ there is a chart (U,ϕ), such that on this chart the two-form σ has the following constant normalform:

$$\sigma|U := \sum_{i=1}^{n} dx_i \wedge dy_i \quad . \tag{1}$$

The above distinguished charts (U,ϕ) will be called in the following symplectic charts.

The two-form σ being non degenerate provides a natural linear isomorphism $\hat{\sigma}: v \to \hat{\sigma}(v) = C_v\sigma$ between the vectorfields and the 1-forms, and allows to define the so-called Hamiltonian vectorfields with Hamilton-function f by $v_f := \hat{\sigma}^{-1}(df)$. With this notation, the Poissonbracket with respect to σ for two functions f , $g \in \Omega^0(M)$ is then defined by

(2) $$\{f,g\} := -\sigma(v_f,v_g) \quad ,$$

and the fact that σ is closed is equivalent to

(3) $$[v_f,v_g] = v_{\{f,g\}} \quad ,$$

for all $f,g \in \Omega^0(M)$. If (M,σ) and (M,τ) are symplectic manifolds, then a map $f \in C^k(M,N)$, $k \geq 1$, is called symplectic, if $f^*\tau = \sigma$. A symplectic map is automatically an immersion, since σ and τ are non degenerate.

Every 2-dimensional, orientable manifold (M,σ) with σ a volumeform, is symplectic. If M is an oriented manifold of dimension $\dim(M) = n$, with volumeform $\nu \in \Omega^n_\infty(M)$, (that is ν simply satisfies $\nu(m) \neq 0$ for all $m \in M$), then a map $f \in C^k(M,M)$, $k \geq 1$ is called volumepreserving if $f^*\nu = \nu$.

b) Results:

We shall prove the following statements in partial answer to a question by F. Takens. Every symplectic C^k-diffeomorphism, $k \geq 1$, can be approximated in the C^k-sense by symplectic C^∞-diffeomorphisms. The proof, based on the generating function technique is elementary and will be carried out in detail. It turns out, that in case of volumepreserving diffeomorphisms, the

Holder topology $C^{p,\alpha}$, p an integer ≥ 1 , $0 < \alpha \leq 1$ is more appropriate. Given a volumepreserving $C^{p,\alpha}$ diffeomorphism f on a compact manifold, then there are C^∞-diffeomorphisms f_n , which are volumepreserving, such that (in local coordinates) $|f_n - f|_{C^p} \to 0$ as $n \to \infty$, and $|f_n|_{C^{p,\alpha}} \leq K$, for some constant K independent of n . The proof makes use of the Hodge decomposition theorem for Holder spaces, which is given in the appendix.

In proving the above approximation statements we also show (Theorem 4) the following. If $\nu_1, \nu_2 \in C^\alpha$, $0 < \alpha \leq 1$ are two volumeforms sufficiently close by (that is to say they lie in a small C^α-neighborhood of a volumeform $\mu \in C^4$), then there is a $C^{1,\alpha}$-diffeomorphism f , $C^{1,\alpha}$-close to the identity and a constant $c > 0$, such that $cf^* \nu_1 = \nu_2$. The corresponding global statement without a loss of derivatives is not known, except in the C^∞ case [1] .

I would like to thank J. Moser and R. Böhme for helpfull discussions. I am very grateful to I.M.P.A. for its hospitality.

2. SYMPLECTIC DIFFEOMORPHISMS

Our aim is to prove the following elementary approximation statement:

Theorem 1. Let (M,σ) and (N,τ) be two symplectic manifolds. The set of symplectic C^∞-diffeomorphisms of M onto N is dense in the C^k-space of symplectic diffeomorphisms of M onto N , for $k \geq 1$.

In the special case of 2-dimensions we obtain:

Corollary. If M is an oriented two-dimensional manifold with volumeform $\nu \in \Omega^2_\infty(M)$, then the volume preserving C^∞-diffeomorphisms of M onto M are dense in the C^k-volume preserving diffeomorphisms of M onto M, for all $k \geq 1$.

The above corollary solves Palis and Pugh's 44. problem in the special case $\dim M = 2$. [6]

The proof of theorem 1 is based on the following well-known Lemma of classical mechanics. It characterizes locally symplectic maps between symplectic manifolds of the same dimension by means of the so-called generating functions.

Lemma 1. Let (M,σ) and (N,τ) be symplectic manifolds such that $\dim(M) = \dim(N) = 2n$. A map $f : M \to N$ of class C^k, $k \geq 1$, is symplectic if and only if the following holds true. For every $m \in M$ there are two symplectic charts, (U,ϕ) on M and (V,ψ) on N, with $m \in U$, $f(m) \in V$ and $f(U) \subset V$, such that for the local representation F of f, $F := \psi \circ f \circ \phi^{-1} \in C^k(\phi(U),\psi(V)) : (x,y) \to (\xi,\eta) := (\pi \circ F(x,y)$, $\pi_2 \circ F(x,y)$ we have

$$\text{(3)} \qquad \begin{aligned} \xi &= (d_\eta S)\,(x,\eta) \\ y &= (d_x S)\,(x,\eta) \end{aligned},$$

with $S \in C^{k+1}(W)$, some open $W \subset \mathbb{R}^{2n}$, satisfying $\mathrm{Det}(d_x d_\eta S(x,\eta)) \neq 0$ for all $(x,\eta) \in W$. The function S is called a local generating function of f. We henceforth abreviate $F := E(S)$.

Proof:

(i) Assume $f \in C^k(M,N)$, $k \geq 1$ is symplectic. If $m \in M$ we can pick by means of the Darboux theorem mentioned above two symplectic charts (U,ϕ) and (V,ψ) with $m \in U$, $f(m) \in V$, and $f(U) \subset V$. We arrange that $\phi(m) = 0$ and $\psi(f(m)) = 0$, hence $F(0) = 0$, for the local representation of f. Explicitly we write $F(x,y) = (u(x,y),v(x,y)) = = (\xi,\eta)$, with $u = (u_1,\ldots,u_n)$, $v = (v_1,\ldots,v_n)$ and $(x,y) \in \phi(U) \subset \subset R^n \times R^n$. Since f is symplectic, we have in our normalform charts $F^* d\theta_2 = d\theta_1$, where these 1-forms θ_2 and θ_1 are defined by $\theta_2(\xi,\eta) = \sum_{k=1}^{n} \xi_k d\eta_k$, and $\theta_1(x,y) = \sum_{k=1}^{n} x_k dy_k$. Therefore the functions $v_1,\ldots,v_n$ are in involution, that is to say $\{v_i,v_j\} = 0$, $1 \leq i$, $j \leq n$, and since they are independent, we can achieve by a further symplectic linear coordinate transformation that $d_y v(x,y)$ is an isomorphism for (x,y) in an open neighborhood of zero. Postponing the proof of this simple remark, we apply the implicit function theorem to get a unique $b \in C^k$, $b(0) = 0$, such that $y = b(x,\eta)$ satisfies $\eta = v(x,y)$. We now define $a \in C^k$ by means of $\xi = u(x,b(x,\eta)) := a(x,\eta))$ and it remains to be shown that $a = d_\eta S$ and $b = d_x S$. To do so we introduce the local maps σ_1 and σ_2 as follows.

$\sigma_1 : (x,\eta) \to (x,b(x,\eta))$ and $\sigma_2 : (x,\eta) \to (a(x,\eta),\eta)$. Apparently $\sigma_1(0) = 0$, $\sigma_2(0) = 0$, moreover σ_1 is a local diffeomorphism and $F = \sigma_2 \circ \sigma_1^{-1}$, the domain of course properly restricted. Thus also σ_2 is a local diffeomorphism. Since F is symplectic, we conclude that the 1-form $\lambda \in \Omega_k(W)$, defined by $\lambda := a d\eta + b dx$ is closed. Indeed with $\theta_3(x,y) := \sum_{i=1}^{n} y_i dx_i$, we can write $\lambda = \sigma_2^* \theta_2 + \sigma_1^* \theta_3$, hence $\sigma_1^{-1*}\lambda = F^* \theta_2 + \theta_3$ and therefore $\sigma_1^{-1*} d\lambda = F^* d\theta_2 + d\theta_3 = F^* d\theta_2 - d\theta_1 = 0$, and we have $d\lambda = 0$. Therefore there is a function $S \in C^{k+1}(W)$ satisfying $a = d_\eta S$ and $b = d_x S$ on W. By construction $d_\eta d_x S(s,\eta) = d_\eta b(x,\eta)$ which is regular for all $(x,\eta) \in W$ if we eventually shrink W, and we have found the required generating function S.

ii) Conversely, if the local representative F is in symplectic charts given by a generating function $S \in C^{k+1}(W)$ according to (3), we have to show $F^* d\theta_2 = d\theta_1$. Apparently $F = \sigma_2 \circ \sigma_1^{-1}$ with the local diffeomorphisms $\sigma_1 : (x,\eta \to (x, d_x S(x,\eta))$ and $\sigma_2 : (x,\eta) \to (d_\eta S(x,\eta),\eta)$. With the previously introduced θ_3 we have $dS = \sigma_2^* \theta_2 + \sigma_1^* \theta_3$, and therefore $F^* d\theta_2 - d\theta_1 = \sigma_1^{-1*} d(\sigma_2^* \theta_2 + \sigma_1^{-1*} \theta_3) = \sigma_1^{-1*} ddS = 0$.

It remains to check, whether symplectic coordinates can be chosen such that $\mathrm{Det}(d_y v(x,y)) \neq 0$, (x,y) in a neighborhood of $0 \in R^{2m}$. Recalling that the functions $v_1, \ldots, v_n$ are independent and in involution, the next simple Lemma says that there is a linear and symplectic change of coordinates, such that $d_y v(0) = id_n$, at $0 \in R^{2m}$.

Lemma 2. Let $(R^{2n}, \sigma = d\theta)$ be the standard linear symplectic vectorspace with the constant symplectic structure (1). Let g_i, $1 \leq i \leq k \leq n$ be k functions locally defined in a neighborhood of $0 \in R^{2n}$ and satisfying the following two conditions:

(i) $\{g_i, g_j\}(0) = 0$, $1 \leq i, j \leq k$,

(ii) $dg_i(0)$, $1 \leq i \leq k$, linearly independent. Then there is a linear symplectic coordinate transformation, such that in the new symplectic coordinates $(x_1, \ldots, x_n, y_1, \ldots, y_n)$ we have $dg_i(0) = dy_i$, $1 \leq i \leq k$.

Proof:

Denote with v_{g_i} , $1 \le i \le k$ the Hamiltonian Vectorfields with Hamilton-function g_i . By (ii), $v_{g_i}(0)$ are linearly independent and by (3) and condition (i) $\sigma(v_{g_i}(0),v_{g_j}(0)) = -\{g_i,g_j\}(0) = 0$ for all $1 \le i , j \le k$. Hence there is a symplectic basis in R^{2n} , (e_i) , $1 \le i \le 2n$, such that $e_i = v_{g_i}(0)$ for all $1 \le i \le k$. Therefore one obtains for all $1 \le i \le k$ and $1 \le j \le 2n$, $dg_i(0)(e_j) = \sigma(v_{g_i}(0),e_j) = \sigma(e_i,e_j) = 1$ for $j = i + n$ and zero otherwise. Thus with respect to this new symplectic basis, the coordinates do the job.

In view of Lemma 1, the idea of the proof of Theorem 1 is obvious, we first smooth out the symplectic diffeomorphism locally by simply smoothing its generating function and afterwards we patch the pieces together following the standard procedure for the smoothing of maps (as f.e. in [2]).

Let $F = E(S)$ be a symplectic C^k-diffeomorphism $k \ge 1$, from $U \subset \mathbb{R}^{2m}$ into $V \subset \mathbb{R}^{2n}$, U simply connected and $\overline{U}$ compact, which is given by a generating function $S \in C^{k+1}(W)$ as in Lemma 1, $F = \sigma_2 \circ \sigma_1^{-1}$, where $\sigma_1 : W \to U$ is a diffeomorphism. We choose $W_3 \subset\subset W_1 = W$, ($\subset\subset$ means W_3 open with $\overline{W}_3$ compact and $\overline{W}_3 \subset W_2$). In order to approximate S on W_3 by a C^∞-function, we pick $\zeta \in C_0^\infty(W_2)$ with $\zeta = 1$ on $\overline{W}_3$ and $\eta \in C_0^\infty(W)$, with $\eta = 1$, on W_2 . For any $\varepsilon > 0$, there is a $\chi_\varepsilon \in C_0^\infty(W)$ such that $|\eta[(\zeta S) - \chi_\varepsilon * (\zeta S)]|_{C^{k+1}(W)} < \varepsilon$. The function $S_1 \in C^{k+1}(W)$ is now defined as follows:

$$S_1 = S - \eta((\zeta S) - \chi_\varepsilon * (\zeta S)) . \tag{4}$$

Apparently $|S_1 - S|_{C^{k+1}(W)} < \varepsilon$, and $S_1 = S$ on $W \setminus \overline{W}_2$, and $S_1 = \chi_\varepsilon * (\zeta S)$ on W_3 , thus $(S_1|W_3) \in C^\infty(W_3)$. For ε sufficiently small we then choose $U_3 \subset\subset U_2 \subset\subset U$, such that $\sigma_1(W_3) \supset U_3$ and $\sigma_1(W_2) \subset U_2$.

Define now $F_1 := E(S_1)$. If ε is sufficiently small we conclude from Lemma 1 and the implicit function theorem that F_1 has the following properties (i)-(v). (i) $(F_1|U_3) \in C^\infty(U_3)$. (ii) F_1 is a symplectic C^k-diffeomorphism from U onto $F(U)$. (iii) $F_1 = F$ on $U\setminus\overline{U}_2$. (iv) $|F_1-F|_{C^k(U)} < \delta(\varepsilon)$, with $\delta(\varepsilon) \to 0$ as $\varepsilon \to 0$. (v) F_1 is of class C^p, $k \le p \le \infty$ on every open set on which F is of class C^p. In addition we remark that there is a symplectic C^k-isotopy, H_t, between F and F_1, such that H_t satisfies (ii)-(v). Indeed, simply define $H_t : E(S_t)$ with the generating function

(5) $$S_t := S - \alpha(t)(S - S_1)$$

where $\alpha \in C^\infty[0,1]$; with $\alpha(t) = 0$ for $0 \le t \le 1/2$ and $\alpha(t) = 1$ for $3/2 \le t \le 1$.

Let now $f \in C^k(M,M)$, $k \ge 1$ be a symplectic diffeomorphism from M onto N , and let Λ be a sufficiently small open neighborhood of f in the C^k-topology. We choose a local finite covering of M , consisting of symplectic charts (U_i,ϕ_i) , $1 \le i \le \infty$, with the following properties. $\overline{U}_i$ is compact, $h(U_i) \subset O_i$ for all $h \in C^k(M,M) \cap \Lambda$, with (O_i,ψ_i) , $1 \le i \le \infty$ being an atlas of symplectic charts on N . Moreover there is a covering $(U_i^{(3)})$, $1 \le i \le \infty$, on M , such that $U_i^{(3)} \subset\subset U_i^{(2)} \subset\subset U_i$, and such that on U_i every symplectic diffeomorphism contained in Λ is given by a generating function as described in Lemma 1, such that the previous local construction can be carried out with respect to $\phi_i(U_i^{(3)}) \subset\subset \phi_i(U_i^{(2)}) \subset\subset \phi_i(U_i)$. The rest of the proof follows the usual procedure, one puts $f_0 = f$, and assumes for $n \ge 2$ that $f_{n-1} \in C^k(M,N) \cap \Lambda$, and such that $f_{n-1}|U_1^{(3)} \cup \ldots \cup U_{n-1}^{(3)}$ is of class C^∞ . In view of the properties of our covering, the local map $F_{n-1} := \psi_n \circ f \circ \phi_n^{-1} \in C^k(\phi_n(U_n), \psi_n(O_n))$ is a symplectic diffeormorphism of $\phi_n(U_n)$ onto its image, which is given by $F_{n-1} := E(S_{n-1})$. Define

$F_n := E(S)$ as the symplectic diffeomorphism from $\phi_n(U_n)$ onto $F_{n-1}(U_n))$ provided by the local construction above. Hence $F_n = F_{n-1}$ outside $\phi_n(U_n^{(2)})$ and $F_n|\phi_n(U_n^{(3)})$ is of class C^∞ and $|F_n - F_{n-1}|C^k(\phi_n(U_n)) < \delta_n$. We then define the next approximation f_n as follows:

$$f_n(m) = f_{n-1}(m) \ , \qquad m \in M \setminus \overline{U_n^{(2)}}$$

(6)

$$f_n(m) = \psi_n^{-1} \circ E(S_n) \circ \phi_n(m) \ , \ m \in U_n \ .$$

If δ_n is chosen sufficiently small, then f is a symplectic diffeomorphism of M onto N , which is contained in the neighborhood Λ of f , and such that f_n is of class C^∞ on $U_1^{(3)} \cup \ldots \cup U_{n-1}^{(3)} \cup U_n^{(3)}$, the latter follows from the local property (v) above. Define finally $g(m) := \lim\limits_{n \to \infty} f_n(m)$, this limit is trivial, since the covering $(U_n^{(3)})$ is locally finite. With the proper choice of (δ_n) we conclude that $g \in C^\infty(M,N) \cap \Lambda$ is a C^∞ symplectic diffeomorphism from M onto N . In addition one obtains from the locally isotopy statement proved above a symplectic C^k-isotopy $h : M \times [0,1] \to N$ connecting f and g . This finishes the proof of Theorem 1.

Remarks

By the same token we have also proved that a C^k Lagrange submanifold $N \subset M$ of a symplectic manifold M can be C^k-approximated by C^∞ Lagrange-submanifolds, $k \geq 1$. The reason is, that such submanifolds are also explicitly described (locally) by means of generating functions: A submanifold $N \subset M$ is called Lagrange, if $T_m(N) = T_m(n)^\sigma$, for all $m \in N$, where $T_m(N)^\sigma \subset T_m(M)$ stands for the σ-complement, thus $\dim(N) = n$ if

$\dim M = 2n$. Let $m \in N$, then we have in a symplectic chart (U,ϕ) , $\phi(N \cap U) = \{(x,y) \in \mathbb{R}^{2n} | f_1(x,y) = \dots = f_n(x,y) = 0\}$. The C^k-functions (f_i) , $1 \leq i \leq n$ are independent and in involution: $\{f_i,f_j\}(m) = 0$, $m \in N \cap U$. This follows simply from the fact that $T_m(N) = \bigcap_{i=1}^{n} \ker df_i(m) = \langle v_{f_i}(m), \dots, v_{f_n}(m)\rangle^{\sigma} =: S_m^{\sigma}$ since by definition $T_m(N)^{\sigma} = T_m(N)$, we conclude $S_m = S_m^{\sigma}$, hence by (3) $0 = \sigma(v_{f_i}(m), v_{f_j}(m)) = -\{f_i,f_j\}(m)$. Therefore, in view of Lemma 3, we may assume that $d_y f(x,y) \neq 0$, where $f = (f_1,\dots,f_n)$, and by the implicit function theorem, there exists a $g \in C^k$ such that $f(x,g(x)) = 0$, so $\phi(N \cap U) = \{(x,g(x))\}$. Since the chart is symplectic, $\sigma = d\theta$, $\theta(x,y) = \sum_{k=1}^{n} y_k dx_k$, therefore the 1-form $\lambda(x) := g(x)dx$ is closed. Indeed with $i : N \to M$ the injection, we have $\lambda = i^*\theta$ and hence $d\lambda = i^* d\theta = 0$, since N is Lagrangian. Therefore there is a C^{k+1} function S with $g(x) = d_x S(x)$ and we obtain:

(7) $$\phi(N \cap U) = \{(x, d_x S(x)) , x \in V \subset \mathbb{R}^n\} .$$

Conversely, such a submanifold in a symplectic chart defines apparently a piece of a C^k-Lagrange submanifold if S is of class C^{k+1} .

3. VOLUME PRESERVING DIFFEOMORPHISMS

a) Holder spaces:

In the following the suitable function-spaces are the Holder spaces $C^{p,\alpha}(M)$, M being a C^{∞}-manifold, p is an integer ≥ 0 and $0 < \alpha \leq 1$. Instead of $C^{p,\alpha}(M)$ we shall simply write $C^a(M)$, where $a = p + \alpha$, $p = [a]$. For completeness sake we add some remarks as to the definition of these Holder spaces.

We consider a fixed convex and compact set $K \subset \mathbb{R}^n$ with interior points. (For all our purpose closed balls will do). For a continuous function $f \in C(K)$ we set

(8) $$\|f\|_\alpha := \sup_{x \neq y} \frac{|f(x) - f(y)|}{|x-y|^\alpha} ,$$

if $0 < \alpha \leq 1$. If instead $p < a \leq p + 1$, where p is a positive integer, we set for $f \in C^p(K)$

(9) $$\|f\|_a := \sup_{|s| = p} \|D^s f\|_{a-p} .$$

For $p < a \leq p + 1$, where p is an integer ≥ 0, the Holder space $C^a(K)$ is the set of all $f \in C^p(K)$, such that $\|f\|_a < \infty$, with the norm:

(10) $$|f|_a := |f|_{C^p(K)} + \|f\|_a .$$

In addition we set $C^0(K) = C(K)$. The same definition applies if the functions f have their values in finite dimensional vector spaces, $f \in C^a(K,\mathbb{R}^m)$. If $f := (f_1,\dots,f_m)$ we set $|f|_a := \sup|f_j|_a$, $1 \leq j \leq m$. The Holder spaces $C^a(K)$ are Banach spaces, and $C^b(K) \subset C^a(K)$ continuously if $0 \leq a \leq b$, the inclusion being compact, if $b > a$.

One of the basic tools in estimating Holdernorms is the following convexity inequality. For $0 \leq a \leq b$ and $0 < \lambda < 1$, there is a constant $C > 0$, C depending on b, such that for all $f \in C^b(K)$

(11) $$|f|_{(1-\lambda)a + \lambda b} \leq C\, |f|_a^{1-\lambda}\, |f|_b^\lambda .$$

This estimate controlls all the intermediate norms. We shall now assemble some well-known estimates which will come in handy later on. Let $0 < a \leq 1$, then clearly for all $f \in C^a(K_1)$, $g \in C^a(K_2)$ with $g(K_2) \subset K_1$ we have

$$(12) \qquad |f \circ g|_a \le |f|_1 |g|_a \quad (\text{resp.}, \le |f|_a |g|_1^a)$$

if $f \in C^1$ resp., $g \in C^1$. More general, the following Lemma about compositions as well as multiplications of Hölder formations holds true:

Lemma 3. (i) For all $f \in C^a(K_1)$ and $g \in C^a(K_2)$, $g(K_2) \subset K_1$, $a \in \mathbb{R}$:

$$|f \circ g|_a \le C(|f|_a \cdot |g|_1^a + |f|_1 |g|_a + |f|_0)$$

with a constant C depending on a.

(ii) For all $f, g \in C^a(K)$, $a \in \mathbb{R}$,

$$|f \cdot g|_a \le C(|f|_a |g|_0 + |f|_0 |g|_a)$$

with some other constant C depending on a only.

The proof is by induction using the estimates (11).

Lemma 4. (i) Let $0 \le a \le 1$, then on $C^a(K)$

$$\frac{1}{2} |f \cdot u - g \cdot v|_a \le |f - g|_a \cdot |u|_a + |u - v|_a |g|_a .$$

(ii) If $0 \le a \le 1$, $u, v \in C^a(K)$, and $f \in C^2(K_1)$ with $|f|_{C^2} \le B$, then

$$|f \circ u - f \circ v|_a \le 2B(|u - v|_a + |u - v|_0 (|u|_a + |v|_a)) .$$

Proof:

The first statement is immediate in view of Lemma 3(ii). In order to verify (ii), we set $\omega = v - u$, and abbreviate $\xi := u(x) + t\omega(x)$, $\eta := u(y) + t\omega(y)$. By the mean value theorem we can write for fixed x,y:

$$\begin{aligned} A(x,y) &= f(u(x)) - f(u(x)) - f(u(y)) + f(v(y)) \\ &= \int_0^1 \frac{d}{dt} (f(\xi) - f(\eta))\, dt \\ &= \int_0^1 (Df(\xi)\omega(x) - Df(\eta)\omega(y))\, dt \end{aligned}$$

$$= \int_0^1 Df(\xi)(\omega(x) - \omega(y))\, dt$$

$$+ \int_0^1 \int_0^1 D^2f(\xi + (1-s)\eta)(\xi-\eta, \omega(y))\, dtds \quad .$$

Consequently, with $|\xi-\eta| = |u(x)-u(y) + t[\omega(x) - \omega(y)]| \leq (|u|_a + t|\omega|_a)|x-y|^a \leq 2(|u|_a + |v|_a)|x-y|^a$, we obtain for $x \neq y$:

$$\frac{A(x,y)}{|x-y|^a} \leq 2B(|u-v|_a + |u-v|_o(|u|_a + |v|_a)) \quad .$$

The statement (ii) follows.

Lemma 3 allows to define $C^a(M)$ if M is any compact C^∞-manifold. To do so simply cover M by finitely many coordinate pachies (U_j, ϕ_j) . A function f on M is then said to belong to $C^a(M)$ if $f_j := f|U_j$ for every j is in C^a as a function of the local coordinates (that is to say $f \circ \phi_j^{-1} \in C^a(K)$, for $\phi_j(U_j) = K \subset \mathbb{R}^n$) . The norm $|f|_a$ is defined as $\sup_j |f_j|_a$. The definition of $C^a(M)$ does not depend on the choice of the covering, while the norm is apparently defined up to equivalence. Similarly, the Hölder-norms of the linear space of sections $C^a(E)$ of a C^∞-vector bundee $E \to M$ over a compact C^∞-manifold M is defined by means of a fixed covering by trivializing charts of E . Also maps of class C^a between two C^∞-manifolds are defined by means of coverings. Two C^a-maps are then close in the C^a-sense, if their local representations with respect to fixed converings are C^a-close.

b) Approximation of volume preserving diffeomorphism

We shall prove the following approximation statement.

Theorem 2. Let M be a compact C^∞-manifold of dimension d with volume element $\mu \in \Omega^d_\infty(M)$. Let f be a volume preserving diffeomorphism of M belonging to the Hölder class $C^{1,\alpha} = C^a$, $a = 1 + \alpha$, $0 < \alpha \leq 1$. Then f can be approximated by C^∞ volume preserving diffeomorphisms in the following sense. There is a sequence f_n of C^∞-diffeomorphisms, with $f_n^*\mu = \mu$, such that

(1) $|f_n - f|_1 \to 0$ as $n \to \infty$

(2) $|f_n|_a \leq K$,

with a constant $K > 0$ depending on $|f|_a$, but independent on n . (The notation (1), (2) is meant with respect to fixed local coordinates in an obvious way as explained in (a)). (It follows from the prove that there is a continuous isotopy of $C^{1,\alpha}$-volumepreserving diffeomorphisms connecting f_n and f).

The converse of theorem 2 is obvious: given a C^1-diffeomorphism f such that (1) and (2) hold true for a sequence f_n of volumepreserving diffeomorphisms, then f is volumepreserving and belongs to C^a .

Proof of Theorem 2:

Since the given f is of class C^a , we can pick a sequence (ϕ_n) of C^∞-diffeomorphisms of M , which are orientation preserving but not necessarily volumepreserving, such that

(13) $$|\phi_n - f|_1 \to 0 \text{ as } n \to \infty , \quad |\phi_n|_a \leq K_1 ,$$

with a constant K_1 depending on f only. For the sequence (ν_n) of volume-forms, defined by

(14) $$\nu_n = \phi_n^*\mu ,$$

the following statements hold true.

Lemma 5. (i) $\nu_n \in \Omega^d_\infty(M)$ and $\int_M \nu_n = \int_M \mu$

(ii) $|\nu_n - \mu|_\beta \to 0$ as $n \to \infty$; $|\nu_n|_\alpha \le K_2$,

for all $0 \le \beta < \alpha$, K_2 is a constant depending on K_1 and $|\mu|_1$ only.

Proof:

(i) follows from the definition. As for (ii), we shall demonstrate, that for all $0 \le \beta \le \alpha$:

(15) $$|\nu_n - \mu|_\beta \le C|\phi_n - f|_{1+\beta} ,$$

where $C > 0$ is a constant independent of n and β . The statement (ii) then follows from (15) and (13) by means of the convexity estimate (11). To get (15) we have to look at the expression $\nu_n - \mu$ in local coordinates. We pick two C^∞-coordinate patches U and V which are compact, such that $f(U) \subset V$ and $\phi_n(U) \subset V$ for all n sufficiently large, which is possible in view of (13). On V we have $\mu|V(y) = m(y)\, dy_1 \wedge \ldots \wedge dy_d$, with some function $m \in C^\infty(V)$. Consequently, the form $\phi_n^* \mu - \mu$ is represented by the function:

(16) $$m(\phi_n(x))\mathrm{Det}(d\phi_n(x)) - m(f(x))\mathrm{Det}(df(x) ,$$

$x \in U$, where the same latter has been used for the local representatives of the mappings as for the mappings themselves. We made use of $f^* \mu = \mu$. The required estimate (15) follows now easily by applying Lemma 4 and (12) to the expression (16), keeping the bounds (13) in mind.

The main step in the proof of Theorem 2 is the construction of a sequence ψ_n of C^∞-diffeomorphisms, with $\psi_n^* \nu_n = \mu$, such that $\psi_n \to \mathrm{id}$ in the C^1-sense and such that $|\psi_n|_a \le K$. Our problem is then solved by the sequence $f_n := \phi_n \circ \psi_n$, since f_n satisfies the required estimates, and $f_n^* \mu = (\phi_n \circ \psi_n)^* \mu = \psi_n^* \nu_n = \mu$. ψ_n will be given later by Theorem 3.

As a side remark we notice, that one is tempted to apply Moser's deformation argument in [1], which easily yields a C^∞-diffeomorphism ψ_n with $\psi_n^* \nu_n = \mu$. This argument runs as follows. Set $\sigma = \nu_n - \mu$, then by Lemma 4, $\sigma \in \Omega_\infty^d(M)$ and $\int \sigma = 0$, hence by the Rham theorem $\sigma = d\gamma$ for some $\gamma \in \Omega_\infty^{d-1}(M)$. Define the path $\nu_t = (1-t)\mu + t\nu_n$ in the space of volume forms connecting μ with ν_n. Since ν_t is a volume form there is a unique C^∞-vector field ξ defined by $C_\xi \nu_t = -\gamma$. For the flow η of ξ, which is defined for all $t \in [0,1]$, M being compact, one then verifies immediately that $\frac{d}{dt}(\eta_t^* \nu_t) = 0$, hence with $\eta_0 = id$, $\eta_t^* \nu_t = \mu$, and consequently $\psi_n^* \nu_n = \mu$, if we define $\psi_n := \eta_1$. Unfortunately in estimating $\psi_n - id$ we consume one derivative, since we can control the size of the C^∞-vectorfield ξ only in the α-norm and not in the $(1+\alpha)$-norm. For in the formula $C_\xi \nu_t = -\gamma$ the unknown from ν_n is involved, which we control by Lemma 4 in the α-norm only. We proceed therefore differently in order to construct ψ_n and solve a local perturbation problem in the neighborhood of the given d-form μ. The statement runs as follows:

<u>Theorem 3</u>. Let $\mu \in C^4$ be a volumeform and let $0 < \alpha \leq 1$. Then there are constants $C > 0$ and $\delta > 0$ such that the following holds true. For every d-form $\nu \in N_\alpha := \{\nu \in C^\alpha \mid |\mu-\nu|_\alpha < \delta\}$ there are a $C^{1+\alpha}$-diffeomorphism f of M and a positive constant c, such that

(i) $c\, f^* \mu = \nu$

(ii) $|f - id|_{1+\alpha} \leq C\, |\mu - \nu|_\alpha$, $|c - 1| \leq C\, |\mu - \nu|_\alpha$.

Moreover, if in addition $\mu \in C^\infty$, then there are constants $C_\ell > 0$, $\ell = 0,1,2,\ldots$, such that for all integers $\ell \geq 0$ and $0 < \beta \leq 1$:

$$\nu \in N_\alpha \cap C^{\ell,\beta} \Rightarrow f \in C^{\ell+1+\beta} \text{ and}$$

$$|f - id|_{\ell+1+\beta} \leq C_\ell\, |\mu - \nu|_{\ell+\beta} .$$

In particular, if $\nu \in N_\alpha \cap C^\infty$, then $f \in C^\infty$.

<u>Remark</u>. If we know for $\nu \in N_\alpha$ that $\int\mu = \int\nu$, then obviously $c = 1$ and consequently $f^*\mu = \nu$. The mapping $\nu \to (c,f)$, $f \in C^{1+\alpha}$, from the open neighborhood N_α into the solutions of the functional equation (i), provided by the theorem, will be continuous.

As an immediate consequence of Theorem 3 we have

<u>Theorem 4</u>. Let $\nu_1,\nu_2 \in N_\alpha$, then there is a $C^{1+\alpha}$-diffeomorphism f , and a constant $c > 0$, such that $cf^* \nu_1 = \nu_2$.

The corresponding global statement for two volumeforms $\nu_1,\nu_2 \in C^\alpha$ is not known. The statement holds true globally however for $\nu_1,\nu_2 \in C^k$ with $f \in C^{k-1}$ for all $3 \le k \le \infty$ [1] .

Postponing the proof of Theorem 3 we finish the proof of Theorem 2. Let $0 < \alpha \le 1$ be as in Theorem 2. By means of Lemma 5 and Theorem 3 (for n sufficiently large) we get a sequence $(g_n$ of C^∞-diffeomorphisms with the properties $g_n^*\mu = \nu_n$, where $\nu_n = \phi_n\mu$, and $|g_n - id|_{1+\beta} \to 0$ as $n \to \infty$, where $0 < \beta < \alpha$, and $|f_n|_{1+\alpha} \le K_3$. In particular $|g_n - id|_1 \to 0$ as $n \to \infty$. For the inverse diffeomorphisms $g_n^{-1} = \psi_n$ one verifies easily using Lemma 4 and (12) that $|\psi_n - id|_1 \to 0$ as $n \to \infty$ and $|\psi_n|_{1+\alpha} \le K_4$ for some other constant $K_4 > 0$ independent of n . Clearly $\psi_n^*\nu_n = \mu$. Finally, the sequence f_n of C^∞-diffeomorphisms, defined as $f_n = \phi_n \circ \psi_n$ satisfies the requirements of Theorem 2. Clearly $f_n^*\mu = \mu$, and from (13) together with the above estimates for ψ_n , re reach by means of Lemma 3(i),

$|f_n|_{1+\alpha} \leq K_5 = K$. To prove $|f_n - f|_1 \to 0$ as $n \to \infty$ we write (in local coordinates) $\phi_n \circ \psi_n - f = (\phi_n - f) + (\phi_n \circ \psi_n - \phi_n)$. By (13), $|\phi_n - f|_1 \to 0$ as $n \to \infty$. Clearly $|\phi_n \circ \psi_n - \phi_n|_o \leq |\phi_n|_1 \, |\psi_n - id|_o \to$ $\to 0$ as $n \to \infty$. Using once more (12), we estimate $\|\phi_n \circ \psi_n - \phi_n\| \leq$ $\leq |(d\phi_n \circ \psi_n)(d\psi_n - 1)|_o + |d\phi_n \circ \psi_n - d\phi_n|_o \leq |\phi_n|_1 \; |\psi_n - id|_1 +$ $+ |\phi_n|_{1+\alpha} |\psi_n - id|_1^\alpha$ which tends to zero as $n \to \infty$, since $0 < \alpha$. This finishes the proof of Theorem 2.

c) Proof of Theorem 3:

Assume the volume form μ belongs to C^4. Fix $0 < \alpha \leq 1$. We want to solve the equation

$$c\, f^* \mu = \nu \qquad (17)$$

for a given ν. If $\nu = \mu$, we have $f = id$ and $c = 1$ as a solution. If $|\nu - \mu|_\alpha < \delta$, δ sufficiently small, we seek a $C^{1+\alpha}$-diffeomorphism f close to the identity and a constant c close to 1 solving (17). We therefore put:

$$f = \exp(v) \;, \quad c = 1 + \hat{c} \;, \quad \nu = \mu + \hat{\nu} \;. \qquad (18)$$

v is a $C^{1+\alpha}$-vector field sufficiently small, and $\hat{c}$ is a small constant. With (18), the equation (17) becomes a nonlinear functional equation for v and c. The Taylor expansion of $(1 + t\hat{c}) \exp(tv)^* \mu$, at $t = 0$, up to order two, yields

$$(1 + \hat{c})\exp(v)^* \mu = \mu + dC_v\mu + \hat{c}\mu + O_2(\hat{c},v) \;, \qquad (19)$$

where $O_2(\hat{c},v)$ stands for the terms of second order in $\hat{c}$ and v. It will be determined explicitely later on (Lemma 7). C_v stands for contraction with the vector field v. Now in order to solve the equation

$$dC_v\mu + \hat{c}\mu = \hat{\nu} - O_2(\hat{c},v) \;, \qquad (20)$$

which follows from (17) and (19), we shall set

$$C_v\mu = \delta\phi \quad , \tag{21}$$

with the Hodge δ-operator. The equation (21) determines v uniquely since μ is a volumeform. Observing that $d\phi = 0$, ϕ being a d-form, we have $dC_v\mu = d\delta\phi = (d\delta + \delta d)\phi =: \Delta\phi$, with the Hodge Δ-operator. It is therefore sufficient to solve the equation

$$\Delta\phi + \hat{c}\mu = \hat{\nu} - O_2(\hat{c},\phi) \quad , \tag{22}$$

for $\hat{\nu}$ given. It turns out, that (22) has a solution $\phi \in C^{2+\alpha}$ which is unique, if $|\phi|_\alpha$ is sufficiently small. For the linearized equation we have:

<u>Lemma 6</u>. Let $\alpha \in \mathbb{R}^+$, and $\mu \in C^\alpha$ be a volumeform, then the linear operator $(\hat{c},\phi) \to \Delta\phi + \hat{c}\mu$ from $\mathbb{R} \times C^{2+\alpha}$ into C^α has a bounded rightinverse, which we denote by $\mathcal{L}$, such that for all $g \in C^\alpha$

$$|\mathcal{L}(g)|_{2+\alpha} \leq C_\alpha |g|_\alpha \quad , \tag{23}$$

where we used the norm $|(\hat{c},\phi)|_{2+\alpha} := \max\{|\hat{c}| , |\phi|_{2+\alpha}\}$.

<u>Proof:</u>

To solve $\Delta\phi + \hat{c}\mu = g$, for $g \in C^\alpha$ given, we first put $\hat{c} := \int g \, (\int\mu)^{-1}$, recalling that $\int\mu \neq 0$. Since $g_1 := (g - \hat{c}\mu)$ satisfies $g_1 \in C^\alpha$ and $\int g_1 = 0$, we conclude by means of the Hodge theory (see the appendix) for Hölder spaces, that $\Delta\phi = g_1$ has a unique solution $\phi \in C^{2+\alpha}$ with $\int\phi = 0$ and $|\phi|_{2+\alpha} \leq C'_\alpha |g_1|_\alpha$. Consequently $|\phi|_{2+\alpha} \leq C_\alpha |g|_\alpha$, with some constant C_α depending on $|\mu|_\alpha$.

Combining the unknown $(\hat{c},\phi) =: u$, we shall write equation (22) as follows:

$$u = \mathcal{L}(\hat{\nu} - O_2(u)) \quad , \tag{24}$$

where the linear operator is defined in Lemma 6. Since, as it turns out, $O_2(u) \in C^\alpha$, if $u \in C^{2+\alpha}$, the above equation makes sense. We shall solve (24) by interation, starting with $u^0 = (0,0)$, $u^1 = \mathcal{L}(\hat{\nu})$ and putting for $n \geq 0$

$$u^{n+1} = \mathcal{L}(\hat{\nu} - O_2(u^n)) \quad . \tag{25}$$

The convergence of this sequence in $C^{2+\alpha}$ will follow from:

<u>Lemma 7</u>. If $\mu \in C^4$ and $u \in C^{2+\alpha}$ then $O_2(u) \in C^\alpha$. Furthermore there is a constant $C > 0$, such that if $u_1, u_2 \in C^{2+\alpha}$ with $|u_1|_{2+\alpha} \leq 1$, $|u_2|_{2+\alpha} \leq 1$, then following estimate holds true:

$$|O_2(u_1) - O_2(u_2)|_\alpha \leq C \cdot (|u_1|_{2+\alpha} + |u_2|_{2+\alpha}) |u_1 - u_2|_{2+\alpha} \quad . \tag{26}$$

<u>Proof</u>:

In coordinates, the equation (17) reads as follows:

$$(1+\hat{c})\ \mu(f(x))\ \mathrm{Det}(df(x)) = \nu(x) \quad . \tag{27}$$

Now, for $(x,v) \in \mathbb{R}^d \times \mathbb{R}^d$, if $|v|$ is small the exponential map is given by $\exp_x(v) = x + v + r(x,v)(v,v)$, where $r(x,v)(.,.)$ is bilinear, and $r \in C^\infty$. Consequently $f(x) = \exp_x(v(x)) = x + v(x) + r(x,v(x))(v(x),v(x))$. Inserting this expression into (27), the Taylor formula yields $\mathrm{div}(\ v) = \hat{\nu} - \hat{c}\mu - O_2(\hat{c},v)$ and with $\mu v = \mathrm{grad}\ \phi$, we have $\Delta\phi = \hat{\nu} - \hat{c}\mu - O_2(\hat{c},\phi)$. The term $O_2(\hat{c},\phi)$ is a finite sum of functions of the form

$$\begin{array}{c} \hat{c}\beta_1(x,\phi,\partial_x\phi)\ P_1(\phi,\partial_x\phi,\partial_x^2\phi) \\ \beta_2(x,\phi,\partial_x\phi)\ P_2(\phi,\partial_x\phi,\partial_x^2\phi) \end{array} \quad , \tag{28}$$

where $\beta_1, \beta_2 \in C^{k-2}$, if $\mu \in C^k$, P_1 is a homogeneous polynomial of degree ≥ 1, P_2 is such a polynomial of degree ≥ 2. Recalling $u := (\hat{c},\phi)$, the required estimate (26) follows readily from (28) by means of the Lemmata 3 and 4. We werely look at a typical term. Assuming $|\phi|_{2+\alpha}, |\psi|_{2+\alpha} \leq 1$ we have to estimate in the α-norm the following function:

$$\beta(x,\partial_x^2\phi)\partial_x^2\phi\,\partial_x^2\phi - \beta(x,\partial_x^2\psi)\,\partial_x^2\psi\,\partial_x^2\psi =$$

$$= \beta(x,\partial_x^2\phi)\;\partial_x^2\phi\;(\partial_x^2\phi - \partial_x^2\psi) + \tag{29}$$

$$+ (\partial_x^2\psi)(\beta(x,\partial_x^2\phi)\;\partial_x^2\phi - \beta(x,\partial_x^2\psi)\;\partial_x^2\psi) \quad .$$

By assumption $\beta \in C^2$. By means of Lemma 3 (ii), the first term is estimated as $\leq 2|\beta(x,\partial_x^2\phi)\;\partial_x^2\phi|_\alpha\;|\partial_x^2\phi - \partial_x^2\psi|_\alpha \leq 8|\beta|_1\;|\phi|_{2+\alpha}\;|\phi-\psi|_{2+\alpha}$. Here we have used that $|\beta(x,\partial_x^2\phi)|_\alpha \leq |\beta|_1(1 + |\phi|_{2+\alpha}) \leq\;|\beta|_1$ by means of (12) and the assumption $|\phi|_{2+\alpha} \leq 1$. By Lemma 3 (ii) the second term in (29) is estimated $\leq 2|\psi|_{2+\alpha}\;|\beta(x,\partial_x^2\phi)\partial_x^2\phi - \beta(x,\partial_x^2\psi)\partial_x^2\psi|_\alpha$. The last factor is $\leq C|\beta|_2\;|\phi-\psi|_{2+\alpha}$ with some numerical constant $C > 0$. For,

$$\beta(x,\partial_x^2\phi)\;\partial_x^2\phi - \beta(x,\partial_x^2\psi)\;\partial_x^2\psi = \beta(x,\partial_x^2\phi)(\partial_x^2\phi - \partial_x^2\psi) + (\beta(x,\partial_x^2\phi) - \beta(x,\partial_x^2\psi))\partial_x^2\psi \,.$$

The first term is treated as above, for the second term we get by means of Lemma 4 (ii) the estimate $|\beta(x,\partial_x^2\phi) - \beta(x,\partial_x^2\psi)|_\alpha \leq C|\beta|_2\;|\phi-\psi|_{2+\alpha}$ if we take $|\psi|_{2+\alpha}, |\phi|_{2+\alpha} \leq 1$ into account. This finishes the proof of the Lemma.

For the sequence u^n, inductively defined by (25) we find by means of (26) and (23),

$$|u^{n+1}-u^n|_{2+\alpha} \leq \frac{1}{2}\;|u^n-u^{n-1}|_{2+\alpha}$$

if $|u^n|_{2+\alpha}, |u^{n-1}|_{2+\alpha} \leq \varepsilon < (4C_\alpha C)^{-1}$. Hence as long as $|u^j|_\alpha \leq \varepsilon$, for $j = 0,1,\dots,n$, we have:

$$|u^{n+1}-u^n|_{2+\alpha} \leq 2^{-n}|u^1|_{2+\alpha} \leq 2^{-n}C_\alpha\;|\hat{v}|_\alpha \tag{30}$$

and

$$|u^{n+1}|_{2+\alpha} \leq 2C_\alpha|\hat{v}|_\alpha \quad . \tag{31}$$

Therefore, if $|\hat{v}|_\alpha < \delta$, with $\delta = (8C_\alpha^2 C)^{-1}$, we get a unique solution $u = \lim_{n\to\infty} u^n$ in $C^{2+\alpha}$ satisfying (24), and $|u|_{2+\alpha} \leq 2C_\alpha|\hat{v}|_\alpha$. Consequently we find for the required vectorfield v , defined by $C_v\mu = \delta\phi$, $u = (c,\phi)$, $|v|_{1+\alpha} \leq K_\alpha|\hat{\nu}|_\alpha$ with some constant K_α depending on $|\mu|_4$, hence $|\exp(v) - id|_{1+\alpha} \leq K_\alpha|\hat{\nu}|_\alpha$. The diffeomorphism $f := \exp(v)$ and the constant c we constructed satisfy (i) and (ii) of Theorem 3.

From now on we assume $\mu \in C^\infty$. There is a constant $K > 0$, such that for $\alpha < \beta \leq 1$ the following holds true. If $\hat{\nu} \in C^\beta$ and $|\hat{\nu}|_\alpha < \delta$, (for δ possibly smaller but independent of β) then the solution u belongs to C^β and

(32) $$|u|_\beta \leq K|\hat{\nu}|_\beta \quad .$$

Indeed, $u^n \in C^\beta$. Recall $|u^n|_{2+\alpha} \leq 2C_\alpha|\hat{\nu}|_\alpha \leq 1$. Use the following estimates for products of functions, which follow by iteration from Lemma 3(ii):

(33) $$|f_1\cdots f_k|_a \leq C \sum_{i=1}^{k} (\prod_{s\neq i} |f_s|_0)|f_i|_a \quad ,$$

where $f_i \in C^a$. It then follows by means of Lemma 6 and (28), that for all $n = 0,1,2,\ldots$, $|u^{n+1}|_\beta \leq K_1|\hat{\nu}|_\beta + K_2|u^n|_2|u^n|_\beta \leq K_1|\hat{\nu}|_\beta + \frac{1}{2}|u^n|_\beta$, if we choose δ sufficiently small. Consequently, for all $n = 0,1,2,\ldots$, $|u^n|_\beta \leq 2K_1|\hat{\nu}|_\beta$, and we have verified (32). Similarly,we shall demonstrate, that if $\hat{\nu} \in C^\infty$ and $|\hat{\nu}|_\alpha < \delta$, for some possibly smaller δ , then

(34) $$|\phi^n|_{\lambda+\beta} \leq K_\lambda$$

for all $n = 0,1,\ldots$, where $u^n = (\hat{c}^n,\phi^n)$. λ runs over the integers ≥ 2 , and $\alpha \leq \beta \leq 1$. Clearly, $\phi^n \in C^\infty$. We already have proved (34) in the case $\lambda = 2$. We proceed by induction, assuming (34) to hold true for $2 \leq \lambda \leq \ell - 1$, $\ell \geq 3$. Differentiating the identity $\Delta\phi^{n+1} = \hat{\nu} - Q_2(\phi^n) - \hat{c}^n\mu$ one gets for $|\rho| = \ell - 2$, $\Delta D^\rho\phi^{n+1} = D^\rho\hat{\nu} + D^\rho Q_2(\phi^n) +$ terms of lower order. Recalling (31), (28) the Leibnizrule and (33) we can estimate the highest order terms in $D^\rho O_2$ (in the β-norm) by $\leq C|\phi^n|_2 \; |D^\rho \partial_x^2 \phi^n|_\beta$ with a constant C independent of n and ρ, while all the remaining terms are of lower order and hence are estimated by an additive constant using the induction hypotheses. We therefore get:

$$(35) \qquad |\Delta D^\rho \phi^{n+1}|_\beta \leq K_1(\ell) + C|\phi^n|_2 \, \|\phi^n\|_{|\rho|+2+\beta} \quad .$$

We apply the interior Schauder estimate, $|f|_{2+\beta} \leq C_1(|\Delta f|_\beta + |f|_o)$ to $f := D^\rho \phi^{n+1}$ and get from (35), $\|\phi^{n+1}\|_{\ell+\beta} \leq K_2(\ell) + CC_1|\phi^n|_2 \, \|\phi^n\|_{\ell+\beta}$.
Consequently for δ sufficiently small, independently of ℓ, we reach for all $n = 0,1,2,...$

$$(36) \qquad \|\phi^{n+1}\|_{\ell+\beta} \leq K_2(\ell) + \frac{1}{2}\|\phi^n\|_{\ell+\beta} \quad .$$

Therefore $\|\phi^n\|_{\ell+\beta} \leq 2K_2(\ell)$, hence (34) holds true for $\lambda = \ell$, hence for all $\lambda \geq 2$.

From (30) and (34) we conclude by means of the inequality (11), that $\phi \in C^\lambda$ and therefore by (34), $\phi \in C^{\lambda+\beta}$. This finishes the proof of Theorem 3.

4. APPENDIX, THE HODGE THEORY FOR HÖLDER SPACES

Let M be a compact oriented C^∞-manifold of dimension d. The Hodge decomposition states that for all $0 \leq j \leq d$, with $\Delta := d\delta + \delta d$,
$\Delta_j = \Delta|\Omega^j_\infty(M)$

$$(37) \qquad \Omega^j_\infty(M) = \Delta(\Omega^j_\infty) \oplus \ker \Delta_j = d(\Omega^{j-1}_\infty) \oplus \delta(\Omega^{j+1}_\infty) \oplus \ker \Delta_j \quad ,$$

where $\oplus$ means orthogonal with respect to the L_2 structure.
Ker $\Delta_j \subset \Omega^1_\infty(M)$ is the finite dimensional space of harmonic j-forms. Each

cohomology class in the de Rham Cohomology has a unique harmonic representative. Moreover, there is a $C > 0$ such that

(38) $$\frac{1}{C}|f|_{L_2} \le |\Delta f|_{L_2}$$

for all $f \in \Omega^j_\infty \cap (\ker \Delta_j)^{\perp}$. (see f.i P.B. Gilkey [3] or F. W. Warner [5]). What we need, however, is a Hodge decomposition for the Banachspace of Hölder sections, with the correspoding estimates. Since this does not seem to be available in the literature, we shall prove it starting with (37) and (38). We shall use a more general setting and consider a smooth vector bundle E over a compact manifold M. We assume a smooth Hermitian inner product denoted by $(.,.)_m$ on E_m, $m \in M$. On $C^\infty(E)$, the vector space of smooth sections of E, we have the L_2 inner product defined by $(f,g) = \int(f(m),g(m))d_\mu(m)$, $f,g \in C^\infty(E)$; is the canonical volume of a Riemanian metric. Assume $L : C^\infty(E) \to C^\infty(E)$ is a selfadjoint elliptic operator of order 2. Then $C^\infty(E) = B^\infty \oplus H^\infty$, where $H^\infty = \ker L \subset C^\infty(E)$ with $\dim(\ker L) < \infty$, and $B^\infty = L(C^\infty(E)) = (\ker L)^\perp \cap C^\infty(E)$. Moreover,

(39) $$\frac{1}{C}|f| \le |Lf| \quad ,$$

for all $f \in B^\infty$, $|\ |$ standing for the L^2-norm. Let $C^a(E)$ be the Banachspace of Hölder sections, $a \in \mathbb{R}^+$. Clearly $C^{a+2}(E) \subset C^a(E) \subset L_2(M)$ for $a \in \mathbb{R}$, since M is compact. The first inclusion is compact, and the second one continuous. Consequently there is for any $\varepsilon > 0$ a constant $C(\varepsilon) > 0$, such that for all $f \in C^{a+2}$, $|f|_a \le \varepsilon|f|_{a+2} + C(\varepsilon)\,|f|$. On the other hand, since L is elliptic of order 2, we conclude by means of the interior Schauder-estimates [4], that for all $a > 0$ $|f|_{a+2} \le C(|Lf|_a + |f|_0)$, with a constant C depending on a. Choosing ε appropriately we therefore obtain $|f|_{a+2} \le C(|Lf|_q + |f|)$, with some other constant $C > 0$. For $f \in B^\infty$ we can estimate by (39), $|f| \le c_1|Lf| \le c_2|Lf|_a$. Therefore we obtain the following estimate for $f \in B^\infty = (\ker L)^\perp \cap C^\infty(E)$ and $a > 0$:

(40) $$\frac{1}{C_a} |f|_{a+2} \le |Lf|_a \le C_a |f|_{a+2} \quad .$$

The second estimate simply follows from the fact, that L is of second order. Now, $\ker L = H^\infty \subset C^\infty(E)$ being finite dimensional, we have, for $a > 0$, $C^a(E) = B^a \oplus H^\infty$, with $B^a = H^{\infty\perp} \cap C^a(E)$. In order to extend the estimate (40) to B^{a+2} , for all $a > 0$, let $f \in B^{a+2}$. It is easily seen that there is a sequence $f_n \in B^\infty$, such that $|f_n|_{a+2} \le K|f|_{a+2}$, and $|f_n - f|_b \to 0$ as $n \to \infty$, for all $b < a$. Clearly $Lf_n \to Lf$ in C^b , since $L : C^{b+2} \to C^b$ is continuous. Consequently, for given $\varepsilon > 0$, we have for $n > N(\varepsilon)$, $|Lf_n|_b \le |Lf|_b + \varepsilon$. Hence by (40) $\frac{1}{C_b} |f_n|_{b+2} \le |Lf|_a + \varepsilon$ for all $b < a$, hence $\frac{1}{C_b} |f_n|_{a+2} \le |Lf|_a + \varepsilon$. Since $f_n \to f$ in C^b , we therefore conclude that $\frac{1}{C_b} |f|_{a+2} \le |Lf|_a + \varepsilon$. ε being arbitrary, we finally get $\frac{1}{C_b} |f|_{a+2} \le |Lf|_a$. We have shown that the estimate (40) holds true for $f \in B^{a+2}$ and $a > 0$. Similarly we shall prove, that for $a > 0$

(41) $$L(B^{a+2}) = B^a \quad .$$

Since L is selfadjoint, $L(C^{a+2}) \subset B^a$. Let $f \in B^a$, pick $f_n \in B^\infty$ with $|f_n|_a \le K\,|f|_a$ and $|f_n - f|_b \to 0$ as $n \to \infty$. Since $L(B^\infty) = B^\infty$, there is a sequence $\phi_n \in B$ with $L\phi_n = f_n$. From (40) we conclude $|\phi_n|_{a+2} \le C|L|_{n\ a} \le C_1 |f|_a$. Moreover, again by (40), $|\phi_n - \phi_m|_{b+2} \le$ $\le C|L\phi_n - L\phi_m|_b \le |f_n - f_m|_b$. Hence $\phi_n \to \phi$ in C^{b+2} , $b < a$, and $\phi \in C^{a+2}$ since $|\phi_n|_{a+2} \le C_1 |f|_a$. From the continuity of $L : C^{b+2} \to C^b$, we conclude, that $L\phi = f$. We have verified (41). Summarizing we have demonstrated that for all $a > 0$ $C^a(E) = L(C^{a+2}(E) \oplus \ker L = B^a \oplus \ker L$, and $L : B^{a+2} \to B^a$ is a linear isomorphism, such that for all $f \in B^{a+2}$, $|f|_{a+2} \le C_a^{-1} |Lf|_a$. In particular, if $L = d\delta + \delta d = \Delta$, we conclude that for $0 \leq j \leq d$, and $a \in \mathbb{R}^+ \setminus \{o\}$

(42) $$\Omega_a^j(M) = \Delta(\Omega_{a+2}^j) \oplus \ker \Delta_j$$
$$= d(\Omega_{a+1}^{j-1}) \oplus \delta(\Omega_{a+1}^{j+1}) \oplus \ker \Delta_j \quad .$$

Moreover, for all $f \in (\ker \Delta)^{\perp} \cap \Omega_{a+2}^j$,

(43) $$\frac{1}{C_a} |f|_{a+2} \leq |\Delta f|_a \leq C_a |f|_{a+2} \quad .$$

For the proof of Lemma 6, we made use of the following criterium which follows readily from (42) together with the de Rham theorem. Let $\dim(M) = d$, then a d-form $f \in \Omega_a^d(M)$, $a > 0$, belongs to B^a if and only if $\int_M f = 0$.

Final remarks. The question, whethereTheorem 2 holds true in the C^1 case (not only $C^{1+\alpha}$, $a > 0$) for $\dim M \geq 3$ remains unanswered. Related to the problem is the following question: given a continuous function $f \in C^o(U)$, $U \subset \mathbb{R}^n$, does there exist a vectorfield $v \in C^1(U,\mathbb{R}^n)$, solving the equation $\operatorname{div} v = f$? Finally, one could ask whether Thereom 3 holds true if M is not compact.

REFERENCES

[1]: J. Moser: "On the volume-elements on a manifold". Trans. of the AMS, 120(1965), 286-294.

[2]: J. R. Munkres: "Elementary Differential Topology". Princeton University Press, (1966).

[3] : F. W. Warner: "Foundations of differentiable manifolds and Lie groups". Scott, Foresman and Co., (1971).

[4]: L. Bers / F. John / M. Schechter: "Partial Differential Equations". Interscience Publishers, (1964).

[5] : P. B. Gilkey: "The Index Theorem and the Heat Equation". Publish and Perish, Inc. (1974).

[6] : J. Palis and C. Pugh: "Fifty problems in dynamical systems". Springer Lecture notes in Mathematics, Vol. 468, (1975), p. 352.

Eduard Zehnder
Department of Mathematics
Ruhr-Universität Bochum
D-4630 Bochum / West-Germany

A simple proof of a generalization of a Theorem by C. L. Siegel

Eduard Zehnder

Our purpose is to give a simple proof of the theorem stated below. The proof illustrates a modification of the Newton iteration method, which has been introduced by H. Rüssmann for a related problem [3]. The technique itself can be successfully applied to more intricate problems involving the so called difficulty of small divisors. The theorem runs as follows.

Theorem: Let $z \to f(z) = \Lambda z + \hat{f}(z)$ be a holomorphic map in a neighborhood of 0 in $\mathbb{C}^n$, $\hat{f}$ contains only terms of order ≥ 2. Assume $\Lambda = (\lambda_1, \ldots, \lambda_n)$ to be diagonal, the eigenvalues λ_k, $1 \leq k \leq n$ satisfying

(1) $$|\lambda^j - \lambda_k| > C_0 \; |j|^{-\nu} \; ,$$

for all integervectors $j = (j_n, \ldots, j_n)$, $j_k \geq 0$ with $|j| = \sum_{k=1}^{n} j_k > 1$. C_0 and ν are two positive constants, and λ^j stands for $\lambda_1^{j_1} \cdot \lambda_2^{j_2} \; \ldots \; \lambda_n^{j_n}$.

Then there is a (unique) holomorphic map $z = u(\zeta) = \zeta + \hat{u}(\zeta)$ in a neighborhood of 0, $\hat{u}$ containing terms of order ≥ 2 only, such that

(2) $$f(u(\zeta)) = u(\Lambda\zeta) \; .$$

The local diffeomorphism u therefore transforms f into its (at 0) linearized map Λ.

<u>Historical comments</u>: The expansion for u, $u(\zeta) = \zeta + \sum_{|\alpha| \geq 2} u_\alpha \zeta^\alpha$ is uniquely determined, if only the eigenvalues λ_k satisfy:

(3) $$\lambda^j - \lambda_k \neq 0$$

for all $|j| \geq 1$. This is easily seen by comparing the coefficients in the deforming relation (2), the coefficients u_α are recursively uniquely determined, if (3) holds true. The main point is to investigate the convergence of the series for u. If we assume in addition to (3), that

(4) $$|\lambda_k| < 1 \quad , \quad 1 \leq k \leq n \quad ,$$

then the convergence is easily established. This was already known to Poincaré in 1879 [11]. Without the restriction (4) on λ_k we are confronted with a small divisor problem. In the special case $n = 1$, the equation (2) is the so called functional equation of Schröder. [1] and also [4] on p. (186 - 198). It is well-known, that the set of λ with $\lambda^j \neq 1$ for $j \geq 1$, for which there is a holomorphic function $f(z) = \lambda z + \hat{f}(z)$ whose series u diverges, form a dense set on the unit circle $|\lambda| = 1$, [2] and [4]. If however λ with $|\lambda| = 1$ satisfies the conditions $|\lambda^j - 1| \geq C_0 |j|^{-\nu}$ for all $j \geq 1$ with some positive constants C_0 and ν, then the series u does converge. This was shown by C. L. Siegel in 1942 [5]. It was actually the first time a small divisor difficulty was overcome. Later, in 1952, C. L. Siegel proved the analogon of the stated theorem for vectorfields near an equilibrium point instead of mappings [6]. His delicate estimate technique of the series u (Cauchy majorants) could not be applied to the more intricate small divisor problems arising in celestial mechanics. To cope with these difficulties, Kolmogorov, Arnold and Moser introduced an entirely different technique for solving certain nonlinear functional equations (As for the literature see in [4]). It consists of a rapidely converging iteration procedure based on a modification of Newton's method, involving infinitely many coordinate transformations. In his book, "Celestial Mechanics II" S. Sternberg proves the theorem for $n > 1$ using the same technique, under the additional assumption however, that

$$|\lambda_k| \leq 1 \quad , \quad 1 \leq k \leq n \quad . \tag{5}$$

This assumption is needed merely for some technical reasons in order to control the domains of the transformations involved, and in his book (p. 96 - 99) Sternberg himself expressed doubt, that the condition (5) is essential (see however [7]). In fact the condition (5) can be dropped, which was noticed by A. Gray who followed Sternberg's exposition in his proof [8] p. 360 - 365. Earlier V. Arnold indicated a possible proof [9] p.(24 - 27) of the stated theorem. He attributes the result to Siegel who proved the case $n = 1$. In this case however the above appearent obstacle which required Sternberg to impose the restriction (5), does not occur.

Nowadays one proves such theorems with a more flexible iteration technique [3] and [10] , as we shall illustrate.

Proof

(a) Idea. We consider the equation (2) with $u = id + \hat{u}$, as a functional equation for holomorphic functions $\hat{u}$ satisfying $\hat{u}(0) = 0$ and $d\hat{u}(0) = 0$. We try to solve $F(u) = 0$, where

$$F(u) := f \circ u - u \circ \Lambda \quad . \tag{6}$$

Since $F(id) = \hat{f}$, which is small in a small neighborhood of zero, we are dealing with a perturbation problem. Indeed, since $\hat{f}$ contains only terms of order ≥ 2 , we may assume that f is holomorphic on $|z_j| < 1$, and that

$$(7) \qquad \sup_{|z_j|<1} |\hat{f}(z)| < \delta_0 ,$$

for δ_0 as small as we want, to be chosen later on. (Otherwise set $z = \varepsilon\zeta$, $\varepsilon > 0$; in the new variables the mapping f then is given by $\zeta \to \Lambda\zeta + \frac{1}{\varepsilon}\hat{f}(\varepsilon\zeta)$).

We shall solve (6) by means of a modification of Newton's iteration method. Assuming $F(u)$ to be small, we are looking for a better approximation $u + v$, which makes $F(u+v)$ smaller. Taylor expansion of $F(u+tv)$ yields

$$(8) \qquad F(u+v) = F(u) + F'(u)v + R(u;v) ,$$

where $F'(u)v := \frac{d}{dt} F(u+tv)/_{t=o}$ is given by

$$(9) \qquad F'(u)v = df \circ u \cdot v - v \circ \Lambda .$$

For the remainder term $R(u;v)$ we have

$$(10) \qquad R(u;v) = \frac{1}{2}\int_0^1 (1-t) \frac{d^2}{dt^2} \hat{f}(u+tv)dt .$$

Following Newton's method, we would have to solve $F(u) + F'(u)v = 0$ achieving $F(u+v) = 0_2(F(u))$. Unfortunately the linear operator $F'(u)$ given by (9) has no rightinverse. But using the special structure of the functional equation (6), we shall be able to construct a sufficiently good approximate rightinverse of $F'(u)$. To do so we follow Rüssmann [3] and differentiate the function $z \to F(u)(z)$. By the chain rule we get:

$$(11) \qquad d\,F(u) = df \circ u \cdot du - du \circ \Lambda \cdot \Lambda .$$

Now put

$$(12) \qquad v := du\ (w) .$$

Comparison of (9) and (11) yields the functional identity $F'(u)v = d\,F(u)\,(w) + du \circ \Lambda\ (\Lambda w - w \circ \Lambda)$. Consequently

$$(13) \qquad F(u+v) = F(u) + du \circ \Lambda\ (\Lambda w - w \circ \Lambda) + R(u;v) + d\,F(u)\,(w) ,$$

where still $v = du\,(w)$. As the next Lemma shows, the linear operator $\Lambda w - w \circ \Lambda$ has a rightinverse. Hence we can solve the equation $F(u) + du \circ \Lambda\ (\Lambda w - w \circ \Lambda)$ in case $du(\Lambda z)$ is invertible. According to (13) we then still have $F(u+v) = O_2(F(u))$. In order to formulate the Lemma we introduce some notations, which will be used also later on.

Let $r > 0$, with $D_r \subset C^n$ we denote the polyzylinder:

$$D_r := \{z \in C^n \mid |z_j| < r\ ,\ 1 \le j \le n\} \quad . \tag{14}$$

If g is holomorphic on D_r , we put

$$|g|_r := \sup_{z \in D_r} |g(z)| \quad . \tag{15}$$

If g is holomorphic on $D_r \cup \Lambda(D_r)$, with the linear isomorphism Λ as intorduced in the theorem, we define

$$\|g\|_r := \sup_{z \in D_r \cup \Lambda(D_r)} |g(z)| \quad . \tag{16}$$

Finally we set

$$\|g\|_{C^1,r} := \max\{\|g\|_r\ ,\ \|dg\|_r\} \quad . \tag{17}$$

For a vectorvalued function g we take the maximum over its components. The following simple Lemma expoits our assumption (1) quantitatively:

<u>Lemma</u>: Assume Λ satisfies (1). Let g be a holomorphic map on D_r into C^n for some $o < r \le 1$, such that $g(o) = o$, $dg(o) = o$ and such that $|g|_r < \infty$. Then the linear equation

$$w(\Lambda z) - \Lambda w(z) = g(z) \tag{18}$$

has a unique solution w with $w(o) = o$, $dw(o) = o$, w is holomorphic on $D_r \cup \Lambda(D_r)$. Moreover the following estimates hold true for all $o < \delta < r$:

$$\|w\|_{C^1,r-\delta} \le C_1\, \delta^{\tau}\, |g|_r \quad , \tag{19}$$

where $\tau = \nu + n + 2$. C_1 is a constant depending on ν , n , C_0 $|\Lambda|$ and $|\Lambda^{-1}|$. In the following we shall write $w = L(g)$ for the above solution, defining so the rightinverse L of the linear operator on the left hand side of (18).

<u>Proof</u>: Let $g = (g^1,\dots,g^n)$. $g^k(z) = \sum_{|\alpha| \geq 2} g^k_\alpha z^\alpha$. Put $w = (w^1,\dots,w^n)$, $w^k(z) = \sum_{|\alpha| \geq 2} w^k_\alpha z^\alpha$. Comparison of the coefficients in (18) yields for the coefficients w^k_α :

$$w^k_\alpha = \frac{g^k_\alpha}{\lambda^\alpha - \lambda_k} \quad .$$

Since $|g|_r < \infty$, we have $|g^k_\alpha| \leq |g|_r \, r^{-|\alpha|}$. Consequently, together with (1) we can estimate

$$|w^k_\alpha| \leq \frac{1}{C_0} |g|_r \quad |\alpha|^\nu \, r^{-|\alpha|} \quad .$$

For the series w^k we therefore get, if $z \in D_{r-\delta}$, $0 < \delta < r$:

$$|w^k(z)| \leq \sum_{|\alpha| \geq 2} |w^k_\alpha| \; |z^\alpha|$$

$$\leq \frac{1}{C_0} |g|_r \sum_{|\alpha| \geq 2} |\alpha|^\nu \left(\frac{r-\delta}{r}\right)^{|\alpha|} \quad .$$

The number of α's with $|\alpha| = k$ is less than k^n . Since $r \leq 1$, we have $\left(\frac{r-\delta}{r}\right)^k = e^{-(\log r - \log(r-\delta))k} \leq e^{-\delta k}$. Consequently .

$$|w^k(z)| \leq \frac{1}{C_0} |g|_r \sum_{k \geq 2} k^{\nu+n} e^{-\delta k}$$

$$\leq \frac{1}{C_0} |g|_r \int_0^\infty x^{\nu+n} e^{-\nu x} \, dx$$

$$= \frac{1}{C_0} |g|_r \; \delta^{-(\nu+n+1)} \int_0^\infty y^{\nu+n} e^{-y} \, dy \quad .$$

Therefore, $|w^k(z)| \leq C_2 \; \delta^{-(\nu+n+1)} |g|_r < \infty$ for $z \in D_{r-\delta}$, and all $0 < \delta < r$. Hence w is holomorphic in D_r . Moreover, since w satisfies the equation (18) we conclude that w is actually holomorphic in $D_r \cup \Lambda(D_r)$, and for $0 < \delta < r$, $|w \circ \Lambda|_{r-\delta} \leq |\Lambda| \, |w|_{r-\delta} + |g|_{r-\delta} \leq (|\Lambda| C_2 + 1)\delta^{-(\nu+n+1)} |g|_r$, since $\delta \leq 1$. By means of the Cauchy estimate we have $|dw|_{r-\delta} \leq 2\delta^{-1}|w|_{r-\delta/2}$, and $|dw \circ \Lambda|_{r-\delta} \leq |d(w \circ \Lambda)|_{r-\delta}|\Lambda^{-1}| \leq 2\delta^{-1} |\Lambda^{-1}| \, |w \circ \Lambda|_{r-\delta/2}$. The required estimate (19) follows with definition (17).

With the linear operator L provided by the Lemma, we shall define inductively the iteration u_k, for $k = 0,1,2,\dots$ as follows. $u_0 = \mathrm{id}$, and for $k \geq 0$

$$u_{k+1} = u_k + v_k \,, \tag{20}$$
$$v_k = du_k\,(w_k)\,, \quad w_k = L((du_k \circ \Lambda)^{-1} \; F(u_k)) \,.$$

By the Lemma and by formula (13) we then have

$$F(u_{k+1}) = d\,F(u_k)\,(w_k) + R(u_k;v_k) \,, \tag{21}$$

which is $O_2(F(u_k))$. In order to prove that the sequence u_k converges to a solution u of $F(u) = 0$, we use the following quantitativ set up.

b) <u>Set up</u>: The domains D_{r_k} are defined by (14) with

$$r_k = \frac{1}{2}\,(1 + 2^{-(k+1)}) \,, \quad k \geq 0 \,. \tag{22}$$

Obviously $\lim_{k\to\infty} r_k = \frac{1}{2}$, and $D_{r_{k+1}} \subset D_{r_k}$ for all $k \geq 0$. The sequence of small numbers, ε_k, is defined by

$$\varepsilon_{k+1} = C^{k+1} \; \varepsilon_k^2 \,, \quad k \geq 0 \,, \tag{23}$$

where $C > 4$ is a large constant depending on ν, C_0, n, $|\Lambda|$ and $|\Lambda^{-1}|$. It will be determined later on. For ε_0 sufficiently small, the sequence ε_k tends rapidely to zero, indeed for all $k \geq 0$

$$\varepsilon_k = C^{-(k+2)} \, (C^2\varepsilon_0)^{(2^k)} \,. \tag{24}$$

We choose $\varepsilon_o = (2C^2)^{-1} \le 16^{-1}$. In particular we then have

(25) $$\varepsilon_{k+1} \le \frac{1}{2}\,\varepsilon_k \le \varepsilon_k - \varepsilon_{k+1} \quad .$$

We shall prove, that if ε_o is sufficiently small (that is to say C sufficiently large), then the following statements for $k = o,1,2,\ldots$ hold true, for the sequence of mappings u_k inductively defined by (20).

(1.k): u_k is holomorphic on $D_{r_k} \cup \Lambda(D_{r_k})$, $u_k(o) = o$, $du_k(o) = 1$, and

$$\|u_k - \mathrm{id}\|_{C^1, r_k} \le \varepsilon_o - \varepsilon_k$$

(2.k): $F(u_k)$ is holomorphic on D_{r_k}, and

$$|F(u_k)|_{r_k} \le \varepsilon_k^2$$

(3.k): v_k is holomorphic on $D_{r_{k+1}} \cup \Lambda(D_{r_{k+1}})$, $v_k(o) = o$, $dv_k(o) = o$, and

$$\|v_k\|_{C^1, r_{k+1}} \le \varepsilon_k - \varepsilon_{k+1} \quad .$$

From (3.k), $k = o,1,2,\ldots$ we conclude, since $v_k = u_{k+1} - u_k$, that $u(z) := \lim_{k\to\infty} u_k(z)$ uniformly for $z \in D_{1/2}$, hence u is a holomorphic map defined on $D_{1/2}$. From (1.k) we have $u(o) = o$ and $du(o) = 1$. As a consequence of (2.k) we have on $D_{1/2}$, $F(u) = \lim_{k\to\infty} F(u_k) = o$. Therefore u solves our problem.

c) <u>Induction sup</u>: The proof of the statements (1.k) - (3.k) is by induction. Statement (1.o) is trivially satisfied, since $u_o = \mathrm{id}$. Statement (2.o) simply exresses the smallnes condition on $\hat{f} = F(\mathrm{id})$, we require:

(26) $$|\hat{f}|_{r_o} \le |\hat{f}|_1 \le \varepsilon_o^2 =: \delta_o \quad .$$

Statement (3.o) follows from (1.o) and (2.o) by means of the Lemma, as we shall see.

We first show, that (1.k) and (2.k) imply (3.k), by means of the Lemma. Indeed, on D_{r_k} we have by (1.k), $\|du_k - 1\|_{r_k} \leq 1/2$, hence $\|du_k\|_{r_k} \leq 2$, and $\|du_k^{-1}\|_{r_k} = \|[1-(du_k-1)]^{-1}\|_{r_k} \leq 2$. Hence by (2.k), $g_k := (du_k \circ \Lambda)^{-1} F(u_k)$ is holomorphic on D_{r_k}, and $|g_k|_{r_k} \leq 2|F(u_k)|_{r_k} \leq 2\varepsilon_k^2$. Since also $g_k(o) = o$, and $dg_k(o) = o$, we can apply the Lemma in order to define $w_k = L(g_k)$. Put $\rho_k = \frac{1}{2}(r_k + r_{k+1})$, hence $r_{k+1} < \rho_k < r_k$. Observing $r_k - \rho_k = \frac{1}{2}(r_k - r_{k+1}) = 2^{-(k+3)}$, we conclude from (19)

$$(27) \qquad \|w_k\|_{\rho_k} \leq C_3^{k+1} \varepsilon_k^2 ,$$

for some constant C_3 (depending on C_1). For $v_k := du_k(w_k)$ we have $\|v_k\|_{\rho_k} \leq 2\|w_k\|_{\rho_k}$. Moreover, by means of the Cauchy estimate, $|dv_k|_{r_{k+1}} \leq (r_{k+1}-\rho_k)^{-1} |v_k|_{r_k}$ and $|dv_k \circ \Lambda|_{r_{k+1}} \leq |d(v_k \circ \Lambda)|_{r_{k+1}} |\Lambda^{-1}| \leq (r_{k+1}-\rho_k)^{-1} |\Lambda^{-1}| |v_k \circ \Lambda|_{\rho_k}$. Consequently, for some constant $C_4 > C_3$ we have

$$(28) \qquad \|v_k\|_{C^1, r_{k+1}} \leq C_4^{k+1} \varepsilon_k^2 \leq C^{k+1} \varepsilon_k^2$$

$$= \varepsilon_{k+1} \leq \varepsilon_k - \varepsilon_{k+1} ,$$

by means of (23) and (25), if we choose $C \geq C_4$. We have proved that (3.k) follows from (1.k) and (2.k).

Trivially the statement 1.(k+1) follows from (1.k) and (3.k). It remains to prove statement 2.(k+1), assuming (1.k+1) and the statements (1.k) - (3.k) to be true. We first observe, that $F(u_{k+1}) = F(u_k + v_k)$ is defined on $D_{r_{k+1}}$. Indeed from (1.k) and (3.k) we get for $z \in D_{r_{k+1}} \subset D_{r_1}$:

$$(29) \quad |u_k(z)+v_k(z)| \le |z| + |u_k(z)-z| + |v_k(z)|$$

$$\le |z| + 2\varepsilon_0 \le r_1 + 2\varepsilon_0 \le 3/4 \quad ,$$

since $\varepsilon_0 \le (16)^{-1}$. Now recall that by (21) we have $F(u_k+v_k) = = dF(u_k)(w_k) + R(u_k;v_k)$. Using (2.k), (27), and the Cauchy estimate, we conclude with some contant $C_5 \ge C_4$, that

$$(30) \quad |dF(u_k)|_{r_{k+1}} |w_k|_{r_{k+1}} \le \frac{1}{2}(C_5^{k+1}\varepsilon_k^2)^2$$

$$\le \frac{1}{2}(C^{k+1}\varepsilon_k^2)^2 = \frac{1}{2}\varepsilon_{k+1}^2 \quad ,$$

if we choose $C \ge C_5$. Finally, from (10) we reach, since by (26), and by the Cauchy estimate, $|d^2\hat{f}|_{r_0} \le 16|\hat{f}|_1 \le (16)\varepsilon_0^2 < 1$,

$$(31) \quad |R(u_k;v_k)|_{r_{k+1}} \le \frac{1}{2}|d^2\hat{f}|_{r_0} |v_k|^2_{r_{k+1}}$$

$$\le \frac{1}{2}\varepsilon_{k+1}^2 \quad ,$$

using (3.k). From (30) and (31) we conclude, that $|F(u_{k+1})|_{r_{k+1}} \le \varepsilon_{k+1}^2$, hence 2.k+1 holds true. The induction step is complet. The Theorem is proved.

Remark: Writing $F(f,u)$ instead of $F(u)$ for (6) we observe that $D_2F(\Lambda,id)(v) = \Lambda v - v \circ \Lambda$. According to the Lemma, the operator $D_2F(\Lambda,id)$ has a rightinverse, which we called L. If now f is as in the theorem, with $df(o) = \Lambda$, then the operator $\eta(u)$, given by $\eta(u)(g) = du \; L((du \circ \Lambda)^{-1}g)$ is an approximate rightinverse of $D_2F(f,u)$ in the technical sense of [10]. Indeed, if $\|u-id\|_{C^1,r} < 1/2$, we have

$\|\eta(u)(g)\|_{C^1,r-\delta} \leq M\ \delta^{-\tau}|g|_r$ for some constant $M > o$. Moreover from $D_2F(f,u) \circ \eta(u)(g) = g + dF(u)L((du \circ \Lambda)^{-1}g)$ (see the identity leading to (13)), we conclude $|(D_2F(f,u) \circ \eta(u) - 1)(g)|_{r-\delta}$ $\leq M\ \delta^{-(\tau+1)}|F(u)|_r\ |g|_r$. Hence the assumptions of the abstract implizit function theorem of [10] are met and hence the theorem follows.

References

[1] E. Schröder: Über iterierte Funktionen.
Math. Ann. 3. (p. 296-322) (1871).

[2] H. Cremer: Über die Häufigkeit der Nichtzentren.
Math. Ann. 115. (p. 573-580) (1938).

[3] H. Rüssmann: Kleine Nenner II: Bemerkungen zur Newton'schen Methode.
Nachr. Akad. Wiss. Göttingen, Math. Phys. Kl. pp. 1-20 1972.

[4] C. L. Siegel / J. K. Moser: "Lectures on Celestial Mechanics"
Springer, 1970.

[5] C. L. Siegel: Iteration of analytic functions.
Ann. Math. 43. 6o7-612 (1942).

[6] C. L. Siegel: Über die Normalform analytischer Differentialgleichungen in der Nähe einer Gleichgewichtslösung. Nachr. Akad. Wiss. Göttingen, Math.-phys. Kl. 21-30 1952.

[7] S. Sternberg: Infinite Lie Groups and the formal Aspects of Dynamical Systems. J. of Math. and Mech. 10 (1961) (451-474) p. 465.

[8] A. Gray: A fixed point theorem for small divisor problems. J. of Diff. eq. 18 (1975) 346-365.

[9] V. Arnold: Singularities of smooth mappings. Russian Math. Surveys Vol. XXIII (1-43), 1968.

[10] E. Zehnder: Generalized implicit function theorem with applications to some small divisor problems. I. Comm. Pure Appl. Math., Vol. 28 (91-140), 1975.

[11] H. Poincaré: Sur les propriétés des fonctions définies par les équations aux différences partielles. Oeuvres, 1, Gauthier-Villars, Paris (1929) (IX 7), pp. XCIX-CX.

Eduard Zehnder
Mathematisches Institut
Ruhr-Universität Bochum
D-4630 Bochum
West-Germany

Vol. 460: O. Loos, Jordan Pairs. XVI, 218 pages. 1975.

Vol. 461: Computational Mechanics. Proceedings 1974. Edited by J. T. Oden. VII, 328 pages. 1975.

Vol. 462: P. Gérardin, Construction de Séries Discrètes p-adiques. »Sur les séries discrètes non ramifiées des groupes réductifs déployés p-adiques«. III, 180 pages. 1975.

Vol. 463: H.-H. Kuo, Gaussian Measures in Banach Spaces. VI, 224 pages. 1975.

Vol. 464: C. Rockland, Hypoellipticity and Eigenvalue Asymptotics. III, 171 pages. 1975.

Vol. 465: Séminaire de Probabilités IX. Proceedings 1973/74. Edité par P. A. Meyer. IV, 589 pages. 1975.

Vol. 466: Non-Commutative Harmonic Analysis. Proceedings 1974. Edited by J. Carmona, J. Dixmier and M. Vergne. VI, 231 pages. 1975.

Vol. 467: M. R. Essén, The Cos $\pi\lambda$ Theorem. With a paper by Christer Borell. VII, 112 pages. 1975.

Vol. 468: Dynamical Systems – Warwick 1974. Proceedings 1973/74. Edited by A. Manning. X, 405 pages. 1975.

Vol. 469: E. Binz, Continuous Convergence on C(X). IX, 140 pages. 1975.

Vol. 470: R. Bowen, Equilibrium States and the Ergodic Theory of Anosov Diffeomorphisms. III, 108 pages. 1975.

Vol. 471: R. S. Hamilton, Harmonic Maps of Manifolds with Boundary. III, 168 pages. 1975.

Vol. 472: Probability-Winter School. Proceedings 1975. Edited by Z. Ciesielski, K. Urbanik, and W. A. Woyczyński. VI, 283 pages. 1975.

Vol. 473: D. Burghelea, R. Lashof, and M. Rothenberg, Groups of Automorphisms of Manifolds. (with an appendix by E. Pedersen) VII, 156 pages. 1975.

Vol. 474: Séminaire Pierre Lelong (Analyse) Année 1973/74. Edité par P. Lelong. VI, 182 pages. 1975.

Vol. 475: Répartition Modulo 1. Actes du Colloque de Marseille-Luminy, 4 au 7 Juin 1974. Edité par G. Rauzy. V, 258 pages. 1975. 1975.

Vol. 476: Modular Functions of One Variable IV. Proceedings 1972. Edited by B. J. Birch and W. Kuyk. V, 151 pages. 1975.

Vol. 477: Optimization and Optimal Control. Proceedings 1974. Edited by R. Bulirsch, W. Oettli, and J. Stoer. VII, 294 pages. 1975.

Vol. 478: G. Schober, Univalent Functions – Selected Topics. V, 200 pages. 1975.

Vol. 479: S. D. Fisher and J. W. Jerome, Minimum Norm Extremals in Function Spaces. With Applications to Classical and Modern Analysis. VIII, 209 pages. 1975.

Vol. 480: X. M. Fernique, J. P. Conze et J. Gani, Ecole d'Eté de Probabilités de Saint-Flour IV–1974. Edité par P.-L. Hennequin. XI, 293 pages. 1975.

Vol. 481: M. de Guzmán, Differentiation of Integrals in R^n. XII, 226 pages. 1975.

Vol. 482: Fonctions de Plusieurs Variables Complexes II. Séminaire François Norguet 1974–1975. IX, 367 pages. 1975.

Vol. 483: R. D. M. Accola, Riemann Surfaces, Theta Functions, and Abelian Automorphisms Groups. III, 105 pages. 1975.

Vol. 484: Differential Topology and Geometry. Proceedings 1974. Edited by G. P. Joubert, R. P. Moussu, and R. H. Roussarie. IX, 287 pages. 1975.

Vol. 485: J. Diestel, Geometry of Banach Spaces – Selected Topics. XI, 282 pages. 1975.

Vol. 486: S. Stratila and D. Voiculescu, Representations of AF-Algebras and of the Group U (∞). IX, 169 pages. 1975.

Vol. 487: H. M. Reimann und T. Rychener, Funktionen beschränkter mittlerer Oszillation. VI, 141 Seiten. 1975.

Vol. 488: Representations of Algebras, Ottawa 1974. Proceedings 1974. Edited by V. Dlab and P. Gabriel. XII, 378 pages. 1975.

Vol. 489: J. Bair and R. Fourneau, Etude Géométrique des Espaces Vectoriels. Une Introduction. VII, 185 pages. 1975.

Vol. 490: The Geometry of Metric and Linear Spaces. Proceedings 1974. Edited by L. M. Kelly. X, 244 pages. 1975.

Vol. 491: K. A. Broughan, Invariants for Real-Generated Uniform Topological and Algebraic Categories. X, 197 pages. 1975.

Vol. 492: Infinitary Logic: In Memoriam Carol Karp. Edited by D. W. Kueker. VI, 206 pages. 1975.

Vol. 493: F. W. Kamber and P. Tondeur, Foliated Bundles and Characteristic Classes. XIII, 208 pages. 1975.

Vol. 494: A Cornea and G. Licea. Order and Potential Resolvent Families of Kernels. IV, 154 pages. 1975.

Vol. 495: A. Kerber, Representations of Permutation Groups II. V, 175 pages. 1975.

Vol. 496: L. H. Hodgkin and V. P. Snaith, Topics in K-Theory. Two Independent Contributions. III, 294 pages. 1975.

Vol. 497: Analyse Harmonique sur les Groupes de Lie. Proceedings 1973–75. Edité par P. Eymard et al. VI, 710 pages. 1975.

Vol. 498: Model Theory and Algebra. A Memorial Tribute to Abraham Robinson. Edited by D. H. Saracino and V. B. Weispfenning. X, 463 pages. 1975.

Vol. 499: Logic Conference, Kiel 1974. Proceedings. Edited by G. H. Müller, A. Oberschelp, and K. Potthoff. V, 651 pages 1975.

Vol. 500: Proof Theory Symposion, Kiel 1974. Proceedings. Edited by J. Diller and G. H. Müller. VIII, 383 pages. 1975.

Vol. 501: Spline Functions, Karlsruhe 1975. Proceedings. Edited by K. Böhmer, G. Meinardus, and W. Schempp. VI, 421 pages. 1976.

Vol. 502: János Galambos, Representations of Real Numbers by Infinite Series. VI, 146 pages. 1976.

Vol. 503: Applications of Methods of Functional Analysis to Problems in Mechanics. Proceedings 1975. Edited by P. Germain and B. Nayroles. XIX, 531 pages. 1976.

Vol. 504: S. Lang and H. F. Trotter, Frobenius Distributions in GL_2-Extensions. III, 274 pages. 1976.

Vol. 505: Advances in Complex Function Theory. Proceedings 1973/74. Edited by W. E. Kirwan and L. Zalcman. VIII, 203 pages. 1976.

Vol. 506: Numerical Analysis, Dundee 1975. Proceedings. Edited by G. A. Watson. X, 201 pages. 1976.

Vol. 507: M. C. Reed, Abstract Non-Linear Wave Equations. VI, 128 pages. 1976.

Vol. 508: E. Seneta, Regularly Varying Functions. V, 112 pages. 1976.

Vol. 509: D. E. Blair, Contact Manifolds in Riemannian Geometry. VI, 146 pages. 1976.

Vol. 510: V. Poènaru, Singularités C^∞ en Présence de Symétrie. V, 174 pages. 1976.

Vol. 511: Séminaire de Probabilités X. Proceedings 1974/75. Edité par P. A. Meyer. VI, 593 pages. 1976.

Vol. 512: Spaces of Analytic Functions, Kristiansand, Norway 1975. Proceedings. Edited by O. B. Bekken, B. K. Øksendal, and A. Stray. VIII, 204 pages. 1976.

Vol. 513: R. B. Warfield, Jr. Nilpotent Groups. VIII, 115 pages. 1976.

Vol. 514: Séminaire Bourbaki vol. 1974/75. Exposés 453 – 470. IV, 276 pages. 1976.

Vol. 515: Bäcklund Transformations. Nashville, Tennessee 1974. Proceedings. Edited by R. M. Miura. VIII, 295 pages. 1976.

Vol. 516: M. L. Silverstein, Boundary Theory for Symmetric Markov Processes. XVI, 314 pages. 1976.

Vol. 517: S. Glasner, Proximal Flows. VIII, 153 pages. 1976.

Vol. 518: Séminaire de Théorie du Potentiel, Proceedings Paris 1972–1974. Edité par F. Hirsch et G. Mokobodzki. VI, 275 pages. 1976.

Vol. 519: J. Schmets, Espaces de Fonctions Continues. XII, 150 pages. 1976.

Vol. 520: R. H. Farrell, Techniques of Multivariate Calculation. X, 337 pages. 1976.

Vol. 521: G. Cherlin, Model Theoretic Algebra – IV, 234 pages. 1976.

Vol. 522: C. O. Bloom and N. D. Kazarinoff, Sho Problems in Inhomogeneous Media: Asymptotic pages. 1976.

Vol. 523: S. A. Albeverio and R. J. Høegh-Kro Theory of Feynman Path Integrals. IV, 139 pages. 1976.

Vol. 524: Séminaire Pierre Lelong (Analyse) Année 1974/75. Edité par P. Lelong. V, 222 pages. 1976.

Vol. 525: Structural Stability, the Theory of Catastrophes, and Applications in the Sciences. Proceedings 1975. Edited by P. Hilton. VI, 408 pages. 1976.

Vol. 526: Probability in Banach Spaces. Proceedings 1975. Edited by A. Beck. VI, 290 pages. 1976.

Vol. 527: M. Denker, Ch. Grillenberger, and K. Sigmund, Ergodic Theory on Compact Spaces. IV, 360 pages. 1976.

Vol. 528: J. E. Humphreys, Ordinary and Modular Representations of Chevalley Groups. III, 127 pages. 1976.

Vol. 529: J. Grandell, Doubly Stochastic Poisson Processes. X, 234 pages. 1976.

Vol. 530: S. S. Gelbart, Weil's Representation and the Spectrum of the Metaplectic Group. VII, 140 pages. 1976.

Vol. 531: Y.-C. Wong, The Topology of Uniform Convergence on Order-Bounded Sets. VI, 163 pages. 1976.

Vol. 532: Théorie Ergodique. Proceedings 1973/1974. Edité par J.-P. Conze and M. S. Keane. VIII, 227 pages. 1976.

Vol. 533: F. R. Cohen, T. J. Lada, and J. P. May, The Homology of Iterated Loop Spaces. IX, 490 pages. 1976.

Vol. 534: C. Preston, Random Fields. V, 200 pages. 1976.

Vol. 535: Singularités d'Applications Differentiables. Plans-sur-Bex. 1975. Edité par O. Burlet et F. Ronga. V, 253 pages. 1976.

Vol. 536: W. M. Schmidt, Equations over Finite Fields. An Elementary Approach. IX, 267 pages. 1976.

Vol. 537: Set Theory and Hierarchy Theory. Bierutowice, Poland 1975. A Memorial Tribute to Andrzej Mostowski. Edited by W. Marek, M. Srebrny and A. Zarach. XIII, 345 pages. 1976.

Vol. 538: G. Fischer, Complex Analytic Geometry. VII, 201 pages. 1976.

Vol. 539: A. Badrikian, J. F. C. Kingman et J. Kuelbs, Ecole d'Eté de Probabilités de Saint Flour V-1975. Edité par P.-L. Hennequin. IX, 314 pages. 1976.

Vol. 540: Categorical Topology, Proceedings 1975. Edited by E. Binz and H. Herrlich. XV, 719 pages. 1976.

Vol. 541: Measure Theory, Oberwolfach 1975. Proceedings. Edited by A. Bellow and D. Kölzow. XIV, 430 pages. 1976.

Vol. 542: D. A. Edwards and H. M. Hastings, Čech and Steenrod Homotopy Theories with Applications to Geometric Topology. VII, 296 pages. 1976.

Vol. 543: Nonlinear Operators and the Calculus of Variations, Bruxelles 1975. Edited by J. P. Gossez, E. J. Lami Dozo, J. Mawhin, and L. Waelbroeck, VII, 237 pages. 1976.

Vol. 544: Robert P. Langlands, On the Functional Equations Satisfied by Eisenstein Series. VII, 337 pages. 1976.

Vol. 545: Noncommutative Ring Theory. Kent State 1975. Edited by J. H. Cozzens and F. L. Sandomierski. V, 212 pages. 1976.

Vol. 546: K. Mahler, Lectures on Transcendental Numbers. Edited and Completed by B. Diviš and W. J. Le Veque. XXI, 254 pages. 1976.

Vol. 547: A. Mukherjea and N. A. Tserpes, Measures on Topological Semigroups: Convolution Products and Random Walks. V, 197 pages. 1976.

Vol. 548: D. A. Hejhal, The Selberg Trace Formula for PSL (2,IR). Volume I. VI, 516 pages. 1976.

Vol. 549: Brauer Groups, Evanston 1975. Proceedings. Edited by D. Zelinsky. V, 187 pages. 1976.

Vol. 550: Proceedings of the Third Japan – USSR Symposium on Probability Theory. Edited by G. Maruyama and J. V. Prokhorov. VI, 722 pages. 1976.

y, Evanston 1976. Proceedings. Edited s. 1976.

K. Wirthmüller, A. A. du Plessis and ical Stability of Smooth Mappings. V,

gories of Algebraic Systems. Vector and Projective Spaces, Semigroups, Rings and Lattices. VIII, 217 pages. 1976.

Vol. 554: J. D. H. Smith, Mal'cev Varieties. VIII, 158 pages. 1976.

Vol. 555: M. Ishida, The Genus Fields of Algebraic Number Fields. VII, 116 pages. 1976.

Vol. 556: Approximation Theory. Bonn 1976. Proceedings. Edited by R. Schaback and K. Scherer. VII, 466 pages. 1976.

Vol. 557: W. Iberkleid and T. Petrie, Smooth S^1 Manifolds. III, 163 pages. 1976.

Vol. 558: B. Weisfeiler, On Construction and Identification of Graphs. XIV, 237 pages. 1976.

Vol. 559: J.-P. Caubet, Le Mouvement Brownien Relativiste. IX, 212 pages. 1976.

Vol. 560: Combinatorial Mathematics, IV, Proceedings 1975. Edited by L. R. A. Casse and W. D. Wallis. VII, 249 pages. 1976.

Vol. 561: Function Theoretic Methods for Partial Differential Equations. Darmstadt 1976. Proceedings. Edited by V. E. Meister, N. Weck and W. L. Wendland. XVIII, 520 pages. 1976.

Vol. 562: R. W. Goodman, Nilpotent Lie Groups: Structure and Applications to Analysis. X, 210 pages. 1976.

Vol. 563: Séminaire de Théorie du Potentiel. Paris, No. 2. Proceedings 1975–1976. Edited by F. Hirsch and G. Mokobodzki. VI, 292 pages. 1976.

Vol. 564: Ordinary and Partial Differential Equations, Dundee 1976. Proceedings. Edited by W. N. Everitt and B. D. Sleeman. XVIII, 551 pages. 1976.

Vol. 565: Turbulence and Navier Stokes Equations. Proceedings 1975. Edited by R. Temam. IX, 194 pages. 1976.

Vol. 566: Empirical Distributions and Processes. Oberwolfach 1976. Proceedings. Edited by P. Gaenssler and P. Révész. VII, 146 pages. 1976.

Vol. 567: Séminaire Bourbaki vol. 1975/76. Exposés 471–488. IV, 303 pages. 1977.

Vol. 568: R. E. Gaines and J. L. Mawhin, Coincidence Degree, and Nonlinear Differential Equations. V, 262 pages. 1977.

Vol. 569: Cohomologie Etale SGA 4½. Séminaire de Géométrie Algébrique du Bois-Marie. Edité par P. Deligne. V, 312 pages. 1977.

Vol. 570: Differential Geometrical Methods in Mathematical Physics, Bonn 1975. Proceedings. Edited by K. Bleuler and A. Reetz. VIII, 576 pages. 1977.

Vol. 571: Constructive Theory of Functions of Several Variables, Oberwolfach 1976. Proceedings. Edited by W. Schempp and K. Zeller. VI, 290 pages. 1977

Vol. 572: Sparse Matrix Techniques, Copenhagen 1976. Edited by V. A. Barker. V, 184 pages. 1977.

Vol. 573: Group Theory, Canberra 1975. Proceedings. Edited by R. A. Bryce, J. Cossey and M. F. Newman. VII, 146 pages. 1977.

Vol. 574: J. Moldestad, Computations in Higher Types. IV, 203 pages. 1977.

Vol. 575: K-Theory and Operator Algebras, Athens, Georgia 1975. Edited by B. B. Morrel and I. M. Singer. VI, 191 pages. 1977.

Vol. 576: V. S. Varadarajan, Harmonic Analysis on Real Reductive Groups. VI, 521 pages. 1977.

Vol. 577: J. P. May, E_∞ Ring Spaces and E_∞ Ring Spectra. IV, 268 pages. 1977.

Vol. 578: Séminaire Pierre Lelong (Analyse) Année 1975/76. Edité par P. Lelong. VI, 327 pages. 1977.

Vol. 579: Combinatoire et Représentation du Groupe Symétrique, Strasbourg 1976. Proceedings 1976. Edité par D. Foata. IV, 339 pages. 1977.